Studies in Systems, Decision and Control

Volume 606

The series "Studies in Systems, Decision and Control" (SSDC) covers both new developments and advances, as well as the state of the art, in the various areas of broadly perceived systems, decision making and control–quickly, up to date and with a high quality. The intent is to cover the theory, applications, and perspectives on the state of the art and future developments relevant to systems, decision making, control, complex processes and related areas, as embedded in the fields of engineering, computer science, physics, economics, social and life sciences, as well as the paradigms and methodologies behind them. The series contains monographs, textbooks, lecture notes and edited volumes in systems, decision making and control spanning the areas of Cyber-Physical Systems, Autonomous Systems, Sensor Networks, Control Systems, Energy Systems, Automotive Systems, Biological Systems, Vehicular Networking and Connected Vehicles, Aerospace Systems, Automation, Manufacturing, Smart Grids, Nonlinear Systems, Power Systems, Robotics, Social Systems, Economic Systems and other. Of particular value to both the contributors and the readership are the short publication timeframe and the world-wide distribution and exposure which enable both a wide and rapid dissemination of research output.

Indexed by SCOPUS, DBLP, WTI Frankfurt eG, zbMATH, SCImago.

All books published in the series are submitted for consideration in Web of Science.

Zahra Eidinejad · Reza Saadati ·
Tofigh Allahviranloo · Chenkuan Li · Javad Vahidi

Optimal Stability Theory and Approximate Solutions of Fractional Systems

New Results on the Analysis of Fractional Equations: Theoretical Insights and Numerical Approximations

Springer

Zahra Eidinejad
Iran University of Science and Technology
Tehran, Iran

Tofigh Allahviranloo
Research Center of Performance
and Productivity Analysis
Istinye University
Istanbul, Türkiye

Javad Vahidi
Faculty of Mathematics and Computer
Science
Iran University of Science and Technology
Tehran, Iran

Reza Saadati
Faculty of Mathematics and Computer
Science
Iran University of Science and Technology
Tehran, Iran

Chenkuan Li
Department of Mathematics and Computer
Science
Brandon University
Brandon, MB, Canada

ISSN 2198-4182 ISSN 2198-4190 (electronic)
Studies in Systems, Decision and Control
ISBN 978-3-031-96703-0 ISBN 978-3-031-96704-7 (eBook)
https://doi.org/10.1007/978-3-031-96704-7

This Springer imprint is published by the registered company Springer Nature Switzerland AG
The registered company address is: Gewerbestrasse 11, 6330 Cham, Switzerland

To my beloved parents:
My most valuable resource while writing, has been unstinting and ungrudging support from parents. Without that, this project might have been postponed indefinitely.

Zahra Eidinejad

Preface

One of the important topics in mathematics and especially in mathematical analysis is fractional calculus. The topic of fractional calculus and its significant applications have appeared to be an appropriate tool in the study of widespread fields of engineering and science. There are a lot of phenomena, and highly accurate modeling is very important in studying them. Fluid mechanics, biological models, electrochemistry, viscoelasticity, and electromagnetics are some examples of these fields. It has been found that the notion of fractional-order differential equations can well describe these models in science and engendering. Here, we can refer to the fractional derivatives of Caputo, Riemann-Liouville, Grinwald-Letnikov, Marchaud, or Hadamard as fractional operators. It should be noted that these operators are the result of changes made to ODEs and PDEs over time. The first approach is the Riemann-Liouville, which is based on the iteration of the classical integral operator for n times and then considers Cauchy's formula where $n!$ is replaced by the Gamma function and hence the fractional integral of noninteger order is defined. Given that solutions of fractional differential equations better evaluate the results in different fields, they are useful in modeling various phenomena. The topic of stability first started with Ulam's famous question about additive mappings in 1940. After that, Hyers and Rassias, by expanding this topic, presented new definitions of stability for additive mappings and continuous maps, which were known as Hyers-Ulam and Hyers-Ulam-Rasias stability, respectively. Since then, many researchers have conducted extensive research on the issue of stability for functional equations in different spaces. For example, the stability of homomorphism equations was studied in 1980. Authors have investigated different equations using fixed point theory in classical and fuzzy spaces. Note that fixed point theory arises in various fields such as dynamical systems, and equilibrium problems. In 1965, the theory of fuzzy sets was introduced. Also, Kramosil and Michalek (1975) introduced the new concept of fuzzy metric space and provided many results. Afterward, many mathematicians studied fixed-point theorems in related spaces. The use of special functions in fractional calculus has always been the focus of many authors. Mittag-Leffler's function (M-LF), which was first discovered by a Swedish mathematician named Mittag-Leffler, is used in many fields of science and engineering and certain areas of physical and applied sciences.

Among the wide applications of this function, it can also be mentioned to solve the problems of physical, biological, and earth sciences. Another important function that is particularly essential in solving partial differential equations and fractional theory is Wright's function (WF). This function, which is a generalization of the exponential function like the Mittag-Leffler function, was first introduced by the British mathematician E. M. Wright. Two other types of functions used in this article are the H-Fox function (H-FF) and the Gauss hypergeometric function (GHF). The H-Fox function introduced by Fox in 1961 has special functions of a very general nature. Also, many special functions we encounter in physics, engineering, and probability theory are special Gauss hypergeometric functions. To determine the optimal control function, we introduce the aggregation function (AF). Aggregation functions have been developing since 1980 because these functions are rooted in different fields. In nonlinear sciences, nonlinear partial equations are widely used in fields such as nonlinear optics, quantum mechanics, molecular biology, surface water waves, and other cases. On the other hand, many physical phenomena that exist in nature are described by exact solutions of nonlinear partial differential equations (PDEs). Nonlinear wave solutions play a fundamental role in various scientific fields such as geochemistry, optical fibers, chemical physics, fluid mechanics, plasma physics, solid-state physics, chemical kinematics, etc. Therefore, for nonlinear models in nature and industry, analytical and semi-analytical solutions have been investigated with many methods including the tanh-coth method, the sine-cosine method, the tanh-function method, the homogeneous balance method, inverse scattering transformation, Darboux transformation, Hirota bilinear method and so on. By converting nonlinear equations to ordinary differential equations (ODEs) with variable coefficients, we can use these techniques to obtain logical solutions. Subjects such as fixed point results, Hyers-Ulam-Rasias Wright stability, Hyers-Ulam-Rasias Kummer stability for fractional integral differential equations, stability analysis using aggregation control functions, investigation of fractional nonlinear problems with integral boundary conditions, analysis of fractional equations using matrix-valued fuzzy controllers, Gauss Hypergeometric stability, Cellular neural networks, pseudo almost periodic solution, pseudo almost automorphic solution, fuzzy measure theory, Z-numbers, Modular stability, *-GENERALIZED-$\mathcal{Z}$-NUMBER-BOREL MEASURES, the multiple exp-function method, the clique polynomials method, the reproducing kernel Hilbert space method, they are thoroughly explored through this book.

This book selects the most significant differential equations and integrodifferential equations introduced by acknowledged mathematical journals. It intends to assist research workers as a text. In this regard, the expectation is that the reader is thoroughly familiar with the basics of analysis. The book is self-contained in the sense that it delivers detailed proofs of all the main theorems and relevant technical lemmas. Thus, graduate students and advanced undergraduates seeking more opportunities to appreciate the elegance of mathematics and enjoy the fun of understanding certain phenomena can take benefit from this book.

This book is arranged into eight chapters as follows. At the end of each section, it presents notes and references to indicate the history of the results and further readings.

Chapter 1 reviews the basic concepts, definitions, and theorems used in the following chapters. Most of the material will be well-known to readers who have had basics of Analysis, Linear Algebra, and Matrix Analysis courses. In this regard, it is presented briefly with no proof. The topics considered less familiar are treated in moderately greater detail.

Chapter 2 presents the alternative fixed point method and applies the Wright function as a control function, to investigate the Hyers-Ulam-Wright stability. In the sequel, we consider the Gauss Hypergeometric function as a control function and stabilize the fractional equations with this control function, which we call the Kummer control function, and apply the Cadariu-Radu method derived from the Diaz-Margolis alternative fixed point theorem to investigate the existence, and uniqueness of solutions for the fractional Volterra integrodifferential equation and the fractional Volterra integral equation. Also, we consider the conformable fractional differential equation (CFD-E) with a constant coefficient. By calculating the optimal control function and considering this function, we obtain the existence of a unique solution and H-U-R-OS_1-multiple stability for the CFD equation. An integral boundary condition problem using Babenko's strategy is also investigated. Then we derive sufficient conditions for the uniqueness and existence of a nonlinear fractional integral differential equation with arbitrary order with the help of o Banach's contractive principle and Laray-Schauder's fixed point theorem. This chapter presents the stability results in classical space for the most important fractional equations.

Chapter 3 introduces the main results in fuzzy spaces. In this chapter, we study fuzzy sets matrix-valued with the generalized t-norms, to define matrix-valued fuzzy k-Banach space (in short MV-k-FB-space) and introduce a new class of matrix-valued fuzzy controllers. We apply the Radu-Miheţ method to get an approximation for a class of ψ–Hilfer fractional Volterra integrodifferential equation in the matrix-valued fuzzy k-normed spaces (MVF-k-N-spaces). The Symmetric Matrix-Valued fuzzy Banach space, the multiple control function, the aggregation function (AG-F), and the optimal control function which means the minimum function, are introduced. The multiple Hyers-Ulam-OS_1 stability of the nonlinear single fractional differential equation, using the desired control function, is proved. In the following, considering a modern class of Matrix-Valued fuzzy control functions based on the approximate Wright functions for the non-homogeneous fractional delay oscillation equation, we obtain a new combined method powered by the Laplace inverse transform and MR approach.

Chapter 4 is the introduction of Picard's special method. The Gauss Hypergeometric stability of the fractional-order differential equation using the Chebyshev norm and the Gauss Hypergeometric stability of the fractional-order differential equation by using the Bielecki norm is proposed.

Chapter 5 evaluates different issues in the field of sustainability. Pseudo almost automorphic (paa) solutions for the integrodifferential equations with reflection, an application of the measure paa, in connection with integrodifferential equations with reflection and delay, a class of non-autonomous cellular neural networks (CNNs) with mixed delays, pseudo almost periodicity, the general measure theory, the stability of the Mittag-Leffler and the attractiveness of the Mittag-Leffler for pseudo almost periodic functions, fuzzy measure theory, ϱ-pseudo almost periodically are discussed.

Chapter 6 introduces a new concept for obtaining a suitable approximation. The idea of the generalized Z-number as a diagonal matrix is proposed. A new space called a Modular space, stochastic fractional Volterra integral differential equation, approximation of the functional inequalities, the concept of Lebesgue-stieltjes-$\star$-Z-Number measure, $\star$-Z-Number differentiation, the Lebesgue-Radon-Nikodyme theorem, the fuzzy measure theory and fuzzy integration considering fuzzy number-valued triangular norm-based decomposable time-stamped, m–dimensional continuous vector-valued mappings are evaluated.

The final chapter describes the methods of solving the problems numerically. The multiple exp-function method for $(2+1)$-dimensional conformable DateJimbo-Kashiwara Miwa (CDJKM) equation is used. The generalized form of the two-dimensional stochastic time-fractional Sine-Gordon equation and the COVID-19 differential system by the new clique polynomial method are evaluated. Approximate solutions of the fractional hybrid integrodifferential equations with Dirichlet boundary conditions using the reproducing kernel Hilbert space numerical method are presented.

Tehran, Iran — Zahra Eidinejad
Tehran, Iran — Reza Saadati
Istanbul, Türkiye — Tofigh Allahviranloo
Brandon, Canada — Chenkuan Li
Tehran, Iran — Javad Vahidi
September 2024

Acknowledgements We would like to express our thanks to referees for reading the manuscript and providing valuable suggestions and comments which have helped to improve the presentation of the book.

Last but not least, it is our pleasure to acknowledge the superb assistance provided by the staff of Springer for the publication of the book.

Contents

Chapter 1
Basic Concepts Review

In this chapter, we review the basic concepts, definitions and theorems which are used throughout the book. The contents presented in this chapter are familiar to experts in this field. Therefore, we avoid giving unnecessary details, and only for the contents that need more explanations, the required details are explained.

1.1 Special Functions

1.1.1 The Gauss Hypergeometric Function

Definition 1.1 ([1]) Let $\mathfrak{d}$ be a real number and consider the generic parameters $\mathfrak{p}, \mathfrak{q}, \mathfrak{r} > 0$. We define the Gauss Hypergeometric function ${}_2F_1 : \mathbb{R}^3 \times \mathcal{T} \longrightarrow (0, \infty)$ by the infinite sum (that is convergent)

$$ {}_2F_1(\mathfrak{p}, \mathfrak{q}, \mathfrak{r}; \mathfrak{d}) = \sum_{k=0}^{\infty} \frac{(\mathfrak{p})_k (\mathfrak{q})_k}{(\mathfrak{r})_k} \frac{\mathfrak{d}^k}{k!} = \frac{\Gamma(\mathfrak{r})}{\Gamma(\mathfrak{p})\Gamma(\mathfrak{q})} \sum_{k=0}^{\infty} \frac{\Gamma(\mathfrak{p}+k)\Gamma(\mathfrak{q}+k)}{\Gamma(\mathfrak{r}+k)} \frac{\mathfrak{d}^k}{k!}. $$

1.1.2 The Mittag-Leffler Function

Definition 1.2 ([2, 3]) One-parameter and two-parameter Mittag-Leffler functions (M-LF) are defined as follows, respectively

$$ E_\alpha(\mathfrak{d}) = \sum_{n=0}^{\infty} \frac{\mathfrak{d}^n}{\Gamma(n\alpha + 1)}, $$

Z. Eidinejad et al., *Optimal Stability Theory and Approximate Solutions of Fractional Systems*, Studies in Systems, Decision and Control 606,
https://doi.org/10.1007/978-3-031-96704-7_1

$$E_{\alpha,\beta}(\mathfrak{d}) = \sum_{n=0}^{\infty} \frac{\mathfrak{d}^n}{\Gamma(n\alpha + \beta)},$$

where $\alpha, \beta \in \mathbb{C}$, $Re(\alpha), Re(\beta) > 0$ and $\Gamma(.)$ used in the above functions is the famous gamma function.

In the following, we will define the multivariate-Mittag-Leffler function.

Definition 1.3 ([6]) The multivariate-Mittag-Leffler function is defined as follows

$$E_{(\alpha_1,\ldots,\alpha_m),\beta}(z_1,\ldots,z_m) = \sum_{k=0}^{\infty} \sum_{k_1+\cdots+k_m=k} \binom{k}{k_1,\ldots,k_m} \frac{z_1^{k_1}\cdots z_m^{k_m}}{\Gamma(\alpha_1 k_1 + \cdots + \alpha_m k_m + \beta)},$$

where $\alpha_i, \beta > 0$ $z_i \in \mathbb{C}$ for $i = 1, 2, \ldots, m$ and

$$\binom{k}{k_1,\ldots,k_m} = \frac{k!}{k_1!\cdots k_m!}.$$

We use two functions called the sine-type Mittag-Leffler and the cosine-type Mittag-Leffler and these two functions are derived from the extension of Mittag-Leffler function.

Definition 1.4 The cosine-type Mittag-Leffler function of a polynomial of degree $p\sigma$, $C_\sigma^\mu(\varrho) : \mathbb{R} \to \mathbb{R}^n$, for $n \in \mathbb{N}$ and $\sigma \in \mathbb{R}^+$ on the interval $[(\mathfrak{p}-1)\,\mu, \mathfrak{p}\mu]$ and for $\varrho = \mathfrak{p}\mu$, $\mathfrak{p} = 0, 1, 2, \ldots, N$ defined by

$$C_\sigma^\mu(\varrho) := \begin{cases} \Theta, & -\infty < \varrho < \mu, \\ I, & -\mu < \varrho < 0, \\ \cdots & \\ I - \vartheta \frac{\varrho^\sigma}{\Gamma(\sigma+1)} + \vartheta^2 \frac{(\varrho-\mu)^{2\sigma}}{\Gamma(2\sigma+1)} + \cdots + (-1)^{\mathfrak{p}} \vartheta^{\mathfrak{p}} \frac{(\varrho-(\mathfrak{p}-1)\varrho)^{\mathfrak{p}\sigma}}{\Gamma(\mathfrak{p}\sigma+1)}, & (\mathfrak{p}-1)\mu \le \varrho < k\mu, \end{cases}.$$

that I denotes a unit matrix and Θ denotes a zero matrix.

Definition 1.5 The sine-type Mittag-Leffler function of a polynomial of degree $p\sigma + 1$, $S_\sigma^\mu(\varrho) : \mathbb{R} \to \mathbb{R}^n$, for $n \in \mathbb{N}$ and $\sigma \in \mathbb{R}^+$ on the interval $[(\mathfrak{p}-1)\mu, \mathfrak{p}\mu)$ and for $\varrho = \mathfrak{p}\mu$, $\mathfrak{p} = 0, 1, 2, \ldots, N$ defined by

$$S_\sigma^\mu(\varrho) := \begin{cases} \Theta, & -\infty < \varrho < \mu, \\ (\varrho - \mu)I, & -\mu < \varrho < 0, \\ \cdots & \\ (\varrho + \mu)I - \vartheta \frac{\varrho^{\sigma+1}}{\Gamma(\sigma+2)} + \vartheta^2 \frac{(\varrho-\mu)^{2\sigma+1}}{\Gamma(2\sigma+2)} + & \\ \cdots + (-1)^{\mathfrak{p}} \vartheta^{\mathfrak{p}} \frac{(\varrho-(\mathfrak{p}-1)\mu)^{\mathfrak{p}\sigma+1}}{\Gamma(\mathfrak{p}\sigma+2)}, & (\mathfrak{p}-1)\mu \le \varrho < \mathfrak{p}\mu, \end{cases}$$

that I denotes a unit matrix and Θ denotes a zero matrix.

1.1.3 The $\mathcal{H}$-Fox Function

Definition 1.6 ([4]) For $0 \leq n \leq p$, $1 \leq m \leq q$, $\{r_j, s_j\} \in \mathbb{C}$, $\{\phi_j, \psi_j\} \in \mathbb{R}^+$, we define the following functions

- $\mathbb{V}(e) = \prod_{j=1}^{m} \Gamma(s_j - \psi_j e)$,
- $\mathbb{W}(e) = \prod_{j=1}^{n} \Gamma(1 - r_j + \phi_j e)$,
- $\mathbb{X}(e) = \prod_{j=m+1}^{q} \Gamma(1 - s_j + \psi_j e)$,
- $\mathbb{Y}(e) = \prod_{j=n+1}^{p} \Gamma(r_j - \phi_j e)$.

In the introduced functions $n = 0$ if and only if $\mathbb{W}(e) = 1$, $m = q$ if and only if $\mathbb{X}(e) = 1$ and $n = p$ if and only if $\mathbb{Y}(e) = 1$. According to the introduced functions, we consider $H_{p,q}^{m,n}(e) = \frac{\mathbb{V}(e)\mathbb{W}(e)}{\mathbb{X}(e)\mathbb{Y}(e)}$. The Mellin–Barnes integral (M-BI) representation of $\mathcal{H}$-Fox function ($\mathcal{H}$-FF) is shown as below

$$\mathcal{H}_{p,q}^{m,n}(\mathfrak{d}) = \frac{1}{2\pi i} \int_{\mathcal{L}} H_{p,q}^{m,n}(e)\mathfrak{d}^e de, \tag{1.1.1}$$

where $\mathfrak{d}^e = \exp\{(\log|\mathfrak{d}| + i \arg \mathfrak{d})\}$ and $\mathcal{L} \in \mathbb{C}$ is a path that is deleted. Also, the symbol $\mathcal{H}_{p,q}^{m,n}(\mathfrak{d}) = \mathcal{H}_{p,q}^{m,n}\left[\mathfrak{d} \middle| \begin{matrix} (r_j, \alpha_j)_{j=1,\ldots,p} \\ (s_j, \beta_j)_{j=1,\ldots,p} \end{matrix}\right]$ is considered for this integral.

1.1.4 The Generalized Bessel Maitland Function

Definition 1.7 ([3, 5]) The generalized Bessel Maitland function (GBMF) or Wright function (WF) of order $1/(1+\sigma)$ is represented by using the series as

$$W_{\sigma,\upsilon}(\mathfrak{d}) = \sum_{m=0}^{\infty} \frac{\mathfrak{d}^m}{m!\Gamma(\sigma m + \upsilon)},$$

for $\sigma > -1$, $\upsilon > 0$, $\mathfrak{d} \in \mathbb{R}$.

1.1.5 The Generalized Aggregation Function

Definition 1.8 ([6, 7]) An n-ary ($n \in \mathbb{N}$) generalized aggregation function (n-AGAF) $A^{(n)} : \mathbb{R}^n \longrightarrow \mathbb{R}$ has the following property

- $a_i \leq b_i \Longrightarrow A^{(n)}(a_1, \ldots, a_n) \leq A^{(n)}(b_1, \ldots, b_n)$, for all $i \in \{1, \ldots, n\}$, and for $(a_1, \ldots, a_n)$, $(b_1, \ldots, b_n) \in \mathbb{R}^n$. For the sake of simplicity, we can remove the n number, which represents the number of variables of the aggregation function, and denote this function as A. Also, when $n = 1$, the aggregation function is shown as $A^{(1)}(x) = x$ for all $x \in \mathbb{R}$.

The following are examples of generalized aggregation functions:

Example 1.1 The arithmetic mean function (AMF) $AM : \mathbb{R}^n \longrightarrow \mathbb{R}$, defined by

$$AM(X) = \frac{1}{n}\sum_{i=1}^{n} x_i.$$

Example 1.2 The geometric mean function (GMF) $GM : \mathbb{R}^n \longrightarrow \mathbb{R}$, define by

$$GM(X) = (\prod_{i=1}^{n} x_i)^{\frac{1}{n}}.$$

Example 1.3 The projection function (PF) $P_k : \mathbb{R}^n \longrightarrow \mathbb{R}$ for $k \in [n]$ and kth argument, define by

$$P_k(x) = x_k,$$

where $x_{(k)}$ is the kth lowest coordinate of x, i.e. $x_{(1)} \leq \cdots \leq x_{(k)} \leq \cdots x_{(n)}$. Also, the following functions show the PF in the first and last coordinates

$$\begin{aligned} P_F(x) &= P_1(x) = x_1, \\ P_L(x) &= P_n(x) = x_n. \end{aligned} \tag{1.1.2}$$

Example 1.4 The order statistic function (OSF) $OS_k : \mathbb{R}^n \longrightarrow \mathbb{R}$ with the kth argument and kth lowest coordinate, define by

$$OS_k(x) = x_{(k)},$$

for any $k \in [n]$.

Example 1.5 The minimum function (MIN-F) and maximum function (MAX-F) are defined as follows respectively

$$\begin{aligned} Min(x) &= OS_1(x) = \min\{x_1, \ldots, x_n\} = \bigwedge_{i=1}^{n} x_i, \\ Max(x) &= OS_n(x) = \max\{x_1, \ldots, x_n\} = \bigvee_{i=1}^{n} x_i. \end{aligned} \tag{1.1.3}$$

Example 1.6 The median function (MF) is defined as follows for odd and even values of $(x_1, \ldots, x_{2k-1})$ and $(x_1, \ldots, x_{2k})$, respectively

$$Med(x_1, \ldots, x_{2k-1}) = x_{(k)}, \tag{1.1.4}$$

$$Med(x_1, \ldots, x_{2k}) = AM(x_{(k)}, x_{(k+1)}) = \frac{x_{(k)} + x_{(k+1)}}{2}.$$

Definition 1.9 $\mathfrak{Z} : \mathbb{I}^n \to \bar{\mathbb{R}}$ is an idempotent function if $\delta_{\mathfrak{Z}} = id$, that is, $\mathfrak{Z}(\ell.w) = w$ for all $w \in \mathbb{I}$. Idempotency is in some areas supposed to be a natural property of AG-Fs, e.g., in multicriteria decision making, where it is commonly accepted that if all criteria are satisfied at the same degree w, implicitly assuming the commensurateness of criteria, then also the overall score should be w.

The AG-F introduced above are idempotent. Here are some examples of non-idempotent AG-Fs.

- The product $\prod(W) = \prod_{\iota=1}^{\ell}(W_\iota)$ $(I \in \{|0, 1|, |0, +\infty|, |1, +\infty|\})$, where $|\iota, \jmath|$ means any of four kinds of intervals, with boundary points ι and $\jmath$, and with convention $0 \times +\infty = 0$.
- The sum function $\sum(W) = \sum_{\iota=1}^{\ell}(W_\iota)$ $(I \in \{|0, +\infty|, |-\infty, 0|, |-\infty, +\infty|\}, +\infty + (-\infty) = -\infty)$.

Tables below gives us good information about the introduced functions so that one can select the appropriate control function. By studying the information in these tables, we choose the minimum function i.e. OS_1 function as the optimal control function (Tables 1.1 and 1.2).

Considering the generalized aggregate function for the values of

$$\Theta = \left(E_{\alpha,\beta}(\mathfrak{d}), W_{\sigma,\upsilon}(\mathfrak{d}), {}_2F_1(\mathfrak{p}, \mathfrak{q}, \mathfrak{r}; \mathfrak{d}), \mathcal{H}_{p,q}^{m,n}(\mathfrak{d}), \exp(\frac{-\mathfrak{d}}{2})\right).$$

The figures below show the three-dimensional representation of the AM and GM aggregation functions (Figs. 1.1 and 1.2).

1.2 Fractional Integrals and Fractional Derivatives

1.2.1 The Conformable Fractional Derivative

The conformable fractional derivative and the Hadamard fractional integral are introduced below [8, 9].

Definition 1.10 For a function $\mathfrak{u} : [a, \infty) \to \mathbb{R}$, the conformable fractional derivative starting from a of order ζ is defined by

$$\mathfrak{D}_\zeta^a \mathfrak{u}(\mathfrak{d}) = \lim_{\iota \to 0} \frac{\mathfrak{u}\left(\mathfrak{d} + \iota(\mathfrak{d} - a)^{1-\zeta}\right) - \mathfrak{u}(\mathfrak{d})}{\iota}, \quad \mathfrak{d} > a, \ 0 < \zeta < 1.$$

If on (a, b), $\mathfrak{D}_\zeta^a \mathfrak{u}(\mathfrak{d})$ exists, then $\mathfrak{D}_\zeta^a \mathfrak{u}(a) = \lim_{\mathfrak{d} \to a^+} \mathfrak{D}_\zeta^a \mathfrak{u}(\mathfrak{d})$.

Table 1.1 In the following table we compare some known control functions

	$\mathfrak{d} = 0.2$	$\mathfrak{d} = 0.5$	$\mathfrak{d} = 5$	$\mathfrak{d} = 10$	$\mathfrak{d} = 100$	$\mathfrak{d} = 200$	$\mathfrak{d} = -1000$	$\mathfrak{d} = 1000$
$E_{\alpha,\beta}(\mathfrak{d})$	0.6361685737	0.7674678689	47.04527171	8189.592028	$3.461016369 \times 10^{13}$	$3.424908139 \times 10^{16}$	$3.214380196 \times 10^{23}$	$3.256719440 \times 10^{23}$
$W_{\sigma,\upsilon}(\mathfrak{d})$	0.6333991537	0.7473103055	4.875407059	22.68741719	$2.030171666 \times 10^{7}$	$1.318507579 \times 10^{10}$	$8.363632094 \times 10^{16}$	$9.533373699 \times 10^{16}$
${}_2F_1(\mathfrak{p}, \mathfrak{q}, \mathfrak{r}; \mathfrak{d})$	11.17112300	0.747310305	$1.023945418 \times 10^{10}$	$2.484080443 \times 10^{12}$	$2.371428685 \times 10^{20}$	$6055550402 \times 10^{22}$	$2.358383184 \times 10^{28}$	$2.360761077 \times 10^{28}$
$\mathcal{H}^{m,n}_{p,q}(\mathfrak{d})$	$2.360761077 \times 10^{28}$	0.00004423988200	0.004423988200	0.01769595280	1.769595280	7.078381120	176.9595280	176.9595280
$\exp(\frac{-\mathfrak{d}}{2})$	0.9355069850	0.8464817249	0.1888756028	0.03567399336	$3.338237806 \times 10^{-15}$	$1.114383154 \times 10^{-29}$	$5.818717687 \times 10^{144}$	$1.718591713 \times 10^{-145}$

Table 1.2 In the following table we compare some known generalized aggregate functions

	$\eth = 0.2$	$\eth = 0.5$	$\eth = 5$	$\eth = 10$	$\eth = 100$	$\eth = 200$	$\eth = -1000$	$\eth = 1000$
$AM(\theta)$	2.675240960	35.86013490	2.047890846×10^9	$4.968160902 \times 10^{11}$	$4.742858062 \times 10^{19}$	$1.211110765 \times 10^{22}$	$1.163743537 \times 10^{144}$	$4.721587288 \times 10^{27}$
$GM(\theta)$	2.675240960	0.3263333716	72.20357457	781.4257743	250396.0891	18482.92633	18482.92633	$1.860480545 \times 10^{-15}$
$OS_n(\theta)$	11.17112300	176.9595280	$1.023945418 \times 10^{10}$	$2.484080443 \times 10^{12}$	$2.371428685 \times 10^{20}$	$6.055550402 \times 10^{22}$	$5.818717687 \times 10^{144}$	$2.360761077 \times 10^{28}$
$OS_1(\theta)$	$7.078381120 \times 10^{-6}$	0.00004423988200	0.004423988200	0.01769595280	$3.338237806 \times 10^{-15}$	$1.114383154 \times 10^{-29}$	176.9595280	$1.718591713 \times 10^{-145}$
$Med(\theta)$	0.6361685737	0.7473103055	4.875407059	22.68741719	1.769595280	$1.318507579 \times 10^{10}$	$3.214380196 \times 10^{23}$	$9.533373699 \times 10^{16}$

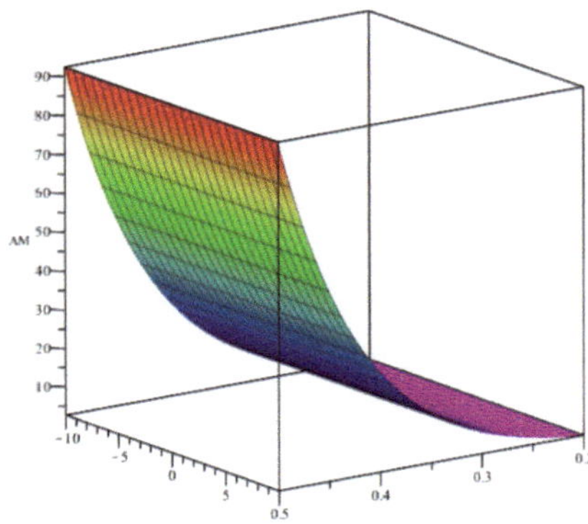

(a) AM function for $\mathfrak{d} \in [0.2, 0.5]$

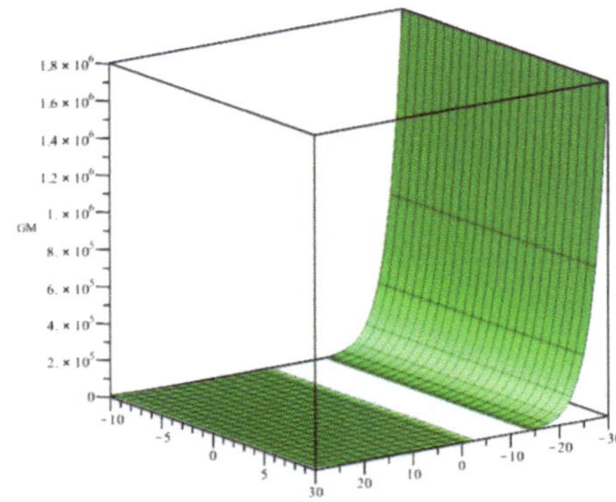

(b) GM function for $\mathfrak{d} \in [-30, 30]$

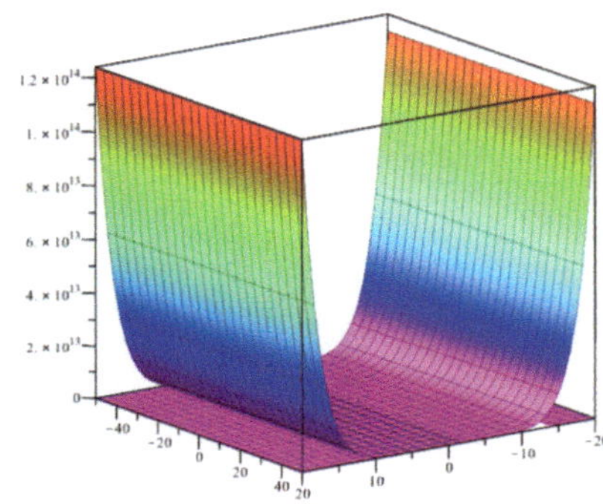

(c) AM and GM functions for $\mathfrak{d} \in [-20, 20]$

Fig. 1.1 3D representation of the AG functions AM and GM for different values

Remark 1.1 If $\mathfrak{D}_{\zeta}^{a}\mathfrak{u}\,(\mathfrak{d}_0)$ exists and is finite, we say that $\mathfrak{u}$ is ζ–differentiable at $\mathfrak{d}_0$. If $\mathfrak{u} \in C^1([a, \infty), \mathbb{R})$, then $\mathfrak{D}_{\zeta}^{a}\mathfrak{u}(\mathfrak{d}) = (\mathfrak{d} - a)^{1-\zeta}\mathfrak{u}'(\mathfrak{d})$.

1.2.2 The Hadamard Fractional Integral

Definition 1.11 For a given function $\mathfrak{u}(\mathfrak{d})$ with respect to $\mathfrak{d}$, the Hadamard fractional integral with order $0 < \zeta < 1$ and parameter $s \in \mathbb{R}$, is defined as

$$ {}_{a}^{H}I_{\mathfrak{d}}^{\zeta,s}\mathfrak{u}(\mathfrak{d}) = \frac{1}{\Gamma(\zeta)} \int_a^{\mathfrak{d}} \left(\frac{J}{x}\right)^{s} \left(\log \frac{x}{J}\right)^{\zeta-1} \frac{\mathfrak{u}(J)}{J}\, \mathrm{d}J, $$

where $\mathfrak{d} \in (a, b)$ and $a \leq b$.

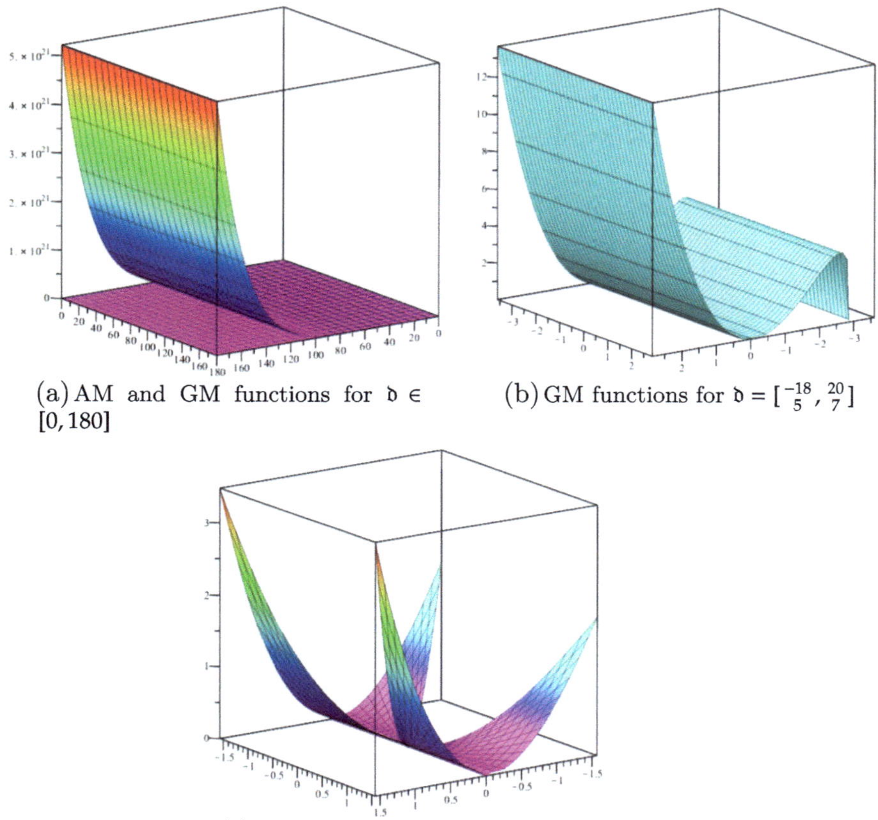

(a) AM and GM functions for $\mathfrak{d} \in [0, 180]$

(b) GM functions for $\mathfrak{d} = [\frac{-18}{5}, \frac{20}{7}]$

(c) GM functions for $\mathfrak{d} \in [-1.5, 1.5]$

Fig. 1.2 3D representation of the AG functions AM and GM for different values

1.2.3 The Riemann-Liouville Fractional Integral

Definition 1.12 ([10]) The Riemann-Liouville fractional integral for function $u(x)$ of order $\alpha \in R^+$ is defined as

$$(I_a^\alpha u)(x) = \frac{1}{\Gamma(\alpha)} \int_a^x (x-t)^{\alpha-1} u(t) dt.$$

In particular,

$$(I_a^0 u)(x) = u(x).$$

Definition 1.13 The Riemann–Liouville fractional integral of order $\Xi^* > 0$ for the function $Q : (0, +\infty) \to \mathbb{R}$ is defined as follows, [11]

$$\mathcal{I}^{\Xi^*} Q(\mathfrak{X}) = \frac{1}{\Gamma(\Xi^*)} \int_0^{\mathfrak{X}} (\mathfrak{X} - \mathrm{k})^{\Xi^*-1} Q(\mathrm{k}) d\mathrm{k},$$

where for Re $(\Xi^*) > 0$, and

$$\Gamma(\Xi^*) = \int_0^{+\infty} \exp(-\mathrm{k}) \mathrm{k}^{\Xi^*-1} d\mathrm{k}.$$

1.2.4 The Caputo Fractional Derivative

Definition 1.14 ([11]) For a continuous function $Q(\mathfrak{X}) : (0, +\infty) \to \mathbb{R}$, the Caputo fractional derivative is defined as follows,

$$\mathcal{D}^{\varrho_0} Q(\mathfrak{X}) = \frac{1}{\Gamma(\lambda - \varrho_0)} \int_0^{\mathfrak{X}} (\mathfrak{X} - \mathrm{k})^{\lambda - \varrho_0 - 1} Q^{(\lambda)}(\mathrm{k}) d\mathrm{k}, \quad (1.2.1)$$

where $\lambda = [\varrho_0]$ (the smallest integer greater than or equal to ϱ_0).

Lemma 1.1 ([11]) *For any* $\Xi \in (\ell - 1, \ell]$ *and* $Q \in C^{\ell-1}$, *the following equation holds*

$$\mathcal{I}^{\Xi} D^{\Xi} Q(\mathfrak{X}) = Q(\mathfrak{X}) + \mathrm{g}_0 + \mathrm{g}_1 \mathfrak{X} + \mathrm{g}_2 \mathfrak{X}^2 + \ldots + \mathrm{g}_{\ell-1} \mathfrak{X}^{\ell-1},$$

where $\mathrm{g}_\lambda \in \mathbb{R}$, $\lambda = 0, 1, 2, \ldots, \ell - 1$.

1.2.5 The Liouville-Caputo Fractional Derivative

Definition 1.15 The Liouville-Caputo fractional derivative for function $u(x)$ of order $\alpha \in R^+$ is defined as

$$({}_C D_a^\alpha u)(x) = I_a^{n-\alpha} \frac{d^n}{dx^n} u(x) = \frac{1}{\Gamma(n - \alpha)} \int_a^x (x - t)^{n-\alpha-1} u^{(n)}(t) dt,$$

for $n \in N = \{1, 2, 3, \ldots\}$ and $n - 1 < \alpha \leq n$.

Lemma 1.2 *Let* $\mathfrak{u} \in C^1([a, \infty), \mathbb{R})$. *For the function* $\mathfrak{u}$ *and* $\mathfrak{d} > a$, $0 < \zeta < 1$, *the following relationship holds:*

$${}_a^H I_{\mathfrak{d}}^{\zeta, s} \mathfrak{D}_\zeta^a \mathfrak{u}(\mathfrak{d}) = \mathfrak{u}(\mathfrak{d}) - \mathfrak{u}(a).$$

1.2.6 The Ψ-Hilfer Fractional Derivative

As one of the aims of this book is to present some types of stabilities involving a class of fractional integrodifferential equations by means of a ψ-Hilfer fractional operator, we introduce such a fractional operator.

Definition 1.16 Consider the functions $\theta, \Psi \in C^n[a,b]$ where $n-1<\tau<n$ and $n \in \mathbb{N}$, the left-sided Ψ-Hilfer fractional derivative of a function θ of order τ on $[a,b]$ and for $0 \leq \kappa \leq 1$ is defined by

$$ {}^{H}\mathcal{D}_{a^+}^{\tau,\kappa;\Psi}\theta(\ell) = \mathcal{I}_{a^+}^{\kappa(n-\tau);\Psi}\left(\frac{1}{\Psi'(\ell)}\frac{d}{d\ell}\right)^n \mathcal{I}_{a^+}^{(1-\kappa)(n-\tau);\Psi}\theta(\ell). \tag{1.2.2} $$

wherein Ψ is an increasing function and $\Psi'(\ell) \neq 0$. We can define the right-sided Ψ-Hilfer fractional derivative in a similar way.

Theorem 1.1 ([12]) *Let $\theta \in C^1[a,b]$, $\tau > 0$ and $0 \leq \kappa \leq 1$, we have*

$$ {}^{H}\mathcal{D}_{0^+}^{\tau,\kappa;\Psi}\mathcal{I}_{0^+}^{\tau;\Psi}\theta(\ell) = \theta(\ell). $$

Definition 1.17 The left-sided fractional integral of a function θ with respect to a function $\Psi(\ell)$ on interval $[a,b]$ for $\tau > 0$ is defined by

$$ \mathcal{I}_{a^+}^{\tau;\Psi}\theta(\ell) = \frac{1}{\Gamma(\tau)}\int_a^{\ell} \Delta_{\Psi}^{\tau}(\ell,\jmath)\theta(\jmath)d\jmath, \tag{1.2.3} $$

that the function $\Psi(\ell)$ is an increasing and positive function with a continuous derivative on (a,b). Also, the $\Delta_{\Psi}^{\tau}(\ell,\jmath)$ function used in the upper integral is $\Delta_{\Psi}^{\tau}(\ell,\jmath) := \Psi'(\jmath)(\Psi(\ell)-\Psi(\jmath))^{\tau-1}$. We can define the right-sided fractional integral in a similar way.

1.2.7 The $\mathcal{ABC}$ Fractional Derivative

Definition 1.18 The $\mathcal{ABC}$ fractional derivative of the function $\mathcal{Z} \in \mathcal{L}^*(i,j)$, where $\Xi^* \in [0,1]$, is defined by [17]

$$ {}^{\mathcal{ABC}}{}_{0}\mathcal{D}_{\zeta}^{\Xi^*}\mathcal{Z}(\zeta) = \frac{\mathcal{W}(\Xi^*)}{1-\varsigma}\int_0^{\zeta} \mathcal{Z}'(\varrho_0)\, E_{\Xi^*}\left[\frac{-\Xi^*(\zeta-\mathrm{k})^{\varsigma}}{1-\Xi^*}\right]d\mathrm{k}, $$

such that for $\mathcal{W}(\Xi^*)$ satisfying $\mathcal{W}(0) = \mathcal{W}(1) = 1$.

1.2.8 The $\mathcal{AB}$-Riemann–Liouville Fractional Derivative

Definition 1.19 $\mathcal{AB}$-Riemann–Liouville fractional derivative of the function $\mathcal{Z} \in \mathcal{L}^*(i, j)$, is described as follows,

$$ {}^{\mathcal{ABC}}{}_0\mathcal{D}_\zeta^{\Xi^*}\mathcal{Z}(\zeta) = \frac{\mathcal{W}(\Xi^*)}{1-\Xi^*}\frac{d}{d\zeta}\int_0^\zeta \mathcal{Z}(\varrho_0)\, E_\Xi^*\left[\frac{-\Xi^*(\zeta-\mathrm{k})^{\Xi^*}}{1-\Xi^*}\right]d\varrho, $$

where $\Xi^* \in [0, 1]$.

1.2.9 The $\mathcal{AB}$-Fractional Integral

Definition 1.20 $\mathcal{AB}$-fractional integral of the function $\mathcal{Z} \in \mathcal{L}^*(i, j)$, $0 < \Xi^* < 1$ is given by [18]:

$$ {}^{\mathcal{AB}}{}_0\mathcal{I}_\zeta^{\Xi^*}\mathcal{Z}(\zeta) = \frac{1-\Xi^*}{\mathcal{W}(\Xi^*)}\mathcal{Z}(\zeta) + \frac{\Xi^*}{\mathcal{W}(\Xi^*)\,\Gamma(\Xi^*)}\int_0^\zeta \mathcal{Z}(\varrho_0)\,(\zeta-\mathrm{k})^{\Xi^*-1}d\varrho. $$

1.2.10 The Newton–Leibniz Formula

Lemma 1.3 ([19, 20]) *The Newton–Leibniz formula for the $\mathcal{ABC}$-fractional derivative and $\mathcal{ABC}$-fractional integral of the function $\mathcal{Z}$ is*

$$ {}^{\mathcal{AB}}{}_0\mathcal{I}_\zeta^{\Xi^*}\left({}^{\mathcal{ABC}}{}_0\mathcal{D}_\zeta^{\Xi^*}\mathcal{Z}(\zeta)\right) = \mathcal{Z}(\zeta) - \mathcal{Z}(0). \tag{1.2.4} $$

This formula has also been proved for the Caputo–Fabrizio derivative in both continuous and discrete states, as well as for the same derivative in the discrete state.

1.2.11 The Classical Laplace Transform

Definition 1.21 The classical Laplace transform is defined by the integral formula

$$ (\pounds g)(s) = \int_0^\infty g(\varrho)e^{-s\varrho}d\varrho, $$

where $g(\varrho)$ is absolutely integrable on $[0, \infty)$. Let J_0^σ denote the Riemann-Liouville fractional integral operator of order $\sigma \in (1, 2)$. Assume the Laplace transforms of

$J_0^\sigma g(\varrho)$, $D_0^\sigma g(\varrho)$ exist for $\varrho \leq 0$ and the Laplace transform of $g(\varrho - \mathrm{b})$ exist for $\varrho \leq \mathrm{b}$. Then we have

- $£\left(J_0^\sigma g(\varrho)\right)(s) = s^{-\sigma}£(g(\varrho))(s)$,
- $£\left(D_0^\sigma g(\varrho)\right)(s) = s^{\sigma}£(g(\varrho))(s) - s^{\sigma-1}g(0) - s^{\sigma-2}g'(0)$,
- $£\left(g_1(\varrho)\right)(s) = e^{-\mathrm{b}s}\int_{-\mathrm{b}}^{0} e^{-\varrho s}g(\varrho)d\varrho + e^{-\mathrm{b}s}£(g(\varrho))(s)$.

$£^{-1}$ is the Laplace inverse transform.

1.2.12 The Generalized Metric Space

Definition 1.22 Let $\mathcal{D}$ be a nonempty set, and let $\delta : \mathcal{D} \times \mathcal{D} \to [0, \infty]$ be a mapping such that for all $\mathfrak{d}, \mathfrak{e}, \mathfrak{c} \in \mathcal{D}$,

1. $\delta(\mathfrak{d}, \mathfrak{e}) = 0$ if and only if $\mathfrak{d} = \mathfrak{e}$;
2. $\delta(\mathfrak{d}, \mathfrak{e}) = \delta(\mathfrak{e}, \mathfrak{d})$;
3. $\delta(\mathfrak{d}, \mathfrak{e}) \leq \delta(\mathfrak{d}, \mathfrak{c}) + \delta(\mathfrak{c}, \mathfrak{e})$.

Then, $(\mathcal{D}, \delta)$ is called a generalized metric space (g.m.s., for short).

1.3 Fuzzy Space

Here, we let $E_1 = [0, p]$, $E_2 = (0, \infty)$, $E_3 = (0, 1]$, $E_4 = [0, \infty]$ and $E_5 = [0, 1]$ (note $E_5^\circ = (0, 1)$ denotes the interior of E_5) and $E_6 = [0, \infty)$.

Let

$$\mathrm{diag}M_n(E_5) = \left\{ \begin{bmatrix} q_1 & & \\ & \ddots & \\ & & q_n \end{bmatrix} = \mathrm{diag}[q_1, \ldots, q_n],\ q_1, \ldots, q_n \in E_5 \right\},$$

where $\mathrm{diag}M_n(E_5)$ is equipped with the partial order relation:

$$\boldsymbol{q} = \mathrm{diag}[q_1, \ldots, q_n],\ \boldsymbol{b} = \mathrm{diag}[b_1, \ldots, b_n] \in \mathrm{diag}M_n(E_5),$$
$$\boldsymbol{q} \preceq \boldsymbol{b} \longrightarrow q_i \leq b_i \text{ for every } i = 1, \ldots, n.$$

Also, $\boldsymbol{q} \prec \boldsymbol{b}$ denotes that $\boldsymbol{q} \preceq \boldsymbol{b}$ and $\boldsymbol{q} \neq \boldsymbol{b}$; $\boldsymbol{q} \ll \boldsymbol{b}$ and $a_i < b_i$ for every $i = 1, \ldots, n$. We define $\boldsymbol{e} = \mathrm{diag}[e, \ldots, e]$ in $\mathrm{diag}M_n(E_5)$ where $e \in E_5$. For instance, $\mathbf{1} = \mathrm{diag}[1, \ldots, 1]$ and $\mathbf{0} = \mathrm{diag}[0, \ldots, 0]$.

Now, we define a class of t-norms [13, 14] on $\mathrm{diag}M_n(E_5)$.

1.3.1 The Generalized T-Norm

Definition 1.23 ([13–15]) Consider the generalized t-norm (in short GTN) $\circledast:$ $\text{diag}M_n(E_5)\times\text{diag}M_n(E_5)\to\text{diag}M_n(E_5)$ which satisfies the following conditions:

(a) $(\forall \boldsymbol{q}\in\text{diag}M_n(E_5))(\boldsymbol{q}\circledast\boldsymbol{1}=\boldsymbol{q})$ (boundary condition);

(b) $(\forall(\boldsymbol{q},\boldsymbol{b})\in(\text{diag}M_n(E_5))^2)(\boldsymbol{q}\circledast\boldsymbol{b}=\boldsymbol{b}\circledast\boldsymbol{q})$ (commutativity);

(c) $(\forall(\boldsymbol{q},\boldsymbol{b},\boldsymbol{c})\in(\text{diag}M_n(E_5))^3)(\boldsymbol{q}\circledast(\boldsymbol{b}\circledast\boldsymbol{c})=(\boldsymbol{q}\circledast\boldsymbol{b})\circledast\boldsymbol{c})$ (associativity);

(d) $(\forall(\boldsymbol{q}_1,\boldsymbol{q}_2,\boldsymbol{b}_1,\boldsymbol{b}_2)\in(\text{diag}M_n(E_5))^4)(\boldsymbol{q}_1\preceq\boldsymbol{q}_2$ and $\boldsymbol{b}_1\preceq\boldsymbol{b}_2\Longrightarrow\boldsymbol{q}_1\circledast\boldsymbol{b}_1\preceq\boldsymbol{q}_2\circledast\boldsymbol{b}_2)$ (monotonicity).

For every $\boldsymbol{q},\boldsymbol{b}\in\text{diag}M_n(E_5)$ and every sequences $\{\boldsymbol{q}_k\}$ and $\{\boldsymbol{b}_k\}$ converging to $\boldsymbol{q}$ and $\boldsymbol{b}$ suppose we have

$$\lim_k(\boldsymbol{q}_k\circledast\boldsymbol{b}_k)=\boldsymbol{q}\circledast\boldsymbol{b},$$

then, $\circledast$ on $\text{diag}M_n(E_5)$ is continuous GTN (in short CGTN).

Now, we present some examples of CGTN.

(1) Define $\circledast_M:\text{diag}M_n(E_5)\times\text{diag}M_n(E_5)\to\text{diag}M_n(E_5)$, such that,

$$\boldsymbol{q}\circledast_M\boldsymbol{b}=\text{diag}[q_1,\ldots,q_n]\circledast_M\text{diag}[b_1,\ldots,b_n]=\text{diag}[\min\{q_1,b_1\},\ldots,\min\{q_n,b_n\}],$$

then $\circledast_M$ is CGTN (minimum CGTN).

(2) Define $\circledast_P:\text{diag}M_n(E_5)\times\text{diag}M_n(E_5)\to\text{diag}M_n(E_5)$, such that,

$$\boldsymbol{q}\circledast_P\boldsymbol{b}=\text{diag}[q_1,\ldots,q_n]\circledast_P\text{diag}[b_1,\ldots,b_n]=\text{diag}[q_1.b_1,\ldots,q_n.b_n],$$

then $\circledast_P$ is CGTN (product CGTN).

(3) Define $\circledast_L:\text{diag}M_n(E_5)\times\text{diag}M_n(E_5)\to\text{diag}M_n(E_5)$, such that,

$$\boldsymbol{q}\circledast_L\boldsymbol{b}=\text{diag}[q_1,\ldots,q_n]\circledast_L\text{diag}[b_1,\ldots,b_n]=\text{diag}[\max\{q_1+b_1-1,0\},\ldots,\max\{q_n+b_n-1,0\}],$$

then $\circledast_P$ is CGTN (Lukasiewicz CGTN).

Now, we present some numerical examples:

$$\text{diag}\left[\frac{1}{2},\frac{3}{4},1,\frac{2}{3},\frac{1}{4}\right]\circledast_M\text{diag}\left[0,\frac{1}{3},\frac{2}{5},\frac{6}{7},1\right]=$$

$$\begin{bmatrix}\frac{1}{2}&&&&\\&\frac{3}{4}&&&\\&&1&&\\&&&\frac{2}{3}&\\&&&&\frac{1}{4}\end{bmatrix}\circledast_M\begin{bmatrix}0&&&&\\&\frac{1}{3}&&&\\&&\frac{2}{5}&&\\&&&\frac{6}{7}&\\&&&&1\end{bmatrix}=\begin{bmatrix}0&&&&\\&\frac{1}{3}&&&\\&&\frac{2}{5}&&\\&&&\frac{2}{3}&\\&&&&\frac{1}{4}\end{bmatrix}=\text{diag}\left[0,\frac{1}{3},\frac{2}{5},\frac{2}{3},\frac{1}{4}\right]$$

$$\operatorname{diag}\left[\frac{1}{2},\frac{3}{4},1,\frac{2}{3},\frac{1}{4}\right]\circledast_P \operatorname{diag}\left[0,\frac{1}{3},\frac{2}{5},\frac{6}{7},1\right]=$$

$$\begin{bmatrix}\frac{1}{2}&&&&\\&\frac{3}{4}&&&\\&&1&&\\&&&\frac{2}{3}&\\&&&&\frac{1}{4}\end{bmatrix}\circledast_P\begin{bmatrix}0&&&&\\&\frac{1}{3}&&&\\&&\frac{2}{5}&&\\&&&\frac{6}{7}&\\&&&&1\end{bmatrix}=\begin{bmatrix}0&&&&\\&\frac{1}{4}&&&\\&&\frac{2}{5}&&\\&&&\frac{4}{7}&\\&&&&\frac{1}{4}\end{bmatrix}=\operatorname{diag}\left[0,\frac{1}{4},\frac{2}{5},\frac{4}{7},\frac{1}{4}\right]$$

$$\operatorname{diag}\left[\frac{1}{2},\frac{3}{4},1,\frac{2}{3},\frac{1}{4}\right]\circledast_L \, diag\left[0,\frac{1}{3},\frac{2}{5},\frac{6}{7},1\right]=$$

$$\begin{bmatrix}\frac{1}{2}&&&&\\&\frac{3}{4}&&&\\&&1&&\\&&&\frac{2}{3}&\\&&&&\frac{1}{4}\end{bmatrix}\circledast_L\begin{bmatrix}0&&&&\\&\frac{1}{3}&&&\\&&\frac{2}{5}&&\\&&&\frac{6}{7}&\\&&&&1\end{bmatrix}=\begin{bmatrix}0&&&&\\&\frac{1}{12}&&&\\&&\frac{2}{5}&&\\&&&\frac{11}{21}&\\&&&&\frac{1}{4}\end{bmatrix}=\operatorname{diag}\left[0,\frac{1}{12},\frac{2}{5},\frac{11}{21},\frac{1}{4}\right].$$

we get

$$\begin{aligned}&\operatorname{diag}\left[\frac{1}{2},\frac{3}{4},1,\frac{2}{3},\frac{1}{4}\right]\circledast_M \operatorname{diag}\left[0,\frac{1}{3},\frac{2}{5},\frac{6}{7},1\right]\\ \succeq\ &\operatorname{diag}\left[\frac{1}{2},\frac{3}{4},1,\frac{2}{3},\frac{1}{4}\right]\circledast_P \operatorname{diag}\left[0,\frac{1}{3},\frac{2}{5},\frac{6}{7},1\right]\\ \succeq\ &\operatorname{diag}\left[\frac{1}{2},\frac{3}{4},1,\frac{2}{3},\frac{1}{4}\right]\circledast_L \operatorname{diag}\left[0,\frac{1}{3},\frac{2}{5},\frac{6}{7},1\right].\end{aligned}$$

1.3.2 Matrix Valued Fuzzy Norm Space

We consider the set of matrix valued fuzzy functions (MVFF) Φ which are left continuous and increasing functions $\varphi : (E_1)^k \times (E_2)^n \to \operatorname{diag}M_n(E_3)$, where $\vec{t} = (t_1, \ldots, t_n) \in (E_2)^n$. Also $\lim_{t\to+\infty} \varphi(p_1, \ldots, p_k, \vec{t}) = \mathbf{1}$ for any $p_1, \ldots, p_k$ in E_1. As an example, the matrix valued fuzzy function $\varphi : (E_1)^k \times (E_2)^3 \to \operatorname{diag}M_n(E_3)$

$$\varphi(p_1,\ldots,p_k,\vec{t})=\operatorname{diag}\left[\exp\left(\frac{-\|p_1\ldots,p_k\|}{t_1}\right),\frac{t_2}{t_2+\|p_1,\ldots,p_k\|},\exp\left(\frac{-\|p_1,\ldots,p_k\|}{t_3}\right)\right],$$

for $\vec{t} \in (E_2)^3$.

In Φ, we define " $\preceq$" as follows:

$$\varphi \precsim \phi \iff \varphi(p_1, \dots, p_k, \vec{t}) \preceq \phi(p_1, \dots, p_k, \vec{t}), \quad \forall \vec{t} \in (E_2)^n \text{ and } p_1, \dots, p_k \in E_1.$$

Definition 1.24 Consider the CGTN $\circledast$, a vector space V and matrix valued fuzzy set (MVFS) $\Theta : V^k \times (E_2)^n \to \text{diag}M_n(E_3)$. In this case, we define a matrix valued fuzzy k-normed space (MVF-K-N-space) $(V, \Theta, \circledast)$ as,

(MVF-K-N1) $\Theta(v_1, \dots, v_k, \vec{t}) = \mathbf{1}$ if and only if $v_1, \dots, v_k$ are linearly dependent and $\vec{t} \in (E_2)^n$;

(MVF-K-N2) $\Theta(\alpha v_1, \dots, v_k, \vec{t}) = \Theta(v_1, \dots, v_k, \frac{\vec{t}}{|\alpha|})$ for all $v_1, \dots, v_k \in V$ and $\alpha \in \mathbb{C}$ with $\alpha \neq 0$;

(MVF-K-N3) $\Theta(v_0 + v_1, v_2, \dots, v_k, \vec{t} + \vec{s}) \succeq \Theta(v_0, v_2, \dots, v_k, \vec{t}) \circledast \Theta(v_1, v_2, \dots, v_k, \vec{s})$ for all $v_1, \dots, v_k \in V$ and any $\vec{t} \in (E_2)^n$ and $\vec{s} \in (E_2)^n$;

(MVF-K-N4) $\lim_{t \to +\infty} \Theta(p_1, \dots, p_k, \vec{t}) = \mathbf{1}$ for any $\vec{t} \in (E_2)^n$.

A complete MVF-k-N-space is called matrix valued fuzzy Banach space(in short MVF-k-B-space).

For an example the matrix valued fuzzy k-norm (MVF-k-N) Θ,

$$\begin{aligned}
&\Theta(v_1, v_2, \dots, v_k, \vec{t}) \\
&= \text{diag}\left[\exp\left(-\frac{\|v_1, v_2, \dots, v_k\|}{t_1}\right), \frac{t_2}{t_2 + \|v_1, v_2, \dots, v_k\|}, \exp\left(-\frac{\|v_1, v_2, \dots, v_k\|}{t_3}\right)\right],
\end{aligned}$$

for $\vec{t} \in (E_2)^3$.

Defines a matrix valued fuzzy k-norm and $(V, \Theta, \circledast_M)$ is an MVF-k-N-space; here $(V, \|\cdot\|)$ is a k-normed vector space. Throughout this book, $\circledast = \circledast_M$.

1.3.3 Matrix Valued Fuzzy Control Function

Similar to the calculations made, we consider the aggregation function of the minimum type to construct the control function. Therefore, we define the control function as follows:

$$\boldsymbol{OS}_1\Big(\boldsymbol{\Upsilon}(\mathfrak{X}, \eta)\Big) = \text{diag}\Big[OS_1\Big(\Upsilon(\mathfrak{X}, \tau)\Big), OS_1\Big(\Upsilon(\mathfrak{X}, \eta)\Big), OS_1\Big(\Upsilon(\mathfrak{X}, \eta)\Big), OS_1\Big(\Upsilon(\mathfrak{X}, \eta)\Big)\Big]. \tag{1.3.1}$$

1.3.3.1 Matrix Valued Fuzzy H-Fox Function

Here, we assume $\mathfrak{F}_1 = [0, p]$, $\mathfrak{F}_2 = (0, +\infty)$, $\mathfrak{F}_3 = (0, 1]$, $\mathfrak{F}_4 = [0, +\infty]$ and $\mathfrak{F}_5 = [0, 1]$ ($\mathfrak{F}_5^\circ = (0, 1)$) and $\mathfrak{F}_6 = [0, +\infty)$. Using the concept of H-Fox function, we define a MVF H-Fox function $\boldsymbol{H}^{m,n}_{\tau,\omega} : (\mathfrak{F}_1)^k \times \mathfrak{F}_2 \to \mathrm{diagM}_n(\mathfrak{F}_3)$ ($\boldsymbol{\tau}, \boldsymbol{\omega} \in \mathfrak{F}_2$) as a control function in the MVFkN-spaces as follows

$$\boldsymbol{H}^{m,n}_{\tau,\omega}\left(-\frac{|\varpi_1 - \vartheta_1, \ldots, \varpi_k - \vartheta_k|}{\mathfrak{y}}\right) =$$
$$\mathrm{diag}\Bigg[H^{m,n}_{\tau,\omega}\left[\left(-\frac{|\varpi_1 - \vartheta_1, \ldots, \varpi_k - \vartheta_k|}{\mathfrak{y}}\right)\Bigg| \begin{matrix}(a_j, \alpha_j)_{j=1,\ldots,p}\\ (b_j, \beta_j)_{j=1,\ldots,p}\end{matrix}\right],$$
$$H^{m,n}_{\tau,\omega}\left[\left(-\frac{|\varpi_1 - \vartheta_1, \ldots, \varpi_k - \vartheta_k|}{\mathfrak{y}}\right)\Bigg| \begin{matrix}(a_j, \alpha_j)_{j=1,\ldots,p}\\ (b_j, \beta_j)_{j=1,\ldots,p}\end{matrix}\right],$$
$$\ldots, H^{m,n}_{\tau,\omega}\left[\left(-\frac{|\varpi_1 - \vartheta_1, \ldots, \varpi_k - \vartheta_k|}{\mathfrak{y}}\right)\Bigg| \begin{matrix}(a_j, \alpha_j)_{j=1,\ldots,p}\\ (b_j, \beta_j)_{j=1,\ldots,p}\end{matrix}\right]\Bigg].$$

For a MVF H-Fox function $\boldsymbol{H}^{m,n}_{\tau,\omega}$, we have

- It is a left continuous and increasing function for positive values.
- $\lim_{\mathfrak{y}\to+\infty} \boldsymbol{H}^{m,n}_{\tau,\omega}\left(-\frac{|\varpi_1-\vartheta_1,\ldots,\varpi_k-\vartheta_k|}{\mathfrak{y}}\right) = \mathbf{1}$.
- For $\boldsymbol{H}^{m,n}_{\tau,\omega}$ and also, for the matrix valued fuzzy function $\boldsymbol{\Phi}_{\tau,\omega}$, we have

$$\boldsymbol{\Phi}_{\tau,\omega} \precsim \boldsymbol{H}^{m,n}_{\tau,\omega} \iff \boldsymbol{\Phi}_{\tau,\omega}(-\frac{|\varpi_1 - \vartheta_1, \ldots, \varpi_k - \vartheta_k|}{\mathfrak{y}}) \preceq \boldsymbol{H}^{m,n}_{\tau,\omega}(-\frac{|\varpi_1 - \vartheta_1, \ldots, \varpi_k - \vartheta_k|}{\mathfrak{y}}),$$

and also, we have

(1) $\boldsymbol{H}^{m,n}_{\tau,\omega}\left(-\frac{|\varpi_1-\vartheta_1,\ldots,\varpi_k-\vartheta_k|}{\mathfrak{y}}\right) \succ 0$.

(2) We can easily show that for $\mathfrak{y} \in \mathfrak{F}_2$, $\boldsymbol{H}^{m,n}_{\tau,\omega}\left(-\frac{|\varpi_1-\vartheta_1,\ldots,\varpi_k-\vartheta_k|}{\mathfrak{y}}\right) = \mathbf{1} \iff \varpi_1 - \vartheta_1, \ldots, \varpi_k - \vartheta_k = 0$.

(3) We show that

$$\boldsymbol{H}^{m,n}_{\tau,\omega}\left(-\frac{|\alpha(\varpi_1 - \vartheta_1, \ldots, \varpi_k - \vartheta_k)|}{\mathfrak{y}}\right) = \boldsymbol{H}^{m,n}_{\tau,\omega}\left(-\frac{|(\varpi_1 - \vartheta_1, \ldots, \varpi_k - \vartheta_k)|}{\frac{\mathfrak{y}}{|\alpha|}}\right).$$

Then

$$\boldsymbol{H}^{m,n}_{\tau,\omega}\left(-\frac{|\alpha(\varpi_1 - \vartheta_1, \ldots, \varpi_k - \vartheta_k)|}{\mathfrak{y}}\right) =$$
$$\mathrm{diag}\Bigg[H^{m,n}_{\tau,\omega}\left[\left(-\frac{-|\alpha(\varpi_1 - \vartheta_1, \ldots, \varpi_k - \vartheta_k)|}{\mathfrak{y}}\right)\Bigg| \begin{matrix}(a_j, \alpha_j)_{j=1,\ldots,p}\\ (b_j, \beta_j)_{j=1,\ldots,p}\end{matrix}\right],$$

$$H^{m,n}_{\tau,\omega}\left[(-\frac{-|\alpha(\varpi_1-\vartheta_1,\ldots,\varpi_k-\vartheta_k)|}{\mathfrak{y}})\middle|\begin{matrix}(a_j,\alpha_j)_{j=1,\ldots,p}\\(b_j,\beta_j)_{j=1,\ldots,p}\end{matrix}\right],$$
$$\ldots,H^{m,n}_{\tau,\omega}\left[\left(-\frac{-|\alpha(\varpi_1-\vartheta_1,\ldots,\varpi_k-\vartheta_k)|}{\mathfrak{y}}\right)\middle|\begin{matrix}(a_j,\alpha_j)_{j=1,\ldots,p}\\(b_j,\beta_j)_{j=1,\ldots,p}\end{matrix}\right]\Bigg]=$$
$$\mathrm{diag}\Bigg[H^{m,n}_{\tau,\omega}\left[\left(-\frac{-|\alpha||(\varpi_1-\vartheta_1,\ldots,\varpi_k-\vartheta_k)|}{\mathfrak{y}}\right)\middle|\begin{matrix}(a_j,\alpha_j)_{j=1,\ldots,p}\\(b_j,\beta_j)_{j=1,\ldots,p}\end{matrix}\right],$$
$$H^{m,n}_{\tau,\omega}\left[\left(-\frac{-|\alpha||(\varpi_1-\vartheta_1,\ldots,\varpi_k-\vartheta_k)|}{\mathfrak{y}}\right)\middle|\begin{matrix}(a_j,\alpha_j)_{j=1,\ldots,p}\\(b_j,\beta_j)_{j=1,\ldots,p}\end{matrix}\right]$$
$$,\ldots,H^{m,n}_{\tau,\omega}\left[\left(-\frac{-|\alpha||(\varpi_1-\vartheta_1,\ldots,\varpi_k-\vartheta_k)|}{\mathfrak{y}}\right)\middle|\begin{matrix}(a_j,\alpha_j)_{j=1,\ldots,p}\\(b_j,\beta_j)_{j=1,\ldots,p}\end{matrix}\right]\Bigg]=$$
$$\mathrm{diag}\Bigg[H^{m,n}_{\tau,\omega}\left[\left(\frac{-|(\varpi_1-\vartheta_1,\ldots,\varpi_k-\vartheta_k)|}{\frac{\mathfrak{y}}{|\alpha|}}\right)\middle|\begin{matrix}(a_j,\alpha_j)_{j=1,\ldots,p}\\(b_j,\beta_j)_{j=1,\ldots,p}\end{matrix}\right],$$
$$H^{m,n}_{\tau,\omega}\left[\left(\frac{-|(\varpi_1-\vartheta_1,\ldots,\varpi_k-\vartheta_k)|}{\frac{\mathfrak{y}}{|\alpha|}}\right)\middle|\begin{matrix}(a_j,\alpha_j)_{j=1,\ldots,p}\\(b_j,\beta_j)_{j=1,\ldots,p}\end{matrix}\right]$$
$$,\ldots,H^{m,n}_{\tau,\omega}\left[\left(\frac{-|(\varpi_1-\vartheta_1,\ldots,\varpi_k-\vartheta_k)|}{\frac{\mathfrak{y}}{|\alpha|}}\right)\middle|\begin{matrix}(a_j,\alpha_j)_{j=1,\ldots,p}\\(b_j,\beta_j)_{j=1,\ldots,p}\end{matrix}\right]\Bigg]=$$
$$\boldsymbol{H}^{\boldsymbol{m,n}}_{\boldsymbol{\tau,\omega}}\left(-\frac{|(\varpi_1-\vartheta_1,\ldots,\varpi_k-\vartheta_k)|}{\frac{\mathfrak{y}}{|\alpha|}}\right).$$

(4) We show that

$$\boldsymbol{H}^{\boldsymbol{m,n}}_{\boldsymbol{\tau,\omega}}\left(-\frac{|((\varpi_1-\vartheta_1)+(\mathfrak{z}_1-\mathfrak{p}_1),\ldots,(\varpi_k-\vartheta_k)+(\mathfrak{z}_k-\mathfrak{p}_k))|}{\mathfrak{y}+\mathfrak{s}}\right)\succeq$$
$$\boldsymbol{H}^{\boldsymbol{m,n}}_{\boldsymbol{\tau,\omega}}\left(-\frac{|((\varpi_1-\vartheta_1),\ldots,(\varpi_k-\vartheta_k))|}{\mathfrak{y}}\right)\circledast\boldsymbol{H}^{\boldsymbol{m,n}}_{\boldsymbol{\tau,\omega}}\left(-\frac{|(\mathfrak{z}_1-\mathfrak{p}_1),\ldots,(\mathfrak{z}_k-\mathfrak{p}_k)|}{\mathfrak{s}}\right).$$

Suppose that

$$H^{m,n}_{\tau,\omega}\left[-\left(\frac{|\varpi_1-\vartheta_1,\ldots,\varpi_k-\vartheta_k|}{\mathfrak{y}}\right)\middle|\begin{matrix}(a_j,\alpha_j)_{j=1,\ldots,p}\\(b_j,\beta_j)_{j=1,\ldots,p}\end{matrix}\right]$$
$$\leq H^{m,n}_{\tau,\omega}\left[-\left(\frac{|\mathfrak{z}_1-\mathfrak{p}_1,\ldots,\mathfrak{z}_k-\mathfrak{p}_k|}{\mathfrak{s}}\right)\middle|\begin{matrix}(a_j,\alpha_j)_{j=1,\ldots,p}\\(b_j,\beta_j)_{j=1,\ldots,p}\end{matrix}\right],$$

then

$$\frac{-|(\varpi_1-\vartheta_1,\ldots,\varpi_k-\vartheta_k)|}{\mathfrak{y}}\leq\frac{-|(\mathfrak{z}_1-\mathfrak{p}_1,\ldots,\mathfrak{z}_k-\mathfrak{p}_k)|}{\mathfrak{s}}\Longrightarrow$$
$$\frac{|(\varpi_1-\vartheta_1,\ldots,\varpi_k-\vartheta_k)|}{\mathfrak{y}}\geq\frac{|(\mathfrak{z}_1-\mathfrak{p}_1,\ldots,\mathfrak{z}_k-\mathfrak{p}_k)|}{\mathfrak{s}}\Longrightarrow$$

$$
\begin{aligned}
&\frac{\mathfrak{s}|(\varpi_1-\vartheta_1,\ldots,\varpi_k-\vartheta_k)|}{\mathfrak{y}} \geq |(\mathfrak{z}_1-\mathfrak{p}_1,\ldots,\mathfrak{z}_k-\mathfrak{p}_k)| \Longrightarrow\\
&\frac{\mathfrak{s}|(\varpi_1-\vartheta_1,\ldots,\varpi_k-\vartheta_k)|}{\mathfrak{y}} + |(\varpi_1-\vartheta_1,\ldots,\varpi_k-\vartheta_k)| \geq\\
&|(\mathfrak{z}_1-\mathfrak{p}_1,\ldots,\mathfrak{z}_k-\mathfrak{p}_k)| + |(\varpi_1-\vartheta_1,\ldots,\varpi_k-\vartheta_k)| \geq\\
&|(\varpi_1-\vartheta_1,\ldots,\varpi_k-\vartheta_k)+(\mathfrak{z}_1-\mathfrak{p}_1,\ldots,\mathfrak{z}_k-\mathfrak{p}_k)| \Longrightarrow\\
&|(\varpi_1-\vartheta_1,\ldots,\varpi_k-\vartheta_k)|\left(\frac{\mathfrak{s}}{\mathfrak{y}}+1\right)\\
&\geq |(\varpi_1-\vartheta_1,\ldots,\varpi_k-\vartheta_k)+(\mathfrak{z}_1-\mathfrak{p}_1,\ldots,\mathfrak{z}_k-\mathfrak{p}_k)| \Longrightarrow\\
&|(\varpi_1-\vartheta_1,\ldots,\varpi_k-\vartheta_k)|\left(\frac{\mathfrak{s}+\mathfrak{y}}{\mathfrak{y}}\right)\\
&\geq |(\varpi_1-\vartheta_1,\ldots,\varpi_k-\vartheta_k)+(\mathfrak{z}_1-\mathfrak{p}_1,\ldots,\mathfrak{z}_k-\mathfrak{p}_k)| \Longrightarrow\\
&\frac{|(\varpi_1-\vartheta_1,\ldots,\varpi_k-\vartheta_k)|}{\mathfrak{y}}\\
&\geq \frac{|(\varpi_1-\vartheta_1,\ldots,\varpi_k-\vartheta_k)+(\mathfrak{z}_1-\mathfrak{p}_1,\ldots,\mathfrak{z}_k-\mathfrak{p}_k)|}{\mathfrak{s}+\mathfrak{y}} \Longrightarrow\\
&\frac{-|(\varpi_1-\vartheta_1,\ldots,\varpi_k-\vartheta_k)|}{\mathfrak{y}}\\
&\leq \frac{-|(\varpi_1-\vartheta_1,\ldots,\varpi_k-\vartheta_k)+(\mathfrak{z}_1-\mathfrak{p}_1,\ldots,\mathfrak{z}_k-\mathfrak{p}_k)|}{\mathfrak{s}+\mathfrak{y}} \Longrightarrow\\
&H^{m,n}_{\tau,\omega}\left[\frac{-|(\varpi_1-\vartheta_1,\ldots,\varpi_k-\vartheta_k)+(\mathfrak{z}_1-\mathfrak{p}_1,\ldots,\mathfrak{z}_k-\mathfrak{p}_k)|}{\mathfrak{s}+\mathfrak{y}}\,\middle|\,\begin{matrix}(a_j,\alpha_j)_{j=1,\ldots,p}\\(b_j,\beta_j)_{j=1,\ldots,p}\end{matrix}\right]\\
&\geq H^{m,n}_{\tau,\omega}\left[\frac{-|(\varpi_1-\vartheta_1,\ldots,\varpi_k-\vartheta_k)|}{\mathfrak{y}}\,\middle|\,\begin{matrix}(a_j,\alpha_j)_{j=1,\ldots,p}\\(b_j,\beta_j)_{j=1,\ldots,p}\end{matrix}\right].
\end{aligned}
$$

Therefore, we have

$$
\begin{aligned}
&\mathrm{diag}\Bigg[H^{m,n}_{\tau,\omega}\left[\frac{-|(\varpi_1-\vartheta_1,\ldots,\varpi_k-\vartheta_k)+(\mathfrak{z}_1-\mathfrak{p}_1,\ldots,\mathfrak{z}_k-\mathfrak{p}_k)|}{\mathfrak{s}+\mathfrak{y}}\,\middle|\,\begin{matrix}(a_j,\alpha_j)_{j=1,\ldots,p}\\(b_j,\beta_j)_{j=1,\ldots,p}\end{matrix}\right],\\
&H^{m,n}_{\tau,\omega}\left[\frac{-|(\varpi_1-\vartheta_1,\ldots,\varpi_k-\vartheta_k)+(\mathfrak{z}_1-\mathfrak{p}_1,\ldots,\mathfrak{z}_k-\mathfrak{p}_k)|}{\mathfrak{s}+\mathfrak{y}}\,\middle|\,\begin{matrix}(a_j,\alpha_j)_{j=1,\ldots,p}\\(b_j,\beta_j)_{j=1,\ldots,p}\end{matrix}\right],\ldots,\\
&H^{m,n}_{\tau,\omega}\left[\frac{-|(\varpi_1-\vartheta_1,\ldots,\varpi_k-\vartheta_k)+(\mathfrak{z}_1-\mathfrak{p}_1,\ldots,\mathfrak{z}_k-\mathfrak{p}_k)|}{\mathfrak{s}+\mathfrak{y}}\,\middle|\,\begin{matrix}(a_j,\alpha_j)_{j=1,\ldots,p}\\(b_j,\beta_j)_{j=1,\ldots,p}\end{matrix}\right]\Bigg] \succeq\\
&\mathrm{diag}\Bigg[H^{m,n}_{\tau,\omega}\left[\frac{-|(\varpi_1-\vartheta_1,\ldots,\varpi_k-\vartheta_k)|}{\mathfrak{y}}\,\middle|\,\begin{matrix}(a_j,\alpha_j)_{j=1,\ldots,p}\\(b_j,\beta_j)_{j=1,\ldots,p}\end{matrix}\right],\\
&H^{m,n}_{\tau,\omega}\left[\frac{-|(\varpi_1-\vartheta_1,\ldots,\varpi_k-\vartheta_k)|}{\mathfrak{y}}\,\middle|\,\begin{matrix}(a_j,\alpha_j)_{j=1,\ldots,p}\\(b_j,\beta_j)_{j=1,\ldots,p}\end{matrix}\right]
\end{aligned}
$$

$$, \dots, H^{m,n}_{\tau,\omega}\left[\frac{-|(\varpi_1 - \vartheta_1, \dots, \varpi_k - \vartheta_k)|}{\mathfrak{y}}\middle| \begin{matrix}(a_j,\alpha_j)_{j=1,\dots,p}\\(b_j,\beta_j)_{j=1,\dots,p}\end{matrix}\right]\Bigg]\circledast$$

$$\text{diag}\Bigg[H^{m,n}_{\tau,\omega}\left[\frac{-|\mathfrak{s}_1 - \mathfrak{p}_1, \dots, \mathfrak{s}_k - \mathfrak{p}_k|}{\mathfrak{s}}\middle| \begin{matrix}(a_j,\alpha_j)_{j=1,\dots,p}\\(b_j,\beta_j)_{j=1,\dots,p}\end{matrix}\right],$$

$$H^{m,n}_{\tau,\omega}\left[\frac{-|\mathfrak{s}_1 - \mathfrak{p}_1, \dots, \mathfrak{s}_k - \mathfrak{p}_k|}{\mathfrak{s}}\middle| \begin{matrix}(a_j,\alpha_j)_{j=1,\dots,p}\\(b_j,\beta_j)_{j=1,\dots,p}\end{matrix}\right],$$

$$\dots, H^{m,n}_{\tau,\omega}\left[\frac{-|\mathfrak{s}_1 - \mathfrak{p}_1, \dots, \mathfrak{s}_k - \mathfrak{p}_k|}{\mathfrak{s}}\middle| \begin{matrix}(a_j,\alpha_j)_{j=1,\dots,p}\\(b_j,\beta_j)_{j=1,\dots,p}\end{matrix}\right]\Bigg].$$

Consequently, if

$$\mathfrak{Y}(\mathfrak{z}_1 - \mathfrak{z}'_1, \dots, \mathfrak{z}_k - \mathfrak{z}'_k, \mathfrak{y}) =$$
$$\text{diag}\Bigg[H^{m,n}_{\tau,\omega}\left(-\frac{|\mathfrak{z}_1 - \mathfrak{z}'_1, \dots, \mathfrak{z}_k - \mathfrak{z}'_k|}{\mathfrak{y}}\right), H^{m,n}_{\tau,\omega}\left(-\frac{|\mathfrak{z}_1 - \mathfrak{z}'_1, \dots, \mathfrak{z}_k - \mathfrak{z}'_k|}{\mathfrak{y}}\right)$$
$$, \dots, H^{m,n}_{\tau,\omega}\left(-\frac{|\mathfrak{z}_1 - \mathfrak{z}'_1, \dots, \mathfrak{z}_k - \mathfrak{z}'_k|}{\mathfrak{y}}\right)\Bigg],$$

for $\mathfrak{y} \in \mathfrak{F}_2$.

1.3.3.2 Matrix Valued Fuzzy Wright Function

Using the concept of Wright function [9], we define a MVF Wright function $\boldsymbol{W}_{\kappa,\varsigma}$ ($\kappa, \varsigma \in \mathfrak{F}_2$) as a control function in the MVFN-spaces by

$$\boldsymbol{W}_{\kappa,\varsigma}\left(-\frac{|\mathfrak{x}|}{\eta}\right) = \text{diag}\left[W_{\kappa,\varsigma}\left(-\frac{|\mathfrak{x}|}{\eta}\right), W_{\kappa,\varsigma}\left(-\frac{|\mathfrak{x}|}{\eta}\right), W_{\kappa,\varsigma}\left(-\frac{|\mathfrak{x}|}{\eta}\right)\right],$$

where, the Wright function $W_{\kappa,\varsigma}$ is define as follows

$$W_{\kappa,\varsigma}\left(-\frac{|\mathfrak{x}|}{\eta}\right) = \sum_{\mathfrak{p}=0}^{\infty}\frac{\left(\frac{-|\mathfrak{x}|}{\eta}\right)^{\mathfrak{p}}}{\mathfrak{p}!\Gamma(\mathfrak{p}\kappa+\varsigma)}.$$

Then, for the MVF Wright function $\boldsymbol{W}_{\kappa,\varsigma}$, we have

- It is left continuous and an increasing function for positive values.
- $\lim_{\eta\to+\infty}\boldsymbol{W}_{\kappa,\varsigma}(-\frac{|\mathfrak{x}|}{\eta}) = \mathbf{1}$.
- For $\boldsymbol{W}_{\kappa,\varsigma}$ and the matrix-valued fuzzy function $\boldsymbol{\Xi}_{\kappa,\varsigma}$, we have

$$\boldsymbol{\Xi}_{\kappa,\varsigma} \precsim \boldsymbol{W}_{\kappa,\varsigma} \Longleftrightarrow \boldsymbol{\Xi}_{\kappa,\varsigma}\left(-\frac{|\mathfrak{x}|}{\eta}\right) \preceq \boldsymbol{W}_{\kappa,\varsigma}\left(-\frac{|\mathfrak{x}|}{\eta}\right),$$

and we further get

(a) $\boldsymbol{W}_{\kappa,\varsigma}\left(-\frac{|\mathfrak{x}|}{\eta}\right) > 0.$

(b) For $\eta \in \mathfrak{F}_2$, $\boldsymbol{W}_{\kappa,\varsigma}\left(-\frac{|\mathfrak{x}|}{\eta}\right) = \mathbf{1} \iff \mathfrak{x} = 0.$

First, we assume that $\boldsymbol{W}_{\kappa,\varsigma}\left(-\frac{|\mathfrak{x}|}{\eta}\right) = \mathbf{1}$. Then

$$\operatorname{diag}\left[W_{\kappa,\varsigma}\left(-\frac{|\mathfrak{x}|}{\eta}\right), W_{\kappa,\varsigma}\left(-\frac{|\mathfrak{x}|}{\eta}\right), W_{\kappa,\varsigma}\left(-\frac{|\mathfrak{x}|}{\eta}\right)\right] =$$

$$\operatorname{diag}\left[\sum_{\mathfrak{p}=0}^{\infty}\frac{\left(\frac{-|\mathfrak{x}|}{\eta}\right)^{\mathfrak{p}}}{\mathfrak{p}!\Gamma(\mathfrak{p}\kappa+\varsigma)}, \sum_{\mathfrak{p}=0}^{\infty}\frac{\left(\frac{-|\mathfrak{x}|}{\eta}\right)^{\mathfrak{p}}}{\mathfrak{p}!\Gamma(\mathfrak{p}\kappa+\varsigma)}, \sum_{\mathfrak{p}=0}^{\infty}\frac{\left(\frac{-|\mathfrak{x}|}{\eta}\right)^{\mathfrak{p}}}{\mathfrak{p}!\Gamma(\mathfrak{p}\kappa+\varsigma)}\right] = \mathbf{1},$$

then $\sum_{\mathfrak{p}=0}^{\infty}\frac{\left(\frac{-|\mathfrak{x}|}{\eta}\right)^{\mathfrak{p}}}{\mathfrak{p}!\Gamma(\mathfrak{p}\kappa+\varsigma)} = 1$. We have

$$1+\sum_{\mathfrak{p}=1}^{\infty}\frac{\left(\frac{-|\mathfrak{x}|}{\eta}\right)^{\mathfrak{p}}}{\mathfrak{p}!\Gamma(\mathfrak{p}\kappa+\varsigma)} = 1 \Longrightarrow \sum_{\mathfrak{p}=1}^{\infty}\frac{\left(\frac{-|\mathfrak{x}|}{\eta}\right)^{\mathfrak{p}}}{\mathfrak{p}!\Gamma(\mathfrak{p}\kappa+\varsigma)} = 0 \Longrightarrow |\mathfrak{x}| = 0 \Longrightarrow \mathfrak{x} = 0.$$

Conversely, suppose that $\mathfrak{x} = 0$, then $\sum_{\mathfrak{p}=1}^{\infty}\frac{\left(\frac{-|\mathfrak{x}|}{\eta}\right)^{\mathfrak{p}}}{\mathfrak{p}!\Gamma(\mathfrak{p}\kappa+\varsigma)} = 0$. As a result

$$\sum_{\mathfrak{p}=0}^{\infty}\frac{\left(\frac{-|\mathfrak{x}|}{\eta}\right)^{\mathfrak{p}}}{\mathfrak{p}!\Gamma(\mathfrak{p}\kappa+\varsigma)} = 1+\sum_{\mathfrak{p}=1}^{\infty}\frac{\left(\frac{-|\mathfrak{x}|}{\eta}\right)^{\mathfrak{p}}}{\mathfrak{p}!\Gamma(\mathfrak{p}\kappa+\varsigma)} = 1 \Longrightarrow \sum_{\mathfrak{p}=0}^{\infty}\frac{\left(\frac{-|\mathfrak{x}|}{\eta}\right)^{\mathfrak{p}}}{\mathfrak{p}!\Gamma(\mathfrak{p}\kappa+\varsigma)} = 1.$$

(c) We shall deduce

$$\boldsymbol{W}_{\kappa,\varsigma}\left(-\frac{|\alpha(\mathfrak{x})|}{\eta}\right) = \boldsymbol{W}_{\kappa,\varsigma}\left(-\frac{|(\mathfrak{x})|}{\frac{\eta}{|\alpha|}}\right).$$

Indeed,

$$\boldsymbol{W}_{\kappa,\varsigma}\left(-\frac{|\mathfrak{x})|}{\eta}\right) =$$

$$\operatorname{diag}\left[\sum_{\mathfrak{p}=0}^{\infty}\frac{\left(\frac{-|\alpha(\mathfrak{x})|}{\eta}\right)^{\mathfrak{p}}}{\mathfrak{p}!\Gamma(\mathfrak{p}\kappa+\varsigma)}, \sum_{\mathfrak{p}=0}^{\infty}\frac{\left(\frac{-|\alpha(\mathfrak{x})|}{\eta}\right)^{\mathfrak{p}}}{\mathfrak{p}!\Gamma(\mathfrak{p}\kappa+\varsigma)}, \sum_{\mathfrak{p}=0}^{\infty}\frac{\left(\frac{-|\alpha(\mathfrak{x})|}{\eta}\right)^{\mathfrak{p}}}{\mathfrak{p}!\Gamma(\mathfrak{p}\kappa+\varsigma)}\right] =$$

$$\operatorname{diag}\left[\sum_{\mathfrak{p}=0}^{\infty}\frac{\left(\frac{-|\alpha||(\mathfrak{x})|}{\eta}\right)^{\mathfrak{p}}}{\mathfrak{p}!\Gamma(\mathfrak{p}\kappa+\varsigma)},\sum_{\mathfrak{p}=0}^{\infty}\frac{\left(\frac{-|\alpha||(\mathfrak{x})|}{\eta}\right)^{\mathfrak{p}}}{\mathfrak{p}!\Gamma(\mathfrak{p}\kappa+\varsigma)},\sum_{\mathfrak{p}=0}^{\infty}\frac{\left(\frac{-|\alpha||(\mathfrak{x})|}{\eta}\right)^{\mathfrak{p}}}{\mathfrak{p}!\Gamma(\mathfrak{p}\kappa+\varsigma)}\right]=$$

$$\operatorname{diag}\left[\sum_{\mathfrak{p}=0}^{\infty}\frac{\left(\frac{-|(\mathfrak{x})|}{\frac{\eta}{|\alpha|}}\right)^{\mathfrak{p}}}{\mathfrak{p}!\Gamma(\mathfrak{p}\kappa+\varsigma)},\sum_{\mathfrak{p}=0}^{\infty}\frac{\left(\frac{-|(\mathfrak{x})|}{\frac{\eta}{|\alpha|}}\right)^{\mathfrak{p}}}{\mathfrak{p}!\Gamma(\mathfrak{p}\tau+\omega)},\sum_{\mathfrak{p}=0}^{\infty}\frac{\left(\frac{-|(\mathfrak{x})|}{\frac{\eta}{|\alpha|}}\right)^{\mathfrak{p}}}{\mathfrak{p}!\Gamma(\mathfrak{p}\kappa+\varsigma)}\right]=$$

$$\boldsymbol{W}_{\kappa,\varsigma}(-\frac{|(\mathfrak{x})|}{\frac{\eta}{|\alpha|}}).$$

(d) Finally, we prove

$$\boldsymbol{W}_{\kappa,\varsigma}\left(-\frac{|(\mathfrak{x}+\epsilon)|}{\eta+\zeta}\right)\succeq\boldsymbol{W}_{\kappa,\varsigma}\left(-\frac{|\mathfrak{x}|}{\eta}\right)\circledast\boldsymbol{W}_{\kappa,\varsigma}\left(-\frac{|\epsilon|}{\zeta}\right).$$

Suppose that $\sum_{\mathfrak{p}=0}^{\infty}\frac{\left(\frac{|(\mathfrak{x})|}{\eta}\right)^{\mathfrak{p}}}{\mathfrak{p}!\Gamma(\mathfrak{p}\kappa+\varsigma)}\leq\sum_{\mathfrak{p}=0}^{\infty}\frac{\left(\frac{|\epsilon|}{\zeta}\right)^{\mathfrak{p}}}{\mathfrak{p}!\Gamma(\mathfrak{p}\kappa+\varsigma)}$, then

$$\frac{-|(\mathfrak{x})|}{\eta}\leq\frac{-|\epsilon|}{\zeta}\Longrightarrow\frac{|(\mathfrak{x})|}{\eta}\geq\frac{|\epsilon|}{\zeta}\Longrightarrow\frac{\zeta|\mathfrak{x}|}{\eta}\geq|\epsilon|\Longrightarrow\frac{\zeta|(\mathfrak{x})|}{\eta}+|(\mathfrak{x})|\geq|(\epsilon)|+|\mathfrak{x}|$$
$$\geq|(\mathfrak{x})+(\epsilon)|\Longrightarrow|(\mathfrak{x})|\left(\frac{\zeta}{\eta}+1\right)\geq|(\mathfrak{x})+(\epsilon)|\Longrightarrow|(\mathfrak{x})|\left(\frac{\zeta+\eta}{\eta}\right)$$
$$\geq|(\mathfrak{x})+(\epsilon)|\Longrightarrow\frac{|(\mathfrak{x})|}{\eta}\geq\frac{|(\mathfrak{x})+(\epsilon)|}{\zeta+\eta}\Longrightarrow\frac{-|(\mathfrak{x})|}{\eta}\leq\frac{-|(\mathfrak{x})+(\epsilon)|}{\zeta+\eta}$$
$$\Longrightarrow\sum_{\mathfrak{p}=0}^{\infty}\frac{\left(\frac{-|(\mathfrak{x})+(\epsilon)|}{\zeta+\eta}\right)^{\mathfrak{p}}}{\mathfrak{p}!\Gamma(\mathfrak{p}\kappa+\varsigma)}\geq\sum_{\mathfrak{p}=0}^{\infty}\frac{\left(\frac{-|(\mathfrak{x})|}{\eta}\right)^{\mathfrak{p}}}{\mathfrak{p}!\Gamma(\mathfrak{p}\kappa+\varsigma)}.$$

Therefore, we have

$$\operatorname{diag}\left[\sum_{\mathfrak{p}=0}^{\infty}\frac{\left(\frac{-|(\mathfrak{x})+(\epsilon)|}{\eta+\zeta}\right)^{\mathfrak{p}}}{\mathfrak{p}!\Gamma(\mathfrak{p}\kappa+\varsigma)},\sum_{\mathfrak{p}=0}^{\infty}\frac{\left(\frac{-|(\mathfrak{x})+(\epsilon)|}{\eta+\zeta}\right)^{\mathfrak{p}}}{\mathfrak{p}!\Gamma(\mathfrak{p}\kappa+\varsigma)},\sum_{\mathfrak{p}=0}^{\infty}\frac{\left(\frac{|\alpha|(-|(\mathfrak{x})+(\epsilon)|)}{\eta+\zeta}\right)^{\mathfrak{p}}}{\mathfrak{p}!\Gamma(\mathfrak{p}\kappa+\varsigma)}\right]\succeq$$

$$\operatorname{diag}\left[\sum_{\mathfrak{p}=0}^{\infty}\frac{\left(\frac{-|(\mathfrak{x})|}{\eta}\right)^{\mathfrak{p}}}{\mathfrak{p}!\Gamma(\mathfrak{p}\kappa+\varsigma)},\sum_{\mathfrak{p}=0}^{\infty}\frac{\left(\frac{-|(\mathfrak{x})|}{\eta}\right)^{\mathfrak{p}}}{\mathfrak{p}!\Gamma(\mathfrak{p}\kappa+\varsigma)},\sum_{\mathfrak{p}=0}^{\infty}\frac{\left(\frac{-|(\mathfrak{x})|}{\eta}\right)^{\mathfrak{p}}}{\mathfrak{p}!\Gamma(\mathfrak{p}\kappa,\varsigma)}\right]\circledast$$

$$\operatorname{diag}\left[\sum_{\mathfrak{p}=0}^{\infty}\frac{\left(\frac{-|\epsilon|}{\zeta}\right)^{\mathfrak{p}}}{\mathfrak{p}!\Gamma(\mathfrak{p}\kappa+\varsigma)},\sum_{\mathfrak{p}=0}^{\infty}\frac{\left(\frac{-|\epsilon|}{\zeta}\right)^{\mathfrak{p}}}{\mathfrak{p}!\Gamma(\mathfrak{p}\kappa+\varsigma)},\sum_{\mathfrak{p}=0}^{\infty}\frac{\left(\frac{-|(\epsilon)|}{\zeta}\right)^{\mathfrak{p}}}{\mathfrak{p}!\Gamma(\mathfrak{p}\kappa+\varsigma)}\right].$$

Consequently, if

$$\mathcal{A}(\epsilon,\eta)=$$
$$\operatorname{diag}\left[W_{\kappa,\varsigma}\left(-\frac{|\epsilon|}{\eta}\right),W_{\kappa,\varsigma}\left(-\frac{|\epsilon|}{\eta}\right),W_{\kappa,\varsigma}\left(-\frac{|\epsilon|}{\eta}\right)\right],$$

for $\eta\in\mathfrak{F}_2$, then $(\mathcal{J},\mathcal{A},\circledast_M)$ is a MVFN-space. From now on, we assume $\circledast=\circledast_M$.

1.4 Alternative Fixed Point Theorem

The following alternative fixed point theorem, was proved by Diaz-Margolis [22].

Theorem 1.2 *Let $\mathfrak{G}$ be a set with a complete $[0,\infty]$ valued metric d. For $m,\bar{m}\in\mathfrak{G}$ consider the self map $\mho$ on $\mathfrak{G}$ such that the following relation holds*

$$d(\mho m,\mho\bar{m})\leq\gamma\, d(\bar{m},m),$$

where $\varrho<1$ is a Lipschitz constant. Let $m\in\mathfrak{G}$. Therefore, one of the following two conditions holds

(i) $d\left(\mho^{\mathbf{e}}m,\mho^{\mathbf{e}+1}m\right)=\infty,\quad\forall\,\mathbf{e}\in\mathbb{N},$

or

(ii) *we can find $\mathbf{e}_0\in\mathbb{N}$ such that:* $d\left(\mho^{\mathbf{e}}m,\mho^{\mathbf{e}+1}m\right)<\infty,\quad\forall\,\mathbf{e}\geq\mathbf{e}_0;$

and if condition (ii) is true, then we always have

(1) *the fixed point $\bar{m}^*$ of $\mho$ is the convergence point of the sequence $\{\mho^{\mathbf{e}}m\}$;*
(2) *in the set, $C^*=\{\bar{m}\in\mathfrak{G}\mid d(\mho^{\mathbf{e}_0}m,\bar{m})<\infty\}$, $\bar{m}^*$ is the unique fixed point of $\mho$;*
(3) $(1-\gamma)d(\bar{m},\bar{m}^*)\leq d(\bar{m},\mho\bar{m})$ *for every $\bar{m}\in\mathfrak{G}$.*

References

1. Khalil, R., Al Horani, M., Yousef, A., Sababheh, M.: A new definition of fractional derivative. J. Comput. Appl. Math. **264**, 65–70 (2014)
2. Kilbas, A.A., Srivastava, H.M., Trujillo, J.J.: Theory and Applications of Fractional Differential Equations. North-Holland Mathematics Studies, vol. 204, pp. xvi+523. Elsevier Science B.V., Amsterdam (2006)
3. Kiryakova, V.: Some special functions related to fractional calculus and fractional (non-integer) order control systems and equations. Facta Univ. Ser. Autom. Control Robot. **7**(1), 79–98 (2008)
4. Mathai, A.M., Saxena, R.K.: The H-function with Applications in Statistics and other Disciplines, pp. xii+192. Halsted Press [John Wiley and Sons], New York, London, Sidney (1978). ISBN: 0-470-26380-6
5. Eidinejad, Z., Saadati, R.: Hyers-Ulam-Rassias-Wright stability for fractional oscillation equation. Discret. Dyn. Nat. Soc. **2022**, (2022)
6. Eidinejad, Z., Saadati, R., Mesiar, R.: Optimum approximation for ς-lie homomorphisms and Jordan ς-lie homomorphisms in ς-lie algebras by aggregation control functions. Mathematics **10**(10), 1704 (2022)
7. Grabisch, M., Marichal, J.L., Mesiar, R., Pap, E.: Aggregation Functions. Encyclopedia of Mathematics and its Applications, vol. 127, pp. xviii+460. Cambridge University Press, Cambridge (2009). ISBN: 978-0-521-51926-7
8. Kilbas, A.A.: Hadamard-type fractional calculus. J. Korean Math. Soc. **38**(6), 1191–1204 (2001)
9. Butzer, P.L., Kilbas, A.A., Trujillo, J.J.: Fractional calculus in the Mellin setting and Hadamard-type fractional integrals. J. Math. Anal. Appl. **269**(1), 1–27 (2002)
10. Li, C.: Several results of fractional derivatives in $\mathcal{D}'(R_+)$. Fract. Calc. Appl. Anal. **18**, 192–207 (2015)
11. Eidinejad, Z., Saadati, S., de la Sen, M.: Radu-Miheţ method for the existence, uniqueness, and approximation of the-Hilfer fractional equations by matrix-valued fuzzy controllers. Axioms **10**(2), 63 (2021)
12. Sousa, J.V.D.C., Capelas de Oliveira, E.: On the Ψ-Hilfer fractional derivative. Commun. Nonlinear Sci. Numer. Simul. **60**, 72–91 (2018)
13. Schweizer, B., Sklar, A.: North-Holland Series in Probability and Applied Mathematics. North-Holland Publishing Co., New York (1983)
14. Hadzic, O., Pap, E.: Fixed Point Theory in Probabilistic Metric Spaces. Mathematics and its Applications, vol. 536. Kluwer Academic Publishers, Dordrecht (2001)
15. Klement, E.P., Mesiar, R., Pap, E.: Triangular Norms. Trends in Logic–Studia Logica Library, vol. 8. Kluwer Academic Publishers, Dordrecht (2000)
16. Sousa, J.V.D.C., Fabio, G.R., de Oliveira, E.C.: Stability of the fractional Volterra integro-differential equation by means of Ψ-Hilfer operator
17. Arqub, O.A., Maayah, B.: Numerical solutions of integrodifferential equations of Fredholm operator type in the sense of the Atangana-Baleanu fractional operator. Chaos Solitons Fractals **117**, 117–124 (2018)
18. Zhang, J., Wang, G., Zhi, X., Zhou, C.: Generalized Euler-Lagrange equations for fuzzy fractional variational problems under gH-Atangana-Baleanu differentiability. J. Funct. Spaces **2018**, 2740678 (2018)
19. Abdeljawad, T., Baleanu, D.: Discrete fractional differences with nonsingular discrete Mittag-Leffler kernels. Adv. Differ. Eq. **2016**, 232 (2016)
20. Abdeljawad, T., Baleanu, D.: On fractional derivatives with exponential kernel and their discrete versions. Rep. Math. Phys. **80**, 11–27 (2017)
21. Liu, L., Dong, Q., Li, G.: Exact solutions and Hyers-Ulam stability for fractional oscillation equations with pure delay. Appl. Math. Lett. **112**, 106666, 7 (2021)
22. Diaz, J.B., Margolis, B.: A fixed point theorem of the alternative, for contractions on a generalized complete metric space. Bull. Am. Math. Soc. **74**, 305–309 (1968)

Chapter 2
Analysis of a New Stability

In this chapter, we consider the non-homogeneous fractional delay oscillation equation with order κ and investigate the existence of a unique solution in matrix-valued fuzzy Banach spaces for this equation using the alternative fixed point theorem. We introduce the Wright controller to investigate the Hyers–Ulam–Wright stability for the NH-FD-O equation with order κ. Using the fractional integral with respect to the Ψ function and the Ψ-Hilfer fractional derivative, we consider Volterra fractional equations. Considering the Gauss Hypergeometric function as a control function, we introduce the concept of the Hyers-Ulam-Rassias-Kummer stability of this fractional equation and study the existence, uniqueness, and approximation for two classes of fractional Volterra integro-differential and fractional Volterra integral. We apply the Cădariu-Radu method derived from the Diaz-Margolis alternative fixed point theorem. Also we investigate a conformable fractional differential equation (CFD-E) with a constant coefficient on a compact interval. Considering the Diaz-Margolis alternative fixed point theorem (D-MAFPT) and using the Cădariu-Radu method (CRM), we prove the existence of a unique solution and the Hyers-Ulam-Rassias-OS_1-multiple stability (H-U-R-OS_1-MS) for the CFD equation. We consider a nonlinear fractional integral differential equation (N-FIDE) with arbitrary order and we show that there is a unique solution for this type of equation according to Babenko's strategy and considering the multivariate-Mittag-Leffler function. All results are proved according to Banach's contractive principle (BCP) and Leray-Schauder's fixed point theorem (LS-FTP).

Z. Eidinejad et al., *Optimal Stability Theory and Approximate Solutions of Fractional Systems*, Studies in Systems, Decision and Control 606,
https://doi.org/10.1007/978-3-031-96704-7_2

2.1 Hyers-Ulam-Rassias-Wright Stability for Fractional Oscillation Equation

2.1.1 *Introduction of Non-homogeneous Fractional Delay Oscillation Equation and Other Preliminaries*

From past to present, fractional calculus has gained considerable popularity and importance due to its many applications in various and wide scientific and engineering fields (see [1, 4]). We consider the non-homogeneous fractional delay oscillation equation as follows

$$D_0^\sigma m(\varrho) = -\vartheta m(\varrho - \mu) + g(\varrho) + \pounds^{-1}(m(s))(\varrho), \quad \varrho \in B := [0, B], \tag{2.1.1}$$

$$m(\varrho) = \psi(\varrho), \quad m'(\varrho) = \psi'(\varrho), \qquad -\mu \leq \varrho \leq 0, \tag{2.1.2}$$

and the concepts and functions used in the equation are as follows

- $g : B \times \mathbb{R}^n \to \mathbb{R}^n$ is an integrable function, $m(\varrho) \in \mathbb{R}^n, \vartheta \in \mathbb{R}^{n\times n}$, $\psi \in C^2([-\mu, 0], \mathbb{R}^n)$ denotes constant matrix, $\mu > 0$ is a fixed time, $B = N\mu$ for a fixed $N \in \{1, 2, \ldots\}$.
- D_0^σ is Caputo fractional derivative with $\mu \in (1, 2)$ defined by

$$D_0^\sigma m(\varrho) = \int_0^\varrho \frac{(\varrho - \beth)^{1-\sigma}}{\Gamma(2-\sigma)} m''(\beth)\mathrm{d}\beth,$$

if the integral exists.

Exact solutions for the differential equation using the exponential function for $\sigma = 1$ and the matrix sine and matrix cosine, for $\sigma = 2$ are obtained in [5, 6]. In 2018, Li and Wang obtained the exact solution of Eq. (2.1.1) for $0 < \sigma < 1$ using the Mittag-Leffler type matrix function [8–10]. We use the alternative fixed point method and apply the Wright function as a control function, to investigate the Hyers-Ulam-Wright stability for Eq. (2.1.1).

For the homogeneous fractional oscillation system

$$D_0^\sigma m(\varrho) = -\vartheta m(\varrho - \mu), \quad \varrho \in B \tag{2.1.3}$$

We present the following Lemma from [7].

Lemma 2.1 *For $C_\sigma^\mu(\varrho)$ and $S_\sigma^\mu(\varrho)$, we find*

(i) $\frac{d}{d\varrho} S_\sigma^\mu(\varrho) = C_\sigma^\mu(\varrho)$ *for all* $\varrho \in \mathbb{R}$.
(ii) $C_\sigma^\mu(\varrho)$ *is a solution of* (2.1.3), $\varrho \in B$.
(iii) $S_\sigma^\mu(\varrho)$ *is a solution of* (2.1.3), $\varrho \in B$.

Theorem 2.1 *If the function m is a solution of Eq.* (2.1.1) *with initial conditions* (2.1.2) *then the function m is defined as follows*

$$m(\varrho) = (s^{\sigma} I + \vartheta e^{-s\mu})\pounds(m(\varrho))(s) - s^{\sigma-1}\psi(0) \\ - s^{\sigma-2}\psi'(0) + \vartheta e^{-s\mu}\int_{-\mu}^{0} e^{-s\mu}\psi(\varrho)\,\mathrm{d}\varrho - \pounds(g(\varrho))(s). \tag{2.1.4}$$

Proof Applying the Laplace transform to both sides of (2.1.1) and according to Laplace's definitions, we derive

$$s^{\sigma}\mathcal{L}(m(\varrho))(s) - s^{\sigma-1}m(0) - s^{\sigma-2}m'(0) = \\ - \vartheta e^{-s\mu}\left(\int_{-\mu}^{0} e^{-sz}y(z)dz + \mathcal{L}(m(\varrho))(s)\right) + \mathcal{L}(g(\varrho))(s) + m(\varrho).$$

We solve this equation for $\mathcal{L}(m(\varrho))(s)$, then we have

$$\left(s^{\sigma} I + \vartheta e^{-s\mu}\right)\mathcal{L}(m(\varrho))(s) = \\ s^{\sigma-1}\psi(0) + s^{\sigma-2}\psi'(0) - \vartheta e^{-s\mu}\int_{-\mu}^{0} e^{-sz}\psi(z)dz + \mathcal{L}(g(\varrho))(s) + m(\varrho),$$

then

$$m(\varrho) = \left(s^{\sigma} I + \vartheta e^{-s\mu}\right)\mathcal{L}(m(\varrho))(s) - s^{\sigma-1}\psi(0) - s^{\sigma-2}\psi'(0) \\ + \vartheta e^{-s\mu}\int_{-\mu}^{0} e^{-sz}\psi(z)dz - \mathcal{L}(g(\varrho))(s).$$

We use the norm $\|s\| = \sum_{i=1}^{\ell}|s_i|\,, s \in \mathbb{R}^{n\times 1}$ and the matrix norm $\|\vartheta\| = \max_{1\le j\le n}\sum_{i=1}^{\ell}\left|\vartheta_{ij}\right|, a \in \mathbb{R}^{n\times n}$.

In the sequel, we provide a definition of Hyers-Ulam-Wright stability.

Definition 2.1 (See [16]). We say that Eq. (2.1.1) is Hyers-Ulam-Wright stable on B with respect to $W_{\sigma,\kappa}(\varrho)$, if there exists $E > 0$ such that for each $\beth > 0$ and for each solution $\bar{m}(\varrho)$ satisfying inequality

$$\left\| D_0^{\sigma}\bar{m}(\varrho) + \vartheta\bar{m}(\varrho - \mu) - g(\varrho) - \pounds^{-1}(\bar{m}(s))(\varrho)\right\| \le \beth W_{\sigma,\kappa}(\varrho), \quad \varrho \in B,$$

there exists a solution $m(\varrho)$ of the problem (2.1.1) such that

$$\|\bar{m}(\varrho) - m(\varrho)\| \le E\beth W_{\sigma,\kappa}(\varrho),$$

for every $\varrho \in B$.

2.1.2 Hyers-Ulam-Wright Stability Property

Theorem 2.2 *Consider the positive constant* $\varkappa$*, such that* $0 < \varkappa < 1$*. Consider the control function* $W_{\sigma,\kappa} : B \longrightarrow \mathbb{R}$*, such that*

$$£(W_{\sigma,\kappa}(s))(\varrho) \leq \varkappa\, W_{\sigma,\kappa}(\varrho), \tag{2.1.5}$$

and

$$\|(s^{\sigma} I + \vartheta e^{-s\mu})\| \leq 1. \tag{2.1.6}$$

If for each $\varrho \in B$*,* $\bar{m} \in C(B, \mathbb{R}^n)$ *is a differentiable function satisfying*

$$\left\| D_0^{\sigma} \bar{m}(\varrho) + \vartheta \bar{m}(\varrho - \mu) - g(\varrho) - £^{-1}(\bar{m}(s))(\varrho) \right\| \leq \beth W_{\sigma,\kappa}(\varrho), \quad \varrho \in B, \tag{2.1.7}$$

then, there is a unique solution $m \in C(B, \mathbb{R}^n)$ *to Eq.* (2.1.1)*, such that*

$$\|\bar{m}(\varrho) - m(\varrho)\| \leq \daleth \beth\, W_{\sigma,\kappa}(\varrho)$$

wherein $\daleth = \frac{\varkappa}{1-\varkappa}$*. Thus, Eq.* (2.1.1)*, has the Hyers-Ulam-Wright stability property.*

Proof To begin the proof, we define the set C as follows

$$C = \{m : B \to \mathbb{R}^n, m \text{ is continuous}\},$$

and consider the mapping $d : C \times C \to [0, \infty]$, which is defined as follows

$$d(m, \bar{m}) = \inf\{A \in [0, \infty] : \|m(\varrho) - \bar{m}(\varrho)\| \leq A\, W_{\sigma,\kappa}(\varrho),\ \varrho \in B\}.$$

Now, we divide the proofs into 3 steps.
Step 1. We show that (C, d) is a $[0, \infty]$-valued complete metric space.

For functions $m, \bar{m} \in C$, we consider $d(m, \bar{m}) > d(m, v) + d(v, \bar{m})$, therefore, there exists $\varrho_0 \in B$ such that $\|m(\varrho_0) - \bar{m}(\varrho_0)\| > d(m, v) + d(v, \bar{m})\ W_{\sigma,\kappa}(\varrho_0)$. Hence, according to the definition d we have

$$\|m(\varrho_0) - \bar{m}(\varrho_0)\| > \|m(\varrho_0) - v(\varrho_0)\| + \|v(\varrho_0) - \bar{m}(\varrho_0)\|,$$

which is a contradiction.

Next, to prove that (C, d) is a $[0, \infty]$-valued complete. Consider a Cauchy sequence and we show that this sequence is uniformly convergent. Now, let $\{m_n\}$ be a Cauchy sequence in (C, d). by using d, $\forall\ \beth > 0$, $\exists\ N_{\beth} \in \mathbb{N}$, $\forall\ m, n \geq N_{\varepsilon}$, $\forall \varrho \in B$ and we have

$$\|m_n(\varrho) - m_m(\varrho)\| < \beth\, W_{\sigma,\kappa}(\varrho). \tag{2.1.8}$$

Since the function $W_{\sigma,\kappa}$ is continuous in the compact set B, due to the inequality (2.1.8), the sequence $\{m_n\}$ uniformly converges to a continuous function of C such as m. Then again, using the definition of d, the following relation is established for us, wherein for $\beth > 0$, $\exists\, N_\beth \in \mathbb{N}$, $n \geq N_\beth$, $\varrho \in C$,

$$\|m_n(\varrho) - m(\varrho)\| < \beth\, W_{\sigma,\kappa}(\varrho).$$

It should be noted that we let $m \to \infty$. As a result, $d(m_n, m) \leq \beth$, and that means (C, d) is $[0, \infty]$-valued complete and the first step is complete.

Now, according to the main equation, we define the mapping $\mho : C \to C$, as follows

$$\mho(m(\varrho)) = \pounds(D_0^\sigma m(\varrho) + \vartheta m(\varrho - \mu) - g(\varrho))(s) = (s^\sigma I + \vartheta e^{-s\mu})\pounds(m(\varrho))(s) - s^{\sigma-1}\psi(0)$$
$$- s^{\sigma-2}\psi'(0) + \vartheta e^{-s\mu} \int_{-\mu}^{0} e^{-s\mu}\psi(\varrho)\, \mathrm{d}\varrho - \pounds(g(\varrho))(s).$$

Step 2. We show that $\mho$ is a contraction. If the functions $m, \bar{m} \in C$ and the coefficient $A_{m\bar{m}} \in [0, \infty]$, we assume that $d(m, \bar{m}) \leq A_{m\bar{m}}$, then the following inequality for $\varrho \in B$ holds

$$\|m(\varrho) - \bar{m}(\varrho)\| < A_{m\bar{m}}\, W_{\sigma,\kappa}(\varrho)$$

and using (2.1.5) and (2.1.6), we have

$$\begin{aligned}
\|\mho m(\varrho) - \mho\bar{m}(\varrho)\| = \Big\| &(s^\sigma I + \vartheta e^{-s\mu})\pounds(m(\varrho))(s) - s^{\sigma-1}\psi(0) \\
&- s^{\sigma-2}\psi'(0) + \vartheta e^{-s\mu} \int_{-\mu}^{0} e^{-s\mu}\psi(\varrho)\, \mathrm{d}\varrho - \pounds(g(\varrho))(s) \\
&- (s^\sigma I + \vartheta e^{-s\mu})\pounds(\bar{m}(\varrho))(s) + s^{\sigma-1}\psi(0) \\
&+ s^{\sigma-2}\psi'(0)_\vartheta e^{-s\mu} \int_{-\mu}^{0} e^{-s\mu}\psi(\varrho)\, \mathrm{d}\varrho + \pounds(g(\varrho))(s)\Big\| \\
= \Big\| &(s^\sigma I + \vartheta e^{-s\mu})\pounds(m(\varrho))(s) - (s^\sigma I + \vartheta e^{-s\mu})\pounds(\bar{m}(\varrho))(s)\Big\| \\
= \Big\| &(s^\sigma I + \vartheta e^{-s\mu})\pounds(m(\varrho) - \bar{m}(\varrho))(s)\Big\| \\
\leq \Big\| &(s^\sigma I + \vartheta e^{-s\mu})\pounds(\|m(\varrho) - \bar{m}(\varrho))\|(s)\Big\|
\end{aligned}$$

$$\leq \left\| (s^{\sigma} I + \vartheta e^{-s\mu}) \pounds (A_{m\bar{m}} W_{\sigma,\kappa}(s))(\varrho) \right\|$$
$$\leq A_{m\bar{m}} \left\| (s^{\sigma} I + \vartheta e^{-s\mu}) \pounds (W_{\sigma,\kappa}(s))(\varrho) \right\|$$
$$\leq A_{m\bar{m}} \left\| (s^{\sigma} I + \vartheta e^{-s\mu}) \varkappa W_{\sigma,\kappa}(\varrho) \right\|$$
$$= A_{m\bar{m}} \varkappa W_{\sigma,\kappa}(\varrho).$$

Thus, we have

$$d(\mho m, \mho \bar{m}) \leq \varkappa d(m, \bar{m}). \tag{2.1.9}$$

Since, $0 < \varkappa < 1$, therefore $\mho$ is a contraction.
Step 3. We will show that $d(\mho \bar{m}, \bar{m}) < \infty$. From inequality (2.1.7), we have

$$\begin{aligned}\|\mho(\bar{m}(\varrho)) - \bar{m}(\varrho)\| &= \left\| \pounds(D_0^{\sigma} \bar{m}(\varrho) + \vartheta \bar{m}(\varrho - \mu) - g(\varrho) - \pounds^{-1}(\bar{m}(s))(\varrho) \right\| \\ &= \Big\| (s^{\sigma} I + \vartheta e^{-s\mu}) \pounds(\bar{m}(\varrho))(s) - s^{\sigma-1} \psi(0) \\ &\quad - s^{\sigma-2} \psi'(0) + \vartheta e^{-s\mu} \int_{-\mu}^{0} e^{-s\mu} \psi(\varrho)\, d\varrho - \pounds(g(\varrho))(s) - \bar{m}(\varrho) \Big\| \\ &\leq \beth \pounds(W_{\sigma,\kappa}(s))(\varrho) \\ &\leq \beth \varkappa W_{\sigma,\kappa}(\varrho).\end{aligned}$$

Consequently,

$$d(\mho \bar{m}, \bar{m}) \leq \beth \varkappa < \infty, \quad \varkappa < 1. \tag{2.1.10}$$

Therefore, all the conditions of Theorem 1.2 are true. Then, we have

(1) The sequence $\{\mho^{\mathfrak{e}} m\}$ converges to a fixed point, say m.
(2) The unique element m is in the set $C^* = \{\bar{m} \in C : d(\mho \bar{m}, m) < \infty\}$ and is the unique fixed point $\mho$, it means $\mho m = m$ or equivalently

$$\begin{aligned} m(\varrho) &= (s^{\sigma} I + \vartheta e^{-s\mu}) \pounds(m(\varrho))(s) - s^{\sigma-1} \psi(0) \\ &\quad - s^{\sigma-2} \psi'(0) + \vartheta e^{-s\mu} \int_{-\mu}^{0} e^{-s\mu} \psi(\varrho)\, d\varrho - \pounds(g(\varrho))(s). \end{aligned} \tag{2.1.11}$$

Since m is a continuous function and according to the Theorem 2.1, we have

$$D_0^{\sigma} m(\varrho) = -\vartheta m(\varrho - \mu) + g(\varrho) + \pounds^{-1} m(\varrho)(s), \quad \varrho \in B := [0, B]. \tag{2.1.12}$$

(3) According to the inequality (2.1.10), we get

$$d(\bar{m}, m) \leq \frac{1}{1-\varkappa} d(\mho \bar{m}, \bar{m}) \leq \beth \frac{\varkappa}{1-\varkappa} W_{\sigma,\kappa}(\varrho).$$

Thus, Eq. (2.1.1) has Hyers-Ulam-Wright stability property.

Now, we show the uniqueness of the obtained point. For convenience, we let

$$d = \frac{\beth \varkappa}{1-\varkappa}.$$

and let t be another differentiable function satisfying the Eq. (2.1.12) and this means that the following equation holds

$$D_0^\sigma t(\varrho) = -\vartheta t(\varrho - \mu) + g(\varrho) + \pounds^{-1}(t(s))(\varrho), \quad \varrho \in B := [0, B]. \tag{2.1.13}$$

Now, we show that t is a fixed point of $\mho$ and $t \in C^*$. Using Eq. (2.1.13), we have $\mho t = t$. In the sequel, we show that $d(\mho \bar{m}, t) < \infty$. Let for $\bar{m} \in C$, $d(\bar{m}, t) < d$. Using Eq. (2.1.13), we have

$$\begin{aligned}
\|\mho(\bar{m}(\varrho)) - t(\varrho)\| &= \left\| \pounds(D_0^\sigma \bar{m}(\varrho) + \vartheta \bar{m}(\varrho - \mu) - g(\varrho))(s) - t(\varrho) \right\| \\
&= \Big\| (s^\sigma I + \vartheta e^{-s\mu})\pounds(\bar{m}(\varrho))(s) - s^{\sigma-1}\psi(0) \\
&\quad - s^{\sigma-2}\psi'(0) + \vartheta e^{-s\mu} \int_{-\mu}^{0} e^{-s\mu}\psi(\varrho)\, d\varrho - \pounds(g(\varrho))(s) - (s^\sigma I + \vartheta e^{-s\mu})\pounds(t(\varrho))(s) + s^{\sigma-1}\psi(0) \\
&\quad + s^{\sigma-2}\psi'(0)_\vartheta e^{-s\mu} \int_{-\mu}^{0} e^{-s\mu}\psi(\varrho)\, d\varrho + \pounds(g(\varrho))(s) \Big\| \\
&= \left\| (s^\sigma I + \vartheta e^{-s\mu})\pounds(\bar{m}(\varrho))(s) - (s^\sigma I + \vartheta e^{-s\mu})\pounds(t(\varrho))(s) \right\| \\
&= \left\| (s^\sigma I + \vartheta e^{-s\mu})\pounds(\bar{m}(\varrho) - t(\varrho))(s) \right\| \\
&\leq \left\| (s^\sigma I + \vartheta e^{-s\mu})\pounds(\|\bar{m}(\varrho) - t(\varrho)\|)(s) \right\| \\
&\leq \left\| (s^\sigma I + \vartheta e^{-s\mu})\pounds(d(W_{\sigma,\kappa}(s))(\varrho) \right\| \\
&\leq d \left\| (s^\sigma I + \vartheta e^{-s\mu})\varkappa W_{\sigma,\kappa}(\varrho) \right\| \\
&\leq d\varkappa W_{\sigma,\kappa}(\varrho)
\end{aligned}$$

that means

$$d(\mho \bar{m}, t) \leq \varkappa d < \infty.$$

2.1.3 Application

Example 2.1 Consider the following the non-homogeneous fractional delay oscillation equation

$$D_0^{\frac{1}{2}} m(\varrho) = -\vartheta m(\varrho - \frac{1}{7}) + \frac{(\varrho - 1)^{-\frac{1}{4}} \sin(\varrho - 1)}{64(1 + \sqrt{\varrho - 1})(1 + |\varrho|)} + £^{-1}(m(s))(\varrho), \tag{2.1.14}$$

$$m(\varrho) = [m_1(\varrho), m_2(\sigma)]^T \quad \psi(\varrho) = [3\varrho, 4\varrho^2]^T, \qquad -\mu \le \varrho \le 0 \tag{2.1.15}$$

where $\varkappa = \frac{1}{4}$, $\sigma = \frac{1}{2}$, $\mu = \frac{1}{7}$, $B = \frac{3}{5}$ and $\vartheta = \begin{bmatrix} \frac{3}{5} & 0 \\ 0 & \frac{2}{5} \end{bmatrix}$. Assume that the following condition is true for the given continuous function

$$£(W_{\frac{1}{2},\kappa}(s))(\varrho) \le \frac{1}{4} W_{\frac{1}{2},\kappa}(\varrho).$$

Let for $\varrho \in B$, $\bar{m} \in C(B, \mathbb{R}^n)$ be a continuous function satisfying

$$\left\| D_0^{\frac{1}{2}} \bar{m}(\varrho) + \vartheta \bar{m}(\varrho - \frac{1}{7}) - \frac{(\varrho - 1)^{-\frac{1}{4}} \sin(\varrho - 1)}{64(1 + \sqrt{\varrho - 1})(1 + |\varrho|)} - £^{-1}(m(s))(\varrho) \right\| \le \beth W_{\frac{1}{2},\kappa}(\varrho), \quad \varrho \in B, \tag{2.1.16}$$

therefore, according to the definitions given, $\bar{m}$ holds in the following inequality

$$\begin{aligned} &\Big\| \bar{m}(\varrho) - (s^{\sigma} I + \vartheta e^{-s\mu}) £(\bar{m}(\varrho))(s) - s^{\sigma-1} \psi(0) \\ &- s^{\sigma-2} \psi'(0) + \vartheta e^{-s\mu} \int_{-\mu}^{0} e^{-s\mu} \psi(\varrho)\, d\varrho - £(\frac{(\varrho - 1)^{-\frac{1}{4}} \sin(\varrho - 1)}{64(1 + \sqrt{\varrho - 1})(1 + |\varrho|)})(s) \Big\| \\ &\le \beth £(W_{\frac{1}{2},\kappa}(s))(\varrho) \\ &\le \frac{1}{4} \beth W_{\frac{1}{2},\kappa}(\varrho) \end{aligned}$$

where $\psi(\varrho) = [3\varrho, 4\varrho^2]^T$. Then, there is a unique solution $m \in C(B, \mathbb{R}^n)$ for Eq. (2.1.1), such that

(1) m is the fixed point of the operator $\mho$, which is defined as follows

$$\begin{aligned} \mho(m(\varrho)) &= (s^{\sigma} I + \vartheta e^{-s\mu}) £(m(\varrho))(s) - s^{\sigma-1} \psi(0) \\ &- s^{\sigma-2} \psi'(0) + \vartheta e^{-s\mu} \int_{-\mu}^{0} e^{-s\mu} \psi(\varrho)\, d\varrho - £(\frac{(\varrho - 1)^{-\frac{1}{4}} \sin(\varrho - 1)}{64(1 + \sqrt{\varrho - 1})(1 + |\varrho|)})(s). \end{aligned}$$

(2) The fixed point m is unique.
(3) m satisfies in Eq. (2.1.14).

Then

$$d(m, \bar{m}) \leq \daleth \beth,$$

or

$$\|\bar{m}(\varrho) - m(\varrho)\| \leq \daleth \beth \, W_{\frac{1}{2},\kappa}(\varrho)$$

where in $\daleth = \frac{1}{3}$.

2.2 Hyers-Ulam-Rassias-Kummer Stability

2.2.1 *Introduction of the Fractional Volterra Equations*

In general, many authors have proposed and proved the existence, uniqueness, and Ulam-Hyers stability of the solution of fractional differential equations using several methods. For example, in 2017, Benchohra and Lazreg investigated the existence of a unique solution and the stability of the following fractional differential equations

$$\begin{cases} {}^{H}\mathfrak{D}^{\tau}\phi(\ell) = \omega(\ell, \phi(\ell), {}^{H}\mathfrak{D}^{\tau}\phi(\ell)), \\ \phi(1) = \phi_1, \end{cases}$$

where ${}^{H}\mathfrak{D}^{\tau}\phi(\ell)$ is the Hadamard fractional derivative of order $0 < \tau \leq 1, \quad \omega : J \times \mathbb{R} \times \mathbb{R} \to \mathbb{R}$ is a given function, $\phi_1 \in \mathbb{R}$ and $t \in J = [1, T], \quad T > 0$. Also, Sousa and Oliveira considered the following fractional Volterra integro-differential equation,

$$\begin{cases} {}^{H}\mathfrak{D}^{\tau,\kappa,\Psi}\mathrm{g}(\ell) = \omega(\ell, \mathrm{g}(\ell)) + \int_0^{\ell} \mathrm{M}(\ell, \jmath, \mathrm{g}(\ell))d\jmath, \\ I_{0_+}^{1-\gamma}\mathrm{g}(0) = \zeta, \end{cases}$$

which is defined using the Hilfer fractional derivative, and invastigate the stability of Hyers-Ulam and Hyers-Ulam Rassias for this equation. In this equation, $\ell \in J = [0, T]$, where $\omega(\ell, u)$ is a continuous function with respect to the variables ℓ and g on $J \times \mathbb{R}$ and $\mathrm{M}(\ell, \jmath, \mathrm{g}(\ell))$ is continuous with respect to $\ell, \ \jmath, \ \mathrm{g}$ on $J \times \mathbb{R} \times \mathbb{R}$ and $I_{0_+}^{1-\gamma}\mathrm{g}(0)$ with $0 < \gamma \leq 1$ is Ψ-Riemann-Liouville fractional integral.

Utilizing the continuous functions $\phi : [0, L] \to Y$, $\mathrm{M} : [0, L] \times [0, L] \times Y \to Y$, $\omega : [0, L] \times Y \to Y$, that Y is a Banach's space and also Ψ-Hilfer fractional derivative ${}^{H}\mathfrak{D}_{a^+}^{\tau,\kappa;\Psi}\phi(\ell)$ that $0 < \tau < 1, 0 \leq \kappa \leq 1$ and $\Delta_{\Psi}^{\tau}(\ell, \jmath) := \Psi'(\jmath)(\Psi(\ell) - \Psi(\jmath))^{\tau-1}$ that the function $\Psi(\ell)$ is an increasing and positive function with a continuous derivative, we define two equations of the fractional Volterra Integro-differential and the fractional Volterra integral as follows

$$^{H}\mathcal{D}_{0^+}^{\tau,\kappa;\Psi}\phi(\ell) = \omega(\ell,\phi(\ell)) + \int_0^{\ell} \mathrm{M}(\ell,\jmath,\phi(\jmath))d\jmath, \tag{2.2.1}$$

and

$$\phi(\ell) = \omega(\ell,\phi(\ell)) + \frac{1}{\Gamma(\tau)}\int_0^{\ell} \Delta_{\Psi}^{\tau}(\ell,\jmath)\mathrm{M}(\ell,\jmath,\phi(\jmath))d\jmath. \tag{2.2.2}$$

We consider the Gauss Hypergeometric function as a control function and stabilize the fractional equations with this control function, which we call the Kummer control function and apply the Cădariu-Radu method derived from the Diaz-Margolis alternative fixed point theorem to investigate the existence, uniqueness of solutions for fractional equations (2.2.1) and (2.2.2).

Now, we present the concept of the Hyers-Ulam-Rassias-Kummer stability of Eqs. (2.2.1) and (2.2.2).

2.2.2 *Preliminaries*

Definition 2.2 ([25–27]) Suppose that ${}_2F_1(\rho,\sigma;\varsigma;\ell)$ is the Kummer function in which $\ell \in [0, L]$. We say that Eq. (2.2.1) has the Hyers-Ulam-Rassias-Kummer stability property if the following inequality holds for the differentiable function $\phi(\ell)$

$$|{}^{H}\mathcal{D}_{0^+}^{\tau,\kappa;\Psi}\phi(\ell) - \omega(\ell,\phi(\ell)) - \int_0^{\ell} \mathrm{M}(\ell,\jmath,\phi(\jmath))d\jmath| \le {}_2F_1(\rho,\sigma;\varsigma;\ell), \tag{2.2.3}$$

then, there exists a solution $\vartheta(\ell)$ of Eq. (2.2.1) such that for some $P > 0$,

$$|{}^{H}\mathcal{D}_{0^+}^{\tau,\kappa;\Psi}\phi(\ell) - {}^{H}\mathcal{D}_{0^+}^{\tau,\kappa;\Psi}\vartheta(\ell)| + |\phi(\ell) - \vartheta(\ell)| \le P\,{}_2F_1(\rho,\sigma;\varsigma;\ell).$$

Definition 2.3 ([25–27]) Suppose that ${}_2F_1(\rho,\sigma;\varsigma;\ell)$ is the Kummer function in which $\ell \in [0, L]$. If the following inequality holds for the differentiable function $\phi(\ell)$

$$|\phi(\ell) - \omega(\ell,\phi(\ell)) - \frac{1}{\Gamma(\tau)}\int_0^{\ell} \Delta_{\Psi}^{\tau}(\ell,\jmath)\mathrm{M}(\ell,\jmath,\phi(\jmath))d\jmath| \le {}_2F_1(\rho,\sigma;\varsigma;\ell), \tag{2.2.4}$$

then, there exists a solution $\vartheta(\ell)$ of Eq. (2.2.2) such that for some $P > 0$

$$|\phi(\ell) - \vartheta(\ell)| \leq P\ {}_2F_1(\rho, \sigma; \varsigma; \ell),$$

then we say that Eq. (2.2.2) is Hyers-Ulam-Rassias-Kummer stable [28–30].

2.2.3 Fractional Volterra Integro-Differential Equations

Here, we have a class of fractional Volterra integrodifferential equations [31–33]. We stabilize the mentioned equation by a Kummer function and apply the Cădariu-Radu Method to guarantee the existence of a unique solution of Eq. (2.2.1). Our results improve and generalize the results of [34], see also [20, 35–43].

Theorem 2.3 *Consider the positive constants* E, E_1 *and* E_2 *such that* $0 < E_1 + (E_1 + E_2 + E_2 E)E < 1$. *Also consider the Banach space* Y, *the continuous functions* $\omega : [0, L] \times Y \to Y$, $\mathrm{M} : [0, L] \times [0, L] \times Y \to Y$ *and* ${}_2F_1 : \mathbb{R}^3 \times [0, L] \to (0, \infty)$ *for all* $\jmath, \ell \in [0, L]$ *and* $\phi, \vartheta \in Y$ *and assume that the following conditions hold*

(1) $|\omega(\ell, \phi) - \omega(\ell, \vartheta)| \leq E_1 |\phi - \vartheta|$,
(2) $|\mathrm{M}(\ell, \jmath, \phi(\jmath)) - \mathrm{M}(\ell, \jmath, \vartheta(\jmath))| \leq E_1 |\phi - \vartheta|$,
(3) $\frac{1}{\Gamma(\tau)} \int_0^\ell \Delta_\Psi^\tau(\ell, \jmath)\ {}_2F_1(\rho, \sigma; \varsigma; \jmath) d\jmath \leq E\ {}_2F_1(\rho, \sigma; \varsigma; \ell)$.

If $\theta : [0, L] \to Y$ *be a differentiable function satisfying* (2.2.3), *then Eq.* (2.2.1) *has a unique solution* $\theta_0 : [0, L] \to Y$. *Therefore, we have*

$$|{}^H\mathcal{D}_{0^+}^{\tau,\kappa;\Psi} \theta(\ell) - {}^H\mathcal{D}_{0^+}^{\tau,\kappa;\Psi} \theta_0(\ell)| + |\theta(\ell) - \theta_0(\ell)| \leq \frac{1 + E}{1 - E_1 + (E_1 + E_2 + E_2 E)E}\ {}_2F_1(\rho, \sigma; \varsigma; \ell),$$

and this means that Eq. (2.2.1), *has Hyers-Ulam-Rassias-Kummer stability property.*

Proof At the first, we define the set

$$\mathrm{W} = \{\phi : [0, L] \to Y, \phi \text{ is differentiable}\}$$

and on this set, consider the mapping $\delta : \mathrm{W} \times \mathrm{W} \to [0, \infty]$ as follows

$$\delta(\phi, \vartheta) = \inf\{P \in [0, \infty] : |{}^H\mathcal{D}_{0^+}^{\tau,\kappa;\Psi} \phi(\ell) - {}^H\mathcal{D}_{0^+}^{\tau,\kappa;\Psi} \vartheta(\ell)| + |\phi(\ell) - \vartheta(\ell)| \leq P\ {}_2F_1(\rho, \sigma; \varsigma; \ell), \ell \in [0, L]\}. \tag{2.2.5}$$

We show that (W, δ) is $[0, \infty]$-valued complete metric space. Therefore, for functions $\phi, \vartheta \in \mathrm{W}$ if we have $\delta(\phi, \vartheta) > \delta(\phi, \nu) + \delta(\nu, \vartheta)$, then there exists $\ell_0 \in [0, L]$ such that

$$|{}^H\mathcal{D}_{0^+}^{\tau,\kappa;\Psi}\phi(\ell_0) - {}^H\mathcal{D}_{0^+}^{\tau,\kappa;\Psi}\vartheta(\ell_0)| + |\phi(\ell_0) - \vartheta(\ell_0)| \\ > (\delta(\phi,\nu) + \delta(\nu,\vartheta))\,{}_2F_1(\rho,\sigma;\varsigma;\ell_0).$$

according to the definition 2.2.5, we have

$$\begin{aligned}&|{}^H\mathcal{D}_{0^+}^{\tau,\kappa;\Psi}\phi(\ell_0) - {}^H\mathcal{D}_{0^+}^{\tau,\kappa;\Psi}\vartheta(\ell_0)| + |\phi(\ell_0) - \vartheta(\ell_0)|\\ &> |{}^H\mathcal{D}_{0^+}^{\tau,\kappa;\Psi}\phi(\ell_0) - {}^H\mathcal{D}_{0^+}^{\tau,\kappa;\Psi}\nu(\ell_0)| + |\phi(\ell_0) - \nu(\ell_0)|\\ &\quad + |{}^H\mathcal{D}_{0^+}^{\tau,\kappa;\Psi}\nu(\ell_0) - {}^H\mathcal{D}_{0^+}^{\tau,\kappa;\Psi}\vartheta(\ell_0)| + |\nu(\ell_0) - \vartheta(\ell_0)|,\end{aligned}$$

which is a contradiction.

In (W, δ), we consider a Cauchy sequence $\{\phi_n\}$. Then for every $\varepsilon > 0$, we can find $N_\varepsilon \in \mathbb{N}$, such that for all $m, n \geq N_\varepsilon$ and $\ell \in [0, L]$, we have

$$|{}^H\mathcal{D}_{0^+}^{\tau,\kappa;\Psi}\phi_n(\ell) - {}^H\mathcal{D}_{0^+}^{\tau,\kappa;\Psi}\phi_m(\ell)| + |\phi_n(\ell) - \phi_m(\ell)| < \varepsilon\,{}_2F_1(\rho,\sigma;\varsigma;\ell). \quad (2.2.6)$$

Due to the continuity of the function ${}_2F_1$, on the compact interval $[0, L]$, the sequence $\{\phi_n\}$ is uniformly convergent to differentiable function $\phi \in$ W and the sequence $\{{}^H\mathcal{D}_{0^+}^{\tau,\kappa;\Psi}\phi_n\}$ also is uniformly convergent to $\{{}^H\mathcal{D}_{0^+}^{\tau,\kappa;\Psi}\phi\}$. Therefore when $m \to \infty$ for $\ell \in [0, L]$ and for $\varepsilon > 0$, we can find $N_\varepsilon \in \mathbb{N}$, such that if $n \geq N_\varepsilon$ we have

$$|{}^H\mathcal{D}_{0^+}^{\tau,\kappa;\Psi}\phi_n(\ell) - {}^H\mathcal{D}_{0^+}^{\tau,\kappa;\Psi}\phi(\ell)| + |\phi_n(\ell) - \phi(\ell)| < \varepsilon\,{}_2F_1(\rho,\sigma;\varsigma;\ell).$$

Thus $\delta(\phi_n, \phi) \leq \varepsilon$. Therefore, we have proved that the space (W, δ) is a $[0, \infty]$-valued complete metric space.

Now, we define the operator $\Omega :$ W $\to$ W as follows

$$\Omega(\phi(\ell)) = \mathcal{I}_{0^+}^{\tau;\Psi}\omega(\ell,\phi(\ell)) + \mathcal{I}_{0^+}^{\tau;\Psi}\left[\int_0^\ell \mathrm{M}(\ell, \jmath, \phi(\jmath))d\jmath\right], \quad (2.2.7)$$

and show that Ω is a contraction operator. For this purpose, suppose that $\phi, \wp \in$ W, $P_{\phi\wp} \in [0, \infty]$ and $\delta(\phi, \vartheta) \leq P_{\phi\vartheta}$. Then

$$|{}^H\mathcal{D}_{0^+}^{\tau,\kappa;\Psi}\phi(\ell) - {}^H\mathcal{D}_{0^+}^{\tau,\kappa;\Psi}\vartheta(\ell)| + |\phi(\ell) - \vartheta(\ell)| < P_{\phi\vartheta}\,{}_2F_1(\rho,\sigma;\varsigma;\ell),$$

for all $\ell \in [0, L]$. Using conditions (1) and (2) in the hypothesis, we have

$$\begin{aligned}&|{}^H\mathcal{D}_{0^+}^{\tau,\kappa;\Psi}(\Omega\phi(\ell) - \Omega\vartheta(\ell))| + |\Omega\phi(\ell) - \Omega\vartheta(\ell)|\\ &\leq |\omega(\ell,\phi(\ell)) - \omega(\ell,\vartheta(\ell))| + \int_0^\ell |\mathrm{M}(\ell,\jmath,\phi(\jmath)) - \mathrm{M}(\ell,\jmath,\vartheta(\jmath))|d\jmath\\ &\quad + |\mathcal{I}_{0^+}^{\tau;\Psi}(\omega(\ell,\phi(\ell)) - \omega(\ell,\vartheta(\ell)))| + |\mathcal{I}_{0^+}^{\tau;\Psi}\left[\int_0^\ell (\mathrm{M}(\ell,\jmath,\phi(\jmath)) - \mathrm{M}(\ell,\jmath,\vartheta(\jmath)))\right]d\jmath|\end{aligned}$$

$$\leq E_1|\phi(\ell)-\vartheta(\ell)|+E_2\int_0^\ell|\phi(\jmath)-\vartheta(\jmath)|d\jmath$$
$$+\mathcal{I}_{0^+}^{\tau;\Psi}(E_1|\phi(\ell)-\vartheta(\ell)|)+\mathcal{I}_{0^+}^{\tau;\Psi}\left[\int_0^\ell E_2|\phi(\ell)-\vartheta(\ell)|d\jmath\right]$$
$$\leq E_1P_{\phi\vartheta}\,{}_2F_1(\rho,\sigma;\varsigma;\ell)+E_2P_{\phi\vartheta}\int_0^\ell {}_2F_1(\rho,\sigma;\varsigma;\jmath)d\jmath+E_1P_{\phi\vartheta}\mathcal{I}_{0^+}^{\tau;\Psi}\,{}_2F_1(\rho,\sigma;\varsigma;\ell)$$
$$+E_2\mathcal{I}_{0^+}^{\tau;\Psi}\left[P_{\phi\vartheta}\int_0^\ell {}_2F_1(\rho,\sigma;\varsigma;\jmath)d\jmath\right]$$
$$\leq[E_1+(E_2+E_1+E_2E)E]P_{\phi\vartheta}\,{}_2F_1(\rho,\sigma;\varsigma;\ell).$$

Or equivalent

$$\delta(\Omega\phi,\Omega\vartheta)\leq[E_1+(E_2+E_1+E_2E)E]\delta(\phi,\vartheta). \tag{2.2.8}$$

From $0<E_1+(E_2+E_1+E_2E)E<1$, we can conclude that Ω is a contractions mapping.

In the sequel, consider the function $\theta\in\mathrm{W}$ and use the inequality (2.2.3), then we have

$$|{}^H\mathcal{D}_{0^+}^{\tau,\kappa;\Psi}(\Omega\theta(\ell)-\theta(\ell)|+|\Omega\theta(\ell)-\theta(\ell)|$$
$$=|\omega(\ell,\theta(\ell))+\int_0^\ell \mathrm{M}(\ell,\jmath,\theta(\jmath))d\jmath-{}^H\mathcal{D}_{0^+}^{\tau,\kappa;\Psi}\theta(\ell)|$$
$$+|\mathcal{I}_{0^+}^{\tau;\Psi}\omega(\ell,\theta(\ell))+\mathcal{I}_{0^+}^{\tau;\Psi}\left[\int_0^\ell \mathrm{M}(\ell,\jmath,\theta(\jmath))d\jmath\right]-\theta(\ell)|$$
$$\leq{}_2F_1(\rho,\sigma;\varsigma;\ell)+|\mathcal{I}_{0^+}^{\tau;\Psi}\omega(\ell,\theta(\ell))+\mathcal{I}_{0^+}^{\tau;\Psi}\left[\int_0^\ell \mathrm{M}(\ell,\jmath,\theta(\jmath))d\jmath\right]-\theta(\ell)|$$
$$\leq(1+E)\,{}_2F_1(\rho,\sigma;\varsigma;\ell).$$

Thus,

$$\delta(\Omega\theta,\theta)\leq 1+E<\infty,\ E<1. \tag{2.2.9}$$

Now all the conditions for the alternative Theorem 1.2 hold. Then

- The mapping Ω has a fixed point like θ_0. It means $\Omega\theta_0=\theta_0$ or equivalently

$$\theta_0(\ell)=\mathcal{I}_{0^+}^{\tau;\Psi}(\omega(\ell,\theta_0(\ell)))+\mathcal{I}_{0^+}^{\tau;\Psi}\left[\int_0^\ell \mathrm{M}(\ell,\jmath,\theta_0(\jmath))d\jmath\right]. \tag{2.2.10}$$

- The fixed point θ_0 is unique in the set $\mathrm{W}^*=\{\vartheta\in\mathrm{W}:\delta(\Omega\theta,\vartheta)<\infty\}$.
- With respect to the continuity of the functions ω, M and the differentiability of the function θ_0 by taking the Ψ-Hilfer fractional derivative of Eq. (2.2.10) and also utilizing Theorem 1.2, we have

$$ {}^{H}\mathcal{D}_{0^+}^{\theta,\kappa;\Psi}\theta_0(\ell) = \omega(\ell,\theta_0(\ell)) + \int_0^{\ell} \mathrm{M}(\ell,\jmath,\theta_0(\jmath))d\jmath, \tag{2.2.11} $$

then, by Eq. (2.2.9), we get

$$ \begin{aligned} \delta(\theta,\theta_0) &\le \frac{1}{1-[E_1+(E_2+E_1+E_2E)E]}\delta(\Omega\theta,\theta) \\ &\le \frac{1+E}{1-[E_1+(E_2+E_1+E_2E)E]}. \end{aligned} $$

Thus, Eq. (2.2.1) has Hyers-Ulam-Rassias-Kummer stability property.

Now, we prove $\mathrm{W}^* = \mathrm{W}$ to show the fixed point of mapping Ω is unique in W. Put

$$ F = \frac{1+E}{1-[E_1+(E_2+E_1+E_2E)E]}, $$

and consider the differentiable function d, which holds in Eq. (2.2.11) and $\theta \in \mathrm{W}$, then we have $\delta(\theta,\mathrm{d}) < F$ and

$$ {}^{H}\mathcal{D}_{0^+}^{\tau,\kappa;\Psi}\mathrm{d}(\ell) = \omega(\ell,\mathrm{d}(\ell)) + \int_0^{\ell} \mathrm{M}(\ell,\jmath,\mathrm{d}(\jmath))d\jmath. \tag{2.2.12} $$

We show that d is a fixed point of Ω and $\mathrm{d} \in \mathrm{W}^*$. From Eq. (2.2.12), we have $\Omega\mathrm{d} = \mathrm{d}$.

Now, we show $\delta(\Omega\theta,\mathrm{d}) < \infty$. By $\delta(\theta,\mathrm{d}) < F$ and Eq. (2.2.12), we get

$$ \begin{aligned} &|{}^{H}\mathcal{D}_{0^+}^{\tau,\kappa;\Psi}(\Omega\theta(\ell)-\mathrm{d}(\ell))| + |\Omega\theta(\ell)-\mathrm{d}(\ell)| \\ &= |\omega(\ell,\theta(\ell)) + \int_0^{\ell}\mathrm{M}(\ell,\jmath,\theta(\jmath))d\jmath - \omega(\ell,\mathrm{d}(\ell)) - \int_0^{\ell}\mathrm{M}(\ell,\jmath,\mathrm{d}(\jmath))d\jmath| \\ &\quad + \left| \mathcal{I}_{0^+}^{\tau;\Psi}\omega(\ell,\theta(\ell)) + \mathcal{I}_{0^+}^{\tau;\Psi}\left[\int_0^{\ell}\mathrm{M}(\ell,\jmath,\theta(\jmath))d\jmath\right] \right. \\ &\qquad \left. - \mathcal{I}_{0^+}^{\tau;\Psi}\omega(\ell,\mathrm{d}(\ell)) - \mathcal{I}_{0^+}^{\tau;\Psi}\left[\int_0^{\ell}\mathrm{M}(\ell,\jmath,\mathrm{d}(\jmath))d\jmath\right]\right| \\ &\le |\omega(\ell,\theta(\ell)) - \omega(\ell,\mathrm{d}(\ell))| + \int_0^{\ell}|\mathrm{M}(\ell,\jmath,\theta(\jmath)) - \mathrm{M}(\ell,\jmath,\mathrm{d}(\jmath))|d\jmath \\ &\quad + |\mathcal{I}_{0^+}^{\tau;\Psi}[\omega(\ell,\theta(\ell)) - \omega(\ell,\mathrm{d}(\ell))]| + |\mathcal{I}_{0^+}^{\tau;\Psi}\left[\int_0^{\ell}[\mathrm{M}(\ell,\jmath,\theta(\jmath)) - \mathrm{M}(\ell,\jmath,\mathrm{d}(\jmath))]d\jmath\right] \\ &\le E_1|\theta(\ell)-\mathrm{d}(\ell)| + E_2\int_0^{\ell}|\theta(\jmath)-\mathrm{d}(\jmath)|d\jmath \\ &\quad + E_1\mathcal{I}_{0^+}^{\tau;\Psi}(|\theta(\ell)-\mathrm{d}(\ell)| + E_2\mathcal{I}_{0^+}^{\tau;\Psi}\left(\int_0^{\ell}|\theta(\jmath)-\mathrm{d}(\jmath)|d\jmath\right) \\ &\le [E_1+(E_2+E_1+E_2E)E]F, \end{aligned} $$

which implies that

$$\delta(\Omega\theta, \mathrm{d}) \leq [E_1 + (E_2 + E_1 + E_2 E)E]F < \infty.$$

2.2.4 Application

Now, we provide an example to illustrate Theorem 2.3.

Example 2.2 Consider the following fractional Volterra integro-differential equation

$$^{H}\mathcal{D}_{0^+}^{\frac{1}{2},\frac{1}{4};\Psi}\mu(\ell) = \frac{1}{100}\mu(\ell)\sin(\mu(\ell)) - 2\ell + \int_0^\ell \frac{1}{200}\cos(\ell + \jmath)\mu(\jmath)\sin(\mu(\jmath))d\jmath, \tag{2.2.13}$$

where μ is a differentiable function. Define $\omega : [0, L] \times Y \longrightarrow Y$ by $\omega(\ell, \mu) = \frac{1}{100}\mu\sin(\mu) - 2\ell$ and $\mathrm{M} : [0, L] \times Y \times Y \longrightarrow Y$ by $\mathrm{M}(\ell, \jmath, \mu) = \frac{1}{200}\mu\sin(\mu)\cos(\ell + \jmath)$ that $a, b, L > 0$ and $\mu(j) = j^2$. Consider the positive coefficients $E = \frac{1}{10}$, $E_1 = \frac{1}{100}$, $E_2 = \frac{1}{200}$ such that $0 < E_1 + (E_1 + E_2 + E_2 E)E < 1$ and for continuous functions ω, M, ${}_2F_1$ we have

(1) $|\omega(\ell, \mu) - \omega(\ell, \mu_0)| = |\frac{1}{100}\mu\sin(\mu) - 2\ell - \frac{1}{100}\mu_0\sin(\mu_0) + 2\ell| \leq |\frac{1}{100}||\mu - \mu_0|$,
(2) $|\mathrm{M}(\ell, \jmath, \mu) - \mathrm{M}(\ell, \jmath, \mu_0)| = |\frac{1}{200}\mu\sin(\mu)\cos(\ell + \jmath) - \frac{1}{200}\mu_0\sin(\mu_0)\cos(\ell + \jmath)| \leq |\frac{1}{200}||\mu - \mu_0|$,
(3) $\frac{1}{\Gamma(\frac{1}{2})}\int_0^\ell \Delta_\Psi^{\frac{1}{2}}(\ell, \jmath)\,{}_2F_1(\rho, \sigma; \varsigma; \jmath)d\jmath \leq \frac{1}{10}\,{}_2F_1(\rho, \sigma; \varsigma; \ell)$.

If the following inequality holds for the differentiable function θ

$$|^{H}\mathcal{D}_{0^+}^{\frac{1}{2},\frac{1}{4};\Psi}\theta(\ell) - \frac{1}{100}\theta(\ell)\sin(\theta(\ell)) + 2\ell - \int_0^\ell \frac{1}{200}\cos(\ell + \jmath)\theta(\jmath)\sin(\theta(\jmath))d\jmath| \leq {}_2F_1(\rho, \sigma; \varsigma; \ell),$$

and $E_1 = |\frac{1}{100}|$ and $E_2 = |\frac{1}{100}|$, Theorem 2.3 implies that, there is a solution θ_0, such that (Fig. 2.1)

- θ_0 is the fixed point of the operator Ω, which is defined as follows

$$\Omega(\theta_0(\ell)) = \mathcal{I}_{0^+}^{\frac{1}{2};\Psi}(\frac{1}{100}\theta_0\sin(\theta_0) - 2\ell) + \mathcal{I}_{0^+}^{\frac{1}{2};\Psi}\left[\int_0^\ell \frac{1}{200}\theta_0\sin(\theta_0)\cos(\ell + \jmath)d\jmath\right].$$

- The fixed point θ_0 is unique.
- θ_0 satisfies in Eq. (2.2.13).
-

$$\delta(\theta, \theta_0) \leq \frac{1 + E}{1 - E_1 + (E_1 + E_2 + E_2 E)E},$$

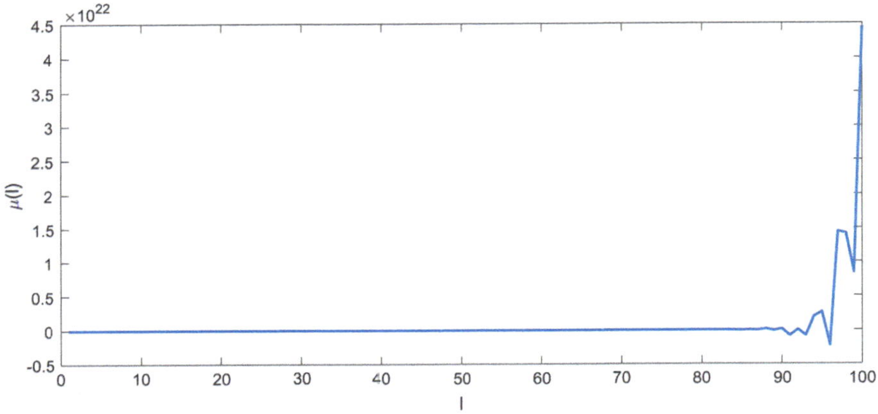

Fig. 2.1 Diagram of the solution of the fractional Volterra integro-differential Eq. 2.2.13

or

$$|{}^{H}\mathcal{D}_{0^+}^{\tau,\kappa;\Psi}\theta(\ell) - {}^{H}\mathcal{D}_{0^+}^{\tau,\kappa;\Psi}\theta_0(\ell)| + |\theta(\ell) - \theta_0(\ell)| \\ \leq \frac{1+E}{1-E_1+(E_1+E_2+E_2E)E}\, {}_2F_1(\rho,\sigma;\varsigma;\ell),$$

where $E_1 + (E_1 + E_2 + E_2E)E = \frac{231}{20000}$ and $\frac{1+E}{1-E_1+(E_1+E_2+E_2E)E} \approx 1.112$. Then Eq. (2.2.13) is Hyers-Ulam-Rassias-Kummer stable.

2.2.5 *Fractional Volterra Integral Equation*

Here, we have a class of fractional Volterra integral equations [31–33]. We stabilize the mentioned equation by a Kummer function and apply the Cădariu-Radu Method to guarantee the existence of a unique solution of Eq. (2.2.2). Our results improve and generalize the results of [34].

Theorem 2.4 *Consider the positive constants E, E_1 and E_2 such that $0 < (E_1 + E_2E) < 1$. On Banach space Y, consider the continuous functions $\omega : [0, L] \times Y \to Y$, $\mathrm{M} : [0, L] \times [0, L] \times Y \to Y$ and ${}_2F_1 : \mathbb{R}^3 \times [0, L] \to (0, \infty)$, for all $\jmath, \ell \in [0, L]$ and $\phi, \vartheta \in Y$ and assume that the following conditions are true for them*

(4) $|\omega(\ell,\phi) - \omega(\ell,\vartheta)| \leq E_1|\phi - \vartheta|$,

(5) $|\mathrm{M}(\ell,\jmath,\phi) - \mathrm{M}(\ell,\jmath,\vartheta)| \leq E_2|\phi - \vartheta|$,

(6) $\frac{1}{\Gamma(\tau)}\int_0^\ell \Delta_\Psi^\tau(\ell,\jmath)\, {}_2F_1(\rho,\sigma;\varsigma;\jmath)d\jmath \leq E\, {}_2F_1(\rho,\sigma;\varsigma;\ell)$.

If $\theta : [0, L] \to Y$ *be a differentiable function satisfying* (2.2.4)*, then Eq.* (2.2.2) *has a unique solution* $\theta_0 : [0, L] \to Y$*. Thus, we have*

$$|\theta(\ell) - \theta_0(\ell)| \leq \frac{1}{1 - (E_1 + E_2 E)} \, {}_2F_1(\rho, \sigma; \varsigma; \ell). \tag{2.2.14}$$

and this means that Eq. (2.2.2) *has Hyers-Ulam-Rassias-Kummer stability property.*

Proof We define the set W as follows

$$\mathrm{W} = \{\phi : [0, L] \to Y : \phi \text{ is continuous}\},$$

Also, we define $\delta : \mathrm{W} \times \mathrm{W} \to [0, \infty]$ as follows

$$\delta(\phi, \vartheta) = \inf\{P \in [0, \infty] : |\phi(\ell) - \vartheta(\ell)| \leq P \, {}_2F_1(\rho, \sigma; \varsigma; \ell), \ \ell \in [0, L]\}.$$

Similar to the argument of the previous theorem, we can easily show that the space (W, δ) is a $[0, \infty]$-valued complete metric space. Since the equation is integral, we consider the functions continuously and define the meter on the set of these functions. In Theorem 2.3, since we have a fractional differential integral equation, we consider the differentiable functions and define the meter on this set of functions.

Now, we define the mapping $\Omega : \mathrm{W} \longrightarrow \mathrm{W}$ by

$$\Omega(\phi(\ell)) = \omega(\ell, \phi(\ell)) + \frac{1}{\Gamma(\tau)} \int_0^{\ell} \Delta_{\Psi}^{\tau}(\ell, \jmath) \mathrm{M}(\ell, \jmath, \phi(\jmath)) d\jmath,$$

and show that Ω is a contraction operator. Suppose that for $\phi, \vartheta \in \mathrm{W}, \delta(\phi, \vartheta) < P_{\phi\vartheta}$ that $P_{\phi\vartheta} > 0$, then we have

$$|\phi(\ell) - \vartheta(\ell)| \leq P_{\phi\vartheta} \, {}_2F_1(\rho, \sigma; \varsigma; \ell),$$

for all $\ell \in [0, L]$. By conditions (4) and (5), we have

$$\begin{aligned}
|\Omega(\phi(\ell)) - \Omega(\vartheta(\ell))| &\leq |\omega(\ell, \phi(\ell)) - \omega(\ell, \vartheta(\ell))| \\
&\quad + \frac{1}{\Gamma(\tau)} \int_0^{\ell} \Delta_{\Psi}^{\tau}(\ell, \jmath) |\mathrm{M}(\ell, \jmath, \phi(\jmath)) - \mathrm{M}(\ell, \jmath, \vartheta(\jmath))| d\jmath \\
&\leq E_1 |\phi(\ell) - \vartheta(\ell)| + E_2 \frac{1}{\Gamma(\tau)} \int_0^{\ell} \Delta_{\Psi}^{\tau}(\ell, \jmath) |\phi(\jmath) - \vartheta(\jmath)| d\jmath \\
&\leq E_1 P_{\phi\vartheta} \, {}_2F_1(\rho, \sigma; \varsigma; \ell) + \frac{P_{\phi\vartheta} E_2}{\Gamma(\tau)} \int_0^{\ell} \Delta_{\Psi}^{\tau}(\ell, \jmath) \, {}_2F_1(\rho, \sigma; \varsigma; \jmath) d\jmath \\
&\leq E_1 P_{\phi\vartheta} \, {}_2F_1(\rho, \sigma; \varsigma; \ell) + E_2 P_{\phi\vartheta} {}_2F_1(\rho, \sigma; \varsigma; \ell) E \\
&\leq (E_1 + E_2 E) P_{\phi\vartheta} \, {}_2F_1(\rho, \sigma; \varsigma; \ell),
\end{aligned}$$

and then

$$\delta(\Omega\phi, \Omega\vartheta) \leq (E_1 + E_2 E) \delta(\phi, \vartheta).$$

Since $0 < (E_1 + E_2 E) < 1$, then Ω is a contraction mapping. In the sequel, considering the function $\theta \in \mathrm{W}$ and utilizing the inequality (2.2.4), we get

$$\delta(\Omega\theta, \theta) \leq 1 < \infty. \tag{2.2.15}$$

Now, all the conditions of Theorem 1.2 hold. Then

- The mapping Ω has a fixed point named by θ_0. It means $\Omega\theta_0 = \theta_0$ or equivalently

$$\theta_0(\ell) = \omega(\ell, \theta_0(\ell)) + \frac{1}{\Gamma(\tau)} \int_0^\ell \Delta_\Psi^\tau(\ell, \jmath)\mathrm{M}(\ell, \jmath, \theta_0(\jmath))d\jmath. \tag{2.2.16}$$

- The fixed point θ_0 is unique in the set $\mathrm{W}^* = \{\vartheta \in \mathrm{W}(\Omega\theta, \vartheta) < \infty\}$.
- By Eq. (2.2.15) and Theorem (1.2), we get

$$\delta(\theta, \theta_0) =\leq \frac{1}{1-(E_1+E_2E)}\delta(\Omega\theta, \theta) \leq \frac{1}{1-(E_1+E_2E)}.$$

Thus, Eq. (2.2.2) is Hyers-Ulam-Rassias-Kummer stable.

Now, we prove $\mathrm{W}^* = \mathrm{W}$ to show the fixed point of mapping Ω is unique in W. Consider another continuous function d that satisfying

$$\mathrm{d}(\ell) = \omega(\ell, \mathrm{d}(\ell)) + \frac{1}{\Gamma(\tau)} \int_0^\ell \Delta_\Psi^\tau(\ell, \jmath)\mathrm{M}(\ell, \jmath, \mathrm{d}(\jmath))d\jmath.$$

In the following, we show that d is a fixed point of Ω and $\mathrm{d} \in \mathrm{W}^*$. From Eq. (2.2.16) we have, $\Omega\mathrm{d} = \mathrm{d}$. Now, we show $\delta(\Omega\theta, \mathrm{d}) < \infty$. From

$$\delta(\theta, \mathrm{d}) \leq \frac{1}{1-(E_1+E_2E)},$$

and conditions (4), (5), and (6), we get

$$\begin{aligned}
&|\Omega\theta(\ell) - \mathrm{d}(\ell)| \\
&= |\omega(\ell, \theta(\ell)) - \frac{1}{\Gamma(\tau)} \int_0^\ell \Delta_\Psi^\tau(\ell, \jmath)\mathrm{M}(\ell, \jmath, \mathrm{d}(\jmath))d\jmath - \omega(\ell, \mathrm{d}(\ell)) \\
&+ \frac{1}{\Gamma(\tau)} \int_0^\ell \Delta_\Psi^\tau(\ell, \jmath)\mathrm{M}(\ell, \jmath, \theta(\jmath))d\jmath| \\
&\leq |\omega(\ell, \theta(\ell)) - \omega(\ell, \mathrm{d}(\ell))| + \frac{1}{\Gamma(\tau)} \int_0^\ell \Delta_\Psi^\tau(\ell, \jmath)|\mathrm{M}(\ell, \jmath, \theta(\jmath)) - \mathrm{M}(\ell, \jmath, \mathrm{d}(\jmath))|d\jmath \\
&\leq E_1|\theta(\ell) - \mathrm{d}(\ell)| + \frac{E_2}{\Gamma(\tau)} \int_0^\ell \Delta_\Psi^\tau(\ell, \jmath)|\theta(\jmath) - \mathrm{d}(\jmath)|d\jmath \\
&\leq E_1 P_{\theta\mathrm{d}}\, {}_2F_1(\rho, \sigma; \varsigma; \ell) + E_2 P_{\theta\mathrm{d}} E\, {}_2F_1(\rho, \sigma; \varsigma; \ell)
\end{aligned}$$

$$\leq (E_1 + E_2 E)\delta(\theta, \mathrm{d})\, {}_2F_1(\rho, \sigma; \varsigma; \ell)$$
$$\leq \frac{(E_1 + E_2 E)}{1 - (E_1 + E_2 E)}\, {}_2F_1(\rho, \sigma; \varsigma; \ell),$$

which implies that $\delta(\Omega\theta, \mathrm{d}) < \infty$.

2.2.6 Application

Example 2.3 Consider the following fractional Volterra integral equation

$$\mu(\ell) = \sqrt{\ell} + \frac{1}{100}\mu(\ell) - \frac{1}{\Gamma(\frac{1}{2})}\int_0^\ell \Delta_\Psi^{\frac{1}{2}}(\ell, \jmath)\frac{\mu(\jmath)}{\sqrt{\ell - \jmath}}d\jmath, \tag{2.2.17}$$

where in μ is a continuous function. Define $\omega : [0, L] \times Y \to Y$ by $\omega(\ell, \mu(\ell)) = \sqrt{\ell} + \frac{1}{100}\mu(\ell)$, and $\mathrm{M} : [0, L] \times Y \times Y \to Y$ by $\mathrm{M}(\ell, \jmath, \mu(\jmath)) = \frac{\mu(\jmath)}{\sqrt{\ell - \jmath}}$ that $L > 0$ and $\mu(j) = j^2$. Consider the positive coefficients E, E_1, E_2, such that $0 < E_1 + E_2 E < 1$ and for continuous functions ω, M, ${}_2F_1$ we have

(4) $|\omega(\ell, \mu(\ell)) - \omega(\ell, \mu_0(\ell))| = |\sqrt{\ell} + \frac{1}{100}\mu - \sqrt{\ell} - \frac{1}{100}\mu_0| \leq \frac{1}{100}|\mu - \mu_0|$,
(5) $|\mathrm{M}(\ell, \jmath, \mu(\jmath)) - \mathrm{M}(\ell, \jmath, \mu_0(\jmath))| = |\frac{\mu}{\sqrt{\ell-\jmath}} - \frac{\mu_0}{\sqrt{\ell-\jmath}}| \leq \frac{1}{\sqrt{\ell-\jmath}}|\mu - \mu_0|$,
(6) $\frac{1}{\Gamma(\frac{1}{2})}\int_0^\ell \Delta_\Psi^{\frac{1}{2}}(\ell, \jmath)\, {}_2F_1(\rho, \sigma; \varsigma; \ell)d\jmath \leq \frac{1}{10}\, {}_2F_1(\rho, \sigma, \varsigma; \ell)$.

If the following inequality holds for the differentiable function θ

$$|\theta(\ell) - \sqrt{\ell} - \frac{1}{100}\theta(\ell) + \int_0^\ell \frac{\theta(\jmath)}{\sqrt{\ell - \jmath}}d\jmath| \leq {}_2F_1(\rho, \sigma, \varsigma; \ell)$$

and $E_1 = \frac{1}{100}$ and $E_2 = \frac{1}{\sqrt{\ell-\jmath}} = \frac{1}{200}$. Theorem (2.4) implies that there is a unique solution θ_0, such that

- θ_0 is the fixed point of the operator Ω, which is defined as follows

$$\Omega(\theta_0) = \sqrt{\ell} + \frac{1}{100}\theta_0(\ell) - \frac{1}{\Gamma(\frac{1}{2})}\int_0^\ell \Delta_\Psi^{\frac{1}{2}}(\ell, \jmath)\frac{\theta_0(\jmath)}{\sqrt{\ell - \jmath}}d\jmath.$$

- The fixed point θ_0 is unique.
- θ_0 satisfies in Eq. (2.2.17).
- $$\delta(\theta, \theta_0) \leq \frac{1}{1 - (E_1 + E_2 E)},$$

or

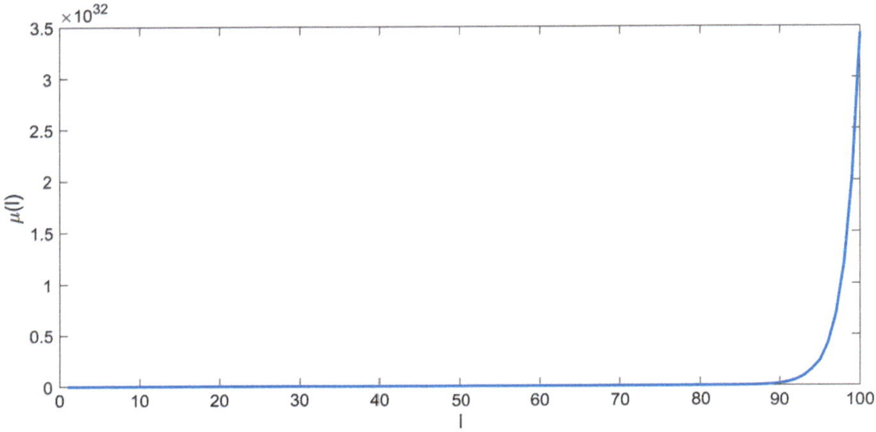

Fig. 2.2 Diagram of the solution of the fractional Volterra integral Eq. 2.2.17

$$|\theta(\ell) - \theta_0(\ell)| \leq \frac{1}{1-(E_1+E_2E)}\, {}_2F_1(\rho, \sigma; \varsigma; \ell),$$

where $E_1 + E_2E = \frac{21}{2000}$ and $\frac{1}{1-(E_1+E_2E)} \approx 1.011$, which implies that Eq. (2.2.17) has Hyers-Ulam-Rassias-Kummer stability property (Fig. 2.2).

2.3 A Class of Conformable Fractional Differential Equations

2.3.1 Introduction of the Conformable Fractional Differential Equations and Other Preliminaries

The incompatibility of Riemann-Liouville and Caputo derivatives in some cases causes high computational complexity in fractional calculations. For this reason, researchers have proposed the conformable fractional derivative equations. To determine the optimal control function, we introduce the aggregation functions (AF). Aggregation functions have been developing since 1980 because these functions are rooted in different fields. Application areas of aggregation functions include applied mathematics, probability, statistics, decision theory, computer science, artificial intelligence, operations research, as well as many applied fields, economics and finance, pattern recognition and image processing, data fusion, multicriteria decision aid, automated reasoning, etc. [62, 63].

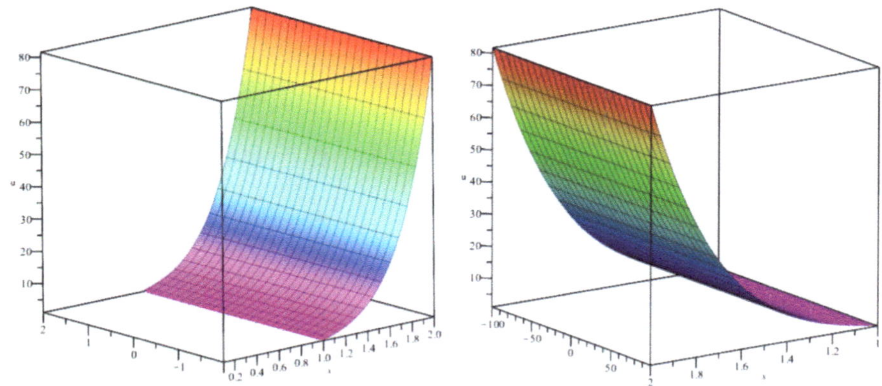

(a) The exact solution of the equation (2.3.15) for $\eth \in [0.1, 2]$

(b) The exact solution of the equation (2.3.15) for $\eth \in (1, 2]$

Fig. 2.3 3D representation to the solution of CFDE

We consider the conformable fractional differential equation (CFD-E) with a constant coefficient, which is defined as follows

$$\begin{cases} \mathfrak{D}_{\zeta}^{a}\mathfrak{u}(\eth) = \lambda\mathfrak{u}(\eth) + \mathfrak{z}(\eth, \mathfrak{u}(\eth)), \ \eth \in (a, b], \ 0 < \zeta < 1, \\ \mathfrak{u}(a) = \mathfrak{u}_{a}, \end{cases} \tag{2.3.1}$$

where $\mathfrak{D}_{\zeta}^{a}\mathfrak{u}$ is called the conformable fractional derivative (CFD) with lower index a of the function $\mathfrak{u}$ and $\mathcal{T} = [a, b]$, $\mathfrak{z} \in C(\mathcal{T} \times \mathbb{R}, \mathbb{R})$. By calculating the optimal control function and considering this function, we obtain the existence of a unique solution and H-U-R-OS_1-multiple stability for CFD Eq. (2.3.1) using the Cădariu-Radu method. We show a type of this equation in the form of a numerical example in which three-dimensional Fig. 2.3 show the exact solution of this equation.

Definition 2.4 For a mapping $\mathfrak{f} : [\mathrm{a}, \infty) \to \mathbb{R}$, the CFD starting from a of order ω is defined by

$$\mathfrak{D}_{\omega}^{\mathrm{a}}\mathfrak{f}(\varpi - \vartheta) = \lim_{\iota \to 0} \frac{\mathfrak{f}\big((\varpi-\vartheta)+\iota((\varpi-\vartheta)-\mathrm{a})^{1-\omega}\big)-\mathfrak{f}(\varpi-\vartheta)}{\iota}, \ (\varpi - \vartheta) > \mathrm{a}, \ 0 < \omega < 1.$$

If on (a, b), $\mathfrak{D}_{\omega}^{\mathrm{a}}\mathfrak{f}(\varpi - \vartheta)$ exists, then $\mathfrak{D}_{\omega}^{\mathrm{a}}\mathfrak{f}(\mathrm{a}) = \lim_{(\varpi-\vartheta)\to \mathrm{a}^{+}} \mathfrak{D}_{\omega}^{\mathrm{a}}\mathfrak{f}(\varpi - \vartheta)$.

Remark 2.1 For a finite given $\mathfrak{D}_{\omega}^{\mathrm{a}}\mathfrak{f}\,((\varpi - \vartheta)_0)$, $\mathfrak{f}$ is ω–differentiable at $(\varpi - \vartheta)_0$. If $\mathfrak{f} \in C^1([\mathrm{a}, \infty), \mathbb{R})$, then $\mathfrak{D}_{\omega}^{\mathrm{a}}\mathfrak{f}(\varpi - \vartheta) = ((\varpi - \vartheta) - \mathrm{a})^{1-\omega}\mathfrak{f}'(\varpi - \vartheta)$.

Definition 2.5 For a mapping $\mathfrak{f}(\varpi - \vartheta)$, the Hadamard fractional integral with order $0 < \omega < 1$ and parameter $s \in \mathbb{R}$, is defined by

$${}_{\mathrm{a}}^{H}I_{(\varpi-\vartheta)}^{\omega,s}\mathfrak{f}(\varpi - \vartheta) = \frac{1}{\Gamma(\omega)} \int_{\mathrm{a}}^{\varpi-\vartheta} \left(\frac{\lambda}{\mathrm{t}}\right)^{s} \left(\log \frac{\mathrm{t}}{\lambda}\right)^{\omega-1} \frac{\mathfrak{f}(\lambda)}{\lambda}\, d\lambda,$$

where $(\varpi - \vartheta) \in (\mathrm{a}, \mathrm{b})$ and $\mathrm{a} \leq \mathrm{b}$ in $\mathbb{R}$.

Lemma 2.2 *Let* $\mathfrak{f} \in C^1([\mathrm{a}, \infty))$. *For the real-valued mapping* $\mathfrak{f}$ *and* $(\varpi - \vartheta) > \mathrm{a},\ 0 < \omega < 1$, *the following relationship holds*

$$ {}_{\mathrm{a}}^{H}I^{\omega,s}_{(\varpi-\vartheta)} \mathfrak{D}^{\mathrm{a}}_{\omega}\mathfrak{f}(\varpi - \vartheta) = \mathfrak{f}(\varpi - \vartheta) - \mathfrak{f}(\mathrm{a}). $$

Theorem 2.5 *Consider the special function*

$$ E_\alpha\left((\frac{(\mathfrak{d} - a)^\zeta}{\zeta})^\lambda \right) = \sum_{k=0}^{\infty} \frac{(\frac{(\mathfrak{d}-a)^\zeta}{\zeta})^{k\lambda}}{\Gamma(1 + \alpha k \lambda)}, \quad \alpha > 0 $$

which E_α *is the Mittag–Leffler function. Suppose that* $\mathfrak{U}(\mathfrak{d}) = E_\alpha\left((\frac{(\mathfrak{d}-a)^\zeta}{\zeta})^\lambda \right)$, *then*

$$ \mathfrak{D}^a_\zeta \mathfrak{U}(\mathfrak{d}) = \lambda \mathfrak{U}(\mathfrak{d}) $$

Proof Using Remark 1.1, we get

$$ \begin{aligned} \mathfrak{D}^a_\zeta \mathfrak{U}(\mathfrak{d}) &= (\mathfrak{d} - a)^{1-\zeta} \frac{\mathrm{d}\mathfrak{U}(\mathfrak{d})}{\mathrm{d}\mathfrak{d}} \\ &= (\mathfrak{d} - a)^{1-\zeta} \lambda (\mathfrak{d} - a)^{\beta - 1} \sum_{k=0}^{\infty} \frac{(\frac{(\mathfrak{d}-a)^\zeta}{\zeta})^{k\lambda}}{\Gamma(1 + \alpha k \lambda)} \\ &= \lambda \sum_{k=0}^{\infty} \frac{(\frac{(\mathfrak{d}-a)^\zeta}{\zeta})^{k\lambda}}{\Gamma(1 + \alpha k \lambda)} \\ &= \lambda \mathfrak{U}(\mathfrak{d}). \end{aligned} $$

For $l \in N$, a subset of $C[a, b]$ is a Banach space $C^{l-1}[a, b]$ defined as follows

$$ C^{l-1}[a, b] = \left\{ u(x) : [a, b] \to R \text{ such that } u^{(l-1)}(x) \text{ is continuous on } [a, b] \right\} $$

with the norm

$$ \|u\| = \max_{x \in [a,b]} |u(x)| < +\infty. $$

In the following theorem, we consider the function $\mathfrak{u}$, in Eq. (2.3.1).

Theorem 2.6 *If the function* $\mathfrak{u} \in C(\mathcal{T}, \mathbb{R})$ *is a solution of Eq. (2.3.1), then the function* $\mathfrak{u}$ *is defined as follows*

$$ \begin{aligned} \mathfrak{u}(\mathfrak{d}) &= \mathfrak{u}_a \sum_{k=0}^{\infty} \frac{(\frac{(\mathfrak{d}-a)^\zeta}{\zeta})^{k\lambda}}{\Gamma(1 + \alpha k \lambda)} \\ &+ \frac{1}{\Gamma(\zeta)} \int_a^{\mathfrak{d}} \sum_{k=0}^{\infty} \frac{(\frac{(\mathfrak{d}-a)^\zeta}{\zeta})^{k\lambda}}{\Gamma(1 + \alpha k \lambda)} \sum_{k=0}^{\infty} \frac{(\frac{(\jmath-a)^\zeta}{\zeta})^{-k\lambda}}{\Gamma(1 + \alpha k \lambda)} \left(\frac{\jmath}{x}\right)^s \left(\log \frac{x}{\jmath}\right)^{\zeta - 1} \frac{\mathfrak{z}(\jmath, \mathfrak{u}(\jmath))}{\jmath}\, \mathrm{d}\jmath. \end{aligned} $$

Proof Any solution of (2.3.1) it should be as follows

$$\mathfrak{u}(\mathfrak{d}) = \sum_{k=0}^{\infty} \frac{(\frac{(\mathfrak{d}-a)^\zeta}{\zeta})^{k\lambda}}{\Gamma(1+\alpha k\lambda)} c(\mathfrak{d}), \tag{2.3.2}$$

where $c(\mathfrak{d})$ is an unknown continuously differentiable function. From (2.3.2) and Remark 2.1, we get

$$\begin{aligned}
\mathfrak{D}_\zeta^a \mathfrak{u}(\mathfrak{d}) &= \mathfrak{D}_\zeta^a \left(\sum_{k=0}^{\infty} \frac{(\frac{(\mathfrak{d}-a)^\zeta}{\zeta})^{k\lambda}}{\Gamma(1+\alpha k\lambda)} c(\mathfrak{d}) \right) \\
&= (\mathfrak{d}-a)^{1-\zeta} \sum_{k=0}^{\infty} \frac{(\frac{(\mathfrak{d}-a)^\zeta}{\zeta})^{k\lambda}}{\Gamma(1+\alpha k\lambda)} (\mathfrak{d}-a)^{\zeta-1} c(\mathfrak{d}) + \sum_{k=0}^{\infty} \frac{(\frac{(\mathfrak{d}-a)^\zeta}{\zeta})^{k\lambda}}{\Gamma(1+\alpha k\lambda)} (\mathfrak{d}-a)^{1-\zeta} c'(\mathfrak{d}) \\
&= \lambda c(\mathfrak{d}) \sum_{k=0}^{\infty} \frac{(\frac{(\mathfrak{d}-a)^\zeta}{\zeta})^{k\lambda}}{\Gamma(1+\alpha k\lambda)} + \sum_{k=0}^{\infty} \frac{(\frac{(\mathfrak{d}-a)^\zeta}{\zeta})^{k\lambda}}{\Gamma(1+\alpha k\lambda)} \mathfrak{D}_\zeta^a c(\mathfrak{d}) \\
&= \lambda \mathfrak{u}(\mathfrak{d}) + \sum_{k=0}^{\infty} \frac{(\frac{(\mathfrak{d}-a)^\zeta}{\zeta})^{k\lambda}}{\Gamma(1+\alpha k\lambda)} \mathfrak{D}_\zeta^a c(\mathfrak{d}).
\end{aligned}$$

As a result

$$\mathfrak{D}_\zeta^a c(\mathfrak{d}) = \sum_{k=0}^{\infty} \frac{(\frac{(\mathfrak{d}-a)^\zeta}{\zeta})^{-k\lambda}}{\Gamma(1+\alpha k\lambda)} \mathfrak{z}(\mathfrak{d}, \mathfrak{u}(\mathfrak{d})). \tag{2.3.3}$$

From (2.3.3) and Lemma 1.2, we get

$$c(\mathfrak{d}) = c(a) + \frac{1}{\Gamma(\zeta)} \int_a^x \sum_{k=0}^{\infty} \frac{(\frac{(\jmath-a)^\zeta}{\zeta})^{-k\lambda}}{\Gamma(1+\alpha k\lambda)} \left(\frac{\jmath}{x}\right)^s \left(\log \frac{x}{\jmath}\right)^{\zeta-1} \frac{\mathfrak{z}(\jmath, \mathfrak{u}(\jmath))}{\jmath} \, d\jmath, \tag{2.3.4}$$

where $c(a) = \mathfrak{u}(a) = \mathfrak{u}_a$. Use (2.3.4) in (2.3.2), the desired result is obtained.

Therefore, the following definition is established.

Definition 2.6 A function $\mathfrak{u} \in C^1(\mathcal{T}, \mathbb{R})$ is called the solution of (2.3.1) if $\mathfrak{u}$ satisfies $\mathfrak{D}_\zeta^a \mathfrak{u}(\mathfrak{d}) = \lambda \mathfrak{u}(\mathfrak{d}) + \mathfrak{z}(\mathfrak{d}, \mathfrak{u}(\mathfrak{d}))$, $\mathfrak{d} \in (a, b]$ and $\mathfrak{u}(a) = \mathfrak{u}_a$. Thus, the function $\mathfrak{u}$ holds in the following integral equation

$$\begin{aligned}
\mathfrak{u}(\mathfrak{d}) &= \mathfrak{u}_a \sum_{k=0}^{\infty} \frac{(\frac{(\mathfrak{d}-a)^\zeta}{\zeta})^{k\lambda}}{\Gamma(1+\alpha k\lambda)} \\
&\quad + \frac{1}{\Gamma(\zeta)} \int_a^{\mathfrak{d}} \sum_{k=0}^{\infty} \frac{(\frac{(\mathfrak{d}-a)^\zeta}{\zeta})^{k\lambda}}{\Gamma(1+\alpha k\lambda)} \sum_{k=0}^{\infty} \frac{(\frac{(\jmath-a)^\zeta}{\zeta})^{-k\lambda}}{\Gamma(1+\alpha k\lambda)} \left(\frac{\jmath}{x}\right)^s \left(\log \frac{x}{\jmath}\right)^{\zeta-1} \frac{\mathfrak{z}(\jmath, \mathfrak{u}(\jmath))}{\jmath} \, d\jmath.
\end{aligned}$$

Definition 2.7 Let $\mathfrak{w}(\mathfrak{d})$ be a given diferentiable function and $OS_1(\Theta)$ be the control function. Consider (2.3.1) and we say that H-U-R-OS_1-stability is established for it, if for $\varepsilon > 0$ the following inequality holds

$$\left|\mathfrak{D}_{\zeta}^{a}\mathfrak{w}(\mathfrak{d}) - \lambda\mathfrak{w}(\mathfrak{d}) - \mathfrak{z}(\mathfrak{d}, \mathfrak{w}(\mathfrak{d}))\right| \leq \varepsilon\ OS_1(\Theta),\quad 0 < \zeta < 1, \mathfrak{d} \in \mathcal{T} \tag{2.3.5}$$

then, exists a solution $\mathfrak{u}(\mathfrak{d})$ of Eq. (2.3.1) such that for $\mathfrak{R} > 0$

$$|\mathfrak{w}(\mathfrak{d}) - \mathfrak{u}(\mathfrak{d})| \leq \mathfrak{R}\ \varepsilon\ OS_1(\Theta),\ \mathfrak{d} \in \mathcal{T}.$$

Next, using the Diaz-Margolis alternative fixed point theorem (D-AMFPT) 1.2, we show that the Hyers- Ulam- Rassias-OS_1-multiple stability (H-U-R-OS_1-MS) for CFD Eq. (2.3.1).

2.3.2 H-U-R-OS_1-MS for CFD Equation

By considering the optimal control function , i.e. the $OS_1(\Theta)$ function, we investigate CFD Eq. (2.3.1), which means the existence of a unique solution, as well as stability using this optimal control function. Our results can be extended to generalize some results presented in [66–79].

Theorem 2.7 *We assume that* $0 < \upsilon = \Im\left(\sum_{k=0}^{\infty} \frac{(\frac{(\mathfrak{d}-a)^{\zeta}}{\zeta})^{k\lambda}}{\Gamma(1+\alpha k\lambda)}\boldsymbol{\Delta}\right) < 1$, *where* λ *is a constant coefficient,* $\boldsymbol{\Delta}$ *and* $\Im$ *are positive coefficients. For the function* $\mathfrak{z} : \mathcal{T} \times \mathbb{R} \longrightarrow \mathbb{R}$ *and the optimal control function* $OS_1 : \mathbb{R}^5 \longrightarrow \mathbb{R}$, *we have*

$$|\mathfrak{z}(\mathfrak{d}, \mathfrak{u}) - \mathfrak{z}(\mathfrak{d}, \mathfrak{w})| \leq \Im|\mathfrak{u} - \mathfrak{w}|, \tag{2.3.6}$$

and

$$\frac{1}{\Gamma(\zeta)}\int_a^{\mathfrak{d}} \sum_{k=0}^{\infty} \frac{(\frac{(\jmath-a)^{\zeta}}{\zeta})^{-k\lambda}}{\Gamma(1+\alpha k\lambda)}\ OS_1(\Theta)\left(\frac{\jmath}{x}\right)^{s}\left(\log\frac{x}{\jmath}\right)^{\zeta-1}\frac{d\jmath}{\jmath} \leq \boldsymbol{\Delta}\ OS_1(\Theta). \tag{2.3.7}$$

If the differentiable function $\mathfrak{w} : \mathcal{T} \to \mathbb{R}$ *satisfy to the inequality condition of*

$$\left|\mathfrak{D}_{\zeta}^{a}\mathfrak{w}(\mathfrak{d}) - \lambda\mathfrak{w}(\mathfrak{d}) - \mathfrak{z}(\mathfrak{d}, \mathfrak{w}(\mathfrak{d}))\right| \leq \varepsilon\ OS_1(\Theta), \tag{2.3.8}$$

and

$$\left|\mathfrak{w}(\mathfrak{d}) - \mathfrak{u}_a \sum_{k=0}^{\infty} \frac{(\frac{(\mathfrak{d}-a)^{\zeta}}{\zeta})^{k\lambda}}{\Gamma(1+\alpha k\lambda)}\right.$$

$$-\frac{1}{\Gamma(\beta)}\int_a^{\mathfrak{d}}\sum_{k=0}^{\infty}\frac{(\frac{(\mathfrak{d}-a)^{\zeta}}{\zeta})^{k\lambda}}{\Gamma(1+\alpha k\lambda)}\sum_{k=0}^{\infty}\frac{(\frac{(\jmath-a)^{\zeta}}{\zeta})^{-k\lambda}}{\Gamma(1+\alpha k\lambda)}\mathfrak{z}(\jmath,\mathfrak{w}(t))\left(\frac{\jmath}{x}\right)^{s}\left(\log\frac{x}{\jmath}\right)^{\zeta-1}\frac{\mathrm{d}\jmath}{\jmath}\Bigg|$$

$$\leq\varepsilon\sum_{k=0}^{\infty}\frac{(\frac{(b-a)^{\zeta}}{\zeta})^{k\lambda}}{\Gamma(1+\alpha k\lambda)}\ \frac{1}{\Gamma(\zeta)}\int_a^{\mathfrak{d}}\sum_{k=0}^{\infty}\frac{(\frac{(\jmath-a)^{\zeta}}{\zeta})^{-k\lambda}}{\Gamma(1+\alpha k\lambda)}OS_1(\Theta)\left(\frac{\jmath}{x}\right)^{s}\left(\log\frac{x}{\jmath}\right)^{\zeta-1}\frac{\mathrm{d}\jmath}{\jmath},$$

for $0<\zeta<1$ *and* $\mathfrak{d}\in\mathcal{T}$, *then, for CFD Eq.* (2.3.1), *there is a unique solution like function* $\mathfrak{u}:\mathcal{T}\to\mathbb{R}$, *such that the following inequality holds*

$$|\mathfrak{w}(\mathfrak{d})-\mathfrak{u}(\mathfrak{d})|\leq\Re\ \varepsilon\ OS_1(\Theta),$$

where $\Re=\frac{\Delta}{1-\upsilon}\sum_{k=0}^{\infty}\frac{(\frac{(b-a)^{\zeta}}{\zeta})^{k\lambda}}{\Gamma(1+\alpha k\lambda)}$. *By establishing this condition, we say that the H-U-R-OS_1-MS for CFD Eq.* (2.3.1) *is satisfied.*

Proof We consider the mapping $d:\mathcal{G}\times\mathcal{G}\to[0,\infty]$ as follows

$$d(\mathfrak{u},\mathfrak{w})=\inf\{Q\in[0,\infty]:|\mathfrak{u}(\mathfrak{d})-\mathfrak{w}(\mathfrak{d})|\leq Q\ OS_1(\Theta),\ \mathfrak{d}\in\mathcal{T}\}.$$

where $\mathcal{G}=\{\mathfrak{u}:\mathcal{T}\to\mathbb{R},\mathfrak{u}\text{ is differentiable}\}$.

In the sequel, we describe the proof step by step.

The first step of the proof is to show that $(\mathcal{G},d)$ is a GCMS. For this purpose, we assume

$$d(\mathfrak{u},\mathfrak{w})>d(\mathfrak{u},\mathfrak{v})+d(\mathfrak{v},\mathfrak{w})$$

for $\mathfrak{u},\mathfrak{w}\in\mathcal{G}$. Then, we have

$$|\mathfrak{u}(\mathfrak{d}_0)-\mathfrak{w}(\mathfrak{d}_0)|>d(\mathfrak{u},\mathfrak{v})+d(\mathfrak{v},\mathfrak{w})\ OS_1(\Theta).$$

Now, with respect to d, we come to a contradiction. That is, we have the following inequality

$$|\mathfrak{u}(\mathfrak{d}_0)-\mathfrak{w}(\mathfrak{d}_0)|>|\mathfrak{u}(\mathfrak{d}_0)-\mathfrak{v}(\mathfrak{d}_0)|+|\mathfrak{v}(\mathfrak{d}_0)-\mathfrak{w}(\mathfrak{d}_0)|,$$

We consider a Cauchy sequence like $\{\mathfrak{u}_n\}$ on $(\mathcal{G},d)$. For $\varepsilon>0$ we can find $N_\varepsilon\in\mathbb{N}$, such that

$$|\mathfrak{u}_n(\mathfrak{d})-\mathfrak{u}_m(\mathfrak{d})|<\varepsilon\ OS_1(\Theta).\tag{2.3.9}$$

for each $m,n\geq N_\varepsilon$ and $\mathfrak{d}\in\mathcal{T}$. Now, using (2.3.9) and taking into account the continuity property of the optimal control function $OS_1(\Theta)$ on $\mathcal{T}$ (which is a compact set), we conclude that the sequence $\{\mathfrak{u}_n\}$ is uniformly convergent to $\mathfrak{u}$ where $\mathfrak{u}\in\mathcal{G}$. Then, for $\varepsilon>0$, we can find $N_\varepsilon\in\mathbb{N}$, such that

$$|\mathfrak{u}_n(\mathfrak{d}) - \mathfrak{u}(\mathfrak{d})| < \varepsilon \; OS_1(\Theta),$$

for $n \geq N_\varepsilon$ and $\mathfrak{d} \in \mathcal{G}$. We assume that $m \to \infty$, then $d(\mathfrak{u}_n, \mathfrak{u}) \leq \varepsilon$. Therefore, we conclude that $(\mathcal{G}, d)$ is a GCMS. In the second step, we define the mapping $\Xi : \mathcal{G} \to \mathcal{G}$ as follows and prove the contractibility of the mapping in the following.

$$\begin{aligned}\Xi(\mathfrak{u}(\mathfrak{d})) &= \mathfrak{u}_a \sum_{k=0}^{\infty} \frac{(\frac{(\mathfrak{d}-a)^\zeta}{\zeta})^{k\lambda}}{\Gamma(1+\alpha k\lambda)} \\ &+ \frac{1}{\Gamma(\beta)} \int_a^{\mathfrak{d}} \sum_{k=0}^{\infty} \frac{(\frac{(\mathfrak{d}-a)^\zeta}{\zeta})^{k\lambda}}{\Gamma(1+\alpha k\lambda)} \sum_{k=0}^{\infty} \frac{(\frac{(\jmath-a)^\zeta}{\zeta})^{-k\lambda}}{\Gamma(1+\alpha k\lambda)} \left(\frac{\jmath}{x}\right)^s \left(\log\frac{x}{\jmath}\right)^{\zeta-1} \frac{\mathfrak{z}(\jmath, \mathfrak{u}(\jmath))}{\jmath}\, d\jmath.\end{aligned}$$

We assume that $d(\mathfrak{u}, \mathfrak{w}) \leq Q_{\mathfrak{u}\mathfrak{w}}$, for $\mathfrak{u}, \mathfrak{w} \in \mathcal{G}$ and the coefficient $Q_{\mathfrak{u}\mathfrak{w}} \in [0, \infty]$. We have

$$|\mathfrak{u}(\mathfrak{d}) - \mathfrak{w}(\mathfrak{d})| < Q_{\mathfrak{u}\mathfrak{w}} \; OS_1(\Theta),$$

for $\mathfrak{d} \in \mathcal{T}$. From Eqs. (2.3.6) and (2.3.7), we get

$$\begin{aligned}|\Xi\mathfrak{u}(\mathfrak{d}) - \Xi\mathfrak{w}(\mathfrak{d})| &= \Bigg|\mathfrak{u}_a \sum_{k=0}^{\infty} \frac{(\frac{(\mathfrak{d}-a)^\zeta}{\zeta})^{k\lambda}}{\Gamma(1+\alpha k\lambda)} \\ &+ \frac{1}{\Gamma(\zeta)} \int_a^{\mathfrak{d}} \sum_{k=0}^{\infty} \frac{(\frac{(\mathfrak{d}-a)^\zeta}{\zeta})^{k\lambda}}{\Gamma(1+\alpha k\lambda)} \sum_{k=0}^{\infty} \frac{(\frac{(\jmath-a)^\zeta}{\zeta})^{-k\lambda}}{\Gamma(1+\alpha k\lambda)} \left(\frac{\jmath}{x}\right)^s \left(\log\frac{x}{\jmath}\right)^{\zeta-1} \frac{\mathfrak{z}(\jmath, \mathfrak{u}(\jmath))}{\jmath}\, d\jmath \\ &- \mathfrak{u}_a \sum_{k=0}^{\infty} \frac{(\frac{(\mathfrak{d}-a)^\zeta}{\zeta})^{k\lambda}}{\Gamma(1+\alpha k\lambda)} \\ &+ \frac{1}{\Gamma(\beta)} \int_a^{\mathfrak{d}} \sum_{k=0}^{\infty} \frac{(\frac{(\mathfrak{d}-a)^\zeta}{\zeta})^{k\lambda}}{\Gamma(1+\alpha k\lambda)} \sum_{k=0}^{\infty} \frac{(\frac{(\jmath-a)^\zeta}{\zeta})^{-k\lambda}}{\Gamma(1+\alpha k\lambda)} \left(\frac{\jmath}{x}\right)^s \left(\log\frac{x}{\jmath}\right)^{\zeta-1} \frac{\mathfrak{z}(\jmath, \mathfrak{w}(\jmath))}{\jmath}\, d\jmath\Bigg| \\ &\leq \frac{1}{\Gamma(\zeta)} \int_a^{\mathfrak{d}} \sum_{k=0}^{\infty} \frac{(\frac{(\mathfrak{d}-a)^\zeta}{\zeta})^{k\lambda}}{\Gamma(1+\alpha k\lambda)} \sum_{k=0}^{\infty} \frac{(\frac{(\jmath-a)^\zeta}{\zeta})^{-k\lambda}}{\Gamma(1+\alpha k\lambda)} \left(\frac{\jmath}{x}\right)^s \left(\log\frac{x}{\jmath}\right)^{\zeta-1} \frac{|\mathfrak{z}(\jmath, \mathfrak{u}(\jmath)) - \mathfrak{z}(\jmath, \mathfrak{w}(\jmath))|}{\jmath}\, d\jmath \\ &\leq \frac{1}{\Gamma(\zeta)} \int_a^{\mathfrak{d}} \sum_{k=0}^{\infty} \frac{(\frac{(\mathfrak{d}-a)^\zeta}{\zeta})^{k\lambda}}{\Gamma(1+\alpha k\lambda)} \sum_{k=0}^{\infty} \frac{(\frac{(\jmath-a)^\zeta}{\zeta})^{-k\lambda}}{\Gamma(1+\alpha k\lambda)} \left(\frac{\jmath}{x}\right)^s \left(\log\frac{x}{\jmath}\right)^{\zeta-1} \frac{\Im|\mathfrak{u}(\jmath) - \mathfrak{w}(\jmath)|}{\jmath}\, d\jmath \\ &\leq L \sum_{k=0}^{\infty} \frac{(\frac{(x-a)^\zeta}{\zeta})^{k\lambda}}{\Gamma(1+\alpha k\lambda)} \frac{1}{\Gamma(\zeta)} \int_a^{\mathfrak{d}} \sum_{k=0}^{\infty} \frac{(\frac{(\jmath-a)^\zeta}{\zeta})^{-k\lambda}}{\Gamma(1+\alpha k\lambda)} Q_{\mathfrak{u}\mathfrak{w}} \; OS_1(\Theta) \left(\frac{\jmath}{x}\right)^s \left(\log\frac{x}{\jmath}\right)^{\zeta-1} \frac{d\jmath}{\jmath} \\ &\leq \Im \; Q_{\mathfrak{u}\mathfrak{w}} \sum_{k=0}^{\infty} \frac{(\frac{(\mathfrak{d}-a)^\zeta}{\zeta})^{k\lambda}}{\Gamma(1+\alpha k\lambda)} \boldsymbol{\Delta} \; OS_1(\Theta) = \left[L \sum_{k=0}^{\infty} \frac{(\frac{(\mathfrak{d}-a)^\zeta}{\zeta})^{k\lambda}}{\Gamma(1+\alpha k\lambda)} \boldsymbol{\Delta}\right] Q_{\mathfrak{u}\mathfrak{w}} \; OS_1(\Theta).\end{aligned}$$

Thus, we have

$$d(\Xi\mathfrak{u}, \Xi\mathfrak{w}) \leq \left[L \sum_{k=0}^{\infty} \frac{(\frac{(\mathfrak{d}-a)^{\zeta}}{\beta})^{k\lambda}}{\Gamma(1+\alpha k\lambda)} \boldsymbol{\Delta} \right] d(\mathfrak{u}, \mathfrak{w}). \tag{2.3.10}$$

Since, $0 < \Im \sum_{k=0}^{\infty} \frac{(\frac{(\mathfrak{d}-a)^{\zeta}}{\zeta})^{k\lambda}}{\Gamma(1+\alpha k\lambda)} \boldsymbol{\Delta} < 1$, therefore Λ is a contraction mapping.
Here and at this step, we show that $d(\Xi\mathfrak{w}, \mathfrak{w}) < \infty$.
Consider the function $\mathfrak{w} \in \mathcal{G}$, we have

$$\begin{aligned}
|\Xi(\mathfrak{w}(\mathfrak{d})) - \mathfrak{w}(\mathfrak{d})| &= \Bigg| \mathfrak{u}_a \sum_{k=0}^{\infty} \frac{(\frac{(\mathfrak{d}-a)^{\zeta}}{\zeta})^{k\lambda}}{\Gamma(1+\alpha k\lambda)} \\
&+ \frac{1}{\Gamma(\zeta)} \int_a^{\mathfrak{d}} \sum_{k=0}^{\infty} \frac{(\frac{(\mathfrak{d}-a)^{\zeta}}{\zeta})^{k\lambda}}{\Gamma(1+\alpha k\lambda)} \sum_{k=0}^{\infty} \frac{(\frac{(\jmath-a)^{\zeta}}{\zeta})^{-k\lambda}}{\Gamma(1+\alpha k\lambda)} \left(\frac{\jmath}{x}\right)^{s} \left(\log\frac{x}{\jmath}\right)^{\zeta-1} \frac{\mathfrak{z}(\jmath, \mathfrak{w}(\jmath))}{\jmath}\, \mathrm{d}\jmath - \mathfrak{w}(\mathfrak{d}) \Bigg| \\
&\leq \varepsilon \sum_{k=0}^{\infty} \frac{(\frac{(b-a)^{\zeta}}{\zeta})^{k\lambda}}{\Gamma(1+\alpha k\lambda)} \frac{1}{\Gamma(\beta)} \int_a^{\mathfrak{d}} \sum_{k=0}^{\infty} \frac{(\frac{(\jmath-a)^{\zeta}}{\zeta})^{-k\lambda}}{\Gamma(1+\alpha k\lambda)} {}_2F_1(\mathfrak{p}, \mathfrak{q}; \mathfrak{r}; \jmath) \left(\frac{\jmath}{x}\right)^{s} \left(\log\frac{x}{\jmath}\right)^{\zeta-1} \frac{\mathrm{d}\jmath}{\jmath} \\
&\leq (\varepsilon \sum_{k=0}^{\infty} \frac{(\frac{(b-a)^{\zeta}}{\zeta})^{k\lambda}}{\Gamma(1+\alpha k\lambda)} \boldsymbol{\Delta})\, OS_1(\Theta).
\end{aligned}$$

Consequently,

$$d(\Xi\mathfrak{w}, \mathfrak{w}) \leq \varepsilon \sum_{k=0}^{\infty} \frac{(\frac{(b-a)^{\zeta}}{\zeta})^{k\lambda}}{\Gamma(1+\alpha k\lambda)} \boldsymbol{\Delta} < \infty, \ \boldsymbol{\Delta} < 1. \tag{2.3.11}$$

Therefore, in this situation, by referring to Theorem 1.2, we find that all the cases of the theorem are true, that is, the sequence $\{\Xi^{\mathbf{e}}\mathfrak{u}\}$ converges to a fixed point $\mathfrak{u}$ of Ξ, where $\mathfrak{u}$ is in the set $\mathcal{G}^* = \{\mathfrak{w} \in \mathcal{G} : d(\Xi\mathfrak{w}, \mathfrak{u}) < \infty\}$. In other words $\Xi\mathfrak{u} = \mathfrak{u}$ or

$$\begin{aligned}
\mathfrak{u}(\mathfrak{d}) &= \mathfrak{u}_a \sum_{k=0}^{\infty} \frac{(\frac{(\mathfrak{d}-a)^{\zeta}}{\zeta})^{k\lambda}}{\Gamma(1+\alpha k\lambda)} \\
&+ \frac{1}{\Gamma(\zeta)} \int_a^{\mathfrak{d}} \sum_{k=0}^{\infty} \frac{(\frac{(\mathfrak{d}-a)^{\zeta}}{\zeta})^{k\lambda}}{\Gamma(1+\alpha k\lambda)} \sum_{k=0}^{\infty} \frac{(\frac{(\jmath-a)^{\zeta}}{\zeta})^{-k\lambda}}{\Gamma(1+\alpha k\lambda)} \left(\frac{\jmath}{x}\right)^{s} \left(\log\frac{x}{\jmath}\right)^{\zeta-1} \frac{\mathfrak{z}(\jmath, \mathfrak{u}(\jmath))}{\jmath}\, \mathrm{d}\jmath.
\end{aligned} \tag{2.3.12}$$

Since $\mathfrak{u}$ is a differentiable function, by the CFD and (2.3.12) and Lemma 1.2, we have

$$\mathfrak{D}_{\zeta}^{a}\mathfrak{u}(\mathfrak{d}) = \lambda\mathfrak{u}(\mathfrak{d}) + \mathfrak{z}(\mathfrak{d}, \mathfrak{u}(\mathfrak{d})). \tag{2.3.13}$$

Also, according to the inequality (2.3.11), we get

$$d(\mathfrak{w}, \mathfrak{u}) \leq \frac{1}{1-\upsilon} d(\Xi\mathfrak{w}, \mathfrak{w}) \leq \varepsilon \frac{\sum_{k=0}^{\infty} \frac{(\frac{(b-a)^{\zeta}}{\zeta})^{k\lambda}}{\Gamma(1+\alpha k\lambda)} \boldsymbol{\Delta}}{1-\upsilon}\, OS_1(\Theta).$$

Thus, H-U-R-OS_1-MS for the CFD Eq. (2.3.1) is established. In the following, to prove the uniqueness of the solution, we assume

$$\nu = \frac{\varepsilon \sum_{k=0}^{\infty} \frac{(\frac{(b-a)^{\zeta}}{\zeta})^{k\lambda}}{\Gamma(1+\alpha k\lambda)} \boldsymbol{\Delta}}{1-\upsilon}.$$

We consider the differentiable function $\mathfrak{x}$ and assume that it applies to Eq. (2.3.13). Then

$$\mathfrak{D}_{\zeta}^{a}\mathfrak{x}(\mathfrak{d}) = \lambda\mathfrak{x}(\mathfrak{d}) + \mathfrak{z}(\mathfrak{d}, \mathfrak{x}(\mathfrak{d})). \tag{2.3.14}$$

To finish proving the uniqueness of the solution, we show that $\mathfrak{x} \in \mathcal{G}^*$ is a fixed point for Ξ. According to (2.3.14), $\Xi\mathfrak{x} = \mathfrak{x}$. In the sequel, we show that $d(\Xi\mathfrak{w}, \mathfrak{x}) < \infty$. Considering (2.3.14) and assuming $\mathfrak{w} \in \mathcal{G}$, $d(\mathfrak{w}, \mathfrak{x}) < \nu$, we have

$$\begin{aligned}
|\Xi(\mathfrak{w}(\mathfrak{d})) - \mathfrak{x}(\mathfrak{d})| &= \Bigg| \mathfrak{u}_a \sum_{k=0}^{\infty} \frac{(\frac{(\mathfrak{d}-a)^{\zeta}}{\zeta})^{k\lambda}}{\Gamma(1+\alpha k\lambda)} \\
&+ \frac{1}{\Gamma(\zeta)} \int_a^{\mathfrak{d}} \sum_{k=0}^{\infty} \frac{(\frac{(\mathfrak{d}-a)^{\zeta}}{\zeta})^{k\lambda}}{\Gamma(1+\alpha k\lambda)} \sum_{k=0}^{\infty} \frac{(\frac{(\jmath-a)^{\zeta}}{\zeta})^{-k\lambda}}{\Gamma(1+\alpha k\lambda)} \left(\frac{\jmath}{x}\right)^s \left(\log\frac{x}{\jmath}\right)^{\zeta-1} \frac{\mathfrak{z}(\jmath, \mathfrak{w}(\jmath))}{\jmath} \,\mathrm{d}\jmath \\
&- \mathfrak{u}_a \sum_{k=0}^{\infty} \frac{(\frac{(\mathfrak{d}-a)^{\zeta}}{\zeta})^{k\lambda}}{\Gamma(1+\alpha k\lambda)} \\
&- \frac{1}{\Gamma(\zeta)} \int_a^{\mathfrak{d}} \sum_{k=0}^{\infty} \frac{(\frac{(\mathfrak{d}-a)^{\zeta}}{\zeta})^{k\lambda}}{\Gamma(1+\alpha k\lambda)} \sum_{k=0}^{\infty} \frac{(\frac{(\jmath-a)^{\zeta}}{\zeta})^{-k\lambda}}{\Gamma(1+\alpha k\lambda)} \left(\frac{\jmath}{x}\right)^s \left(\log\frac{x}{\jmath}\right)^{\zeta-1} \frac{\mathfrak{z}(\jmath, \mathfrak{x}(\jmath))}{\jmath} \,\mathrm{d}\jmath \Bigg| \\
&\le \frac{1}{\Gamma(\zeta)} \int_a^{\mathfrak{d}} \sum_{k=0}^{\infty} \frac{(\frac{(\mathfrak{d}-a)^{\zeta}}{\zeta})^{k\lambda}}{\Gamma(1+\alpha k\lambda)} \sum_{k=0}^{\infty} \frac{(\frac{(\jmath-a)^{\zeta}}{\zeta})^{-k\lambda}}{\Gamma(1+\alpha k\lambda)} \left(\frac{\jmath}{x}\right)^s \left(\log\frac{x}{\jmath}\right)^{\zeta-1} \frac{|\mathfrak{z}(\jmath, \mathfrak{w}(\jmath)) - \mathfrak{z}(\jmath, \mathfrak{x}(\jmath))|}{\jmath} \,\mathrm{d}\jmath \\
&\le \frac{1}{\Gamma(\zeta)} \int_a^{\mathfrak{d}} \sum_{k=0}^{\infty} \frac{(\frac{(\mathfrak{d}-a)^{\zeta}}{\zeta})^{k\lambda}}{\Gamma(1+\alpha k\lambda)} \sum_{k=0}^{\infty} \frac{(\frac{(\jmath-a)^{\zeta}}{\zeta})^{-k\lambda}}{\Gamma(1+\alpha k\lambda)} \left(\frac{\jmath}{x}\right)^s \left(\log\frac{x}{\jmath}\right)^{\zeta-1} \frac{\Im|\mathfrak{w}(\jmath) - \mathfrak{x}(\jmath)|}{\jmath} \,\mathrm{d}\jmath \\
&\le \Im \sum_{k=0}^{\infty} \frac{(\frac{(\mathfrak{d}-a)^{\zeta}}{\zeta})^{k\lambda}}{\Gamma(1+\alpha k\lambda)} \frac{1}{\Gamma(\zeta)} \int_a^{\mathfrak{d}} \sum_{k=0}^{\infty} \frac{(\frac{(\jmath-a)^{\zeta}}{\zeta})^{-k\lambda}}{\Gamma(1+\alpha k\lambda)} \nu\, OS_1(\Theta) \left(\frac{\jmath}{x}\right)^s \left(\log\frac{x}{\jmath}\right)^{\zeta-1} \frac{\mathrm{d}\jmath}{\jmath} \\
&\le \Im \sum_{k=0}^{\infty} \frac{(\frac{(\mathfrak{d}-a)^{\zeta}}{\zeta})^{k\lambda}}{\Gamma(1+\alpha k\lambda)} \nu \boldsymbol{\Delta}\, OS_1(\Theta) = [\Im \sum_{k=0}^{\infty} \frac{(\frac{(\mathfrak{d}-a)^{\zeta}}{\zeta})^{k\lambda}}{\Gamma(1+\alpha k\lambda)} \boldsymbol{\Delta}]\nu\, OS_1(\Theta),
\end{aligned}$$

that means

$$d(\Xi\mathfrak{w}, \mathfrak{x}) \le \left[\Im \sum_{k=0}^{\infty} \frac{(\frac{(\mathfrak{d}-a)^{\zeta}}{\zeta})^{k\lambda}}{\Gamma(1+\alpha k\lambda)} \boldsymbol{\Delta} \right] \nu < \infty.$$

2.3.3 Application

Example 2.4 We consider the following CFD-E (Fig. 2.3)

$$\begin{cases} \mathfrak{D}^1_{\frac{1}{2}} \mathfrak{u}(\mathfrak{d}) = \mathfrak{u}(\mathfrak{d}) + \frac{1}{5+x^2}(\frac{\mathfrak{u}^2(\mathfrak{d})}{\mathfrak{u}(\mathfrak{d})+1}) \cos(\mathfrak{u}(\mathfrak{d})), \mathfrak{d} \in (1, 2] \\ \mathfrak{u}(1) = 1 \end{cases} \tag{2.3.15}$$

where $\zeta = \frac{1}{2}$, $\lambda = 1$, $\Im = \frac{1}{8}$, $\boldsymbol{\Delta} = \frac{1}{2}$, $\upsilon = \frac{1}{16} \sum_{k=0}^{\infty} \frac{(2(x-1)^{\frac{1}{2}})^k}{\Gamma(1+\alpha k)} < 1$(for $\mathfrak{d} = 2, \alpha = 1$ we have $\upsilon = \frac{1}{16} \exp(2) = 0/4618 < 1$) and $\mathfrak{u}_1 = 1$. Also in this equation, $\mathfrak{z}(\mathfrak{d}, \mathfrak{u}(\mathfrak{d})) = \frac{1}{5+\mathfrak{d}^2}(\frac{\mathfrak{u}^2(\mathfrak{d})}{\mathfrak{u}(\mathfrak{d})+1}) \cos(\mathfrak{u}(\mathfrak{d}))$. We consider the following inequalities

$$|\mathfrak{z}(\mathfrak{d}, \mathfrak{u}(\mathfrak{d})) - \mathfrak{z}(\mathfrak{d}, \mathfrak{w}(\mathfrak{d}))| \leq \frac{1}{8}|\mathfrak{u}(\mathfrak{d}) - \mathfrak{w}(\mathfrak{d})|, \tag{2.3.16}$$

and

$$\frac{1}{\Gamma(\frac{1}{2})} \int_a^{\mathfrak{d}} \sum_{k=0}^{\infty} \frac{(2(\jmath - a)^{\frac{1}{2}})^{-k\lambda}}{\Gamma(1+\alpha k\lambda)} OS_1(\Theta) \left(\frac{\jmath}{x}\right)^s \left(\log \frac{x}{\jmath}\right)^{-\frac{1}{2}} \frac{d\jmath}{\jmath} \leq \frac{1}{2} OS_1(\Theta). \tag{2.3.17}$$

Let $\mathfrak{w} \in C([1, 2], \mathbb{R})$ be a differentiable function such that

$$\left| \mathfrak{D}^1_{\frac{1}{2}} \mathfrak{w}(\mathfrak{d}) - \mathfrak{w}(\mathfrak{d}) - \frac{1}{5+x^2}(\frac{\mathfrak{u}^2(\mathfrak{d})}{\mathfrak{u}(\mathfrak{d})+1}) \cos(\mathfrak{u}(\mathfrak{d})) \right| \leq \varepsilon\, OS_1(\Theta), \mathfrak{d} \in (1, 2], \tag{2.3.18}$$

then, $\mathfrak{w}$ is a solution of the inequality

$$\begin{aligned} &\Bigg| \mathfrak{w}(\mathfrak{d}) - \sum_{k=0}^{\infty} \frac{(2(\mathfrak{d}-1)^{\frac{1}{2}})^k}{\Gamma(1+\alpha k)} \\ &- \frac{1}{\Gamma(\frac{1}{2})} \int_a^{\mathfrak{d}} \sum_{k=0}^{\infty} \frac{(2(\mathfrak{d}-1)^{\frac{1}{2}})^k}{\Gamma(1+\alpha k)} \sum_{k=0}^{\infty} \frac{\left(2(\jmath-1)^{\frac{1}{2}}\right)^{-k}}{\Gamma(1+\alpha k)} \frac{1}{5+\mathfrak{d}^2}(\frac{\mathfrak{w}^2(\mathfrak{d})}{\mathfrak{w}(\mathfrak{d})+1}) \cos(\mathfrak{w}(\mathfrak{d})) \left(\frac{\jmath}{x}\right)^s \left(\log \frac{x}{\jmath}\right)^{-\frac{1}{2}} \frac{d\jmath}{\jmath} \Bigg| \\ &\leq \varepsilon \sum_{k=0}^{\infty} \frac{2^k}{\Gamma(1+\alpha k)} \frac{1}{\Gamma(\frac{1}{2})} \int_a^{\mathfrak{d}} \sum_{k=0}^{\infty} \frac{\left(2(\jmath-1)^{\frac{1}{2}}\right)^{-k}}{\Gamma(1+\alpha k)} OS_1(\Theta) \left(\frac{\jmath}{x}\right)^s \left(\log \frac{x}{\jmath}\right)^{-\frac{1}{2}} \frac{d\jmath}{\jmath}, \end{aligned}$$

also, there exists a unique differentiable function $\mathfrak{u} \in C([1, 2], \mathbb{R})$ from (2.3.15) such that for each $\mathfrak{d} \in [1, 2]$ we have

$$\begin{aligned} \mathfrak{u}(\mathfrak{d}) &= \sum_{k=0}^{\infty} \frac{(2(\mathfrak{d}-1)^{\frac{1}{2}})^k}{\Gamma(1+k\alpha)} \\ &+ \frac{1}{\Gamma(\frac{1}{2})} \int_1^{\mathfrak{d}} \sum_{k=0}^{\infty} \frac{(2(\mathfrak{d}-1)^{\frac{1}{2}})^k}{\Gamma(1+k\alpha)} \sum_{k=0}^{\infty} \frac{(2(\jmath-1)^{\frac{1}{2}})^{-k}}{\Gamma(1+k\alpha)} \left(\frac{\jmath}{x}\right)^s \left(\log \frac{x}{\jmath}\right)^{-1/2} \frac{\frac{1}{5+x^2}(\frac{\mathfrak{u}^2(\mathfrak{d})}{\mathfrak{u}(\mathfrak{d})+1}) \cos(\mathfrak{u}(\mathfrak{d}))}{\jmath} d\jmath, \end{aligned}$$

therefore, for $\mathfrak{R} = \frac{2}{1-\upsilon} \sum_{k=0}^{\infty} \frac{2^k}{\Gamma(1+\alpha k)}$

$$d(\mathfrak{w}, \mathfrak{u}) \leq \mathfrak{R}\, \varepsilon,$$

which implies that

$$|\mathfrak{w}(\mathfrak{d}) - \mathfrak{u}(\mathfrak{d})| \leq \mathfrak{R}\, \varepsilon\, OS_1(\Theta).$$

2.4 Fractional Nonlinear Problem with Integral Boundary Condition

2.4.1 *Introduction of the Fractional Integral Boundary Condition Problems*

Cabada and his colleague Hamdi [85], Young, and his colleagues [88] were among those who have investigated two models of IBCPs. These two fractional differential equations with boundary conditions are defined as follows:

$$\begin{cases} {}_{RL}D^{\beta}_{0+}u(x) + f(x, u(x)) = 0, & x \in (0, 1) \\ u(0) = u'(0) = 0, \quad u(1) = \rho \int_0^1 u(x)dx, \end{cases} \tag{2.4.1}$$

for the continuous function $f(x, u)$ and $2 < \beta \leq 3,\ 0 < \rho < \beta$ and a more general problem

$$\begin{cases} -({}_{RL}D^{\beta}_{0+})u(x) = f(x, u(x)) + \gamma g(x), & x \in (0, 1) \\ u(0) = u'(0) = \theta, \quad u(1) = \rho \int_0^1 u(x)dx, \end{cases}$$

for the continuous functions $f(x, u)\ : [0, 1] \times K \to K,\ g\ : [0, 1] \to K$ and $2 < \beta \leq 3,\ 0 < \rho < \beta, \gamma \in R$. K and θ are the normal cone and a zero element in the Banach space E, respectively. $\gamma \in R$ is also a variable sign parameter. Cabada and Hamdi proved their results by considering Guo-Krasnoselskii's FPT for Eq. (2.4.1). The following N-FDE is another IBCP that has been investigated by Wang et al., using the monotonic iterative method [89]:

$$\begin{cases} ({}_{RL}D^{q}_{0+})u(x) = f(x, u(x)), & x \in [0, T], \quad T > 0, \\ u(0) = \lambda \int_0^T u(x)dx + d, \quad d \in R, \end{cases}$$

for the continuous function $f : [0, T] \times R \to R$ and $\lambda \geq 0$, $0 < q < 1$. Here, for $l, m \in N$ and $0 \leq \alpha_n < \cdots < \alpha_1 \leq l-1 < \alpha \leq l$, $0 < \beta_1 < \cdots < \beta_m$, $0 \leq a < b < +\infty$, we consider the following N-FIDE, which is an IBCP:

$$\begin{cases} {}_C D_a^{\alpha} u(x) - \lambda_1\, {}_C D_a^{\alpha_1} u(x) - \cdots - \lambda_n\, {}_C D_a^{\alpha_n} u(x) - \lambda_{n+1} I_a^{\beta_1} u(x) - \cdots - \lambda_{n+m} I_a^{\beta_m} u(x) \\ \quad = f(x, u(x)), \quad x \in [a, b] \\ u(a) = u'(a) = \cdots = u^{(l-2)}(a) = 0, \\ u(b) = \lambda_1 I_a^{\alpha-\alpha_1} u(b) + \cdots + \lambda_n I_a^{\alpha-\alpha_n} u(b) + \lambda_{n+1} I_a^{\alpha+\beta_1} u(b) + \cdots + \lambda_{n+m} I_a^{\alpha+\beta_m} u(b), \end{cases} \tag{2.4.2}$$

where $f : [a, b] \times \mathbb{R} \to \mathbb{R}$ is a mapping satisfying certain conditions and λ_j is a constant for all $j = 1, 2, \ldots, n+m$. In particular for $l = 1$, Eq. (2.4.2) turns out to be

$$\begin{cases} {}_C D_a^{\alpha} u(x) - \lambda_n\, u(x) - \lambda_{n+1} I_a^{\beta_1} u(x) - \cdots - \lambda_{n+m} I_a^{\beta_m} u(x) \\ \quad = f(x, u(x)), \quad x \in [a, b] \\ u(b) = \lambda_n I_a^{\alpha} u(b) + \lambda_{n+1} I_a^{\alpha+\beta_1} u(b) + \cdots + \lambda_{n+m} I_a^{\alpha+\beta_m} u(b). \end{cases}$$

One of the important tools for solving differential and integral equations with initial conditions is Babenko's strategy. This method is generally the same as the Laplace transform, which works on equations with constant coefficients. Also, this method is used for differential and integral equations with continuous and limited variable coefficients [94]. In the following, to show the applications of this approach, we will investigate the solutions of two problems in the space $C^{l-1}[a, b]$. These equations are defined as follows:

$$\begin{cases} {}_C D_0^{\alpha} u(x) + a(x) I_0^{\beta} u(x) = g(x), \quad x \in [0, T], \quad l-1 < \alpha \leq l, \quad \beta \geq 0, \\ u(0) = u'(0) = \cdots = u^{(l-1)}(0) = 0, \end{cases} \tag{2.4.3}$$

where $a(x),\ g(x) \in C[0, T]$, and

$$\begin{cases} {}_C D_a^{\alpha} u(x) - \lambda_1\, {}_C D_a^{\alpha_1} u(x) - \cdots - \lambda_n\, {}_C D_a^{\alpha_n} u(x) - \lambda_{n+1} I_a^{\beta_1} u(x) - \cdots - \lambda_{n+m} I_a^{\beta_m} u(x) \\ \quad = f(x), \quad x \in [a, b] \\ u(a) = u'(a) = \cdots = u^{(l-2)}(a) = 0, \\ u(b) = \lambda_1 I_a^{\alpha-\alpha_1} u(b) + \cdots + \lambda_n I_a^{\alpha-\alpha_n} u(b) + \lambda_{n+1} I_a^{\alpha+\beta_1} u(b) + \cdots + \lambda_{n+m} I_a^{\alpha+\beta_m} u(b). \end{cases} \tag{2.4.4}$$

where $f \in C[a, b]$.

2.4.2 Preliminaries

Lemma 2.3 *Assume that* $a(x),\ g(x) \in C[0, T]$. *The Eq.* (2.4.3) *has a unique solution.*

Proof First, we can apply the operator I_0^{α} to both sides of equation

$${}_C D_0^{\alpha} u(x) + a(x) I_0^{\beta} u(x) = g(x),$$

then using the initial condition $u(0) = u'(0) = \cdots = u^{(l-1)}(0) = 0$, we have

$$u(x) + I_0^\alpha a(x) I_0^\beta u(x) = I_0^\alpha g(x).$$

Then

$$(1 + I_0^\alpha a(x) I_0^\beta) u(x) = u(x) + I_0^\alpha a(x) I_0^\beta u(x) = I_0^\alpha g(x).$$

Due to the boundedness of the variable $I_0^\alpha a(x) I_0^\beta$ on $C[a,b]$ and considering that $1 + I_0^\alpha a(x) I_0^\beta$ is a normal variable, we use Babenko's strategy. Then using $\left(I_0^\alpha a(x) I_0^\beta\right)^k I_0^\alpha = I_0^\alpha \left(a(x) I_0^{\alpha+\beta}\right)^k$, we have

$$\begin{aligned} u(x) = &\left(1 + I_0^\alpha a(x) I_0^\beta\right)^{-1} I_0^\alpha g(x) = \sum_{k=0}^{\infty} (-1)^k \left(I_0^\alpha a(x) I_0^\beta\right)^k I_0^\alpha g(x) \\ = &\sum_{k=0}^{\infty} (-1)^k I_0^\alpha \left(a(x) I_0^{\alpha+\beta}\right)^k g(x), \end{aligned}$$

In the following, we show that in the space $C^{l-1}[a,b]$, the obtained series is convergent on the right side of the above equation. Clearly

$$\left\| I_0^\alpha g(x) \right\| = \max_{x \in [0,T]} \frac{1}{\Gamma(\alpha)} \left| \int_0^x (x-t)^{\alpha-1} g(t) dt \right| \leq \frac{T^\alpha}{\Gamma(\alpha+1)} \|g\|,$$

and

$$\left\| I_0^\alpha a(x) I_0^\beta g(x) \right\| \leq M \frac{T^{\alpha+\beta}}{\Gamma(\alpha+\beta+1)} \|g\|,$$

where $M = \max_{x \in [0,T]} |a(x)|$. Furthermore,

$$\begin{aligned} \|u\| \leq &\sum_{k=0}^{\infty} \left\| \left(I_0^\alpha a(x) I_0^\beta\right)^k I_0^\alpha g(x) \right\| \leq \|g\| \sum_{k=0}^{\infty} M^k \frac{T^{(\alpha+\beta)k+\alpha}}{\Gamma((\alpha+\beta)k+\alpha+1)} \\ = &\|g\| T^\alpha E_{\alpha+\beta,\,\alpha+1}\left(M T^{\alpha+\beta}\right), \end{aligned}$$

which infers that $u \in C[0,T]$. Moreover, from the identity

$$u(x) = \sum_{k=0}^{\infty} (-1)^k I_0^\alpha \left(a(x) I_0^{\alpha+\beta}\right)^k g(x) = I_0^\alpha \sum_{k=0}^{\infty} (-1)^k \left(a(x) I_0^{\alpha+\beta}\right)^k g(x),$$

we deduce that $u \in C^{l-1}[a,b]$ due to the factor I_0^α.

Lemma 2.4 *Assume that* $0 \le \alpha_n < \cdots < \alpha_1 \le l-1 < \alpha \le l \in N$, $0 < \beta_1 < \cdots < \beta_m$, $0 \le a < b < +\infty$, λ_j *is a constant for all* $j = 1, 2, \ldots, n+m$ *and* f *is a continuous mapping from* $[a, b]$ *to* R. *The Eq.* (2.4.4) *has a unique solution*

$$u(x) = \sum_{k=0}^{\infty} \sum_{k_1+\cdots+k_{n+m}=k} \binom{k}{k_1, \ldots, k_{n+m}} \lambda_1^{k_1} \cdots \lambda_{n+m}^{k_{n+m}} \cdot I_a^{\alpha+(\alpha-\alpha_1)k_1+\cdots+(\alpha+\beta_m)k_{n+m}} f(x)$$
$$- \frac{1}{(b-a)^{l-1}\Gamma(\alpha)} \int_a^b (b-x)^{\alpha-1} f(x)dx \cdot \sum_{k=0}^{\infty} \sum_{k_1+\cdots+k_{n+m}=k} \binom{k}{k_1, \ldots, k_{n+m}} \cdot$$
$$\lambda_1^{k_1} \cdots \lambda_{n+m}^{k_{n+m}} \cdot I_a^{(\alpha-\alpha_1)k_1+\cdots+(\alpha+\beta_m)k_{n+m}} (x-a)^{l-1}$$

in the space $C^{l-1}[a, b]$.

Proof Like the previous lemma, here we start the proof by applying I_a^{α} to both sides of the following equation. Then, we have

$$ {}_C D_a^{\alpha} u(x) - \lambda_1 \, {}_C D_a^{\alpha_1} u(x) - \cdots - \lambda_n \, {}_C D_a^{\alpha_n} u(x) - \lambda_{n+1} I_a^{\beta_1} u(x) - \cdots - \lambda_{n+m} I_a^{\beta_m} u(x)$$
$$= f(x),$$

to have

$$u(x) - \lambda_1 \, I_a^{\alpha-\alpha_1} u(x) - \cdots - \lambda_n I_a^{\alpha-\alpha_n} u(x) - \lambda_{n+1} I_a^{\alpha+\beta_1} u(x) - \cdots - \lambda_{n+m} I_a^{\alpha+\beta_m} u(x)$$
$$= I_a^{\alpha} f(x) + c(x-a)^{l-1}, \tag{2.4.5}$$

by utilizing the condition

$$u(a) = u'(a) = \cdots = u^{(l-2)}(a) = 0,$$

and $0 \le \alpha_n < \cdots < \alpha_1 \le l-1 < \alpha \le l \in N$, where c is a constant to be determined by the boundary condition.

Setting $x = b$ in Eq. (2.4.5)and using the condition

$$u(b) = \lambda_1 I_a^{\alpha-\alpha_1} u(b) + \cdots + \lambda_n I_a^{\alpha-\alpha_n} u(b) + \lambda_{n+1} I_a^{\alpha+\beta_1} u(b) + \cdots + \lambda_{n+m} I_a^{\alpha+\beta_m} u(b),$$

we get

$$I_a^{\alpha} f(b) + c(b-a)^{l-1} = 0,$$

which implies that

$$c = -\frac{1}{(b-a)^{l-1}} I_a^{\alpha} f(b).$$

Hence, Eq. (2.4.5) turns out to be

$$\left(1 - \lambda_1 \, I_a^{\alpha-\alpha_1} - \cdots - \lambda_n I_a^{\alpha-\alpha_n} - \lambda_{n+1} I_a^{\alpha+\beta_1} - \cdots - \lambda_{n+m} I_a^{\alpha+\beta_m}\right) u(x)$$
$$= I_a^{\alpha} f(x) - \frac{(x-a)^{l-1}}{(b-a)^{l-1}\Gamma(\alpha)} \int_a^b (b-x)^{\alpha-1} f(x)dx.$$

By Babenko's approach, we come to

$$u(x) = \left(1 - \lambda_1\, I_a^{\alpha-\alpha_1} - \cdots - \lambda_n I_a^{\alpha-\alpha_n} - \lambda_{n+1} I_a^{\alpha+\beta_1} - \cdots - \lambda_{n+m} I_a^{\alpha+\beta_m}\right)^{-1} \cdot \left[I_a^\alpha f(x) - \frac{(x-a)^{l-1}}{(b-a)^{l-1}\Gamma(\alpha)} \int_a^b (b-x)^{\alpha-1} f(x)dx\right]$$

$$= \sum_{k=0}^{\infty} \left(\lambda_1\, I_a^{\alpha-\alpha_1} + \cdots + \lambda_n I_a^{\alpha-\alpha_n} + \lambda_{n+1} I_a^{\alpha+\beta_1} + \cdots + \lambda_{n+m} I_a^{\alpha+\beta_m}\right)^{k} \cdot \left[I_a^\alpha f(x) - \frac{(x-a)^{l-1}}{(b-a)^{l-1}\Gamma(\alpha)} \int_a^b (b-x)^{\alpha-1} f(x)dx\right].$$

It follows from the multinomial theorem that

$$\left(\lambda_1\, I_a^{\alpha-\alpha_1} + \cdots + \lambda_n I_a^{\alpha-\alpha_n} + \lambda_{n+1} I_a^{\alpha+\beta_1} + \cdots + \lambda_{n+m} I_a^{\alpha+\beta_m}\right)^{k}$$
$$= \sum_{k_1+\cdots+k_{n+m}=k} \binom{k}{k_1, \ldots, k_{n+m}} \left(\lambda_1\, I_a^{\alpha-\alpha_1}\right)^{k_1} \cdots \left(\lambda_{n+m} I_a^{\alpha+\beta_m}\right)^{k_{n+m}}$$
$$= \sum_{k_1+\cdots+k_{n+m}=k} \binom{k}{k_1, \ldots, k_{n+m}} \lambda_1^{k_1} \cdots \lambda_{n+m}^{k_{n+m}}\, I_a^{(\alpha-\alpha_1)k_1+\cdots+(\alpha+\beta_m)k_{n+m}}.$$

Thus,

$$u(x) = \sum_{k=0}^{\infty} \sum_{k_1+\cdots+k_{n+m}=k} \binom{k}{k_1, \ldots, k_{n+m}} \lambda_1^{k_1} \cdots \lambda_{n+m}^{k_{n+m}} \cdot I_a^{\alpha+(\alpha-\alpha_1)k_1+\cdots+(\alpha+\beta_m)k_{n+m}} f(x)$$
$$- \frac{1}{(b-a)^{l-1}\Gamma(\alpha)} \int_a^b (b-x)^{\alpha-1} f(x)dx \cdot \sum_{k=0}^{\infty} \sum_{k_1+\cdots+k_{n+m}=k} \binom{k}{k_1, \ldots, k_{n+m}} \cdot$$
$$\lambda_1^{k_1} \cdots \lambda_{n+m}^{k_{n+m}} \cdot I_a^{(\alpha-\alpha_1)k_1+\cdots+(\alpha+\beta_m)k_{n+m}} (x-a)^{l-1}.$$

Evidently, $u^{(l-1)}(x)$ is continuous due to the two factors I_a^α and $(x-a)^{l-1}$. Further, we consider

$$\|u\| \leq \|f\|\,(b-a)^\alpha \sum_{k=0}^{\infty} \sum_{k_1+\cdots+k_{n+m}=k} \binom{k}{k_1, \ldots, k_{n+m}} |\lambda_1|^{k_1} \cdots |\lambda_{n+m}|^{k_{n+m}} \cdot \frac{(b-a)^{(\alpha-\alpha_1)k_1+\cdots+(\alpha+\beta_m)k_{n+m}}}{\Gamma((\alpha-\alpha_1)k_1 + \cdots + (\alpha+\beta_m)k_{n+m} + \alpha + 1)} +$$
$$+ \frac{\|f\|\,(b-a)^\alpha}{\Gamma(\alpha+1)} \sum_{k=0}^{\infty} \sum_{k_1+\cdots+k_{n+m}=k} \binom{k}{k_1, \ldots, k_{n+m}} |\lambda_1|^{k_1} \cdots |\lambda_{n+m}|^{k_{n+m}} \cdot \frac{(b-a)^{(\alpha-\alpha_1)k_1+\cdots+(\alpha+\beta_m)k_{n+m}}}{\Gamma((\alpha-\alpha_1)k_1 + \cdots + (\alpha+\beta_m)k_{n+m} + 1)}$$
$$= \|f\|\,(b-a)^\alpha \cdot \left(E_{(\alpha-\alpha_1,\ldots,\alpha+\beta_m,\, \alpha+1)}\left(|\lambda_1|(b-a)^{\alpha-\alpha_1}, \ldots, |\lambda_{n+m}|(b-a)^{\alpha+\beta_m}\right)\right.$$
$$\left. + \frac{1}{\Gamma(\alpha+1)} E_{(\alpha-\alpha_1,\ldots,\alpha+\beta_m,\, 1)}\left(|\lambda_1|(b-a)^{\alpha-\alpha_1}, \ldots, |\lambda_{n+m}|(b-a)^{\alpha+\beta_m}\right)\right) < +\infty,$$

which claims that $u \in C^{l-1}[a,b]$. Given that the following system

$$\begin{cases} {}_CD_a^{\alpha}u(x) - \lambda_1\, {}_CD_a^{\alpha_1}u(x) - \cdots - \lambda_n\, {}_CD_a^{\alpha_n}u(x) - \lambda_{n+1}I_a^{\beta_1}u(x) - \cdots - \lambda_{n+m}I_a^{\beta_m}u(x) \\ \quad = 0, \quad x \in [a,b] \\ u(a) = u'(a) = \cdots = u^{(l-2)}(a) = 0, \\ u(b) = \lambda_1 I_a^{\alpha-\alpha_1}u(b) + \cdots + \lambda_n I_a^{\alpha-\alpha_n}u(b) + \lambda_{n+1}I_a^{\alpha+\beta_1}u(b) + \cdots + \lambda_{n+m}I_a^{\alpha+\beta_m}u(b), \end{cases}$$

has only zero solution in $C^{l-1}[a,b]$, therefore uniqueness is proved.

2.4.3 Existence and Uniqueness

Theorem 2.8 *We consider the following conditions for the continuous function* $f : [a,b] \times R \to R$ *and* θ*:*

- f *satisfies in the the Lipschitz condition with constant* $L \geq 0$*, that is*

$$|f(x,y_1) - f(x,y_2)| \leq L|y_1 - y_2|,$$

  for $y_1, y_2 \in R$.
- *we suppose*

$$\theta = L(b-a)^{\alpha}\,[E_{(\alpha-\alpha_1,\ldots,\alpha+\beta_m,\ \alpha+1)}\left(|\lambda_1|(b-a)^{\alpha-\alpha_1}, \ldots, |\lambda_{n+m}|(b-a)^{\alpha+\beta_m}\right)$$
$$+\frac{1}{\Gamma(\alpha+1)}E_{(\alpha-\alpha_1,\ldots,\alpha+\beta_m,\ 1)}\left(|\lambda_1|(b-a)^{\alpha-\alpha_1}, \ldots, |\lambda_{n+m}|(b-a)^{\alpha+\beta_m}\right)] < 1.$$

Then the boundary value problem (2.4.2) has a unique solution in the space $C^{l-1}[a,b]$.

Proof Define a nonlinear mapping T over $C^{l-1}[a,b]$ by

$$(Tu)(x) =$$
$$\sum_{k=0}^{\infty}\sum_{k_1+\cdots+k_{n+m}=k}\binom{k}{k_1,\ldots,k_{n+m}}\lambda_1^{k_1}\cdots\lambda_{n+m}^{k_{n+m}}\cdot I_a^{\alpha+(\alpha-\alpha_1)k_1+\cdots+(\alpha+\beta_m)k_{n+m}}f(x,u(x))$$
$$-\frac{1}{(b-a)^{l-1}\Gamma(\alpha)}\int_a^b (b-x)^{\alpha-1}f(x,u(x))dx\cdot\sum_{k=0}^{\infty}\sum_{k_1+\cdots+k_{n+m}=k}\binom{k}{k_1,\ldots,k_{n+m}}\cdot$$
$$\lambda_1^{k_1}\cdots\lambda_{n+m}^{k_{n+m}}\cdot I_a^{(\alpha-\alpha_1)k_1+\cdots+(\alpha+\beta_m)k_{n+m}}(x-a)^{l-1}.$$

Clearly,

$$\begin{aligned}\max_{x\in[a,b]}|f(x,u(x))| &= \max_{x\in[a,b]}|f(x,u(x)) - f(x,0) + f(x,0)| \\ &\leq \max_{x\in[a,b]}L|u(x)| + \max_{x\in[a,b]}|f(x,0)| \\ &= L\,\|u\| + \max_{x\in[a,b]}|f(x,0)| < +\infty,\end{aligned}$$

using the Lipschitz condition and noting that $f(x,0)$ is a continuous function over $[a,b]$.

It follows from the proof of Lemma 2.4 that $(Tu)(x) \in C^{l-1}[a,b]$. We need to prove that T is contractive. Indeed for $u,\ v \in C^{l-1}[a,b]$,

$$
\begin{aligned}
&|(Tu)(x)-(Tv)(x)| \le \\
&\sum_{k=0}^{\infty}\sum_{k_1+\cdots+k_{n+m}=k}\binom{k}{k_1,\ldots,k_{n+m}}|\lambda_1|^{k_1}\cdots|\lambda_{n+m}|^{k_{n+m}}\cdot \\
&\frac{1}{\Gamma((\alpha-\alpha_1)k_1+\cdots+(\alpha+\beta_m)k_{n+m}+\alpha)}\cdot \\
&\left|\int_a^x (x-t)^{(\alpha-\alpha_1)k_1+\cdots+(\alpha+\beta_m)k_{n+m}+\alpha-1}(f(t,u(t))-f(t,v(t))dt\right| \\
&+\frac{1}{(b-a)^{l-1}\Gamma(\alpha)}\left|\int_a^b (b-x)^{\alpha-1}(f(x,u(x))-f(x,v(x)))dx\right|\cdot \\
&\sum_{k=0}^{\infty}\sum_{k_1+\cdots+k_{n+m}=k}\binom{k}{k_1,\ldots,k_{n+m}}\cdot \\
&|\lambda_1|^{k_1}\cdots|\lambda_{n+m}|^{k_{n+m}}\cdot I_a^{(\alpha-\alpha_1)k_1+\cdots+(\alpha+\beta_m)k_{n+m}}(x-a)^{l-1} \\
&\le L(b-a)^{\alpha}\|u-v\|\sum_{k=0}^{\infty}\sum_{k_1+\cdots+k_{n+m}=k}\binom{k}{k_1,\ldots,k_{n+m}}|\lambda_1|^{k_1}\cdots|\lambda_{n+m}|^{k_{n+m}}\cdot \\
&\frac{(b-a)^{(\alpha-\alpha_1)k_1+\cdots+(\alpha+\beta_m)k_{n+m}}}{\Gamma((\alpha-\alpha_1)k_1+\cdots+(\alpha+\beta_m)k_{n+m}+\alpha+1)}+ \\
&L(b-a)^{\alpha}\|u-v\|\frac{1}{\Gamma(\alpha+1)}\sum_{k=0}^{\infty}\sum_{k_1+\cdots+k_{n+m}=k}\binom{k}{k_1,\ldots,k_{n+m}}|\lambda_1|^{k_1}\cdots|\lambda_{n+m}|^{k_{n+m}}\cdot \\
&\frac{(b-a)^{\Gamma((\alpha-\alpha_1)k_1+\cdots+(\alpha+\beta_m)k_{n+m}}}{\Gamma((\alpha-\alpha_1)k_1+\cdots+(\alpha+\beta_m)k_{n+m}+1)} \\
&=L(b-a)^{\alpha}\|u-v\|\,E_{(\alpha-\alpha_1,\ldots,\alpha+\beta_m,\,\alpha+1)}\left(|\lambda_1|(b-a)^{\alpha-\alpha_1},\ldots,|\lambda_{n+m}(b-a)^{\alpha+\beta_m}\right) \\
&+L(b-a)^{\alpha}\|u-v\|\frac{1}{\Gamma(\alpha+1)}\cdot \\
&E_{(\alpha-\alpha_1,\ldots,\alpha+\beta_m,\,1)}\left(|\lambda_1|(b-a)^{\alpha-\alpha_1},\ldots,|\lambda_{n+m}|(b-a)^{\alpha+\beta_m}\right) \\
&=\theta\,\|u-v\|.
\end{aligned}
$$

Since $\theta<1$, Eq. (2.4.2) has a unique solution in the space $C^{l-1}[a,b]$ by BCP.

Regarding the existence of solutions to Eq. (2.4.2), we have the following theorem.

Theorem 2.9 *We consider the following conditions*

- *For $M_1>0$, $M_2>0$ and $y\in R$, the continuous function $f:[a,b]\times R\to R$ satisfying*
$$|f(x,y)|\le M_1+M_2|y|,$$
- *suppose that*
$$\Theta = \ M_2(b-a)^{\alpha}(E_{(\alpha-\alpha_1,\ldots,\alpha+\beta_m,\,\alpha+1)}\left(|\lambda_1|(b-a)^{\alpha-\alpha_1},\ldots,|\lambda_{n+m}|(b-a)^{\alpha+\beta_m}\right)$$

$$+\frac{1}{\Gamma(\alpha+1)}E_{(\alpha-\alpha_1,\ldots,\alpha+\beta_m,\,1)}\left(|\lambda_1|(b-a)^{\alpha-\alpha_1},\ldots,|\lambda_{n+m}|(b-a)^{\alpha+\beta_m}\right)) < 1.$$

Then Eq. (2.4.2) has at least one solution in the space $C^{l-1}[a,b]$.

Proof We again consider the nonlinear mapping T over $C^{l-1}[a,b]$ given by

$$(Tu)(x) =$$
$$\sum_{k=0}^{\infty}\sum_{k_1+\cdots+k_{n+m}=k}\binom{k}{k_1,\ldots,k_{n+m}}\lambda_1^{k_1}\cdots\lambda_{n+m}^{k_{n+m}}\cdot I_a^{\alpha+(\alpha-\alpha_1)k_1+\cdots+(\alpha+\beta_m)k_{n+m}}f(x,u(x))$$
$$-\frac{1}{(b-a)^{l-1}\Gamma(\alpha)}\int_a^b (b-x)^{\alpha-1}f(x,u(x))dx\cdot\sum_{k=0}^{\infty}\sum_{k_1+\cdots+k_{n+m}=k}\binom{k}{k_1,\ldots,k_{n+m}}\cdot$$
$$\lambda_1^{k_1}\cdots\lambda_{n+m}^{k_{n+m}}\cdot I_a^{(\alpha-\alpha_1)k_1+\cdots+(\alpha+\beta_m)k_{n+m}}(x-a)^{l-1}.$$

Obviously,

$$\|Tu\| \le (M_1+M_2\|u\|)(b-a)^{\alpha}\cdot$$
$$(E_{(\alpha-\alpha_1,\ldots,\alpha+\beta_m,\,\alpha+1)}\left(|\lambda_1|(b-a)^{\alpha-\alpha_1},\ldots,|\lambda_{n+m}|(b-a)^{\alpha+\beta_m}\right)$$
$$+\frac{1}{\Gamma(\alpha+1)}E_{(\alpha-\alpha_1,\ldots,\alpha+\beta_m,\,1)}\left(|\lambda_1|(b-a)^{\alpha-\alpha_1},\ldots,|\lambda_{n+m}|(b-a)^{\alpha+\beta_m}\right)) < +\infty,$$

which claims that the mapping T is from the space $C^{l-1}[a,b]$ to itself. We are going to prove (i) T is continuous. Indeed,

$$|(Tu)(x)-(Tv)(x)| \le \max_{x\in[a,b]}|f(x,u(x))-f(x,v(x))|(b-a)^{\alpha}\cdot$$
$$(E_{(\alpha-\alpha_1,\ldots,\alpha+\beta_m,\,\alpha+1)}\left(|\lambda_1|(b-a)^{\alpha-\alpha_1},\ldots,|\lambda_{n+m}|(b-a)^{\alpha+\beta_m}\right)$$
$$+\frac{1}{\Gamma(\alpha+1)}E_{(\alpha-\alpha_1,\ldots,\alpha+\beta_m,\,1)}\left(|\lambda_1|(b-a)^{\alpha-\alpha_1},\ldots,|\lambda_{n+m}|(b-a)^{\alpha+\beta_m}\right)) < +\infty,$$

which according to the continuity of f, the continuity of v is the result.

(ii) Here, we need to prove that T in $C^{l-1}[a,b]$ is a mapping from bounded sets to bounded sets. In fact, let W be a bounded set in $C^{l-1}[a,b]$, then there exists a constant $Z \in R^+$ such that

$$|f(x,u(x))| \le M_1 + M_2|u(x)| \le Z,$$

for all $u \in W$. This implies that

$$\|Tu\| \le Z(b-a)^{\alpha}\cdot$$
$$(E_{(\alpha-\alpha_1,\ldots,\alpha+\beta_m,\,\alpha+1)}\left(|\lambda_1|(b-a)^{\alpha-\alpha_1},\ldots,|\lambda_{n+m}|(b-a)^{\alpha+\beta_m}\right)$$
$$+\frac{1}{\Gamma(\alpha+1)}E_{(\alpha-\alpha_1,\ldots,\alpha+\beta_m,\,1)}\left(|\lambda_1|(b-a)^{\alpha-\alpha_1},\ldots,|\lambda_{n+m}|(b-a)^{\alpha+\beta_m}\right)) < +\infty.$$

Therefore, the set

$$\{Tu \ : \ u \in W\} = TW$$

is uniformly bounded in $C^{l-1}[a, b]$.

(iii) T is completely continuous from $C^{l-1}[a, b]$ to itself, which is a subspace of $C[a, b]$. Now, we show that T is equicontinuous on every bounded set such as W in $C^{l-1}[a, b]$. For this purpose, we consider Arzela–Ascoli theorem and assume that $a \leq \tau_1 < \tau_2 \leq b$ and $u \in W$. Then, we have

$$\begin{aligned}
&|(Tu)(\tau_2) - (Tu)(\tau_1)| \\
&\leq \sum_{k=0}^{\infty} \sum_{k_1+\cdots+k_{n+m}=k} \binom{k}{k_1, \ldots, k_{n+m}} |\lambda_1|^{k_1} \cdots |\lambda_{n+m}|^{k_{n+m}} \cdot \\
&\frac{1}{\Gamma(\gamma)} \left| \int_a^{\tau_2} (\tau_2 - t)^{\gamma-1} f(t, u(t)) dt - \int_a^{\tau_1} (\tau_1 - t)^{\gamma-1} f(t, u(t)) dt \right| \\
&+ \frac{Z(b-a)^{\alpha}}{(b-a)^{l-1}\Gamma(\alpha+1)} \sum_{k=0}^{\infty} \sum_{k_1+\cdots+k_{n+m}=k} \binom{k}{k_1, \ldots, k_{n+m}} |\lambda_1|^{k_1} \cdots |\lambda_{n+m}|^{k_{n+m}} \cdot \\
&\frac{1}{\Gamma(\gamma-\alpha)} \left| \int_a^{\tau_2} (\tau_2 - t)^{\gamma-\alpha-1} (t-a)^{l-1} dt - \int_a^{\tau_1} (\tau_1 - t)^{\gamma-\alpha-1} (t-a)^{l-1} dt \right| \\
&= I_1 + I_2,
\end{aligned}$$

where

$$\gamma = \alpha + (\alpha - \alpha_1)k_1 + \cdots + (\alpha + \beta_m)k_{n+m} \geq \alpha.$$

Regarding I_1, we have

$$\begin{aligned}
&\int_a^{\tau_2} (\tau_2 - t)^{\gamma-1} f(t, u(t)) dt \\
&= \int_a^{\tau_1} (\tau_2 - t)^{\gamma-1} f(t, u(t)) dt + \int_{\tau_1}^{\tau_2} (\tau_2 - t)^{\gamma-1} f(t, u(t)) dt.
\end{aligned}$$

Then,

$$\begin{aligned}
&\int_a^{\tau_2} (\tau_2 - t)^{\gamma-1} f(t, u(t)) dt - \int_a^{\tau_1} (\tau_1 - t)^{\gamma-1} f(t, u(t)) dt \\
&= \int_a^{\tau_1} [(\tau_2 - t)^{\gamma-1} - (\tau_1 - t)^{\gamma-1}] f(t, u(t)) dt + \int_{\tau_1}^{\tau_2} (\tau_2 - t)^{\gamma-1} f(t, u(t)) dt,
\end{aligned}$$

and by the mean value theorem,

$$\left| \int_a^{\tau_1} [(\tau_2 - t)^{\gamma-1} - (\tau_1 - t)^{\gamma-1}] f(t, u(t)) dt \right|$$

$$\leq Z\left[\frac{(\tau_2-a)^\gamma}{\gamma}-\frac{(\tau_1-a)^\gamma}{\gamma}\right]\leq Z(\tau_2-\tau_1)(b-a)^{\gamma-1}.$$

Furthermore,

$$\left|\int_{\tau_1}^{\tau_2}(\tau_2-t)^{\gamma-1}f(t,u(t))dt\right|\leq Z\frac{(\tau_2-\tau_1)^\gamma}{\gamma}.$$

In summary,

$$\begin{aligned}
I_1 \leq\ & Z(\tau_2-\tau_1)(b-a)^{\alpha-1}\sum_{k=0}^{\infty}\sum_{k_1+\cdots+k_{n+m}=k}\binom{k}{k_1,\ldots,k_{n+m}}|\lambda_1|^{k_1}\cdots|\lambda_{n+m}|^{k_{n+m}}\cdot\\
& \frac{(b-a)^{(\alpha-\alpha_1)k_1+\cdots+(\alpha+\beta_m)k_{n+m}}}{\Gamma(\alpha+(\alpha-\alpha_1)k_1+\cdots+(\alpha+\beta_m)k_{n+m})}\\
& + Z(\tau_2-\tau_1)^\alpha\sum_{k=0}^{\infty}\sum_{k_1+\cdots+k_{n+m}=k}\binom{k}{k_1,\ldots,k_{n+m}}|\lambda_1|^{k_1}\cdots|\lambda_{n+m}|^{k_{n+m}}\cdot\\
& \frac{(b-a)^{(\alpha-\alpha_1)k_1+\cdots+(\alpha+\beta_m)k_{n+m}}}{\Gamma(\alpha+1+(\alpha-\alpha_1)k_1+\cdots+(\alpha+\beta_m)k_{n+m})}\\
& = Z(\tau_2-\tau_1)(b-a)^{\alpha-1}E_{(\alpha-\alpha_1,\ldots,\alpha+\beta_m,\,\alpha)}\left(|\lambda_1|(b-a)^{\alpha-\alpha_1},\ldots,|\lambda_{n+m}|(b-a)^{\alpha+\beta_m}\right)\\
& + Z(\tau_2-\tau_1)^\alpha E_{(\alpha-\alpha_1,\ldots,\alpha+\beta_m,\,\alpha+1)}\left(|\lambda_1|(b-a)^{\alpha-\alpha_1},\ldots,|\lambda_{n+m}|(b-a)^{\alpha+\beta_m}\right).
\end{aligned}$$

As for I_2,

$$\begin{aligned}
& \frac{1}{\Gamma(\gamma-\alpha)}\left|\int_a^{\tau_2}(\tau_2-t)^{\gamma-\alpha-1}(t-a)^{l-1}dt-\int_a^{\tau_1}(\tau_1-t)^{\gamma-\alpha-1}(t-a)^{l-1}dt\right|\\
& \leq\frac{(b-a)^{l-1}}{\Gamma(\gamma-\alpha)}\left[\left|\int_a^{\tau_1}((\tau_2-t)^{\gamma-\alpha-1}-(\tau_1-t)^{\gamma-\alpha-1})dt\right|+\int_{\tau_1}^{\tau_2}(\tau_2-t)^{\gamma-\alpha-1}dt\right]\\
& \leq\frac{(b-a)^{l-1}}{\Gamma(\gamma+1-\alpha)}\left[(\tau_2-a)^{\gamma-\alpha}-(\tau_1-a)^{\gamma-\alpha}\right]+\\
& \quad\frac{(b-a)^{l-1}}{\Gamma(\gamma+1-\alpha)}(\tau_2-\tau_1)^{\gamma-\alpha},
\end{aligned}$$

for $\gamma>\alpha$.

Clearly,

$$\begin{aligned}
I_2\leq\ & \frac{Z(b-a)^\alpha}{(b-a)^{l-1}\Gamma(\alpha+1)}\left[(\tau_2-a)^{l-1}-(\tau_1-a)^{l-1}\right]+\\
& \frac{Z(b-a)^\alpha}{\Gamma(\alpha+1)}\sum_{k=1}^{\infty}\sum_{k_1+\cdots+k_{n+m}=k}\binom{k}{k_1,\ldots,k_{n+m}}|\lambda_1|^{k_1}\cdots|\lambda_{n+m}|^{k_{n+m}}\cdot\\
& \frac{1}{\Gamma(\gamma+1-\alpha)}\left[(\tau_2-a)^{\gamma-\alpha}-(\tau_1-a)^{\gamma-\alpha}\right]+
\end{aligned}$$

$$\frac{Z(b-a)^{\alpha}}{\Gamma(\alpha+1)} \sum_{k=1}^{\infty} \sum_{k_1+\cdots+k_{n+m}=k} \binom{k}{k_1, \ldots, k_{n+m}} |\lambda_1|^{k_1} \cdots |\lambda_{n+m}|^{k_{n+m}} \cdot$$
$$\frac{1}{\Gamma(\gamma+1-\alpha)} (\tau_2 - \tau_1)^{\gamma-\alpha}.$$

Obviously the second term above contains the factor $\tau_2 - \tau_1$ from the mean value theorem and third includes the factor $(\tau_2 - \tau_1)^{\alpha-\alpha_1}$, by noting that the index k starts from 1 rather than zero. Thus, T is completely continuous.

(iv) As a final step, we shall show that for some $0 < \lambda < 1$ the set

$$X = \left\{u \in C^{l-1}[a,b] \;:\; u = \lambda T u\right\}$$

is bounded. Indeed,

$$\begin{aligned}\|u\| \le\ & \|Tu\| \le (M_1 + M_2 \|u\|)(b-a)^{\alpha} \cdot \\ & (E_{(\alpha-\alpha_1,\ldots,\alpha+\beta_m,\, \alpha+1)} \left(|\lambda_1|(b-a)^{\alpha-\alpha_1}, \ldots, |\lambda_{n+m}|(b-a)^{\alpha+\beta_m}\right) \\ & + \frac{1}{\Gamma(\alpha+1)} E_{(\alpha-\alpha_1,\ldots,\alpha+\beta_m,\, 1)} \left(|\lambda_1|(b-a)^{\alpha-\alpha_1}, \ldots, |\lambda_{n+m}|(b-a)^{\alpha+\beta_m}\right)) < +\infty,\end{aligned}$$

which deduces that

$$\begin{aligned}(1-\Theta)\,\|u\| \le\ & M_1 (E_{(\alpha-\alpha_1,\ldots,\alpha+\beta_m,\, \alpha+1)} \left(|\lambda_1|(b-a)^{\alpha-\alpha_1}, \ldots, |\lambda_{n+m}|(b-a)^{\alpha+\beta_m}\right) \\ & + \frac{1}{\Gamma(\alpha+1)} E_{(\alpha-\alpha_1,\ldots,\alpha+\beta_m,\, 1)} \left(|\lambda_1|(b-a)^{\alpha-\alpha_1}, \ldots, |\lambda_{n+m}|(b-a)^{\alpha+\beta_m}\right)).\end{aligned}$$

Hence,

$$\begin{aligned}\|u\| \le\ & \frac{M_1}{1-\Theta} (E_{(\alpha-\alpha_1,\ldots,\alpha+\beta_m,\, \alpha+1)} \left(|\lambda_1|(b-a)^{\alpha-\alpha_1}, \ldots, |\lambda_{n+m}|(b-a)^{\alpha+\beta_m}\right) \\ & + \frac{1}{\Gamma(\alpha+1)} E_{(\alpha-\alpha_1,\ldots,\alpha+\beta_m,\, 1)} \left(|\lambda_1|(b-a)^{\alpha-\alpha_1}, \ldots, |\lambda_{n+m}|(b-a)^{\alpha+\beta_m}\right)).\end{aligned}$$

This completes the proof of Theorem 2.9.

Theorem 2.10 *Assume $f : [a,b] \times R \to R$ is a continuous and bounded function. Then Eq. (2.4.2) has at least one solution in the space $C^{l-1}[a,b]$.*

Proof It follows immediately from the proof of Theorem 2.9 by setting $M_2 = 0$.

Remark 2.2 It would be interesting and challenging to consider the following integral boundary value problem with variable coefficients:

$$\begin{cases} {}_C D_a^{\alpha} u(x) - \lambda_1(x)\, {}_C D_a^{\alpha_1} u(x) - \cdots - \lambda_n(x)\, {}_C D_a^{\alpha_n} u(x) \\ -\lambda_{n+1}(x) I_a^{\beta_1} u(x) - \cdots - \lambda_{n+m}(x) I_a^{\beta_m} u(x) = f(x, u(x)), \quad x \in [a,b] \\ u(a) = u'(a) = \cdots = u^{(l-2)}(a) = 0, \\ u(b) = I_a^{\alpha} \lambda_1(b)\, {}_C D_a^{\alpha_1} u(b) + \cdots + I_a^{\alpha} \lambda_n(b)\, {}_C D_a^{\alpha_n} u(b) \\ +I_a^{\alpha} \lambda_{n+1}(b) I_a^{\beta_1} u(b) + \cdots + I_a^{\alpha} \lambda_{n+m}(b) I_a^{\beta_m} u(b), \end{cases}$$

where $\lambda_j(x) \in C[a, b]$ for all $j = 1, 2, \ldots, n+m$ and $l - 1 < \alpha \le l$.

2.4.4 Application

Example 2.5 We consider the following IBCP

$$\begin{cases} {}_C D_0^{1.5} u(x) - 2\, {}_C D_0^{0.5} u(x) + I_0^{0.1} u(x) = \frac{1}{50} \sin(x u(x)) + x^3, \quad x \in [0, 1] \\ u(0) = 0, \\ u(1) = 2 I_0 u(1) - I_0^{1.6} u(1) = 2 \int_0^1 u(x) dx - \frac{1}{\Gamma(1.6)} \int_0^1 (1-x)^{0.6} u(x) dx, \end{cases} \tag{2.4.6}$$

According to the proven results, Eq. (2.4.6) has a unique solution in the space $C^1[0, 1]$.

Proof Clearly,

$$f(x, u(x)) = \frac{1}{50} \sin(x u(x)) + x^3,$$

and

$$|f(x, u(x)) - f(x, v(x))| \le \frac{1}{50} |x u(x) - x v(x)| \le \frac{1}{50} |u(x) - v(x)|,$$

using the fact $0 \le x \le 1$, which implies that $L = \frac{1}{50}$. In addition, from Theorem 2.8

$$\theta = \frac{1}{50} \left[E_{(1, 1.6,\, 2.5)}(2, 1) + \frac{1}{\Gamma(2.5)} E_{(1, 1.6,\, 1)}(2, 1) \right].$$

By the definition,

$$E_{(1, 1.6,\, 2.5)}(2, 1) = \sum_{k=0}^{\infty} \sum_{k_1 + k_2 = k} \binom{k}{k_1, k_2} \frac{2^{k_1}}{\Gamma(k_1 + 1.6 k_2 + 2.5)}.$$

Evidently,

$$\sum_{k_1 + k_2 = k} \binom{k}{k_1, k_2} = 2^k,$$

$$\frac{2^{k_1}}{\Gamma(k_1 + 1.6 k_2 + 2.5)} \le \frac{2^k}{\Gamma(k + 2.5)},$$

then, according to the calculations, we have

$$E_{(1,1.6,\ 2.5)}(2,1) \leq \sum_{k=0}^{\infty} \frac{4^k}{\Gamma(k+2.5)} \approx 6.51075,$$

On the other hand,

$$\frac{1}{\Gamma(2.5)} E_{(1,1.6,\ 1)}(2,1) \leq 0.752253 * \sum_{k=0}^{\infty} \frac{4^k}{\Gamma(k+1)} \approx 41.0716.$$

In summary,

$$\theta = \frac{1}{50}[6.51075 + 41.0716] < 1.$$

By Theorem 2.8, the integral boundary value problem has a unique solution in the space $C^1[0,1]$.

Example 2.6 Consider the following FDE which is an IBCP.

$$\begin{cases} {}_C D_0^{2.3} u(x) - 4\ {}_C D_0^{1.3} u(x) = \cos^2(x + u(x)) + x^2 + \arctan u^2(x), \\ u(0) = u'(0) = 0, \\ u(1) = 4 I_0 u(1) = 4 \int_0^1 u(x) dx, \end{cases} \tag{2.4.7}$$

According to the proven results, Eq. (2.4.7) has at least one solution in the space $C^2[0,1]$

Proof The function

$$f(x, u(x)) = \cos^2(x + u(x)) + x^2 + \arctan u^2(x),$$

is clearly bounded. From Theorem 2.10, the above equation with the IBCP has at least one solution in the space $C^2[0,1]$.

References

1. Diethelm, K.: The Analysis of Fractional Differential Equations. An Application-oriented Exposition Using Differential Operators of Caputo Type. Lecture Notes in Mathematics, vol. 2004. Springer, Berlin (2010)
2. Ilyas Butt, R., ur Rehman, M.: Ulam-Hyers-Mittag-Leffler stability of fractional difference equations with delay. Rocky Mountain J. Math. **51**(3), 891–901 (2021)
3. Heydari, M. H.; Atangana, A. *An accurate approach based on the orthonormal shifted discrete Legendre polynomials for variable-order fractional Sobolev equation.* Adv. Difference Equ. 2021, Paper No. 272, 12 pp
4. Yang, D; Bai, Ch. *Existence of ground state solutions for an asymptotically 2-linear fractional Schrodinger-Poisson system.* Bound. Value Probl. 2020, Paper No. 5, 9 pp
5. Khusainov, D. Ya.; Shuklin, G. V. *On relative controllability in systems with pure delay.* (Russian) ; translated from Prikl. Mekh. 41 (2005), no. 2, 118–130 Internat. Appl. Mech. 41 (2005), no. 2, 210–221

6. Khusainov, D. Ya.; Diblik, l; Ruzhichkova, M.; Lukacheva, Ya. *A representation of the solution of the Cauchy problem for an oscillatory system with pure delay*. Nonlinear Oscil. (N. Y.) 11 (2008), no. 2, 276–285
7. Liu, L.; Dong, Q.; Li, G. *Exact solutions and Hyers-Ulam stability for fractional oscillation equations with pure delay*. Appl. Math. Lett. 112 (2021), 106666, 7 pp
8. Li, M., Wang, J.: Finite time stability and relative controllability of Riemann-Liouville fractional delay differential equations. Math. Methods Appl. Sci. **42**(18), 6607–6623 (2019)
9. Li, M., Wang, J.: Exploring delayed Mittag-Leffler type matrix functions to study finite time stability of fractional delay differential equations. Appl. Math. Comput. **324**, 254–265 (2018)
10. Liang, Ch., Wang, J., O'Regan, D.: Representation of a solution for a fractional linear system with pure delay. Appl. Math. Lett. **77**, 72–78 (2018)
11. Kiryakova, V.: Some special functions related to fractional calculus and fractional (non-integer) order control systems and equations. Facta Univ. Ser. Autom. Control Robot. **7**(1), 79–98 (2008)
12. Podlubny, I. *Fractional differential equations. An introduction to fractional derivatives, fractional differential equations, to methods of their solution and some of their applications*. Mathematics in Science and Engineering, 198. Academic Press, Inc., San Diego, CA, 1999
13. Shiri, B., Wu, G.-C., Baleanu, D.: Terminal value problems for the nonlinear systems of fractional differential equations. Appl. Numer. Math. **170**, 162–178 (2021)
14. Benkerrouche, A.; Baleanu, D.; Said Souid, M.; Hakem, A.; Inc, M. *Boundary value problem for nonlinear fractional differential equations of variable order via Kuratowski MNC technique*. Adv. Difference Equ. 2021, Paper No. 365
15. Diaz, J.B., Margolis, B.: A fixed point theorem of the alternative, for con- tractions on a generalized complete metric space. Bull. Amer. Math. Soc. **74**, 305–309 (1968)
16. Wang, Ch; Xu, T-Zh. *Hyers-Ulam stability of fractional linear differential equations involving Caputo fractional derivatives*. Appl. Math. 60 (2015), no. 4, 383–393
17. Liu, K., Feckan, M., Wang, J.: A fixed-point approach to the hyers-ulam stability of caputo-fabrizio fractional differential equations. Mathematics **8**(4), 647 (2020)
18. Kilbas, A. A; S, H. M.; Trujillo, J. J. *Theory and applications of fractional differential equations. North-Holland Mathematics Studies, 204*. Elsevier Science B.V., Amsterdam, (2006), 1–523
19. Oldham, K. B.; Spanier, J. *The fractional calculus. Theory and applications of differentiation and integration to arbitrary order*. With an annotated chronological bibliography by Bertram Ross. *Mathematics in Science and Engineering, Vol. 111*. Academic Press [A subsidiary of Harcourt Brace Jovanovich, Publishers], New York-London, (1974), 45–66
20. Podlubny, I. *Fractional differential equations. An introduction to fractional derivatives, fractional differential equations, to methods of their solution and some of their applications. Mathematics in Science and Engineering,* 198. Academic Press, Inc., San Diego, CA, (1999), 223–242
21. Sousa, J. V. d. C.; Capelas de Oliveira, E. *On the Ψ-Hilfer fractional derivative. Communications in Nonlinear Science and Numerical Simulation* 60 (2018), 72–91
22. Amann, H. *Ordinary differential equations. An introduction to nonlinear analysis*. Translated from the German by Gerhard Metzen. *De Gruyter Studies in Mathematics, 13*. Walter de Gruyter & Co., Berlin, (1990), 136–197
23. Chicone, C. *Ordinary differential equations with applications. Second edition*. Texts in *Applied Mathematics, 34*. Springer, New York, (2006). 145–322
24. Abramowitz, M.; Stegun, I. A. *Handbook of mathematical functions with formulas, graphs, and mathematical tables. National Bureau of Standards Applied Mathematics Series, 55* For sale by the Superintendent of Documents, U.S. Government Printing Office, Washington, D.C. (1964), (1 page) on 239–239
25. Sousa, J. V. d. C.; Capelas de Oliveira, E. *Two new fractional derivatives of variable order with non-singular kernel and fractional differential equation. Computational and Applied Mathematics*. 37 (2018), no. 4, 5375–5394

26. Eidinejad, Z; Saadati, R; De La Sen, M. *Picard Method for Existence, Uniqueness, and Gauss Hypergeomatric Stability of the Fractional-Order Differential Equations. Mathematical Problems in Engineering.* (2021), no. 2–3
27. Abdo, Mohammed S.; Panchal, Satish K.; Wahash, Hanan A. *Ulam-Hyers-Mittag-Leffler stability for a ψ-Hilfer problem with fractional order and infinite delay.Results Appl. Math.* 7 (2020), Paper No. 100115, 12 pp
28. Ravikumar K, Ramkumar K, Chalishajar D. *Existence and Stability Results for Second-Order Neutral Stochastic Differential Equations With Random Impulses and Poisson Jumps.European Journal of Mathematical Analysis.* (2021), no. 1–8
29. Agarwal R, Jain S, Agarwal RP. *Analytic solution of generalized space time advection-dispersion equation with fractional Laplace operator.J Nonlinear Sci Appl.* (2016), no. 3545–54
30. Ramkumar K, Ravikumar K. *Controllability of Neutral Impulsive Stochastic Integrodifferential Equations Driven by a Rosenblatt Process and Unbounded Delay.Discontinuity, Nonlinearity, and Complexity.* (2021), no. 02, 311–21
31. Abdo, Mohammed S.; Panchal, Satish K. *Fractional integro-differential equations involving ψ-Hilfer fractional derivative.Adv. Appl. Math. Mech.* 11 (2019), no. 2, 338–359
32. Abdo, M.S., Panchal, S.K., Hussien, H.S.: Fractional integro-differential equations with non-local conditions and ψ -Hilfer fractional derivative. Math. Model. Anal. **24**(4), 564–584 (2019)
33. Wahash, H.A., Abdo, M.S., Panchal, S.K.: Fractional integrodifferential equations with non-local conditions and generalized Hilfer fractional derivative. Ufa Math. J. **11**(4), 151–171 (2019)
34. Sousa, J. V. d. C.; Rodrigues, F. G.; Capelas de Oliveira, E. *Stability of the fractional Volterra integro-differential equation by means of ψ-Hilfer operator. Math. Methods Appl. Sci.* 42 (2019), no. 9, 3033–3043
35. Zhou, Y.: Existence and uniqueness of solutions for a system of fractional differential equations. Fract. Calc. Appl. Anal. **12**(2), 195–204 (2009)
36. Zhou, Y.; Wang, J.; Zhang, L. *Basic theory of fractional differential equations.* Second edition [of MR3287248]. *World Scientific Publishing Co. Pte. Ltd.*, Hackensack, NJ, (2017), 1–380
37. Huang, J., Li, Y.: Hyers-Ulam stability of delay differential equations of first order. Math. Nachr. **289**(1), 60–66 (2016)
38. Graily, E.; Vaezpour, S. M.; Saadati, R.; Cho, Y. J. *Generalization of fixed point theorems in ordered metric spaces concerning generalized distance.Fixed Point Theory Appl.* 2011, 2011:30, 8 pp
39. Shakeri, S.; Ciric, L. J. B. *Common fixed point theorem in partially ordered L-fuzzy metric spaces. Fixed Point Theory Appl.* 2010, Art. ID 125082, 13 pp
40. Ciric, L., Abbas, M., Damjanovic, B.: Common fuzzy fixed point theorems in ordered metric spaces. Math. Comput. Modelling. **53**(9–10), 1737–1741 (2011)
41. Cho, Y. J. *Lattictic non-Archimedean random stability of ACQ functional equation. Adv. Difference Equ.* (2011), 2011:31, 12 pp
42. Mihet, D., Saadati, R., Vaezpour, S.M.: The stability of an additive functional equation in Menger probabilistic ϕ -normed spaces. Math. Slovaca. **61**(5), 817–826 (2011)
43. Cho, Y.J., Park, C., Rassias, Th.M.: Stability of functional equations in Banach algebras. Springer, Cham (2015)
44. Sousa, J. V. d. C.; Capelas de Oliveira, E. *Mittag–Leffler Functions and the Truncated $\mathcal{V}$-fractional Derivative. Mediterranean Journal of Mathematics.* 14 (2017), no. 61–26
45. Lakshmikantham, V.: Theory of fractional functional differential equations. Nonlinear Anal. **69**(10), 3337–3343 (2008)
46. Phuong, N. D.; Tuan, N. A.; Kumar, D.; Tuan, N. H. *Initial value problem for fractional Volterra integrodifferential pseudo-parabolic equations. Math. Model. Nat. Phenom.* 16 (2021), Paper No. 27, 14 pp
47. Salim, A.; Benchohra, M.; Karapinar, E.; Lazreg, J. E. *Existence and Ulam stability for impulsive generalized Hilfer-type fractional differential equations. Adv. Difference Equ.* (2020), Paper No. 601, 21 pp

48. Wang, Z., Cen, D., Mo, Y.: Sharp error estimate of a compact L 1-ADI scheme for the two-dimensional time-fractional integro-differential equation with singular kernels. Appl. Numer. Math. **159**, 190–203 (2021)
49. Janfada, M., Sadeghi, Gh.: Stability of the Volterra integrodifferential equation. Folia Math. **18**(1), 11–20 (2013)
50. Diaz, J.B., Margolis, B.: A fixed point theorem of the alternative, for contractions on a generalized complete metric space. Bull. Amer. Math. Soc. **74**, 305–309 (1968)
51. Kilbas, AA.; Srivastava, Hari M.; Trujillo, Juan J. *Theory and applications of fractional differential equations*. North-Holland Mathematics Studies, 204. Elsevier Science B.V., Amsterdam, 2006. xvi+523 pp
52. Agarwal, Ravi P.; Benchohra, M; Hamani, S.*A survey on existence results for boundary value problems of nonlinear fractional differential equations and inclusions*. Acta Appl. Math. 109 (2010), no. 3, 973–1033
53. Zhou, Y., Zhang, L.: Existence and multiplicity results of homoclinic solutions for fractional Hamiltonian systems. Comput. Math. Appl. **73**(6), 1325–1345 (2017)
54. Kilbas, Anatoly A. *Hadamard-type fractional calculus*. J. Korean Math. Soc. 38 (2001), no. 6, 1191–1204
55. Butzer, P.L., Kilbas, A.A., Trujillo, J.J.: Fractional calculus in the Mellin setting and Hadamard-type fractional integrals. J. Math. Anal. Appl. **269**(1), 1–27 (2002)
56. Khalil, R., Al Horani, M., Yousef, A., Sababheh, M.: A new definition of fractional derivative. J. Comput. Appl. Math. **264**, 65–70 (2014)
57. Horani, M., R. Khalil and T. Abdeljawad. *Conformable Fractional Semigroups of Operators*. Functional Analysis, (2014)
58. Li, M., Wang, J.R., O'Regan, D.: Existence and Ulam's stability for conformable fractional differential equations with constant coefficients. Bull. Malays. Math. Sci. Soc. **42**(4), 1791–1812 (2019)
59. Eidinejad.; Z, Saadati.; R. *Hyers-Ulam-Rassias-Wright stability for fractional oscillation equation.* Discrete Dynamics in Nature and Society 2022 (2022)
60. M.; Abramowitz. *Handbook of mathematical functions with formulas, graphs, and mathematical tables.* J. Appl. Mech., 32 (1965), 239–239
61. Eidinejad, Z., Saadati, R.: Hyers-Ulam-Rassias-Kummer stability of the fractional integro-differential equations. Math. Biosci. Eng. **19**(7), 6536–6550 (2022)
62. Fox, Ch.: The G and H functions as symmetrical Fourier kernels. Trans. Amer. Math. Soc. **98**, 395–429 (1961)
63. Eidinejad, Z., Saadati, R., Mesiar, R.: Optimum Approximation for ς -Lie Homomorphisms and Jordan ς -Lie Homomorphisms in ς -Lie Algebras by Aggregation Control Functions. Mathematics **10**(10), 1704 (2022)
64. Mathai, A. M.; Saxena, R. K. *The H-function with applications in statistics and other disciplines.* Halsted Press [John Wiley and Sons], New York-London-Sidney, 1978. xii+192 pp. ISBN: 0-470-26380-6
65. Grabisch, M; Marichal, JL; Mesiar, R; Pap, E. *Aggregation functions.* Encyclopedia of Mathematics and its Applications, 127. Cambridge University Press, Cambridge, 2009. xviii+460 pp. ISBN: 978-0-521-51926-7
66. Issa, M.S., Bani, H., Ahmed, A., Ghadle, K.P.: Numerical solutions of fuzzy integro-differential equations of the second kind. J. Math. Comput. Sci. **23**, 67–74 (2021)
67. Park, C., O'Regan, D., Saadati, R.: Stability of some set-valued functional equations. Appl. Math. Lett. **24**, 1910–1914 (2011)
68. Yan, H., Li, J.: SIQR dynamics in a random network with heterogeneous connections with infection age. J. Nonlinear Sci. Appl. **14**, 196–211 (2021)
69. Harir, A., Melliani, S., Chadli, L.S.: Solving fuzzy Burgers equation by variational iteration method. J. Math. Comput. Sci. **21**, 136–149 (2020)
70. Allahviranloo, T., Ahmady, E., Ahmady, N.: A method for solving nth order fuzzy linear differential equations. International Journal Computer Math. **86**, 730–742 (2009)

71. L-Sayed, A.M.A.; El-Owaidy, H.; Ahmed, R.G. Solvability of a boundary value problem of self-reference functional differential equation with infinite point and integral conditions. *J. Math. Comput. Sci.***2020**, *21*, 296–308
72. Demma, M., Saadati, R., Vetro, P.: Fixed point results on b-metric space via Picard sequences and b-simulation functions. Iran. J. Math. Sci. Inform. **11**, 123–136 (2016)
73. Heilat, A.S., Hailat, R.S.: Extended cubic B-spline method for solving a system of non-linear second-order boundary value problems. J. Math. Comput. Sci. **21**, 231–242 (2020)
74. Ezadi, S., Allahviranloo, T., Mohammadi, S.: Two new methods for ranking of Z-numbers based on sigmoid function and sign method. International Journal of Intelligent Systems **33**, 1476–1487 (2018)
75. Hieu, L.M., Hanh, T.T.H., Thanh, D.N.H.: A finite-difference scheme for initial boundary value problem of the Gamma equation in the pricing of financial derivatives. J. Math. Comput. Sci. **20**, 283–291 (2020)
76. Saadati, Reza; Park, Choonkil. A note on IFN-spaces. *Adv. Difference Equ.***2014**, Paper No. 63, 3 pp
77. Mehreen, N., Anwar, M.: Hermite-Hadamard and Hermite-Hadamard-Fejer type inequalities for p-convex functions via new fractional conformable integral operators. J. Math. Comput. Sci. **19**, 230–240 (2019)
78. Khalil, R.; AL Horani, M.; Abu Hammad, M. Geometric meaning of conformable derivative via fractional cords. *J. Math. Comput. Sci.***2019**, *19*, 241–245
79. Abbasbandy, S.: Allah Viranloo T. Numerical solution of fuzzy differential equation, Mathematical and Computational Applications **7**, 41–52 (2002)
80. Sudsutad, W, Tariboon, J: Existence results of fractional integro-differential equations with m-point multi-term fractional order integral boundary conditions. Bound. Value Probl. 2012, 94 (2012). https://doi.org/10.1186/1687-2770-2012-94
81. Zhao, K: Triple positive solutions for two classes of delayed nonlinear fractional FDEs with nonlinear integral boundary value conditions. Bound. Value Probl. 2015, 181 (2015). https://doi.org/10.1186/s13661-015-0445-y
82. Ahmad, B., Sivasundaram, S.: On four-point nonlocal boundary value problems of nonlinear integro-differential equations of fractional order. Appl. Math. Comput. **217**, 480–487 (2010)
83. Ntouyas, SK, Al-Sulami, HH: A study of coupled systems of mixed order fractional differential equations and inclusions with coupled integral fractional boundary conditions. Adv Differ Equ 2020, 73 (2020). https://doi.org/10.1186/s13662-020-2539-9
84. S. Meng, S, Cui, Y: Multiplicity results to a conformable fractional differential equations involving integral boundary condition. Complexity 2019 (2019), Article ID 8402347
85. Cabada, A., Hamdi, Z.: Nonlinear fractional differential equations with integral boundary value conditions. Appl. Math. Comput. **228**, 251–257 (2014)
86. Sun, Y., Zhao, M.: Positive solutions for a class of fractional differential equations with integral boundary conditions. Appl. Math. Lett. **34**, 17–21 (2014)
87. Ahmad, B, Nieto, JJ: Existence results for nonlinear boundary value problems of fractional integro-differential equations with integral boundary conditions. Bound. Value Probl. 2009 (2009), Article ID 708576
88. Yang, C, Guo, Y, Zhai, C: An integral boundary value problem of fractional differential equations with a sign-changed parameter in Banach spaces. Complexity 2021, Article ID 9567931. https://doi.org/10.1155/2021/9567931
89. Wang, X.H., Wang, L.P., Zeng, Q.H.: Fractional differential equations with integral boundary conditions. J. Nonlinear Sci. Appl. **8**, 309–314 (2015)
90. Li, C.: Several results of fractional derivatives in $\mathcal{D}'(R_+)$. Fract. Calc. Appl. Anal. **18**, 192–207 (2015)
91. Eidinejad, Z., Saadati, S., de la Sen, M.: Radu-Miheţ Method for the Existence, Uniqueness, and Approximation of the -Hilfer Fractional Equations by Matrix-Valued Fuzzy Controllers. Axioms **10**(2), 63 (2021)
92. Hadid, S.B., Luchko, Y.F.: An operational method for solving fractional differential equations of an arbitrary real order. Panamer. Math. J. **6**, 57–73 (1996)

93. Eidinejad, Z., Saadati, R., Mesiar, R.: Optimum Approximation for ς-Lie Homomorphisms and Jordan ς-Lie Homomorphisms in ς-Lie Algebras by Aggregation Control Functions. Mathematics **10**(10), 1704 (2022)
94. Babenkos, Y.I.: Heat and mass transfer. Khimiya, Leningrad (1986). (in Russian)

Chapter 3
Analysis of a New Stability on Matrix Valued Fuzzy Spaces

We apply the Radu-Miheţ method which is derived from an alternative fixed point theorem with a class of matrix-valued fuzzy controllers to approximate a fractional Volterra integrodifferential equation with ψ-Hilfer fractional derivative in matrix-valued fuzzy k-normed spaces to obtain an approximation for this type of fractional equations. Using a class of aggregation control functions, we introduce the concept of multiple-HU-OS_1-stability and get an optimum approximation for a nonlinear single fractional differential equation (NS-ABC-FDE) with a Mittag–Leffler kernel. Considering the matrix-valued fuzzy k-normed spaces and matrix-valued fuzzy H-Fox function as a control function, we investigate the existence of a unique solution and Hyers-Ulam-H-Fox stability for the conformable fractional differential equation with a constant coefficient. In a fuzzy environment, we introduce a matrix-valued fuzzy Wright controller class to investigate the Hyers–Ulam–Wright stability for the non-homogeneous fractional delay oscillation equation (NH-FD-O) with order κ.

3.1 Radu-Miheţ Method Using Matrix-Valued Fuzzy Controllers

3.1.1 Introduction of the ψ-Hilfer Fractional Volterra Integro-Differential Equation

Consider the ψ-Hilfer fractional Volterra integro-differential equation, defined by

$$^{H}\mathcal{D}_{0^+}^{\kappa,\tau,\psi}\Omega(p) = \boldsymbol{S}(p, \omega(p)) + \int_0^p \mathcal{M}(p, \sigma, \Omega(\sigma))d\sigma, \tag{3.1.1}$$

Z. Eidinejad et al., *Optimal Stability Theory and Approximate Solutions of Fractional Systems*, Studies in Systems, Decision and Control 606,
https://doi.org/10.1007/978-3-031-96704-7_3

where $\boldsymbol{S}(p, \omega(p))$ is a continuous function (in short CF) with respect to the variables p, ω and also $\mathcal{M}(p, \sigma, \Omega(\sigma))$ is a CF with respect to p, σ and Ω on E_1, $\kappa \in E_5^\circ$, $\tau \in E_5$, $\boldsymbol{S} : E_1 \times V \to V$ and $\mathcal{M} : E_1 \times E_1 \times V \to V$.

Let function $\phi : (E_1)^k \times (E_2)^n \to \mathrm{diagM_n(E_3)}$ be a matrix valued fuzzy function. The Eq. (3.1.1) is said to be Hyers-Ulam-Rassias stable, if $\Omega(p_1), \ldots, \Omega(p_k)$ is a given differentiable function, satisfying

$$\Theta\Big({}_0\mathcal{D}_{p_1}^{\kappa,\tau,\psi}\Omega(p_1) - \boldsymbol{S}(p_1, \Omega(p_1)) - \int_0^{p_1} \mathcal{M}(p_1, \sigma_1, \Omega(\sigma_1))d\sigma_1, \ldots,$$
$$ {}_0\mathcal{D}_{p_k}^{\kappa,\tau,\psi}\Omega(p_k) - \boldsymbol{S}(p_k, \Omega(p_k)) - \int_0^{p_k} \mathcal{M}(p_k, \sigma_k, \Omega(\sigma_k))d\sigma_k, \vec{\mu}\Big) \succeq \phi(p_1, \ldots, p_k, \vec{\mu}),$$

for $p_1, \ldots, p_k \in E_1$, and we can find a solution $\Upsilon(p_1) \ldots, \Upsilon(p_k)$ of Eq. (3.1.1) such that for some $r > 0$,

$$\Theta(\Omega(p_1) - \Upsilon(p_1), \ldots, \Omega(p_k) - \Upsilon(p_k), \vec{\mu}) \succeq \phi\left(p_1, \ldots, p_k, \frac{\vec{\mu}}{r}\right).$$

We study fuzzy sets matrix valued with the generalized t-norms, to define matrix valued fuzzy k-Banach space (in short MV-k-FB-space) and introduce a new class of matrix valued fuzzy controllers.

3.1.2 Best Approximation ψ−Hilfer Fractional Volterra Integro-Differential Equation

We apply the Radu-Miheţ method derived from Theorem 1.2 to study Hyers-Ulam-Rassias stability of functional equation (3.1.1) for more details we refer to [4, 13]. Consider the MVF-k-B-space $(V, \Theta, \circledast)$ and matrix valued fuzzy function (MVFF) $\phi : (E_1)^k \times (E_2)^n \to \mathrm{diagM_n(E_3)}$. We set

$$B := \{\Omega : E_1 \to V, \Omega \text{ is differentiable}\}$$

and define a mapping d from $B \times B$ to E_4 by

$$d(\Omega, \Upsilon) = \inf\Big\{C \in E_6 : \Theta\Big({}_0\mathcal{D}_{p_1}^{\kappa,\tau,\psi}\Omega(p_1) - {}_0\mathcal{D}_{p_1}^{\kappa,\tau,\psi}\Upsilon(p_1), \ldots, {}_0\mathcal{D}_{p_k}^{\kappa,\tau,\psi}\Omega(p_k) - {}_0\mathcal{D}_{p_k}^{\kappa,\tau,\psi}\Upsilon(p_k), \vec{\mu}\Big) \circledast$$
$$\Theta\Big(\Omega(p_1) - \Upsilon(p_1), \ldots, \Omega(p_k) - \Upsilon(p_k), \vec{\mu}\Big), \succeq \phi\Big(p_1, \ldots, p_k, \frac{\vec{\mu}}{C}\Big),$$
$$\forall \;\; \Omega, \Upsilon \in B, \;\; p_1, \ldots, p_k \in E_1, \;\; \vec{\mu} \in (E_2)^n\Big\}.$$

Theorem 3.1 *(B, d) is a complete E_4-valued metric fuzzy space.*

Proof First, we show that (B, d) is a E_4-valued metric fuzzy space.

We show that $d(\Omega, \Upsilon) = 0$ if and only if $\Omega = \Upsilon$. Let $d(\Omega, \Upsilon) = 0$, we have

$$\inf\left\{C \in E_6 : \Theta\left({}_0\mathcal{D}_{p_1}^{\kappa,\tau,\psi}\Omega(p_1) - {}_0\mathcal{D}_{p_1}^{\kappa,\tau,\psi}\Upsilon(p_1), \ldots, {}_0\mathcal{D}_{p_k}^{\kappa,\tau,\psi}\Omega(p_k) - {}_0\mathcal{D}_{p_k}^{\kappa,\tau,\psi}\Upsilon(p_k), \vec{\mu}\right)\circledast \right.$$
$$\Theta(\Omega(p_1) - \Upsilon(p_1), \ldots, \Omega(p_k) - \Upsilon(p_k), \vec{\mu}) \succeq \phi\left(p_1, \ldots, p_k, \frac{\vec{\mu}}{C}\right),$$
$$\left. \forall\, \Omega, \Upsilon \in B,\ p_1, \ldots, p_k \in E_1,\ \vec{\mu} \in (E_2)^n\right\} = 0,$$

and thus

$$\Theta\left({}_0\mathcal{D}_{p_1}^{\kappa,\tau,\psi}\Omega(p_1) - {}_0\mathcal{D}_{p_1}^{\kappa,\tau,\psi}\Upsilon(p_1), \ldots, {}_0\mathcal{D}_{p_k}^{\kappa,\tau,\psi}\Omega(p_k) - {}_0\mathcal{D}_{p_k}^{\kappa,\tau,\psi}\Upsilon(p_k), \vec{\mu}\right)\circledast$$
$$\Theta(\Omega(p_1) - \Upsilon(p_1), \ldots, \Omega(p_k) - \Upsilon(p_k), \vec{\mu}) \succeq \phi\left(p_1, \ldots, p_k, \frac{\vec{\mu}}{C}\right),$$

for all $C \in E_6$. Tend C to zero in the above inequality, we get

$$\Theta\left({}_0\mathcal{D}_{p_1}^{\kappa,\tau,\psi}\Omega(p_1) - {}_0\mathcal{D}_{p_1}^{\kappa,\tau,\psi}\Upsilon(p_1), \ldots, {}_0\mathcal{D}_{p_k}^{\kappa,\tau,\psi}\Omega(p_k) - {}_0\mathcal{D}_{p_k}^{\kappa,\tau,\psi}\Upsilon(p_k), \vec{\mu}\right) = \mathbf{1},$$

and then

$$\Theta(\Omega(p_1) - \Upsilon(p_1), \ldots, \Omega(p_k) - \Upsilon(p_k), \vec{\mu}) = \mathbf{1},$$

thus $\Omega(p_1, \ldots, p_k) = \Upsilon(p_1, \ldots, p_k)$ for every $p_1, \ldots, p_k \in E_1$, and vice versa. It is straight forward to show $d(\Omega, \Upsilon) = d(\Omega, \Upsilon)$ for every $\Omega, \Upsilon \in B$. Now, let $d(\Omega, \Upsilon) = e_1 \in E_2$ and $d(\Omega, \Upsilon) = e_2 \in E_2$. Then, we have

$$\Theta\left({}_0\mathcal{D}_{p_1}^{\kappa,\tau,\psi}\Omega(p_1) - {}_0\mathcal{D}_{p_1}^{\kappa,\tau,\psi}\Upsilon(p_1), \ldots, {}_0\mathcal{D}_{p_k}^{\kappa,\tau,\psi}\Omega(p_k) - {}_0\mathcal{D}_{p_k}^{\kappa,\tau,\psi}\Upsilon(p_k), \vec{\mu}\right)\circledast$$
$$\Theta(\Omega(p_1) - \Upsilon(p_1), \ldots, \Omega(p_k) - \Upsilon(p_k), \vec{\mu}) \succeq \phi\left(p_1, \ldots, p_k, \frac{\vec{\mu}}{e_1}\right),$$

and

$$\Theta\left({}_0\mathcal{D}_{p_1}^{\kappa,\tau,\psi}\Upsilon(p_1) - {}_0\mathcal{D}_{p_1}^{\kappa,\tau,\psi}\omega(p_1), \ldots, {}_0\mathcal{D}_{p_k}^{\kappa,\tau,\psi}\Upsilon(p_k) - {}_0\mathcal{D}_{p_k}^{\kappa,\tau,\psi}\omega(p_k), \vec{\mu}\right)\circledast$$
$$\Theta(\Upsilon(p_1) - \omega(p_1), \ldots, \Upsilon(p_k) - \omega(p_k), \vec{\mu}) \succeq \phi\left(p_1, \ldots, p_k, \frac{\vec{\mu}}{e_2}\right),$$

for every $\vec{\mu} \in (E_2)^n$. Then, we have

$$
\begin{aligned}
&\Theta\left({}_0\mathcal{D}^{\kappa,\tau,\psi}_{p_1}\Omega(p_1) - {}_0\mathcal{D}^{\kappa,\tau,\psi}_{p_1}\omega(p_1), \ldots, {}_0\mathcal{D}^{\kappa,\tau,\psi}_{p_k}\Omega(p_k) - {}_0\mathcal{D}^{\kappa,\tau,\psi}_{p_k}\omega(p_k), (e_1+e_2)\vec{\mu}\right) \circledast\\
&\Theta\left(\Omega(p_1)-\omega(p_1), \ldots, \Omega(p_k)-\omega(p_k), (e_1+e_2)\vec{\mu}\right)\\
&\succeq\Big[\Theta\left({}_0\mathcal{D}^{\kappa,\tau,\psi}_{p_1}\Omega(p_1) - {}_0\mathcal{D}^{\kappa,\tau,\psi}_{p_1}\Upsilon(p_1), \ldots, {}_0\mathcal{D}^{\kappa,\tau,\psi}_{p_k}\Omega(p_k) - {}_0\mathcal{D}^{\kappa,\tau,\psi}_{p_k}\Upsilon(p_k), e_1\vec{\mu}\right) \circledast\\
&\Theta\left({}_0\mathcal{D}^{\kappa,\tau,\psi}_{p_1}\Upsilon(p_1) - {}_0\mathcal{D}^{\kappa,\tau,\psi}_{p_1}\omega(p_1), \ldots, {}_0\mathcal{D}^{\kappa,\tau,\psi}_{p_k}\Upsilon(p_k) - {}_0\mathcal{D}^{\kappa,\tau,\psi}_{p_k}\omega(p_k), e_2\vec{\mu}\right)\Big]\\
&\circledast\Big[\Theta\left(\Omega(p_1)-\Upsilon(p_1), \ldots, \Omega(p_k)-\Upsilon(p_k), e_1\vec{\mu}\right) \circledast\ \Theta\left(\Upsilon(p_1)-\omega(p_1), \ldots, \Upsilon(p_k)-\omega(p_k), e_2\vec{\mu}\right)\Big]\\
&\succeq \phi(p_1, \ldots, p_k, \vec{\mu}) \circledast \phi(p_1, \ldots, p_k, \vec{\mu})\\
&= \phi(p_1, \ldots, p_k, \vec{\mu}),
\end{aligned}
$$

and then $d(\Omega, \omega) \leq e_1 + e_2$. Thus, $d(\Omega, \omega) \leq d(\Omega, \Upsilon) + d(\Upsilon, \omega)$. Now, we are ready to prove (B, d) is complete. Suppose that $\{\Omega_k\}_k$ be a Cauchy sequence in (B, d). Let $p_1, \ldots, p_k \in E_1$. Assume that $\vec{v} \in (E_2)^n$ and $\lambda \in E_5{}^\circ$ are arbitrary and consider $\vec{\mu} \in (E_2)^n$ such that $\phi(p_1, \ldots, p_k, \vec{\mu}) \succ 1 - \lambda$ for each λ in $(0, 1)$. For $e\vec{\mu} < \vec{v}$ choose $k_0 \in \mathbb{N}$ such that

$$
d(\Omega_k, \Omega_\ell) < e \quad \forall\, k, \ell \geq k_0.
$$

Then

$$
\begin{aligned}
&\Theta\left({}_0\mathcal{D}^{\kappa,\tau,\psi}_{p_1}\Omega_k(p_1) - {}_0\mathcal{D}^{\kappa,\tau,\psi}_{p_1}\Omega_l(p_1), \ldots, {}_0\mathcal{D}^{\kappa,\tau,\psi}_{p_k}\Omega_k(p_k) - {}_0\mathcal{D}^{\kappa,\tau,\psi}_{p_k}\Omega_l(p_k), \vec{v}\right) \circledast\\
&\Theta\left(\Omega_k(p_1)-\Omega_l(p_1), \ldots, \Omega_k(p_k)-\Omega_l(p_k), \vec{v}\right) \succeq\\
&\Theta\left({}_0\mathcal{D}^{\kappa,\tau,\psi}_{p_1}\Omega_k(p_1) - {}_0\mathcal{D}^{\kappa,\tau,\psi}_{p_1}\Omega_l(p_1), \ldots, {}_0\mathcal{D}^{\kappa,\tau,\psi}_{p_k}\Omega_k(p_k) - {}_0\mathcal{D}^{\kappa,\tau,\psi}_{p_k}\Omega_l(p_k), e\vec{\mu}\right) \circledast\\
&\Theta(\Omega_k(p_1)-\Omega_l(p_1), \ldots, \Omega_k(p_k)-\Omega_l(p_k), e\vec{\mu})\\
&\succeq \phi(p_1, \ldots, p_k, \vec{\mu})\\
&\succ 1-\lambda.
\end{aligned}
$$

Then

$$
\Theta\left({}_0\mathcal{D}^{\kappa,\tau,\psi}_{p_1}\Omega_k(p_1) - {}_0\mathcal{D}^{\kappa,\tau,\psi}_{p_1}\Omega_l(p_1), \ldots, {}_0\mathcal{D}^{\kappa,\tau,\psi}_{p_k}\Omega_k(p_k) - {}_0\mathcal{D}^{\kappa,\tau,\psi}_{p_k}\Omega_l(p_k), \vec{v}\right) \succ 1-\lambda,
$$

and $\Theta(\Omega_k(p_1) - \Omega_l(p_1), \ldots, \Omega_k(p_k) - \Omega_l(p_k), \vec{v}) \succ 1 - \lambda$ and i.e., the sequence both $\{\Omega_k(p_1), \Omega_k(p_2), \ldots, \Omega_k(p_k)\}_k$ and $\{{}_0\mathcal{D}^{\kappa,\tau,\psi}_{p_1}\Omega_k(p_1), {}_0\mathcal{D}^{\kappa,\tau,\psi}_{p_2}\Omega_k(p_2), \ldots, {}_0\mathcal{D}^{\kappa,\tau,\psi}_{p_k}\Omega_k(p_k)\}_k$ are Cauchy in complete space $(V, \Theta, \circledast)$ on compact set E_1, then, they are uniformly convergent to the mapping $\Omega : E_1 \to V$ and ${}_0\mathcal{D}^{\kappa,\tau,\psi}_{p_1}\Omega, {}_0\mathcal{D}^{\kappa,\tau,\psi}_{p_2}\Omega, \ldots, {}_0\mathcal{D}^{\kappa,\tau,\psi}_{p_k}\Omega$ respectively. The uniform convergence leads us to the fact that Ω is differentiable, i.e., an element of B and then (B, d) is complete.

Now, we are ready to study Hyers-Ulam-Rassias stability of the $\psi-$Hilfer fractional Volterra integro-differential equation (3.1.1) and get the best approximation with better estimate for the $\psi-$Hilfer fractional Volterra integro-differential equation.

Theorem 3.2 *Let* $(V, \Theta, \circledast)$ *be an MVFB-space and* L_1, L_2, L_3, L_4 *and* P *be positive constant such that* $N_1 > 2$ *where* $N_1 = \min[L_1, L_2L_3, L_1L_4, L_2L_3L_4]$. *Assume that the continuous mappings* $\boldsymbol{S} : E_1 \times V \to V$, $\mathcal{M} : E_1 \times E_1 \times V \to V$ *with matrix valued fuzzy function* $\phi : (E_1)^k \times (E_2)^n \to \mathrm{diagM_n(E_3)}$ *satisfying*

$$\Theta(\boldsymbol{S}(p_1, \Omega(p_1)) - \boldsymbol{S}(p_1, \Upsilon(p_1)), \ldots, \boldsymbol{S}(p_k, \Omega(p_k)) - \boldsymbol{S}(p_k, \Upsilon(p_k)), \vec{\mu}) \succeq \Theta\left(\Omega(p_1) - \Upsilon(p_1), \ldots, \Omega(p_k) - \Upsilon(p_k), L_1\vec{\mu}\right), \tag{3.1.2}$$

$$\Theta(\mathcal{M}(p_1, \sigma_1, \Omega(\sigma_1)) - \mathcal{M}(p_1, \sigma_1, \Upsilon(\sigma_1)), \ldots, \mathcal{M}(p_k, \sigma_k, \Omega(\sigma_k)) - \mathcal{M}(p_k, \sigma_k, \Upsilon(\sigma_k), \vec{\mu}) \succeq \Theta\left(\Omega(\sigma_1) - \Upsilon(\sigma_1), \ldots, \Omega(\sigma_k) - \Upsilon(\sigma_k), L_2\vec{\mu}\right), \tag{3.1.3}$$

$\sigma_i \preceq p_i \ \ (i = 1, 2, \ldots, k)$.

$$\inf_{\xi_1, \ldots, \xi_k \in E_1} \phi(\xi_1, \ldots, \xi_k, \vec{\mu}) \succeq \phi\left(p_1, \ldots, p_k, L_3 P \vec{\mu}\right), \tag{3.1.4}$$

and

$$\Theta(\Omega(p_1), \Omega(p_2), \ldots, \Omega(p_k), \vec{\mu}) \succeq \phi(p_1, \ldots, p_k, \vec{\mu}), \ \ \textit{implies that} \tag{3.1.5}$$
$$\Theta\left({}_0\mathcal{I}^{\kappa,\psi}_{p_1} \Omega(\sigma_1) d\sigma_1, \ldots, {}_0\mathcal{I}^{\kappa,\psi}_{p_k} \Omega(\sigma_k) d\sigma_K, \vec{\mu}\right) \succeq \phi\left(p_1, \ldots, p_k, L_4\vec{\mu}\right),$$

for every $p_1, \ldots, p_k \in E_1$, $\Omega, \Upsilon : E_1 \to V$ *and* $\vec{\mu} \in (E_2)^n$.

Let $\omega : E_1 \to V$ *be a differentiable function satisfying*

$$\Theta\Bigg({}_0\mathcal{D}^{\kappa,\tau,\psi}_{p_1} \omega(p_1) - \boldsymbol{S}(p_1, \omega(p_1)) - \int_0^{p_1} \mathcal{M}(p_1, \sigma_1, \omega(\sigma_1)) d\sigma_1, \ldots, \tag{3.1.6}$$
$$ {}_0\mathcal{D}^{\kappa,\tau,\psi}_{p_k} \omega(p_k) - \boldsymbol{S}(p_k, \omega(p_k)) - \int_0^{p_k} \mathcal{M}(p_k, \sigma_k, \omega(\sigma_k)) d\sigma_k, \vec{\mu}\Bigg)$$
$$\succeq \phi(p_1, \ldots, p_k, \vec{\mu}),$$

for every $p_1, \ldots, p_k \in E_1$ *and* $\vec{\mu} \in (E_2)^n$. *Therefore, we are to be able find a unique differentiable function* $\omega_0 : E_1 \to V$ *such that*

$${}_0\mathcal{D}^{\kappa,\tau,\psi}_{p} \omega_0(p) = \boldsymbol{S}(p, \omega_0(p)) + \int_0^{p} \mathcal{M}(p, \sigma, \omega_0(\sigma)) d\sigma, \tag{3.1.7}$$

and

$$\Theta\left({}_0\mathcal{D}^{\kappa,\tau,\psi}_{p_1} \omega(p_1) - {}_0\mathcal{D}^{\kappa,\tau,\psi}_{p_1} \omega_0(p_1), \ldots, {}_0\mathcal{D}^{\kappa,\tau,\psi}_{p_k} \omega(p_k) - {}_0\mathcal{D}^{\kappa,\tau,\psi}_{p_k} \omega_0(p_k), \vec{\mu}\right) \circledast \tag{3.1.8}$$

$$\Theta(\omega(p_1) - \omega_0(p_1), \ldots, \omega(p_k) - \omega_0(p_k), \vec{\mu}) \succeq \phi\left(p_1, \ldots, p_k, \frac{N_2}{N_3}\vec{\mu}\right),$$

for every $p_1, \ldots, p_k \in E_1$ *and* $\vec{\mu} \in (E_2)^n$ *and* $N_2 = \min[1, L_4]$ *and* $N_3 = \frac{\min[L_1, L_2L_3, L_1L_4, L_2L_3L_4]}{\min[L_1, L_2L_3, L_1L_4, L_2L_3L_4]-2}$.

Proof We set

$$B := \{\Omega : E_1 \to V, \Omega \text{ is differentiable}\}$$

and introduce the E_4-valued metric on B as,

$$d(\Omega, \Upsilon) = \inf\Big\{C \in E_6 : \Theta\left({}_0\mathcal{D}^{\kappa,\tau,\psi}_{p_1}\Omega(p_1) - {}_0\mathcal{D}^{\kappa,\tau,\psi}_{p_1}\Upsilon(p_1), \ldots, {}_0\mathcal{D}^{\kappa,\tau,\psi}_{p_k}\Omega(p_k) - {}_0\mathcal{D}^{\kappa,\tau,\psi}_{p_k}\Upsilon(p_k), \vec{\mu}\right)$$
$$\circledast\ \Theta(\Omega(p_1) - \Upsilon(p_1), \ldots, \Omega(p_k) - \Upsilon(p_k), \vec{\mu}) \succeq \phi\left(p_1, \ldots, p_k, \frac{\vec{\mu}}{C}\right),$$
$$\forall\ \Omega, \Upsilon \in B,\ \ p_1, \ldots, p_k \in E_1,\ \ \vec{\mu} \in (E_2)^n\Big\}.$$

By Theorem 3.1, we have (B, d) is a complete E_4-valued metric space. Now, we define the mapping Λ from B to B by

$$\Lambda(\Omega(p_i)) = {}_0\mathcal{I}^{\kappa,\psi}_{p_i}(\mathbf{S}(\sigma, \Omega(\sigma))) + {}_0\mathcal{I}^{\kappa,\psi}_{p_i}\left(\int_0^\sigma \mathcal{M}(\sigma, \varepsilon, \Omega(\varepsilon))d\varepsilon\right), \tag{3.1.9}$$

where $\kappa \in E_5{}^\circ$, $\mathbf{S} : E_1 \times V \to V$, $\mathcal{M} : E_1 \times E_1 \times V \to V$ and $p_i \in E_1 (i = 1, 2, \ldots, k)$. We prove Λ is a strictly contractive mapping. Let $\Omega, \Upsilon \in B$, $C \in E_2$ and $d(\Omega, \Upsilon) < \vartheta$, then we have

$$\Theta\left({}_0\mathcal{D}^{\kappa,\tau,\psi}_{p_1}\Omega(p_1) - {}_0\mathcal{D}^{\kappa,\tau,\psi}_{p_1}\Upsilon(p_1), \ldots, {}_0\mathcal{D}^{\kappa,\tau,\psi}_{p_k}\Omega(p_k) - {}_0\mathcal{D}^{\kappa,\tau,\psi}_{p_k}\Upsilon(p_k), \vartheta\vec{\mu}\right) \circledast$$
$$\Theta(\Omega(p_1) - \Upsilon(p_1), \vartheta\vec{\mu}) \succeq \phi(p_1, \ldots, p_k, \vec{\mu}),$$
$$\forall\, \Omega, \Upsilon \in B,\ p_1, \ldots, p_k \in E_1,\ \vec{\mu} \in (E_2)^n.$$

Using properties (MVF-K-N2) and (MVF-K-N3) of Definition 1.24 and (3.1.9), we have

$$\Theta\left({}_0\mathcal{D}^{\kappa,\tau,\psi}_{p_1}\Lambda(\Omega(p_1)) - {}_0\mathcal{D}^{\kappa,\tau,\psi}_{p_1}\Lambda(\Upsilon(p_1)), \ldots, {}_0\mathcal{D}^{\kappa,\tau,\psi}_{p_k}\Lambda(\Omega(p_k)) - {}_0\mathcal{D}^{\kappa,\tau,\psi}_{p_k}\Lambda(\Upsilon(p_k)), 2\vartheta\vec{\mu}\right) \tag{3.1.10}$$
$$\circledast\ \Theta\left(\Lambda(\Omega(p_1)) - \Lambda(\Upsilon(p_1)), \ldots, \Lambda(\Omega(p_k)) - \Lambda(\Upsilon(p_k)), 2\vartheta\vec{\mu}\right)$$
$$= \Theta\Big([\mathbf{S}(p_1, \Omega(p_1)) - \mathbf{S}(p_1, \Upsilon(p_1))] + \int_0^{p_1}[\mathcal{M}(p_1, \sigma_1, \Omega(\sigma_1)) - \mathcal{M}(p_1, \sigma_1, \Upsilon(\sigma_1))]d\sigma_1,$$
$$\ldots, [\mathbf{S}(p_k, \Omega(p_k)) - \mathbf{S}(p_k, \Upsilon(p_k))] + \int_0^{p_k}[\mathcal{M}(p_k, \sigma_k, \Omega(\sigma_k)) - \mathcal{M}(p_k, \sigma_k, \Upsilon(\sigma_k))]d\sigma_k, 2\vartheta\vec{\mu}\Big)$$

$$\circledast\, \Theta\Big({}_0\mathcal{I}^{\kappa,\psi}_{p_1}\Big(\mathbf{S}(p_1, \Omega(p_1)) - \mathbf{S}(p_1, \Upsilon(p_1))\Big) + {}_0\mathcal{I}^{\kappa,\psi}_{p_1}\Big(\int_0^{p_1}[\mathcal{M}(p_1, \sigma_1, \Omega(\sigma_1)) - \mathcal{M}(p_1, \sigma_1, \Upsilon(\sigma_1))]d\sigma_1\Big),$$
$$\cdots, {}_0\mathcal{I}^{\kappa,\psi}_{p_k}\Big(\mathbf{S}(p_k, \Omega(p_k)) - \mathbf{S}(p_k, \Upsilon(p_k))\Big) +$$
$${}_0\mathcal{I}^{\kappa,\psi}_{p_k}\Big(\int_0^{p_k}[\mathcal{M}(p_k, \sigma_k, \Omega(\sigma_k)) - \mathcal{M}(p_k, \sigma_k, \Upsilon(\sigma_k))]d\sigma_k\Big), 2\vartheta\vec{\mu}\Big)$$
$$\succeq \Theta\left(\mathbf{S}(p_1, \Omega(p_1)) - \mathbf{S}(p_1, \Upsilon(p_1)), \ldots, \mathbf{S}(p_k, \Omega(p_k)) - \mathbf{S}(p_k, \Upsilon(p_k)), \vartheta\vec{\mu}\right)$$
$$\circledast\, \Theta\Big(\int_0^{p_1}[\mathcal{M}(p_1, \sigma_1, \Omega(\sigma_1)) - \mathcal{M}(p_1, \sigma_1, \Upsilon(\sigma_1))]d\sigma_1,$$
$$\cdots, \int_0^{p_k}[\mathcal{M}(p_k, \sigma_k, \Omega(\sigma_k)) - \mathcal{M}(p_k, \sigma_k, \Upsilon(\sigma_k))]d\sigma_k, \vartheta\vec{\mu}\Big)$$
$$\circledast\, \Theta\left({}_0\mathcal{I}^{\kappa,\psi}_{p_1}[\mathbf{S}(p_1, \Omega(p_1)) - \mathbf{S}(p_1, \Upsilon(p_1))], \ldots, {}_0\mathcal{I}^{\kappa,\psi}_{p_k}[\mathbf{S}(p_k, \Omega(p_k)) - \mathbf{S}(p_k, \Upsilon(p_k))], \vartheta\vec{\mu}\right)$$
$$\circledast\, \Theta\Big({}_0\mathcal{I}^{\kappa,\psi}_{p_1}\int_0^{p_1}[\mathcal{M}(p_1, \sigma_1, \Omega(\sigma_1)) - \mathcal{M}(p_1, \sigma_1, \Upsilon(\sigma_1))]d\sigma_1,$$
$$\cdots, {}_0\mathcal{I}^{\kappa,\psi}_{p_k}\int_0^{p_k}[\mathcal{M}(p_k, \sigma_k, \Omega(\sigma_k)) - \mathcal{M}(p_k, \sigma_k, \Upsilon(\sigma_k))]d\sigma_k, \vartheta\vec{\mu}\Big).$$

In the last part of (3.1.10) there are four formulas, in the next steps we work on them to get new formulas derived from the control function ϕ. Let $0 = \bar{y}_1 < \bar{y}_2 < \cdots < \bar{y}_k = p_i$, $\Delta\bar{y}_i = \bar{y}_i - \bar{y}_{i-1} = \frac{p_i}{k}$, $i = 1, 2, \ldots, k$ and $\|\Delta\bar{y}\| = \max_{1\le i\le k}(\Delta\bar{y}_i)$.

Step 1. From (3.1.2) we have

$$\Theta\left(\mathbf{S}(p_1, \Omega(p_1)) - \mathbf{S}(p_1, \Upsilon(p_1)), \ldots, \mathbf{S}(p_k, \Omega(p_k)) - \mathbf{S}(p_k, \Upsilon(p_k)), \vartheta\vec{\mu}\right) \succeq \tag{3.1.11}$$
$$\Theta\left(\Omega(p_1) - \Upsilon(p_1), \ldots, \Omega(p_k) - \Upsilon(p_k), L_1\vartheta\vec{\mu}\right) \succeq \phi\left(p_1, \ldots, p_k, L_1\vec{\mu}\right).$$

Step 2. Using (MVF-K-N2) and (MVF-K-N3) of Definition 1.24, continuity property of MVFF Θ, (3.1.3) and (3.1.4), we get

$$\Theta\Big(\int_0^{p_1}\big[\mathcal{M}(p_1, \sigma_1, \Omega(\sigma_1)) - \mathcal{M}(p_1, \sigma_1, \Upsilon(\sigma_1))\big]d\sigma_1, \tag{3.1.12}$$
$$\cdots, \int_0^{p_k}\big[\mathcal{M}(p_k, \sigma_k, \Omega(\sigma_k)) - \mathcal{M}(p_k, \sigma_k, \Upsilon(\sigma_k))\big]d\sigma_k, \vartheta\vec{\mu}\Big)$$
$$= \Theta\Big(\lim_{\|\Delta\bar{y}\|\to 0}\sum_{j=1}^{k}[\mathcal{M}(p_1, \bar{y}_j, \Omega(\bar{y}_j)) - \mathcal{M}(p_1, \bar{y}_j, \Upsilon(\bar{y}_j))]\Delta\bar{y}_i,$$
$$\cdots, \lim_{\|\Delta\bar{y}\|\to 0}\sum_{j=1}^{k}[\mathcal{M}(p_k, \bar{y}_j, \Omega(\bar{y}_j)) - \mathcal{M}(p_k, \bar{y}_j, \Upsilon(\bar{y}_j))]\Delta\bar{y}_i, \vartheta\vec{\mu}\Big)$$
$$= \lim_{\|\Delta\bar{y}\|\to 0}\Theta\Big(\sum_{j=1}^{k}[\mathcal{M}(p_1, \bar{y}_j, \omega(\bar{y}_j)) - \mathcal{M}(p_1, \bar{y}_j, \Upsilon(\bar{y}_j))]\Delta\bar{y}_i,$$

$$\cdots, \sum_{j=1}^{k}[\mathcal{M}(p_k, \bar{y}_j, \omega(\bar{y}_j)) - \mathcal{M}(p_k, \bar{y}_j, \Upsilon(\bar{y}_j))]\Delta\bar{y}_i, \vartheta\vec{\mu}\Big)$$

$$\succeq \lim_{\|\Delta\bar{y}\|\to 0} \circledast_M \Theta\Big([\mathcal{M}(p_1, \bar{y}_j, \Omega(\bar{y}_j)) - \mathcal{M}(p_1, \bar{y}_j, \Upsilon(\bar{y}_j))]\Delta\bar{y}_i,$$

$$\cdots, [\mathcal{M}(p_k, \bar{y}_j, \Omega(\bar{y}_j)) - \mathcal{M}(p_k, \bar{y}_j, \Upsilon(\bar{y}_j))]\Delta\bar{y}_i, \frac{\vartheta\vec{\mu}}{k}\Big)$$

$$\succeq \inf_{\xi_1,\ldots,\xi_k\in E_1} \Theta\Big(\mathcal{M}(p_1, \xi_1, \Omega(\xi_1)) - \mathcal{M}(p_1, \xi_1, \Upsilon(\xi_1)),$$

$$\cdots, \mathcal{M}(p_k, \xi_k, \Omega(\xi_k)) - \mathcal{M}(p_k, \xi_k, \Upsilon(\xi_k)), \frac{k\vartheta\vec{\mu}}{kP}\Big)$$

$$\succeq \inf_{\xi_1,\ldots,\xi_k\in E_1} \Theta\left(\Omega(\xi_1) - \Upsilon(\xi_1), \ldots, \Omega(\xi_k) - \Upsilon(\xi_k), \frac{kL_2\vartheta\vec{\mu}}{kP}\right)$$

$$\succeq \inf_{\xi_1,\ldots,\xi_k\in E_1} \phi\left(\xi_1, \ldots, \xi_k, \frac{P\vec{\mu}}{L_2}\right)$$

$$\succeq \phi\left(p_1, \ldots, p_k, L_2L_3\vec{\mu}\right).$$

Step 3. Using (3.1.5) and (3.1.11), we get

$$\Theta\Big({}_0\mathcal{I}_{p_1}^{\kappa,\psi}\Big(\boldsymbol{S}(\sigma_1, \Omega(\sigma_1)) - \boldsymbol{S}(\sigma_1, \Upsilon(\sigma_1))\Big), \quad (3.1.13)$$

$$\cdots, {}_0\mathcal{I}_{p_k}^{\kappa,\psi}\left(\boldsymbol{S}(\sigma_k, \Omega(\sigma_k)) - \boldsymbol{S}(\sigma_k, \Upsilon(\sigma_k))\right), \vartheta\vec{\mu}\Big) \succeq \phi\left(p_1, \ldots, p_k, L_1L_4\vec{\mu}\right).$$

Step 4. Using (3.1.5) and (3.1.12), we get

$$\Theta\Big({}_0\mathcal{I}_{p_1}^{\kappa,\psi}\Big(\int_0^{\sigma_1}[\mathcal{M}(\sigma_1, \varepsilon_1, \Omega(\varepsilon_1)) - \mathcal{M}(\sigma_1, \varepsilon_1, \Upsilon(\varepsilon_1))]d\varepsilon_1\Big), \quad (3.1.14)$$

$$\cdots, {}_0\mathcal{I}_{p_k}^{\kappa,\psi}\Big(\int_0^{\sigma_k}[\mathcal{M}(\sigma_k, \varepsilon_k, \Omega(\varepsilon_k)) - \mathcal{M}(\sigma_k, \varepsilon_k, \Upsilon(\varepsilon_k))]d\varepsilon_k\Big), \vartheta\vec{\mu}\Big) \succeq \phi\left(p, L_2L_3L_4\vec{\mu}\right).$$

Form (3.1.10), (3.1.11), (3.1.12), (3.1.13) and (3.1.14), we have

$$\Theta\left({}_0\mathcal{D}_{p_1}^{\kappa,\tau,\psi}\Lambda(\Omega(p_1)) - {}_0\mathcal{D}_{p_1}^{\kappa,\tau,\psi}\Lambda(\Upsilon(p_1)), \cdots, {}_0\mathcal{D}_{p_k}^{\kappa,\tau,\psi}\Lambda(\Omega(p_k)) - {}_0\mathcal{D}_{p_k}^{\kappa,\tau,\psi}\Lambda(\Upsilon(p_k)), 2\vartheta\vec{\mu}\right) \circledast \quad (3.1.15)$$

$$\Theta\left(\Lambda(\Omega(p_1)) - \Lambda(\Upsilon(p_1)), \ldots, \Lambda(\Omega(p_k)) - \Lambda(\Upsilon(p_k)), 2\vartheta\vec{\mu}\right)$$

$$\succeq \phi(p_1, \ldots, p_k, L_1\vec{\mu}) \circledast \phi(p_1, \ldots, p_k, L_2L_3\vec{\mu}) \circledast \phi(p_1, \ldots, p_k, L_1L_4\vec{\mu}) \circledast \phi(p_1, \ldots, p_k, L_2L_3L_4\vec{\mu})$$
$$\succeq \phi(p_1, \ldots, p_k, N_1\vec{\mu})$$

where in $N_1 = \min[L_1, L_2L_3, L_1L_4, L_2L_3L_4]$.
Then

$$\begin{aligned}\Theta\Big({}_0\mathcal{D}_{p_1}^{\kappa,\tau,\psi}\Lambda(\Omega(p_1)) - {}_0\mathcal{D}_{p_1}^{\kappa,\tau,\psi}\Lambda(\Upsilon(p_1)), \ldots, {}_0\mathcal{D}{p_k}^{\kappa,\tau,\psi}\Lambda(\Omega(p_k)) - {}_0\mathcal{D}_{p_k}^{\kappa,\tau,\psi}\Lambda(\Upsilon(p_k)), 2\vartheta\vec{\mu}\Big)\\ \circledast\ \Theta(\Lambda(\Omega(p_1)) - \Lambda(\Upsilon(p_1)), \ldots, \Lambda(\Omega(p_k)) - \Lambda(\Upsilon(p_k)), 2\vartheta\vec{\mu})\\ \succeq \phi(p_1, \ldots, p_k, N_4\vec{\mu}),\end{aligned} \tag{3.1.16}$$

which $N_4 = \frac{1}{2}\min[L_1, L_2L_3, L_1L_4, L_2L_3L_4]$, and implies that

$$d(\Lambda(\Omega), \Lambda(\Upsilon)) \leq \frac{\vartheta}{N_4}, \tag{3.1.17}$$

and then

$$d(\Lambda(\Omega), \Lambda(\Upsilon)) \leq \frac{1}{N_4}d(\Omega, \Upsilon). \tag{3.1.18}$$

Thus, Λ with Lipschitz constant $\frac{1}{N_4}$ is a strictly contractive mapping.

Let $\omega \in B$, we show that $d(\Lambda(\omega), \omega) < \infty$. Using (3.1.5) and (3.1.6) we get

$$\begin{aligned}&\Theta\Big({}_0\mathcal{D}_{p_1}^{\kappa,\tau,\psi}[\Lambda(\omega(p_1)) - \omega(p_1)], \ldots, {}_0\mathcal{D}_{p_k}^{\kappa,\tau,\psi}[\Lambda(\omega(p_k)) - \omega(p_k)], \vec{\mu}\Big) \qquad (3.1.19)\\ &\circledast\ \Theta(\Lambda(\omega(p_1)) - \omega(p_1), \ldots, \Lambda(\omega(p_k)) - \omega(p_k), \vec{\mu})\\ &= \Theta\Big(\boldsymbol{S}(p_1, \omega(p_1)) + \int_0^{p_1}\mathcal{M}(p_1, \sigma_1, \omega(\sigma_1))d\sigma_1 - {}_0\mathcal{D}_{p_1}^{\kappa,\tau,\psi}\omega(p_1),\\ &\ldots, \boldsymbol{S}(p_k, \omega(p_k)) + \int_0^{p_k}\mathcal{M}(p_k, \sigma_k, \omega(\sigma_k))d\sigma_k - {}_0\mathcal{D}_{p_k}^{\kappa,\tau,\psi}\omega(p_k), \vec{\mu}\Big)\\ &\circledast\ \Theta\Big({}_0\mathcal{I}_{p_1}^{\kappa,\psi}(\boldsymbol{S}(p_1, \omega(p_1))) + {}_0\mathcal{I}_{p_1}^{\kappa,\psi}(\int_0^{p_1}\mathcal{M}(p_1, \sigma_1, \omega(\sigma_1))d\sigma_1) - {}_0\mathcal{I}_{p_1}^{\kappa,\psi}\ {}_0\mathcal{D}_{p_1}^{\kappa,\tau,\psi}\omega(p_1),\\ &\ldots, {}_0\mathcal{I}_{p_k}^{\kappa,\psi}(\boldsymbol{S}(p_k, \omega(p_k))) + {}_0\mathcal{I}_{p_k}^{\kappa,\psi}(\int_0^{p_k}\mathcal{M}(p_k, \sigma_k, \omega(\sigma_k))d\sigma_k) - {}_0\mathcal{I}_{p_k}^{\kappa,\psi}\ {}_0\mathcal{D}_{p_k}^{\kappa,\tau,\psi}\omega(p_k), \vec{\mu}\Big)\\ &= \Theta\Big(\boldsymbol{S}(p_1, \omega(p_1)) + \int_0^{p_1}\mathcal{M}(p_1, \sigma_1, \omega(\sigma_1))d\sigma_1 - {}_0\mathcal{D}_{p_1}^{\kappa,\tau,\psi}\omega(p_1),\\ &\ldots, \boldsymbol{S}(p_k, \omega(p_k)) + \int_0^{p_k}\mathcal{M}(p_k, \sigma_k, \omega(\sigma_k))d\sigma_k - {}_0\mathcal{D}_{p_k}^{\kappa,\tau,\psi}\omega(p_k), \vec{\mu}\Big)\\ &\circledast\ \Theta\Big({}_0\mathcal{I}_{p_1}^{\kappa,\psi}\Big[\boldsymbol{S}(p_1, \omega(p_1)) + \Big(\int_0^{p_1}\mathcal{M}(p_1, \sigma_1, \omega(\sigma_1))d\sigma_1\Big) - {}_0\mathcal{D}_{p_1}^{\kappa,\tau,\psi}\omega(p_1)\Big],\\ &\ldots, {}_0\mathcal{I}_{p_k}^{\kappa,\psi}\Big[\boldsymbol{S}(p_k, \omega(p_k)) + \Big(\int_0^{p_k}\mathcal{M}(p_k, \sigma_k, \omega(\sigma_k))d\sigma_k\Big) - {}_0\mathcal{D}_{p_k}^{\kappa,\tau,\psi}\omega(p_k)\Big], \vec{\mu}\Big)\\ &\succeq \phi(p_1, \ldots, p_k, \vec{\mu}) \circledast \phi(p_1, \ldots, p_k, L_4\vec{\mu})\\ &\succeq \phi(p_1, \ldots, p_k, N_2\vec{\mu}),\end{aligned}$$

for every $\vec{\mu} \in (E_2)^n$ and $N_2 = \min[1, L_4]$. Then we have $d(\Lambda(\omega), \omega) < \frac{1}{N_2} < \infty$.

Now, Theorem 1.2 enables us to find an element ω_0 in B satisfying the following:
(1) ω_0 is a fixed point of Λ, i.e.,

$$\begin{aligned}\omega_0(p) =& \Lambda(\omega_0(p)) \\ =& {}_0\mathcal{I}_p^{\kappa,\psi}(\boldsymbol{S}(\sigma, \omega_0(\sigma))) +_0 \mathcal{I}_p^{\kappa,\psi}\left(\int_0^\sigma \mathcal{M}(\sigma, \varepsilon, \omega_0(\varepsilon))d\varepsilon\right),\end{aligned} \tag{3.1.20}$$

which is unique in the set

$$B^* = \{\Omega \in B : d(\Lambda(\omega), \Omega) < \infty\}.$$

Take ${}_0\mathcal{D}_p^{\kappa,\tau,\psi}$ from (3.1.20) we get

$${}_0\mathcal{D}_p^{\kappa,\tau,\psi}\omega_0(p) = \boldsymbol{S}(p, \omega_0(p)) + \int_0^p \mathcal{M}(p, \sigma, \omega_0(\sigma))d\sigma, \tag{3.1.21}$$

where $\kappa \in E_5{}^\circ$, $\boldsymbol{S} : E_1 \times V \to V$, $\mathcal{M} : E_1 \times E_1 \times V \to V$.
(2) $d(\Lambda^k(\omega), \omega_0) \to 0$ as $n \to \infty$;
(3) $d(\omega, \omega_0) \leq N_3 d(\Lambda(\omega), w) \leq \frac{N_3}{N_2}$,

which $N_2 = \min[1, L_4]$ and $N_3 = \frac{\min[L_1, L_2L_3, L_1L_4, L_2L_3L_4]}{\min[L_1, L_2L_3, L_1L_4, L_2L_3L_4]-2}$ and implies that

$$\begin{aligned}&\Theta\left({}_0\mathcal{D}_{p_1}^{\kappa,\tau,\psi}\omega(p_1) - {}_0\mathcal{D}_{p_1}^{\kappa,\tau,\psi}\omega_0(p_1), \ldots, {}_0\mathcal{D}_{p_k}^{\kappa,\tau,\psi}\omega(p_k) - {}_0\mathcal{D}_{p_k}^{\kappa,\tau,\psi}\omega_0(p_k), \vec{\mu}\right) \circledast \\ &\qquad \Theta\left(\omega(p_1) - \omega_0(p_1), \ldots, \omega(p_k) - \omega_0(p_k), \vec{\mu}\right) \\ &\qquad \succeq \phi\left(p_1, \ldots, p_k, \frac{N_2}{N_3}\vec{\mu}\right),\end{aligned} \tag{3.1.22}$$

for every $p_1, \ldots, p_k \in E_1$ and $\vec{\mu} \in (E_2)^n$.

Now, we prove that the fixed point in B^* is unique. Suppose that ς_0 be an element of B satisfying (3.1.7) and (3.1.8), we prove that $\varsigma_0 = \omega_0$ and $\varsigma_0 \in B^*$. From (3.1.7) we get

$${}_0\mathcal{D}_p^{\kappa,\tau,\psi}\varsigma_0(p) = \boldsymbol{S}(p, \varsigma_0(p)) + \int_0^p \mathcal{M}(p, \sigma, \varsigma_0(\sigma))d\sigma, \tag{3.1.23}$$

and then

$$\begin{aligned}\varsigma_0(p) =& {}_0\mathcal{I}_p^{\kappa,\psi}\boldsymbol{S}(\sigma, \varsigma_0(\sigma)) +_0 \mathcal{I}_p^{\kappa,\psi}\int_0^\sigma \mathcal{M}(\sigma, \varepsilon, \varsigma_0(\varepsilon))d\varepsilon \\ =& \Lambda(\varsigma_0(p)),\end{aligned} \tag{3.1.24}$$

where $\kappa \in E_5{}^\circ$, $\boldsymbol{S} : E_1 \times V \to V$, $\mathcal{M} : E_1 \times E_1 \times V \to V$.

Now, we show that

$$\varsigma_0 \in \{\Omega \in B : d(\Lambda(\omega), \Omega) < \infty\},$$

i.e., $d(\Lambda(\omega), \varsigma_0) < \infty$. We set $\jmath = \frac{\min[1,L_4](\min[L_1,L_2L_3,L_1L_4,L_2L_3L_4]-2)}{\min[L_1,L_2L_3,L_1L_4,L_2L_3L_4]}$, from (3.1.8) we get

$$\begin{aligned} &\Theta\left({}_0\mathcal{D}_{p_1}^{\kappa,\tau,\psi}\omega(p_1) - {}_0\mathcal{D}_{p_1}^{\kappa,\tau,\psi}\varsigma_0(p_1), \ldots, {}_0\mathcal{D}_{p_k}^{\kappa,\tau,\psi}\omega(p_k) - {}_0\mathcal{D}_{p_k}^{\kappa,\tau,\psi}\varsigma_0(p_k), \vec{\mu}\right) \circledast \\ &\quad \Theta\left(\omega(p_1) - \varsigma_0(p_1), \ldots, \omega(p_k) - \varsigma_0(p_k), \vec{\mu}\right) \succeq \phi\left(p_1, \ldots, p_k, \jmath\vec{\mu}\right), \end{aligned} \tag{3.1.25}$$

for every $p_1, \ldots, p_k \in E_1$ and $\vec{\mu} \in (E_2)^n$.

Form (3.1.2) and (3.1.25) we get

$$\begin{aligned} &\Theta\left(\boldsymbol{S}(p_1, \omega(p_1)) - \boldsymbol{S}(p_1, \varsigma_0(p_1)), \ldots, \boldsymbol{S}(p_k, \omega(p_k)) - \boldsymbol{S}(p_k, \varsigma_0(p_k)), \vec{\mu}\right) \\ &\quad \succeq \Theta\left(\omega(p_1) - \varsigma_0(p_1), \ldots, \omega(p_k) - \varsigma_0(p_k), L_1\vec{\mu}\right) \\ &\quad \succeq \phi\left(p_1, \ldots, p_k, L_1\jmath\vec{\mu}\right), \end{aligned} \tag{3.1.26}$$

also, from (3.1.3) and (3.1.25) we get

$$\begin{aligned} &\Theta(\mathcal{M}(p_1, \sigma_1, \omega(\sigma_1)) - \mathcal{M}(p_1, \sigma_1, \varsigma_0(\sigma_1)), \ldots, \\ &\quad \mathcal{M}(p_k, \sigma_k, \omega(\sigma_k)) - \mathcal{M}(p_k, \sigma_k, \varsigma_0(\sigma_k)), \vec{\mu}) \\ &\succeq \Theta\left(\omega(\sigma_1) - \varsigma_0(\sigma_1), \ldots, \omega(\sigma_k) - \varsigma_0(\sigma_k), L_2\vec{\mu}\right) \\ &\quad \succeq \phi\left(p_1, \ldots, p_k, L_2\jmath\vec{\mu}\right), \end{aligned} \tag{3.1.27}$$

for every $p_1, \ldots, p_k \in E_1, \sigma_i \preceq p_i \ \ (i = 1, 2, \ldots, k)$ and $\vec{\mu} \in (E_2)^n$.

Now, using step 2 and (3.1.27) we get

$$\begin{aligned} &\Theta\Bigg(\int_0^{p_1}[\mathcal{M}(p_1, \sigma_1, \omega(\sigma_1)) - \mathcal{M}(p_1, \sigma_1, \varsigma_0(\sigma_1))]d\sigma_1, \ldots, \\ &\int_0^{p_k}[\mathcal{M}(p_k, \sigma_k, \omega(\sigma_k)) - \mathcal{M}(p_k, \sigma_k, \varsigma_0(\sigma_k))]d\sigma_k, \vec{\mu}\Bigg) \\ &\quad \succeq \phi\left(p_1, \ldots, p_k, L_2L_3\vec{\mu}\right) \\ &\quad \succeq \phi\left(p_1, \ldots, p_k, L_2L_3\jmath\vec{\mu}\right). \end{aligned} \tag{3.1.28}$$

Using triangular inequality (MVF-K-N3), (3.1.26) and (3.1.28) we get

$$\begin{aligned}&\Theta\Big(S(p_1,\omega(p_1))-S(p_1,\varsigma_0(p_1))+\int_0^{p_1}[\mathcal{M}(p_1,\sigma_1,\omega(\sigma_1))-\mathcal{M}(p_1,\sigma_1,\varsigma_0(\sigma_1))]d\sigma_1,\ldots,\\&S(p_k,\omega(p_k))-S(p_k,\varsigma_0(p_k))+\int_0^{p_k}[\mathcal{M}(p_k,\sigma_k,\omega(\sigma_k))-\mathcal{M}(p_k,\sigma_k,\varsigma_0(\sigma_k))]d\sigma_k,2\vec{\mu}\Big)\\&\succeq\Theta\left(S(p_1,\omega(p_1))-S(p_1,\varsigma_0(p_1)),\ldots,S(p_k,\omega(p_k))-S(p_k,\varsigma_0(p_k)),\vec{\mu}\right)\\&\circledast\Theta\Big(\int_0^{p_1}[\mathcal{M}(p_1,\sigma_1,\omega(\sigma_1))-\mathcal{M}(p_1,\sigma_1,\varsigma_0(\sigma_1))]d\sigma_1,\ldots,\\&\int_0^{p_k}[\mathcal{M}(p_k,\sigma_k,\omega(\sigma_k))-\mathcal{M}(p_k,\sigma_k,\varsigma_0(\sigma_k))]d\sigma_k,\vec{\mu}\Big)\\&\succeq\phi\left(p_1,\ldots,p_k,L_1\jmath\vec{\mu}\right)\circledast\phi\left(p_1,\ldots,p_k,L_2L_3\jmath\vec{\mu}\right)\\&\succeq\phi\left(p_1,\ldots,p_k,N_5\jmath\vec{\mu}\right),\end{aligned}\tag{3.1.29}$$

where in $N_5=\min[L_1,L_2L_3]$ and then

$$\begin{aligned}&\Theta\Big(S(p_1,\omega(p_1))-S(p_1,\varsigma_0(p_1))+\int_0^{p_1}\left[\mathcal{M}(p_1,\sigma_1,\omega(\sigma_1))-\mathcal{M}(p_1,\sigma_1,\varsigma_0(\sigma_1))\right]d\sigma_1,\ldots,\\&S(p_k,\omega(p_k))-S(p_k,\varsigma_0(p_k))+\int_0^{p_k}\left[\mathcal{M}(p_k,\sigma_k,\omega(\sigma_k))-\mathcal{M}(p_k,\sigma_k,\varsigma_0(\sigma_k))\right]d\sigma_k,\vec{\mu}\Big)\\&\succeq\phi\left(p_1,\ldots,p_k,N_6\jmath\vec{\mu}\right),\end{aligned}\tag{3.1.30}$$

where in $N_6=\frac{1}{2}\min[L_1,L_2L_3]$. We apply (3.1.5) and get

$$\begin{aligned}&\Theta\Big({}_0\mathcal{I}_{p_1}^{\kappa,\psi}\left[S(\sigma_1,\omega(\sigma_1))-S(\sigma_1,\varsigma_0(\sigma_1))\right]+{}_0\mathcal{I}_{p_1}^{\kappa,\psi}\int_0^{\sigma_1}[\mathcal{M}(\sigma_1,\varepsilon_1,\omega(\varepsilon_1))-\mathcal{M}(\sigma_1,\varepsilon_1,\varsigma_0(\varepsilon_1))]d\varepsilon_1,\\&\cdots,{}_0\mathcal{I}_{p_k}^{\kappa,\psi}\left[S(\sigma,\omega(\sigma))-S(\sigma,\varsigma_0(\sigma))\right]+{}_0\mathcal{I}_{p_k}^{\kappa,\psi}\int_0^{\sigma}[\mathcal{M}(\sigma,\varepsilon,\omega(\varepsilon))-\mathcal{M}(\sigma,\varepsilon,\varsigma_0(\varepsilon))]d\varepsilon,\vec{\mu}\Big)\\&\succeq\phi\left(p_1,\ldots,p_k,L_4N_6\jmath\vec{\mu}\right),\end{aligned}\tag{3.1.31}$$

for every $p_1,\ldots,p_k\in E_1$ and $\vec{\mu}\in(E_2)^n$ and $N_6=\frac{1}{2}\min[L_1,L_2L_3]$.

Using (3.1.30), (3.1.30) and (3.1.31) we get

$$\begin{aligned}&\Theta\left({}_0\mathcal{D}_{p_1}^{\kappa,\tau,\psi}[\Lambda(\omega(p_1))-\varsigma_0(p_1)],\ldots,{}_0\mathcal{D}_{p_k}^{\kappa,\tau,\psi}[\Lambda(\omega(p_k))-\varsigma_0(p_k)],\vec{\mu}\right)\\&\circledast\Theta\left(\Lambda(\omega(p_1))-\varsigma_0(p_1),\ldots,\Lambda(\omega(p_k))-\varsigma_0(p_k),\vec{\mu}\right)\\&=\Theta\Big(S(p_1,\omega(p_1))-S(p_1,\varsigma_0(p_1))+\textstyle\int_0^{p_1}[\mathcal{M}(p_1,\sigma_1,\omega(\sigma_1))-\mathcal{M}(p_1,\sigma_1,\varsigma_0(\sigma_1))]d\sigma_1,\ldots,\\&S(p_k,\omega(p_k))-S(p_k,\varsigma_0(p_k))+\textstyle\int_0^{p_k}[\mathcal{M}(p_k,\sigma_k,\omega(\sigma_k))-\mathcal{M}(p_k,\sigma_k,\varsigma_0(\sigma_k))]d\sigma_k,\vec{\mu}\Big)\\&\circledast\Theta\Big({}_0\mathcal{I}_{p_1}^{\kappa,\psi}[S(\sigma_1,\omega(\sigma_1))-S(\sigma_1,\varsigma_0(\sigma_1))]+{}_0\mathcal{I}_{p_1}^{\kappa,\psi}\int_0^{\sigma_1}[\mathcal{M}(\sigma_1,\varepsilon_1,\omega(\varepsilon_1))-\mathcal{M}(\sigma_1,\varepsilon_1,\varsigma_0(\varepsilon_1))]d\varepsilon_1,\ldots,\\&{}_0\mathcal{I}_{p_k}^{\kappa,\psi}[S(\sigma_k,\omega(\sigma_k))-S(\sigma_k,\varsigma_0(\sigma_k))]+{}_0\mathcal{I}_{p_k}^{\kappa,\psi}\int_0^{\sigma_k}[\mathcal{M}(\sigma_k,\varepsilon_k,\omega(\varepsilon_k))-\mathcal{M}(\sigma_k,\varepsilon_k,\varsigma_0(\varepsilon_k))]d\varepsilon_k,\vec{\mu}\Big)\\&\succeq\phi\left(p_1,\ldots,p_k,N_6\jmath\vec{\mu}\right)\circledast\phi\left(p_1,\ldots,p_k,L_4N_6\jmath\vec{\mu}\right)\\&\succeq\phi\left(p_1,\ldots,p_k,\frac{(L_4+1)N_6\jmath\vec{\mu}}{L_4}\right),\end{aligned}\tag{3.1.32}$$

which $N_6=\frac{1}{2}\min[L_1,L_2L_3]$, which implies that

$$d(\Lambda(\omega),\varsigma_0)\leq\frac{L_4}{\jmath(L_4+1)N_6}<\infty$$

then $\varsigma_0\in B^*$.

3.1.3 Best Approximation of $\psi-$ Hilfer Fractional Volterra Integral Equation

Now, we are ready to study Hyers-Ulam-Rassias stability of the $\psi-$ Hilfer fractional Volterra integral equation

$$\Omega(p) = \boldsymbol{S}(p, \Omega(p)) + {}_0\mathcal{I}_p^{\kappa,\psi} \mathcal{M}(p, \sigma, \Omega(\sigma)), \tag{3.1.33}$$

where $\kappa \in E_5{}^{\circ}, \boldsymbol{S} : E_1 \times V \to V, \mathcal{M} : E_1 \times E_1 \times V \to V$ and get the best approximation with better estimate for the pseudo $\psi-$ Hilfer fractional Volterra integral equation. Our method can be use for new problems for more problems and details we refer to [23–33].

Theorem 3.3 *Let $(V, \Theta, \circledast)$ be an MVFB-space and L_1, L_2, L_3, L_4 and T be positive constant such that $N_1 > 2$ which $N_1 = \min[L_1, L_2L_3, L_1L_4, L_2L_3L_4]$. Assume that the continuous mappings $\boldsymbol{S} : E_1 \times V \to V$, $\mathcal{M} : E_1 \times E_1 \times V \to V$ with MVFF $\phi : (E_1)^k \times (E_2)^n \to \mathrm{diag}(\mathrm{M_n}(\mathrm{E_3}))$ satisfying (3.1.2), (3.1.3), (3.1.4) and (3.1.5).*

Let $\omega : E_1 \to V$ be a differentiable function satisfying

$$\begin{aligned}\Theta\Big(&\omega(p_1) - \boldsymbol{S}(p_1, \omega(p_1)) - {}_0\mathcal{I}_{p_1}^{\kappa,\psi} \mathcal{M}(p_1, \sigma_1, \omega(\sigma_1)), \ldots, \\ &\omega(p_k) - \boldsymbol{S}(p_k, \omega(p_k)) - {}_0\mathcal{I}_{p_k}^{\kappa,\psi} \mathcal{M}(p_k, \sigma_k, \omega(\sigma_k)), \vec{\mu}\Big) \\ &\succeq \phi(p_1, \ldots, p_k, \vec{\mu}),\end{aligned} \tag{3.1.34}$$

for every $p_1, \ldots, p_k \in E_1$ and $\vec{\mu} \in (E_2)^n$. Then we are to be able find a unique differentiable function $\omega_0 : E_1 \to V$ such that

$$\omega_0(p) = \boldsymbol{S}(p, \omega_0(p)) + {}_0\mathcal{I}_p^{\kappa,\psi} \mathcal{M}(p, \sigma, \omega_0(\sigma)), \tag{3.1.35}$$

and

$$\begin{aligned}\Theta(\omega(p_1) - \omega_0(p_1), \ldots, \omega(p_k) - \omega_0(p_k), \vec{\mu}) \succeq \\ \phi\left(p_1, \ldots, p_k, \frac{N_2}{N_3}\vec{\mu}\right),\end{aligned} \tag{3.1.36}$$

for every $p_1, \ldots, p_k \in E_1$ and $\vec{\mu} \in (E_2)^n$ and $N_2 = \min[1, L_4]$ and $N_3 = \frac{\min[L_1, L_2L_3, L_1L_4, L_2L_3L_4]}{\min[L_1, L_2L_3, L_1L_4, L_2L_3L_4]-2}$.

Proof We set

$$B := \{\Omega : E_1 \to V, \Omega \text{ is differentiable}\}$$

and introduce the E_4-valued metric on B as,

$$d(\Omega, \Upsilon) = \inf\left\{C \in E_6 : \Theta\left(\Omega(p_1) - \Upsilon(p_1), \ldots, \Omega(p_k) - \Upsilon(p_k), \vec{\mu}\right) \succeq \phi\left(p_1, \ldots, p_k, \frac{\vec{\mu}}{C}\right), \right.$$
$$\left. \forall\, \Omega, \Upsilon \in B,\ p_1, \ldots, p_k \in E_1,\ \vec{\mu} \in (E_2)^n\right\}.$$

By Theorem 3.1, we have (B, d) is a complete E_4-valued metric space.

Now, we define the mapping Λ from B to B by

$$\Lambda(\Omega(p_i)) = \boldsymbol{S}(\sigma, \Omega(\sigma)) + {}_0\mathcal{I}^{\kappa,\psi}_{p_i}\left(\int_0^{\sigma} \mathcal{M}(\sigma, \varepsilon, \Omega(\varepsilon))d\varepsilon\right), \tag{3.1.37}$$

where $\kappa \in (E_5)^{\circ}$, $\boldsymbol{S} : E_1 \times V \to V$, $\mathcal{M} : E_1 \times E_1 \times V \to V$ and $p_i \in E_1 (i = 1, 2, \ldots, k)$. We prove Λ is a strictly contractive mapping. Let $\Omega, \Upsilon \in B$, $C \in E_6$ and $d(\Omega, \Upsilon) < \vartheta$, then we have

$$\Theta(\Omega(p_1) - \Upsilon(p_1), \ldots, \Omega(p_k) - \Upsilon(p_k), \vartheta\vec{\mu}) \succeq \phi(p_1, \ldots, p_k, \vec{\mu}),$$
$$\forall\, \Omega, \Upsilon \in B,\ p_1, \ldots, p_k \in E_1,\ \vec{\mu} \in (E_2)^n.$$

Using properties (MVF-K-N2) and (MVF-K-N3) of Definition 1.24, (3.1.2), (3.1.3), (3.1.4), (3.1.5) and (3.1.37), we have

$$\begin{aligned}
&\Theta\left(\Lambda(\Omega(p_1)) - \Lambda(\Upsilon(p_1)), \ldots, \Lambda(\Omega(p_k)) - \Lambda(\Upsilon(p_k)), 2\vartheta\vec{\mu}\right)\\
&= \Theta\Big([\boldsymbol{S}(p_1, \Omega(p_1)) - \boldsymbol{S}(p_1, \Upsilon(p_1))] + {}_0\mathcal{I}^{\kappa,\psi}_{p_1}[\mathcal{M}(p_1, \sigma_1, \Omega(\sigma_1)) - \mathcal{M}(p_1, \sigma_1, \Upsilon(\sigma_1))]d\sigma_1, \ldots,\\
&[\boldsymbol{S}(p_k, \Omega(p_k)) - \boldsymbol{S}(p_k, \Upsilon(p_k))] + {}_0\mathcal{I}^{\kappa,\psi}_{p_k}[\mathcal{M}(p_k, \sigma_k, \Omega(\sigma_k)) - \mathcal{M}(p_k, \sigma_k, \Upsilon(\sigma_k))]d\sigma_k, 2\vartheta\vec{\mu}\Big)
\end{aligned} \tag{3.1.38}$$

$$\begin{aligned}
&\succeq \Theta\left(\boldsymbol{S}(p_1, \Omega(p_1)) - \boldsymbol{S}(p_1, \Upsilon(p_1)), \ldots, \boldsymbol{S}(p_k, \Omega(p_k)) - \boldsymbol{S}(p_k, \Upsilon(p_k)), \vartheta\vec{\mu}\right)\\
&\circledast \Theta\Big(\int_0^{p_1}[\mathcal{M}(p_1, \sigma_1, \Omega(\sigma_1)) - \mathcal{M}(p_1, \sigma_1, \Upsilon(\sigma_1))]d\sigma_1, \ldots,\\
&\int_0^{p_k}[\mathcal{M}(p_k, \sigma_k, \Omega(\sigma_k)) - \mathcal{M}(p_k, \sigma_k, \Upsilon(\sigma_k))]d\sigma_k, \vartheta\vec{\mu}\Big)\\
&\succeq \phi\left(p_1, \ldots, p_k, L_1\vec{\mu}\right) \circledast \phi\left(p_1, \ldots, p_k, L_2L_4\vec{\mu}\right)\\
&\succeq \phi\left(p_1, \ldots, p_k, N_7\vec{\mu}\right),
\end{aligned} \tag{3.1.39}$$

where in $N_7 = \min[L_1, L_2L_4]$. Then,

$$\begin{aligned}
&\Theta\left(\Lambda(\Omega(p_1)) - \Lambda(\Upsilon(p_1)), \ldots, \Lambda(\Omega(p_k)) - \Lambda(\Upsilon(p_k)), \vartheta\vec{\mu}\right)\\
&\succeq \phi\left(p_1, \ldots, p_k, N_8\vec{\mu}\right),
\end{aligned} \tag{3.1.40}$$

for every $p_1, \ldots, p_k \in E_1$ and $\vec{\mu} \in (E_2)^n$ and $N_8 = \frac{1}{2}\min[L_1, L_2L_4]$.

Then

$$d(\Lambda(\Omega), \Lambda(\Upsilon)) \leq \frac{\vartheta}{N_8}, \tag{3.1.41}$$

and

$$d(\Lambda(\Omega), \Lambda(\Upsilon)) \leq \frac{1}{N_8} d(\Omega, \Upsilon). \tag{3.1.42}$$

Then, Λ is a strictly contractive mapping with Lipschitz constant $\frac{1}{N_8}$.

Let $\omega \in B$, we show that $d(\Lambda(\omega), \omega) < \infty$. Using (3.1.34) we get

$$\begin{aligned} &\Theta\left(\Lambda(\omega(p_1)) - \omega(p_1), \ldots, \Lambda(\omega(p_k)) - \omega(p_k), \vec{\mu}\right) \\ &= \Theta\Big(\boldsymbol{S}(p_1, \omega(p_1)) + {}_0\mathcal{I}_{p_1}^{\kappa,\psi} \mathcal{M}(p_1, \sigma_1, \omega(\sigma_1)) d\sigma_1 - \omega(p_1), \\ &\cdots, \boldsymbol{S}(p_k, \omega(p_k)) + {}_0\mathcal{I}_{p_k}^{\kappa,\psi} \mathcal{M}(p_k, \sigma_k, \omega(\sigma_k)) d\sigma_k - \omega(p_k), \vec{\mu}\Big) \\ &\succeq \phi(p_1, \ldots, p_k, \vec{\mu}) \end{aligned} \tag{3.1.43}$$

for every $\vec{\mu} \in (E_2)^n$. Then we have $d(\Lambda(\omega), \omega) < 1$.

Now Theorem 1.2, enables us to find an element ω_0 in B satisfying the following:
(1) ω_0 is a fixed point of Λ, i.e.,

$$\begin{aligned} \omega_0(p) &= \Lambda(\omega_0(p)) \\ &= \boldsymbol{S}(\sigma, \omega_0(\sigma)) + {}_0\mathcal{I}_p^{\kappa,\psi} \left(\mathcal{M}(\sigma, \varepsilon, \omega_0(\varepsilon)) d\varepsilon\right), \end{aligned} \tag{3.1.44}$$

which is unique in the set

$$B^* = \{\Omega \in B : d(\Lambda(\omega), \Omega) < \infty\}.$$

(2) $d(\Lambda^k(\omega), \omega_0) \to 0$ as $n \to \infty$;
(3) $d(\omega, \omega_0) \leq N_9 d(\Lambda(\omega), \omega) \leq N_9$,

where in $N_9 = \frac{\min[L_1, L_2 L_4]}{\min[L_1, L_2 L_4] - 2}$ and implies that

$$\Theta(\omega(p_1) - \omega_0(p_1), \ldots, \omega(p_k) - \omega_0(p_k), \vec{\mu}) \succeq \phi(p_1, \ldots, p_k, \frac{\vec{\mu}}{N_9}), \tag{3.1.45}$$

for every $p_1, \ldots, p_k \in E_1$ and $\vec{\mu} \in (E_2)^n$. We use the same method of proving Theorem 3.2, and show that the fixed point in B^* is unique.

3.1.4 Applications

We apply the main results to solve some examples.

Example 3.1 Let $(\mathbb{R}, \Theta, \circledast)$ be an matrix valued fuzzy Banach space. Let $\Omega, \Upsilon : E_1 \to \mathbb{R}$ such that $\Omega(p_i) = {p_i}^4$ and $\Upsilon(p_i) = {p_i}^2$ and define $\boldsymbol{S}(p_i, \Omega(p_i)) =$

$\frac{1}{L_1}\Omega(p_i)$. Define $\mathcal{M}: E_1 \times E_1 \times \mathbb{R} \to \mathbb{R}$ as $\mathcal{M}(p_i, \sigma_i, \Omega(\sigma_i)) = \exp(p_i{}^2 - \sigma_i{}^2)\Omega(\sigma_i)$ for every $p_i \in E_1$ and $\sigma_i \preceq p_i \quad (i = 1, 2, \ldots, k)$.

Then we have

$$\begin{aligned}
&\Theta\left(\boldsymbol{S}(p_1, \omega(p_1)) - \boldsymbol{S}(p_1, \Upsilon(p_1)), \ldots, \boldsymbol{S}(p_1, \omega(p_k)) - \boldsymbol{S}(p_k, \Upsilon(p_k)), \vec{\mu}\right) \\
&= \Theta(\frac{1}{L_1}\Omega(p_1) - \frac{1}{L_1}\Upsilon(p_1), \ldots, \frac{1}{L_1}\Omega(p_k) - \frac{1}{L_1}\Upsilon(p_k), \vec{\mu}) \\
&= \Theta(\frac{1}{L_1}p_1^4 - \frac{1}{L_1}p_1^2, \ldots, \frac{1}{L_1}p_k^4 - \frac{1}{L_1}p_k^2, \vec{\mu}) \\
&= \operatorname{diag}\Bigg[\\
&\exp\left(\frac{-\frac{1}{L_1}|p_1^4 - p_1^2, \ldots, p_k^4 - p_k^2|}{\mu_1}\right), \frac{\mu_2}{\mu_2 + \frac{1}{L_1}|p_1^4 - p_1^2, \ldots, p_k^4 - p_k^2|}, \exp\left(\frac{-\frac{1}{L_1}|p_1^4 - p_1^2, \ldots, p_k^4 - p_k^2|}{\mu_3}\right)\Bigg] \\
&\succeq \operatorname{diag}\Bigg[\\
&\exp\left(\frac{-|p_1^4 - p_1^2, \ldots, p_k^4 - p_k^2|}{L_1\mu_1}\right), \frac{L_1\mu_2}{L_1\mu_2 + |p_1^4 - p_1^2, \ldots, p_k^4 - p_k^2|}, \exp\left(\frac{-|p_1^4 - p_1^2, \ldots, p_k^4 - p_k^2|}{L_1\mu_3}\right)\Bigg] \\
&= \Theta\left(p_1^4 - p_1^2, \ldots, p_k^4 - p_k^2, L_1\vec{\mu}\right) \\
&= \Theta\left(\Omega(p_1) - \Upsilon(p_1), \ldots, \Omega(p_k) - \Upsilon(p_k), L_1\vec{\mu}\right),
\end{aligned} \tag{3.1.46}$$

$$\begin{aligned}
&\Theta\left(\mathcal{M}(p_1, \sigma_1, \Omega(\sigma_1)) - \mathcal{M}(p_1, \sigma_1, \Upsilon(\sigma_1)), \ldots, \mathcal{M}(p_k, \sigma_k, \Omega(\sigma_k)) - \mathcal{M}(p_k, \sigma_k, \Upsilon(\sigma_k)), \vec{\mu}\right) \\
&= \Theta(\exp(p_1^2 - \sigma_1{}^2)\sigma_1{}^4 - \exp(p_1^2 - \sigma_1{}^2)\sigma_1{}^2, \ldots, \exp\left(p_k^2 - \sigma_k{}^2\right)\sigma_k{}^4 - \exp(p_k^2 - \sigma_k{}^2)\sigma_k{}^2, \vec{\mu}) \\
&= \operatorname{diag}\Bigg[\\
&\exp\left(\frac{-|\exp(p_1^2 - \sigma_1{}^2)\sigma_1{}^4 - \exp(p_1^2 - \sigma_1{}^2)\sigma_1{}^2, \ldots, \exp(p_k^2 - \sigma_k{}^2)\sigma_k{}^4 - \exp(p_k^2 - \sigma_k{}^2)\sigma_k{}^2|}{\mu_1}\right), \\
&\frac{\mu_2}{\mu_2 + |\exp(p_1^2 - \sigma_1{}^2)\sigma_1{}^4 - \exp(p_1^2 - \sigma_1{}^2)\sigma_1{}^2, \ldots, \exp(p_k^2 - \sigma_k{}^2)\sigma_k{}^4 - \exp(p_k^2 - \sigma_k{}^2)\sigma_k{}^2|}, \\
&\exp\left(\frac{-|\exp(p_1^2 - \sigma_1{}^2)\sigma_1{}^4 - \exp(p_1^2 - \sigma_1{}^2)\sigma_1{}^2, \ldots, \exp(p_k^2 - \sigma_k{}^2)\sigma_k{}^4 - \exp(p_k^2 - \sigma_k{}^2)\sigma_k{}^2|}{\mu_3}\right)\Bigg] \\
&\succeq \operatorname{diag}\Bigg[\\
&\exp\left(\frac{-|\sigma_1{}^4 - \sigma_1{}^2, \ldots, \sigma_k{}^4 - \sigma_k{}^2|}{\frac{\mu_1}{\exp(p_1^2 - \sigma_1{}^2), \ldots, \exp(p_k^2 - \sigma_k{}^2)}}\right), \frac{\frac{\mu_2}{\exp(p_1^2 - \sigma_1{}^2), \ldots, \exp(p_k^2 - \sigma_k{}^2)}}{\frac{\mu_2}{\exp(p_1^2 - \sigma_1{}^2), \ldots, \exp(p_k^2 - \sigma_k{}^2)} + |\sigma_1{}^4 - \sigma_1{}^2, \ldots, \sigma_k{}^4 - \sigma_k{}^2|}, \\
&\exp\left(\frac{-|\sigma_1{}^4 - \sigma_1{}^2, \ldots, \sigma_k{}^4 - \sigma_k{}^2|}{\frac{\mu_3}{\exp(p_1^2 - \sigma_1{}^2), \ldots, \exp(p_k^2 - \sigma_k{}^2)}}\right)\Bigg] \\
&= \Theta\left(\Omega(\sigma_1) - \Upsilon(\sigma_1), \ldots, \Omega(\sigma_k) - \Upsilon(\sigma_k), \frac{\vec{\mu}}{|\exp(p_1^2 - \sigma_1{}^2), \ldots, \exp(p_k^2 - \sigma_k{}^2)|}\right) \\
&\succeq \Theta\left(\Omega(\sigma_1) - \Upsilon(\sigma_1), \ldots, \Omega(\sigma_k) - \Upsilon(\sigma_k), \frac{\vec{\mu}}{K}\right),
\end{aligned} \tag{3.1.47}$$

for some $K \in E_6$.

Let matrix valued fuzzy function $\phi : (E_1)^k \times (E_2)^3 \to \mathrm{diagM_n(E_3)}$ satisfying (3.1.4) and (3.1.5).

Let $\omega : E_1 \to \mathbb{R}$ be a differentiable function satisfying

$$\begin{aligned}&\mathrm{diag}\Big[\\&\exp\left(\frac{-|{}_0\mathcal{D}_{p_1}^{\kappa,\tau,\psi}\omega(p_1)-\frac{1}{L_1}\omega(p_1)-\int_0^{p_1}\exp(p_1^2-\sigma_1{}^2)\omega(\sigma_1)d\sigma_1,\ldots,{}_0\mathcal{D}_{p_k}^{\kappa,\tau,\psi}\omega(p_k)-\frac{1}{L_1}\omega(p_k)-\int_0^{p_k}\exp(p_k^2-\sigma_k{}^2)\omega(\sigma_k)d\sigma_k|}{\mu_1}\right),\\&\frac{\mu_2}{\mu_2+|{}_0\mathcal{D}_{p_1}^{\kappa,\tau,\psi}\omega(p_1)-\frac{1}{L_1}\omega(p_1)-\int_0^{p_1}\exp(p_1^2-\sigma_1{}^2)\omega(\sigma_1)d\sigma_1,\ldots,{}_0\mathcal{D}_{p_k}^{\kappa,\tau,\psi}\omega(p_k)-\frac{1}{L_1}\omega(p_k)-\int_0^{p_k}\exp(p_k^2-\sigma_k{}^2)\omega(\sigma_k)d\sigma_k|},\\&\exp\left(\frac{-|{}_0\mathcal{D}_{p_1}^{\kappa,\tau,\psi}\omega(p_1)-\frac{1}{L_1}\omega(p_1)-\int_0^{p_1}\exp(p_1^2-\sigma_1{}^2)\omega(\sigma_1)d\sigma_1,\ldots,{}_0\mathcal{D}_{p_k}^{\kappa,\tau,\psi}\omega(p_k)-\frac{1}{L_1}\omega(p_k)-\int_0^{p_k}\exp(p_k^2-\sigma_k{}^2)\omega(\sigma_k)d\sigma_k|}{\mu_3}\right)\Big]\\&\succeq \mathrm{diag}\Big[\\&\exp\left(\frac{-|p_1,\ldots,p_k|}{\mu_1}\right),\frac{\mu_2}{\mu_2+|p_1,\ldots,p_k|},\exp\left(\frac{-|p_1,\ldots,p_k|}{\mu_3}\right)\Big].\end{aligned}\tag{3.1.48}$$

for every $p_1,\ldots,p_k \in E_1$ and $\vec{\mu} \in (E_2)^3$. Now, Theorem 3.2 implies that, if $H_1 > 2$, which $H_1 = \min[L_1, \frac{L_3}{K}, L_1L_4, \frac{L_3L_4}{K}]$. we are to be able find a unique differentiable function $\omega_0 : E_1 \to \mathbb{R}$ such that

$${}_0\mathcal{D}_p^{\kappa,\tau,\psi}\omega_0(p) = \frac{1}{L_1}\omega_0(p) + \int_0^p \exp(p^2-\sigma^2)\omega(\sigma)d\sigma, \tag{3.1.49}$$

and

$$\begin{aligned}&\mathrm{diag}\Big[\\&\exp\left(\frac{-|{}_0\mathcal{D}_{p_1}^{\kappa,\tau,\psi}\omega(p_1)-{}_0\mathcal{D}_{p_1}^{\kappa,\tau,\psi}\omega_0(p_1),\ldots,{}_0\mathcal{D}_{p_k}^{\kappa,\tau,\psi}\omega(p_k)-{}_0\mathcal{D}_{p_k}^{\kappa,\tau,\psi}\omega_0(p_k)|}{\mu_1}\right),\\&\frac{\mu_2}{\mu_2+|{}_0\mathcal{D}_{p_1}^{\kappa,\tau,\psi}\omega(p_1)-{}_0\mathcal{D}_{p_1}^{\kappa,\tau,\psi}\omega_0(p_1),\ldots,{}_0\mathcal{D}_{p_k}^{\kappa,\tau,\psi}\omega(p_k)-{}_0\mathcal{D}_{p_k}^{\kappa,\tau,\psi}\omega_0(p_k)|},\\&\exp\left(\frac{-|{}_0\mathcal{D}_{p_1}^{\kappa,\tau,\psi}\omega(p_1)-{}_0\mathcal{D}_{p_1}^{\kappa,\tau,\psi}\omega_0(p_1),\ldots,{}_0\mathcal{D}_{p_k}^{\kappa,\tau,\psi}\omega(p_k)-{}_0\mathcal{D}_{p_k}^{\kappa,\tau,\psi}\omega_0(p_k)|}{\mu_3}\right)\Big]\\&\circledast\ \mathrm{diag}\Big[\\&\exp\left(\frac{-|\omega(p_1)-\omega_0(p_1),\ldots,\omega(p_k)-\omega_0(p_k)|}{\mu_1}\right),\frac{\mu_2}{\mu_2+|\omega(p_1)-\omega_0(p_1),\ldots,\omega(p_k)-\omega_0(p_k)|},\\&\exp\left(\frac{-|\omega(p_1)-\omega_0(p_1),\ldots,\omega(p_k)-\omega_0(p_k)|}{\mu_3}\right)\Big]\end{aligned}\tag{3.1.50}$$

$$\begin{aligned}&\succeq \mathrm{diag}\Big[\\&\exp\left(\frac{(-|p_1,\ldots,p_k|)}{\frac{N_1}{H_2}\mu_1}\right),\frac{\frac{N_1}{H_2}\mu_2}{\frac{N_1}{H_2}\mu_2+|p_1,\ldots,p_k|},\exp\left(\frac{(-|p_1,\ldots,p_k|)}{\frac{N_1}{H_2}\mu_3}\right)\Big].\end{aligned}\tag{3.1.51}$$

for every $p_1, \ldots, p_k \in E_1$ and $\vec{\mu} \in (E_2)^3$ and $N_1 = \min[1, L_4]$ and $H_2 = \frac{\min[L_1, \frac{L_3}{K}, L_1L_4, \frac{L_3L_4}{K}]}{\min[L_1, \frac{L_3}{K}, L_1L_4, \frac{L_3L_4}{K}]-2}$.

Example 3.2 Let $(\mathbb{R}, \Theta, \circledast)$ be an matrix valued fuzzy Banach space. Consider $\Omega, \Upsilon : E_1 \to \mathbb{R}$ such that $\Omega(p_i) = {p_i}^4$ and $\Upsilon(p_i) = {p_i}^2$ and define $S(p_i, \Omega(p_i)) = \frac{1}{L_1}\Omega(p_i)$. Define $\mathcal{M} : E_1 \times E_1 \times \mathbb{R} \to \mathbb{R}$ as $\mathcal{M}(p_i, \sigma_i, \Omega(\sigma_i)) = \exp({p_i}^2 - {\sigma_i}^2)\Omega(\sigma_i)$ for every $p_1, \ldots, p_k \in E_1$ and $\sigma_i \preceq p_i \quad (i = 1, 2, \ldots, k)$ satisfying (3.1.47).

Let matrix valued fuzzy function $\phi : (E_1)^k \times (E_2)^3 \to \mathrm{diagM_n(E_3)}$ satisfying (3.1.4) and (3.1.5).

Let $\omega : E_1 \to \mathbb{R}$ be a differentiable function satisfying

$$\begin{aligned}
&\mathrm{diag}\Bigg[\\
&\exp\left(\frac{-|\omega(p_1)-\frac{1}{L_1}\omega(p_1)-\ {}_0\mathcal{I}_{p_1}^{\kappa,\psi}\exp(p_1^2-{\sigma_1}^2)\omega(\sigma_1)d\sigma_1,\ldots,\omega(p_k)-\frac{1}{L_1}\omega(p_k)-\ {}_0\mathcal{I}_{p_k}^{\kappa,\psi}\exp(p_k^2-{\sigma_k}^2)\omega(\sigma_k)d\sigma_k|}{\mu_1}\right),\\
&\frac{\mu_2}{\mu_2+|\omega(p_1)-\frac{1}{L_1}\omega(p_1)-\ {}_0\mathcal{I}_{p_1}^{\kappa,\psi}\exp(p_1^2-{\sigma_1}^2)\omega(\sigma_1)d\sigma_1,\ldots,\omega(p_k)-\frac{1}{L_1}\omega(p_k)-\ {}_0\mathcal{I}_{p_k}^{\kappa,\psi}\exp(p_k^2-{\sigma_k}^2)\omega(\sigma_k)d\sigma_k|},\\
&\exp\left(\frac{-|\omega(p_1)-\frac{1}{L_1}\omega(p_1)-\ {}_0\mathcal{I}_{p_1}^{\kappa,\psi}\exp(p_1^2-{\sigma_1}^2)\omega(\sigma_1)d\sigma_1,\ldots,\omega(p_k)-\frac{1}{L_1}\omega(p_k)-\ {}_0\mathcal{I}_{p_k}^{\kappa,\psi}\exp(p_k^2-{\sigma_k}^2)\omega(\sigma_k)d\sigma_k|}{\mu_3}\right)\Bigg]\\
&\succeq \mathrm{diag}\Bigg[\\
&\exp\left(\frac{-|p_1,\ldots,p_k|}{\mu_1}\right), \frac{\mu_2}{\mu_2+|p_1,\ldots,p_k|}, \exp\left(\frac{-|p_1,\ldots,p_k|}{\mu_3}\right)\Bigg]
\end{aligned} \tag{3.1.52}$$

for every $p_1, \ldots, p_k \in E_1$ and $\vec{\mu} \in (E_2)^3$.

Now, Theorem 3.3 implies that, if $H_3 > 2$, which $H_3 = \min[L_1, \frac{L_4}{K}]$ we are to be able find a unique differentiable function $\omega_0 : E_1 \to \mathbb{R}$ such that

$$\omega_0(p) = \frac{1}{L_1}\omega_0(p) + {}_0\mathcal{I}_p^{\kappa,\psi}\exp(p^2 - \sigma^2)\omega(\sigma)d\sigma, \tag{3.1.53}$$

and

$$\begin{aligned}
&\mathrm{diag}\Bigg[\\
&\exp\left(\frac{-|\omega(p_1)-\omega_0(p_1),\ldots,\omega(p_k)-\omega_0(p_k)|}{\mu_1}\right), \frac{\mu_2}{\mu_2+|\omega(p_1)-\omega_0(p_1),\ldots,\omega(p_k)-\omega_0(p_k)|},\\
&\exp\left(\frac{-|\omega(p_1)-\omega_0(p_1),\ldots,\omega(p_k)-\omega_0(p_k)|}{\mu_3}\right)\Bigg]\\
&\succeq \mathrm{diag}\Bigg[\\
&\exp\left(\frac{-|p_1,\ldots,p_k|}{\frac{N_1}{H_4}\mu_1}\right), \frac{\frac{N_1}{H_4}\mu_2}{\frac{N_1}{H_4}\mu_2+|p_1,\ldots,p_k|}, \exp\left(\frac{-|p_1,\ldots,p_k|}{\frac{N_1}{H_4}\mu_3}\right)\Bigg]
\end{aligned} \tag{3.1.54}$$

for every $p_1, \dots, p_k \in E_1$ and $\vec{\mu} \in (E_2)^3$ and $N_2 = \min[1, L_4]$ and $H_4 = \frac{\min[L_1, \frac{L_4}{K}]}{\min[L_1, \frac{L_4}{K}]-2}$.

3.2 New Stability Results of the ABC Fractional Differential Equation

3.2.1 *Introduction of the ABC Fractional Differential Equation and Other Preliminaries*

We consider the NS-ABC-FD equation, which is as follows, and then investigate its multiple-HU-OS_1-stability using a multiple fuzzy controller. We have

$$\begin{cases} {}^{\mathcal{ABC}}{}_0\mathcal{D}^{\epsilon}\left[\Delta_\theta\left[{}^{ABC}{}_0\mathcal{D}^{\varrho_0}\psi(\mathfrak{X})\right]\right] = -\phi_1^*(\mathfrak{X}, \psi(\mathfrak{X})), \\ \Phi_p\left[{}^{\mathcal{ABC}}{}_0\mathcal{D}^{\varrho_0}\psi(\mathfrak{X})\right]\big|_{\mathfrak{X}=0} = 0,\ \psi(1) = 0, \end{cases} \tag{3.2.1}$$

for $\epsilon, \varrho_0 \in (0, 1]$ and the continuous function $\phi_1^*(\mathfrak{X}, \psi(\mathfrak{X})) \in C[i, j],\ i = 0,\ j = 1$. $\mathcal{D}^{\epsilon}$ and $\mathcal{D}^{\varrho_0}$ are $\mathcal{ABC}$-fractional derivative operators. $\Delta_\theta(\mathrm{m})$ is the nonlinear operator such that $\Delta_\theta(\mathrm{m}) = |\mathrm{m}|^{\theta-2}\mathrm{m}$, $\frac{1}{\theta} + \frac{1}{\eta} = 1$ and $\Delta_\theta^{-1} = \Delta_\eta$. Also, for $\mathfrak{X} \in (0, 1]$, $\psi(\mathfrak{X}) > 0$.

$$\begin{cases} {}^{\mathcal{A}}\mathcal{BC}\mathbb{D}^{\varrho_0}\psi(\mathfrak{X}) = \phi_1^*(\mathfrak{X}, \psi(\mathfrak{X})), \\ \psi(0) = \psi_0, \end{cases} \tag{3.2.2}$$

where ${}^{\mathcal{ABC}}{}_0\mathcal{D}^{\epsilon}$ is the $\mathcal{ABC}$-fractional derivative operator and $\varrho_0 \in (0, 1)$, ${}^{\mathcal{ABC}}{}_0\mathcal{D}^{\epsilon}\psi(\mathfrak{X})$, $\phi_1^*(\mathfrak{X}, \psi(\mathfrak{X})) \in C[i, j]$ [21]. The function $\psi(\mathfrak{X})$ in Eq. (3.2.2) is the same function in Eq. (3.2.1).

Definition 3.1 Let function $\boldsymbol{OS_1}\Big(\boldsymbol{\Upsilon}(\mathfrak{X}, \eta)\Big)$ be a MVFF. The Eq. (3.2.1) is said to be multiple-HU-OS_1 stable, if $\psi(\mathfrak{X})$ is a given differentiable function satisfying

$$\mathcal{N}\Big({}^{\mathcal{A}}\mathcal{BC}_0\mathcal{D}^{\epsilon}\left[\Delta_\theta\left[{}^{\mathcal{ABC}}{}_0\mathcal{D}^{\varrho_0}\psi(\mathfrak{X})\right]\right] + \phi_1^*(\mathfrak{X}, \psi(\mathfrak{X})), \eta\Big) \succeq \boldsymbol{OS_1}\Big(\boldsymbol{\Upsilon}(\mathfrak{X}, \eta)\Big), \tag{3.2.3}$$

for $\mathfrak{X} \in [0, j]$, and we can find a solution $\wp(\mathfrak{X})$ of Eq. (3.2.1) such that for some $\gamma > 0$,

$$\mathcal{N}\Big(\wp(\mathfrak{X}) - \psi(\mathfrak{X}), \eta\Big) \succeq \boldsymbol{OS_1}\Big(\boldsymbol{\Upsilon}(\mathfrak{X}, \frac{\eta}{\gamma})\Big).$$

Our method can be used to get new results from [41–44].

Remark 3.1 $\psi(\mathfrak{X})$ is a solution of (3.2.1) if and only if for $\sigma, \varrho_0 \in (0, 1]$, $\phi_1^*(\mathfrak{X}, \psi(\mathfrak{X})) \in C[0, 1]$ and $\phi_1^*(0, \psi(0)) = 0$, we have

$$\psi(\mathfrak{X}) = \int_0^1 \mathfrak{G}^{\varrho_0}(\mathfrak{X}, \mathrm{k})\Delta_\eta \left({}^{\mathcal{AB}}\mathcal{I}_0^\epsilon \left[\phi_1^*(\mathfrak{X}, \phi(\mathfrak{X}))\right]\right) d\mathfrak{X},$$

where

$$\mathfrak{G}^{\varrho_0}(\mathfrak{X}, \mathrm{k}) = \begin{cases} \frac{\varrho_0}{\mathcal{W}(\varrho_0)} \frac{(1-\mathrm{k})^{\varrho_0 - 1}}{\Gamma(\varrho_0)} - \frac{\varrho_0}{\mathcal{W}(\varrho_0)} \frac{(\mathfrak{X}-\mathrm{k})^{\varrho_0 - 1}}{\Gamma(\varrho_0)} & \mathrm{k} \le \mathfrak{X}, \\ \frac{\varrho_0}{\mathcal{W}(\varrho_0)} \frac{(1-\mathrm{k})^{\varrho_0 - 1}}{\Gamma(\varrho_0)} & \mathrm{k} \ge \mathfrak{X}, \end{cases} \tag{3.2.4}$$

for the function $\mathfrak{G}^{\varrho_0}(\mathfrak{X}, \mathrm{k})$ defined by Eq. (3.2.4), we have

- $\mathfrak{G}^{\varrho_0}(\mathfrak{X}, \mathrm{k}) > 0$ for all $\mathrm{k}, \mathfrak{X} \in (0, 1)$;
- the function $\mathfrak{G}^{\varrho_0}(\mathfrak{X}, \mathrm{k})$ is a decreasing multivalued function and $\mathfrak{G}^{\varrho_0}(0, \mathrm{k}) = \max_{\mathfrak{X}\in[0,1]} \mathfrak{G}^{\varrho_0}(\mathfrak{X}, \mathrm{k})$; and
- with an assumption of $0 \le \mathfrak{X}^{\varrho_0 - 1} \le 0.5$, $\mathfrak{G}^{\varrho_0}(\mathfrak{X}, \mathrm{k}) \ge \mathfrak{X}^{\epsilon - 1} \max_{\mathfrak{X}\in[0,1]} \mathfrak{G}^{\varrho_0}(\mathfrak{X}, \mathrm{k})$ for $\mathrm{k}, \mathfrak{X} \in (0, 1)$.

3.2.2 *Multiple-HU-$\mathcal{O}S_1$ Stability for NS-ABC-FDE*

Now, we use the FPT based on the Theorem 1.2 to show (3.2.1) is multiple-HU-$\mathcal{O}S_1$ stable in SMVFBS $(\mathcal{A}, \mathcal{N}, \circledast)$ with MVFF $\boldsymbol{\mathcal{O}S_1}\left(\boldsymbol{\Upsilon}(\mathfrak{X}, \eta)\right)$.

We define the set $\mathcal{F}$ as follows

$$\mathcal{F} = \{\psi : \mathfrak{L} \to \mathbb{R}^n, \psi \text{ is differentiable}\},$$

and $\delta : \mathcal{F} \times \mathcal{F} \to [0, +\infty]$, is given by

$$\delta(\psi, \wp) = \inf\left\{\wp \in [0, +\infty) : \ \mathcal{N}\Big(\psi(\mathfrak{X}) - \wp(\mathfrak{X}), \eta\Big) \succeq \boldsymbol{\mathcal{O}S_1}\Big(\boldsymbol{\Upsilon}(\mathfrak{X}, \frac{\eta}{\wp})\Big), \forall \ \psi, \wp \in \mathcal{F}, \ \mathfrak{X} \in [0, j], \ \ \eta \in (0, +\infty)\right\}.$$

Theorem 3.4 *$(\mathcal{F}, \delta)$ is a complete $[0, +\infty]$-valued metric space.*

Proof We first have $\delta(\psi, \wp) = 0$ if and only if $\psi = \wp$. Assume that $\delta(\psi, \wp) = 0$, then

$$\inf\left\{\vartheta \in [0, +\infty) : \mathcal{N}\Big(\psi(\mathfrak{X}) - \wp(\mathfrak{X}), \eta\Big) \succeq \boldsymbol{\mathcal{O}S_1}\Big(\boldsymbol{\Upsilon}(\mathfrak{X}, \frac{\eta}{\vartheta})\Big), \forall \ \psi, \wp \in \mathcal{F},\right.$$

$$\mathfrak{x} \in [0, j], \quad \eta \in (0, +\infty)\Big\} = 0,$$

and then

$$\mathcal{N}\Big(\psi(\mathfrak{X}) - \wp(\mathfrak{X}), \eta\Big) \succeq \boldsymbol{OS}_1\Big(\boldsymbol{\Upsilon}(\mathfrak{x}, \frac{\eta}{\vartheta})\Big),$$

for all $\vartheta \in [0, +\infty)$. Let ϑ tend to zero in the above inequality, we get

$$\mathcal{N}\Big(\psi(\mathfrak{X}) - \wp(\mathfrak{X}), \eta\Big) = \mathbf{1}.$$

Thus $\psi(\mathfrak{X}) = \wp(\mathfrak{X})$ for every $\mathfrak{X} \in [0, j]$, and vice versa. Moreover, we have $\delta(\psi, \wp) = \delta(\wp, \psi)$ for every $\psi, \wp \in \mathcal{F}$. Now, let $\delta(\psi, \wp) = \alpha_1 \in (0, +\infty)$ and $\delta(\wp, \mathrm{w}) = \alpha_2 \in (0, +\infty)$. Then, we have

$$\mathcal{N}\Big(\psi(\mathfrak{X}) - \wp(\mathfrak{X}), \eta\Big) \succeq \boldsymbol{OS}_1\Big(\boldsymbol{\Upsilon}\Big(\mathfrak{X}, \frac{\eta}{\alpha_1}\Big)\Big),$$

and

$$\mathcal{N}\Big(\wp(\mathfrak{X}) - \mathrm{w}(\mathfrak{X}), \eta\Big) \succeq \boldsymbol{OS}_1\Big(\boldsymbol{\Upsilon}\Big(\mathfrak{X}, \frac{\eta}{\alpha_2}\Big)\Big),$$

for each $\eta \in (0, +\infty)$. Then we have

$$\begin{aligned}
\mathcal{N}\Big((\psi(\mathfrak{X}) - \mathrm{w}(\mathfrak{X}), (\alpha_1 + \alpha_2)\eta\Big) &\succeq \Big[\mathcal{N}\Big((\psi(\mathfrak{X}) - \wp(\mathfrak{X}), (\alpha_1)\eta\Big) \circledast \mathcal{N}\Big((\wp(\mathfrak{X}) - \mathrm{w}(\mathfrak{X}), (\alpha_2)\eta\Big)\Big] \\
&\succeq \boldsymbol{OS}_1\Big(\boldsymbol{\Upsilon}(\mathfrak{X}, \eta)\Big) \circledast \boldsymbol{OS}_1\Big(\boldsymbol{\Upsilon}(\mathfrak{X}, \eta)\Big) \\
&= \boldsymbol{OS}_1\Big(\boldsymbol{\Upsilon}(\mathfrak{X}, \eta)\Big),
\end{aligned}$$

and then, $\delta(\wp, \mathrm{w}) \le \alpha_1 + \alpha_2$ and $\delta(\wp, \mathrm{w}) \le \delta(\psi, \wp) + \delta(\wp, \mathrm{w})$. Now we show that $(\mathcal{F}, \delta)$ is complete. For this purpose, we consider a Cauchy sequence like $\{\wp_k\}$ and assume that $\mathfrak{X} \in [0, j], \ \tau \in (0, +\infty), \ \eta \in (0, +\infty)$. We consider

$$\boldsymbol{OS}_1\Big(\boldsymbol{\Upsilon}(\mathfrak{X}, \eta)\Big) \succ \mathbf{1} - \boldsymbol{\Im}.$$

For $\alpha\beta < \varsigma$ choose $q_0 \in \mathbb{N}$ where

$$\delta(\wp_q, \wp_p) < \alpha \quad \text{for all } q, p \ge q_0.$$

Consequently,

$$\mathcal{N}\Big(\wp_q(\mathfrak{X}) - \wp_p(\mathfrak{X}), \eta\Big) \succeq \mathcal{N}\Big(\wp_q(\mathfrak{X}) - \wp_p(\mathfrak{X}), \alpha\eta\Big) \succeq \boldsymbol{OS_1}\Big(\boldsymbol{\Upsilon}(\boldsymbol{\mathfrak{X}}, \boldsymbol{\eta})\Big) \succ \mathbf{1} - \boldsymbol{\mathfrak{I}},$$

and then

$$\mathcal{N}\Big(\wp_q(\mathfrak{X}) - \wp_p(\mathfrak{X}), \eta\Big) \succ \mathbf{1} - \boldsymbol{\mathfrak{I}},$$

and then $\{\wp_q(\mathfrak{X})\}_k$ is a Cauchy sequence in complete space $(\mathcal{A}, \mathcal{N}, \circledast)$ on compact set $[0, j]$ and uniformly converges to $\wp : [0, j] \to \mathcal{A}$. Therefore, taking into account the uniformly convergent property, $\wp \in \mathcal{F}$, that is, $\wp$ is a differentiable function. Therefore, the completeness of $(\mathcal{F}, \delta)$ is the result.

Here, we are ready to study multiple-HU-OS_1 stability and approximate NS-ABC-FDEs (3.2.1).

Theorem 3.5 *Let $(\mathcal{A}, \mathcal{N}, \circledast)$ be a SMVFBS and consider the constants ϵ, γ and ϱ where $0 < \epsilon\gamma\varrho < 1$. Let*

- $$\mathcal{N}\Big(\mathfrak{G}^{\varrho_0}(\mathfrak{X}, \xi)\Delta_\kappa\Big(\phi_1^*(\mathfrak{X}, \psi(\mathfrak{X})) - \phi_1^*(\mathfrak{X}, \wp(\mathfrak{X}))\Big), \eta\Big) \succeq \mathcal{N}\Big(\psi(\mathfrak{X}) - \wp(\mathfrak{X}), \frac{\eta}{\gamma}\Big), \tag{3.2.5}$$

- $$\mathcal{N}\Big(\psi(\mathfrak{X}), \eta\Big) \succeq \boldsymbol{OS_1}\Big(\boldsymbol{\Upsilon}(\boldsymbol{\mathfrak{X}}, \boldsymbol{\eta})\Big)$$

  implies that

  $$\mathcal{N}\Big(I\psi(\mathfrak{X}), \eta\Big) \succeq \boldsymbol{OS_1}\Big(\boldsymbol{\Upsilon}(\boldsymbol{\mathfrak{X}}, \frac{\boldsymbol{\eta}}{\boldsymbol{\epsilon}})\Big). \tag{3.2.6}$$

- *By considering the MVFF $\boldsymbol{OS_1} : [0, 1] \longrightarrow [0, 1]$ as the control function, we have*

  $$\inf_{\xi \in E_1} \boldsymbol{OS_1}\Big(\boldsymbol{\Upsilon}(\boldsymbol{\xi}, \boldsymbol{\eta})\Big) \succeq \boldsymbol{OS_1}\Big(\boldsymbol{\Upsilon}(\boldsymbol{\mathfrak{X}}, \frac{\boldsymbol{\eta}}{\boldsymbol{P\varrho}})\Big). \tag{3.2.7}$$

Let $\psi : [0, j] \to \mathcal{A}$ be a differentiable function satisfying

$$\mathcal{N}\Big({}^{\mathcal{ABC}}{}_0\mathcal{D}^{\epsilon}\left[\Delta_\theta\left[{}^{\mathcal{ABC}}{}_0\mathcal{D}^{\varrho_0}\psi(\mathfrak{X})\right]\right] + \phi_1^*(\mathfrak{X}, \psi(\mathfrak{X})), \eta\Big) \succeq \boldsymbol{OS_1}\Big(\boldsymbol{\Upsilon}(\boldsymbol{\mathfrak{X}}, \boldsymbol{\eta})\Big), \tag{3.2.8}$$

and then, there is a unique solution $\wp : [0, j] \to \mathcal{A}$ *for Eq. (3.2.1) such that*

$$\mathcal{N}\Big(\psi(\mathfrak{X}) - \wp(\mathfrak{X}), \eta\Big) \succeq \boldsymbol{OS}_1\Big(\boldsymbol{\Upsilon}(\mathfrak{X}, \frac{\eta}{F})\Big),$$

where $F = \frac{\epsilon\gamma\varsigma}{1-\epsilon\gamma\varsigma}$, $\mathfrak{X} \in [0, j]$ *and* $\eta \in (0, +\infty)$.

Proof We set

$$\mathcal{F} := \{\wp : [0, j] \to \mathcal{A}, \wp \text{ is differentiable}\},$$

and introduce the $[0, +\infty]$-valued metric on $\mathcal{F}$ as

$$\inf\Big\{B \in [0, +\infty) : \mathcal{N}\Big(\psi(\mathfrak{X}) - \wp(\mathfrak{X})), \eta\Big) \succeq \boldsymbol{OS}_1\Big(\boldsymbol{\Upsilon}(\mathfrak{X}, \frac{\eta}{B})\Big), \forall\ \wp, \psi \in \mathcal{F}, \\ \mathfrak{X} \in [0, j],\quad \eta \in (0, +\infty)\Big\} = 0.$$

By Theorem 3.4, we have $(\mathcal{F}, \delta)$ is a complete $[0, +\infty]$-valued metric space.

Step 1. Now, we define the mapping $\mathcal{R}^* : \mathcal{F} \to \mathcal{F}$, as follows

$$\mathcal{R}^*(\psi(\mathfrak{X})) = \int_0^1 \mathfrak{G}^{\varrho_0}(\mathfrak{X}, \mathrm{k})\Delta_\kappa\left({}^{\mathcal{AB}}\mathcal{I}_0^\epsilon\left[\phi_1^*(\mathfrak{X}, \psi(\mathfrak{X}))\right]\right) d\mathrm{k}, \tag{3.2.9}$$

for $\mathfrak{X} \in [0, j]$. Let $\psi, \wp \in \mathcal{F}$ and consider the coefficient $B_{\psi\wp} \in [0, +\infty]$ with $\delta(\psi, \wp) \leq B_{\psi\wp}$, thus

$$\mathcal{N}\Big(\psi(\mathfrak{X}) - \wp(\mathfrak{X}), B_{\psi\wp}\eta\Big) \succeq \boldsymbol{OS}_1\Big(\boldsymbol{\Upsilon}(\mathfrak{X}, \eta)\Big),$$

for all $\psi, \wp \in \mathcal{F}$, $\mathfrak{X} \in [0, j]$ and $\eta \in (0, +\infty)$. Applying $(\aleph_2)$ and $(\aleph_3)$, we imply that

$$\mathcal{N}\Big(\mathcal{R}^*(\psi(\mathfrak{X})) - \mathcal{R}^*(\wp(\mathfrak{X})), B_{\psi\wp}\eta\Big) \tag{3.2.10}$$
$$= \mathcal{N}\Big(\int_0^1 \mathfrak{G}^{\varrho_0}(\mathfrak{X}, \mathrm{k})\Delta_\kappa\left({}^{\mathcal{AB}}\mathcal{I}_0^\epsilon\left[\phi_1^*(\mathfrak{X}, \psi(\mathfrak{X}))\right]\right) d\mathrm{k} - \int_0^1 \mathfrak{G}^{\varrho_0}(\mathfrak{X}, \mathrm{k})\Delta_\kappa\left({}^{\mathcal{AB}}\mathcal{I}_0^\epsilon\left[\phi_1^*(\mathfrak{X}, \wp(\mathfrak{X}))\right]\right) d\mathrm{k}, B_{\psi\wp}\eta\Big)$$
$$= \mathcal{N}\Big(\int_0^1 \Big(\mathfrak{G}^{\varrho_0}(\mathfrak{X}, \mathrm{k})\Delta_\kappa\left({}^{\mathcal{AB}}\mathcal{I}_0^\epsilon\left[\phi_1^*(\mathfrak{X}, \psi(\mathfrak{X}))\right]\right) - \mathfrak{G}^{\varrho_0}(\mathfrak{X}, \mathrm{k})\Delta_\kappa\left({}^{\mathcal{AB}}\mathcal{I}_0^\epsilon\left[\phi_1^*(\mathfrak{X}, \wp(\mathfrak{X}))\right]\right)\Big) d\mathrm{k}, B_{\psi\wp}\eta\Big)$$
$$= \mathcal{N}\Big(\int_0^1 \Big(\mathfrak{G}^{\varrho_0}(\mathfrak{X}, \mathrm{k})\Delta_\kappa\left({}^{\mathcal{AB}}\mathcal{I}_0^\epsilon\left[\phi_1^*(\mathfrak{X}, \psi(\mathfrak{X})) - \phi_1^*(\mathfrak{X}, \wp(\mathfrak{X}))\right]\right) d\mathrm{k}, B_{\psi\wp}\eta\Big).$$

In the following, we have

(1)

$$\mathcal{N}\Big(\mathfrak{G}^{\varrho_0}(\mathfrak{X},\xi)\Delta_\kappa\Big(\Big[\phi_1^*(\mathfrak{X},\psi(\mathfrak{X}))-\phi_1^*(\mathfrak{X},\wp(\mathfrak{X}))\Big]\Big),B_{\psi\wp}\eta\Big) \tag{3.2.11}$$
$$\succeq \mathcal{N}\Big(\Big(\mathfrak{G}^{\varrho_0}(\mathfrak{X},\xi)\Delta_\kappa\Big(\Big[\psi(\mathfrak{X})-\wp(\mathfrak{X})\Big]\Big),\frac{B_{\psi\wp}\eta}{\gamma}\Big)$$
$$\succeq \boldsymbol{OS_1}\Big(\boldsymbol{\Upsilon}(\boldsymbol{\mathfrak{X}},\frac{\boldsymbol{\eta}}{\boldsymbol{\gamma}})\Big),$$

(2)

$$\mathcal{N}\Big({}^{\mathcal{A}\mathcal{B}}\mathcal{I}_0^\epsilon\int_0^1\Big(\mathfrak{G}^{\varrho_0}(\mathfrak{X},\mathrm{k})\Delta_\kappa\left(\left[\phi_1^*(\mathfrak{X},\psi(\mathfrak{X}))-\phi_1^*(\mathfrak{X},\wp(\mathfrak{X}))\right]\right)d\mathrm{k},B_{\psi\wp}\eta\Big) \tag{3.2.12}$$
$$=\mathcal{N}\Big({}^{\mathcal{A}\mathcal{B}}\mathcal{I}_0^\epsilon\lim_{\|\bar{y}\|\to 0}\sum_{j=1}^{n}\Big(\mathfrak{G}^{\varrho_0}(\mathfrak{X},\bar{y}_j)\Delta_\kappa\Big(\Big[\phi_1^*(\mathfrak{X},\psi(\mathfrak{X}))-\phi_1^*(\mathfrak{X},\wp(\mathfrak{X}))\Big]\Big)\Delta\bar{y}_j,B_{\psi\wp}\eta\Big)$$
$$=\lim_{\|\bar{y}\|\to 0}\mathcal{N}\Big({}^{\mathcal{A}\mathcal{B}}\mathcal{I}_0^\epsilon\sum_{j=1}^{n}\Big(\mathfrak{G}^{\varrho_0}(\mathfrak{X},\bar{y}_j)\Delta_\kappa\Big(\Big[\phi_1^*(\mathfrak{X},\psi(\mathfrak{X}))-\phi_1^*(\mathfrak{X},\wp(\mathfrak{X}))\Big]\Big)\Delta\bar{y}_j,B_{\psi\wp}\eta\Big)$$
$$\succeq\lim_{\|\bar{y}\|\to 0}\circledast_M\mathcal{N}\Big({}^{\mathcal{A}\mathcal{B}}\mathcal{I}_0^\epsilon\Big(\mathfrak{G}^{\varrho_0}(\mathfrak{X},\bar{y}_j)\Delta_\kappa\Big(\Big[\phi_1^*(\mathfrak{X},\psi(\mathfrak{X}))-\phi_1^*(\mathfrak{X},\wp(\mathfrak{X}))\Big]\Big)\Delta\bar{y}_j,\frac{B_{\psi\wp}\eta}{k}\Big)$$
$$\succeq\inf_{\xi\in E_1}\mathcal{N}\Big({}^{\mathcal{A}\mathcal{B}}\mathcal{I}_0^\epsilon\Big(\mathfrak{G}^{\varrho_0}(\mathfrak{X},\xi)\Delta_\kappa\Big(\Big[\phi_1^*(\mathfrak{X},\psi(\mathfrak{X}))-\phi_1^*(\mathfrak{X},\wp(\mathfrak{X}))\Big]\Big),\frac{kB_{\psi\wp}\eta}{kP}\Big)$$
$$\succeq\inf_{\xi\in E_1}\mathcal{N}\Big(\mathfrak{G}^{\varrho_0}(\mathfrak{X},\xi)\Delta_\kappa\Big(\Big[\phi_1^*(\mathfrak{X},\psi(\mathfrak{X}))-\phi_1^*(\mathfrak{X},\wp(\mathfrak{X}))\Big]\Big),\frac{kB_{\psi\wp}\eta}{kP\epsilon}\Big)$$
$$\succeq\inf_{\xi\in E_1}\mathcal{N}\Big(\mathfrak{G}^{\varrho_0}(\mathfrak{X},\xi)\Delta_\kappa\Big(\Big[\psi(\mathfrak{X})-\wp(\mathfrak{X})\Big]\Big),\frac{kB_{\psi\wp}\eta}{kP\epsilon\gamma}\Big)$$
$$\succeq\inf_{\xi\in E_1}\boldsymbol{OS_1}\Big(\boldsymbol{\Upsilon}(\boldsymbol{\xi},\frac{\boldsymbol{\eta}}{\boldsymbol{P\epsilon\gamma}})\Big)$$
$$\succeq\boldsymbol{OS_1}\Big(\boldsymbol{\Upsilon}(\boldsymbol{\mathfrak{X}},\frac{\boldsymbol{\eta}}{\boldsymbol{\epsilon\gamma\varrho}})\Big).$$

From (3.2.10)–(3.2.12), we have

$$\mathcal{N}\Big(\int_0^1\mathfrak{G}^{\varrho_0}(\mathfrak{X},\mathrm{k})\Delta_\kappa\left({}^{\mathcal{A}\mathcal{B}}\mathcal{I}_0^\epsilon\left[\phi_1^*(\mathfrak{X},\psi(\mathfrak{X}))\right]\right)d\mathrm{k}-\int_0^1\mathfrak{G}^{\varrho_0}(\mathfrak{X},\mathrm{k})\Delta_\kappa\left({}^{\mathcal{A}\mathcal{B}}\mathcal{I}_0^\epsilon\left[\phi_1^*(\mathfrak{X},\wp(\mathfrak{X}))\right]\right)d\mathrm{k},B_{\psi\wp}\eta\Big) \tag{3.2.13}$$
$$\succeq\boldsymbol{OS_1}\Big(\boldsymbol{\Upsilon}(\boldsymbol{\mathfrak{X}},\frac{\boldsymbol{\eta}}{\boldsymbol{\epsilon\gamma\varrho}})\Big),$$

which implies that

$$\delta(\mathcal{R}^*(\psi(\mathfrak{X})),\mathcal{R}^*(\wp(\mathfrak{X})))\leq B_{\psi\wp}\epsilon\gamma\varrho,$$

and then

$$\delta(\mathcal{R}^*(\psi(\mathfrak{X})), \mathcal{R}^*(\wp(\mathfrak{X}))) \leq \epsilon\gamma\varrho\delta(\psi, \wp),$$

where $0 < \epsilon\gamma\varrho < 1$; therefore, $\mathcal{R}^*$ is a contractional mapping.
Step 2. We will show that $\delta(\mathcal{R}^*(\psi(\mathfrak{X})), \psi(\mathfrak{X})) < +\infty$. Let $\wp \in \mathcal{F}$, and we have

$$\begin{aligned}
&\mathcal{N}\Big(\mathcal{R}^*(\psi(\mathfrak{X})) - \psi(\mathfrak{X}), \eta\Big) \qquad (3.2.14)\\
&= \mathcal{N}\Big(\int_0^1 \mathfrak{G}^{\varrho_0}(\mathfrak{X}, \mathrm{k})\Delta_\kappa\left({}^{\mathcal{AB}}\mathcal{I}_0^\epsilon\left[\phi_1^*(\mathfrak{X}, \psi(\mathfrak{X}))\right]\right)d\mathrm{k} - {}^{\mathcal{AB}}\mathcal{I}_0^{\epsilon\,\mathcal{ABC}}{}_0\mathcal{D}^\epsilon\psi(\mathfrak{X}), \eta\Big)\\
&= \mathcal{N}\Big(\Big[\frac{\varrho_0}{\mathcal{W}(\varrho_0)}\mathcal{I}^{\varrho_0}_{\mathfrak{X}=1} - \frac{\varrho_0}{\mathcal{W}(\varrho_0)}\mathcal{I}^{\varrho_0}_{\mathfrak{X}}\Big]\Delta_\kappa\Big({}^{\mathcal{AB}}\mathcal{I}_0^\epsilon\left[\phi_1^*(\mathfrak{X}, \psi(\mathfrak{X}))\right]\Big) - {}^{\mathcal{AB}}\mathcal{I}_0^{\epsilon\,\mathcal{ABC}}{}_0\mathcal{D}^\epsilon\psi(\mathfrak{X}), \eta\Big)\\
&= \mathcal{N}\Big(\Big[\frac{1-\varrho_0}{\mathcal{W}(\varrho_0)} + \frac{\varrho_0}{\mathcal{W}(\varrho_0)}\mathcal{I}^{\varrho_0}_{\mathfrak{X}=1}\Big]\Delta_\kappa\Big({}^{\mathcal{AB}}\mathcal{I}_0^\epsilon\left[\phi_1^*(\mathfrak{X}, \psi(\mathfrak{X}))\right]\Big) - \Big[\frac{1-\varrho_0}{\mathcal{W}(\varrho_0)} + \frac{\varrho_0}{\mathcal{W}(\varrho_0)}\mathcal{I}^{\varrho_0}_{\mathfrak{X}}\Big]\\
&\Delta_\kappa\Big({}^{\mathcal{AB}}\mathcal{I}_0^\epsilon\left[\phi_1^*(\mathfrak{X}, \psi(\mathfrak{X}))\right]\Big) - {}^{\mathcal{AB}}\mathcal{I}_0^{\epsilon\,\mathcal{ABC}}{}_0\mathcal{D}^\epsilon\psi(\mathfrak{X}), \eta\Big)\\
&= \mathcal{N}\Big({}^{\mathcal{AB}}\mathcal{I}_0^\epsilon\Big(\Delta_\kappa\Big({}^{\mathcal{AB}}\mathcal{I}_0^\epsilon\left[\phi_1^*(\mathfrak{X}, \psi(\mathfrak{X}))\right]\Big)|_{\mathfrak{X}=1} - {}^{\mathcal{AB}}\mathcal{I}_0^\epsilon\Big(\Delta_\kappa\Big({}^{\mathcal{AB}}\mathcal{I}_0^\epsilon\left[\phi_1^*(\mathfrak{X}, \psi(\mathfrak{X}))\right]\Big) -\\
&{}^{\mathcal{AB}}\mathcal{I}_0^{\epsilon\,\mathcal{ABC}}{}_0\mathcal{D}^\epsilon\psi(\mathfrak{X}), \eta\Big)\\
&= \mathcal{N}\Big({}^{\mathcal{AB}}\mathcal{I}_0^\epsilon\Big(-\Delta_\kappa\Big({}^{\mathcal{AB}}\mathcal{I}_0^\epsilon\left[\phi_1^*(\mathfrak{X}, \psi(\mathfrak{X}))\right]\Big) - {}^{\mathcal{AB}}\mathcal{I}_0^{\epsilon\,\mathcal{ABC}}{}_0\mathcal{D}^\epsilon\psi(\mathfrak{X}), \eta\Big)\\
&= \mathcal{N}\Big({}^{\mathcal{AB}}\mathcal{I}_0^\epsilon\Big[\Big(-\Delta_\kappa\Big({}^{\mathcal{AB}}\mathcal{I}_0^\epsilon\left[\phi_1^*(\mathfrak{X}, \psi(\mathfrak{X}))\right]\Big) - {}^{\mathcal{ABC}}{}_0\mathcal{D}^\epsilon\psi(\mathfrak{X})\Big], \eta\Big)\\
&= \mathcal{N}\Big({}^{\mathcal{AB}}\mathcal{I}_0^\epsilon\Big[\Big(-\Delta_\kappa\Big[{}^{\mathcal{ABC}}{}_0\mathcal{D}^\epsilon\psi(\mathfrak{X})\Big] - {}^{\mathcal{AB}}\mathcal{I}_0^\epsilon\left[\phi_1^*(\mathfrak{X}, \psi(\mathfrak{X}))\right]\Big)\eta\Big)\Big]\\
&= \mathcal{N}\Big({}^{\mathcal{AB}}\mathcal{I}_0^\epsilon\Big[{}^{\mathcal{ABC}}{}_0\mathcal{D}^\epsilon\left[\Delta_\theta\left[{}^{\mathcal{ABC}}{}_0\mathcal{D}^{\varrho_0}\psi(\mathfrak{X})\right]\right] + \phi_1^*(\mathfrak{X}, \psi(\mathfrak{X}))\Big]\\
&\succeq \boldsymbol{OS_1}\Big(\boldsymbol{\Psi}(\boldsymbol{\mathfrak{X}}, \frac{\boldsymbol{\eta}}{\boldsymbol{\epsilon}})\Big),
\end{aligned}$$

for every $\eta \in (0, +\infty)$. Then we have $\delta(\mathcal{R}^*(\psi(\mathfrak{X})), \psi(\mathfrak{X})) < +\infty$. Therefore, all conditions of Theorem 1.2 are satisfied. Thus,

(1) $\{\mathcal{R}^{*\kappa}\psi(\mathfrak{X})\} \to \psi$ (a FP); and
(2) in $\mathcal{F}^* = \{\wp \in \mathcal{F} : d(\mathcal{R}^*\wp, \wp) < +\infty\}$ we get $\mathcal{R}^*\psi(\mathfrak{X}) = \psi(\mathfrak{X})$ or equivalently

$$\psi(\mathfrak{X}) = \int_0^1 \mathfrak{G}^{\varrho_0}(\mathfrak{X}, \mathrm{k})\Delta_\kappa\left({}^{\mathcal{AB}}\mathcal{I}_0^\epsilon\left[\phi_1^*(\mathfrak{X}, \psi(\mathfrak{X}))\right]\right)d\mathrm{k}. \qquad (3.2.15)$$

By using (3.2.15), we get

$${}^{\mathcal{ABC}}{}_0\mathcal{D}^\epsilon\left[\Delta_\theta\left[{}^{\mathcal{ABC}}{}_0\mathcal{D}^{\varrho_0}\psi(\mathfrak{X})\right]\right] = -\phi_1^*(\mathfrak{X}, \psi(\mathfrak{X})), \quad \mathfrak{X} \in \mathfrak{L} := [0, j]. \qquad (3.2.16)$$

(3) By using Inequality (3.2.14), we get

$$\delta(\psi, \psi) \le \frac{1}{1-\epsilon\gamma\varrho}\delta(\mathcal{R}^*\psi, \psi) \le \frac{\epsilon}{1-\epsilon\gamma\varrho}.$$

Hence, the Eq. (3.2.1) is multiple-HU-OS_1 stable. Assume that

$$\beth = \frac{\epsilon}{1-\epsilon\gamma\varrho},$$

and consider $\mathfrak{T}$ satisfying (3.2.16), then

$$^{\mathcal{ABC}}{}_0\mathcal{D}^{\epsilon}\left[\Delta_\theta\left[^{\mathcal{ABC}}{}_0\mathcal{D}^{\varrho_0}\mathfrak{T}(\mathfrak{X})\right]\right] = -\phi_1^*(\mathfrak{X}, \mathfrak{T}(\mathfrak{X})), \quad \mathfrak{X} \in \mathfrak{L} := [0, j]. \tag{3.2.17}$$

To show $\mathfrak{T}$ is a FP of $\mathcal{F}^*$ and $\mathfrak{T} \in \mathcal{F}^*$, we apply (3.2.17), and get $\mathcal{R}^*\mathfrak{T} = \mathfrak{T}$.

Now, we show that $d(\mathcal{R}^*\psi(\mathfrak{X}), \mathfrak{T}) < \infty$. Let $\psi(\mathfrak{X}) \in \mathcal{F}$, $\delta(\psi(\mathfrak{X}), \mathfrak{T}(\mathfrak{X})) < \beth$, i.e.,

$$\mathcal{N}\Big(\psi(\mathfrak{X}) - \mathfrak{T}(\mathfrak{X}), \eta\Big) \succeq \boldsymbol{OS_1}\Big(\boldsymbol{\Upsilon}(\boldsymbol{\mathfrak{X}}, \frac{\eta}{\beth})\Big), \tag{3.2.18}$$

From (3.2.5), (3.2.18) we have

$$\mathcal{N}\Big(\mathfrak{G}^{\varrho_0}(\mathfrak{X}, \mathrm{k})\Delta_\kappa\Big(\phi_1^*(\mathfrak{X}, \psi(\mathfrak{X})) - \phi_1^*(\mathfrak{X}, \mathfrak{T}(\mathfrak{X}))\Big), \eta\Big) \succeq \mathcal{N}\Big(\psi(\mathfrak{X}) - \mathfrak{T}(\mathfrak{X}), \frac{\eta}{\gamma}\Big) \succeq \boldsymbol{OS_1}\Big(\boldsymbol{\Upsilon}(\boldsymbol{\mathfrak{X}}, \frac{\eta}{\gamma\beth})\Big), \tag{3.2.19}$$

and using Eq. (3.2.17), we get

$$\begin{aligned}
&\mathcal{N}\Big(\mathcal{R}^*(\psi(\mathfrak{X})) - \mathfrak{T}(\mathfrak{X}), \eta\Big)\\
&= \mathcal{N}\Big(\int_0^1 \mathfrak{G}^{\varrho_0}(\mathfrak{X}, \mathrm{k})\Delta_\kappa\left(^{\mathcal{AB}}\mathcal{I}_0^{\epsilon}\left[\phi_1^*(\mathfrak{X}, \psi(\mathfrak{X}))\right]\right)d\mathrm{k} - \int_0^1 \mathfrak{G}^{\varrho_0}(\mathfrak{X}, \mathrm{k})\Delta_\kappa\left(^{\mathcal{AB}}\mathcal{I}_0^{\epsilon}\left[\phi_1^*(\mathfrak{X}, \mathfrak{T}(\mathfrak{X}))\right]\right)d\mathrm{k}, \eta\Big)\\
&= \mathcal{N}\Big(\int_0^1 \mathfrak{G}^{\varrho_0}(\mathfrak{X}, \mathrm{k})\Delta_\kappa\left(^{\mathcal{AB}}\mathcal{I}_0^{\epsilon}\left[\phi_1^*(\mathfrak{X}, \phi(\mathfrak{X}) - \phi_1^*(\mathfrak{X}, \mathfrak{T}(\mathfrak{X}))\right]\right)d\mathrm{k}, \eta\Big).
\end{aligned}$$

In the sequel, considering Eq. (3.2.18) and using step 1, we have

$$\begin{aligned}
&\mathcal{N}\Big(\int_0^1 \mathfrak{G}^{\varrho_0}(\mathfrak{X}, \mathrm{k})\Delta_\kappa\left(^{\mathcal{AB}}\mathcal{I}_0^{\epsilon}\left[\phi_1^*(\mathfrak{X}, \phi(\mathfrak{X}) - \phi_1^*(\mathfrak{X}, \mathfrak{T}(\mathfrak{X}))\right]\right)d\mathrm{k}, \eta\Big)\\
&\succeq \boldsymbol{OS_1}\Big(\boldsymbol{\Upsilon}(\boldsymbol{\mathfrak{X}}, \frac{\eta}{\epsilon\gamma\varrho}\Big)
\end{aligned} \tag{3.2.20}$$

$$\succeq \boldsymbol{OS}_1\left(\boldsymbol{\Upsilon}(\mathfrak{X}, \frac{\eta}{\beth\epsilon\gamma\varrho}\right),$$

and then

$$\delta(\mathcal{R}^*(\psi(\mathfrak{X})), \mathfrak{T}) \leq \epsilon\gamma\varrho\beth < +\infty.$$

3.2.3 Application

Example 3.3 For the NS-ABC-FDE

$$\begin{cases} \mathcal{D}^{\frac{3}{4}}\left[\Delta_5\left[\mathcal{D}^{\frac{1}{4}}\psi(\mathfrak{X})\right]\right] + \left[\sqrt{\psi} + \frac{1}{\psi^{\frac{3}{33}}}\right] = 0, \\ \left(\Delta_5\left[\mathcal{D}^{\frac{3}{4}}\psi(\mathfrak{X})\right)\right]\right) \mid_{\mathfrak{X}=0} = 0 = \left(\Delta_5\left(\mathcal{D}^{\frac{1}{4}}\psi(\mathfrak{X})\right)\right)'\Big|_{\mathfrak{X}=0}, \psi(1) = 0 = \psi'(0), \end{cases} \tag{3.2.21}$$

in which $\theta = 5, \kappa = \frac{5}{9}, j = 1, \epsilon = \frac{3}{4}, \gamma = \frac{1}{8}, \varrho = \frac{5}{8}, \varrho_0 = \frac{1}{4}, i = \frac{2}{100}$, and $\phi_1^*(\mathfrak{X}, \psi(\mathfrak{X})) = \sqrt{\psi} + \frac{1}{\psi^{\frac{3}{33}}}$ where $\phi_1^* \in C((0, 1) \times (0, +\infty))$, we have that

►

$$\mathcal{N}\left(\mathfrak{G}^{\frac{1}{4}}(\mathfrak{X}, \xi)\Delta_{\frac{5}{9}}\left(\sqrt{\psi} + \frac{1}{\psi^{\frac{3}{33}}} - \sqrt{\wp} + \frac{1}{\wp^{\frac{3}{33}}}, \eta\right) \succeq \mathcal{N}\left(\psi(\mathfrak{X}) - \wp(\mathfrak{X})\right), 8\eta\right). \tag{3.2.22}$$

► By considering the MVFF $\boldsymbol{OS}_1 : [0, 1] \longrightarrow [0, 1]$ as the control function, we have

$$\inf_{\xi\in E_1} \boldsymbol{OS}_1\left(\boldsymbol{\Upsilon}(\xi, \eta)\right) \succeq \boldsymbol{OS}_1\left(\boldsymbol{\Upsilon}(\mathfrak{X}, \frac{\eta}{\frac{5}{8}\boldsymbol{P}})\right). \tag{3.2.23}$$

If $\psi \in C(\mathfrak{L}, \mathbb{R}^n)$ satisfying

$$\mathcal{N}\left(\mathcal{D}^{\frac{3}{4}}\left[\Delta_5\left[\mathcal{D}^{\frac{1}{4}}\psi(\mathfrak{X})\right]\right] + \left[\sqrt{\psi} + \frac{1}{\psi^{\frac{3}{33}}}\right], \eta\right) \succeq \boldsymbol{OS}_1\left(\boldsymbol{\Upsilon}(\mathfrak{X}, \eta)\right), \tag{3.2.24}$$

then, there is $\wp \in C(\mathfrak{L}, \mathbb{R}^n)$ from (3.2.21), such that

$$\mathcal{R}^*(\psi(\mathfrak{X})) = \int_0^1 \mathfrak{G}^{\frac{1}{4}}(\mathfrak{X}, \mathrm{k})\Delta_{\frac{5}{9}}\left({}^{\mathcal{AB}}\mathcal{I}_0^{\frac{3}{4}}\left[\phi_1^*(\mathfrak{X}, \psi(\mathfrak{X}))\right]\right) d\mathrm{k}.$$

Therefore,

$$\delta(\psi, \wp) \leq F,$$

and

$$\mathcal{N}\Big(\psi(\mathfrak{X}) - \wp(\mathfrak{X}), \eta\Big) \succeq \boldsymbol{O}\,\boldsymbol{S}_1\Big(\boldsymbol{\Upsilon}(\mathfrak{X}, \frac{\eta}{F})\Big),$$

where in $F = \frac{\epsilon\gamma\varrho}{1-\epsilon\gamma\varrho} = 0.0622406639$.

3.3 Best Approximation for the Conformable Fractional Differential Equation

3.3.1 Introduction of the Conformable Fractional Differential Equations and Other Preliminaries

Our goal is to obtain an approximation for the conformable fractional differential equation using the alternative fixed point theorem in MVkFB-space. The fuzzy control functions presented have a dynamic situation and can model new events, such as the COVID-19 disease, explained in [55]. Using fuzzy controllers, the stability analysis of differential equations and integral equations can be studied.

We consider the following conformable FDE with constant coefficient:

$$\begin{cases} \mathfrak{D}^{\mathrm{a}}_{\omega}\mathfrak{f}(\varpi - \vartheta) = \varrho\mathfrak{f}(\varpi - \vartheta) + \mathrm{K}(\varpi - \vartheta, \mathfrak{f}(\varpi - \vartheta)), \ \ \varpi, \vartheta \in (\mathrm{a}, \mathrm{b}], \ \ 0 < \omega < 1, \\ \mathfrak{f}(\mathrm{a}) = \mathfrak{f}_{\mathrm{a}} \end{cases} \tag{3.3.1}$$

where $\mathfrak{D}^{\mathrm{a}}_{\omega}\mathfrak{f}$ is called the conformable fractional derivative (CFD) with lower index a of the function $\mathfrak{f}$ and $\mathcal{T} = [\mathrm{a}, \mathrm{b}]$, $\mathrm{K} \in C(\mathcal{T} \times \mathbb{R}, \mathbb{R})$ [45, 46].

Theorem 3.6 *Consider the Mittag-Leffler map*

$$E_\tau\left(\left(\frac{((\varpi - \vartheta) - \mathrm{a})^{\omega}}{\omega}\right)^{\varrho}\right) = \sum_{\varsigma=0}^{\infty} \frac{\left(\frac{((\varpi-\vartheta)-\mathrm{a})^{\omega}}{\omega}\right)^{\varsigma\varrho}}{\Gamma(1 + \tau\varsigma\varrho)}, \ \ \tau > 0.$$

Suppose that $\mathcal{Z}(\varpi - \vartheta) = E_\tau\left(\left(\frac{((\varpi-\vartheta)-\mathrm{a})^{\omega}}{\omega}\right)^{\varrho}\right)$, *then*

$$\mathfrak{D}^{\mathrm{a}}_{\omega}\mathcal{Z}(\varpi - \vartheta) = \varrho\mathcal{Z}(\varpi - \vartheta).$$

Proof Using Remark 2.1, we have

$$\begin{aligned}
\mathfrak{D}_{\omega}^{\mathrm{a}}\mathcal{Z}(\varpi-\vartheta) &= ((\varpi-\vartheta)-\mathrm{a})^{1-\omega}\frac{(\varpi-\vartheta)\mathcal{Z}(\varpi-\vartheta)}{\mathrm{b}(\varpi-\vartheta)} \\
&= ((\varpi-\vartheta)-\mathrm{a})^{1-\omega}\varrho((\varpi-\vartheta)-\mathrm{a})^{\omega-1}\sum_{\varsigma=0}^{\infty}\frac{(\frac{((\varpi-\vartheta)-\mathrm{a})^{\omega}}{\omega})^{\varsigma\varrho}}{\Gamma(1+\tau\varsigma\varrho)} \\
&= \lambda\sum_{\varsigma=0}^{\infty}\frac{(\frac{((\varpi-\vartheta)-\mathrm{a})^{\omega}}{\omega})^{\varsigma\varrho}}{\Gamma(1+\tau\varsigma\varrho)} \\
&= \varrho\mathcal{Z}(\varpi-\vartheta).
\end{aligned}$$

Next, we study the mapping $\mathfrak{f}$.

Theorem 3.7 *If for Eq. (3.3.1), the mapping $\mathfrak{f}\in C(\mathcal{T},\mathbb{R})$ is a solution, thus*

$$\begin{aligned}
\mathfrak{f}(\varpi-\vartheta) &= \mathfrak{f}_{\mathrm{a}}\sum_{\varsigma=0}^{\infty}\frac{(\frac{((\varpi-\vartheta)-\mathrm{a})^{\omega}}{\omega})^{\varsigma\varrho}}{\Gamma(1+\tau\varsigma\varrho)} \\
&\quad+\frac{1}{\Gamma(\omega)}\int_{\mathrm{a}}^{\varpi-\vartheta}\sum_{\varsigma=0}^{\infty}\frac{(\frac{((\varpi-\vartheta)-\mathrm{a})^{\omega}}{\omega})^{\varsigma\varrho}}{\Gamma(1+\tau\varsigma\varrho)}\sum_{\varsigma=0}^{\infty}\frac{(\frac{(\lambda-\mathrm{a})^{\omega}}{\omega})^{-\varsigma\varrho}}{\Gamma(1+\tau\varsigma\varrho)}\left(\frac{\lambda}{\mathrm{t}}\right)^{s}\left(\log\frac{\mathrm{t}}{\lambda}\right)^{\omega-1}\frac{\mathrm{K}(\lambda,\mathfrak{f}(\lambda))}{\lambda}\,\mathrm{d}\lambda.
\end{aligned}$$

Proof Any solution of (3.3.1) it should be as

$$\mathfrak{f}(\varpi-\vartheta)=\sum_{\varsigma=0}^{\infty}\frac{(\frac{((\varpi-\vartheta)-\mathrm{a})^{\omega}}{\omega})^{\varsigma\varrho}}{\Gamma(1+\tau\varsigma\varrho)}c(\varpi-\vartheta), \tag{3.3.2}$$

where $c(\varpi-\vartheta)$ is an unknown continuously differentiable function. From (3.3.2) and by Remark 2.1, we get

$$\begin{aligned}
\mathfrak{D}_{\omega}^{\mathrm{a}}\mathfrak{f}(\varpi-\vartheta) &= \mathfrak{D}_{\omega}^{\mathrm{a}}\left(\sum_{\varsigma=0}^{\infty}\frac{(\frac{((\varpi-\vartheta)-\mathrm{a})^{\omega}}{\omega})^{\varsigma\varrho}}{\Gamma(1+\tau\varsigma\varrho)}c(\varpi-\vartheta)\right) \\
&= ((\varpi-\vartheta)-\mathrm{a})^{1-\omega}\sum_{\varsigma=0}^{\infty}\frac{(\frac{((\varpi-\vartheta)-\mathrm{a})^{\omega}}{\omega})^{k\varrho}}{\Gamma(1+\tau\varsigma\varrho)}((\varpi-\vartheta)-\mathrm{a})^{\omega-1}c(\varpi-\vartheta) \\
&\quad+\sum_{\varsigma=0}^{\infty}\frac{(\frac{((\varpi-\vartheta)-\mathrm{a})^{\omega}}{\omega})^{\varsigma\varrho}}{\Gamma(1+\tau\varsigma\varrho)}((\varpi-\vartheta)-\mathrm{a})^{1-\omega}c'(\varpi-\vartheta) \\
&= \varrho c(\varpi-\vartheta)\sum_{\varsigma=0}^{\infty}\frac{(\frac{((\varpi-\vartheta)-\mathrm{a})^{\omega}}{\omega})^{\varsigma\varrho}}{\Gamma(1+\tau\varsigma\varrho)}+\sum_{\varsigma=0}^{\infty}\frac{(\frac{((\varpi-\vartheta)-\mathrm{a})^{\omega}}{\omega})^{\varsigma\varrho}}{\Gamma(1+\tau\varsigma\varrho)}\mathfrak{D}_{\omega}^{\mathrm{a}}c(\varpi-\vartheta) \\
&= \varrho\mathfrak{f}(\varpi-\vartheta)+\sum_{\varsigma=0}^{\infty}\frac{(\frac{((\varpi-\vartheta)-\mathrm{a})^{\omega}}{\omega})^{\varsigma\varrho}}{\Gamma(1+\tau\varsigma\varrho)}\mathfrak{D}_{\omega}^{\mathrm{a}}c(\varpi-\vartheta).
\end{aligned}$$

As a result

$$\mathfrak{D}^{\mathrm{a}}_{\omega} c(\varpi - \vartheta) = \sum_{\varsigma=0}^{\infty} \frac{((\frac{(\varpi-\vartheta)-\mathrm{a})^{\omega}}{\omega})^{-\varsigma\varrho}}{\Gamma(1+\tau\varsigma\varrho)} \mathrm{K}(\varpi - \vartheta, \mathfrak{f}(\varpi - \vartheta)). \tag{3.3.3}$$

From (3.3.3) and Lemma 2.2, we get

$$c(\varpi - \vartheta) = c(\mathrm{a}) + \frac{1}{\Gamma(\omega)} \int_{\mathrm{a}}^{\varpi-\vartheta} \sum_{\varsigma=0}^{\infty} \frac{(\frac{(\lambda-\mathfrak{t})^{\omega}}{\omega})^{-\varsigma\varrho}}{\Gamma(1+\tau\varsigma\varrho)} \left(\frac{\lambda}{\mathrm{t}}\right)^{s} \left(\log\frac{\mathrm{t}}{\lambda}\right)^{\omega-1} \frac{\mathrm{K}(\lambda, \mathfrak{f}(\lambda))}{\lambda} \, \mathrm{d}\lambda, \tag{3.3.4}$$

where $c(\mathrm{a}) = \mathfrak{f}(\mathrm{a}) = \mathfrak{f}_{\mathrm{a}}$. Use (3.3.4) and (3.3.2), the desired result is obtained.

Definition 3.2 A mapping $\mathfrak{f} \in C^{1}(\mathcal{T}, \mathbb{R})$ is said to be the solution of (3.3.1) if $\mathfrak{f}$ satisfies $\mathfrak{D}^{\mathrm{a}}_{\omega}\mathfrak{f}(\varpi - \vartheta) = \varrho\mathfrak{f}(\varpi - \vartheta) + \mathrm{K}(\varpi - \vartheta, \mathfrak{f}(\varpi - \vartheta))$, $\varpi - \vartheta \in (\mathrm{a}, \mathfrak{d}]$ and $\mathfrak{f}(\mathrm{a}) = \mathfrak{f}_{\mathrm{a}}$. Thus,

$$\begin{aligned}\mathfrak{f}(\varpi - \vartheta) &= \mathfrak{f}_{\mathrm{a}} \sum_{\varsigma=0}^{\infty} \frac{(\frac{((\varpi-\vartheta)-\mathrm{a})^{\omega}}{\omega})^{\varsigma\varrho}}{\Gamma(1+\tau\varsigma\varrho)} \\ &+ \frac{1}{\Gamma(\omega)} \int_{\mathrm{a}}^{\varpi-\vartheta} \sum_{\varsigma=0}^{\infty} \frac{(\frac{((\varpi-\vartheta)-\mathrm{a})^{\omega}}{\omega})^{\varsigma\varrho}}{\Gamma(1+\tau\varsigma\varrho)} \sum_{\varsigma=0}^{\infty} \frac{(\frac{(\lambda-\mathrm{a})^{\omega}}{\omega})^{-\varsigma\varrho}}{\Gamma(1+\tau\varsigma\varrho)} \left(\frac{\lambda}{\mathrm{t}}\right)^{s} \left(\log\frac{\mathrm{t}}{\lambda}\right)^{\omega-1} \frac{\mathrm{K}(\lambda, \mathfrak{f}(\lambda))}{\lambda} \, \mathrm{d}\lambda.\end{aligned}$$

Definition 3.3 Let function $\boldsymbol{H}^{m,n}_{\tau,\omega}$ be a MVF function. The Eq. (3.3.1) is said to be Hyers-Ulam-H-Fox stable, if $\mathfrak{h}(\varpi_1 - \vartheta_1), \ldots, \mathfrak{h}(\varpi_k - \vartheta_k)$ is a given differentiable function, satisfying

$$\begin{aligned}&\mathcal{N}\Big(\mathfrak{D}^{\mathrm{a}}_{\omega}\mathfrak{h}(\varpi_1 - \vartheta_1) - \varrho\mathfrak{h}(\varpi_1 - \vartheta_1) - \mathrm{K}(\varpi_1 - \vartheta_1, \mathfrak{h}(\varpi_1 - \vartheta_1)), \ldots, \\ &\mathfrak{D}^{\mathrm{a}}_{\omega}\mathfrak{h}(\varpi_k - \vartheta_k) - \varrho\mathfrak{h}(\varpi_k - \vartheta_k) - \mathrm{K}(\varpi_k - \vartheta_k, \mathfrak{h}(\varpi_k - \vartheta_k)), \aleph\Big) \\ &\succeq \boldsymbol{H}^{m,n}_{\tau,\omega}(-\frac{|\varpi_1 - \vartheta_1, \ldots, \varpi_k - \vartheta_k|}{\aleph}),\end{aligned} \tag{3.3.5}$$

for $(\varpi_1 - \vartheta_1, \ldots, \varpi_k - \vartheta_k) \in \mathfrak{F}_1$, and we can find a solution $\mathfrak{f}(\varpi_1 - \vartheta_1) \cdots, \mathfrak{f}(\varpi_k - \vartheta_k)$ of Eq. (3.3.1) such that for some $\eta > 0$,

$$\mathcal{N}\Big(\mathfrak{h}(\varpi_1 - \vartheta_1) - \mathfrak{f}(\varpi_1 - \vartheta_1), \ldots, \mathfrak{h}(\varpi_k - \vartheta_k) - \mathfrak{f}(\varpi_k - \vartheta_k), \aleph\Big) \succeq \boldsymbol{H}^{m,n}_{\tau,\omega}\left(-\frac{|\varpi_1 - \vartheta_1, \ldots, \varpi_k - \vartheta_k|}{\frac{\aleph}{\eta}}\right).$$

Remark 3.2 Let $\mathfrak{h}$ be a solution of the inequality (3.3.5). Then, $\mathfrak{h}$ is a solution of the following integral inequality:

$$\mathcal{N}\Big(\mathfrak{h}(\varpi_1 - \vartheta_1) - \mathfrak{h}_{\mathrm{a}} \sum_{\varsigma=0}^{\infty} \frac{(\frac{((\varpi_1-\vartheta_1)-\mathrm{a})^{\omega}}{\omega})^{\varsigma\varrho}}{\Gamma(1+\tau\varsigma\varrho)}$$

$$-\frac{1}{\Gamma(\omega)}\int_{a}^{\varpi-\vartheta}\sum_{\varsigma=0}^{\infty}\frac{(\frac{((\varpi_1-\vartheta_1)-a)^{\omega}}{\omega})^{\varsigma\varrho}}{\Gamma(1+\tau\varsigma\varrho)}\sum_{\varsigma=0}^{\infty}\frac{(\frac{((\lambda_1-\lambda'_1)-a)^{\omega}}{\omega})^{-\varsigma\varrho}}{\Gamma(1+\tau\varsigma\varrho)}\mathrm{K}((\lambda_1-\lambda'_1),\mathfrak{h}(\varpi_1-\vartheta_1))$$
$$\left(\frac{(\lambda_1-\lambda'_1)}{t}\right)^{s}\left(\log\frac{t}{(\lambda_1-\lambda'_1)}\right)^{\omega-1}\frac{d(\lambda_1-\lambda'_1)}{(\lambda_1-\lambda'_1)},\ldots,\mathfrak{h}(\varpi_k-\vartheta_k)-\mathfrak{h}_a\sum_{\varsigma=0}^{\infty}\frac{(\frac{((\varpi_k-\vartheta_k)-a)^{\omega}}{\omega})^{\varsigma\varrho}}{\Gamma(1+\tau\varsigma\varrho)}$$
$$-\frac{1}{\Gamma(\omega)}\int_{a}^{\varpi-\vartheta}\sum_{\varsigma=0}^{\infty}\frac{(\frac{((\varpi_k-\vartheta_k)-a)^{\omega}}{\omega})^{\varsigma\varrho}}{\Gamma(1+\tau\varsigma\varrho)}\sum_{\varsigma=0}^{\infty}\frac{(\frac{((\lambda_k-\lambda'_k)-a)^{\omega}}{\omega})^{-\varsigma\varrho}}{\Gamma(1+\tau\varsigma\varrho)}\mathrm{K}((\lambda_k-\lambda'_k),\mathfrak{h}(\varpi_k-\vartheta_k))$$
$$\left.\left(\frac{(\lambda_k-\lambda'_k)}{t}\right)^{s}\left(\log\frac{t}{(\lambda_k-\lambda'_k)}\right)^{\omega-1}\frac{d(\lambda_k-\lambda'_k)}{(\lambda_k-\lambda'_k)},\aleph\right)$$
$$\succeq \boldsymbol{H}^{m,n}_{\tau,\omega}\left(\left|\frac{1}{\Gamma(\omega)}\int_{a}^{\varpi-\vartheta}\sum_{\varsigma=0}^{\infty}\frac{(\frac{((\lambda_1-\lambda'_1)-a)^{\omega}}{\omega})^{-\varsigma\varrho}}{\Gamma(1+\tau\varsigma\varrho)}\right.\right.$$
$$\left(\frac{(\lambda_1-\lambda'_1)}{t}\right)^{s}\left(\log\frac{t}{(\lambda_1-\lambda'_1)}\right)^{\omega-1}\frac{d(\lambda_1-\lambda'_1)}{(\lambda_1-\lambda'_1)},\ldots,\frac{1}{\Gamma(\omega)}\int_{a}^{\varpi-\vartheta}\sum_{\varsigma=0}^{\infty}\frac{(\frac{((\lambda_k-\lambda'_k)-a)^{\omega}}{\omega})^{-\varsigma\varrho}}{\Gamma(1+\tau\varsigma\varrho)}$$
$$\left.\left(\frac{(\lambda_k-\lambda'_k)}{t}\right)^{s}\left(\log\frac{t}{(\lambda_k-\lambda'_k)}\right)^{\omega-1}\frac{d(\lambda_k-\lambda'_k)}{(\lambda_k-\lambda'_k)}\right|\frac{1}{\frac{\aleph}{F\sum_{\varsigma=0}^{\infty}\frac{(\frac{(b-a)^{\omega}}{\omega})^{\varsigma\varrho}}{\Gamma(1+\tau\varsigma\varrho)}}}\Bigg).$$

for every $\varpi_1-\vartheta_1,\ldots,\varpi_k-\vartheta_k\in\mathfrak{F}_1$ and $\aleph\in\mathfrak{F}_2$.

3.3.2 Hyers-Ulam-H-Fox Stability for Conformable Fractional Differential Equation

Now, we use the FPT based on the Theorem 1.2 to show (3.3.1) is Hyers-Ulam-H-Fox stable [61] in MVFkB-space $(\mathcal{J},\mathcal{N},\circledast)$ with MVFF $\boldsymbol{H}^{m,n}_{\tau,\omega}$ [63–65].

We set the set $\mathcal{H}$ as follows

$$\mathcal{H}=\{\mathfrak{f}:\mathfrak{F}_1\to\mathcal{J},\mathfrak{f}\text{ is differentiable}\},$$

and consider the mapping $d:\mathcal{H}\times\mathcal{H}\to\mathfrak{F}_4$ as

$$d(\mathfrak{f},\mathfrak{h})=\inf\Bigg\{\Re\in\mathfrak{F}_6:\mathcal{N}\Big(\mathfrak{h}(\varpi_1-\vartheta_1)-\mathfrak{f}(\varpi_1-\vartheta_1),\ldots,\mathfrak{h}(\varpi_k-\vartheta_k)-\mathfrak{f}(\varpi_k-\vartheta_k),\aleph\Big)$$
$$\succeq\boldsymbol{H}^{m,n}_{\tau,\omega}\left(-\frac{|\varpi_1-\vartheta_1,\ldots,\varpi_k-\vartheta_k|}{\frac{\aleph}{\Re}}\right),$$
$$\forall\ \mathfrak{h},\mathfrak{f}\in\mathcal{H},\quad(\varpi_1-\vartheta_1,\ldots,\varpi_k-\vartheta_k)\in\mathfrak{F}_1,\quad\aleph\in\mathfrak{F}_2\Bigg\}.$$

Theorem 3.8 *$(\mathcal{H},d)$ is a complete $\mathfrak{F}_4$-valued metric space.*

Proof We have $d(\mathfrak{h},\mathfrak{f})=0$ if and only if $\mathfrak{h}=\mathfrak{f}$. Assume that $d(\mathfrak{h},\mathfrak{f})=0$, then

$$\inf\Bigg\{\mathfrak{R}\in\mathfrak{F}_6:\mathcal{N}\Big(\mathfrak{h}(\varpi_1-\vartheta_1)-\mathfrak{f}(\varpi_1-\vartheta_1),\ldots,\mathfrak{h}(\varpi_k-\vartheta_k)-\mathfrak{f}(\varpi_k-\vartheta_k),\aleph\Big)$$
$$\succeq\boldsymbol{H}^{m,n}_{\tau,\omega}\left(-\frac{|\varpi_1-\vartheta_1,\ldots,\varpi_k-\vartheta_k|}{\frac{\aleph}{\mathfrak{R}}}\right),$$
$$\forall\ \mathfrak{h},\mathfrak{f}\in\mathcal{H},\ \ (\varpi_1-\vartheta_1,\ldots,\varpi_k-\vartheta_k)\in\mathfrak{F}_1,\ \ \aleph\in\mathfrak{F}_2\Bigg\}=0,$$

and, then

$$\mathcal{N}\Big(\mathfrak{h}(\varpi_1-\vartheta_1)-\mathfrak{f}(\varpi_1-\vartheta_1),\ldots,\mathfrak{h}(\varpi_k-\vartheta_k)-\mathfrak{f}(\varpi_k-\vartheta_k),\aleph\Big)$$
$$\succeq\boldsymbol{H}^{m,n}_{\tau,\omega}\left(-\frac{|\varpi_1-\vartheta_1,\ldots,\varpi_k-\vartheta_k|}{\frac{\aleph}{\mathfrak{R}}}\right),$$

for all $\mathfrak{R}\in\mathfrak{F}_6$. Tend $\mathfrak{R}$ to zero in the above inequality, we get

$$\mathcal{N}\Big(\mathfrak{h}(\varpi_1-\vartheta_1)-\mathfrak{f}(\varpi_1-\vartheta_1),\ldots,\mathfrak{h}(\varpi_k-\vartheta_k)-\mathfrak{f}(\varpi_k-\vartheta_k),\aleph\Big)=\mathbf{1}.$$

Thus $\mathfrak{h}(\varpi_1-\vartheta_1,\ldots,\varpi_k-\vartheta_k)=\mathfrak{f}(\varpi_1-\vartheta_1,\ldots,\varpi_k-\vartheta_k)$ for every $(\varpi_1-\vartheta_1,\ldots,\varpi_k-\vartheta_k)\in\mathfrak{F}_1$, and vice versa. Also we have $d(\mathfrak{h},\mathfrak{f})=d(\mathfrak{f},\mathfrak{h})$ for every $\mathfrak{h},\mathfrak{f}\in\mathcal{H}$. Now, let $d(\mathfrak{h},\mathfrak{f})=\alpha_1\in\mathfrak{F}_2$ and $d(\mathfrak{h},\mathfrak{f})=\alpha_2\in\mathfrak{F}_2$. Then, we have

$$\mathcal{N}\Big(\mathfrak{h}(\varpi_1-\vartheta_1)-\mathfrak{f}(\varpi_1-\vartheta_1),\ldots,\mathfrak{h}(\varpi_k-\vartheta_k)-\mathfrak{f}(\varpi_k-\vartheta_k),\aleph\Big)\succeq\boldsymbol{H}^{m,n}_{\tau,\omega}\left(-\frac{|\varpi_1-\vartheta_1,\ldots,\varpi_k-\vartheta_k|}{\frac{\aleph}{\alpha_1}}\right),$$

and

$$\mathcal{N}\Big(\mathfrak{h}(\varpi_1-\vartheta_1)-\mathfrak{f}(\varpi_1-\vartheta_1),\ldots,\mathfrak{h}(\varpi_k-\vartheta_k)-\mathfrak{f}(\varpi_k-\vartheta_k),\aleph\Big)\succeq\boldsymbol{H}^{m,n}_{\tau,\omega}\left(-\frac{|\varpi_1-\vartheta_1,\ldots,\varpi_k-\vartheta_k|}{\frac{\aleph}{\alpha_2}}\right),$$

for every $\aleph\in\mathfrak{F}_2$. Then we have

$$\mathcal{N}\Big(\mathfrak{h}(\varpi_1-\vartheta_1)-\mathfrak{p}(\varpi_1-\vartheta_1),\ldots,\mathfrak{h}(\varpi_k-\vartheta_k)-\mathfrak{p}(\varpi_k-\vartheta_k),(\alpha_1+\alpha_2)\aleph\Big)$$
$$\succeq\Big[\mathcal{N}\Big(\mathfrak{h}(\varpi_1-\vartheta_1)-\mathfrak{f}(\varpi_1-\vartheta_1),\ldots,\mathfrak{h}(\varpi_k-\vartheta_k)-\mathfrak{f}(\varpi_k-\vartheta_k),(\alpha_1)\aleph\Big)$$
$$\circledast\,\mathcal{N}\Big(\mathfrak{f}(\varpi_1-\vartheta_1)-\mathfrak{p}(\varpi_1-\vartheta_1),\ldots,\mathfrak{f}(\varpi_k-\vartheta_k)-\mathfrak{p}(\varpi_k-\vartheta_k),(\alpha_2)\aleph\Big)\Big]$$

$$\succeq \boldsymbol{H}^{m,n}_{\tau,\omega}\left(-\frac{|\varpi_1-\vartheta_1,\ldots,\varpi_k-\vartheta_k|}{\aleph}\right)\circledast \boldsymbol{H}^{m,n}_{\tau,\omega}\left(-\frac{|\varpi_1-\vartheta_1,\ldots,\varpi_k-\vartheta_k|}{\aleph}\right)$$
$$= \boldsymbol{H}^{m,n}_{\tau,\omega}\left(-\frac{|\varpi_1-\vartheta_1,\ldots,\varpi_k-\vartheta_k|}{\aleph}\right),$$

and then $d(\mathfrak{h},(\mathfrak{p}))\leq \alpha_1+\alpha_2$. Thus, $d(\mathfrak{h},(\mathfrak{p}))\leq d(\mathfrak{h},(\mathfrak{f}))+d(\mathfrak{f},(\mathfrak{p}))$. To show the completeness of $(\mathcal{H},d)$, we suppose that $\{\mathfrak{h}_k\}_k$ is a Cauchy sequence in $(\mathcal{H},d)$. Let $\varpi_1-\vartheta_1,\ldots,\varpi_k-\vartheta_k\in\mathfrak{F}_1$. Assume that $\gamma\in\mathfrak{F}_2$ and $\theta\in\mathfrak{F}_5{}^{\circ}$ are arbitrary and consider $\aleph\in\mathfrak{F}_2$ such that $\boldsymbol{H}^{m,n}_{\tau,\omega}\left(-\frac{|\varpi_1-\vartheta_1,\ldots,\varpi_k-\vartheta_k|}{\aleph}\right)\succ \mathbf{1}-\boldsymbol{\theta}$. For $\alpha\aleph<\gamma$ choose $k_0\in\mathbb{N}$ such that

$$d(\mathfrak{h}_k,\mathfrak{h}_\ell)<\alpha\quad \forall\, k,\ell\geq k_0.$$

Then

$$\mathcal{N}\bigg(\mathfrak{h}_k(\varpi_1-\vartheta_1)-\mathfrak{h}_\ell(\varpi_1-\vartheta_1),\ldots,\mathfrak{h}_k(\varpi_k-\vartheta_k)-\mathfrak{h}_\ell(\mathrm{x}_k-\mathrm{y}_k),\gamma\bigg)$$
$$\succeq\mathcal{N}\bigg(\mathfrak{h}_k(\varpi_1-\vartheta_1)-\mathfrak{h}_\ell(\varpi_1-\vartheta_1),\ldots,\mathfrak{h}_k(\varpi_k-\vartheta_k)-\mathfrak{h}_\ell(\varpi_k-\vartheta_k),\alpha\aleph\bigg)$$
$$\succeq \boldsymbol{H}^{m,n}_{\tau,\omega}\left(-\frac{|\varpi_1-\vartheta_1,\ldots,\varpi_k-\vartheta_k|}{\aleph}\right)$$
$$\succ \mathbf{1}-\boldsymbol{\theta}.$$

Therefore

$$\mathcal{N}\bigg(\mathfrak{h}(\varpi_1-\vartheta_1)-\mathfrak{h}(\varpi_1-\vartheta_1),\ldots,\mathfrak{h}(\varpi_k-\vartheta_k)-\mathfrak{h}(\varpi_k-\vartheta_k),\gamma\bigg)\succ \mathbf{1}-\boldsymbol{\theta},$$

which implies that, the sequence $\{\mathfrak{h}_k(\varpi_1-\vartheta_1),\mathfrak{h}_k(\varpi_2-\vartheta_2),\ldots,\mathfrak{h}_k(\varpi_k-\vartheta_k)\}_k$ is Cauchy in complete space $(\mathcal{J},\mathcal{N},\circledast)$ on compact set $\mathfrak{F}_1$. Then, it is uniformly convergent to the mapping $\mathfrak{h}:\mathfrak{F}_1\to\mathcal{J}$. By uniform convergence property we conclude that $\mathfrak{h}$ is differentiable, i.e., an element of $\mathcal{H}$ and then $(\mathcal{H},d)$ is complete.

Now, we can investigate Hyers–Ulam–H–Fox stability and get an approximation for the solution of conformable FDE (3.3.1). In [5, 17, 41, 66–88] there are new stability problem that ones can prove them by our method.

Theorem 3.9 *Let $(\mathcal{J},\mathcal{N},\circledast)$ be an MVFkB-space and consider the constant coefficient ϱ, E and $\wp$. Let $\beth=\wp\left(\sum_{\varsigma=0}^{\infty}\sum_{\varsigma_1\cdots\varsigma_k=\varsigma}^{\infty}\binom{\varsigma}{\varsigma_1,\ldots,\varsigma_k}\frac{(\frac{((\varpi_1-\vartheta_1)-\mathrm{a})^{\omega}}{\omega})^{\varsigma_1\varrho}\ldots(\frac{((\varpi_k-\vartheta_k)-\mathrm{a})^{\omega}}{\omega})^{\varsigma_k\varrho}}{\Gamma(\varsigma_1\cdots\varsigma_k+\tau\varsigma\varrho)}\right)$ $\mathrm{E}<1$.*

Soppose that the following conditions hold:

- *For continuous function $\mathrm{K}:\mathfrak{F}_1\times\mathcal{J}\to\mathcal{J}$,*

$$\begin{aligned} &\mathcal{N}\Big(\mathrm{K}(\varpi_1-\vartheta_1,\mathfrak{f}(\varpi_1-\vartheta_1))-\mathrm{K}(\varpi_1-\vartheta_1,\mathfrak{h}(\varpi_1-\vartheta_1)),\\ &\quad\cdots,\mathrm{K}(\varpi_k-\vartheta_k,\mathfrak{f}(\varpi_k-\vartheta_k))-\mathrm{K}(\varpi_k-\vartheta_k,\mathfrak{h}(\varpi_k-\vartheta_k)),\aleph\Big) \qquad (3.3.6)\\ &\succeq\mathcal{N}\Big(\mathfrak{h}(\varpi_1-\vartheta_1)-\mathfrak{f}(\varpi_1-\vartheta_1),\ldots,\mathfrak{h}(\varpi_k-\vartheta_k)-\mathfrak{f}(\varpi_1-\vartheta_1),\frac{\aleph}{\Re}\Big), \end{aligned}$$

► *MVFF* $\boldsymbol{H}^{m,n}_{\tau,\omega}$ *satisfying*

$$\begin{aligned} &\boldsymbol{H}^{m,n}_{\tau,\omega}\Bigg(\Bigg|\frac{1}{\Gamma(\omega)}\Bigg[\int_{\mathrm{a}}^{(\varpi_1-\vartheta_1)}\sum_{\varsigma=0}^{\infty}\frac{(\frac{((\lambdabar_1-\lambdabar'_1)-\mathrm{a})^{\omega}}{\omega})^{-\varsigma\varrho}}{\Gamma(1+\tau\varsigma\varrho)}\left(\frac{(\lambdabar_1-\lambdabar'_1)}{\mathrm{t}}\right)^{s}\left(\log\frac{\mathrm{t}}{(\lambdabar_1-\lambdabar'_1)}\right)^{\omega-1}\frac{\mathrm{d}(\lambdabar_1-\lambdabar'_1)}{(\lambdabar_1-\lambdabar'_1)}, \qquad (3.3.7)\\ &\cdots,\int_{\mathrm{a}}^{(\varpi_k-\vartheta_k)}\sum_{\varsigma=0}^{\infty}\frac{(\frac{((\lambdabar_k-\lambdabar'_k)-\mathrm{a})^{\omega}}{\omega})^{-\varsigma\varrho}}{\Gamma(1+\tau\varsigma\varrho)}\left(\frac{(\lambdabar_k-\lambdabar'_k)}{\mathrm{t}}\right)^{s}\left(\log\frac{\mathrm{t}}{(\lambdabar_k-\lambdabar'_k)}\right)^{\omega-1}\frac{\mathrm{d}(\lambdabar_k-\lambdabar'_k)}{(\lambdabar_k-\lambdabar'_k)}\Bigg]\Bigg|\frac{1}{\aleph}\Bigg)\\ &\succeq\boldsymbol{H}^{m,n}_{\tau,\omega}\left(-\frac{|\varpi_1-\vartheta_1,\ldots,\varpi_k-\vartheta_k|}{\frac{\aleph}{\mathrm{E}}}\right). \end{aligned}$$

Let $\mathfrak{h}:\mathfrak{F}_1\to\mathcal{J}$ *be a differentiable function satisfying*

$$\begin{aligned} &\mathcal{N}\Big(\mathfrak{D}^{\mathrm{a}}_{\omega}\mathfrak{h}(\varpi_1-\vartheta_1)-\varrho\mathfrak{h}(\varpi_1-\vartheta_1)-\mathfrak{k}(\varpi_1-\vartheta_1,\mathfrak{h}(\varpi_1-\vartheta_1)), \qquad (3.3.8)\\ &\quad\cdots,\mathfrak{D}^{\mathrm{a}}_{\omega}\mathfrak{h}(\varpi_k-\vartheta_k)-\lambda\mathfrak{h}(\varpi_k-\vartheta_k)-\mathrm{K}(\varpi_k-\vartheta_k,\mathfrak{h}(\varpi_k-\vartheta_k)),\aleph\Big)\\ &\succeq\boldsymbol{H}^{m,n}_{\tau,\omega}\left(-\frac{|\varpi_1-\vartheta_1,\ldots,\varpi_k-\vartheta_k|}{\frac{\aleph}{F}}\right), \end{aligned}$$

Then, there is a unique solution $\mathfrak{f}:\mathfrak{F}_1\to\mathcal{J}$ *for Eq. (3.3.1) such that*

$$\begin{aligned} &\mathfrak{Y}\Big(\mathfrak{h}(\varpi_1-\vartheta_1)-\mathfrak{f}(\varpi_1-\vartheta_1),\ldots,\mathfrak{h}(\varpi_k-\vartheta_k)-\mathfrak{f}(\varpi_k-\vartheta_k),\aleph\Big)\succeq\\ &\boldsymbol{H}^{m,n}_{\tau,\omega}\left(-\frac{|\varpi_1-\vartheta_1,\ldots,\varpi_k-\vartheta_k|}{\frac{\aleph(1-\beth)}{\mathrm{E}\sum_{\varsigma=0}^{\infty}\frac{(\frac{(\mathrm{b}-\mathrm{a})^{\omega}}{\omega})^{\varsigma\varrho}}{\Gamma(1+\tau\varsigma\varrho)}F}}\right), \end{aligned}$$

for every $\varpi_1-\vartheta_1,\ldots,\varpi_k-\vartheta_k\in\mathfrak{F}_1$ *and* $\aleph\in\mathfrak{F}_2$.

Proof We set

$$\mathcal{H}:=\{\mathfrak{f}:\mathfrak{F}_1\to\mathcal{J},\ \mathfrak{f}\text{ is differentiable}\},$$

and introduce the $\mathfrak{F}_4$-valued metric on $\mathcal{H}$ as,

$$\inf\Big\{\Re\in\mathfrak{F}_6:\mathcal{N}\Big(\mathfrak{h}(\varpi_1-\vartheta_1)-\mathfrak{f}(\varpi_1-\vartheta_1),\ldots,\mathfrak{h}(\varpi_k-\vartheta_k)-\mathfrak{f}(\varpi_k-\vartheta_k),\aleph\Big)$$

$$\succeq \boldsymbol{H}^{m,n}_{\tau,\omega}\left(-\frac{|\varpi_1-\vartheta_1,\ldots,\varpi_k-\vartheta_k|}{\frac{\aleph}{\Re}}\right),$$

$$\forall\ \mathfrak{f},\mathfrak{h}\in\mathcal{H},\quad \varpi_1-\vartheta_1,\ldots,\varpi_k-\vartheta_k\in\mathfrak{F}_1,\quad \aleph\in\mathfrak{F}_2\Big\}=0.$$

By Theorem 3.8, we have $(\mathcal{H}, d)$ is a complete $\mathfrak{F}_4$-valued metric space.

Step 1. We define Π from $\mathcal{H}$ to $\mathcal{H}$ by

$$\Pi(\mathfrak{f}(\varpi_i-\vartheta_i))=\mathfrak{f}_{\mathrm{a}}\sum_{\varsigma=0}^{\infty}\frac{(\frac{((\varpi_i-\vartheta_i)-\mathrm{a})^{\omega}}{\omega})^{\varsigma\varrho}}{\Gamma(1+\tau\varsigma\varrho)}+\frac{1}{\Gamma(\omega)}\int_{\mathrm{a}}^{(\varpi_i-\vartheta_i)}\sum_{\varsigma=0}^{\infty}\frac{(\frac{((\varpi_i-\vartheta_i)-\mathrm{a})^{\omega}}{\omega})^{\varsigma\varrho}}{\Gamma(1+\tau\varsigma\varrho)} \tag{3.3.9}$$
$$\sum_{\varsigma=0}^{\infty}\frac{(\frac{(\lambda_i-\lambda'_i)-\mathrm{a})^{\omega}}{\omega})^{-k\varrho}}{\Gamma(1+\tau\varsigma\varrho)}\left(\frac{(\lambda_i-\lambda'_i)}{\mathrm{t}}\right)^{s}\left(\log\frac{\mathrm{t}}{(\lambda_i-\lambda'_i)}\right)^{\omega-1}\frac{\mathrm{K}((\lambda_i-\lambda'_i),\mathfrak{f}((\lambda_i-\lambda'_i)))}{(\lambda_i-\lambda'_i)}\,\mathrm{d}(\lambda_i-\lambda'_i),$$

for $\varpi_i-\vartheta_i\in\mathfrak{F}_1 (i=1,2,\ldots,k)$ and we show Π is a strictly contractive mapping. Let $\mathfrak{f},\mathfrak{h}\in\mathcal{H}$ and consider the coefficient $\Re_{\mathfrak{f}\mathfrak{h}}\in\mathfrak{F}_4$ with $d(\mathfrak{f},\mathfrak{h})\le\Re_{\mathfrak{f}\mathfrak{h}}$, thus

$$\mathcal{N}\Big(\mathfrak{f}(\varpi_1-\vartheta_1)-\mathfrak{h}(\varpi_1-\vartheta_1),\ldots,\mathfrak{f}(\varpi_k-\vartheta_k)-\mathfrak{h}(\varpi_k-\vartheta_k),\Re_{\mathfrak{f}\mathfrak{h}}\aleph\Big)$$
$$\succeq \boldsymbol{H}^{m,n}_{\tau,\omega}\left(-\frac{|\varpi_1-\vartheta_1,\ldots,\varpi_k-\vartheta_k|}{\aleph}\right)$$

for all $\mathfrak{f},\mathfrak{h}\in\mathcal{H},\ \varpi_1-\vartheta_1,\ldots,\varpi_k-\vartheta_k\in\mathfrak{F}_1$ and $\aleph\in\mathfrak{F}_2$. Applying (MVFkN2) and (MVFkN3), imply that

$$\mathcal{N}\Big(\Pi\mathfrak{f}(\varpi_1-\vartheta_1)-\Pi\mathfrak{h}(\varpi_1-\vartheta_1),\ldots,\Pi\mathfrak{f}(\varpi_k-\vartheta_k)-\Pi\mathfrak{h}(\varpi_k-\vartheta_k),\Re_{\mathfrak{f}\mathfrak{h}}\aleph\Big) \tag{3.3.10}$$
$$=\mathcal{N}\Big(\mathfrak{f}_{\mathrm{a}}\sum_{\varsigma=0}^{\infty}\frac{(\frac{((\varpi_1-\vartheta_1)-\mathrm{a})^{\omega}}{\omega})^{\varsigma\varrho}}{\Gamma(1+\tau\varsigma\varrho)}+\frac{1}{\Gamma(\omega)}\int_{\mathrm{a}}^{(\varpi_1-\vartheta_1)}\sum_{\varsigma=0}^{\infty}\frac{(\frac{((\varpi_1-\vartheta_1)-\mathrm{a})^{\omega}}{\omega})^{\varsigma\varrho}}{\Gamma(1+\tau\varsigma\varrho)}\sum_{\varsigma=0}^{\infty}\frac{(\frac{(\lambda_1-\lambda'_1)-\mathrm{a})^{\omega}}{\omega})^{-\varsigma\varrho}}{\Gamma(1+\tau\varsigma\varrho)}$$
$$\left(\frac{(\lambda_1-\lambda'_1)}{\mathrm{t}}\right)^{s}\left(\log\frac{\mathrm{t}}{(\lambda_1-\lambda'_1)}\right)^{\omega-1}\frac{\mathfrak{k}((\lambda_1-\lambda'_1),\mathfrak{f}((\lambda_1-\lambda'_1)))}{(\lambda_1-\lambda'_1)}\,\mathrm{d}(\lambda_1-\lambda'_1)$$
$$-\ \mathfrak{h}_{\mathrm{a}}\sum_{\varsigma=0}^{\infty}\frac{(\frac{((\varpi_1-\vartheta_1)-\mathrm{a})^{\omega}}{\omega})^{\varsigma\varrho}}{\Gamma(1+\tau\varsigma\varrho)}+\frac{1}{\Gamma(\omega)}\int_{\mathrm{a}}^{(\varpi_1-\vartheta_1)}\sum_{\varsigma=0}^{\infty}\frac{(\frac{((\varpi_1-\vartheta_1)-\mathrm{a})^{\omega}}{\omega})^{\varsigma\varrho}}{\Gamma(1+\tau\varsigma\varrho)}\sum_{\varsigma=0}^{\infty}\frac{(\frac{(\lambda_1-\lambda'_1)-\mathrm{a})^{\omega}}{\omega})^{-\varsigma\varrho}}{\Gamma(1+\tau\varsigma\varrho)}$$
$$\left(\frac{(\lambda_1-\lambda'_1)}{\mathrm{t}}\right)^{s}\left(\log\frac{\mathrm{t}}{(\lambda_1-\lambda'_1)}\right)^{\omega-1}\frac{\mathrm{K}((\lambda_1-\lambda'_1),\mathfrak{h}((\lambda_1-\lambda'_1)))}{(\lambda_1-\lambda'_1)}\,\mathrm{d}(\lambda_1-\lambda'_1),$$
$$\ldots,\ \mathfrak{f}_{\mathrm{a}}\sum_{\varsigma=0}^{\infty}\frac{(\frac{((\varpi_k-\vartheta_k)-\mathrm{a})^{\omega}}{\omega})^{\varsigma\varrho}}{\Gamma(1+\tau\varsigma\varrho)}+\frac{1}{\Gamma(\omega)}\int_{\mathrm{a}}^{(\varpi_k-\vartheta_k)}\sum_{\varsigma=0}^{\infty}\frac{(\frac{((\varpi_k-\vartheta_k)-\mathrm{a})^{\omega}}{\omega})^{\varsigma\varrho}}{\Gamma(1+\tau\varsigma\varrho)}\sum_{\varsigma=0}^{\infty}\frac{(\frac{(\lambda_k-\lambda'_k)-\mathrm{a})^{\omega}}{\omega})^{-\varsigma\varrho}}{\Gamma(1+\tau\varsigma\varrho)}$$
$$\left(\frac{(\lambda_k-\lambda'_k)}{\mathrm{t}}\right)^{s}\left(\log\frac{\mathrm{t}}{(\lambda_k-\lambda'_k)}\right)^{\omega-1}\frac{\mathrm{K}((\lambda_k-\lambda'_k),\mathfrak{f}((\lambda_k-\lambda'_k)))}{(\lambda_k-\lambda'_k)}\,\mathrm{d}(\lambda_k-\lambda'_k)$$

$$- \quad \mathfrak{h}_{\mathrm{a}} \sum_{\varsigma=0}^{\infty} \frac{(\frac{((\varpi_k-\vartheta_k)-\mathrm{a})^{\omega}}{\omega})^{\varsigma\varrho}}{\Gamma(1+\tau\varsigma\varrho)} + \frac{1}{\Gamma(\omega)} \int_{\mathrm{a}}^{(\varpi_k-\vartheta_k)} \sum_{\varsigma=0}^{\infty} \frac{(\frac{((\varpi_k-\vartheta_k)-\mathrm{a})^{\omega}}{\omega})^{\varsigma\varrho}}{\Gamma(1+\tau\varsigma\varrho)} \sum_{\varsigma=0}^{\infty} \frac{(\frac{(\lambdabar_k-\lambdabar'_k)-\mathrm{a})^{\omega}}{\omega})^{-\varsigma\varrho}}{\Gamma(1+\tau\varsigma\varrho)}$$

$$\left(\frac{(\lambdabar_k-\lambdabar'_k)}{\mathrm{t}}\right)^{s} \left(\log \frac{\mathrm{t}}{(\lambdabar_k-\lambdabar'_k)}\right)^{\omega-1} \frac{\mathrm{K}((\lambdabar_k-\lambdabar'_k), \mathfrak{h}((\lambdabar_k-\lambdabar'_k)))}{(\lambdabar_k-\lambdabar'_k)} \,\mathrm{d}(\lambdabar_k-\lambdabar'_k), \Re_{\mathfrak{f}\mathfrak{h}}\aleph \Big)$$

$$\succeq \mathcal{N}\Big(\frac{1}{\Gamma(\omega)} \int_{\mathrm{a}}^{(\varpi_1-\vartheta_1)} \sum_{\varsigma=0}^{\infty} \frac{(\frac{((\varpi_1-\vartheta_1)-\mathrm{a})^{\omega}}{\omega})^{\varsigma\varrho}}{\Gamma(1+\tau\varsigma\varrho)} \sum_{\varsigma=0}^{\infty} \frac{(\frac{(\lambdabar_1-\lambdabar'_1)-\mathrm{c})^{\omega}}{\omega})^{-\varsigma\varrho}}{\Gamma(1+\tau\varsigma\varrho)}$$

$$\left(\frac{(\lambdabar_1-\lambdabar'_1)}{\mathrm{t}}\right)^{s} \left(\log \frac{\mathrm{t}}{(\lambdabar_1-\lambdabar'_1)}\right)^{\omega-1} \frac{\mathrm{K}((\lambdabar_1-\lambdabar'_1), \mathfrak{f}((\lambdabar_1-\lambdabar'_1))) - \mathrm{K}((\lambdabar_1-\lambdabar'_1), \mathfrak{h}((\lambdabar_1-\lambdabar'_1)))}{(\lambdabar_1-\lambdabar'_1)}$$

$$\mathrm{d}(\lambdabar_1-\lambdabar'_1),$$

$$\ldots, \frac{1}{\Gamma(\omega)} \int_{\mathrm{a}}^{(\varpi_k-\vartheta_k)} \sum_{\varsigma=0}^{\infty} \frac{(\frac{((\varpi_k-\vartheta_k)-\mathrm{a})^{\omega}}{\omega})^{\varsigma\varrho}}{\Gamma(1+\tau\varsigma\varrho)} \sum_{\varsigma=0}^{\infty} \frac{(\frac{(\lambdabar_k-\lambdabar'_k)-\mathrm{a})^{\omega}}{\omega})^{-\varsigma\varrho}}{\Gamma(1+\tau\varsigma\varrho)} \left(\frac{(\lambdabar_k-\lambdabar'_k)}{\mathrm{t}}\right)^{s} \left(\log \frac{\mathrm{t}}{(\lambdabar_k-\lambdabar'_k)}\right)^{\omega-1}$$

$$\frac{\mathrm{K}((\lambdabar_k-\lambdabar'_k), \mathfrak{f}((\lambdabar_k-\lambdabar'_k))) - \mathrm{K}((\lambdabar_k-\lambdabar'_k), \mathfrak{h}((\lambdabar_k-\lambdabar'_k)))}{(\lambdabar_k-\lambdabar'_k)} \,\mathrm{d}(\lambdabar_k-\lambdabar'_k), \Re_{\mathfrak{f}\mathfrak{h}}\aleph \Big)$$

$$\succeq \mathcal{N}\Big(\frac{1}{\Gamma(\omega)} \int_{\mathrm{a}}^{(\varpi_1-\vartheta_1)} \sum_{\varsigma=0}^{\infty} \frac{(\frac{((\varpi_1-\vartheta_1)-\mathrm{a})^{\omega}}{\omega})^{\varsigma\varrho}}{\Gamma(1+\tau\varsigma\varrho)} \sum_{\varsigma=0}^{\infty} \frac{(\frac{(\lambdabar_1-\lambdabar'_1)-\mathrm{a})^{\omega}}{\omega})^{-\varsigma\varrho}}{\Gamma(1+\tau\varsigma\varrho)}$$

$$\left(\frac{(\lambdabar_1-\lambdabar'_1)}{\mathrm{t}}\right)^{s} \left(\log \frac{\mathrm{t}}{(\lambdabar_1-\lambdabar'_1)}\right)^{\omega-1} \frac{\mathfrak{f}(\lambdabar_1-\lambdabar'_1) - \mathfrak{h}(\lambdabar_1-\lambdabar'_1)}{(\lambdabar_1-\lambdabar'_1)} \,\mathrm{d}(\lambdabar_1-\lambdabar'_1),$$

$$\ldots, \frac{1}{\Gamma(\zeta)} \int_{\mathrm{a}}^{(\varpi_k-\vartheta_k)} \sum_{\varsigma=0}^{\infty} \frac{(\frac{((\varpi_k-\vartheta_k)-\mathrm{a})^{\omega}}{\omega})^{\varsigma\varrho}}{\Gamma(1+\tau\varsigma\varrho)} \sum_{\varsigma=0}^{\infty} \frac{(\frac{(\lambdabar_k-\lambdabar'_k)-\mathrm{a})^{\omega}}{\omega})^{-\varsigma\varrho}}{\Gamma(1+\tau\varsigma\varrho)} \left(\frac{(\lambdabar_k-\lambdabar'_k)}{\mathrm{t}}\right)^{s} \left(\log \frac{\mathrm{t}}{(\lambdabar_k-\lambdabar'_k)}\right)^{\omega-1}$$

$$\frac{\mathfrak{f}(\lambdabar_k-\lambdabar'_k) - \mathfrak{h}(\lambdabar_k-\lambdabar'_k)}{(\lambdabar_k-\lambdabar'_k)} \,\mathrm{d}(\lambdabar_k-\lambdabar'_k), \frac{\Re_{\mathfrak{f}\mathfrak{h}}\mu}{\wp} \Big)$$

$$\succeq \boldsymbol{H}^{m,n}_{\tau,\omega}\Big(\Big| \sum_{\varsigma=0}^{\infty} \frac{(\frac{((\varpi_1-\vartheta_1)-\mathrm{a})^{\omega}}{\omega})^{\varsigma\lambda}}{\Gamma(1+\alpha\varsigma\varrho)} \frac{1}{\Gamma(\omega)} \int_{\mathrm{a}}^{(\varpi_1-\vartheta_1)} \sum_{\varsigma=0}^{\infty} \frac{(\frac{(\lambdabar_1-\lambdabar'_1)-\mathrm{a})^{\omega}}{\omega})^{-\varsigma\varrho}}{\Gamma(1+\tau\varsigma\varrho)}$$

$$\left(\frac{(\lambdabar_1-\lambdabar'_1)}{\mathrm{t}}\right)^{s} \left(\log \frac{\mathrm{t}}{(\lambdabar_1-\lambdabar'_1)}\right)^{\omega-1} \frac{\mathrm{d}(\lambdabar_1-\lambdabar'_1)}{\lambdabar_1-\lambdabar'_1},$$

$$\ldots, \sum_{\varsigma=0}^{\infty} \frac{(\frac{((\varpi_k-\vartheta_k)-\mathrm{a})^{\omega}}{\omega})^{\varsigma\varrho}}{\Gamma(1+\tau\varsigma\varrho)} \frac{1}{\Gamma(\omega)} \int_{\mathrm{a}}^{(\varpi_k-\vartheta_k)} \sum_{\varsigma=0}^{\infty} \frac{(\frac{(\lambdabar_k-\lambdabar'_k)-\mathrm{a})^{\omega}}{\omega})^{-\varsigma\varrho}}{\Gamma(1+\tau\varsigma\varrho)} \left(\frac{(\lambdabar_k-\lambdabar'_k)}{\mathrm{t}}\right)^{s} \left(\log \frac{\mathrm{t}}{(\lambdabar_k-\lambdabar'_k)}\right)^{\omega-1}$$

$$\frac{\mathrm{d}(\lambdabar_k-\lambdabar'_k)}{(\lambdabar_k-\lambdabar'_k)}\Big| \frac{1}{\frac{\aleph}{\wp}} \Big)$$

$$\succeq \boldsymbol{H}^{m,n}_{\tau,\omega}\Big(\Big| \sum_{\varsigma=0}^{\infty} \frac{(\frac{((\varpi_1-\vartheta_1)-\mathrm{a})^{\omega}}{\omega})^{\varsigma\varrho}}{\Gamma(1+\tau\varsigma\varrho)}, \ldots, \sum_{\varsigma=0}^{\infty} \frac{(\frac{((\varpi_k-\vartheta_k)-\mathrm{a})^{\omega}}{\omega})^{\varsigma\varrho}}{\Gamma(1+\tau\varsigma\varrho)} \frac{1}{\Gamma(\omega)} \Big[\int_{\mathrm{a}}^{(\varpi_1-\vartheta_1)} \sum_{\varsigma=0}^{\infty} \frac{(\frac{(\lambdabar_1-\lambdabar'_1)-\mathrm{a})^{\omega}}{\omega})^{-\varsigma\varrho}}{\Gamma(1+\tau\varsigma\varrho)}$$

$$, \ldots, \int_{\mathrm{a}}^{(\varpi_k-\vartheta_k)} \sum_{\varsigma=0}^{\infty} \frac{(\frac{(\lambdabar_k-\lambdabar'_k)-\mathrm{a})^{\omega}}{\omega})^{-\varsigma\varrho}}{\Gamma(1+\tau\varsigma\varrho)} \left(\frac{(\lambdabar_k-\lambdabar'_k)}{\mathrm{t}}\right)^{s} \left(\log \frac{\mathrm{t}}{(\lambdabar_k-\lambdabar'_k)}\right)^{\omega-1}$$

$$\frac{\mathrm{d}(\lambdabar_k-\lambdabar'_k)}{(\lambdabar_k-\lambdabar'_k)} \Big]\Big| \frac{1}{\frac{\aleph}{\wp}} \Big)$$

$$\succeq H^{m,n}_{\tau,\omega}\left(-\frac{|\varpi_1-\vartheta_1,\ldots,\varpi_k-\vartheta_k|}{\frac{\aleph}{\wp\left(\sum_{\varsigma=0}^{\infty}\sum_{\varsigma_1\cdots\varsigma_k=\varsigma}^{\infty}\binom{\varsigma}{\varsigma_1,\ldots,\varsigma_k}\frac{(\frac{((\varpi_1-\vartheta_1)-\mathrm{a})^{\omega}}{\omega})^{\varsigma_1\varrho}\ldots(\frac{((\varpi_k-\vartheta_k)-\mathrm{a})^{\omega}}{\omega})^{\varsigma_k\varrho}}{\Gamma(\varsigma_1\cdots\varsigma_k+\tau\varsigma\varrho)}\right)\mathrm{E}}}\right),$$

which implies that

$$d(\Pi(\mathrm{f}),\Pi(\mathfrak{h}))\leq\wp\left(\sum_{\varsigma=0}^{\infty}\sum_{\varsigma_1\cdots\varsigma_k=\varsigma}^{\infty}\binom{\varsigma}{\varsigma_1,\ldots,\varsigma_k}\frac{(\frac{((\varpi_1-\vartheta_1)-\mathrm{a})^{\omega}}{\omega})^{\varsigma_1\varrho}\ldots(\frac{((\varpi_k-\vartheta_k)-\mathrm{a})^{\omega}}{\omega})^{\varsigma_k\varrho}}{\Gamma(\varsigma_1\cdots\varsigma_k+\tau\varsigma\varrho)}\right)\mathrm{E}\Re_{\mathrm{f}\mathfrak{h}},$$

and then

$$d(\Pi(\mathrm{f}),\Pi(\mathfrak{h}))\leq\wp\left(\sum_{\varsigma=0}^{\infty}\sum_{\varsigma_1\cdots\varsigma_k=\varsigma}^{\infty}\binom{\varsigma}{\varsigma_1,\ldots,\varsigma_k}\frac{(\frac{((\varpi_1-\vartheta_1)-\mathrm{a})^{\omega}}{\omega})^{\varsigma_1\varrho}\ldots(\frac{((\varpi_k-\vartheta_k)-\mathrm{a})^{\omega}}{\omega})^{\varsigma_k\varrho}}{\Gamma(\varsigma_1\cdots\varsigma_k+\tau\varsigma\varrho)}\right)\mathrm{E}d(\mathrm{f},\mathfrak{h}),$$

where $\quad 0<\wp\left(\sum_{\varsigma=0}^{\infty}\sum_{\varsigma_1\cdots\varsigma_k=\varsigma}^{\infty}\binom{\varsigma}{\varsigma_1,\ldots,\varsigma_k}\frac{(\frac{((\varpi_1-\vartheta_1)-\mathfrak{c})^{\omega}}{\omega})^{\varsigma_1\varrho}\ldots(\frac{((\varpi_k-\vartheta_k)-\mathfrak{c})^{\omega}}{\omega})^{\varsigma_k\varrho}}{\Gamma(\varsigma_1\cdots\varsigma_k+\tau\varsigma\varrho)}\right)\mathrm{E}<1,$

therefore Π is a contractional mapping.

Step 2. We will show that $d(\Pi(\mathfrak{h}),\mathfrak{h})<\infty$. Let $\mathfrak{h}\in\mathcal{H}$, we have

$$\mathcal{N}\Big(\Pi(\mathfrak{h}(\varpi_1-\vartheta_1))-\mathfrak{h}(\varpi_1-\vartheta_1),\ldots,\Pi(\varpi_k-\vartheta_k))-\mathfrak{h}(\varpi_k-\vartheta_k),\aleph\Big) \tag{3.3.11}$$

$$=\mathcal{N}\Bigg(\mathfrak{h}_{\mathrm{a}}\sum_{\varsigma=0}^{\infty}\frac{(\frac{((\varpi_1-\vartheta_1)-\mathrm{a})^{\omega}}{\omega})^{\varsigma\varrho}}{\Gamma(1+\tau\varsigma\varrho)}+\frac{1}{\Gamma(\omega)}\int_{\mathrm{a}}^{(\varpi_1-\vartheta_1)}\sum_{\varsigma=0}^{\infty}\frac{(\frac{((\varpi_1-\vartheta_1)-\mathrm{a})^{\omega}}{\omega})^{\varsigma\varrho}}{\Gamma(1+\tau\varsigma\varrho)}\sum_{\varsigma=0}^{\infty}\frac{(\frac{(\lambda_1-\lambda'_1)-\mathrm{a})^{\omega}}{\omega})^{-\varsigma\varrho}}{\Gamma(1+\tau\varsigma\varrho)}$$

$$\left(\frac{(\lambda_1-\lambda'_1)}{\mathrm{t}}\right)^{s}\left(\log\frac{\mathrm{t}}{(\lambda_1-\lambda'_1)}\right)^{\omega-1}\frac{\mathrm{K}((\lambda_1-\lambda'_1),\mathfrak{h}((\lambda_1-\lambda'_1)))}{(\lambda_1-\lambda'_1)}\,d(\lambda_1-\lambda'_1)-\mathfrak{h}(\varpi_1-\vartheta_1),\ldots,$$

$$\mathfrak{h}_{\mathrm{a}}\sum_{\varsigma=0}^{\infty}\frac{(\frac{((\varpi_k-\vartheta_k)-\mathrm{a})^{\omega}}{\omega})^{\varsigma\varrho}}{\Gamma(1+\tau\varsigma\varrho)}+\frac{1}{\Gamma(\omega)}\int_{\mathrm{a}}^{(\varpi_k-\vartheta_k)}\sum_{\varsigma=0}^{\infty}\frac{(\frac{((\varpi_k-\vartheta_k)-\mathrm{a})^{\omega}}{\omega})^{\varsigma\varrho}}{\Gamma(1+\tau\varsigma\varrho)}\sum_{\varsigma=0}^{\infty}\frac{(\frac{(\lambda_k-\lambda'_k)-\mathrm{a})^{\omega}}{\omega})^{-\varsigma\varrho}}{\Gamma(1+\tau\varsigma\varrho)}$$

$$\left(\frac{(\lambda_k-\lambda'_k)}{\mathrm{t}}\right)^{s}\left(\log\frac{\mathrm{t}}{(\lambda_k-\lambda'_k)}\right)^{\omega-1}\frac{\mathrm{K}((\lambda_k-\lambda'_k),\mathfrak{h}((\lambda_k-\lambda'_k)))}{(\lambda_k-\lambda'_k)}\,d(\lambda_k-\lambda'_k)-\mathfrak{h}(\varpi_k-\vartheta_k),\aleph\Bigg)$$

$$\succeq H^{m,n}_{\tau,\omega}\Bigg(\Bigg|\sum_{\varsigma=0}^{\infty}\frac{(\frac{(\mathrm{b}-\mathrm{a})^{\omega}}{\omega})^{\varsigma\varrho}}{\Gamma(1+\tau\varsigma\varrho)}\Bigg[\frac{1}{\Gamma(\beta)}\int_{\mathrm{a}}^{(\varpi_1-\vartheta_1)}\sum_{\varsigma=0}^{\infty}\frac{(\frac{((\lambda_1-\lambda'_1)-\mathrm{a})^{\omega}}{\omega})^{-\varsigma\varrho}}{\Gamma(1+\tau\varsigma\varrho)}\left(\frac{(\lambda_1-\lambda'_1)}{\mathrm{t}}\right)^{s}\left(\log\frac{\mathrm{t}}{(\lambda_1-\lambda'_1)}\right)^{\omega-1}$$

$$\frac{d(\lambda_1-\lambda'_1)}{(\lambda_1-\lambda'_1)},\ldots,\frac{1}{\Gamma(\omega)}\int_{\mathrm{a}}^{(\varpi_k-\vartheta_k)}\sum_{\varsigma=0}^{\infty}\frac{(\frac{((\lambda_k-\lambda'_k)-\mathrm{a})^{\omega}}{\omega})^{-\varsigma\varrho}}{\Gamma(1+\tau\varsigma\varrho)}\left(\frac{(\lambda_k-\lambda'_k)}{\mathrm{t}}\right)^{s}\left(\log\frac{\mathrm{t}}{(\lambda_k-\lambda'_k)}\right)^{\omega-1}$$

$$\frac{d(\lambda_k-\lambda'_k)}{(\lambda_k-\lambda'_k)}\Bigg]\Bigg|\frac{1}{\aleph}\Bigg)$$

$$\succeq H^{m,n}_{\tau,\omega}\left(-\frac{|\varpi_1-\vartheta_1,\ldots,\varpi_k-\vartheta_k|}{\frac{\aleph}{F\sum_{\varsigma=0}^{\infty}\frac{(\frac{(\mathrm{b}-\mathrm{a})^{\omega}}{\omega})^{\varsigma\varrho}}{\Gamma(1+\tau\varsigma\varrho)}\mathrm{E}}}\right).$$

Consequently,

$$d(\Pi\mathfrak{h}, \mathfrak{h}) \leq F \sum_{\varsigma=0}^{\infty} \frac{(\frac{(\mathrm{b}-\mathrm{a})^{\omega}}{\omega})^{\varsigma\varrho}}{\Gamma(1+\tau\varsigma\varrho)} \mathrm{E} < \infty, \tag{3.3.12}$$

for every $\aleph \in \mathfrak{F}_2$ and $\mathrm{E} < 1$. Then, we have $d(\Pi(\mathfrak{h}), \mathfrak{h}) < \infty$. Therefore, all the conditions of Theorem 1.2 hold. Then, we have

(1) The sequence $\{\Pi^{\mathrm{e}}\mathfrak{f}\}$ converges to a fixed point such as $\mathfrak{f}$.
(2) The unique element $\mathfrak{f}$ is in the set $\mathcal{H}^* = \{\mathfrak{h} \in \mathcal{H} : d(\Pi\mathfrak{h}, \mathfrak{f}) < \infty\}$ and is the unique fixed point of Π, it means $\Pi\mathfrak{f} = \mathfrak{f}$ or equivalently

$$\mathfrak{f}(\varpi_i - \vartheta_i) = \mathfrak{f}_{\mathrm{a}} \sum_{\varsigma=0}^{\infty} \frac{(\frac{((\varpi_i-\vartheta_i)-\mathrm{a})^{\omega}}{\omega})^{\varsigma\varrho}}{\Gamma(1+\tau\varsigma\varrho)} + \frac{1}{\Gamma(\omega)} \int_{\mathrm{a}}^{(\varpi_i-\vartheta_i)} \sum_{\varsigma=0}^{\infty} \frac{(\frac{((\varpi_i-\vartheta_i)-\mathrm{a})^{\omega}}{\omega})^{\varsigma\varrho}}{\Gamma(1+\tau\varsigma\lambda)} \sum_{\varsigma=0}^{\infty} \frac{(\frac{(\varpi_i-\vartheta_i)-\mathrm{a})^{\omega}}{\omega})^{-\varsigma\varrho}}{\Gamma(1+\tau\varsigma\varrho)} \tag{3.3.13}$$

$$\left(\frac{(\lambda_i - \lambda'_i)}{\mathrm{t}}\right)^{s} \left(\log \frac{\mathrm{t}}{(\lambda_i - \lambda'_i)}\right)^{\omega-1} \frac{\mathrm{K}((\lambda_i - \lambda'_i), \mathfrak{f}((\lambda_i - \lambda'_i)))}{(\lambda_i - \lambda'_i)} \mathrm{d}(\lambda_i - \lambda'_i),$$

where $i = 1, \ldots, k$.

Since $\mathfrak{f}$ is a differentiable function, by the CFD and according to the Eq. (3.3.13) and Lemma 2.2, we have

$$\mathfrak{D}_{\omega}^{\mathrm{a}}\mathfrak{f}(\varpi - \vartheta) = \varrho\mathfrak{f}(\varpi - \vartheta) + \mathrm{K}(\varpi - \vartheta, \mathfrak{f}(\varpi - \vartheta)). \tag{3.3.14}$$

(3) Using Inequality (3.3.12), we get

$$d(\mathfrak{w}, \mathfrak{f}) \leq \frac{1}{1-\beth} d(\Pi\mathfrak{h}, \mathfrak{h}) \leq \frac{F \sum_{\varsigma=0}^{\infty} \frac{(\frac{(\mathrm{b}-\mathrm{a})^{\omega}}{\omega})^{\varsigma\varrho}}{\Gamma(1+\tau\varsigma\varrho)} \mathrm{E}}{1-\beth}$$

thus, the Eq. (3.3.1) has Hyers–Ulam–H–Fox stability property.

Now, we show the uniqueness of the obtained point. For convenience, we consider

$$\epsilon = \frac{F \sum_{\varsigma=0}^{\infty} \frac{(\frac{(\mathrm{b}-\mathrm{a})^{\omega}}{\omega})^{\varsigma\varrho}}{\Gamma(1+\tau\varsigma\varrho)} \mathrm{E}}{1-\beth},$$

and let $\mathfrak{g}$ be another differentiable function satisfying the Eq. (3.3.14) and this means that the following equation holds

$$\mathfrak{D}_{\omega}^{\mathrm{a}}\mathfrak{g}(\varpi - \vartheta) = \varrho\mathfrak{g}(\varpi - \vartheta) + \mathrm{K}(\varpi - \vartheta, \mathfrak{g}(\varpi - \vartheta)). \tag{3.3.15}$$

We are ready to prove that $\mathfrak{g}$ is a fixed point of Π and $\mathfrak{g} \in \mathcal{H}^*$. Using Eq. (3.3.15), we get $\Pi\mathfrak{g} = \mathfrak{g}$. Now, we show that $d(\Pi\mathfrak{h}, \mathfrak{g}) < \infty$. Let $\mathfrak{h} \in \mathcal{H}$, $d(\mathfrak{h}, \mathfrak{g}) < \epsilon$ and use Eq. (3.3.15), we get

$$\mathcal{N}\Big(\Pi(\mathfrak{h}(\varpi_1 - \vartheta_1)) - \mathfrak{g}(\varpi_1 - \vartheta_1), \ldots, \Pi(\mathfrak{h}(\varpi_k - \vartheta_k)) - \mathfrak{g}(\varpi_k - \vartheta_k), \aleph\Big)$$

$$= \mathcal{N}\Big(\mathfrak{h}_a \sum_{\varsigma=0}^{\infty} \frac{(\frac{((\varpi_1-\vartheta_1)-a)^{\omega}}{\omega})^{\varsigma\varrho}}{\Gamma(1+\tau\varsigma\varrho)} + \frac{1}{\Gamma(\omega)} \int_a^{(\varpi_1-\vartheta_1)} \sum_{\varsigma=0}^{\infty} \frac{(\frac{((\varpi_1-\vartheta_1)-a)^{\omega}}{\omega})^{\varsigma\varrho}}{\Gamma(1+\tau\varsigma\varrho)} \sum_{\varsigma=0}^{\infty} \frac{(\frac{((\lambda_1-\lambda'_1)-a)^{\omega}}{\omega})^{-\varsigma\varrho}}{\Gamma(1+\tau\varsigma\varrho)}$$

$$\left(\frac{(\lambda_1-\lambda'_1)}{t}\right)^s \left(\log\frac{t}{(\lambda_1-\lambda'_1)}\right)^{\omega-1} \frac{K((\lambda_1-\lambda'_1), \mathfrak{h}((\lambda_1-\lambda'_1)))}{(\lambda_1-\lambda'_1)} d(\lambda_1-\lambda'_1) - \mathfrak{g}_a \sum_{\varsigma=0}^{\infty} \frac{(\frac{((\varpi_1-\vartheta_1)-a)^{\omega}}{\omega})^{\varsigma\varrho}}{\Gamma(1+\tau\varsigma\varrho)}$$

$$- \frac{1}{\Gamma(\omega)} \int_a^{(\varpi_1-\vartheta_1)} \sum_{\varsigma=0}^{\infty} \frac{(\frac{((\varpi_1-\vartheta_1)-a)^{\omega}}{\omega})^{\varsigma\varrho}}{\Gamma(1+\tau\varsigma\varrho)} \sum_{\varsigma=0}^{\infty} \frac{(\frac{((\lambda_1-\lambda'_1)-a)^{\omega}}{\omega})^{-\varsigma\varrho}}{\Gamma(1+\tau\varsigma\varrho)}$$

$$\left(\frac{(\lambda_1-\lambda'_1)}{t}\right)^s \left(\log\frac{t}{(\lambda_1-\lambda'_1)}\right)^{\omega-1} \frac{K((\lambda_1-\lambda'_1), \mathfrak{g}((\lambda_1-\lambda'_1)))}{(\lambda_1-\lambda'_1)} d(\lambda_1-\lambda'_1), \ldots,$$

$$\mathfrak{h}_a \sum_{\varsigma=0}^{\infty} \frac{(\frac{((\varpi_k-\vartheta_k)-a)^{\omega}}{\omega})^{\varsigma\varrho}}{\Gamma(1+\tau\varsigma\varrho)} + \frac{1}{\Gamma(\omega)} \int_a^{(\varpi_k-\vartheta_k)} \sum_{\varsigma=0}^{\infty} \frac{(\frac{((\varpi_k-\vartheta_k)-a)^{\omega}}{\omega})^{\varsigma\varrho}}{\Gamma(1+\alpha\varsigma\varrho)} \sum_{\varsigma=0}^{\infty} \frac{(\frac{((\lambda_k-\lambda'_k)-a)^{\omega}}{\omega})^{-\varsigma\varrho}}{\Gamma(1+\tau\varsigma\varrho)}$$

$$\left(\frac{(\lambda_k-\lambda'_k)}{t}\right)^s \left(\log\frac{t}{(\lambda_k-\lambda'_k)}\right)^{\omega-1} \frac{K((\lambda_k-\lambda'_k), \mathfrak{h}((\lambda_k-\lambda'_k)))}{(\lambda_k-\lambda'_k)} d(\lambda_k-\lambda'_k) -$$

$$\mathfrak{g}_a \sum_{\varsigma=0}^{\infty} \frac{(\frac{((\varpi_k-\vartheta_k)-a)^{\omega}}{\omega})^{\varsigma\varrho}}{\Gamma(1+\tau\varsigma\varrho)} - \frac{1}{\Gamma(\omega)} \int_a^{(\varpi_k-\vartheta_k)} \sum_{\varsigma=0}^{\infty} \frac{(\frac{((\varpi_k-\vartheta_k)-a)^{\omega}}{\omega})^{\varsigma\varrho}}{\Gamma(1+\alpha\varsigma\varrho)} \sum_{\varsigma=0}^{\infty} \frac{(\frac{((\lambda_k-\lambda'_k)-a)^{\omega}}{\omega})^{-\varsigma\varrho}}{\Gamma(1+\alpha\varsigma\varrho)}$$

$$\left(\frac{(\lambda_k-\lambda'_k)}{t}\right)^s \left(\log\frac{t}{(\lambda_k-\lambda'_k)}\right)^{\omega-1} \frac{K((\lambda_k-\lambda'_k), \mathfrak{g}((\lambda_k-\lambda'_k)))}{(\lambda_k-\lambda'_k)} d(\lambda_k-\lambda'_k), \aleph\Big)$$

$$\succeq \mathcal{N}\Big(\frac{1}{\Gamma(\omega)} \int_a^{(\varpi_1-\vartheta_1)} \sum_{\varsigma=0}^{\infty} \frac{(\frac{((\varpi_1-\vartheta_1)-a)^{\omega}}{\omega})^{\varsigma\varrho}}{\Gamma(1+\tau\varsigma\varrho)} \sum_{\varsigma=0}^{\infty} \frac{(\frac{((\lambda_1-\lambda'_1)-a)^{\zeta}}{\zeta})^{-\varsigma\varrho}}{\Gamma(1+\tau\varsigma\varrho)}$$

$$\left(\frac{(\lambda_1-\lambda'_1)}{t}\right)^s \left(\log\frac{t}{(\lambda_1-\lambda'_1)}\right)^{\omega-1} \frac{K((\lambda_1-\lambda'_1), \mathfrak{h}((\lambda_1-\lambda'_1))) - K((\lambda_1-\lambda'_1), \mathfrak{g}((\lambda_1-\lambda'_1)))}{(\lambda_1-\lambda'_1)} d(\lambda_1-\lambda'_1),$$

$$\ldots, \frac{1}{\Gamma(\omega)} \int_a^{(\varpi_k-\vartheta_k)} \sum_{\varsigma=0}^{\infty} \frac{(\frac{((\varpi_k-\vartheta_k)-a)^{\omega}}{\omega})^{\varsigma\varrho}}{\Gamma(1+\tau\varsigma\varrho)} \sum_{\varsigma=0}^{\infty} \frac{(\frac{((\lambda_k-\lambda'_k)-a)^{\omega}}{\omega})^{-\varsigma\varrho}}{\Gamma(1+\tau\varsigma\varrho)}$$

$$\left(\frac{(\lambda_k-\lambda'_k)}{t}\right)^s \left(\log\frac{t}{(\lambda_k-\lambda'_k)}\right)^{\omega-1} \frac{k((\lambda_k-\lambda'_k), \mathfrak{h}((\lambda_k-\lambda'_k))) - k((\lambda_k-\lambda'_k), \mathfrak{g}((\lambda_k-\lambda'_k)))}{(\lambda_k-\lambda'_k)} d(\lambda_k-\lambda'_k), \aleph\Big)$$

$$\succeq \mathfrak{Y}\Big(\frac{1}{\Gamma(\omega)} \int_a^{(\varpi_1-\vartheta_1)} \sum_{\varsigma=0}^{\infty} \frac{(\frac{((\varpi_1-\vartheta_1)-a)^{\omega}}{\omega})^{\varsigma\varrho}}{\Gamma(1+\tau\varsigma\varrho)} \sum_{\varsigma=0}^{\infty} \frac{(\frac{((\lambda_1-\lambda'_1)-a)^{\zeta}}{\zeta})^{-\varsigma\varrho}}{\Gamma(1+\tau\varsigma\varrho)}$$

$$\left(\frac{(\lambda_1-\lambda'_1)}{t}\right)^s \left(\log\frac{t}{(\lambda_1-\lambda'_1)}\right)^{\zeta-1} \frac{\mathfrak{h}((\lambda_1-\lambda'_1))) - \mathfrak{g}((\lambda_1-\lambda'_1)))}{(\lambda_1-\lambda'_1)} d(\lambda_1-\lambda'_1), \ldots,$$

$$\frac{1}{\Gamma(\omega)} \int_a^{(\varpi_k-\vartheta_k)} \sum_{\varsigma=0}^{\infty} \frac{(\frac{((\varpi_k-\vartheta_k)-a)^{\omega}}{\omega})^{\varsigma\varrho}}{\Gamma(1+\tau\varsigma\varrho)} \sum_{\varsigma=0}^{\infty} \frac{(\frac{((\lambda_k-\lambda'_k)-a)^{\omega}}{\omega})^{-\varsigma\varepsilon}}{\Gamma(1+\alpha\varsigma\varrho)}$$

$$\left(\frac{(\lambda_k-\lambda'_k)}{t}\right)^s \left(\log\frac{t}{(\lambda_k-\lambda'_k)}\right)^{\omega-1} \frac{\mathfrak{h}((\lambda_k-\lambda'_k))) - \mathfrak{g}((\lambda_k-\lambda'_k)))}{(\lambda_k-\lambda'_k)} d(\lambda_k-\lambda'_k), \frac{\aleph}{\wp}\Big)$$

$$\succeq \boldsymbol{H}^{m,n}_{\tau,\omega}\Big(\Big\|\Big[\sum_{\varsigma=0}^{\infty} \frac{(\frac{((\varpi_1-\vartheta_1)-a)^{\omega}}{\omega})^{\varsigma\varrho}}{\Gamma(1+\alpha\varsigma\varrho)}, \ldots, \sum_{\varsigma=0}^{\infty} \frac{(\frac{((\varpi_k-\vartheta_k)-a)^{\omega}}{\omega})^{\varsigma\varrho}}{\Gamma(1+\alpha\varsigma\varrho)}\Big]\epsilon\Big(\frac{1}{\Gamma(\omega)} \int_a^{(\varpi_1-\vartheta_1)} \sum_{\varsigma=0}^{\infty} \frac{(\frac{((\lambda_1-\lambda'_1)-a)^{\omega}}{\omega})^{-\varsigma\varrho}}{\Gamma(1+\alpha\varsigma\varrho)}$$

$$\left(\frac{(\lambda_1-\lambda'_1)}{t}\right)^s\left(\log\frac{t}{(\lambda_1-\lambda'_1)}\right)^{\omega-1}\frac{d(\lambda_1-\lambda'_1)}{(\lambda_1-\lambda'_1)},\ldots,$$

$$\frac{1}{\Gamma(\omega)}\int_a^{(\varpi_k-\vartheta_k)}\sum_{\varsigma=0}^{\infty}\frac{(\frac{((\lambda_k-\lambda'_k)-a)^{\omega}}{\omega})^{-\varsigma\varrho}}{\Gamma(1+\tau\varsigma\varrho)}\left(\frac{(\lambda_k-\lambda'_k)}{t}\right)^s\left(\log\frac{t}{(\lambda_k-\lambda'_k)}\right)^{\omega-1}\frac{d(\lambda_k-\lambda'_k)}{(\lambda_k-\lambda'_k)}\Bigg|\frac{1}{\frac{\aleph}{\wp}}\Bigg)$$

$$\succeq \boldsymbol{H}^{m,n}_{\tau,\omega}\Bigg(\Bigg|\frac{1}{\Gamma(\omega)}\int_a^{(\varpi_1-\vartheta_1)}\sum_{\varsigma=0}^{\infty}\frac{(\frac{((\lambda_1-\lambda'_1)-a)^{\omega}}{\omega})^{-\varsigma\varrho}}{\Gamma(1+\tau\varsigma\varrho)}\left(\frac{(\lambda_1-\lambda'_1)}{t}\right)^s\left(\log\frac{t}{(\lambda_1-\lambda'_1)}\right)^{\omega-1}\frac{d(\lambda_1-\lambda'_1)}{(\lambda_1-\lambda'_1)},\ldots,$$

$$\frac{1}{\Gamma(\omega)}\int_a^{(\varpi_k-\vartheta_k)}\sum_{\varsigma=0}^{\infty}\frac{(\frac{((\lambda_k-\lambda'_k)-a)^{\omega}}{\omega})^{-\varsigma\varrho}}{\Gamma(1+\tau\varsigma\varrho)}\left(\frac{(\lambda_k-\lambda'_k)}{t}\right)^s\left(\log\frac{t}{(\lambda_k-\lambda'_k)}\right)^{\omega-1}\frac{d(\lambda_k-\lambda'_k)}{(\lambda_k-\lambda'_k)}\Bigg|$$

$$\frac{1}{\dfrac{\aleph}{\wp\left(\sum_{\varsigma=0}^{\infty}\sum_{\varsigma_1\cdots\varsigma_k=\varsigma}^{\infty}\binom{\varsigma}{\varsigma_1,\ldots,\varsigma_k}\frac{(\frac{((\varpi_1-\vartheta_1)-a)^{\omega}}{\omega})^{\varsigma_1\varrho}\ldots(\frac{((\varpi_k-\vartheta_k)-a)^{\omega}}{\omega})^{\varsigma_k\varrho}}{\Gamma(\varsigma_1\cdots\varsigma_k+\tau\varsigma\varrho)}\right)\epsilon}}\Bigg)$$

$$\succeq \boldsymbol{H}^{m,n}_{\tau,\omega}\left(-\frac{|\varpi_1-\vartheta_1,\ldots,\varpi_k-\vartheta_k|}{\dfrac{\aleph}{\wp\left(\sum_{\varsigma=0}^{\infty}\sum_{\varsigma_1\cdots\varsigma_k=\varsigma}^{\infty}\binom{\varsigma}{\varsigma_1,\ldots,\varsigma_k}\frac{(\frac{((\varpi_1-\vartheta_1)-a)^{\omega}}{\omega})^{\varsigma_1\varrho}\ldots(\frac{((\varpi_k-\vartheta_k)-a)^{\omega}}{\omega})^{\varsigma_k\varrho}}{\Gamma(\varsigma_1\cdots\varsigma_k+\tau\varsigma\varrho)}\right)\epsilon E}}\right),$$

then

$$d(\Pi\mathfrak{h},\mathfrak{g})\leq\left[\wp\left(\sum_{\varsigma=0}^{\infty}\sum_{\varsigma_1\cdots\varsigma_k=\varsigma}^{\infty}\binom{\varsigma}{\varsigma_1,\ldots,\varsigma_k}\frac{(\frac{((\varpi_1-\vartheta_1)-a)^{\omega}}{\omega})^{\varsigma_1\varrho}\ldots(\frac{((\varpi_k-\vartheta_k)-a)^{\omega}}{\omega})^{\varsigma_k\varrho}}{\Gamma(\varsigma_1\cdots\varsigma_k+\tau\varsigma\varrho)}\right)E\right]\epsilon<\infty.$$

3.3.3 Applications

Now, we provide numerical examples according to the results obtained.

Example 3.4 We consider the following conformable FDE

$$\begin{cases}\mathfrak{D}^1_{\frac{1}{2}}\mathfrak{f}(\varpi-\vartheta)=\mathfrak{f}(\varpi-\vartheta)+\frac{1}{5+(\varpi-\vartheta)^2}\left(\frac{\mathfrak{f}^2(\varpi-\vartheta)}{\mathfrak{f}(\varpi-\vartheta)+1}\right)\cos(\mathfrak{f}(\varpi-\vartheta)),\ (\varpi-\vartheta)\in(1,2]\\ \mathfrak{f}(1)=1\end{cases}\tag{3.3.16}$$

where $\omega=\frac{1}{2}$, $\varrho=1$, $\wp=\frac{1}{8}$, for $\varsigma=1$, $(\varpi_1-\vartheta_1)=2,\tau=1$ we have $\beth=0/4618<1$. Also in this equation, $K(\varpi-\vartheta,\mathfrak{f}(\varpi-\vartheta))=\frac{1}{5+(\varpi-\vartheta)^2}\left(\frac{\mathfrak{f}^2(\varpi-\vartheta)}{\mathfrak{f}(\varpi-\vartheta)+1}\right)$ $\cos(\mathfrak{f}(\varpi-\vartheta))$. Let for mapping $\mathfrak{f}$ and the MVF control function $\boldsymbol{H}^{m,n}_{\tau,\omega}$ we have

▶

$$\mathcal{N}\Big(K((\varpi_1-\vartheta_1),\mathfrak{f}((\varpi_1-\vartheta_1)))-K((\varpi_1-\vartheta_1),\mathfrak{h}((\varpi_1-\vartheta_1))),\ldots,$$

$$K((\varpi_k-\vartheta_k),\mathfrak{f}((\varpi_k-\vartheta_k)))-K((\varpi_k-\vartheta_k),\mathfrak{h}((\varpi_k-\vartheta_k))),\aleph\Big)$$

$$\succeq \mathcal{N}\left(\mathfrak{f}((\varpi_1-\vartheta_1))-\mathfrak{h}((\varpi_1-\vartheta_1)),\ldots,\mathfrak{f}((\varpi_k-\vartheta_k))-\mathfrak{h}((\varpi_k-\vartheta_k)),\frac{\aleph}{\frac{1}{8}}\right)$$

▶

$$\boldsymbol{H}^{m,n}_{\tau,\omega}\left(\left|\frac{1}{\Gamma(\frac{1}{2})}\left[\int_{\mathrm{a}}^{(\varpi_1-\vartheta_1)}\sum_{\varsigma=0}^{\infty}\frac{((2(\lambda_1-\lambda'_1)-\mathrm{a})^{\frac{1}{2}})^{-\varsigma\varrho}}{\Gamma(1+\tau\varsigma\varrho)}\left(\frac{(\lambda_1-\lambda'_1)}{\mathrm{t}}\right)^{s}\left(\log\frac{\mathrm{t}}{(\lambda_1-\lambda'_1)}\right)^{-\frac{1}{2}}\frac{\mathrm{d}(\lambda_1-\lambda'_1)}{(\lambda_1-\lambda'_1)},\right.\right.\right.$$

$$\left.\left.\ldots,\int_{\mathrm{a}}^{(\varpi_k-\vartheta_k)}\sum_{\varsigma=0}^{\infty}\frac{(2(\lambda_k-\lambda'_k)-\mathrm{a})^{\frac{1}{2}})^{-\varsigma\varrho}}{\Gamma(1+\tau\varsigma\varrho)}\left(\frac{(\lambda_k-\lambda'_k)}{\mathrm{t}}\right)^{s}\left(\log\frac{\mathrm{t}}{(\lambda_k-\lambda'_k)}\right)^{\omega-1}\frac{\mathrm{d}(\lambda_k-\lambda'_k)}{(\lambda_k-\lambda'_k)}\right]\frac{1}{\aleph}\right|\right)$$

$$\succeq \boldsymbol{H}^{m,n}_{\tau,\omega}\left(-\frac{|\varpi_1-\vartheta_1,\ldots,\varpi_k-\vartheta_k|}{\frac{\aleph}{\frac{1}{2}}}\right).$$

If $\mathfrak{h}\in C([1,2],\mathbb{R})$ be a differentiable function such that (Figs. 3.1 and 3.2)

$$\mathcal{N}\left(\mathfrak{D}^1_{\frac{1}{2}}\mathfrak{h}(\varpi_1-\vartheta_1)-\mathfrak{h}(\varpi_1-\vartheta_1)-\frac{1}{5+(\varpi_1-\vartheta_1)^2}\left(\frac{\mathfrak{h}^2(\varpi_1-\vartheta_1)}{\mathfrak{h}(\varpi_1-\vartheta_1)+1}\right)\cos(\mathfrak{h}(\varpi_1-\vartheta_1)),\ldots,\right.$$

$$\left.\mathfrak{D}^1_{\frac{1}{2}}\mathfrak{h}(\varpi_k-\vartheta_k)-\mathfrak{h}(\varpi_k-\vartheta_k)-\frac{1}{5+(\varpi_k-\vartheta_k)^2}\left(\frac{\mathfrak{h}^2(\varpi_k-\vartheta_k)}{\mathfrak{h}(\varpi_k-\vartheta_k)+1}\right)\cos(\mathfrak{h}(\varpi_k-\vartheta_k)),\aleph\right)$$

$$\succeq \boldsymbol{H}^{m,n}_{\tau,\omega}\left(-\frac{|\varpi_1-\vartheta_1,\ldots,\varpi_k-\vartheta_k|}{\frac{\aleph}{F}}\right), \tag{3.3.17}$$

then, $\mathfrak{h}$ is a solution of the inequality

$$\mathcal{N}\left(\mathfrak{h}(\varpi_1-\vartheta_1)-\mathfrak{h}_{\mathrm{a}}\sum_{\varsigma=0}^{\infty}\frac{(2(\varpi_1-\vartheta_1)-\mathrm{a})^{\frac{1}{2}})^{\varsigma}}{\Gamma(1+\tau\varsigma)}\right.$$

$$-\frac{1}{\Gamma(\frac{1}{2})}\int_{\mathrm{a}}^{(\varpi_1-\vartheta_1)}\sum_{\varsigma=0}^{\infty}\frac{(2(\varpi_1-\vartheta_1)-\mathrm{a})^{\frac{1}{2}})^{\varsigma}}{\Gamma(1+\tau\varsigma)}\sum_{\varsigma=0}^{\infty}\frac{((2(\lambda_1-\lambda'_1)-\mathrm{a})^{\frac{1}{2}})^{-\varsigma}}{\Gamma(1+\tau\varsigma)}$$

$$\frac{1}{5+(\lambda_1-\lambda'_1)^2}\left(\frac{\mathfrak{h}^2(\lambda_1-\lambda'_1)}{\mathfrak{h}(\lambda_1-\lambda'_1)+1}\right)\cos(\mathfrak{h}(\lambda_1-\lambda'_1))\left(\frac{(\lambda_1-\lambda'_1)}{\mathrm{t}}\right)^{s}\left(\log\frac{\mathrm{t}}{(\lambda_1-\lambda'_1)}\right)^{-\frac{1}{2}}\frac{\mathrm{d}(\lambda_1-\lambda'_1)}{(\lambda_1-\lambda'_1)},\ldots,$$

$$\mathfrak{h}(\varpi_k-\vartheta_k)-\mathfrak{f}_{\mathrm{a}}\sum_{\varsigma=0}^{\infty}\frac{((2(\varpi_k-\vartheta_k)-\mathrm{a})^{\frac{1}{2}})^{\varsigma}}{\Gamma(1+\tau\varsigma)}-\frac{1}{\Gamma(\frac{1}{2})}\int_{\mathrm{a}}^{(\varpi_k-\vartheta_k)}\sum_{\varsigma=0}^{\infty}\frac{((2(\varpi_k-\vartheta_k)-\mathrm{a})^{\frac{1}{2}})^{\varsigma}}{\Gamma(1+\tau\varsigma)}$$

$$\sum_{\varsigma=0}^{\infty}\frac{((2(\lambda_k-\lambda'_k)-\mathrm{a})^{\frac{1}{2}})^{-\varsigma}}{\Gamma(1+\tau\varsigma)}\frac{1}{5+(\lambda_k-\lambda'_k)^2}\left(\frac{\mathfrak{h}^2(\lambda_k-\lambda'_k)}{\mathfrak{h}(\lambda_k-\lambda'_k)+1}\right)\cos(\mathfrak{h}(\lambda_k-\lambda'_k))$$

$$\left.\left(\frac{(\lambda_k-\lambda'_k)}{\mathrm{t}}\right)^{s}\left(\log\frac{\mathrm{t}}{(\lambda_k-\lambda'_k)}\right)^{-\frac{1}{2}}\frac{\mathrm{d}(\lambda_k-\lambda'_k)}{(\lambda_k-\lambda'_k)},\aleph\right)$$

$$\succeq \boldsymbol{H}^{m,n}_{\tau,\omega}\left(\left|\frac{1}{\Gamma(\frac{1}{2})}\int_{\mathrm{a}}^{(\varpi_1-\vartheta_1)}\sum_{\varsigma=0}^{\infty}\frac{((2(\lambda_1-\lambda'_1)-\mathrm{a})^{\frac{1}{2}})^{-\varsigma}}{\Gamma(1+\tau\varsigma)}\left(\frac{(\lambda_1-\lambda'_1)}{\mathrm{t}}\right)^{s}\left(\log\frac{\mathrm{t}}{(\lambda_1-\lambda'_1)}\right)^{-\frac{1}{2}}\right.\right.$$

$$\frac{\mathrm{d}(\lambda_1-\lambda'_1)}{(\lambda_1-\lambda'_1)},\ldots,$$

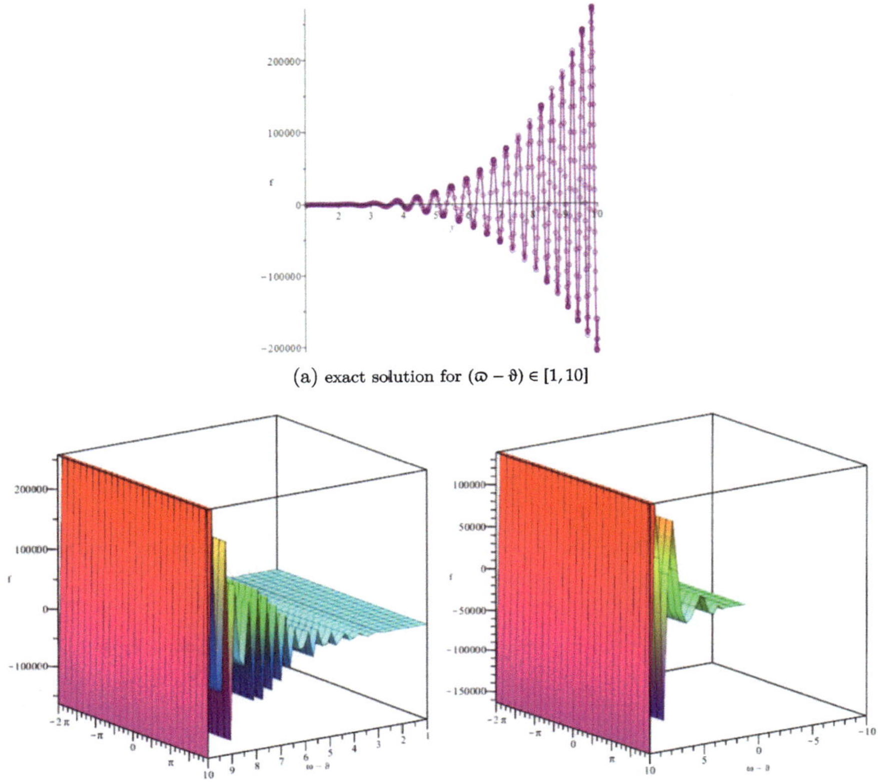

(a) exact solution for $(\varpi-\vartheta)\in[1,10]$

(b) exact solution for $(\varpi-\vartheta)\in[1,10]$

(c) exact solution for $(\varpi-\vartheta)\in[-10,10]$

Fig. 3.1 Graphs related to the exact solution of the conformable FDE (3.3.16) for $\tau=\frac{1}{2}, s=1, t=\frac{1}{4}$

$$\frac{1}{\Gamma(\frac{1}{2})}\int_{a}^{(\varpi_k-\vartheta_k)} \sum_{\varsigma=0}^{\infty} \frac{((2(\lambda_k-\lambda'_k)-a)^{\frac{1}{2}})^{-\varsigma}}{\Gamma(1+\tau\varsigma)}\left(\frac{(\lambda_k-\lambda'_k)}{t}\right)^{s}\left(\log\frac{t}{(\lambda_k-\lambda'_k)}\right)^{-\frac{1}{2}}$$
$$\frac{d(\lambda_k-\lambda'_k)}{(\lambda_k-\lambda'_k)}\Bigg|\frac{1}{\frac{\aleph}{F\sum_{\varsigma=0}^{\infty}\frac{(2)^{\varsigma}}{\Gamma(1+\tau\varsigma)}}}\Bigg),$$

thus we can find a unique differentiable function $\mathfrak{f}\in C([1,2],\mathbb{R})$ from (Theorem 3.9) such that for each $(\varpi-\vartheta)\in[1,2]$ we have

$$\mathfrak{f}(\varpi_i-\vartheta_i)=\sum_{\varsigma=0}^{\infty}\frac{(2((\varpi_i-\vartheta_i)-1)^{\frac{1}{2}})^{\varsigma}}{\Gamma(1+\varsigma\tau)}$$
$$+\frac{1}{\Gamma(\frac{1}{2})}\int_{1}^{(\varpi_i-\vartheta_i)}\sum_{\varsigma=0}^{\infty}\frac{(2((\varpi_i-\vartheta_i)-1)^{\frac{1}{2}})^{\varsigma}}{\Gamma(1+\varsigma\tau)}\sum_{\varsigma=0}^{\infty}\frac{(2(\lambda_i-\lambda'_i-1)^{\frac{1}{2}})^{-\varsigma}}{\Gamma(1+\varsigma\tau)}\left(\frac{\lambda_i-\lambda'_i}{t}\right)^{s}\left(\log\frac{t}{\lambda_i-\lambda'_i}\right)^{-1/2}$$

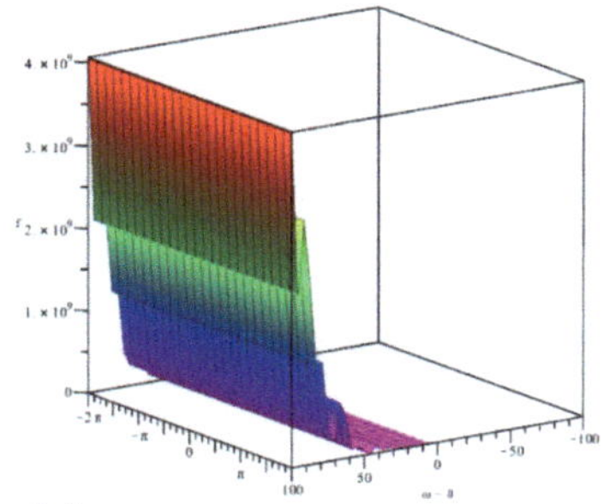

(a) exact solution for $(\varpi - \vartheta) \in [-100, 100]$

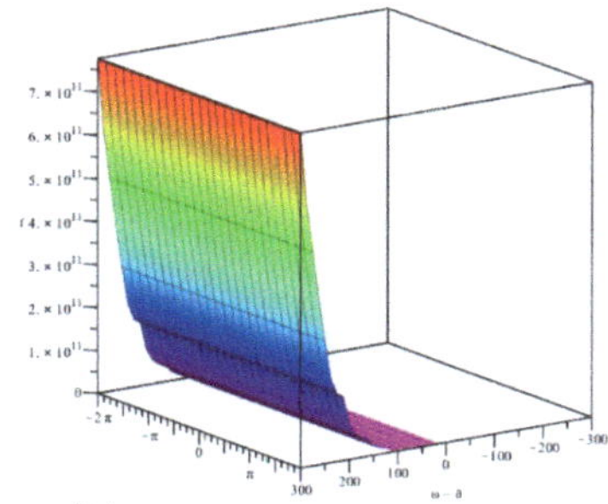

(b) exact solution for $(\varpi - \vartheta) \in [-300, 300]$

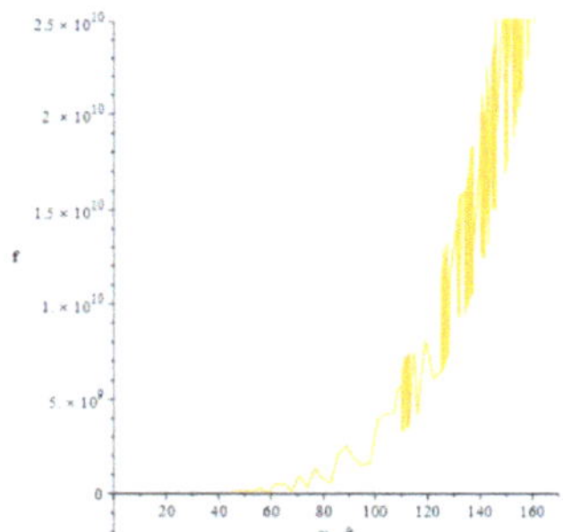

(c) exact solution for $(\varpi - \vartheta) \in [-300, 300]$

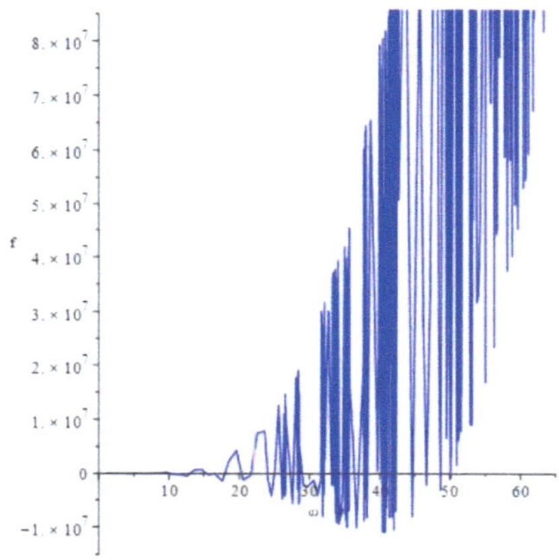

(d) exact solution for $(\varpi - \vartheta) \in [-100, 100]$

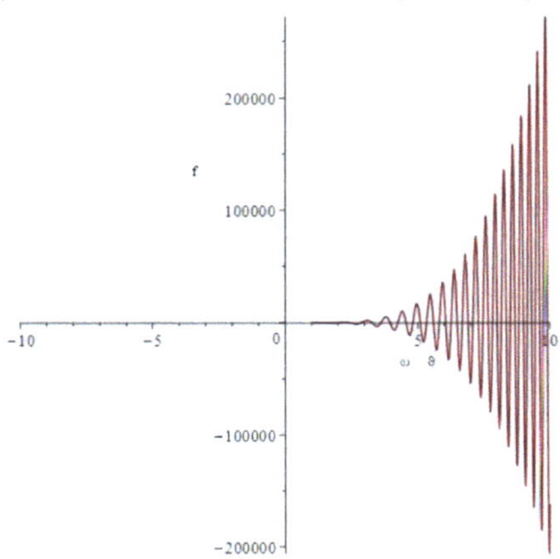

(e) exact solution for $(\varpi - \vartheta) \in [-10, 10]$

Fig. 3.2 Graphs related to the exact solution of the conformable FDE (3.3.16) for $\tau = \frac{1}{2}, s = 1, t = \frac{1}{4}$

$$\frac{\frac{1}{5+(\varpi_i-\vartheta_i)^2}\left(\frac{\mathfrak{f}^2(\varpi_i-\vartheta_i)}{\mathfrak{f}((\varpi_i-\vartheta_i)+1}\right)\cos(\mathfrak{f}(\varpi_i-\vartheta_i))}{(\lambda_i-\lambda'_i)}\,\mathrm{d}(\lambda_i-\lambda'_i),$$

for any $i = 1, \ldots, k$. Therefore

$$d(\mathfrak{h}, \mathfrak{f}) \leq \frac{2}{1-\beth} \sum_{\varsigma=0}^{\infty} \frac{2^{\varsigma}}{\Gamma(1+\tau\varsigma)} F,$$

and

$$\mathcal{N}\Big(\mathfrak{h}(\varpi_1-\vartheta_1)-\mathfrak{f}(\varpi_1-\vartheta_1), \ldots, \mathfrak{h}(\varpi_k-\vartheta_k)-\mathfrak{f}(\varpi_k-\vartheta_k), \aleph\Big)$$
$$\succeq \boldsymbol{H}^{m,n}_{\tau,\omega}\left(-\frac{|\varpi_1-\vartheta_1, \ldots, \varpi_k-\vartheta_k|}{\frac{\aleph(1-\beth)}{2\sum_{\varsigma=0}^{\infty}\frac{2^{\varsigma}}{\Gamma(1+\tau\varsigma)}F}}\right).$$

Example 3.5 We consider the following conformable FDE

$$\begin{cases} \mathfrak{D}^1_{\frac{1}{4}}\mathfrak{f}(\varpi-\vartheta) = \mathfrak{f}(\varpi-\vartheta) + \frac{\sqrt{(\varpi-\vartheta)+1}}{6(\varpi-\vartheta)^2+6}\sin(\mathfrak{f}(\varpi-\vartheta)), (\varpi-\vartheta) \in (1, 2] \\ \mathfrak{f}(1) = 1 \end{cases} \tag{3.3.18}$$

where $\omega = \frac{1}{4}$, $\varrho = 1$, $\wp = \frac{1}{100}$, for $\varsigma = 2$, $(\varpi_1 - \vartheta_1) = 4$, $\tau = 2$ we have $\beth = 0.004274 < 1$. Also in this equation, $\mathfrak{k}(\varpi-\vartheta, \mathfrak{f}(\varpi-\vartheta)) = \frac{\sqrt{(\varpi-\vartheta)+1}}{6(\varpi-\vartheta)^2+6}\sin(\mathfrak{f}(\varpi-\vartheta))$. Let for mapping $\mathfrak{k}$ and the MVF control function $\boldsymbol{H}^{m,n}_{\tau,\omega}$ we have

▶

$$\mathcal{N}\Big(\mathrm{K}((\varpi_1-\vartheta_1), \mathfrak{f}((\varpi_1-\vartheta_1))) - \mathrm{K}((\varpi_1-\vartheta_1), \mathfrak{h}((\varpi_1-\vartheta_1))), \ldots,$$
$$\mathrm{K}((\varpi_k-\vartheta_k), \mathfrak{f}((\varpi_k-\vartheta_k))) - \mathrm{K}((\varpi_k-\vartheta_k), \mathfrak{h}((\varpi_k-\vartheta_k))), \aleph\Big)$$
$$\succeq \mathfrak{Y}\Big(\mathfrak{f}((\varpi_1-\vartheta_1)) - \mathfrak{h}((\varpi_1-\vartheta_1)), \ldots, \mathfrak{f}((\varpi_k-\vartheta_k)) - \mathfrak{h}((\varpi_k-\vartheta_k)), 100\aleph\Big).$$

▶

$$\boldsymbol{H}^{m,n}_{\tau,\omega}\Bigg(\Bigg|\frac{1}{\Gamma(\frac{1}{4})}\Bigg[\int_{\mathfrak{c}}^{(\varpi_1-\vartheta_1)} \sum_{\varsigma=0}^{\infty} \frac{((4(\lambda_1-\lambda'_1)-\mathrm{a})^{\frac{1}{4}})^{-\varsigma\varrho}}{\Gamma(1+\tau\varsigma\varrho)} \left(\frac{(\lambda_1-\lambda'_1)}{\mathrm{t}}\right)^{s} \left(\log\frac{\mathrm{t}}{(\xi_1-\xi'_1)}\right)^{-\frac{3}{4}}$$
$$\frac{\mathrm{d}(\lambda_1-\lambda'_1)}{(\lambda_1-\lambda'_1)},$$
$$\ldots, \int_{\mathrm{a}}^{(\varpi_k-\vartheta_k)} \sum_{\varsigma=0}^{\infty} \frac{(4(\lambda_k-\lambda'_k)-\mathrm{a})^{\frac{1}{4}})^{-\varsigma\varrho}}{\Gamma(1+\tau\varsigma\varrho)} \left(\frac{(\lambda_k-\lambda'_k)}{\mathrm{t}}\right)^{s} \left(\log\frac{\mathrm{t}}{(\lambda_k-\lambda'_k)}\right)^{-\frac{3}{4}} \frac{\mathrm{d}(\lambda_k-\lambda'_k)}{(\lambda_k-\lambda'_k)}\Bigg]\frac{1}{\aleph}\Bigg|\Bigg)$$

$$\succeq H^{m,n}_{\tau,\omega}\left(-\frac{|\varpi_1-\vartheta_1,\ldots,\varpi_k-\vartheta_k|}{4\aleph}\right).$$

If $\mathfrak{h}\in C([1,2],\mathbb{R})$ be a differentiable function such that

$$\mathcal{N}\Bigg(\mathfrak{D}^1_{\frac{1}{4}}\mathfrak{h}(\varpi_1-\vartheta_1)-\mathfrak{h}(\varpi_1-\vartheta_1)-\frac{\sqrt{(\varpi_1-\vartheta_1)+1}}{6(\varpi_1-\vartheta_1)^2+6}\sin(\mathfrak{h}(\varpi_1-\vartheta_1)),\ldots, \tag{3.3.19}$$

$$\mathfrak{D}^1_{\frac{1}{4}}\mathfrak{h}(\varpi_k-\vartheta_k)-\mathfrak{h}(\varpi_k-\vartheta_k)-\frac{\sqrt{(\varpi_k-\vartheta_k)+1}}{6(\varpi_k-\vartheta_k)^2+6}\sin(\mathfrak{h}(\varpi_k-\vartheta_k)),\aleph\Bigg)$$
$$\succeq H^{m,n}_{\tau,\omega}\left(-\frac{|\varpi_1-\vartheta_1,\ldots,\varpi_k-\vartheta_k|}{3\aleph}\right),$$

then, $\mathfrak{h}$ is a solution of the inequality

$$\mathcal{N}\Bigg(\mathfrak{h}(\varpi_1-\vartheta_1)-\mathfrak{h}_{\mathrm{a}}\sum_{\varsigma=0}^{\infty}\frac{(4(\varpi_1-\vartheta_1)-\mathrm{a})^{\frac{1}{4}})^{\varsigma}}{\Gamma(1+\tau\varsigma)}$$
$$-\frac{1}{\Gamma(\frac{1}{4})}\int_{\mathrm{a}}^{(\varpi_1-\vartheta_1)}\sum_{\varsigma=0}^{\infty}\frac{(4(\varpi_1-\vartheta_1)-\mathrm{a})^{\frac{1}{4}})^{\varsigma}}{\Gamma(1+\tau\varsigma)}\sum_{\varsigma=0}^{\infty}\frac{((4(\lambda_1-\lambda'_1)-\mathrm{a})^{\frac{1}{4}})^{-\varsigma}}{\Gamma(1+\tau\varsigma)}$$
$$\frac{\sqrt{(\lambda_1-\lambda'_1)+1}}{6(\lambda_1-\lambda'_1)^2+6}\sin(\mathfrak{h}(\lambda_1-\lambda'_1))\left(\frac{(\lambda_1-\lambda'_1)}{\mathrm{t}}\right)^s\left(\log\frac{\mathrm{t}}{(\lambda_1-\lambda'_1)}\right)^{-\frac{3}{4}}\frac{\mathrm{d}(\lambda_1-\lambda'_1)}{(\lambda_1-\lambda'_1)},\ldots,$$
$$\mathfrak{h}(\varpi_k-\vartheta_k)-\mathfrak{f}_{\mathrm{a}}\sum_{\varsigma=0}^{\infty}\frac{((4(\varpi_k-\vartheta_k)-\mathrm{a})^{\frac{1}{4}})^{\varsigma}}{\Gamma(1+\tau\varsigma)}-\frac{1}{\Gamma(\frac{1}{4})}\int_{\mathrm{a}}^{(\varpi_k-\vartheta_k)}\sum_{\varsigma=0}^{\infty}\frac{((4(\varpi_k-\vartheta_k)-\mathrm{a})^{\frac{1}{4}})^{\varsigma}}{\Gamma(1+\tau\varsigma)}$$
$$\sum_{\varsigma=0}^{\infty}\frac{((4(\lambda_k-\lambda'_k)-\mathfrak{c})^{\frac{1}{4}})^{-\varsigma}}{\Gamma(1+\tau\varsigma)}\frac{\sqrt{(\lambda_k-\lambda'_k)+1}}{6(\lambda_k-\lambda'_k)^2+6}\sin(\mathfrak{h}(\lambda_k-\lambda'_k))$$
$$\left(\frac{(\lambda_k-\lambda'_k)}{\mathrm{t}}\right)^s\left(\log\frac{\mathrm{t}}{(\lambda_k-\lambda'_k)}\right)^{-\frac{3}{4}}\frac{\mathrm{d}(\lambda_k-\lambda'_k)}{(\lambda_k-\lambda'_k)},\aleph\Bigg)$$
$$\succeq H^{m,n}_{\tau,\omega}\Bigg(\Bigg|\frac{1}{\Gamma(\frac{1}{4})}\int_{\mathrm{a}}^{(\varpi_1-\vartheta_1)}\sum_{\varsigma=0}^{\infty}\frac{((4(\lambda_1-\lambda'_1)-\mathrm{a})^{\frac{1}{4}})^{-\varsigma}}{\Gamma(1+\tau\varsigma)}\left(\frac{(\lambda_1-\lambda'_1)}{\mathrm{t}}\right)^s\left(\log\frac{\mathrm{t}}{(\lambda_1-\lambda'_1)}\right)^{-\frac{3}{4}}\frac{\mathrm{d}(\lambda_1-\lambda'_1)}{(\lambda_1-\lambda'_1)},\ldots,$$
$$\frac{1}{\Gamma(\frac{1}{4})}\int_{\mathrm{a}}^{(\varpi_k-\vartheta_k)}\sum_{\varsigma=0}^{\infty}\frac{((4(\lambda_k-\lambda'_k)-\mathrm{a})^{\frac{1}{4}})^{-\varsigma}}{\Gamma(1+\tau\varsigma)}\left(\frac{(\lambda_k-\lambda'_k)}{\mathrm{t}}\right)^s\left(\log\frac{\mathrm{t}}{(\lambda_k-\lambda'_k)}\right)^{-\frac{3}{4}}\frac{\mathrm{d}(\lambda_k-\lambda'_k)}{(\lambda_k-\lambda'_k)}\Bigg|$$
$$\frac{1}{\frac{\aleph}{\frac{1}{3}\sum_{\varsigma=0}^{\infty}\frac{(4)^{\varsigma}}{\Gamma(1+\tau\varsigma)}}}\Bigg),$$

thus we can find a unique differentiable function $\mathfrak{f}\in C([1,2],\mathbb{R})$ from (3.3.18) such that for each $(\varpi-\vartheta)\in[1,2]$ we have (Fig. 3.3)

$$\mathfrak{f}(\varpi_i-\vartheta_i)=\sum_{\varsigma=0}^{\infty}\frac{(4((\varpi_i-\vartheta_i)-1)^{\frac{1}{4}})^{\varsigma}}{\Gamma(1+\varsigma\tau)}$$

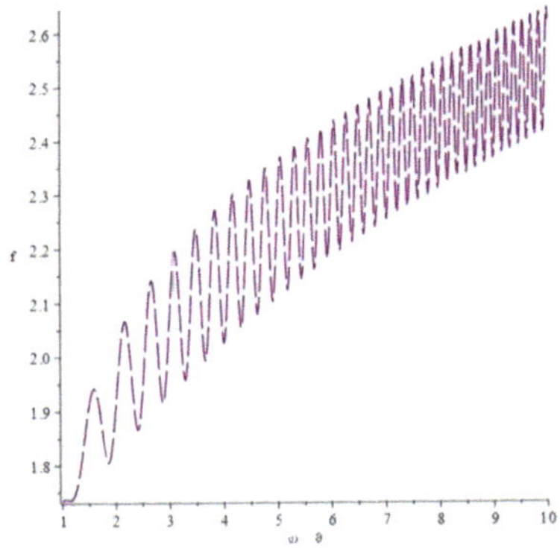

(a) exact solution for $(\varpi - \vartheta) \in [1, 10]$

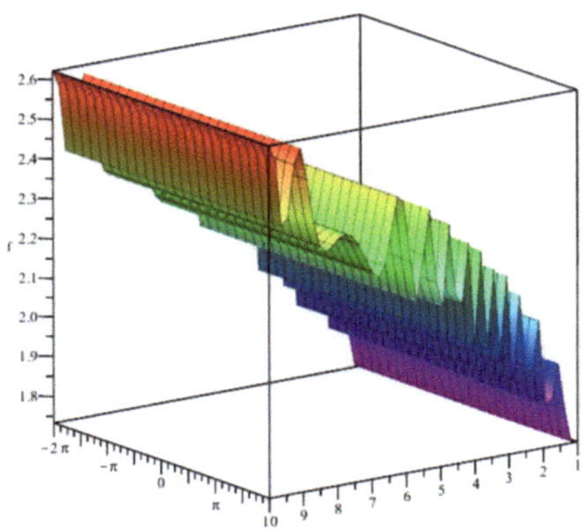

(b) exact solution for $(\varpi - \vartheta) \in [1, 10]$

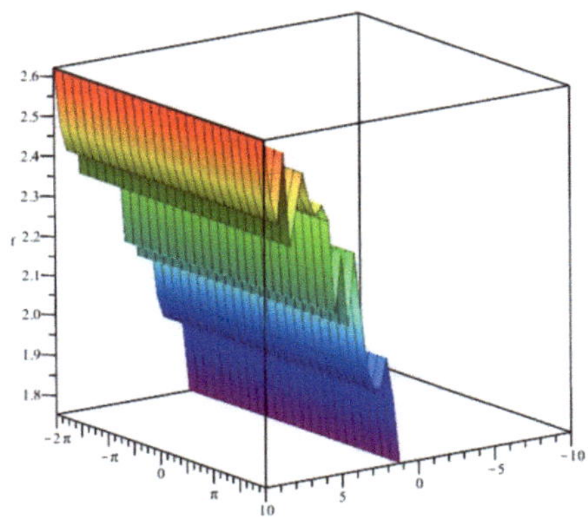

(c) exact solution for $(\varpi - \vartheta) \in [-10, 10]$

Fig. 3.3 Graphs related to the exact solution of the conformable FDE (3.3.18) for $\tau = 2$, $s = 1$, $t = \frac{1}{4}$

$$+ \frac{1}{\Gamma(\frac{1}{4})} \int_1^{(\varpi_i - \vartheta_i)} \sum_{\varsigma=0}^{\infty} \frac{(4((\varpi_i - \vartheta_i) - 1)^{\frac{1}{4}})^{\varsigma}}{\Gamma(1+\varsigma\tau)} \sum_{\varsigma=0}^{\infty} \frac{(2(\lambda\!\!\!^{-}_i - \lambda\!\!\!^{-\prime}_i - 1)^{\frac{1}{4}})^{-\varsigma}}{\Gamma(1+\varsigma\tau)} \left(\frac{\lambda\!\!\!^{-}_i - \lambda\!\!\!^{-\prime}_i}{\mathfrak{f}}\right)^{s} \left(\log \frac{\mathfrak{f}}{\lambda\!\!\!^{-}_i - \lambda\!\!\!^{-\prime}_i}\right)^{-\frac{3}{4}}$$
$$\frac{\frac{\sqrt{(\varpi_i - \vartheta_i)+1}}{6(\varpi_i - \vartheta_i)^2+6} \sin(\mathfrak{h}(\varpi_i - \vartheta_i))}{(\lambda\!\!\!^{-}_i - \lambda\!\!\!^{-\prime}_i)} \, \mathrm{d}(\lambda\!\!\!^{-}_i - \lambda\!\!\!^{-\prime}_i),$$

for any $i = 1, \ldots, k$. Therefore

$$d(\mathfrak{h}, \mathfrak{f}) \leq \frac{4}{1 - \gimel} \sum_{\varsigma=0}^{\infty} \frac{4^{\varsigma}}{\Gamma(1+\tau\varsigma)} \times \frac{1}{3},$$

and

$$\mathcal{N}\Big(\mathfrak{h}(\varpi_1-\vartheta_1)-\mathfrak{f}(\varpi_1-\vartheta_1),\ldots,\mathfrak{h}(\varpi_k-\vartheta_k)-\mathfrak{f}(\varpi_k-\vartheta_k),\aleph\Big)$$
$$\succeq \boldsymbol{H}^{m,n}_{\tau,\omega}\left(-\frac{|\varpi_1-\vartheta_1,\ldots,\varpi_k-\vartheta_k|}{\frac{\aleph(1-\beth)}{4\sum_{\varsigma=0}^{\infty}\frac{2^{\varsigma}}{\Gamma(1+\tau\varsigma)}\times\frac{1}{3}}}\right).$$

3.4 Best Approximation for the Non-homogeneous Fractional Delay Oscillation Equations

3.4.1 Introduction of the Non-homogeneous Fractional Delay Oscillation Equation and Other Preliminaries

Here, our goal is to obtain an approximation for the NH-FDO-E by a new combined method powered by the Laplace inverse transform and MR approach in the MVFB-space [65, 95–98]. We consider the following NH-FDO-E:

$$D_0^{\kappa}\mathrm{u}(\mathfrak{x}) = -\rho\mathrm{u}(\mathfrak{x}-\xi)+\mathrm{k}(\mathfrak{x})+\mathcal{L}^{-1}(\mathrm{u}(\mathfrak{r}))(\mathfrak{x}),\quad \mathfrak{x}\in\mathfrak{L}:=[0,L] \tag{3.4.1}$$

$$\mathrm{u}(\mathfrak{x}) = \Upsilon(\mathfrak{x}),\quad \mathrm{u}'(\mathfrak{x})=\Upsilon'(\mathfrak{x}),\qquad -\xi\le\mathfrak{x}\le 0, \tag{3.4.2}$$

where

- $\mathrm{k}:\mathfrak{L}\times\mathbb{R}^n\to\mathbb{R}^n$ is an integrable function, $\mathrm{u}(\mathfrak{x})\in\mathbb{R}^n$, $\rho\in\mathbb{R}^{n\times n}$ denotes constant matrix, $\Upsilon\in C^2([-\xi,0],\mathbb{R}^n)$ for $\xi>0$ being a fixed time.
- D_0^{κ} is the standard Caputo fractional derivative with $\kappa\in(1,2)$, defined by

$$D_0^{\kappa}\mathrm{u}(\mathfrak{x}) = \int_0^{\mathfrak{x}}\frac{(\mathfrak{x}-t)^{1-\kappa}}{\Gamma(2-\kappa)}\mathrm{u}''(t)\mathrm{d}t,$$

if the integral exists.

Definition 3.4 Consider the MVF $\boldsymbol{W}_{\kappa,\varsigma}$. We say (3.4.1) has Hyers-Ulam-Wright stability, when a given differentiable map $\mathrm{v}(\mathfrak{x})$ satisfies

$$\mathcal{A}\Big(D_0^{\kappa}\mathrm{v}(\mathfrak{x})+\rho\mathrm{v}(\mathfrak{x}-\xi)-\mathrm{k}(\mathfrak{x})-\mathcal{L}^{-1}(\mathrm{v}(\mathfrak{r}))(\mathfrak{x}),\eta\Big)\succeq\boldsymbol{W}_{\kappa,\varsigma}\Big(-\frac{|\mathfrak{x}|}{\eta}\Big), \tag{3.4.3}$$

for $\mathfrak{x}\in\mathfrak{F}_1$ and there is a solution $\mathrm{u}(\mathfrak{x})$ of (3.4.1) such that for some $\lambda>0$,

$$\mathcal{A}\left(\mathrm{u}(\mathfrak{x}) - \mathrm{v}(\mathfrak{x}), \eta\right) \succeq \boldsymbol{W}_{\kappa,\varsigma}\left(-\frac{|\mathfrak{x}|}{\frac{\eta}{\lambda}}\right).$$

3.4.2 *Hyers-Ulam-Wright Stability for Non-homogeneous Fractional Delay Oscillation Equation*

Now, we use a new method, the Laplace-Mihet-Radu method, to show (3.4.1) is Hyers-Ulam-Wright stable [62] in MVFB-space $(\mathcal{J}, \mathcal{A}, \circledast)$ with MVFF $\boldsymbol{W}_{\kappa,\varsigma}$.

We choose the set $\mathcal{Z}$ as follows

$$\mathcal{Q} = \{\mathrm{u} : \mathfrak{L} \to \mathbb{R}^n, \text{ u is differentiable}\},$$

and define the mapping $d : \mathcal{Q} \times \mathcal{Q} \to [0, \infty]$ as

$$d(\mathrm{u}, \mathrm{v}) = \inf\left\{\wp \in \mathfrak{F}_6 : \mathcal{A}\left(\mathrm{u}(\mathfrak{x}) - \mathrm{v}(\mathfrak{x}), \eta\right) \succeq \boldsymbol{W}_{\kappa,\varsigma}\left(-\frac{|\mathfrak{x}|}{\frac{\eta}{\wp}}\right), \forall\ \mathrm{u}, \mathrm{v} \in \mathcal{Z},\ \ \mathfrak{x} \in \mathfrak{F}_1,\ \ \eta \in \mathfrak{F}_2\right\}.$$

Theorem 3.10 *The $\mathfrak{F}_4$-valued metric space $(\mathcal{Q}, d)$ is complete.*

Proof We have $d(\mathrm{u}, \mathrm{v}) = 0$ iff $\mathrm{u} = \mathrm{v}$. Assuming that $d(\mathrm{u}, \mathrm{v}) = 0$, then we come to

$$\inf\left\{\wp \in \mathfrak{F}_6 : \mathcal{A}\left(\mathrm{u}(\mathfrak{x}) - \mathrm{v}(\mathfrak{x}), \eta\right) \succeq \boldsymbol{W}_{\kappa,\varsigma}\left(-\frac{|\mathfrak{x}|}{\frac{\eta}{\wp}}\right), \forall\ \mathrm{u}, \mathrm{v} \in \mathcal{Q},\ \ \mathfrak{x} \in \mathfrak{F}_1,\ \ \eta \in \mathfrak{F}_2\right\} = 0,$$

and then

$$\mathcal{A}\left(\mathrm{u}(\mathfrak{x}) - \mathrm{v}(\mathfrak{x}), \eta\right) \succeq \boldsymbol{W}_{\kappa,\varsigma}\left(-\frac{|\mathfrak{x}|}{\frac{\eta}{\wp}}\right), \tag{3.4.4}$$

for all $\wp \in \mathfrak{F}_6$. Let $\wp \to 0$ in (3.4.4), we get

$$\mathcal{A}\left(\mathrm{u}(\mathfrak{x}) - \mathrm{v}(\mathfrak{x}), \eta\right) = \mathbf{1}.$$

Thus $\mathrm{u}(\mathfrak{x}) = \mathrm{v}(\mathfrak{x})$ for every $\mathfrak{x} \in \mathfrak{F}_1$, and vice versa. Also we have $d(\mathrm{u}, \mathrm{v}) = d(\mathrm{v}, \mathrm{u})$ for every $\mathrm{u}, \mathrm{v} \in \mathcal{Q}$. Assuming $d(\mathrm{u}, \mathrm{v}) = \alpha_1 \in \mathfrak{F}_2$ and $d(\mathrm{v}, \mathrm{w}) = \alpha_2 \in \mathfrak{F}_2$, then we have

$$\mathcal{A}\left(\mathrm{u}(\mathfrak{x}) - \mathrm{v}(\mathfrak{x}), \eta\right) \succeq \boldsymbol{W}_{\kappa,\varsigma}\left(-\frac{|\mathfrak{x}|}{\frac{\eta}{\alpha_1}}\right),$$

and

$$\mathcal{A}\left(\mathrm{v}(\mathfrak{x}) - \mathrm{w}(\mathfrak{x}), \eta\right) \succeq \boldsymbol{W}_{\kappa,\varsigma}\left(-\frac{|\mathfrak{x}|}{\frac{\eta}{\alpha_2}}\right),$$

for every $\eta \in \mathfrak{F}_2$. Therefore, we get

$$\begin{aligned}\mathcal{A}\left((\mathrm{u}(\mathfrak{x}) - \mathrm{w}(\mathfrak{x}), (\alpha_1 + \alpha_2)\eta\right) &\succeq \left[\mathcal{A}\left((\mathrm{u}(\mathfrak{x}) - \mathrm{v}(\mathfrak{x}), (\alpha_1)\eta\right) \circledast \mathcal{A}\left((\mathrm{v}(\mathfrak{x}) - \mathrm{w}(\mathfrak{x}), (\alpha_2)\eta\right)\right] \\ &\succeq \boldsymbol{W}_{\kappa,\varsigma}\left(-\frac{|\mathfrak{x}|}{\eta}\right) \circledast \boldsymbol{W}_{\kappa,\varsigma}\left(-\frac{|\mathfrak{x}|}{\eta}\right) \\ &= \boldsymbol{W}_{\kappa,\varsigma}\left(-\frac{|\mathfrak{x}|}{\eta}\right).\end{aligned}$$

This infers that $d(\mathrm{v}, \mathrm{w}) \preceq d(\mathrm{u}, \mathrm{v}) + d(\mathrm{v}, \mathrm{w})$. To show the completeness of $(\mathcal{Q}, d)$, we suppose that $\{\mathrm{v}_{\mathbf{q}}\}_k$ is a Cauchy sequence in $(\mathcal{Q}, d)$. Assume that $\mathfrak{x} \in \mathfrak{F}_1$, $\tau \in \mathfrak{F}_2$, $\mathfrak{I} \in \mathfrak{F}_5{}^{\circ}$ and $\eta \in \mathfrak{F}_2$ in which $\boldsymbol{W}_{\kappa,\varsigma}\left(-\frac{|\mathfrak{x}|}{\eta}\right) \succ \mathbf{1} - \mathfrak{I}$. For $\alpha\eta < \tau$ choose $\mathbf{q}_0 \in \mathbb{N}$ such that

$$d(\mathrm{v}_{\mathbf{q}}, \mathrm{v}_{\mathbf{p}}) < \alpha \qquad \forall\, \mathbf{q}, \mathbf{p} \geq \mathbf{q}_0.$$

Then

$$\mathcal{A}\left(\mathrm{v}_{\mathbf{q}}(\mathfrak{x}) - \mathrm{v}_{\mathbf{p}}(\mathfrak{x}), \tau\right) \succeq \mathcal{A}\left(\mathrm{v}_{\mathbf{q}}(\mathfrak{x}) - \mathrm{v}_{\mathbf{p}}(\mathfrak{x}), \alpha\eta\right) \succeq \boldsymbol{W}_{\kappa,\varsigma}\left(-\frac{|\mathfrak{x}|}{\eta}\right) \succ \mathbf{1} - \mathfrak{I}.$$

and then

$$\mathcal{A}\left(\mathrm{v}_{\mathbf{q}}(\mathfrak{x}) - \mathrm{v}(\mathfrak{x})_{\mathbf{p}}, \tau\right) \succ \mathbf{1} - \mathfrak{I},$$

which implies that, the sequence $\{\mathrm{v}_{\mathbf{q}}(\mathfrak{x})\}_k$ is Cauchy in the complete space $(\mathcal{J}, \mathcal{A}, \circledast)$ on compact set $\mathfrak{F}_1$. Then, it is uniformly convergent to the mapping $\mathrm{v} : \mathfrak{F}_1 \to \mathcal{J}$. By uniform convergence property we conclude that $\mathrm{v} \in \mathcal{Q}$ and then $(\mathcal{Q}, d)$ is complete.

Now, we can investigate Hyers–Ulam–Wright stability and get an approximation for the non-homogeneous fractional delay oscillation Eq. (3.4.1).

Theorem 3.11 *Consider the MVFB-space $(\mathcal{J}, \mathcal{A}, \circledast)$ and the constant θ, such that $0 < \theta < 1$. Suppose that the following conditions hold:*

- ► *By considering the MVFF* $\boldsymbol{W}_{\kappa,\varsigma} : \mathfrak{L} \longrightarrow \mathbb{R}$ *as the control function, we have*

$$\boldsymbol{W}_{\kappa,\varsigma}\left(\frac{-|\mathcal{L}((\mathfrak{r}))(\mathfrak{x})|}{\eta}\right)\succeq\boldsymbol{W}_{\kappa,\varsigma}\left(-\frac{|\mathfrak{x}|}{\frac{\eta}{\theta}}\right). \tag{3.4.5}$$

Let $\mathrm{v}:\mathfrak{F}_1\to\mathcal{J}$ *be a differentiable function satisfying*

$$\mathcal{A}\left(D_0^{\kappa}\mathrm{v}(\mathfrak{x})+\rho\mathrm{v}(\mathfrak{x}-\xi)-\mathfrak{k}(\mathfrak{x})-\mathcal{L}^{-1}(\mathrm{v}(\mathfrak{r}))(\mathfrak{x}),\eta\right)\succeq\boldsymbol{W}_{\kappa,\varsigma}\left(-\frac{|\mathfrak{x}|}{\frac{\eta}{\varpi}}\right). \tag{3.4.6}$$

Then we can find a unique solution $\mathrm{u}:\mathfrak{F}_1\to\mathcal{J}$ *for Eq. (3.4.1) such that*

$$\mathcal{A}\left(\mathrm{u}(\mathfrak{x})-\mathrm{v}(\mathfrak{x}),\eta\right)\succeq\boldsymbol{W}_{\kappa,\varsigma}\left(-\frac{|\mathfrak{x}|}{\frac{\eta}{F\varpi}}\right),$$

where $F=\frac{\theta}{1-\theta}$, *and for every* $\mathfrak{x}\in\mathfrak{F}_1$ *and* $\eta\in\mathfrak{F}_2$.

Proof Consider complete $\mathfrak{F}_4$-valued metric space $(\boldsymbol{Q},d)$ defined at Theorem 3.10.

Step 1. According to the main equation, we define the mapping $\Delta:\boldsymbol{Q}\to\boldsymbol{Q}$, as follows

$$\Delta(\mathrm{u}(\mathfrak{x}))=\mathcal{L}(D_0^{\kappa}\mathrm{u}(\mathfrak{x})+\rho\mathrm{u}(\mathfrak{x}-\xi)-\mathrm{k}(\mathfrak{x}))(\mathfrak{r})=(\mathfrak{r}^{\kappa}+\rho e^{-\mathfrak{r}\xi})\mathcal{L}\mathrm{u}(\mathfrak{x}))(\mathfrak{r})-\mathfrak{r}^{\kappa-1}\Upsilon(0) \tag{3.4.7}$$

$$-\mathfrak{r}^{\kappa-2}\Upsilon'(0)+\rho e^{-\mathfrak{r}\xi}\int_{-\xi}^{0}e^{-\mathfrak{r}\xi}\Upsilon(\mathfrak{x})\,d\mathfrak{x}-\mathcal{L}(\mathrm{k}(\mathfrak{x}))(\mathfrak{r}),$$

for $\mathfrak{x}\in\mathfrak{F}_1$ and we show Δ is a strictly contractive mapping.

Let $\mathrm{u},\mathrm{v}\in\boldsymbol{Q}$ and consider the coefficient $B_{\mathrm{uv}}\in\mathfrak{F}_4$ with $d(\mathrm{u},\mathrm{v})\preceq B_{\mathrm{uv}}$, thus

$$\mathcal{A}\left(\mathrm{u}(\mathfrak{x})-\mathrm{v}(\mathfrak{x}),B_{\mathrm{uv}}\eta\right)\succeq\boldsymbol{W}_{\kappa,\varsigma}\left(-\frac{|\mathfrak{x}|}{\eta}\right)$$

for all $\mathrm{u},\mathrm{v}\in\boldsymbol{Q}$, $\mathfrak{x}\in\mathfrak{F}_1$ and $\eta\in\mathfrak{F}_2$. Applying (MVF-1-N2) and (MVF-1-N2), we have

$$\mathcal{A}\left(\Delta\mathrm{u}(\mathfrak{x})-\Delta\mathrm{v}(\mathfrak{x}),B_{\mathrm{uv}}\eta\right)=\mathcal{A}\Big((\mathfrak{r}^{\kappa}+\rho e^{-\mathfrak{r}\xi})\mathcal{L}(\mathrm{u}(\mathfrak{x})(\mathfrak{r})-\mathfrak{r}^{\kappa-1}\Upsilon(0) \tag{3.4.8}$$

$$-\mathfrak{r}^{\kappa-2}\Upsilon'(0)+\rho e^{-\mathfrak{r}\xi}\int_{-\xi}^{0}e^{-\mathfrak{r}\xi}\Upsilon(\mathfrak{x})\,d\mathfrak{x}-\mathcal{L}(\mathrm{k}(\mathfrak{x}))(\mathfrak{r})$$

$$-(\mathfrak{r}^{\kappa}+\rho e^{-\mathfrak{r}\xi})\mathcal{L}(\mathrm{v}(\mathfrak{x}))(\mathfrak{r})+\mathfrak{r}^{\kappa-1}\Upsilon(0)$$

$$+\mathfrak{r}^{\kappa-2}\Upsilon'(0)_{\rho}e^{-\mathfrak{r}\xi}\int_{-\xi}^{0}e^{-\mathfrak{r}\xi}\Upsilon(\mathfrak{x})\,d\mathfrak{x}+\mathcal{L}(\mathrm{k}(\mathfrak{x}))(\mathfrak{r}),\eta\Big)$$

$$
\begin{aligned}
&= \mathcal{A}\left((\mathrm{r}^{\kappa} + \rho e^{-\mathrm{r}\xi})\mathcal{L}(\mathrm{u}(\mathfrak{x}))(\mathrm{r}) - (\mathrm{r}^{\kappa} + \rho e^{-\mathrm{r}\xi})\mathcal{L}(\mathrm{v}(\mathfrak{x}))(\mathrm{r}), \eta\right) \\
&= \mathcal{A}\left((\mathrm{r}^{\kappa} + \rho e^{-\mathrm{r}\xi})\mathcal{L}(\mathrm{u}(\mathfrak{x}) - \mathrm{v}(\mathfrak{x}))(\mathrm{r}), \eta\right) \\
&= \mathcal{A}\left(\mathcal{L}(|\mathrm{u}(\mathfrak{x}) - \mathrm{v}(\mathfrak{x})|)(\mathrm{r}), \frac{\eta}{|(\mathrm{r}^{\kappa} + \rho e^{-\mathrm{r}\xi})|}\right) \\
&\succeq \boldsymbol{W}_{\kappa,\varsigma}\left(\frac{-\mathcal{L}(\mathfrak{x})(\mathrm{r})|}{\frac{\eta}{B_{\mathrm{uv}}|(\mathrm{r}^{\kappa}+\rho e^{-\mathrm{r}\xi})|}}\right) \\
&\succeq \boldsymbol{W}_{\kappa,\varsigma}\left(\frac{-|\mathfrak{x}|}{\frac{\eta}{B_{\mathrm{uv}}\theta|(\mathrm{r}^{\kappa}+\rho e^{-\mathrm{r}\xi})|}}\right) \\
&\succeq \boldsymbol{W}_{\kappa,\varsigma}\left(\frac{-|\mathfrak{x}|}{\frac{\eta}{B_{\mathrm{uv}}\theta}}\right),
\end{aligned}
$$

which implies that

$$d(\Delta\mathrm{u}, \Delta\mathrm{v}) \leq \theta B_{\mathrm{uv}},$$

and then

$$d(\Delta\mathrm{u}, \Delta\mathrm{v}) \leq \theta d(\mathrm{u}, \mathrm{v}),$$

where $0 < \theta < 1$, therefore Δ is a contractive mapping.

Step 2. We will show that $d(\Delta(\mathrm{v}), \mathrm{v}) < \infty$. Let $\mathrm{v} \in \mathfrak{L}$, we have

$$
\begin{aligned}
&\mathcal{A}\left(\Delta(\mathrm{v}(\mathfrak{x})) - \mathrm{v}(\mathfrak{x}), \eta\right) \qquad (3.4.9)\\
&= \mathcal{A}\left(\mathcal{L}(D_0^{\kappa}\mathrm{v}(\mathfrak{x}) + \rho\mathrm{v}(\mathfrak{x} - \xi) - \mathrm{k}(\mathfrak{x}) - \mathcal{L}^{-1}(\mathrm{v}(\mathrm{r}))(\mathfrak{x}), \eta\right) \\
&= \mathcal{A}\left((\mathrm{r}^{\kappa} + \rho e^{-\mathrm{r}\xi})\mathcal{L}(\mathrm{v}(\mathfrak{x}))(\mathrm{r}) - \mathrm{r}^{\kappa-1}\Upsilon(0) - \mathrm{r}^{\kappa-2}\Upsilon'(0) + \rho e^{-\mathrm{r}\xi}\int_{-\xi}^{0} e^{-\mathrm{r}\xi}\Upsilon(\mathfrak{x})\,d\mathfrak{x} - \mathcal{L}(\mathrm{k}(\mathfrak{x}))(\mathrm{r}) - \mathrm{v}(\mathfrak{x}), \eta\right) \\
&\succeq \boldsymbol{W}_{\kappa,\varsigma}\left(-\frac{\mathcal{L}(|\mathrm{r}|)(\mathfrak{x})}{\frac{\eta}{\varpi}}\right) \\
&\succeq \boldsymbol{W}_{\kappa,\varsigma}\left(-\frac{|\mathfrak{x}|}{\frac{\eta}{\varpi\theta}}\right).
\end{aligned}
$$

Consequently,

$$d(\Delta\mathrm{v}, \mathrm{v}) \leq \varpi\theta < +\infty, \quad \theta < 1, \qquad (3.4.10)$$

for every $\eta \in \mathfrak{F}_2$. Then we have $d(\Delta\mathrm{v}, \mathrm{v}) < +\infty$.

Therefore, all the conditions of Theorem 1.2 hold. Then, we have

(1) The sequence $\{\Delta^{q}\mathrm{v}\}$ converges to a fixed point such as $\mathfrak{m}$.
(2) The unique element v is in the set $\mathcal{Q}^* = \{\mathrm{v} \in \mathcal{Q} : d(\Delta\mathrm{v}, \mathrm{v}) < \infty\}$ and is the unique fixed point of Δ, it means that $\Delta\mathrm{v} = \mathrm{v}$ or equivalently

$$\begin{aligned}\mathrm{v}(\mathfrak{x}) &= (\mathfrak{r}^{\kappa} + \rho e^{-\mathfrak{r}\xi})\mathcal{L}(\mathrm{u}(\mathfrak{x}))(\mathfrak{r}) - \mathfrak{r}^{\kappa-1}\Upsilon(0) \\ &- \mathfrak{r}^{\kappa-2}\Upsilon'(0) + \rho e^{-\mathfrak{r}\xi}\int_{-\xi}^{0} e^{-\mathfrak{r}\xi}\Upsilon(\mathfrak{x})\,d\mathfrak{x} - \mathcal{L}(\mathrm{k}(\mathfrak{x}))(\mathfrak{r}),\end{aligned} \tag{3.4.11}$$

Since u is a differentiable function, by the NH-FDO-E and according to the Eq. (3.4.11), we have

$$D_0^{\kappa}\mathrm{u}(\mathfrak{x}) = -\rho\mathrm{u}(\mathfrak{x} - \xi) + \mathrm{k}(\mathfrak{x}) + \mathcal{L}^{-1}\mathrm{u}(\mathfrak{x})(\mathfrak{r}), \quad \mathfrak{x} \in \mathfrak{L} := [0, L]. \tag{3.4.12}$$

(3) Using Inequality (3.4.10), we get

$$d(\mathrm{u}, \mathrm{v}) \le \frac{1}{1-\theta} d(\Delta\mathrm{v}, \mathrm{v}) \le \frac{\varpi\theta}{1-\theta},$$

thus, Eq. (3.4.1) has Hyers–Ulam–Wright stability property. Now, we show the uniqueness of the

$$\mathfrak{d} = \frac{\varpi\theta}{1-\theta}.$$

and let h be another differentiable function satisfying Eq. (3.4.12) and this means that the following equation holds

$$D_0^{\kappa}\mathrm{h}(\mathfrak{x}) = -\rho\mathrm{h}(\mathfrak{x} - \xi) + \mathrm{k}(\mathfrak{x}) + \mathcal{L}^{-1}(\mathrm{h}(\mathfrak{r}))(\mathfrak{x}), \quad \mathfrak{x} \in \mathfrak{L} := [0, L]. \tag{3.4.13}$$

We are ready to prove that h is a fixed point of Δ and $\mathrm{h} \in \mathcal{Q}^*$. Using Eq. (3.4.13), we get $\Delta\mathrm{h} = \mathrm{h}$. Now, we show that $d(\Delta\mathrm{v}, \mathrm{h}) < \infty$. Let $\mathrm{v} \in \mathcal{Q}$, $d(\mathrm{v}, \mathrm{h}) < \mathfrak{d}$ and use Eq. (3.4.13), we get

$$\begin{aligned}&\mathcal{A}\Big(\Delta(\mathrm{v}(\mathfrak{x})) - \mathrm{h}(\mathfrak{x}), \eta\Big) \\ &= \Big(\mathcal{L}(D_0^{\kappa}\mathrm{v}(\mathfrak{x}) + \rho\mathrm{v}(\mathfrak{x} - \xi) - \mathrm{k}(\mathfrak{x}))(\mathfrak{r}) - \mathrm{h}(\mathfrak{x}), \eta\Big) \\ &= \mathcal{A}\Big((\mathfrak{r}^{\kappa} + \rho e^{-\mathfrak{r}\xi})\mathcal{L}(\mathrm{v}(\mathfrak{x}))(\mathfrak{r}) - \mathfrak{r}^{\kappa-1}\Upsilon(0) \\ &- s^{\kappa-2}\Upsilon'(0) + \rho e^{-\mathfrak{r}\xi}\int_{-\xi}^{0} e^{-\mathfrak{r}\xi}\Upsilon(\mathfrak{x})\,d\mathfrak{x} - \mathcal{L}(\mathrm{k}(\mathfrak{x}))(\mathfrak{r}) - (\mathfrak{r}^{\kappa} + \rho e^{-\mathfrak{r}\xi})\mathcal{L}(\mathrm{h}(\mathfrak{x}))(\mathfrak{r}) + \mathfrak{r}^{\kappa-1}\Upsilon(0)\end{aligned}$$

$$+ \mathfrak{r}^{\kappa-2}\Upsilon'(0)_{\rho}e^{-\mathfrak{r}\xi}\int_{-\xi}^{0} e^{-\mathfrak{r}\xi}\Upsilon(\mathfrak{x})\,d\mathfrak{x} + \mathcal{L}(\mathrm{k}(\mathfrak{x}))(\mathfrak{r}), \eta\Big)$$
$$= \mathcal{A}\Big((\mathfrak{r}^{\kappa} + \rho e^{-\mathfrak{r}\xi})\mathcal{L}(\mathrm{v}(\mathfrak{x}))(\mathfrak{r}) - (\mathfrak{r}^{\kappa} + \rho e^{-\mathfrak{r}\xi})\mathcal{L}(\mathrm{h}(\mathfrak{x}))(\mathfrak{r}), \eta\Big)$$
$$= \mathcal{A}\Big((\mathfrak{r}^{\kappa} + \rho e^{-\mathfrak{r}\xi})\mathcal{L}(\mathrm{v}(\mathfrak{x}) - \mathrm{h}(\mathfrak{x}))(\mathfrak{r}), \eta\Big)$$
$$\succeq \mathbf{W}_{\kappa,\varsigma}\left(\frac{-|\mathcal{L}(\mathfrak{x})(\mathfrak{r})|}{\frac{\eta}{\mathfrak{d}|(\mathfrak{r}^{\kappa}+\rho e^{-\mathfrak{r}\xi})|}}\right)$$
$$\succeq \mathbf{W}_{\kappa,\varsigma}\left(\frac{-|\mathfrak{x}|}{\frac{\eta}{\mathfrak{d}\theta|(\mathfrak{r}^{\kappa}+\rho e^{-\mathfrak{r}\xi})|}}\right)$$
$$\succeq \mathbf{W}_{\kappa,\varsigma}\left(\frac{-|\mathfrak{x}|}{\frac{\eta}{\mathfrak{d}\theta}}\right),$$

then

$$d(\Delta \mathrm{v}, \mathrm{h}) \leq \theta\mathfrak{d} < +\infty.$$

3.4.3 Application

Now, we provide a numerical example to demonstrate the main results obtained.

Example 3.6 Consider the following the NH-FDO-E

$$D_0^{\frac{1}{2}}\mathrm{u}(\mathfrak{x}) = -\rho\mathrm{u}(\mathfrak{x} - \frac{1}{7}) + \frac{(\mathfrak{x}-1)^{-\frac{1}{4}}\sin(\mathfrak{x}-1)}{64(1+\sqrt{\mathfrak{x}-1})(1+|\mathfrak{x}|)} + \mathcal{L}^{-1}(\mathrm{u}(\mathfrak{r}))(\mathfrak{x}), \tag{3.4.14}$$
$$\mathrm{u}(\mathfrak{x}) = [\mathrm{u}_1(\mathfrak{x}), \mathrm{u}_2(\mathfrak{x})]^T \quad \Upsilon(\mathfrak{x}) = [3\mathfrak{x}, 4\mathfrak{x}^2]^T, \qquad -\xi \leq \mathfrak{x} \leq 0 \tag{3.4.15}$$

where $\kappa = \frac{1}{2}$, $\xi = \frac{1}{7}$ and $\rho = \begin{bmatrix} \frac{3}{5} & 0 \\ 0 & \frac{2}{5} \end{bmatrix}$. Assume that the following condition is true for the given continuous function

▶

$$\mathbf{W}_{\frac{1}{2},\varsigma}\Big(-\frac{|\mathcal{L}((s))(\mathfrak{x})|}{\eta}\Big) \succeq \mathbf{W}_{\frac{1}{2},\varsigma}\Big(-\frac{|\mathfrak{x}|}{4\eta}\Big).$$

Let $\mathrm{v} \in C(\mathfrak{L}, \mathbb{R}^n)$
be a differentiable function such that

$$\mathcal{A}\Big(D_0^{\frac{1}{2}}\mathrm{v}(\mathfrak{x}) + \rho\mathrm{v}(\mathfrak{x} - \frac{1}{7}) - \frac{(\mathfrak{x}-1)^{-\frac{1}{4}}\sin(\mathfrak{x}-1)}{64(1+\sqrt{\mathfrak{x}-1})(1+|\mathfrak{x}|)} - \mathcal{L}^{-1}(\mathrm{u}(s))(\mathfrak{x}), \eta\Big) \tag{3.4.16}$$

$$= \operatorname{diag}\Bigg[W_{\frac{1}{2},\varsigma}\Bigg(-\frac{-|D_0^{\frac{1}{2}}\mathrm{v}(\mathfrak{x})+\rho\mathrm{v}(\mathfrak{x}-\frac{1}{7})-\frac{(\mathfrak{x}-1)^{-\frac{1}{4}}\sin(\mathfrak{x}-1)}{64(1+\sqrt{\mathfrak{x}-1})(1+|\mathfrak{x}|)}-\mathcal{L}^{-1}(\mathrm{u}(\mathfrak{r}))(\mathfrak{x})|}{\eta}\Bigg),$$
$$W_{\frac{1}{2},\varsigma}\Bigg(-\frac{-|D_0^{\frac{1}{2}}\mathrm{v}(\mathfrak{x})+\rho\mathrm{v}(\mathfrak{x}-\frac{1}{7})-\frac{(\mathfrak{x}-1)^{-\frac{1}{4}}\sin(\mathfrak{x}-1)}{64(1+\sqrt{\mathfrak{x}-1})(1+|\mathfrak{x}|)}-\mathcal{L}^{-1}(\mathrm{v}(\mathfrak{r}))(\mathfrak{x})|}{\eta}\Bigg)\Bigg] \tag{3.4.17}$$
$$\succeq W_{\frac{1}{2},\varsigma}\Bigg(-\frac{|\mathfrak{x}|}{\frac{\eta}{\varpi}}\Bigg),$$

then, v is a solution of the inequality

$$\mathcal{A}\Big(\mathrm{v}(\mathfrak{x}) - (\mathfrak{r}^{\kappa}+\rho e^{-\mathfrak{r}\xi})\mathcal{L}(\mathrm{v}(\mathfrak{x}))(\mathfrak{r}) - \mathfrak{r}^{\kappa-1}\Upsilon(0)$$
$$-s^{\kappa-2}\Upsilon'(0) + \rho e^{-\mathfrak{r}\xi}\int_{-\xi}^{0} e^{-\mathfrak{r}\xi}\Upsilon(\mathfrak{x})\,d\mathfrak{x} - \mathcal{L}\big(\frac{(\mathfrak{x}-1)^{-\frac{1}{4}}\sin(\mathfrak{x}-1)}{64(1+\sqrt{\mathfrak{x}-1})(1+|\mathfrak{x}|)}\big)(\mathfrak{r}), \eta\Big)$$
$$= \operatorname{diag}\Bigg[W_{\frac{1}{2},\varsigma}\Bigg(-\frac{|\mathrm{v}(\mathfrak{x})-(\mathfrak{r}^{\kappa}+\rho e^{-\mathfrak{r}\xi})\mathcal{L}(\mathrm{v}(\mathfrak{x}))(\mathfrak{r})-\mathfrak{r}^{\kappa-1}\Upsilon(0)-\mathfrak{r}^{\kappa-2}\Upsilon'(0)+\eta e^{-\mathfrak{r}\xi}\int_{-\xi}^{0}e^{-\mathfrak{r}\xi}\Upsilon(\mathfrak{x})\,d\mathfrak{x}-\mathfrak{L}\big(\frac{(\mathfrak{x}-1)^{-\frac{1}{4}}\sin(\mathfrak{x}-1)}{64(1+\sqrt{\mathfrak{x}-1})(1+|\mathfrak{x}|)}\big)(\mathfrak{r})|}{\eta}\Bigg),$$
$$W_{\frac{1}{2},\varsigma}\Bigg(-\frac{|\mathrm{v}(\mathfrak{x})-(\mathfrak{r}^{\kappa}+\rho e^{-\mathfrak{r}\xi})\mathcal{L}(\mathrm{v}(\mathfrak{x}))(\mathfrak{r})-\mathfrak{r}^{\kappa-1}\Upsilon(0)-\mathfrak{r}^{\kappa-2}\Upsilon'(0)+\rho e^{-\mathfrak{r}\xi}\int_{-\xi}^{0}e^{-\mathfrak{r}\xi}\Upsilon(\mathfrak{x})\,d\mathfrak{x}-\mathcal{L}\big(\frac{(\mathfrak{x}-1)^{-\frac{1}{4}}\sin(\mathfrak{x}-1)}{64(1+\sqrt{\mathfrak{x}-1})(1+|\mathfrak{x}|)}\big)(\mathfrak{r})|}{\eta}\Bigg)\Bigg]$$
$$\succeq \boldsymbol{W}_{\frac{1}{2},\varsigma}\Bigg(-\frac{|\mathcal{L}(\mathfrak{r})(\mathfrak{x})|}{\frac{\eta}{\varpi}}\Bigg)$$
$$\succeq \boldsymbol{W}_{\kappa,\varsigma}\Bigg(-\frac{|\mathfrak{x}|}{\frac{4\eta}{\varpi}}\Bigg),$$

where $\Upsilon(\mathfrak{x}) = [3\mathfrak{x}, 4\mathfrak{x}^2]^T$. Thus we can find a unique differentiable function $\mathrm{u} \in C(\mathfrak{L}, \mathbb{R}^n)$ from (3.4.14) such that for each $\varrho \in [1, 2]$ we have (Fig. 3.4)

$$\Delta(\mathrm{u}(\mathfrak{x})) = (\mathfrak{r}^{\kappa}+\rho e^{-\mathfrak{r}\xi})\mathcal{L}(\mathrm{u}(\mathfrak{x}))(\mathfrak{r}) - \mathfrak{r}^{\kappa-1}\Upsilon(0)$$
$$- \mathfrak{r}^{\kappa-2}\Upsilon'(0) + \rho e^{-\mathfrak{r}\xi}\int_{-\xi}^{0} e^{-\mathfrak{r}\xi}\Upsilon(\mathfrak{x})\,d\mathfrak{x} - \mathcal{L}\big(\frac{(\mathfrak{x}-1)^{-\frac{1}{4}}\sin(\mathfrak{x}-1)}{64(1+\sqrt{\mathfrak{x}-1})(1+|\mathfrak{x}|)}\big)(\mathfrak{r}).$$

Therefore,

$$d(\mathrm{u}, \mathrm{v}) \leq F\,\varpi,$$

and

$$\mathcal{A}\Big(\mathrm{v}(\mathfrak{x}) - \mathrm{u}(\mathfrak{x}), \eta\Big) \succeq \operatorname{diag}\Big[\mathrm{W}_{\kappa,\varsigma}\big(-\frac{|\mathrm{v}(\mathfrak{x})-\mathrm{u}(\mathfrak{x})|}{\eta}\big), \mathrm{W}_{\kappa,\varsigma}\big(-\frac{|\mathrm{v}(\mathfrak{x})-\mathrm{u}(\mathfrak{x})|}{\eta}\big), \mathrm{W}_{\kappa,\varsigma}\big(-\frac{|\mathrm{v}(\mathfrak{x})-\mathrm{u}(\mathfrak{x})|}{\eta}\big)\Big]$$
$$\succeq \boldsymbol{W}_{\kappa,\varsigma}\Bigg(-\frac{|\mathfrak{x}|}{\frac{\eta}{F\varpi}}\Bigg).$$

where in $F = \frac{1}{3}$.

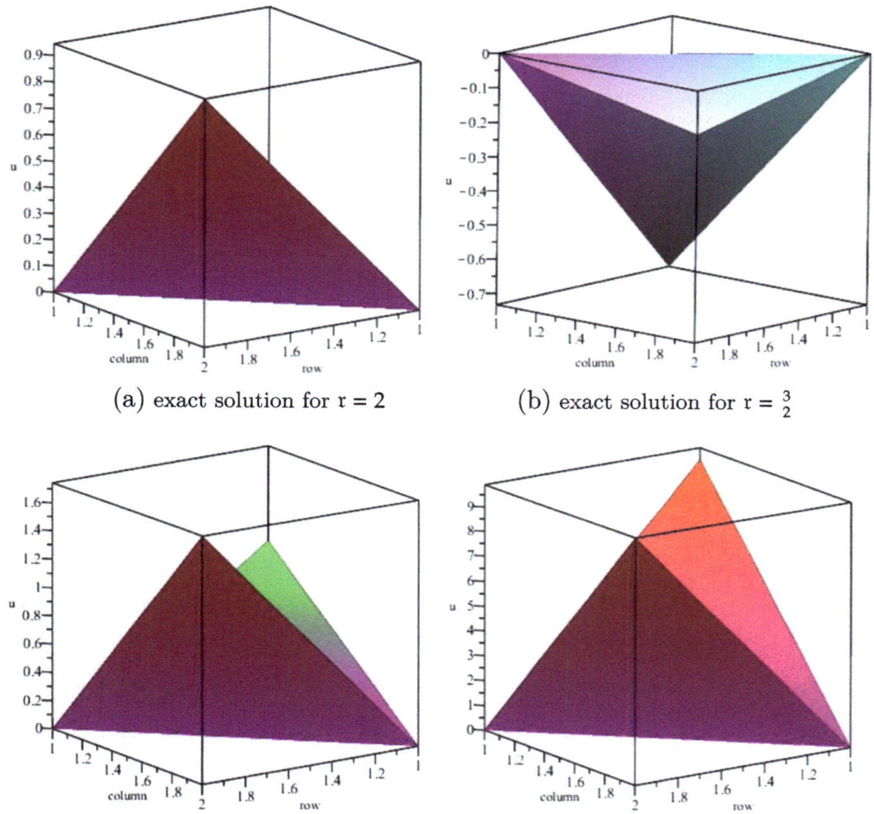

(a) exact solution for $\mathfrak{r} = 2$

(b) exact solution for $\mathfrak{r} = \frac{3}{2}$

(c) exact solution for $\mathfrak{r} = \frac{5}{2}$

(d) exact solution for $\mathfrak{r} = 10$

Fig. 3.4 Graphs related to the exact solution of the NH-FDO-E (3.4.14) for $\xi = \frac{1}{7}, \kappa = \frac{1}{2}$

References

1. Agarwal, R.P., Hristova, S., O'Regan, D.: Lyapunov functions to Caputo reaction-diffusion fractional neural networks with time-varying delays. J. Math. Comput. Sci. **18**(3), 328–345 (2018)
2. Almeida, R., Tavares, D., Torres, D.F.M.: The Variable-Order Fractional Calculus of Variations. Springer Briefs in Applied Sciences and Technology, pp. xiv+124. Springer, Cham (2019)
3. Cădariu, L., Radu, V.: Fixed points and the stability of Jensen's functional equation. JIPAM. J. Inequalities Pure Appl. Math. **4**(1), Article 4, 7 (2003)
4. Cădariu, L., Radu, V.: On the stability of the Cauchy functional equation: a fixed point approach. Grazer Math. Ber. **346**, 43–52 (2004)
5. Chaharpashlou, R., Saadati, R., Atangana, A.: Ulam-Hyers-Rassias stability for nonlinear Ψ-Hilfer stochastic fractional differential equation with uncertainty. Adv. Differ. Equ. **2020**, Paper No. 339, 10
6. Diaz, J.B., Margolis, B.: A fixed point theorem of the alternative, for contractions on a generalized complete metric space. Bull. Am. Math. Soc. **74**, 305–309 (1968)

7. El-Sayed, A.M.A., Al-Issa, S.M.: Existence of integrable solutions for integro-differential inclusions of fractional order; coupled system approach. J. Nonlinear Sci. Appl. **13**(4), 180–186 (2020)
8. El-Moneam, M.A., Ibrahim, T.F., Elamody, S.: Stability of a fractional difference equation of high order. J. Nonlinear Sci. Appl. **12**(2), 65–74 (2019)
9. Golet, I.: Some remarks on functions with values in probabilistic normed spaces. Math. Slovaca **57**(3), 259–270 (2007)
10. Hadzic, O., Pap, E.: Fixed Point Theory in Probabilistic Metric Spaces. Mathematics and its Applications, vol. 536. Kluwer Academic Publishers, Dordrecht (2001)
11. Khan, O., Araci, S., Saif, M.: Fractional calculus formulas for Mathieu-type series and generalized Mittag-Leffler function. J. Math. Comput. Sci. **20**(2), 122–130 (2020)
12. Klement, E.P., Mesiar, R., Pap, E.: Triangular Norms. Trends in Logic–Studia Logica Library, vol. 8. Kluwer Academic Publishers, Dordrecht (2000)
13. Miheţ, D., Radu, V.: On the stability of the additive Cauchy functional equation in random normed spaces. J. Math. Anal. Appl. **343**(1), 567–572 (2008)
14. Saadati, R.: Nonlinear contraction and fuzzy compact operator in fuzzy Banach algebras. Fixed Point Theory **20**(1), 289–297 (2019)
15. Schweizer, B., Sklar, A.: North-Holland Series in Probability and Applied Mathematics. North-Holland Publishing Co., New York (1983)
16. Šerstnev, A.N.: On the notion of a random normed space. Dokl. Akad. Nauk SSSR **149**, 280–283 (1963) (in Russian)
17. Sene, N.: Global asymptotic stability of the fractional differential equations. J. Nonlinear Sci. Appl. **13**(3), 171–175 (2020)
18. Sousa, J.V.D.C., Fabio, G.R., de Oliveira, E.C.: Stability of the fractional Volterra integro-differential equation by means of Ψ-Hilfer operator. Math. Meth. Appl. Sci. **42**, 3033–3043 (2019)
19. Sousa, J.V.D.C., de Oliveira, E.C.: On a new operator in fractional calculus and applications. J. Fixed Point Theory Appl. **20**(3), Paper No. 96, 21 (2018)
20. Wang, J.R., Zhou, Y.: Mittag-Leffler-Ulam stabilities of fractional evolution equations. Appl. Math. Lett. **25**(4), 723–728 (2012)
21. Muniyappan, P., Rajan, S.: Stability of a class of fractional integro-differential equation with nonlocal initial conditions. Acta Math. Univ. Comenian. (N.S.) **87**(1), 85–95 (2018)
22. Selvam, A.G.M., Baleanu, D., Alzabut, J., Vignesh, D., Abbas, S.: On Hyers-Ulam Mittag-Leffler stability of discrete fractional Duffing equation with application on in- verted pendulum. Adv. Differ. Equ. 456 (2020)
23. Alzabut, J., Ahmad, B., Etemad, S., Rezapour, S., Zada, A.: Novel existence techniques on the generalized ϕ-Caputo fractional inclusion boundary problem. Adv. Differ. Equ. 135 (2021)
24. Matar, M.M., Abbas, M.I., Alzabut, J., Kaabar, M.K.A., Etemad, S., Rezapour, S.: Investigation of the p-Laplacian nonperiodic nonlinear boundary value problem via generalized Caputo fractional derivatives. Adv. Differ. Equ. Paper No. 68, 18 (2021)
25. Zhou, H., Alzabut, J., Rezapour, S., Samei, M.E.: Uniform persistence and almost periodic solutions of a nonautonomous patch occupancy model. Adv. Differ. Equ. Paper No. 143, 12 (2020)
26. Rezapour, S., Imran, A., Hussain, A., Martinez, F., Etemad, S., Kaabar, M.K.A.: Condensing functions and approximate endpoint criterion for the existence analysis of quantum integro-difference FBVPs. Symmetry **13**, 469 (2021)
27. Korovina, M.V., Matevossian, H., Smirnov, I.N.: On the asymptotics of solutions of the wave operator with meromorphic coefficients. (Russian) Mat. Zametki **109**(2), 312–317 (2021)
28. Matevossian, H.A.: On the biharmonic problem with the Steklov-type and farwig boundary conditions. Lobachevskii J. Math. **41**(10), 2053–2059 (2020)
29. Al-Issa, S.M., Mawed, N.M.: Results on solvability of nonlinear quadratic integral equations of fractional orders in Banach algebra. J. Nonlinear Sci. Appl. **14**(4), 181–195 (2021)
30. Satsanit, W.: On the solution linear and nonlinear fractional beam equation. J. Nonlinear Sci. Appl. **14**(3), 139–147 (2021)

31. Chaharpashlou, R., Saadati, R.: Best approximation of a nonlinear fractional Volterra integro-differential equation in matrix MB-space. Adv. Differ. Equ. **2021**, 118
32. Asaduzzaman, M., Kilicman, A., Ali, M.Z.: Presence and diversity of positive solutions for a Caputo-type fractional order nonlinear differential equation with an advanced argument. J. Math. Comput. Sci. **23**(3), 230–244 (2021)
33. Ababneh, O.: Adaptive synchronization and anti-synchronization of fractional order chaotic optical systems with uncertain parameters. J. Math. Comput. Sci. **23**(4), 302–314 (2021)
34. Jarad, F., Abdeljawad, T., Hammouch, Z.: On a class of ordinary differential equations in the frame of Atangana-Baleanu fractional derivative. Chaos Solitons Fractals **117**, 16–20 (2018)
35. Arqub, O.A., Maayah, B.: Numerical solutions of integrodifferential equations of Fredholm operator type in the sense of the Atangana-Baleanu fractional operator. Chaos Solitons Fractals **117**, 117–124 (2018)
36. Zhang, J., Wang, G., Zhi, X., Zhou, C.: Generalized Euler-Lagrange equations for fuzzy fractional variational problems under gH-Atangana-Baleanu differentiability. J. Funct. Spaces **2018**, 2740678 (2018)
37. Abdeljawad, T., Baleanu, D.: Discrete fractional differences with nonsingular discrete Mittag-Leffler kernels. Adv. Differ. Eq. **2016**, 232 (2016)
38. Abdeljawad, T., Baleanu, D.: On fractional derivatives with exponential kernel and their discrete versions. Rep. Math. Phys. **80**, 11–27 (2017)
39. Eidinejad, Z., Saadati, R., de la Sen, M.: Radu-mihet method for the existence, uniqueness, and approximation of the ψ-hilfer fractional equations by matrix-valued fuzzy controllers. Axioms **10**, 63 (2021)
40. Eidinejad, Z., Saadati, R., Mesiar, R.: Optimum approximation for ς-lie homomorphisms and jordan ς-lie homomorphisms in ς-lie algebras by aggregation control functions. Mathematics **10**, 1704 (2022)
41. Bayour, B., Torres, D.F.M.: Existence of solution to a local fractional nonlinear differential equation. J. Comput. Appl. Math. **312**, 127–133 (2017)
42. Stojiljković, V., Radojević, S., Çetin, E., Čavić, V.S., Radenović, S.: Sharp bounds for trigonometric and hyperbolic functions with application to fractional calculus. Symmetry **14**, 1260 (2022)
43. Todorčević, V.: Harmonic Quasiconformal Mappings and Hyperbolic Type Metrics. Springer International Publishing, Berlin, Heidelberg, Germany (2019)
44. Gambo, Y.Y., Ameen, R., Jarad, F., Abdeljawad, T.: Existence and uniqueness of solutions to fractional differential equations in the frame of generalized Caputo fractional derivatives. Adv. Differ. Eq. **2018**, 134 (2018)
45. Kilbas, A.A.: Hadamard-type fractional calculus. J. Korean Math. Soc. **38**(6), 1191–1204 (2001)
46. Butzer, P.L., Kilbas, A.A., Trujillo, J.J.: Fractional calculus in the Mellin setting and Hadamard-type fractional integrals. J. Math. Anal. Appl. **269**(1), 1–27 (2002)
47. Khalil, R., Al Horani, M., Yousef, A., Sababheh, M.: A new definition of fractional derivative. J. Comput. Appl. Math. **264**, 65–70 (2014)
48. Abdeljawad, T.: On conformable fractional calculus. J. Comput. Appl. Math. **279**, 57–66 (2015)
49. Kumar, P., Govindaraj, V., Khan, Z.A.: Some novel mathematical results on the existence and uniqueness of generalized Caputo-type initial value problems with delay. AIMS Math. **7**(6), 10483–10494 (2022)
50. Odibat, Z., Erturk, V.S., Kumar, P., Makhlouf, A.B., Govindaraj, V.: An implementation of the generalized differential transform scheme for simulating impulsive fractional differential equations. Math. Probl. Eng. **2022**, (2022)
51. Chung, W.S.: Fractional Newton mechanics with conformable fractional derivative. J. Comput. Appl. Math. **290**, 150–158 (2015)
52. Anderson, D.R., Ulness, D.J.: Newly defined conformable derivatives. Adv. Dyn. Syst. Appl. **10**(2), 109–137 (2015)

53. Al-Refai, M., Abdeljawad, T.: Fundamental results of conformable Sturm-Liouville eigenvalue problems. Complexity **2017**, Art. ID 3720471, 7
54. Eidinejad, Z., Saadati, R., de la Sen, M.: Radu-Mihet method for the existence, uniqueness, and approximation of the ψ-Hilfer fractional equations by matrix-valued fuzzy controllers. Axioms **10**(2), 63 (2021)
55. Ghaffari, A., Saadati, R.: $*$-fuzzy measure model for COVID-19 disease. Adv. Differ. Equ. **2021**, Paper No. 202, 18
56. Hadzic, O., Pap, E., Budincevic, M.: Countable extension of triangular norms and their applications to the fixed point theory in probabilistic metric spaces. Uncertainty modelling, 2001 (Bratislava). Kybernetika (Prague) **38**(3), 363–382 (2002)
57. Klement, E.P., Mesiar, R., Pap, E.: Triangular Norms. Trends in Logic Studia Logica Library, vol. 8. Kluwer Academic Publishers, Dordrecht (2000)
58. Schweizer, B., Sklar, A.: Probabilistic Metric Spaces. North-Holland Series in Probability and Applied Mathematics, pp. xvi+275. North-Holland Publishing Co., New York (1983). ISBN: 0-444-00666-4
59. Hadid, S.B., Luchko, Y.: An operational method for solving fractional differential equations of an arbitrary real order. Panamer. Math. J. **6**, 57–73 (1996)
60. El-Shahed, M., Salem, A.: An extension of Wright function and its properties. J. Math. **2015**, Art. ID 950728, 11
61. Cadariu, L., Radu, V.: Fixed points and the stability of Jensen's functional equation. JIPAM. J. Inequalities Pure Appl. Math. **4**(1), Article 4, 7 (2003)
62. Kiryakova, V.: Some special functions related to fractional calculus and fractional (non-integer) order control systems and equations. Facta Univ. Ser. Autom. Control Robot. **7**(1), 79–98 (2008)
63. Ullah, A., Ullah, Z., Abdeljawad, T., Hammouch, Z., Shah, K.: A hybrid method for solving fuzzy Volterra integral equations of separable type kernels. J. King Saud Univ. Sci. **33**(1), 101246 (2021)
64. Arfan, M., Shah, K., Ullah, A., Abdeljawad, T.: Study of fuzzy fractional order diffusion problem under the Mittag-Leffler Kernel Law. Phys. Scr. **96**(7), 074002 (2021)
65. Arfan, M., Shah, K., Abdeljawad, T., Hammouch, Z.: An efficient tool for solving two-dimensional fuzzy fractional-ordered heat equation. Numer. Methods Part. Differ. Equ. **37**(2), 1407–1418 (2021)
66. Diethelm, K.: The Analysis of Fractional Differential Equations. An Application-Oriented Exposition Using Differential Operators of Caputo Type. Lecture Notes in Mathematics, vol. 2004, pp. viii+247. Springer, Berlin (2010)
67. Kilbas, A.A., Srivastava, H.M., Trujillo, J.J.: Theory and Applications of Fractional Differential Equations. North-Holland Mathematics Studies, vol. 204. Elsevier Science B.V., Amsterdam (2006)
68. Agarwal, R.P., Benchohra, M., Hamani, S.: A survey on existence results for boundary value problems of nonlinear fractional differential equations and inclusions. Acta Appl. Math. **109**(3), 973–1033 (2010)
69. Zhou, Y., Zhang, L.: Existence and multiplicity results of homoclinic solutions for fractional Hamiltonian systems. Comput. Math. Appl. **73**(6), 1325–1345 (2017)
70. Horani, M., Khalil, R., Abdeljawad, T.: Conformable fractional semigroups of operators. Funct. Anal. (2014)
71. Abramowitz, M., Stegun, I.A.: Handbook of Mathematical Functions with Formulas, Graphs, and Mathematical Tables. National Bureau of Standards Applied Mathematics Series, vol. 55 For sale by the Superintendent of Documents, U.S. Government Printing Office, Washington, D.C. (1964)
72. Zhou, Y., Wang, J., Zhang, L.: Basic Theory of Fractional Differential Equations, 2nd edn. [of MR3287248]. World Scientific Publishing Co. Pte. Ltd., Hackensack, NJ (2017)
73. Agarwal, R., Hristova, S., O'Regan, D.: A survey of Lyapunov functions, stability and impulsive Caputo fractional differential equations. Fract. Calc. Appl. Anal. **19**(2), 290–318 (2016)

74. Wang, J., Feckan, M., Zhou, Y.: A survey on impulsive fractional differential equations. Fract. Calc. Appl. Anal. **19**(4), 806–831 (2016)
75. Wang, J., Feckan, M., Zhou, Y.: Center stable manifold for planar fractional damped equations. Appl. Math. Comput. **296**, 257–269 (2017)
76. Wang, J., Feckan, M., Zhou, Y.: Ulam's type stability of impulsive ordinary differential equations. J. Math. Anal. Appl. **395**(1), 258–264 (2012)
77. Khan, O., Araci, S., Saif, M.: Fractional calculus formulas for Mathieu-type series and generalized Mittag-Leffler function. J. Math. Comput. Sci. **20**(2), 122–130 (2020)
78. El-Ajou, A.: Taylor's expansion for fractional matrix functions: theory and applications. J. Math. Comput. Sci. **21**(1), 1–17 (2020)
79. Wang, J., Li, X.: A uniformed method to UlamHyers stability for some linear fractional equations. Mediterr. J. Math. **13**, 625–635 (2016)
80. Li, M., Wang, J.: Finite time stability and relative controllability of Riemann-Liouville fractional delay differential equations. Math. Methods Appl. Sci. **42**(18), 6607–6623 (2019)
81. Zhou, Y., Peng, L.: On the time-fractional Navier-Stokes equations. Comput. Math. Appl. **73**(6), 874–891 (2017)
82. Asaduzzaman, M., Kilicman, A., Ali, M.Z.: Presence and diversity of positive solutions for a Caputo-type fractional order nonlinear differential equation with an advanced argument. J. Math. Comput. Sci. **23**, 230–244 (2021)
83. Horani, M.A.L., Hammad, M.A., Khalilb, R.: Variation of parameters for local fractional nonhomogenous lineardifferential equations. J. Math. Comput. Sci. **16**, 147–153 (2016)
84. Abdeljawad, T., Alzabut, J., Jarad, F.: A generalized Lyapunov-type inequality in the frame of conformable derivatives. Adv. Differ. Equ. **2017**, Paper No. 321, 10
85. Pospisil, M., Pospisiilova, S.L.: Sturm's theorems for conformable fractional differential equations. Math. Commun. **21**(2), 273–281 (2016)
86. Hammad, M.A., Khalil, R.: Abels formula and Wronskian for conformable fractional differential equations. Int. J. Differ. Equ. Appl. **13**, 177–183 (2014)
87. Zheng, A., Feng, Y., Wang, W.: The Hyers Ulam stability of the conformable fractional differential equation. Math. Aeterna **5**, 485–492 (2015)
88. Iyiola, O.S., Nwaeze, E.R.: Some new results on the new conformable fractional calculus with application using D'Alambert approach. Prog. Fract. Differ. Appl. **2**, 17 (2016)
89. Khusainov, D.Y., Shuklin, G.V.: On relative controllability in systems with pure delay. (Russian); translated from Prikl. Mekh. **41**(2), 118–130 (2005). Int. Appl. Mech. **41**(2), 210–221 (2005)
90. Liu, L., Dong, Q., Li, G.: Exact solutions and Hyers-Ulam stability for fractional oscillation equations with pure delay. Appl. Math. Lett. **112**, 106666, 7 (2021)
91. Khusainov, D.Y., Diblik, l., Ruzhichkova, M., Lukacheva, Y.: A representation of the solution of the Cauchy problem for an oscillatory system with pure delay. Nonlinear Oscil. (N. Y.) **11**(2), 276–285 (2008)
92. Li, M., Wang, J.: Finite time stability and relative controllability of Riemann- Liouville fractional delay differential equations. Math. Methods Appl. Sci. **42**(18), 6607–6623 (2019)
93. Li, M., Wang, J.: Exploring delayed Mittag-Leffler type matrix functions to study finite time stability of fractional delay differential equations. Appl. Math. Comput. **324**, 254–265 (2018)
94. Eidinejad, Z., Saadati, R., de la Sen, M.: Radu-Mihet Method for the Existence, Uniqueness, and Approximation of the ψ-Hilfer Fractional Equations by Matrix-Valued Fuzzy Controllers. Axioms **10**(2), 63 (2021)
95. Younus, A., Muhammad, A., Usama, A., Tehmina, B., Abdeljawad, T.: Some fundamental results on fuzzy conformable differential calculus. J. Fract. Calc. Nonlinear Syst. **2**(2), 31–61 (2021)
96. Arfan, M., Shah, K., Ullah, A., Abdeljawad, T.: Study of fuzzy fractional order diffusion problem under the Mittag-Leffler Kernel Law. Phys. Scr. **96**(7), 074002 (2021)
97. Younus, A., Asif, M., Atta, U., Bashir, T., Abdeljawad, T.: Analytical solutions of fuzzy linear differential equations in the conformable setting. J. Fract. Calc. Nonlinear Syst. **2**(2), 13–30 (2021)

98. Arfan, M., Shah, K., Abdeljawad, T., Hammouch, Z.: An efficient tool for solving two-dimensional fuzzy fractional-ordered heat equation. Numer. Methods Part. Differ. Equ. **37**(2), 1407–1418 (2021)
99. Podlubny, I.: Fractional Differential Equations. An Introduction to Fractional Derivatives, Fractional Differential Equations, to Methods of their Solution and Some of their Applications. Mathematics in Science and Engineering, vol. 198. Academic Press, Inc., San Diego, CA (1999)

Chapter 4
Picard Method

4.1 Picard Method for Gauss Hypergeometric Stability

4.1.1 Introduction of the Fractional-Order Differential Equation and Other Preliminaries

We consider the following fractional-order differential equation

$$ {}^{c}D_{\jmath}^{\alpha}\rho(\jmath) = k(\jmath, \rho(\jmath), \rho(h(\jmath))), \quad \jmath \in J \subset \mathbb{R}, \quad \alpha \in (0,1), \tag{4.1.1}$$

we investigate the existence of a unique solution and the Gauss Hypergeometric stability of this fractional-order differential equation (for more detail we refer to [1, 10, 17, 20, 22, 23, 26]). In the above equation, ${}^{c}D_{\jmath}^{\alpha}$ for a function ρ given on the interval $J = (0, q]$, $q\in\mathbb{R}_{+}$ is the Caputo fractional derivative of order α and the functions $k \in C(J \times \mathbb{R}^2, \mathbb{R})$, $h \in C(\bar{J}, \bar{J})$ ($\bar{J}$ is the colure of J) with $h(\jmath) \leq \jmath$. In the first theorem, we investigate the Gauss Hypergeometric stability of the fractional-order differential equation using the Chebyshev norm and in the second theorem, we have proved the Gauss Hypergeometric stability of the equation by using Bielecki norm.

Definition 4.1 Let $\tau > 0$, $k \in C(J \times \mathbb{R}^2, \mathbb{R})$ and $\Gamma(\tau)$ is Gamma function. On an interval $[0, \jmath]$, the Riemann–Liouville fractional integral of order τ is defined by

$$I_{\jmath}^{\tau}k(\jmath) = \frac{1}{\Gamma(\tau)}\int_{0}^{\jmath}\frac{k(\ell)}{(\jmath-\ell)^{1-\tau}}d\ell, \quad \jmath > 0, \ \tau > 0,$$

Definition 4.2 Assume that $n-1 < \tau < n$, $k : [0, \infty) \to \mathbb{R}$. The Riemann–Liouville derivative of order τ and the Caputo derivative of order τ respectively are defined by

Z. Eidinejad et al., *Optimal Stability Theory and Approximate Solutions of Fractional Systems*, Studies in Systems, Decision and Control 606,
https://doi.org/10.1007/978-3-031-96704-7_4

$$ {}^{L}D_{J}^{\tau}k(J) = \frac{1}{\Gamma(n-\tau)} \frac{d^n}{dJ^n} \int_0^J \frac{k(\ell)}{(J-\ell)^{\tau} + 1 - n} d\ell, \quad J > 0. $$

and

$$ {}^{c}D_{J}^{\tau}k(J) ={}^{L} D_{J}^{\tau}\left(k(J) - \sum_{k=0}^{n-1} \frac{J^k}{k!} k^{(k)}(0)\right), \quad J > 0. $$

Remark 4.1 We consider fractional-order differential equation (4.1.1). If w is a differential function, satisfying

$$ |{}^{c}D_{J}^{\alpha}w(t) - k(J, w(J), w(h(J)))| \leq \varepsilon F(\aleph, \beta; \wp; J^{\alpha}) \tag{4.1.2} $$

for $k \in C(J \times \mathbb{R}^2, \mathbb{R})$ and $\varepsilon > 0$, then w also satisfying the following integral inequality:

$$ \begin{aligned} &\left| w(J) - w(0) - \frac{1}{\Gamma(\alpha)} \int_0^J (J-\ell)^{\alpha-1} k(\ell, w(\ell), w(h(\ell))) d\ell \right| \\ &\leq \frac{\varepsilon}{\Gamma(\alpha)} \int_0^J (J-\ell)^{\alpha-1} F(\aleph, \beta; \wp, \ell^{\alpha}) d\ell \\ &\leq \frac{\varepsilon}{\Gamma(\alpha)} \int_0^J (J-\ell)^{\alpha-1} \frac{\Gamma(c)}{\Gamma(\aleph)\Gamma(\beta)} \sum_{k=0}^{\infty} \frac{\Gamma(\aleph+k)\Gamma(\beta+k)}{\Gamma(\wp+k)} \frac{\ell^{\alpha k}}{k!} d\ell \\ &= \frac{\varepsilon}{\Gamma(\alpha)} \frac{\Gamma(\wp)}{\Gamma(\aleph)\Gamma(\beta)} \sum_{k=0}^{\infty} \frac{\Gamma(\aleph+k)\Gamma(\beta+k)}{\Gamma(\wp+k)} \frac{1}{k!} \int_0^J (J-\ell)^{\alpha-1} \ell^{\alpha k} d\ell \\ &= \frac{\varepsilon}{\Gamma(\alpha)} \frac{\Gamma(c)}{\Gamma(a)\Gamma(b)} \sum_{k=0}^{\infty} \frac{\Gamma(\aleph+k)\Gamma(\beta+k)}{\Gamma(\wp+k)} \frac{J^{(k+1)\alpha}\Gamma(k\alpha+1)\Gamma(\alpha)}{\Gamma((k+1)\alpha+1)k!} \\ &\leq \varepsilon \frac{\Gamma(\wp)}{\Gamma(\aleph)\Gamma(\beta)} \sum_{n=0}^{\infty} \frac{\Gamma(\aleph+n)\Gamma(\beta+n)}{\Gamma(\wp+n)} \frac{J^{n\alpha}\Gamma(n\alpha+1)}{\Gamma((n+1)\alpha+1)n!} \\ &\leq \varepsilon \frac{\Gamma(\wp)}{\Gamma(\aleph)\Gamma(\beta)} \sum_{n=0}^{\infty} \frac{\Gamma(\aleph+n)\Gamma(\beta+n)}{\Gamma(\wp+n)} \frac{1}{n!} \frac{J^{n\alpha}\Gamma(n\alpha+1)}{\Gamma(n\alpha+1)} \\ &= \varepsilon F(\aleph, \beta; \wp, J^{\alpha}). \end{aligned} \tag{4.1.3} $$

where F is the Guass hypergeometric series [7] determined by

$$ F(\aleph, \beta; \wp; z) = \sum_{k=0}^{\infty} \frac{(\aleph)_k (\beta)_k}{(\wp)_k} \frac{z^k}{k!} = \frac{\Gamma(\wp)}{\Gamma(\aleph)\Gamma(\beta)} \sum_{k=0}^{\infty} \frac{\Gamma(\aleph+n)\Gamma(\beta+n)}{\Gamma(\wp+n)} \frac{z^k}{k!} $$

Remark 4.2 Assume that w be a function such that $w \in C(J, \mathbb{R})$. Then w is a solution of the inequality (4.1.2) if and only if we can find a function $\tilde{f}_w \in C(J, \mathbb{R})$ such that

- $|\tilde{f}(\jmath)| \leq \varepsilon F(\aleph, \beta; \wp; \jmath^{\alpha})$ for all $\jmath \in J$;
- ${}^{c}D_{\jmath}^{\alpha} w(\jmath) = k(\jmath, w(\jmath), w(h(\jmath))) + \tilde{f}(\jmath)$ for all $\jmath \in J$.

Definition 4.3 We say that the Eq. (4.1.1) is Gauss Hypergeometric stable w.r.t $F(\aleph, \beta; \wp; \jmath^{\alpha})$ if for each $\varepsilon > 0$, there exists a constant value $c_F > 0$, such that for a differentiable function w satisfying (4.1.2), we can find a ρ such that the following holds

$$
{}^{c}D_{\jmath}^{\alpha} \rho(\jmath) = k(\jmath, \rho(\jmath), \rho(h(\jmath))), \quad \jmath \in J \subset \mathbb{R}, \quad \alpha \in (0, 1).
$$

then, we have

$$
|w(\jmath) - \rho(\jmath)| \leq c_F \varepsilon F(\aleph, \beta; \wp; \jmath^{\alpha}), \quad \jmath \in J. \tag{4.1.4}
$$

Definition 4.4 Consider the metric space (Y, δ) and the operator T on Y. Now, T is a Picard operator if there exists $y^* \in Y$ such that

- $E_T = y^*$ wherever $E_T = \{y \in Y : T(y) = y\}$ is the fixed point set of T;
- The sequence $(T^n(y_0))_{n \in \mathbb{N}}$ converges to y^* for all $y_0 \in Y$.

Lemma 4.1 *If we consider an increasing Picard operator $T : Y \longrightarrow Y$, that $(Y, \delta, \leq)$ is an ordered metric space. Therefore, for each $y \in Y$, $y \leq T(y)$ we have $y \leq y_T^*$.*

In the sequel, we express the Gronwall lemma (see Lemma 7.1.1, [9]).

Lemma 4.2 *We consider the continuous functions $\mu, \nu : J \to [0, \infty)$. For the nondecreasing function ν, there are positive constants κ and $0 < \alpha < 1$ such that*

$$
\mu(\jmath) \leq \nu(\jmath) + \kappa \int_0^{\jmath} (\jmath - \ell)^{\alpha - 1} \mu(\ell) d\ell, \quad \jmath \in J,
$$

therefore

$$
\mu(\jmath) \leq \nu(\jmath) + \int_0^{\jmath} \left(\sum_{n=1}^{\infty} \frac{(\kappa \Gamma(\alpha))^n}{\Gamma(n\alpha)} (\jmath - \ell)^{n\alpha - 1} \nu(\ell) \right) d\ell, \quad \jmath \in J. \tag{4.1.5}
$$

Remark 4.3 For continuous and nondecreasing functions $\nu(j)$ and $\mu(\jmath)$ on J, We have:

$$
\mu(\jmath) \leq \nu(\jmath) F(\aleph, \beta; \wp; \kappa \Gamma(\alpha) \jmath^{\alpha}).
$$

4.1.2 Gauss Hypergeometric Stability Using the Chebyshev Norm

Here, we prove the Gauss Hypergeometric stability of Eq. (4.1.1). Before proving the Gauss Hypergeometric stability, we first investigate the existence of a unique solution to the fractional-order differential equation. We prove the first theorem utilizing Chebyshev norm $\|.\|_C$ ($\|\rho\|_C = \max_{\jmath \in J} |\rho(\jmath)|$).

Theorem 4.1 *We assume that the functions* $k \in C(J \times \mathbb{R}^2, \mathbb{R})$ *and* $h \in C(\bar{J}, [a, q])$ *exist, such that* $h(\jmath) \leq \jmath$. *If*

(i) There exists $\mathbf{L}_k > 0$ *such that* $|k(\jmath, u_1, u_2) - k(\jmath, v_1, v_2)| \leq \mathbf{L}_k \sum_{i=1}^{2} |u_i - v_i|$ *for all* $\jmath \in J$, u_i, $v_i \in \mathbb{R}$, $i = 1, 2$

(ii) $\frac{2\mathbf{L}_k q^\alpha}{\Gamma(\alpha+1)} < 1$.

are satisfied, therefore, (4.1.1) has a unique solution in $C(J, \mathbb{R})$ *and is Gauss Hypergeometric stable.*

Proof According to the Eq. (4.1.1), we have

$$\rho(\jmath) = \varphi(0) + \frac{1}{\Gamma(\alpha)} \int_0^J (\jmath - \ell)^{\alpha-1} k(\ell, \rho(\ell), \rho(h(\ell))) d\ell, \quad \jmath \in J. \tag{4.1.6}$$

It is easy to see that (4.1.1) is equivalent to (4.1.6).

We first prove the existence of solution for the Eq. (4.1.6). We define the mapping Λ_k in $Y := C(J, \mathbb{R})$ as

$$\Lambda_k \rho(\jmath) = \varphi(0) + \frac{1}{\Gamma(\alpha)} \int_0^J (\jmath - \ell)^{\alpha-1} k(\ell, \rho(\ell), \rho(h(\ell))) d\ell, \quad \jmath \in J. \tag{4.1.7}$$

Now, we require to demonstration that Λ_k determined in (4.1.7) is a contraction mapping on $Y := C(J, \mathbb{R})$.

According (i), we have

$$\begin{aligned}
&|\Lambda_k \rho(\jmath) - \Lambda_k w(\jmath)| \\
&\quad \leq \frac{1}{\Gamma(\alpha)} \int_0^J (\jmath - \ell)^{\alpha-1} |k(\ell, \rho(\ell), \rho(h(\ell))) - k(\ell, w(\ell), w(h(\ell)))| d\ell \\
&\quad \leq \frac{\mathbf{L}_k}{\Gamma(\alpha)} \int_0^J (\jmath - \ell)^{\alpha-1} \left(\max_{0 \leq \ell \leq q} |\rho(\ell) - w(\ell)| + \max_{0 \leq \ell \leq q} |\rho(h(\ell)) - w(h(\ell))| \right) d\ell \\
&\quad \leq \frac{2\mathbf{L}_k}{\Gamma(\alpha)} \|\rho - w\|_c \int_0^J (\jmath - \ell)^{\alpha-1} d\ell \\
&\quad \leq \frac{2\mathbf{L}_k q^\alpha}{\Gamma(\alpha + 1)} \|\rho - w\|_c,
\end{aligned}$$

for all $\jmath \in J$, which implies that

$$\|\Lambda_k(\rho) - \Lambda_k(w)\|_C \leq \frac{2\mathbf{L}_k q^\alpha}{\Gamma(\alpha + 1)} \|\rho - w\|_C, \quad \rho, w \in C(J, \mathbb{R}).$$

Thus, Λ_k is a contraction mapping w.r.t the Chebyshev norm $\|.\|_C$ on Y. Now, to continue the proof, we utilize the Banach contraction principle.

Now we prove Gauss Hypergeometric stability. We assume that ρ is a unique solution to (4.1.1). Then, for $w \in C(J, \mathbb{R})$ that is a given differentiable function, satisfying (4.1.2), we have

$$\rho(\jmath) = w(0) + \frac{1}{\Gamma(\alpha)} \int_0^{\jmath} (\jmath - \ell)^{\alpha-1} k(\ell, \rho(\ell), \rho(h(\ell)))d\ell, \quad \jmath \in J.$$

Obviously, Remark 4.1 implies that

$$\left| w(\jmath) - w(0) - \frac{1}{\Gamma(\alpha)} \int_0^{\jmath} (\jmath - \ell)^{\alpha-1} k(\ell, w(\ell), w(h(\ell)))d\ell \right| \leq \varepsilon F(\aleph; \beta; \wp; \jmath^{\alpha})$$

for $\jmath \in J$.
Also according to the (i) we have

$$\begin{aligned}
&|w(\jmath) - \rho(\jmath)| \\
&\leq \left| w(\jmath) - w(0) - \frac{1}{\Gamma(\alpha)} \int_0^{\jmath} (\jmath - \ell)^{\alpha-1} k(\ell, w(\ell), w(h(\ell)))d\ell \right| \\
&+ \left| \frac{1}{\Gamma(\alpha)} \int_0^{\jmath} (\jmath - \ell)^{\alpha-1} k(\ell, w(\ell), w(h(\ell)))d\ell \right. \\
&\left. - \frac{1}{\Gamma(\alpha)} \int_0^{\jmath} (\jmath - \ell)^{\alpha-1} k(\ell, \rho(\ell), \rho(h(\ell)))d\ell \right| \\
&\leq \varepsilon F(\aleph, \beta; \wp; \jmath^{\alpha}) + \frac{\mathbf{L}_k}{\Gamma(\alpha)} \left(\int_0^{\jmath} (\jmath - \ell)^{\alpha-1} [|w(\ell) - \rho(\ell)|d\ell + |w(h(\ell)) - \rho(h(\ell))|]d\ell \right),
\end{aligned} \tag{4.1.8}$$

for all $\jmath \in J$. Now, we consider the operator $T : C(J, \mathbb{R}_+) \longrightarrow C(J, \mathbb{R}_+)$, for each $\sigma \in C(J, \mathbb{R}_+)$ as follows

$$T\sigma(\jmath) = \varepsilon F(\aleph, \beta; \wp; \jmath^{\alpha}) + \frac{\mathbf{L}_k}{\Gamma(\alpha)} \left(\int_0^{\jmath} (\jmath - \ell)^{\alpha-1} \sigma(\ell)d\ell + \int_0^{\jmath} (\jmath - \ell)^{\alpha-1} \sigma(h(\ell))d\ell \right), \quad \jmath \in J.$$

In the continuation, we show that T is a Picard operator. By (ii), we have

$$\begin{aligned}
&|T\sigma(\jmath) - T\vartheta(\jmath)| \\
&\leq \frac{\mathbf{L}_k}{\Gamma(\alpha)} \left(\int_0^{\jmath} (\jmath - \ell)^{\alpha-1} [|\sigma(\ell) - \vartheta(\ell)|d\ell + |\sigma(h(\ell)) - \vartheta(h(\ell))|]d\ell \right) \\
&\leq \frac{2\mathbf{L}_k q^{\alpha}}{\Gamma(\alpha+1)} \|\sigma - \vartheta\|_C,
\end{aligned}$$

for all $\jmath \in J$ and for every $\sigma, \vartheta \in C(J, \mathbb{R}_+)$, which implies that

$$\|T(\sigma) - T(\vartheta)\|_C \leq \frac{2\mathbf{L}_k q^{\alpha}}{\Gamma(\alpha+1)} \|\sigma - \vartheta\|_C.$$

Consequently, T w.r.t the Chebyshev norm $\|.\|_C$ is a contraction mapping on $C(J, \mathbb{R}_+)$. Thus, utilizing the Banach contraction principle we obtain that T is a Picard operator and $E_T = \{\sigma^*\}$. Now for $\jmath \in J$

$$\sigma^*(\jmath) = \varepsilon F(\aleph, \beta; \wp; \jmath^\alpha) + \frac{\mathbf{L}_k}{\Gamma(\alpha)}\left(\int_0^\jmath (\jmath - \ell)^{\alpha-1}\sigma^*(\ell)d\ell + \int_0^\jmath (\jmath - \ell)^{\alpha-1}\sigma^*(h(\ell))d\ell\right).$$

In the sequel, we show that the solution σ^* is increasing. Suppose that $\jmath_1, \jmath_2 \in J$ and denote $m := \min_{\ell\in J}[\sigma^*(\ell) + \sigma^*(h(\ell))] \in \mathbb{R}_+$, we have

$$\begin{aligned}
\sigma^*(\jmath_2) - \sigma^*(\jmath_1) &= \varepsilon[F(\aleph, \beta; \wp; \jmath_2^\alpha) - F(\aleph. \beta, \wp; \jmath_1^\alpha)] \\
&\quad \frac{\mathbf{L}_k}{\Gamma(\alpha)}\int_0^{\jmath_1} [(\jmath_2 - \ell)^{\alpha-1} - (\jmath_1 - \ell)^{\alpha-1}][\sigma^*(\ell) + \sigma^*(h(\ell))]d\ell \\
&\quad + \frac{\mathbf{L}_k}{\Gamma(\alpha)}\int_{\jmath_1}^{\jmath_2} (\jmath_2 - \ell)^{\alpha-1}[\sigma^*(\ell) + \sigma^*(h(\ell))]d\ell \\
&\geq \varepsilon[F(\aleph, \beta; \wp; \jmath_2^\alpha) - F(\aleph, \beta; \wp; \jmath_1^\alpha)] + \frac{m\mathbf{L}_k}{\Gamma(\alpha)}\int_0^{\jmath_1} [(\jmath_2 - \ell)^{\alpha-1} - (\jmath_1 - \ell)^{\alpha-1}]d\ell \\
&\quad + \frac{m\mathbf{L}_k}{\Gamma(\alpha)}\int_{\jmath_1}^{\jmath_2} (\jmath_2 - \ell)^{\alpha-1}d\ell \\
&= \varepsilon[F(\aleph, \beta; \wp; \jmath_2^\alpha) - F(\aleph, \beta; \wp; \jmath_1^\alpha)] + \frac{m\mathbf{L}_k}{\Gamma(\alpha+1)}(\jmath_2^\alpha - \jmath_1^\alpha) > 0.
\end{aligned}$$

Then, σ^* is increasing. From $h(\jmath) \leq \jmath$, we obtain

$$\sigma^*(h(\jmath)) \leq \sigma^*(\jmath)$$

and

$$\sigma^*(\jmath) \leq \varepsilon F(\aleph, \beta; \wp; \jmath^\alpha) + \frac{2\mathbf{L}_k}{\Gamma(\alpha)}\int_0^\jmath (\jmath - \ell)^{\alpha-1}\sigma^*(\ell)d\ell.$$

Now, utilizing Lemma 4.2, Remark 4.2 and $c_F := F(\aleph, \beta; \wp; 2\mathbf{L}_k q^\alpha)$, we get

$$\sigma^*(\jmath) \leq c_F \varepsilon F(\aleph, \beta; \wp; \jmath^\alpha), \quad \jmath \in J.$$

Since T is a Picard operator and increasing, by Lemma 4.1 and from (4.1.8), we conclude that $\sigma \leq \sigma^*$. If we consider $\sigma = |w - \rho|$, then

$$|w(\jmath) - \rho(\jmath)| \leq c_F \varepsilon F(\aleph, \beta; wp; \jmath^\alpha), \quad \jmath \in J.$$

Therefore, the Eq. (4.1.1) is Gauss Hypergeometric stable.

4.1.3 Application

In the following, we give an example to illustrate the previous theorem.

Example 4.1 Consider a differential equation for $\alpha = \frac{1}{2}$ and for any $\jmath \in (0, 1]$ $(q = 1)$ as follows

$$ {}^{c}D_{\jmath}^{\frac{1}{2}}\rho(\jmath) = 0.01\rho^{\frac{5}{2}}\sqrt{\pi}(1+\jmath)^{-\frac{3}{2}}, \quad \jmath \in (0, 1]. \tag{4.1.9}$$

According to Eq. (4.1.2) and using the fractional-order differential equation, the following inequality is established for the differentiable function w

$$ |{}^{c}D_{\jmath}^{\frac{1}{2}} w(\jmath) - k(\jmath, w(\jmath), w((1+\jmath)^{-\frac{3}{2}})| \le \varepsilon F(\aleph, \beta; \wp; \jmath^{\frac{1}{2}}).$$

The functions k and h used in the above inequality are considered as follows

$$ h(\Im) = (1+\Im)^{-\frac{3}{2}}, \quad k(\Im, \rho(\Im), \rho(h(\Im))) = 0.01\rho^{\frac{5}{2}}\sqrt{\pi}(1+\Im)^{-\frac{3}{2}}.$$

To be Ulam–Hyers–Hypergeometric stable, we need a positive constant $\mathbf{L}_k$ and a coefficient of Lipschitz $\frac{2\mathbb{L}_k q^{\alpha}}{\Gamma(\alpha+1)}$. Then we consider $\mathbf{L}_k = \frac{2}{5}$ and as a result $\frac{2\mathbb{L}_k q^{\alpha}}{\Gamma(\alpha+1)} = \frac{8}{5\sqrt{\pi}} \approx 0.9 < 1$.

According to Theorem 4.1, all the conditions for the existence of the solution and the Gauss Hypergeometric stability are established. Thus the coefficient $c_F = F(\aleph, \beta; \wp; \frac{4}{5})$ is exist to us such that

$$ |w(\jmath) - \rho(\jmath)| \le c_F \varepsilon F(\aleph, \beta; \wp; \jmath^{\frac{1}{2}}), \quad \jmath \in (0, 1].$$

4.1.4 Gauss Hypergeometric Stability Using the Bielecki's Norm

In the next theorem, we prove the Gauss Hypergeometric stability utilizing the Bielecki's norm $\|.\|_B (\|\rho\|_B := \max_{\jmath \in I} |\rho(\jmath)| \exp(-\eta \jmath), \ \ \eta > 0, \ \ I \subset \mathbb{R}_+)$.

Theorem 4.2 *Assume that the functions $k \in C(J \times \mathbb{R}^2, \mathbb{R})$, $h \in C(\bar{J}, [a, q])$ exist, such that $h(\jmath) \le \jmath$. If*

(i) There exists $\mathbf{L}_k > 0$ such that $|k(\jmath, u_1, u_2) - k(\jmath, v_1, v_2)| \le \mathbf{L}_k \sum_{i=1}^{2} |u_i - v_i|$ for all $\jmath \in J$, u_i, $v_i \in \mathbb{R}$, $i = 1, 2$;

(ii) $\frac{2\mathbf{L}_k q^{\alpha} \exp(\eta q)}{\Gamma(\alpha)\sqrt{2(2\alpha-1)\eta}} < 1$ for some $\alpha \in (\frac{1}{2}, 1)$ and $\eta > 0$

are satisfied, therefore, the Eq. (4.1.1) has a unique solution in $C(J, \mathbb{R})$ and is Gauss Hypergeometric stable.

Proof Here, we prove that, $\Lambda_k : Y \longrightarrow Y$ defined in the Theorem 4.1 with the Bielecki's norm $\|.\|_B$, is a contraction operator. Also, we ignore from saying the similar arguments that expressed in Theorem 4.1. Then, for all $\jmath \in J$, we have

$$\begin{aligned}
&|\Lambda_k\rho(\jmath) - \Lambda_k w(\jmath)| \\
&\leq \frac{1}{\Gamma(\alpha)} \int_0^{\jmath} (\jmath - \ell)^{\alpha-1} |k(\ell, \rho(\ell), \rho(h(\ell))) - k(\ell, w(\ell), w(h(\ell)))| d\ell \\
&\leq \frac{\mathbf{L}_k}{\Gamma(\alpha)} \int_0^{\jmath} (\jmath - \ell)^{\alpha-1} \exp(\eta\ell) \Big(\max_{0\leq\ell\leq q} |\rho(\ell) - w(\ell)| \exp(-\eta\ell) \\
&\quad + \max_{0\leq\ell\leq q} |\rho(h(\ell)) - w(h(\ell))| \exp(-\eta\ell) \Big) d\ell \\
&\leq \frac{2\mathbf{L}_k}{\Gamma(\alpha)} \|\rho - w\|_B \int_0^{\jmath} (\jmath - \ell)^{\alpha-1} \exp(\eta\ell) d\ell.
\end{aligned}$$

Now, we utilize Holder's inequality for $\alpha \in (\frac{1}{2}, 1)$. Then

$$\begin{aligned}
\int_0^{\jmath} (\jmath - \ell)^{\alpha-1} \exp(\eta\ell) &\leq \left(\int_0^{\jmath} (\jmath - \ell)^{2(\alpha-1)} d\ell \right)^{\frac{1}{2}} \left(\int_0^{\jmath} \exp(2\eta\ell) dl \right)^{\frac{1}{2}} \\
&\leq \frac{1}{\sqrt{2\eta}} \frac{q^\alpha}{\sqrt{2\alpha - 1}} \exp(\eta q).
\end{aligned}$$

Therefore,

$$\|\Lambda_k(\rho) - \Lambda_k(w)\|_B \leq \frac{2\mathbf{L}_k q^\alpha \exp(\eta q)}{\Gamma(\alpha)\sqrt{2(2\alpha - 1)\eta}} \|\rho - w\|_B.$$

Consequently, we have

$$\|\Lambda_k(\rho) - \Lambda_k(w)\|_B \leq \frac{2\mathbf{L}_k q^\alpha \exp(\eta q)}{\Gamma(\alpha)\sqrt{2(2\alpha - 1)\eta}} \|\rho - w\|_B, \quad \rho, w \in C(J, \mathbb{R}).$$

Therefore, Λ_k on Y is a contraction mapping w.r.t the Bielecki's norm $\|.\|_B$. Now, we apply the Banach contraction principle to prove (ii). To prove that T in Theorem 4.1 is a Picard operator, we show that T is a contraction. For all $\jmath \in J$, $\alpha \in (\frac{1}{2}, 1)$ and for every $\sigma, \vartheta \in C(J, \mathbb{R}_+)$, we have

$$|T\sigma(\jmath) - T\vartheta(\jmath)| \leq \frac{2\mathbf{L}_k q^\alpha \exp(\eta q)}{\Gamma(\alpha)\sqrt{2(2\alpha - 1)\eta}} \|\rho - w\|_B,$$

then,

$$\|T(\sigma) - T(\vartheta)\|_B \frac{2\mathbf{L}_k q^\alpha \exp(\eta q)}{\Gamma(\alpha)\sqrt{2(2\alpha - 1)\eta}} \|\rho - w\|_B,$$

Consequently, T is a contraction mapping on $C(J, \mathbb{R}_+)$ with Bielecki's norm $\|.\|_B$. The Gauss Hypergeometric stability proof, is exactly the same as (4.1).

4.1.5 Application

Example 4.2 Consider a fractional-order differential equation as

$$ {}^{c}D_{\jmath}^{\frac{2}{3}}\rho(\jmath) = \frac{1}{27}\Gamma(\frac{2}{3})(\jmath - h), \quad \jmath\in(0, \frac{1}{4}], \tag{4.1.10}$$

for $\alpha = \frac{2}{3}$ and for any $\jmath\in(0, \frac{1}{4}]$. Let $\mathbf{L}_k = \frac{1}{2}$ and $\eta = \frac{1}{2}$. Consequently

$$\frac{2\mathbf{L}_k q^{\alpha}\exp(\eta q)}{\Gamma(\alpha)\sqrt{2(2\alpha-1)\eta}} = \frac{(\frac{1}{4})^{\frac{2}{3}}\exp(\frac{1}{8})}{\Gamma(\frac{2}{3})\sqrt{\frac{1}{3}}} < 0.58 < 1.$$

According to Eq. (4.1.2) and using the fractional-order differential equation, for the differentiable function w, we have

$$|{}^{c}D_{\jmath}^{\frac{2}{3}} w(\jmath) - k(\jmath, w(\jmath), w(\jmath - h))| \leq \varepsilon F(\aleph, \beta; \wp; \jmath^{\frac{2}{3}}).$$

All the conditions for the existence of the solution and the Gauss Hypergeometric stability are established By Theorem 4.2. Therefore, the coefficient $c_F = F(\aleph, \beta; \wp; (\frac{1}{4})^{\frac{2}{3}})$ is exist and we have

$$|w(\jmath) - \rho(\jmath)| \leq c_F \varepsilon F(\aleph, \beta; \wp; \jmath^{\frac{2}{3}}), \quad \jmath\in(0, \frac{1}{4}].$$

Remark 4.4 In Theorem 4.2 we examined our problem for $\alpha \in (\frac{1}{2}, 1)$ by utilizing Holder's inequality. At present, we re-examine the problem for $\alpha \in (0, 1)$.

Let $t \in I$, therefore, we have

$$\begin{aligned} |\Lambda_k\rho(\jmath) - \Lambda_k w(\jmath)| &\leq \frac{2\mathbf{L}_k}{\Gamma(\alpha)}\|\rho - w\|_B \int_0^{\jmath} (\jmath - \ell)^{\alpha-1}\exp(\eta\ell)d\ell \\ &\leq \frac{2\mathbf{L}_k q^{\alpha}\exp(\eta q)}{\Gamma(\alpha+1)}\|\rho - w\|_B. \end{aligned}$$

For some $\alpha\in(0, 1)$ and $\eta > 0$, we have $\frac{2\mathbf{L}_k q^{\alpha}\exp(\eta q)}{\Gamma(\alpha+1)} < 1$. Now, we be able demonstrate that Λ_k is a contraction via the Bielecki's norm $\|.\|_B$ on Y as well.

References

1. András, Sz., Kolumbán, J.J.: On the Ulam-Hyers stability of first order differential systems with nonlocal initial conditions. Nonlinear Anal. **82**, 1–11 (2013)
2. András, S., Mészáros, A.R.: Ulam-Hyers stability of dynamic equations on time scales via Picard operators. Appl. Math. Comput. **219**(9), 4853–4864 (2013)
3. Burger, M., Ozawa, N., Thom, A.: On Ulam stability. Isr. J. Math. **193**, 109–129 (2013)
4. Cadariu, L.: Stabilitatea Ulam-Hyers-Bourgin Pentru Ecuatii Functionale. Timiisara: Ed. Univ. Vest Timiisoara (2007)
5. Cimpean, D.S., Popa, D.: Hyers-Ulam stability of Euler's equation. Appl. Math. Lett. **24**(9), 1539–1543 (2011)
6. Gradshteyn, I.S., Ryzhik, I.M.: Table of integrals, series, and products. Translated from the Russian. Sixth edition. Translation edited and with a preface by Alan Jeffrey and Daniel Zwillinger. Academic, Inc., San Diego, CA (2000). xlvii+1163 pp. ISBN: 0-12-294757-6
7. Handbook of mathematical functions with formulas, graphs, and mathematical tables. Edited by Milton Abramowitz and Irene A. Stegun. Reprint of the 1972 edition. Dover Publications, Inc., New York (1992). xiv+1046 pp. ISBN: 0-486-61272-4
8. Hegyi, B., Jung, S.-M.: On the stability of Laplace's equation. Appl. Math. Lett. **26**(5), 549–552 (2013). MR3027761
9. Hyers, D.H., Isac, G., Rassias, T.M.: Stability of functional equations in several variables, vol. 34. Progress in Nonlinear Differential Equations Their Applications. Birkhäuser, Boston (MA) (1998)
10. Ibrahim, R.W.: Generalized Ulam-Hyers stability for fractional differential equations. Internat. J. Math. **23**(5), 1250056, 9 pp (2012)
11. Ibrahim, R.W.: Ulam stability for fractional differential equation in complex domain. Abstr. Appl. Anal. Art. ID 649517, 8 pp (2012)
12. Ibrahim, R.W.: Ulam-Hyers stability for Cauchy fractional differential equation in the unit disk. Abstr. Appl. Anal. Art. ID 613270, 10 pp (2012)
13. Jung, S.-M.: Hyers-Ulam-Rassias stability of functional equations in mathematical analysis. Hadronic Press, Inc., Palm Harbor, FL, 2001. ix+256 pp. ISBN: 1-57485-051-2
14. Jung, S.M.: Hyers-Ulam stability of linear differential equations of first order. Appl. Math. Lett. **17**, 1135–1140 (2004)
15. Kilbas, A.A., Srivastava, H.M., Trujillo, J.J.: Theory and applications of fractional differential equations, vol. 204. Elsevier, North-Holland mathematics studies. Amsterdam (2006)
16. Lungu, N., Popa, D.: Hyers-Ulam stability of a first order partial differential equation. J. Math. Anal. Appl. **385**(1), 86–91 (2012)
17. Otrocol, D., llea, V.: Ulam stability for a delay differential equation. Cent. Eur. J. Math. **11**, 1296–1303 (2013)
18. Rus, I.A.: Gronwall lemma approach to the Hyers-Ulam-Rassias stability of an integral equation. Nonlinear analysis and variational problems, pp. 147–152. Springer Optimization and Its Applications, 35. Springer, New York (2010)
19. Rus, I.A.: Ulam stabilities of ordinary differential equations in a Banach space. Carpathian J. Math. **26**(1), 103–107 (2010)
20. Rus, I.A.: Ulam stability of ordinary differential equations. Studia Univ. Babe's Bolyai Math. **54**, 125–133 (2009)
21. Rus, I.A.: Gronwall lemmas: ten open problems. Sci. Math. Jpn. **70**, 221–228 (2009)
22. Wang, J., Li, X.E.: Ulam type stability of fractional order ordinary differential equations. J. Appl. Math. Comput. (2013). https://doi.org/10.1007/s12190-013-0731-8
23. Wang, J.R., Lv, L., Zhou, Y.: Ulam stability and data dependence for fractional differential equations with Caputo derivative. Electron. J. Qual. Theory Differ. Equ. (63), 10 pp (2011)

24. Wang, J.R., Fečkan, M., Zhou, Y.: Ulam's type stability of impulsive ordinary differential equations. J. Math. Anal. Appl. **395**(1), 258–264 (2012)
25. Wang, J.R., Zhou, Y., Fečkan, M.: Nonlinear impulsive problems for fractional differential equations and Ulam stability. Comput. Math. Appl. **64**(10), 3389–3405 (2012)
26. Wang, J.R., Lv, L.L., Zhou, Y.: New concepts and results in stability of fractional differential equations. Commun. Nonlinear Sci. Numer. Simul. **17**(6), 2530–2538 (2012)

Chapter 5
Cellular Neural Networks, Pseudo Almost Automorphic Solution

In this chapter, using almost automorphic functions, we study the solutions of the integro-differential equations with reflection, which are of pseudo almost automorphic ($\mathcal{PAA}$) type, by introducing the Mittag-Lefler function. In the sequel, using the general measure theory and considering the Mittag-Leffler function, we first obtain the existence of a unique solution for the cellular neural equations and then investigate the stability of the Mittag-Leffler and the attractiveness of the Mittag-Leffler for pseudo almost periodic functions. Considering fuzzy measure theory and matrix-valued fuzzy norm spaces, we study a differential system of non-autonomous cellular neural networks with mixed delays (NA-CNN-MD).

5.1 On the Existence and Uniqueness of Pseuso Almost Automorphic Solutions

5.1.1 Introduction of the Integro Differential Equations with Reflection

Here, the main purpose is to investigate the existence of solutions for the IDE-R, which is defined as follows (considering the continuous functions of $k, \varphi : \mathbb{R}^3 \to \mathbb{R}$ and $\mathcal{L} : \mathbb{R}^+ \to \mathbb{R}^+$):

$$\begin{aligned} \psi'(w) &= \tau\psi(w) + \kappa\psi(-w) + h(w) + k(w, \psi(\vartheta(w)), \psi(\vartheta(-w))) \\ &+ \int_w^{+\infty} \mathcal{L}(z-w)\varphi(z, \psi(\vartheta(w)), \psi(\vartheta(-w)))dw \\ &+ \int_{-w}^{+\infty} \mathcal{L}(z+w)\varphi(z, \psi(\vartheta(z)), \psi(\vartheta(-z)))dz, \quad w \in \mathbb{R}, \end{aligned} \tag{5.1.1}$$

Z. Eidinejad et al., *Optimal Stability Theory and Approximate Solutions of Fractional Systems*, Studies in Systems, Decision and Control 606,
https://doi.org/10.1007/978-3-031-96704-7_5

for $\tau \in \mathbb{R}, \quad \kappa \in \mathbb{R}^*$. Because of the importance of IDE-R in different fields, the stability of these equations has been studied; we refer the reader to [6, 7]. Also, Gupta studied the existence of a unique bounded almost periodic solution for certain kinds of differential equation [8] and in addition two equations

$$\psi'(w) = \tau\psi(w) + \kappa\psi(-w) + h(w), \quad \kappa \neq 0, \quad w \in \mathbb{R}, \tag{5.1.2}$$

and

$$\psi'(w) = \tau\psi(w) + \kappa\psi(-w) + h(w) + k(w, \psi(w), \psi(-w)) \;\; \kappa \neq 0, \quad w \in \mathbb{R}, \tag{5.1.3}$$

were investigated [9] and properties of the solutions which are of weighted pseudo almost periodic type were presented. Also by considering two measures η and μ, the authors studied the solutions of Eqs. (5.1.2) and (5.1.3), which are of η-pseudo almost periodic (η-$\mathcal{PAP}$) type and μ-pseudo almost periodic (μ-$\mathcal{PAP}$) [9].

5.1.2 Preliminaries

Definition 5.1 ([10]) The specific two parametric M-LF is defined as follows

$$E_{\alpha,\theta}(v) = \sum_{\ell=0}^{\infty} \frac{v^{\ell}}{\Gamma(\ell\alpha + \theta)}, \qquad Re(\alpha) > 0, \theta \in \mathbb{C}.$$

Definition 5.2 ([4]) For Banach space $\mathcal{Y}$ and every $k \in C(\mathbb{R}, \mathcal{Y})$, suppose that (z_ℓ) is a real sequence. If there is a sub-sequence $\left(z_{\ell_k}\right)$ such that

$$\lim_{\ell_k \to \infty} k\left(r + z_{\ell_k}\right) = k(r) \text{ and } \lim_{\ell_k \to \infty} h\left(r - z_{\ell_k}\right) = k(r),$$

then, k is said to be almost automorphic or $k \in \mathcal{AA}(\mathbb{R}, \mathcal{Y})$, for every $r \in \mathbb{R}$.

Definition 5.3 ([11]) We consider a ζ-field $\mathcal{Z}$ as type Lebesque of $\mathbb{R}$ and suppose $\mathcal{M}$ is the space of all positive measures on $\mathcal{Z}$. Then $\eta \in \mathcal{M}$ if

(1) $\eta([\tau, \kappa]) < \infty$, for all $\tau \leq \kappa \in \mathbb{R}$,
(2) $\eta(\mathbb{R}) = +\infty$.

Definition 5.4 ([31]) Given the Banach space $\mathcal{Y}$ and the positive measure $\eta \in \mathcal{M}$, a function $k : \mathbb{R} \to \mathcal{Y}$ that is bounded continuous is called η-ergodic, $k \in \mathcal{E}(\mathbb{R}, \mathcal{Y}, \eta)$, if

$$\lim_{s \to \infty} \frac{1}{\eta([-s, s])} \int_{[-s,s]} \|k(w)\| d\eta(w) = 0,$$

where

$$\eta([-\mathrm{s},\mathrm{s}]) := \int_{-\mathrm{s}}^{\mathrm{s}} d\eta(r).$$

Definition 5.5 ([13]) Given the Banach space $\mathcal{Y}$ and the positive measure $\eta \in \mathcal{M}$, a function $k : \mathbb{R} \to \mathcal{Y}$ that is continuous is called η-$\mathcal{PAA}$ if

$$k = h_1 + \mathrm{u}_1,$$

where h_1 is an almost automorphic function ($h_1 \in \mathcal{AA}(\mathbb{R}, \mathcal{Y})$) and u_1 is an ergodic function.

To prove the main results of this paper, we consider the following hypotheses:

($\mathcal{N}_1$) There exist a continuous and increasing function $\vartheta : \mathbb{R} \to \mathbb{R}$ such that for all $v \in \mathcal{AA}(\mathbb{R}, \mathbb{R})$, we have $vo\vartheta \in \mathcal{AA}(\mathbb{R}, \mathbb{R})$.

($\mathcal{N}_2$) For every $\gamma \in \mathbb{R}$, there exist $\vartheta > 0$ and a bounded interval J such that for positive measure η, we have $\eta(\{\tau + \gamma : \tau \in \mathcal{U}\}) \leq \vartheta\eta(\mathcal{U})$, whenever $\mathcal{U} \in \mathcal{Z}$ satisfies $\mathcal{U} \cap J = \emptyset$.

($\mathcal{N}_3$) There exist $\jmath, \ell > 0$ such that for all $\mathcal{U} \in \mathcal{Z}$,

$$\eta(-\mathcal{U}) \leq \jmath + \ell\eta(\mathcal{U}).$$

($\mathcal{N}_4$) There is a function $\rho : \mathbb{R} \to \mathbb{R}^+$ such that for all $\mathrm{E} \in \mathbb{B}(\mathbb{R})$, $\eta \in \mathcal{M}$ and $\eta_\vartheta(\mathrm{E}) = \eta\left(\vartheta^{-1}(\mathrm{E})\right)$ we have $d\eta_\vartheta(r) \leq \rho(r)d\eta(r)$, ρ is also continuous, strictly increasing and

$$\limsup \frac{\eta[-\mathrm{P(a)}, \mathrm{P(a)}]}{\eta[-\mathrm{a}, \mathrm{a}]} \mathrm{Q(P(a))} < +\infty,$$

where $\mathrm{P(a)} = |\vartheta(\mathrm{a})| + |\vartheta(-\mathrm{a})|$ and $\mathrm{Q(P(a))} = \sup_{r \in [-\mathrm{P(a)}, \mathrm{P(a)}]} \rho(\mathrm{P})$.

($\mathcal{N}_5$) Given $\rho = \sqrt{\tau^2 - \kappa^2}$, where $\tau > \kappa$, the following holds

$$\mathrm{D}_1(\rho, \eta) := \sup_{\mathrm{s}>0} \left\{ \int_{-\mathrm{s}}^{\mathrm{s}} \sum_{\ell=0}^{\infty} \frac{(-\rho(r+\mathrm{s}))^\ell}{\Gamma(\ell\alpha + \theta)} d\eta(r) \right\} < \infty,$$

and

$$\mathrm{D}_2(\rho, \eta) := \sup_{\mathrm{s}>0} \left\{ \int_{-\mathrm{s}}^{\mathrm{s}} \sum_{\ell=0}^{\infty} \frac{(-\rho(-r+\mathrm{s}))^\ell}{\Gamma(\ell\alpha + \theta)} d\eta(r) \right\} < \infty.$$

($\mathcal{N}_6$) $k : \mathbb{R} \times \mathbb{R}^2 \to \mathbb{R}$ has a Lipschitz coefficient $\mathrm{H}_k > 0$ such that

$$|k(r, v_1, w_1) - k(r, v_2, w_2)| \leq \mathrm{H}_k (|v_1 - v_2| + |w_1 - w_2|),$$

for all $(v_1, w_1), (v_2, w_2) \in \mathbb{R}^2$.

($\mathcal{N}_7$) $\varphi : \mathbb{R} \times \mathbb{R}^2 \to \mathbb{R}$ has a Lipschitz coefficient $H_\varphi > 0$ such that

$$|\varphi(r, \psi_1, \psi_2) - \psi(r, F_1, F_2)| < H_\varphi(|\psi_1 - F_1| + |\psi_2 - F_2|),$$

for all $\psi_1, \psi_2, F_1, F_2 \in \mathbb{R}$.

($\mathcal{N}_8$) There exists $\mathcal{L} : \mathbb{R}^+ \to \mathbb{R}^+$ such that $g = \int_0^{+\infty} \mathcal{L}(w)dw < \infty$.

To prove the results we consider two states for the Lipschitz coefficients of the functions above. In one state (above) these coefficients are constant and in the second state (below) they are not constant. In the following, we express the necessary conditions according to the second state.

($\mathcal{N}_9$) $k : \mathbb{R} \times \mathbb{R}^2 \to \mathbb{R}$ has a Lipschitz function $H_k \in L^p(\mathbb{R}, \mathbb{R}, dv) \cap L^p(\mathbb{R}, \mathbb{R}, d\eta)$ such that

$$|k(r, v_1, w_1) - k(r, v_2, w_2)| \leq H_k(r)(|v_1 - v_2| + |w_1 - w_2|),$$

where $\eta \in \mathcal{M}$, $p > 1$ and for all $(v_1, w_1), (v_2, w_2) \in \mathbb{R}^2$.

($\mathcal{N}_{10}$) $\varphi : \mathbb{R} \times \mathbb{R}^2 \to \mathbb{R}$ has a Lipschitz function $H_\varphi \in L^p(\mathbb{R}, \mathbb{R}, dv) \cap L^p(\mathbb{R}, \mathbb{R}, d\eta)$ such that

$$|\varphi(r, v_1, w_1) - \varphi(r, v_2, w_2)| \leq H_\varphi(r)(|v_1 - v_2| + |w_1 - w_2|),$$

where $\eta \in \mathcal{M}$, $p > 1$ and for all $(v_1, w_1), (v_2, w_2) \in \mathbb{R}^2$.

($\mathcal{N}_{11}$) There exists $\mathcal{L} : \mathbb{R}^+ \to \mathbb{R}^+$, such that

$$\int_0^{+\infty} (\mathcal{L}(w))^\gamma dw < +\infty, \text{ for all } \gamma > 1.$$

Theorem 5.1 ([12]) *If we assume that condition ($\mathcal{N}_2$) holds, then $\mathcal{PAA}(\mathbb{R}, \mathcal{Y}, \eta)$ ($\eta - \mathcal{PAA}$) is translation invariant and $(\mathcal{PAA}(\mathbb{R}, \mathcal{Y}, \eta), \|.\|_\infty)$ is a Banach space, where $\eta \in \mathcal{M}$.*

Lemma 5.1 ([9]) *If we assume that condition ($\mathcal{N}_3$) holds, then for $h \in \mathcal{PAA}(\mathbb{R}, \mathcal{Y}, \eta)$, we have*

$$h(-r) \in \mathcal{PAA}(\mathbb{R}, \mathcal{Y}, \eta),$$

for every $r \in \mathbb{R}$.

Lemma 5.2 ([9]) *Suppose $\eta \in \mathcal{M}$ is a measure that holds in condition ($\mathcal{N}_2$), then for all $p \geq 1$,*

$$L^p(\mathbb{R}, \mathcal{Y}, d\eta) \subset \mathcal{E}(\mathbb{R}, \mathcal{Y}, \eta).$$

Lemma 5.3 ([14]) *For every $F \in \mathcal{PAA}(\mathbb{R}, \mathbb{R}, \eta)$, if we assume that conditions ($\mathcal{N}_1$) and ($\mathcal{N}_4$) are true, then we have*

$$F(\vartheta(r)) \in (\mathcal{PAA}(\mathbb{R}, \mathbb{R}, \eta)),$$

where $r \in \mathbb{R}$.

Lemma 5.4 ([14]) *For every* $k \in \mathcal{PAA}(\mathbb{R}^3, \mathbb{R}, \eta)$ *and* $F \in \mathcal{PAA}(\mathbb{R}, \mathbb{R}, \eta)$, *if we assume that conditions* $(\mathcal{N}_4)$,$(\mathcal{N}_6)$ *and* $(\mathcal{N}_1)$-$(\mathcal{N}_3)$*are true, then we have*

$$k(r, F(\vartheta(r)), F(\vartheta(-r)) \in (\mathcal{PAA}(\mathbb{R}, \mathbb{R}, \eta)),$$

where $r \in \mathbb{R}$.

Lemma 5.5 ([14]) *For every* $\varphi \in \mathcal{PAA}(\mathbb{R}^3, \mathbb{R}, \eta)$ *and* $F \in \mathcal{PAA}(\mathbb{R}, \mathbb{R}, \eta)$, *if we assume that conditions* $(\mathcal{N}_4)$,$(\mathcal{N}_6)$, $(\mathcal{N}_8)$*and* $(\mathcal{N}_1)$-$(\mathcal{N}_3)$*are true, then we have*

$$\int_r^{\infty} \mathcal{L}(z-r)\varphi(z, F(\vartheta(z)), F(\vartheta(-z))dz \in \mathcal{PAA}(\mathbb{R}, \mathbb{R}, \eta),$$

where $r \in \mathbb{R}$.

Lemma 5.6 ([14]) *For every* $h \in \mathcal{PAA}(\mathbb{R}, \mathbb{R}^2, \eta)$, $\varphi \in \mathcal{PAA}(\mathbb{R} \times \mathbb{R}^2, \mathbb{R}, \eta)$, *if we assume that conditions* $(\mathcal{N}_2)$ *and* $(\mathcal{N}_{10})$ *are true, then*

$$\varphi(r, h(r)) \in \mathcal{PAA}(\mathbb{R}, \mathbb{R}, \eta),$$

where $r \in \mathbb{R}$ *and* $\eta \in \mathcal{M}$.

5.1.3 η-𝒫𝒜𝒜 Solution for Eq. (5.1.1) *in Two-States*

5.1.3.1 The First Stage

Theorem 5.2 *Let* $k, \varphi \in \mathcal{PAP}(\mathbb{R}, \mathbb{R}, \eta)$ *and assume that* $(\mathcal{N}_4)$-$(\mathcal{N}_8)$, $(\mathcal{N}_1)$-$(\mathcal{N}_3)$ *are satisfied. Then Eq.* (5.1.1) *has a unique* η-$\mathcal{PAA}$ *solution if*

$$\frac{|\rho-\tau|+|\rho+\tau|+2|\kappa|}{\rho(\sum_{\ell=0}^{\infty}\frac{(\rho r)^{\ell}}{\Gamma(\ell\alpha+\theta)}\sum_{\ell=0}^{\infty}\frac{(-\rho)^{\ell}r^{\ell+1}}{(\ell+1)\Gamma(\ell\alpha+\theta)})}(\mathrm{H}_k+2g\mathrm{H}_{\varphi})<1.$$

Proof For any $k, \varphi \in \mathcal{PAA}(\mathbb{R}, \mathbb{R}, \eta)$, we consider the following equation:

$$\begin{aligned}\Xi v(r) = &-\frac{1}{2\rho}\left[\sum_{\ell=0}^{\infty}\frac{(\rho r)^{\ell}}{\Gamma(\ell\alpha+\theta)}\int_r^{\infty}\sum_{\ell=0}^{\infty}\frac{(-\rho w)^{\ell}}{\Gamma(\ell\alpha+\theta)}((\rho-\tau)k(w, v(w), v(-w))+\kappa k(-w, v(-w), v(w)))dw\right] \\ &+\frac{1}{2\rho}\left[\sum_{\ell=0}^{\infty}\frac{(-\rho r)^{\ell}}{\Gamma(\ell\alpha+\theta)}\int_{-\infty}^{r}\sum_{\ell=0}^{\infty}\frac{(\rho w)^{\ell}}{\Gamma(\ell\alpha+\theta)}((\rho+\tau)k(w, v(w), v(-w))-\kappa k(-w, v(-w), v(w)))dw\right]\end{aligned} \tag{5.1.4}$$

$$-\frac{1}{2\rho}\left[\sum_{\ell=0}^{\infty}\frac{(\rho r)^{\ell}}{\Gamma(\ell\alpha+\theta)}\int_{r}^{\infty}\sum_{\ell=0}^{\infty}\frac{(-\rho w)^{\ell}}{\Gamma(\ell\alpha+\theta)}((\rho-\tau)h(w)+\kappa h(-w))dw\right]$$
$$+\frac{1}{2\rho}\left[\sum_{\ell=0}^{\infty}\frac{(-\rho r)^{\ell}}{\Gamma(\ell\alpha+\theta)}\int_{-\infty}^{r}\sum_{\ell=0}^{\infty}\frac{(\rho w)^{\ell}}{\Gamma(\ell\alpha+\theta)}((\rho+\tau)h(w)-\kappa h(-w))dw\right],$$

where

$$h(w)=\int_{w}^{+\infty}\mathcal{L}(z-w)\varphi(z,\psi(\vartheta(z)),\psi(\vartheta(-z)))dz + \int_{-w}^{+\infty}\mathcal{L}(z+w)\varphi(z,\psi(\vartheta(z)),\psi(\vartheta(-z)))dz.$$

Now, considering Lemmas 5.1, 5.4 and 5.5 and for each $w\in\mathbb{R}$, we have $\int_{w}^{+\infty}\mathcal{L}(z-w)\varphi(z,\psi(\vartheta(z)),\psi(\vartheta(-z)))dz\in\mathcal{PAA}(\mathbb{R},\mathbb{R},\eta)$, and $\int_{-w}^{+\infty}\mathcal{L}(z+w)\varphi(z,\psi(\vartheta(z)),\psi(\vartheta(-z)))dz\in\mathcal{PAA}(\mathbb{R},\mathbb{R},\eta)$ so this means that $h\in\mathcal{PAA}(\mathbb{R},\mathbb{R},\eta)$. Thus, according to the basic lemmas and theorems presented in the previous section, Ξ is a mapping from $\mathcal{PAA}(\mathbb{R},\mathbb{R},\eta)$ to $\mathcal{PAA}(\mathbb{R},\mathbb{R},\eta)$. Now, we consider the following equation

$$\mathcal{G}(r,F(\vartheta(r)),F(\vartheta(-r))) = k(r,F(\vartheta(r)),F(\vartheta(-r))) + \int_{w}^{+\infty}\mathcal{L}(z-w)\varphi(z,\psi(\vartheta(z)),\psi(\vartheta(-z)))dz + \int_{-w}^{+\infty}\mathcal{L}(z+w)\varphi(z,\psi(\vartheta(z)),\psi(\vartheta(-z)))dz. \tag{5.1.5}$$

We will show that $\Xi:\mathcal{PAA}(\mathbb{R},\mathbb{R},\eta)\to\mathcal{PAA}(\mathbb{R},\mathbb{R},\eta)$ is a contraction. According to $(\boldsymbol{\mathcal{N}_1})$, it follows that $\vartheta:\mathbb{R}\to\mathbb{R}$ is bijective. Then, we have

$$\begin{aligned}
&|\,\mathcal{G}(r,F(\vartheta(r)),F(\vartheta(-r)))-\mathcal{G}(r,\psi(\vartheta(r)),\psi(\vartheta(-r)))\,|\\
&=|k(r,F(\vartheta(r)),F(\vartheta(-r)))-k(r,\psi(\vartheta(r)),\psi(\vartheta(-r)))|\\
&\quad+\int_{r}^{+\infty}\mathcal{L}(z-r)(\varphi(z,F(\vartheta(z)),F(\vartheta(-z)))-\varphi(z,\psi(\vartheta(z)),\psi(\vartheta(-z))))dz\\
&\quad+\int_{-r}^{+\infty}\mathcal{L}(r+z)(\varphi(z,F(\vartheta(z)),F(\vartheta(-z)))-\varphi(z,\psi(\vartheta(z)),\psi(\vartheta(-z))))dz\\
&\le|k(r,F(\vartheta(r)),F(\vartheta(-r)))-k(r,\psi(\vartheta(r)),\psi(\vartheta(-r)))|\\
&\quad+\int_{0}^{+\infty}\mathcal{L}(z)(\varphi((z+r),F(\vartheta(z+r)),F(\vartheta(-(z+r))))-\varphi((z+r,\psi(\vartheta(z+r)),\psi(\vartheta(-(z+r)))))dz\\
&\quad+\int_{0}^{+\infty}\mathcal{L}(z)(\varphi(z-r,F(\varphi(z-r)),F(\vartheta(-(z-r))))-\varphi(z-r,\psi(\vartheta(z-r)),\psi(\vartheta(-(z-r)))))dr\\
&\le 2\left(\mathrm{H}_k+2g\mathrm{H}_\varphi\right)\|F-\psi\|_\infty,
\end{aligned}$$

for all $\psi, F\in\mathcal{PAA}(\mathbb{R},\mathbb{R},\eta)$. Therefore

$$|\Xi F(r)-\Xi\psi(r)|\le\frac{|\rho-\tau|+|\rho+\tau|+2|\kappa|}{\rho(\sum_{\ell=0}^{\infty}\frac{(\rho r)^{\ell}}{\Gamma(\ell\alpha+\theta)}\sum_{\ell=0}^{\infty}\frac{(-\rho)^{\ell}r^{\ell+1}}{(\ell+1)\Gamma(\ell\alpha+\theta)})}(\mathrm{H}_k+2g\mathrm{H}_\varphi)\|F-\psi\|_\infty.$$

Since

$$\frac{|\rho-\tau|+|\rho+\tau|+2|\kappa|}{\rho(\sum_{\ell=0}^{\infty}\frac{(\rho t)^{\ell}}{\Gamma(\ell\alpha+\theta)}\sum_{\ell=0}^{\infty}\frac{(-\rho)^{\ell}r^{\ell+1}}{(\ell+1)\Gamma(\ell\alpha+\theta)})}(\mathrm{H}_k+2g\mathrm{H}_{\varphi})\|F-\psi\|_{\infty}<1,$$

we have that $\Xi : \mathcal{PAA}(\mathbb{R},\mathbb{R},\eta) \to \mathcal{PAA}(\mathbb{R},\mathbb{R},\eta)$ is a contraction and according to the Banach fixed point theorem Ξ has a unique fixed point in $\mathcal{PAA}(\mathbb{R},\mathbb{R},\eta)$. We conclude that (5.1.1) has a unique η-$\mathcal{PAA}$ solution.

5.1.3.2 The Second Stage

Theorem 5.3 *Consider $k,\varphi \in \mathcal{PAP}(\mathbb{R}\times\mathbb{R}^2,\mathbb{R},\eta)$ and assume that conditions ($\mathcal{N}_1$)-($\mathcal{N}_5$) and ($\mathcal{N}_9$)-($\mathcal{N}_{10}$) are satisfied. Then, Eq.* (5.1.1) *has a unique η-$\mathcal{PAA}$ solution if*

$$\|\mathrm{H}_k\|_{L^p(\mathbb{R},\mathbb{R},dv)}+2\Big(\int_0^{+\infty}(\mathcal{L}(w))^q\Big)^{\frac{1}{q}}\|\mathrm{H}_{\varphi}\|_{L^p(\mathbb{R},\mathbb{R},dv)}<\frac{\sum_{\ell=0}^{\infty}\frac{((-\rho)^{\frac{1}{q}}r)^{\ell}}{\rho q^{\frac{1}{q}}\Gamma(\ell\alpha+\theta)}\sum_{\ell=0}^{\infty}\frac{(-\rho)^{\frac{\ell}{q}}r^{\ell+1}}{(\ell+1)\Gamma(\ell\alpha+\theta)}}{|\rho-\tau|+|\rho+\tau|+2|\kappa|},$$

where $\frac{1}{p}+\frac{1}{q}=1$.

Proof (*Here we are considering Lipschitz functions*) Similar to the proof of the above theorem, we consider the Ξ function. As we have proved, Ξ is a mapping from $\mathcal{PAA}(\mathbb{R},\mathbb{R},\eta)$ to $\mathcal{PAA}(\mathbb{R},\mathbb{R},\eta)$. Now, to complete the proof of this theorem we show that Ξ is a contraction. In the following, we use Eq. (5.1.5) introduced in the previous theorem. Then, we have

$$\begin{aligned}
&|\,\mathcal{G}(r,F(\vartheta(r)),F(\vartheta(-r)))-\mathcal{G}(r,\psi(\vartheta(r)),\psi(\vartheta(-r)))\,|\\
&\quad\le|k(r,F(\vartheta(r)),F(\vartheta(-r)))-k(r,\psi(\vartheta(r)),\psi(\vartheta(-r)))|\\
&\qquad+\int_0^{+\infty}\mathcal{L}(z)(\varphi((z+r),F(\vartheta(z+r)),F(\vartheta(-(z+r))))-\varphi((z+r,\psi(\vartheta(z+r)),\psi(\vartheta(-(z+r)))))dz\\
&\qquad+\int_0^{+\infty}\mathcal{L}(z)(\varphi(z-r,F(\vartheta(z-r)),F(\vartheta(-(z-r))))-\varphi(z-r,\psi(\vartheta(z-r)),\psi(\vartheta(-(z-r)))))dz\\
&\quad\le\left[2\mathrm{H}_k(r)+4\Big(\int_0^{+\infty}(\mathcal{L}(w))^q dk\Big)^{\frac{1}{q}}\|\mathrm{H}_{\varphi}\|_{L^p(\mathbb{R},\mathbb{R},dv)}\right]\|F-\psi\|_{\infty},
\end{aligned}$$

where $\psi,F\in\mathcal{PAA}(\mathbb{R},\mathbb{R},\eta)$, and hence

$$\begin{aligned}
|\Xi F(r)-\Xi\psi(r)|\le&\frac{|\rho-\tau|+|\rho+\tau|+2|\kappa|}{\sum_{\ell=0}^{\infty}\frac{((-\rho)^{\frac{1}{q}}r)^{\ell}}{\rho q^{\frac{1}{q}}\Gamma(\ell\alpha+\theta)}\sum_{\ell=0}^{\infty}\frac{(-\rho)^{\frac{\ell}{q}}r^{\ell+1}}{(\ell+1)\Gamma(\ell\alpha+\theta)}}\Big[\|\mathrm{H}_k\|_{L^p(\mathbb{R},\mathbb{R},dv)}\\
&+2\Big(\int_0^{+\infty}(\mathcal{L}(w))^q dw\Big)^{\frac{1}{q}}\|\mathrm{H}_{\varphi}\|_{L^p(\mathbb{R},\mathbb{R},dv)}\Big]\|F-\psi\|_{\infty}.
\end{aligned}$$

Since

$$\frac{|\rho-\tau|+|\rho+\tau|+2|\kappa|}{\sum_{\ell=0}^{\infty}\frac{((-\rho)^{\frac{1}{q}}r)^{\ell}}{\rho q^{\frac{1}{q}}\Gamma(\ell\alpha+\theta)}\sum_{\ell=0}^{\infty}\frac{(-\rho)^{\frac{\ell}{q}}r^{\ell+1}}{(\ell+1)\Gamma(\ell\alpha+\theta)}}\left[\|\mathrm{H}_k\|_{L^p(\mathbb{R},\mathbb{R},dv)}+2\left(\int_0^{+\infty}(\mathcal{L}(w))^q dw\right)^{\frac{1}{q}}\|\mathrm{H}_\varphi\|_{L^p(\mathbb{R},\mathbb{R},dv)}\right]<1,$$

then the operator $\Xi : \mathcal{PAA}(\mathbb{R}, \mathbb{R}, \eta) \to \mathcal{PAA}(\mathbb{R}, \mathbb{R}, \eta)$ is a contraction and according to the Banach fixed point theorem Ξ has a unique fixed point in $\mathcal{PAP}(\mathbb{R}, \mathbb{R}, \eta)$. We conclude that Eq. (5.1.1) has a unique η-$\mathcal{PAA}$ solution.

Proposition 5.1 *If we use the exponential function instead of Mittag-Leffler function, all the conditions and theorems raised are also valid, refer to [14] for details.*

5.1.4 Application

We consider a measure that is defined as follows

$$d\eta(r) = \varpi(r)dr,$$

where $\varpi(r) = \sum_{\ell=0}^{\infty}\frac{(\sin r)^{\ell}}{\Gamma(\alpha\ell+\theta)}$ and $r \in \mathbb{R}$. According to the measure we have defined $\eta \in \mathcal{M}$, then condition $(\boldsymbol{N_2})$ is true. Let $J = [\tau, \kappa]$, Given that $2 + \sin r \geq \sin(-r)$, we have

$$\sum_{\ell=0}^{\infty}\frac{(-1)^{\ell}}{\Gamma(\alpha\ell+\theta)}dr \leq d\eta(r) = \sum_{\ell=0}^{\infty}\frac{(\sin r)^{\ell}}{\Gamma(\alpha\ell+\theta)}dr \leq \sum_{\ell=0}^{\infty}\frac{1}{\Gamma(\alpha\ell+\theta)}dr \Longrightarrow \tag{5.1.6}$$

$$\sum_{\ell=0}^{\infty}\frac{(-1)^{\ell}}{\Gamma(\alpha\ell+\theta)}(\kappa-\tau) \leq \eta(r) \leq \sum_{\ell=0}^{\infty}\frac{1}{\Gamma(\alpha\ell+\theta})(\kappa-\tau) \Longrightarrow$$

$$\sum_{\ell=0}^{\infty}\frac{(-1)^{\ell}}{\Gamma(\alpha\ell+\theta)}(\kappa-\tau) \leq \eta(r) \leq \sum_{\ell=0}^{\infty}\frac{1}{\Gamma(\alpha\ell+\theta)}(\kappa-\tau) \Longrightarrow$$

$$1+\sum_{\ell=0}^{\infty}\frac{2^{\ell}}{\Gamma(\alpha\ell+\theta)}(\sum_{\ell=0}^{\infty}\frac{(-1)^{\ell}}{\Gamma(\alpha\ell+\theta)}(\kappa-\tau)) \leq 1+\sum_{\ell=0}^{\infty}\frac{2^{\ell}}{\Gamma(\alpha\ell+\theta)}\eta(r) \leq$$

$$1+\sum_{\ell=0}^{\infty}\frac{2^{\ell}}{\Gamma(\alpha\ell+\theta)}(\sum_{\ell=0}^{\infty}\frac{1}{\Gamma(\alpha\ell+\theta)}(\kappa-\tau)),$$

then, $1+\sum_{\ell=0}^{\infty}\frac{2^{\ell}}{\Gamma(\alpha\ell+\theta)}\eta(J)\geq\eta(-J)$. Therefore, we conclude that condition $(\mathcal{N}_3)$ holds. Given the main differential integral equation of this paper, i.e. Eq. (5.1.1), we consider the following equation

$$\begin{aligned}v'(r)=&4v(r)+\sqrt{7}v(-r)+\frac{\sum_{\ell=0}^{\infty}\frac{(-|r|)^{\ell}}{\Gamma(\alpha\ell+\theta)}}{15}[\sin v(\frac{1}{r+1}-e)+\cos v(-(\frac{1}{r+1})+e)] \qquad (5.1.7)\\&+\int_{r}^{+\infty}\mathcal{L}(z-r)\frac{\sum_{\ell=0}^{\infty}\frac{(-|z|)^{\ell}}{\Gamma(\alpha\ell+\theta)}}{15}[\sin v(\frac{1}{z+1}-e)+\cos v(-(\frac{1}{z+1})+e)]dz\\&+\int_{-r}^{+\infty}\mathcal{L}(z+r)\frac{\sum_{\ell=0}^{\infty}\frac{(-|z|)^{\ell}}{\Gamma(\alpha\ell+\theta)}}{15}[\sin v(\frac{1}{z+1}-e)+\cos v(-(\frac{1}{z+1})+e)]dz,\end{aligned}$$

where e represents the delay and a positive real number, $\mathcal{L}(z)=\sum_{\ell=0}^{\infty}\frac{(-|z|)^{\ell}}{\Gamma(\alpha\ell+\theta)}$ for all $z\in\mathbb{R}^+$. In order for condition $(\mathcal{N}_1)$ to be true, we set $\vartheta(r)=\frac{1}{r+1}-e$. Next, we put

$$\begin{aligned}\tau=4,\kappa=\sqrt{7},\rho=\sqrt{\tau^2-\kappa^2}=3\text{ and }k(r,v,w)=\varphi(r,v,w)\\=\frac{1}{15}\sum_{\ell=0}^{\infty}\frac{(-|r|)^{\ell}}{\Gamma(\alpha\ell+\theta)}[\sin v+\cos w].\end{aligned}$$

Let $p=q=\frac{1}{2}$. Then

$$|k(r,v_1,w_1)-k(r,v_2,w_2)|\leq \mathrm{H}_k(r)\left(|v_1-v_2|+|y_1-w_2|\right),$$

for all $(v_1,w_1),(v_2,w_2)\in\mathbb{R}^2$, and

$$|\varphi(r,v_1,w_1)-\varphi(r,v_2,w_2)|\leq \mathrm{H}_{\varphi}(r)\left(|v_1-v_2|+|w_1-w_2|\right),$$

for all $(v_1,w_1),(v_2,w_2)\in\mathbb{R}^2$, where

$$\left[r\rightarrow \mathrm{H}_k(r)=\mathrm{H}_{\varphi}(r)=\frac{1}{15}\sum_{\ell=0}^{\infty}\frac{(-|r|)^{\ell}}{\Gamma(\alpha\ell+\theta)}\right]\in L^2(\mathbb{R},\mathbb{R},dv)\cap L^2(\mathbb{R},\mathbb{R},d\eta).$$

We consider $\alpha=1,\theta=1$, and

$$\|\mathrm{H}_k\|_{L^2(\mathbb{R},\mathbb{R},dv)}=\|\mathrm{H}_{\varphi}\|_{L^2(\mathbb{R},\mathbb{R},dv)}=\frac{1}{15}\text{ and }\|\mathrm{H}_k\|_{L^2(\mathbb{R},\mathbb{R},d\eta)}=\|\mathrm{H}_{\varphi}\|_{L^2(\mathbb{R},\mathbb{R},d\eta)}\leq\frac{1}{15}\sqrt{\exp(1)}.$$

This implies that hypothesis $(\mathcal{N}_7)$ is satisfied. Note

$$\begin{aligned}&\|\mathrm{H}_k\|_{L^2(\mathbb{R},\mathbb{R},dv)}+2\left(\int_0^{+\infty}(\mathcal{L}(w))^2dw\right)^{\frac{1}{2}}\|\mathrm{H}_{\varphi}\|_{L^2(\mathbb{R},\mathbb{R},dv)}\\&=\frac{\sqrt{2}+1}{15}<\frac{e^{18}}{|\rho-\tau|+|\rho+\tau|+2|\kappa|}=\frac{e^{18}}{8+2\sqrt{7}}.\end{aligned}$$

Thus, all the conditions of theorem 5.3 are satisfied. Then Eq. (5.1.7) has a unique η-$\mathcal{PAP}$ solution.

5.2 Mittag-Leffler Stability and Attractiveness of Pseudo Almost Periodic Solutions

5.2.1 Introduction of the Delayed Cellular Neural Networks

We shall introduce a neural cellular system and investigate the solutions of this differential equation, which are of the ϕ-pseudo almost periodic type (for more details see [24–26]). Assume that ϕ is a measure, η is a positive measurable function in $\mathbb{R}$, and ϕ_1 is singular Lebesgue measure. Here, ϕ measure is defined as follows

$$d\phi(\mathrm{y}) = \eta(\mathrm{y})d\mathrm{y} + d\phi_1(\mathrm{y}).$$

The cellular neural system with mixed delay is described by

$$\begin{aligned} \mathrm{z}_p'(\mathrm{y}) =& -\mathrm{e}_p(\mathrm{y})\mathrm{z}_p(\mathrm{y}) + \sum_{q=1}^{n} \bar{\vartheta}_{pq}(\mathrm{y})\bar{\Theta}_q\left(\mathrm{z}_q(\mathrm{y})\right) + \sum_{q=1}^{n} \vartheta_{pq}(\mathrm{y})\Theta_q\left(\mathrm{z}_q\left(\mathrm{y}-\zeta_{pq}\right)\right) \\ &+ \sum_{q=1}^{n} \mathrm{h}_{pq}(\mathrm{y}) \int_0^{\infty} \psi_{pq}(\mathrm{r})\widehat{\Theta}_q\left(\mathrm{z}_q(\mathrm{y}-\mathrm{r})\right) d\mathrm{r} + \mathrm{L}_p(\mathrm{y}), \text{ for } \mathrm{y} \in \mathbb{R}. \end{aligned} \tag{5.2.1}$$

This system with the initial value is expressed as follows

$$\mathrm{z}_p'(\mathrm{y}) = -\mathrm{e}_p(\mathrm{y})\mathrm{z}_p(\mathrm{y}) + \sum_{q=1}^{n} \bar{\vartheta}_{pq}(\mathrm{y})\bar{\Theta}_q\left(\mathrm{z}_q(\mathrm{y})\right) + \sum_{q=1}^{n} \vartheta_{pq}(\mathrm{y})\Theta_q\left(\mathrm{z}_q\left(\mathrm{y}-\zeta_{\ell_J}\right)\right) \tag{5.2.2}$$

$$+ \sum_{q=1}^{n} \mathrm{h}_{pq}(\mathrm{y}) \int_0^{\infty} \psi_{pq}(\mathrm{r})\widehat{\Theta}_q\left(\mathrm{z}_q(\mathrm{y}-\mathrm{r})\right) d\mathrm{r} + \mathrm{L}_p(\mathrm{y}), \text{ for } \mathrm{y} \geqslant 0,$$

$$\mathrm{z}_p(\mathrm{y}) = \xi_p(\mathrm{y}), \text{ for } \mathrm{y} \leqslant 0. \tag{5.2.3}$$

The parameters in this equation are

- $\mathrm{z}_p(\mathrm{y})$ is the i-th neuron state,
- $\mathrm{e}_p(\mathrm{y})$ represents the rate of decay,
- $\bar{\Theta}_p$, Θ_p, $\widehat{\Theta}_p$ are the activation functions of the i-th neuron and are real functions,
- $\mathrm{L}_p(\mathrm{y})$ is the input,
- ζ_{pq} are the delays that are constant,
- ψ_{pq} is the transmission delay kernel.

Considering a special case of the stated measure, i.e. $d\phi(\mathrm{y}) = \eta(\mathrm{y})d(\mathrm{y}),\ \phi_1 = 0$, the ϕ-pseudo almost periodic solutions of the above system are of the weighted pseudo almost periodic functions type.

We shall derive some sufficient conditions on existence and uniqueness results for the cellular neural equation [17, 27–29]. We first obtain the unique solution for Eq. (5.2.1). In the sequel, we shall prove our main results, i.e., the Mittag-Leffler stability and the Mittag-Leffler attractiveness for the ϕ-pseudo almost periodic solutions of Eq. (5.2.2), which improve upon and extend [30–36].

We denote the space of all positive measures on Lebesgue ψ-field $\mathcal{A}$ with $\mathcal{N}$. If μ is a positive measure, then we have

- $\phi(\mathbb{R}) = +\infty$,
- $\phi([\ell, \jmath]) < \infty$, for all $\ell, \jmath \in \mathbb{R}(\ell \leqslant \jmath)$.

Considering $\mathcal{BC}(\mathbb{R}, \mathcal{Y})$ as the space of all continuous and bounded functions, as well as the supremum norm $\|\mathrm{g}\|_\infty = \sup_{\mathrm{y}\in\mathbb{R}} \|\mathrm{g}(\mathrm{y})\|$, we have a Banach space.

5.2.2 Preliminaries

Definition 5.6 The Mittag-Leffler function is defined by

$$E_\lambda(\mathrm{g}) = \sum_{\varsigma=0}^{\infty} \frac{\mathrm{g}^\varsigma}{\Gamma(\varsigma\lambda + 1)},$$

where λ is a real number, $\lambda \leq 0$, and g is a complex variable. The generalization of $E_\lambda(\mathrm{g})$ is defined as

$$E_{\lambda,\mu}(\mathrm{g}) = \sum_{\varsigma=0}^{\infty} \frac{\mathrm{g}^\varsigma}{\Gamma(\varsigma\lambda + \mu)},$$

where $\lambda, \mu \in \mathbb{C},\quad Re(\lambda) > 0,\quad Re(\mu) > 0$.

Definition 5.7 If $0 < \lambda \leq 1$ and ν is a complex number, then

$$cosh_\lambda(\nu x^\lambda) = \sum_{\varsigma=0}^{\infty} \frac{\nu^{2\varsigma} x^{2\varsigma\lambda}}{\Gamma(2\varsigma\lambda + 1)}$$

is called λ-order fractional hyperbolic cosine function and

$$sinh_\lambda(\nu x^\lambda) = \sum_{\varsigma=0}^{\infty} \frac{\nu^{2\varsigma+1} x^{(2\varsigma+1)\lambda}}{\Gamma((2\varsigma + 1)\lambda + 1)}$$

is called λ-order fractional hyperbolic sine function.

Proposition 5.2 *Assume that* $0 < \lambda \leq 1$. *Then*

$$cosh_{\lambda}(\nu x^{\lambda}) = \frac{E_{\lambda}(\nu x^{\lambda}) + E_{\lambda}(-\nu x^{\lambda})}{2},$$
$$sinh_{\lambda}(\nu x^{\lambda}) = \frac{E_{\lambda}(\nu x^{\lambda}) - E_{\lambda}(-\nu x^{\lambda})}{2}.$$

Definition 5.8 A continuous function $\mathrm{g} : \mathbb{R} \to \mathcal{Y}$ is an almost periodic function if

$$\|\mathrm{g}(\mathrm{y} + \zeta) - \mathrm{g}(\mathrm{y})\| < \varpi,$$

for all $\mathrm{y} \in \mathbb{R}$ and every $\varpi > 0$ and $\zeta \in [\ell, \jmath]$.

Definition 5.9 Let $\phi \in \mathcal{N}$. A bounded continuous function $\mathrm{g} : \mathbb{R} \to \mathcal{Y}$ is said to be ϕ-ergodic if

$$\lim_{\mathrm{u} \to +\infty} \frac{1}{\phi([-\mathrm{u}, \mathrm{u}])} \int_{-\mathrm{u}}^{\mathrm{u}} \|\mathrm{g}(\mathrm{y})\| d\phi(\mathrm{y}) = 0.$$

Definition 5.10 Suppose that $\phi \in \mathcal{N}$, k and ω are almost periodic and ϕ-ergodic functions, respectively. Then $\mathrm{g} : \mathbb{R} \to \mathcal{Y}$ is a ϕ-pseudo almost periodic function, if $\mathrm{g} = \mathrm{k} + \omega$.

We denote the space of all almost periodically functions by $\mathcal{AP}(\mathbb{R}, \mathcal{Y})$, the space of all ϕ-ergodic functions by $\mathcal{E}(\mathbb{R}, \mathcal{Y}, \phi)$, and the space of all ϕ-pseudo almost periodic functions by $\mathcal{PAP}(\mathbb{R}, \mathcal{Y}, \phi)$. All these spaces, equipped with the supremum norm, are Banach spaces. Also, we have $\mathcal{AP}(\mathbb{R}, \mathcal{Y}) \subset \mathcal{PAP}(\mathbb{R}, \mathcal{Y}, \phi) \subset \mathcal{BC}(\mathbb{R}, \mathcal{Y})$, for more details see [24].

Definition 5.11 Let $\mathrm{z}^*(\mathrm{y}) = \left\{\mathrm{z}_p^*(\mathrm{y})\right\}_{p=1}^{n}$ be a solution of Eq. (5.2.2), with initial value $\left\{\mathrm{z}_p^*(\mathrm{y}) : \mathrm{y} \leqslant 0\right\}$. Suppose that for every solution $\mathrm{z}(\mathrm{y}) = \left\{\mathrm{z}_p(\mathrm{y})\right\}_{p=1}^{n}$ of Eq. (5.2.2) with initial value $\xi = \left\{\xi_p(\mathrm{y})\right\}$, there exist constants $\gamma > 0$ and $W_{\xi} > 1$ such that

$$\left|\mathrm{z}_p(\mathrm{y}) - \mathrm{z}_y^*(\mathrm{y})\right| \leqslant W_{\xi} \left\|\xi - \mathrm{z}^*\right\|_1 \sum_{\varsigma=0}^{\infty} \frac{(-\gamma \mathrm{y})^{\varsigma}}{\Gamma(\varsigma\lambda + 1)},$$

for all $\mathrm{y} > 0, \ p = 1, 2, 3, \ldots, n,$ where

$$\left\|\xi - \mathrm{z}^*\right\|_1 = \sup_{-\infty \leq c \leq 0} \max_{p=1,2,3,\ldots,n} \left|\xi_p(c) - \mathrm{z}_p^*(c)\right|.$$

Then the property of Mittag-Leffler stability is established for z^*.

We can derive the Mittag-Leffler attractiveness from the Mittag-Leffler stability, for more details see [37–43].

Definition 5.12 Let $z^*(y) = \{z_p^*(y)\}_{p=1}^n$ be a solution of Eq. (5.2.2), with initial value $\{z_p^*(y) : y \leqslant 0\}$. Suppose that there exists $\varrho > 0$ such that

$$\lim_{y\to+\infty} \sum_{\varsigma=0}^{\infty} \frac{(\varrho y)^{\varsigma}}{\Gamma(\varsigma\lambda+1)} \left\|z(y) - z^*(y)\right\| = 0,$$

for any solution $z(y) = \{z_p(y)\}_{p=1}^n$ of Eq. (5.2.2). Then the property of Mittag-Leffler attractive is established for z^*.

If the Mittag-Leffler stability for any solution of Eq. (5.2.2) is established, then z depends on its initial value $\{z(y) : -\infty < y \leqslant 0\}$.

Definition 5.13 The convolution of functions v and x from $\mathbb{R}$ to $\mathbb{R}$, if any, is defined as follows

$$(v \star x)(y) := \int_{-\infty}^{+\infty} v(r)x(y-r)dr,$$

where $\phi \in \mathcal{N}$ and for $p, q = 1, 2, 3, \ldots, n$, $\bar{\vartheta}_{pq}, \vartheta_{pq}, h_{pq}, L_p \in \mathcal{PAP}(R, \mathbb{R}, \phi)$ and $e_p \in \mathcal{AP}(R, \mathbb{R})$.

Definition 5.14 ([44]) Let $(G, \mathcal{A})$ be a Borel space. If ϕ and τ are two measures defined on $(G, \mathcal{A})$, we say that ϕ and τ are mutually singular, if there are disjoint sets R and D in $\mathcal{A}$ such that $G = R \cup D$ and $\tau(R) = \phi(D) = 0$.

Definition 5.15 ([44]) Assume that ϕ and τ are the two measures defined on the Borel space $(G, \mathcal{A})$. We say that τ is absolute continuous relative to ϕ if we have

$$(\phi(R) = 0) \Rightarrow (\tau(R) = 0),$$

for each $R \in \mathcal{A}$.

Following Lebesgue-Radon-Nikodym (L-R-N) [44], we assume that $d\phi(y) = \eta(y)dy + d\phi_1$. We impose the following assumptions for every $p = 1, 2, 3, \ldots, n$:

(I₁) $\bar{\Theta}_q, \Theta_q, \Theta_q$ are globaly Lipschitzian with Lipschitz constants $\mathcal{T}_q^{\bar{\Theta}}, \mathcal{T}_q^{\Theta}$ and $\mathcal{T}_q^{\hat{\Theta}}$, respectively.

(I₂) $\psi_{pq} : \mathbb{R}^+ \longrightarrow \mathbb{R}$ is bounded and continuous.

(I₃) There exists $n > 0$ such that

$$\left|\psi_{pq}(y)\right| \sum_{\varsigma=0}^{\infty} \frac{(ny)^{\varsigma}}{\Gamma(\varsigma\lambda+1)}$$

is integrable on $\mathbb{R}^+$.

(I₄) For the bounded interval L and for all $\zeta \in \mathbb{R}$, there is $\epsilon > 0$ such that $\phi(R+\zeta) \leqslant \epsilon\phi(R)$, when $R \in \mathcal{A}$ satisfies $R \cap L = \varnothing$.

($\mathbf{I}_5$) There are $\widehat{\mathtt{e}}_p \in BC(\mathbb{R}, [0, +\infty))$, $\mathtt{o}_p > 0$, such that

$$\sum_{\varsigma=0}^{\infty} \frac{(-\int_{\mathtt{b}}^{\mathtt{y}} \mathtt{e}_p(\mathtt{r})d\mathtt{r})^{\varsigma}}{\Gamma(\varsigma\lambda+1)} \leqslant \mathtt{o}_p \sum_{\varsigma=0}^{\infty} \frac{(-\int_{\mathtt{b}}^{\mathtt{y}} \widehat{\mathtt{e}}_p(\mathtt{r})d\mathtt{r})^{\varsigma}}{\Gamma(\varsigma\lambda+1)}, \quad \forall \mathtt{y}, \mathtt{b} \in \mathbb{R}, \quad \mathtt{y} \geqslant \mathtt{b}$$

and

$$\hat{\mathtt{e}}_p^+ = \sup_{\mathtt{y}\in\mathbb{R}} \hat{\mathtt{e}}_p(\mathtt{y}), \quad \widehat{\mathtt{e}}_p^- = \inf_{\mathtt{y}\in R} \hat{\mathtt{e}}_p(\mathtt{y}) > 0.$$

($\mathbf{I}_6$) There are $\widehat{\mathtt{e}}_p \in \mathcal{BC}(\mathbb{R}, [0, +\infty))$, $\mathtt{o}_p > 0$, $\kappa_p > 0$ and $\sigma_p > 0$ such that

$$\sup_{\mathtt{r}\in\mathbb{R}} \left\{-\hat{\mathtt{e}}_p(\mathtt{y}) + \mathtt{o}_p \left[\sigma_p^{-1} \sum_{q=1}^{n} \left(|\bar{\vartheta}_p(\mathtt{y})| \mathcal{T}_q^{\bar{\Theta}} + |\vartheta_{pq}(\mathtt{y})| \mathcal{T}_q^{\sigma} + \hat{\mathcal{T}}_q^{\hat{\Theta}} |\mathtt{h}_p(\mathtt{y})| \int_0^{\infty} |\psi_p(\mathtt{r})| d\mathtt{r}\right) \sigma_q\right]\right\} \ll -\kappa_p < 0.$$

($\mathbf{I}_7$) There are $\widehat{\mathtt{e}}_p \in \mathcal{BC}(\mathbb{R}, [0, +\infty))$, $\mathtt{o}_p > 0$, $\kappa_p > 0$ and $\sigma_p > 0$ such that

$$0 < \max_{p=1,2,3,\ldots,n} \left\{ \frac{\widehat{\mathtt{e}}_p^+}{\widehat{\mathtt{e}}_p^-} - \frac{\kappa_p}{\widehat{\mathtt{e}}_p^+} \right\} < 1.$$

($\mathbf{I}_8$) For all $\ell, \jmath, \imath \in \mathbb{R}$ such that $0 \leqslant \ell < \jmath \leqslant \imath$, there exist $\zeta_0 \geqslant 0$ and $\lambda_0 > 0$ such that

$$|\zeta| \geqslant \zeta_0 \Rightarrow \phi((\ell+\zeta, \jmath+\zeta)) \geqslant \lambda_0 \phi([\zeta, \imath+\zeta]).$$

Hypothesis ($\mathbf{I}_4$) implies ($\mathbf{I}_8$) and the converse assertion is not true. Also if assumption ($\mathbf{I}_8$) holds, then $(\mathcal{PAP}(\mathbb{R}, \mathbb{R}^n, \phi), \|\cdot\|_{\infty})$ is a Banach space. If ($\mathbf{I}_4$) holds, for $\mathtt{g} \in \mathcal{PAP}(\mathbb{R}, \mathbb{R}^n, \phi)$ and every $\zeta \in \mathbb{R}$, $\mathtt{g}(\mathtt{y}-\zeta) \in \mathcal{PAP}(\mathbb{R}, \mathbb{R}^n, \phi)$ and for $\mathtt{U} \in \mathcal{E}(\mathbb{R}, \mathbb{R}, \phi)$ and $\mathtt{E} \in L^1(\mathbb{R})$ we have $\mathtt{E} \star \mathtt{U} \in \mathcal{E}(\mathbb{R}, \mathbb{R}, \phi)$. The proofs can be found in [24].

Theorem 5.4 ([45]) *For any integrable function* $g : \mathbb{R} \longrightarrow \mathbb{R}$ *such that* $\mathtt{g} \in \mathcal{AP}(\mathbb{R}, \mathbb{R})$*, we have* $\mathtt{g} \star \mathtt{k} \in \mathcal{AP}(\mathbb{R}, \mathbb{R})$.

Theorem 5.5 ([46]) *For any* ζ *on the interval with positive length* l_{ϱ}*, and any* $\varrho > 0$,

$$\|\mathtt{g}(\mathtt{y}-\zeta) - \mathtt{g}(\mathtt{y})\| < \varrho,$$

and

$$\|\mathtt{k}(\mathtt{y}-\zeta) - \mathtt{k}(\mathtt{y})\| < \varrho,$$

where $\mathtt{g}, \mathtt{k} \in \mathcal{AP}(\mathbb{R}, \mathbb{R})$ *and for all* $\mathtt{y} \in \mathbb{R}$*. In particular* $\mathtt{kg} \in \mathcal{AP}(\mathbb{R}, \mathbb{R})$.

Remark 5.1 For a globally Lipschitzian mapping $Q : \mathcal{Y} \to \mathcal{Z}$ such that $\mathcal{Y}$ and $\mathcal{Z}$ are two Banach spaces and every almost periodic functions ω, we have $Q \circ \omega \in \mathcal{AP}(\mathbb{R}, \mathcal{Y})$, which means that $Q \circ \omega$ is a almost periodic function.

5.2.3 *Measures with Conditions* (**I**$_4$) *and* (**I**$_8$)

Next, we shall introduce three examples of measures that satisfy conditions (**I**$_4$) and (**I**$_8$).

Example 5.1 ([24]) Due to the measure of the Lebesgue type, we consider a measure $\phi \in \mathcal{N}$ which is not absolutely continuous and satisfying (**I**$_8$). This measure is defined as $d\phi(\mathrm{y}) = d\mathrm{y} + d\mathfrak{x}$, where $d\mathrm{y}$ is a measure of the Lebesgue type. Also, $\mathfrak{x}$ is the measure on ($\mathbb{R}$, $\mathcal{A}$), which in $\mathcal{A}$ is the ψ-field of the Lebesgue type. This measure is defined as follows

$$\mathfrak{x}(\mathrm{R}) = \begin{cases} \mathrm{card}(\mathrm{R} \cap \mathbb{Z}) & \text{if } \mathrm{R} \cap \mathbb{Z} \text{ is infinite,} \\ \infty & \text{if } \mathrm{R} \cap \mathbb{Z} \text{ is infinite.} \end{cases}$$

Example 5.2 Consider the following measure

$$d\phi_{\psi,\varkappa}(\mathrm{y}) = \sum_{\varsigma=0}^{\infty} \frac{(\psi \mathrm{y})^{\varsigma}}{\Gamma(\varsigma\lambda+1)} d\mathrm{y} + \varkappa \sum_{n=-\infty}^{\infty} \left(\sum_{\varsigma=0}^{\infty} \frac{(\psi n)^{\varsigma}}{\Gamma(\varsigma\lambda+1)} \right) \delta_n,$$

where $\psi \geqslant 0$, $\varkappa > 0$ and according to the integer n, δ_n is a Dirac measure (DM), and $\sum_{n=-\infty}^{\infty} \left(\sum_{\varsigma=0}^{\infty} \frac{(\psi n)^{\varsigma}}{\Gamma(\varsigma\lambda+1)} \right) \delta_n$ is a generalized Dirac comb (GDC). In DCM when $\psi = 0$, this measure is called Dirac comb (DC). Since $[\zeta, \mathrm{b}+\zeta] \subset [\zeta, [\mathrm{b}]+1+\mathrm{b}]$, we shall show that (**I**$_8$) is established for $\mathrm{b} > 0$, such that $\mathrm{b} \geq \jmath$.

$$\begin{aligned}
\phi_{\psi,\varkappa}([\zeta, \mathrm{b}+\zeta]) &= \int_{\zeta}^{\iota+\zeta} \sum_{\varsigma=0}^{\infty} \frac{(\psi \mathrm{y})^{\varsigma}}{\Gamma(\varsigma\lambda+1)} d\mathrm{y} + \varkappa \sum_{n\in[\zeta,\mathrm{b}+\zeta]} \left(\sum_{\varsigma=0}^{\infty} \frac{(\psi n)^{\varsigma}}{\Gamma(\varsigma\lambda+1)} \right) \\
&\leqslant \sum_{\varsigma=0}^{\infty} \frac{\psi^{\varsigma}\left(\frac{(\mathrm{b}+\zeta)^{\varsigma+1}}{\varsigma+1} - \frac{\zeta^{\varsigma+1}}{\varsigma+1} \right)}{\Gamma(\varsigma\lambda+1)} + \varkappa \sum_{n=[\zeta]}^{\mathrm{b}+\zeta} \left(\sum_{\varsigma=0}^{\infty} \frac{(\psi n)^{\varsigma}}{\Gamma(\varsigma\lambda+1)} \right) \\
&= \sum_{\varsigma=0}^{\infty} \frac{\psi^{\varsigma}\left(\frac{(\mathrm{b}+\zeta)^{\varsigma+1}}{\varsigma+1} - \frac{\zeta^{\varsigma+1}}{\varsigma+1} \right)}{\Gamma(\varsigma\lambda+1)} + \varkappa \sum_{\varsigma=0}^{\infty} \frac{(\psi\zeta)^{\varsigma}}{\Gamma(\varsigma\lambda+1)} \sum_{z=0}^{\mathrm{b}} \left(\sum_{\varsigma=0}^{\infty} \frac{(\psi z)^{\varsigma}}{\Gamma(\varsigma\lambda+1)} \right) \\
&= \sum_{\varsigma=0}^{\infty} \frac{(\psi)^{\varsigma}}{\Gamma(\varsigma\lambda+1)} \left[\left(\frac{(\mathrm{b}+\zeta)^{\varsigma+1}}{\varsigma+1} - \frac{\zeta}{\zeta+1} \right) + \varkappa(\zeta)^{\varsigma} \sum_{z=0}^{\mathrm{b}} \left(\sum_{\varsigma=0}^{\infty} \frac{(\psi z)^{\varsigma}}{\Gamma(\varsigma\lambda+1)} \right) \right].
\end{aligned}$$

We also have that

$$\phi_{\psi,\varkappa}(\ell+\zeta, \jmath+\zeta) \geqslant \int_{\ell+\zeta}^{\jmath+\zeta} \sum_{\varsigma=0}^{\infty} \frac{(\psi \mathrm{y})^{\varsigma}}{\Gamma(\varsigma\lambda+1)} d\mathrm{y} = \sum_{\varsigma=0}^{\infty} \frac{(\psi)^{\varsigma}}{\Gamma(\varsigma\lambda+1)} \left(\frac{(\jmath+\zeta)^{\varsigma+1}}{\varsigma+1} - \frac{(\ell+\zeta)^{\varsigma+1}}{\varsigma+1} \right).$$

The conclusion follows with $\zeta_0 = 0$ and

$$\varrho_0 = \sum_{\varsigma=0}^{\infty} \frac{(\psi)^{\varsigma}}{\Gamma(\varsigma\lambda+1)} \left(\frac{(\jmath+\zeta)^{\varsigma+1}}{\varsigma+1} - \frac{(\ell+\zeta)^{\varsigma+1}}{\varsigma+1} \right)$$
$$/ \left(\sum_{\varsigma=0}^{\infty} \frac{(\psi)^{\varsigma}}{\Gamma(\varsigma\lambda+1)} \left[\left(\frac{(\mathrm{b}+\zeta)^{\varsigma+1}}{\varsigma+1} - \frac{\zeta}{\zeta+1} \right) + \varkappa(\zeta)^{\varsigma} \sum_{z=0}^{\mathrm{b}} \left(\sum_{\varsigma=0}^{\infty} \frac{(\psi z)^{\varsigma}}{\Gamma(\varsigma\lambda+1)} \right) \right] \right).$$

This means that $\left(\mathcal{PAP}\left(\mathbb{R}, \mathbb{R}^n, \phi_{\psi,\varkappa}\right), \|.\|_\infty\right)$ is a Banach space. The measure $\phi_{\psi,\varkappa}$ does not apply to $(\mathbf{I}_4)$. In the following, we shall prove this. Let

$$\mathrm{D} = \cup_{n=-\infty}^{\infty} \left(n - \frac{1}{2} - \varrho_n, n - \frac{1}{2} + \varrho_n \right),$$

with

$$\varrho = \frac{1}{\psi} \sinh^{-1} \left(\frac{1}{2^{|n|+1} \sum_{\varsigma=0}^{\infty} \frac{(\psi n)^{\varsigma}}{\Gamma(\varsigma\lambda+1)}} \right)$$

and $\zeta = \frac{1}{2}$. Then

$$\mathrm{D} + \zeta = \cup_{n=-\infty}^{\infty} (n - \varrho_n, n + \varrho_n)$$

contains $\mathbb{Z}$ and

$$\phi_{\psi,\varkappa}(\mathrm{D}+\zeta) = \sum_{n=-\infty}^{\infty} \int_{n-\varrho_n}^{n+\varrho_n} \sum_{\varsigma=0}^{\infty} \frac{(\psi \mathrm{y})^{\varsigma}}{\Gamma(\varsigma\lambda+1)} d\mathrm{y} + \sum_{n=-\infty}^{\infty} \sum_{\varsigma=0}^{\infty} \frac{(\psi n)^{\varsigma}}{\Gamma(\varsigma\lambda+1)} = \sum_{n=-\infty}^{\infty} \frac{1}{2^{|n|+1}} + \infty = \infty,$$

provided that $\mathrm{D} \cap \mathbb{Z} = \varnothing$. Now,

$$\phi_{\psi,\varkappa}(\mathrm{D}) = \sum_{n=-\infty}^{\infty} \int_{n-\frac{1}{2}+\varrho_n}^{n+\frac{1}{2}+\varrho_n} \sum_{\varsigma=0}^{\infty} \frac{(\psi \mathrm{y})^{\varsigma}}{\Gamma(\varsigma\lambda+1)} d\mathrm{y} + 0 = \sum_{\varsigma=0}^{\infty} \frac{(\frac{\psi}{2})^{\varsigma}}{\Gamma(\varsigma\lambda+1)} \sum_{n=-\infty}^{\infty} \frac{1}{2^{|n|+1}} < \infty.$$

Therefore, if $\mathrm{R} = \mathrm{D}/\mathrm{L}$, where H is a bounded interval, we obtain

$$\phi_{\psi,\varkappa}(\mathrm{R}+\zeta) \geqslant \sum_{n\in\mathbb{Z}/\mathrm{L}} \sum_{\varsigma=0}^{\infty} \frac{(\psi n)^{\varsigma}}{\Gamma(\varsigma\lambda+1)} = \infty,$$

provided that $\phi_{\psi,\varkappa}(\mathrm{R}) \leqslant \phi_{\psi,\varkappa}(\mathrm{D}) < \infty$.

Example 5.3 We consider the following measure for $\phi \in \mathcal{N}$, $\psi \geqslant 0$, $\varkappa > 0$,

$$d\phi_{\psi,\varkappa}(\mathrm{y}) = \sum_{\varsigma=0}^{\infty} \frac{(\psi \mathrm{y})^{\varsigma}}{\Gamma(\varsigma\lambda+1)} d\mathrm{y} + \varkappa \sum_{n=1}^{\infty} \frac{1}{n^2} \delta_{\frac{1}{n}},$$

where $\delta_{\frac{1}{n}}$ is the Dirac measure at $\frac{1}{n}$ and satisfying $(\mathbf{I}_4)$. For $\zeta \in \mathbb{R}$, let $\mathtt{L} = (-1 - |\zeta|, 1 + \zeta]$. We can easily see that $\mathtt{R} \cap \mathtt{L} = \emptyset$, $\mathtt{R} \cap [0, 1] = \emptyset$ and also $(\mathtt{R} + \zeta) \cap [0, 1] = \emptyset$. Then,

$$\begin{aligned}\phi(\mathtt{R}+\zeta) &= \int_{\mathtt{R}+\zeta} \sum_{\varsigma=0}^{\infty} \frac{(\psi \mathtt{y})^{\varsigma}}{\Gamma(\varsigma\lambda+1)} d\mathtt{y} + 0 = \sum_{\varsigma=0}^{\infty} \frac{(\psi)^{\varsigma}}{\Gamma(\varsigma\lambda+1)} \left(\int_{\mathtt{R}+\zeta} \mathtt{y}^{\varsigma} d\mathtt{y} \right) \\ &= \sum_{\varsigma=0}^{\infty} \frac{(\psi)^{\varsigma}}{\Gamma(\varsigma\lambda+1)} \frac{(\mathtt{R}+\zeta)^{\varsigma+1}}{\varsigma+1} = \sum_{\varsigma=0}^{\infty} \frac{(\psi)^{\varsigma}}{\Gamma(\varsigma\lambda+1)} \frac{(\mathtt{R}+\zeta)^{\varsigma+1}}{(\varsigma+1)(\mathtt{R}+\zeta)^{\varsigma+1}\zeta^{-\varsigma}\mathtt{R}^{-(\varsigma+1)}} \\ &= \sum_{\varsigma=0}^{\infty} \frac{(\psi\zeta)^{\varsigma}}{\Gamma(\varsigma\lambda+1)} \frac{\mathtt{R}^{\varsigma+1}}{\varsigma+1} = \phi(\mathtt{R}).\end{aligned}$$

The conclusion follows with $\vartheta = 1$.

5.2.4 On the Integral Solution of Eq. (5.2.4)

Proposition 5.3 *(i) Assume that* $(\mathbf{I}_1)$ *and* $(\mathbf{I}_2)$ *hold. If* $\mathtt{z}_q \in \mathcal{BC}(\mathbb{R}, \mathbb{R})$*, then*

$$\vartheta_{pq}(\mathtt{y})\Theta_q\left(\mathtt{z}_q\left(\mathtt{y}-\zeta_{pq}\right)\right) \in \mathcal{BC}(\mathbb{R}, \mathbb{R}),\ \bar{\vartheta}_{pq}(\mathtt{y})\bar{\Theta}_q\left(\mathtt{z}_q(\mathtt{y})\right) \in \mathcal{BC}(\mathbb{R}, \mathbb{R})$$

and

$$\mathtt{h}_{pq}(\mathtt{y}) \int_0^{\infty} \psi_{pq}(\mathtt{r})\widehat{\Theta}_q\left(\mathtt{z}_q(\mathtt{y}-\mathtt{r})\right) d\mathtt{r} \in \mathcal{BC}(\mathbb{R}, \mathbb{R}),$$

for $p, q = 1, 2, 3 \ldots, n$.

(ii) Assume further that $(\mathbf{I}_1)$, $(\mathbf{I}_2)$ *and* $(\mathbf{I}_4)$ *hold. If* $\mathtt{z}_q \in \mathcal{PAP}(\mathbb{R}, \mathbb{R}, \phi)$*, then*

$$\vartheta_{pq}(\mathtt{y})\Theta_q\left(\mathtt{z}_q\left(\mathtt{y}-\zeta_{pq}\right)\right) \in \mathcal{PAP}(\mathbb{R}, \mathbb{R}, \phi),\ \bar{\vartheta}_{pq}(\mathtt{y})\bar{\Theta}_q\left(\mathtt{z}_q(\mathtt{y})\right) \in \mathcal{PAP}(\mathbb{R}, \mathbb{R}, \phi)$$

and

$$\mathtt{h}_{pq}(\mathtt{y}) \int_0^{\infty} \psi_{pq}(\mathtt{r})\widehat{\Theta}_q\left(\mathtt{z}_q(\mathtt{y}-\mathtt{r})\right) d\mathtt{r} \in \mathcal{PAP}(\mathbb{R}, \mathbb{R}, \phi).$$

Proof This follows from [30, Theorem 4.1].

In the sequel, we set $\bar{\mathtt{z}}_p(\mathtt{y}) = \sigma_{\mathtt{y}}^{-1}\mathtt{z}_{(\mathtt{y})}$. Then, Eq. (5.2.1) is transformed to the following system

$$\bar{\mathtt{z}}_p(\mathtt{y}) = -\mathtt{e}_p(\mathtt{y})\bar{\mathtt{z}}_p(\mathtt{y}) + \sigma_p^{-1} \sum_{q=1}^{n} \vartheta_p(\mathtt{y})\bar{\Theta}_q\left(\sigma\,\bar{\mathtt{z}}_p(\mathtt{y})\right) \tag{5.2.4}$$

$$+\sigma_p^{-1}\sum_{q=1}^{n}\vartheta_{pq}(\mathrm{y})\Theta_q\left(\sigma;\bar{\mathrm{z}}_q\left(\mathrm{y}-\zeta_{pq}\right)\right)$$
$$+\sigma_p^{-1}\sum_{q=1}^{n}\mathrm{h}_{pq}(\mathrm{y})\int_0^{\infty}\psi_{pq}(v)\hat{\Theta}_q\left(\sigma_p\bar{\mathrm{z}}_q(\mathrm{y}-v)\right)dv+\sigma_p^{-1}\mathrm{L}_p(\mathrm{y}).$$

Now, we show that the integral solutions of Eq. (5.2.4) are mappings of $\mathcal{PAP}(\mathbb{R},\mathbb{R}^n,\phi)$ to itself.

Theorem 5.6 *Assuming conditions* $(\mathbf{I}_1)$, $(\mathbf{I}_2)$ *and* $(\mathbf{I}_5)$, *we define the nonlinear mapping of* $\mathfrak{P}$ *on* $\mathcal{BC}(\mathbb{R},\mathbb{R}^n)$ *for* $p=1,2,3,\ldots,n$ *as follows:*

$$(\mathfrak{P}\mathrm{U})_p(\mathrm{y})=\int_0^{\mathrm{y}}\sum_{\varsigma=0}^{\infty}\frac{(-\int_{\mathrm{b}}^{\mathrm{y}}\mathrm{e}_p(\mathrm{r})d\mathrm{r})^{\varsigma}}{\Gamma(\varsigma\lambda+1)} \tag{5.2.5}$$
$$\left[\sigma^{-1}\sum_{q=1}^{n}\bar{\vartheta}_{pq}(\mathrm{b})\bar{\Theta}_q\left(\sigma_q\mathrm{U}_q(\mathrm{b})\right)+\sigma_p^{-1}\sum_{q=1}^{n}\vartheta_{pq}(\mathrm{b})\Theta_q\left(\sigma_q\mathrm{U}_q\left(\mathrm{b}-\zeta_q\right)\right)\right.$$
$$\left.+\sigma_p^{-1}\sum_{p=1}^{n}\mathrm{h}_{pq}(\mathrm{b})\int_0^{\infty}\psi_{pq}(v)\hat{\Theta}_q\left(\sigma_q\mathrm{U}_q(\mathrm{b}-v)\right)dv+\sigma_p^{-1}\mathrm{L}_p(\mathrm{b})\right]d\mathrm{b},$$

If we assume condition $(\mathbf{I}_4)$ *along with the other three conditions, then* $\mathfrak{P}\in\mathcal{PAP}(\mathbb{R},\mathbb{R}^n,\phi)$.

Proof We have $\mathfrak{P}\mathrm{U}\in\mathcal{BC}(\mathbb{R},\mathbb{R}^n)$ (see [45]). According to Theorem 5.3, for $p=1,2,3,\ldots,n$, there exist $\Delta_p\in\mathcal{AP}(\mathbb{R},\mathbb{R})$ and $\Psi_p\in\mathcal{E}(\mathbb{R},\mathbb{R},\phi)$ such that

$$\sigma_p^{-1}\sum_{q=1}^{n}\bar{\vartheta}_{pq}(\mathrm{y})\bar{\Theta}_q\left(\sigma_q\mathrm{U}_q(\mathrm{y})\right)+\sigma_p^{-1}\sum_{q=1}^{n}\vartheta_{pq}(\mathrm{y})\Theta_q\left(\sigma_q\mathrm{U}_q\left(\mathrm{y}-\zeta_{pq}\right)\right)+\sigma_p^{-1}\sum_{q=1}^{n}\mathrm{h}_{pq}(\mathrm{y})$$
$$\times\int_0^{\infty}\psi_{pq}(v)\widehat{\Theta}_q\left(\sigma_q\mathrm{U}_q(\mathrm{y}-v)\right)dv+\sigma_p^{-1}\mathrm{L}_p(\mathrm{y})=\Delta_p+\Psi_p\in\mathcal{PAP}(\mathbb{R},\mathbb{R},\phi).$$

(1) We claim that

$$\int_0^{\mathrm{y}}\sum_{\varsigma=0}^{\infty}\frac{(-\int_{\mathrm{b}}^{\mathrm{y}}\mathrm{e}_p(\mathrm{r})d\mathrm{r})}{\Gamma(\varsigma\lambda+1)}\Delta_p(\mathrm{b})d\mathrm{b}\in\mathcal{AP}(\mathbb{R},\mathbb{R}),\quad p=1,2,3,\ldots,n. \tag{5.2.6}$$

In fact,

$$\Lambda_p(\mathrm{y})=:\int_0^{\mathrm{y}}\sum_{\varsigma=0}^{\infty}\frac{(-\int_{\mathrm{b}}^{\mathrm{y}}\mathrm{e}_p(\mathrm{r})d\mathrm{r})^{\varsigma}}{\Gamma(\varsigma\lambda+1)}\Delta_p(\mathrm{b})d\mathrm{b},\quad p=1,2,3,\ldots,n.$$

According to Theorem 5.5, and since $\Delta_p, \mathrm{e}_p \in \mathcal{AP}(\mathbb{R}, \mathbb{R})$, for every $\varpi > 0$, there exists a number such as ζ belonging to an interval of positive length l_ϖ such that $\left|\Delta_p(\mathrm{y}+\zeta) - \Delta_p(\mathrm{y})\right| < \varpi$, and

$$\left|\sum_{\varsigma=0}^{\infty} \frac{-\int_{\mathrm{b}+\zeta}^{\alpha+\zeta} \mathrm{e}_p(\mathrm{r})d\mathrm{r}}{\Gamma(\varsigma\lambda+1)} \mathrm{e}_p(\mathrm{y}+\zeta) - \sum_{\varsigma=0}^{\infty} \frac{-\int_{\alpha}^{\mathrm{y}} \mathrm{e}_p(\mathrm{r})d\mathrm{r}}{\Gamma(\varsigma\lambda+1)} \mathrm{e}_p(\mathrm{y})\right| < \varpi,$$

for all $\mathrm{y} \in \mathbb{R}$ (see [46]). Then

$$\begin{aligned}
&\left|\Lambda_p(\mathrm{y}+\zeta) - \Lambda_p(\mathrm{y})\right| \qquad (5.2.7)\\
&= \left|\int_0^{\mathrm{y}+\zeta} \sum_{\varsigma=0}^{\infty} \frac{(-\int_{\mathrm{b}}^{\mathrm{y}+\zeta} \mathrm{e}_p(\mathrm{r})d\mathrm{r})^\varsigma}{\Gamma(\varsigma\lambda+1)} \Delta_p(\mathrm{b})d\mathrm{b} - \int_0^{\mathrm{y}} \sum_{\varsigma=0}^{\infty} \frac{(-\int_{\mathrm{b}}^{\mathrm{y}} \mathrm{e}_p(\mathrm{r})d\mathrm{r})^\varsigma}{\Gamma(\varsigma\lambda+1)} \Delta_p(\mathrm{b})d\mathrm{b}\right|\\
&= \left|\int_0^{\mathrm{y}} \sum_{\varsigma=0}^{\infty} \frac{(-\int_{\mathrm{b}+\zeta}^{\mathrm{y}+\zeta} \mathrm{e}_p(\mathrm{r})d\mathrm{r})^\varsigma}{\Gamma(\varsigma\lambda+1)} \Delta_p(\mathrm{b}+\zeta)d\mathrm{b} - \int_0^{\mathrm{y}} \sum_{\varsigma=0}^{\infty} \frac{(-\int_{\mathrm{b}}^{\mathrm{y}} \mathrm{e}_p(\mathrm{r})d\mathrm{r})^\varsigma}{\Gamma(\varsigma\lambda+1)} \Delta_p(\mathrm{b})d\mathrm{b}\right|\\
&= | \int_0^{\mathrm{y}} \sum_{\varsigma=0}^{\infty} \frac{(-\int_{\mathrm{b}+\zeta}^{\mathrm{y}+\zeta} \mathrm{e}_p(\mathrm{r})d\mathrm{r})^\varsigma}{\Gamma(\varsigma\lambda+1)} \Delta p(\mathrm{b}+\zeta)d\mathrm{b} - \int_0^{\mathrm{y}} \sum_{\varsigma=0}^{\infty} \frac{(-\int_{\mathrm{b}}^{\mathrm{y}} \mathrm{e}_p(\mathrm{r})du\mathrm{r})^\varsigma}{\Gamma(\varsigma\lambda+1)} \Delta_p(\mathrm{b}+\zeta)d\mathrm{b}\\
&\quad + \int_0^{\mathrm{y}} \sum_{\varsigma=0}^{\infty} \frac{(-\int_{\mathrm{b}}^{\mathrm{y}} \mathrm{e}_p(\mathrm{r})d\mathrm{r})^\varsigma}{\Gamma(\varsigma\lambda+1)} \Delta_p(\mathrm{b}+\zeta)d\zeta - \int_0^{\mathrm{y}} \sum_{\varsigma=0}^{\infty} \frac{(-\int_{\mathrm{b}}^{\mathrm{y}} \mathrm{e}_p(\mathrm{r})d\mathrm{r})^\varsigma}{\Gamma(\varsigma\lambda+1)} \Delta p(\mathrm{b})d\mathrm{b} \,|\\
&\leqslant \int_0^{\mathrm{y}} \left|\sum_{\varsigma=0}^{\infty} \frac{(-\int_{\mathrm{b}+\zeta}^{\mathrm{y}+\zeta} \mathrm{e}_p(\mathrm{r})d\mathrm{r})^\varsigma}{\Gamma(\varsigma\lambda+1)} - \sum_{\varsigma=0}^{\infty} \frac{(-\int_{\mathrm{b}}^{\mathrm{y}} \mathrm{e}_p(\mathrm{r})d\mathrm{r})^\varsigma}{\Gamma(\varsigma\lambda+1)}\right| \left|\Delta_p(\mathrm{b}+\zeta)\right| d\mathrm{b}\\
&\quad + \int_0^{\mathrm{y}} \sum_{\varsigma=0}^{\infty} \frac{(-\int_{\mathrm{b}}^{\mathrm{y}} \mathrm{e}_p(\mathrm{r})d\mathrm{r})^\varsigma}{\Gamma(\varsigma\lambda+1)} \left|\Delta_p(\mathrm{b}+\zeta) - \Delta_p(\mathrm{r})\right| d\mathrm{r}.
\end{aligned}$$

Let

$$\mathcal{J} = \left|\sum_{\varsigma=0}^{\infty} \frac{(-\int_{\mathrm{b}+\varpi}^{\mathrm{y}+\zeta} \mathrm{e}_p(\mathrm{r})d\mathrm{r})^\varsigma}{\Gamma(\varsigma\lambda+1)} - \sum_{\varsigma=0}^{\infty} \frac{(-\int_{\mathrm{b}}^{\mathrm{y}} \mathrm{e}_p(\mathrm{r})d\mathrm{r})^\varsigma}{\Gamma(\varsigma\lambda+1)}\right|.$$

Since

$$\left(\sum_{\varsigma=0}^{\infty} \frac{(-\int_{\mathrm{b}}^{\mathrm{y}} \mathrm{e}_p(\mathrm{r})d\mathrm{r})^\varsigma}{\Gamma(\varsigma\lambda+1)}\right)'_{\mathrm{y}} = \varsigma\left(-\int_{\mathrm{b}}^{\mathrm{y}} \mathrm{e}_p(\mathrm{r})d\mathrm{r}\right)^{\varsigma-1} (-\mathrm{e}_p(\mathrm{r})),\ \mathrm{e}_p \in \mathcal{AP}(\mathbb{R}, \mathbb{R}),$$

using assumption (I_8), we obtain that

$$\mathcal{J} = \left|\left[\sum_{\varsigma=0}^{\infty} \frac{(-\int_{\alpha}^{\mathrm{y}} \mathrm{e}_p(\mathrm{r})d\mathrm{r})^\varsigma}{\Gamma(\varsigma\lambda+1)} \sum_{\varsigma=0}^{\infty} \frac{(-\int_{\mathrm{b}+\zeta}^{\alpha+\zeta} \mathrm{e}_p(\mathrm{r})d\mathrm{r})^\varsigma}{\Gamma(\varsigma\lambda+1)}\right]_{\alpha+\mathrm{b}}^{\mathrm{y}}\right| \qquad (5.2.8)$$

$$
\begin{aligned}
=&\Bigg|\Bigg[\int_{b}^{y}\sum_{\varsigma=0}^{\infty}\frac{(-\int_{\alpha}^{y}e_p(r)dr)^{\varsigma}}{\Gamma(\varsigma\lambda+1)}\left(\sum_{\varsigma=0}^{\infty}\frac{(-\int_{b+\zeta}^{\alpha+\zeta}e_p(r)dr)^{\varsigma}}{\Gamma(\varsigma\lambda+1)}\right)'_{\alpha}d\alpha\\
&+\int_{s}^{t}\left(\sum_{\varsigma=0}^{\infty}\frac{(-\int_{\alpha}^{y}e_p(r)dr)^{\varsigma}}{\Gamma(\varsigma\lambda+1)}\right)'_{\alpha}\sum_{\varsigma=0}^{\infty}\frac{(-\int_{b+\zeta}^{\alpha+\zeta}e_p(r)dr)^{\varepsilon}}{\Gamma(\varsigma\lambda+1)}d\alpha\Bigg]\Bigg|\\
=&\Bigg|\Bigg[\int_{b}^{y}\sum_{\varsigma=0}^{\infty}\frac{(-\int_{\alpha}^{y e_p(r)dr})^{\varsigma}}{\Gamma(\varsigma\lambda+1)}\left(\sum_{\varsigma=0}^{\infty}\frac{\varsigma(-\int_{b+\zeta)}^{\alpha+\zeta}e_p(r)dr)^{\varsigma-1}(-e_p(\alpha+\zeta))}{\Gamma(\varsigma\lambda+1)}d\alpha\right)\\
&+\int_{b}^{y}\sum_{\varsigma=0}^{\infty}\frac{\varsigma(-\int_{\alpha}^{y}e_p(r)dr)^{\varsigma-1}(-e_p(\alpha))}{\Gamma(\varsigma\lambda+1)}\sum_{\varsigma=0}^{\infty}\frac{-\int_{b+\zeta}^{\alpha+\zeta}e_p(r)dr}{\Gamma(\varsigma\lambda+1)}d\alpha\Bigg]\Bigg|\\
=&\Bigg|\int_{b}^{y}\sum_{\varsigma=0}^{\infty}\frac{(-\int_{\alpha}^{y}e_p(r)dr)^{\varsigma}}{\Gamma(\varsigma\lambda+1)}\sum_{\varsigma=0}^{\infty}\frac{(-\int_{b+\zeta}^{\alpha+\zeta}e_p(r)dr)^{\varsigma}}{\Gamma(\varsigma\lambda+1)}\left(\sum_{\varsigma=0}^{\infty}\frac{-\int_{b+\zeta}^{\alpha+\zeta}e_p(r)dr}{\Gamma(\varsigma\lambda+1)}(e_p(\alpha+\zeta))\right.\\
&\left.-\sum_{\varsigma=0}^{\infty}\frac{-\int_{\alpha}^{y}e_p(r)dr}{\Gamma(\varsigma\lambda+1)}(e_p(\alpha))\right)d\alpha\Bigg|\\
\le&\int_{b}^{y}\sum_{\varsigma=0}^{\infty}\frac{(-\int_{\alpha}^{y}e_p(r)dr)^{\varsigma}}{\Gamma(\varsigma\lambda+1)}\sum_{\varsigma=0}^{\infty}\frac{(-\int_{b+\zeta}^{\alpha+\zeta}e_p(r)dr)^{\varsigma}}{\Gamma(\varsigma\lambda+1)}\Bigg|\left(\sum_{\varsigma=0}^{\infty}\frac{-\int_{b+\zeta}^{\alpha+\zeta}e_p(r)dr}{\Gamma(\varsigma\lambda+1)}(e_p(\alpha+\zeta))\right.\\
&\left.-\sum_{\varsigma=0}^{\infty}\frac{-\int_{\alpha}^{y}e_p(r)dr}{\Gamma(\varsigma\lambda+1)}(e_p(\alpha))\right)\Bigg|d\alpha\\
\leqslant&(o_p)^2\in\int_{b}^{y}\sum_{\varsigma=0}^{\infty}\frac{(-\widehat{e}_p^{-(y-b)})^{\varsigma}}{\Gamma(\varsigma\lambda+1)}d\alpha\\
\leqslant&o_p^2\varrho\sum_{\varsigma=0}^{\infty}\frac{(-\widehat{e}_p^{-(y-b)})^{\varsigma}}{\Gamma(\varsigma\lambda+1)}(y-b).
\end{aligned}
$$

Then Δ_p is a continuous and bounded function, given that $\Delta_p\in\mathcal{AP}(\mathbb{R},\mathbb{R})$. Now, by putting Eqs. (5.2.8) in (5.2.7), we have

$$
\begin{aligned}
&|\Lambda_p(y+\zeta)-\Lambda_p(y)|\leqslant o_p^2\varrho\int_{0}^{y}\sum_{\varsigma=0}^{\infty}\frac{(-\widehat{e}_p^{-}(y-b))^{\varsigma}}{\Gamma(\varsigma\lambda+1)}(y-b)\left|\Delta_p(b+\zeta)\right|db\\
&+\varrho\int_{0}^{y}\sum_{\varsigma=0}^{\infty}\frac{(-\int_{b}^{y}e_p(r)dr)^{\varsigma}}{\Gamma(\varsigma\lambda+1)}db\leqslant o_p^2\varrho\left\|\Delta_p\right\|_{\infty}\int_{0}^{y}\sum_{\varsigma=0}^{\infty}\frac{(-\widehat{e}_p^{-}(y-b))^{\varsigma}}{\Gamma(\varsigma\lambda+1)}(y-b)db\\
&+o_p\varrho\int_{-\infty}^{y}\sum_{\varsigma=0}^{\infty}\frac{(-\int_{b}^{y}\widehat{e}_p(r)dr)^{\varsigma}}{\Gamma(\varsigma\lambda+1)}db\leqslant o_p^2\varrho\left\|\Delta_p\right\|_{\infty}\int_{0}^{y}\sum_{\varsigma=0}^{\infty}\frac{(-\widehat{e}_p^{-}(y-b))^{\varsigma}}{\Gamma(\varsigma\lambda+1)}(y-b)db\\
&+o_p\varrho\int_{0}^{y}\sum_{\varsigma=0}^{\infty}\frac{(-\widehat{e}_p^{-}(y-b))^{\varsigma}}{\Gamma(\varsigma\lambda+1)}d\lambda\leqslant(o_p)^2\varrho\left\|\Delta_p\right\|_{\infty}\sum_{\varsigma=0}^{\infty}\frac{(\widehat{e}_p^{-})^{\varsigma}y^{\varsigma+2}}{\Gamma(\varsigma\lambda+1)(\varsigma+2)}+o_p\varrho\sum_{\varsigma=0}^{\infty}\frac{(\widehat{e}_p^{-})^{\varsigma}y^{\varsigma+1}}{\Gamma(\varsigma\lambda+1)(\varsigma+1)}.
\end{aligned}
$$

We obtain that

$$
\int_{0}^{y}\sum_{\varsigma=0}^{\infty}\frac{(-\int_{b}^{y}e_p(r)dr)^{\varsigma}}{\Gamma(\varsigma\lambda+1)}\Delta_p(b)db\in\mathcal{AP}(\mathbb{R},\mathbb{R}).
$$

(2) Let us show that

$$\int_0^{y}\sum_{\varsigma=0}^{\infty}\frac{(-\int_b^y e_p(r)dr)^{\varsigma}}{\Gamma(\varsigma\lambda+1)}\Psi_p(b)db \in \mathcal{E}(\mathbb{R},\mathbb{R},\phi).$$

According to hypothesis (I_4), for $\Psi_p \in \mathcal{E}(\mathbb{R},\mathbb{R},\phi)$ and for $p=1,\ldots,n$, we have

$$\begin{aligned}0 &= \lim_{u\longrightarrow+\infty}\frac{1}{\phi((|-u,u))}\int_{-u}^{u}\int_0^{y}\sum_{\varsigma=0}^{\infty}\frac{(-\widehat{e}_p^{-}r)^{\varsigma}}{\Gamma(\varsigma\lambda+1)}\big|\Psi_p(y-r)\big|drd\phi(y)\\ &= \lim_{u\longrightarrow+\infty}\frac{1}{\phi(|-uu)}\int_{-u}^{u}\int_0^{y}\sum_{\varsigma=0}^{\infty}\frac{(-\widehat{e}_p^{-}(y-b))^{\varsigma}}{\Gamma(\varsigma\lambda+1)}\big|\Psi_p(b)\big|dbd\phi(y).\end{aligned}$$

Then, by (I_5),

$$\begin{aligned}0 &\leqslant \lim_{u\longrightarrow+\infty}\frac{1}{\phi(|-u,u))}\int_{-u}^{u}\int_0^{y}\sum_{\varsigma=0}^{\infty}\frac{(-\int_b^y e_p(\xi)d\xi)^{\varsigma}}{\Gamma(\varsigma\lambda+1)}\big|\Psi_p(b)\big|dsd\phi(y)\\ &\leqslant O_p\lim_{u\longrightarrow+\infty}\frac{1}{\phi(|-u,u])}\int_{-u}^{u}\int_0^{y}\sum_{\varsigma=0}^{\infty}\frac{(-\widehat{e}_p^{-}(y-b))^{\varsigma}}{\Gamma(\varsigma\lambda+1)}\big|\Psi_p(b)\big|dbd\phi(y)\\ &=0.\end{aligned}$$

Hence,

$$\int_0^{y}\sum_{\varsigma=0}^{\infty}\frac{(-\int_b^y e_p(r)dr)^{\varsigma}}{\Gamma(\varsigma\lambda+1)}\Psi_p(b)db \in \mathcal{E}(\mathbb{R},\mathbb{R},\phi),$$

for $p=1,2,3,\ldots,n$. Combined with (5.2.6), we have

$$\begin{aligned}(\mathfrak{P}U)_p(y) &= \int_0^{y}\sum_{\varsigma=0}^{\infty}\frac{(-\int_b^y e_p(r)dr)^{\varsigma}}{\Gamma(\varsigma\lambda+1)}\Delta_p(b)db+\int_0^{y}\sum_{\varsigma=0}^{\infty}\frac{(-\int_b^y e_p(r)dr)^{\varsigma}}{\Gamma(\varsigma\lambda+1)}\Psi_p(b)db\\ &\in \mathcal{PAP}(\mathbb{R},\mathbb{R},\phi).\end{aligned}$$

5.2.5 *ϕ-Pseudo Almost Periodic Solutions*

Assuming that the solution of Eq. (5.2.1) is theϕ-pseudo almost periodic type, we shall prove the existence and uniqueness of these solutions.

Theorem 5.7 *(1) Given assumptions* ($\mathbf{I}_1$), ($\mathbf{I}_2$), *and* ($\mathbf{I}_5$), *there is a unique solution* $z^* \in \mathcal{BC}(\mathbb{R},\mathbb{R}^n)$ *for Eq.* (5.2.1).

(2) Given assumptions $(\mathbf{I}_1), (\mathbf{I}_2), (\mathbf{I}_5), (\mathbf{I}_4), (\mathbf{I}_6)$, *and* $(\mathbf{I}_7)$, *we have* $\mathrm{z}^* \in \mathcal{PAP}(\mathbb{R}, \mathbb{R}^n, \phi)$.

Proof Let $\mathrm{U}, \mathrm{G} \in \mathcal{BC}(\mathbb{R}, \mathbb{R}^n)$ (resp. $\mathrm{U}, \mathrm{G} \in \mathcal{PAP}(\mathbb{R}, \mathbb{R}^n, \phi)$). Then, in view of Theorem 5.6, we have that $\mathfrak{P}\mathrm{U}, \mathfrak{P}\mathrm{G} \in \mathcal{BC}(\mathbb{R}, \mathbb{R}^n)$ (resp. $\mathfrak{P}\mathrm{U}, \mathfrak{P}\mathrm{G} \in \mathcal{PAP}(\mathbb{R}, \mathbb{R}^n, \phi)$). Let

$$\mathcal{S} = \left|(\mathfrak{P}\mathrm{U})_p(\mathrm{y}) - (\mathfrak{G}\mathrm{G})_p(\mathrm{y})\right|.$$

Using $(\mathbf{I}_1)$ and $(\mathbf{I}_5)$, $(\mathbf{I}_6)$, we get that

$$\begin{aligned}
\mathcal{S} = |\int_0^{\mathrm{y}} \sum_{\varsigma=0}^{\infty} & \frac{(-\int_{\mathrm{b}}^{\mathrm{y}} \mathrm{e}_p(\mathrm{r})d\mathrm{r})^{\varsigma}}{\Gamma(\varsigma\lambda+1)} \sigma_p^{-1} \sum_{q=1}^{n} \left(\bar{\vartheta}_{pq}(\mathrm{b}) \left(\bar{\Theta}_q\left(\sigma_q \mathrm{U}_q(\mathrm{b})\right) - \bar{\Theta}_q\left(\sigma_q \mathrm{G}_q(\mathrm{b})\right)\right)\right. \\
& + \sigma_p^{-1} \sum_{q=1}^{n} \vartheta_{pq}(\mathrm{b}) \left(\Theta_q\left(\sigma_q \mathrm{U}_q\left(\mathrm{b}-\zeta_{pq}\right)\right) - \Theta_q\left(\sigma_q \mathrm{G}_q\left(\mathrm{b}-\zeta_{pq}\right)\right)\right) \\
& \left. + \sigma_p^{-1} \sum_{q=1}^{n} \mathrm{h}_{pq}(\mathrm{b}) \int_0^{\infty} \psi_{pq}(v) \left(\widehat{\Theta}_q\left(\sigma_q \mathrm{U}_q(\mathrm{b}-v)\right) - \widehat{\Theta}_q\left(\sigma_q \mathrm{U}_q(\mathrm{b}-v)\right)\right) dv \right) d\mathrm{b}\, | \\
\leqslant \mathrm{o}_p \sigma_p^{-1} & \int_0^{\mathrm{y}} \sum_{\varsigma=0}^{\infty} \frac{(-\int_{\mathrm{b}}^{\mathrm{y}} \hat{\mathrm{e}}_p(\mathrm{r})d\mathrm{r})^{\varsigma}}{\Gamma(\varsigma\lambda+1)} \sum_{q=1}^{n} \left(\left|\bar{\vartheta}_{pq}(\mathrm{b})\right| \mathcal{T}_q^{\bar{\Theta}} + \left|\vartheta_{pq}(\mathrm{b})\right| \mathcal{T}_q^{\Theta}\right. \\
& \left.\left. + \mathcal{T}_q^{\hat{\Theta}} \left|\mathrm{h}_{pq}(\mathrm{s})\right| \int_0^{\infty} \left|\psi_{pq}(\mathrm{r})\right| d\mathrm{r}\right) \sigma_q d\mathrm{b}\right) \|\mathrm{U}-\mathrm{G}\|_{\infty} \\
\leqslant \int_0^{\mathrm{y}} \sum_{\varsigma=0}^{\infty} & \frac{(-\int_{\mathrm{b}}^{\mathrm{y}} \widehat{\mathrm{e}}_p(\mathrm{r})d\mathrm{r})^{\varsigma}}{\Gamma(\varsigma\lambda+1)} \left[\widehat{\mathrm{e}}_p(\mathrm{b}) - \kappa_p\right] d\mathrm{b} \|\mathrm{U}-\mathrm{G}\|_{\infty} \\
\leqslant \Bigg[\widehat{\mathrm{e}}_p^{+} & \int_0^{\mathrm{y}} \sum_{\varsigma=0}^{\infty} \frac{(-\int_{\mathrm{b}}^{\mathrm{y}} \widehat{\mathrm{e}}_p(\mathrm{r})d\mathrm{r})^{\varsigma}}{\Gamma(\varsigma\lambda+1)} d\mathrm{b} - \kappa_p \int_0^{\mathrm{y}} \sum_{\varsigma=0}^{\infty} \frac{(-\int_{\mathrm{b}}^{\mathrm{y}} \widehat{\mathrm{e}}_p(\mathrm{r})d\mathrm{r})^{\varsigma}}{\Gamma(\varsigma\lambda+1)} d\mathrm{b} \Bigg] \|\mathrm{U}-\mathrm{G}\|_{\infty} \\
\leqslant \max_{p=1,2,3,n} & \left\{ \frac{\hat{\mathrm{e}}_p^{+}}{\hat{\mathrm{e}}_p^{-}} - \frac{\kappa_p}{\hat{\mathrm{e}}_p^{+}} \right\} \|\mathrm{U}-\mathrm{G}\|_{\infty},
\end{aligned}$$

so

$$\left|(\mathfrak{P}\mathrm{U})_p(t) - (\mathfrak{P}\mathrm{G})_p(\mathrm{y})\right| \leqslant \max_{p=1,2,3,\ldots,n} \left\{ \frac{\widehat{\mathrm{e}}_p^{+}}{\widehat{\mathrm{e}}_p^{-}} - \frac{\kappa_p}{\widehat{\mathrm{e}}_p^{+}} \right\} \|\mathrm{U}-\mathrm{G}\|_{\infty}.$$

Considering condition $(\mathbf{I}_7)$ and due to the $0 < \max_{p=1,2,3,\ldots,n} \left\{ \frac{\widehat{\mathrm{e}}_+^{+}}{\mathrm{e}_p^{-}} - \frac{\kappa_1}{\mathrm{e}_p^{+}} \right\} < 1$, we conclude that $\mathfrak{P} \in \mathcal{BC}(\mathbb{R}, \mathbb{R}^n)$ is a contraction (since $\mathcal{PAP}(\mathbb{R}, \mathbb{R}^n, \mu)$ is a Banach space according to condition $(\mathbf{I}_3)$, it is similarly contraction on this space). Therefore, we conclude that $\bar{\mathrm{z}}^* \in \mathcal{BC}(\mathbb{R}, \mathbb{R}^n)$ (or $\bar{\mathrm{z}}^* \in \mathcal{PAP}(\mathbb{R}, \mathbb{R}^n, \phi)$) is a unique fixed point for $\mathfrak{P}$. Also, given (5.2.4), $\mathrm{z}^* \in \mathcal{BC}(\mathbb{R}, \mathbb{R}^n)$ (or $\mathrm{z}^* \in \mathcal{PAP}(\mathbb{R}, \mathbb{R}^n, \phi)$) is a unique solution of type ϕ-$\mathcal{PAP}$ for Eq. (5.2.1).

In the sequel, we investigate the Mittag-Leffler stability and the Mittag-Leffler attractiveness for the unique solution of Eq. (5.2.2), which is of type ϕ-$\mathcal{PAP}$. First, we state the following lemma.

Lemma 5.7 *Assuming conditions* ($\mathbf{I}_3$) *and* ($\mathbf{I}_6$), *we define* $\mathfrak{Z} = \{\mathfrak{Z}_p\}_{p=1}^n : [0, m] \longrightarrow \mathbb{R}^n$ *as follows*

$$\mathfrak{Z}_p(\mathfrak{w}) = \sup_{\mathrm{y}\in\mathbb{R}} \mathfrak{Z}_p(\mathfrak{w}, \mathrm{y}),$$

where

$$\mathfrak{Z}_p(\mathfrak{w}, \mathrm{y}) = \mathfrak{w} - \widehat{\mathrm{e}}_p(\mathrm{y}) + \mathrm{o}_p\sigma_p^{-1} \sum_{q=1}^n$$
$$\left(|\bar{\vartheta}_{pq}(\mathrm{y})|\mathcal{T}_q^{\bar{\Theta}} + |\vartheta_{pq}(\mathrm{y})|\mathcal{T}_q^{\Theta} \sum_{\varsigma=0}^{\infty} \frac{(\mathfrak{w}\zeta_{pq})^{\varsigma}}{\Gamma(\varsigma\lambda+1)} + \mathcal{T}_q^{\widehat{\Theta}} |\mathrm{h}_{pq}(\mathrm{y})| \int_0^{\infty} |\psi_{pq}(\mathrm{r})| \sum_{\varsigma=0}^{\infty} \frac{(\mathfrak{w}\mathrm{r})^{\varsigma}}{\Gamma(\varsigma\lambda+1)} d\mathrm{r}\right) \sigma_q.$$

Then $\mathfrak{Z}_p(\nu) < 0$ *for* $0 < \nu_0 < m$ *and for all* $0 < \nu < \nu_0$.

Proof According to condition ($\mathbf{I}_3$), function $\mathrm{y} \mapsto \mathfrak{Z}_p(\mathfrak{w}, \mathrm{y})$ is defined on the interval $[0, m]$. Then according to condition ($\mathbf{I}_6$), we have

$$\begin{aligned}\mathfrak{Z}_p(0) &= \sup_{\mathrm{y}\in\mathbb{R}} \mathfrak{Z}_p(0, \mathrm{y})\\ &= \sup_{\mathrm{y}\in\mathbb{R}} \left\{-\hat{\mathrm{e}}_p(\mathrm{y}) + \mathrm{o}_p\sigma_p^{-1} \sum_{q=1}^n \left(|\bar{\vartheta}_{pq}(\mathrm{y})|\mathcal{T}_q^{\bar{\Theta}} + |\vartheta_{pq}(\mathrm{y})|\mathcal{T}_q^{\Theta} + \mathcal{T}_q^{\widehat{\Theta}} |\mathrm{h}_{pq}(\mathrm{y})| \int_0^{\infty} |\psi_{pq}(\mathrm{r})|\, d\mathrm{r}\right) \sigma_q\right\}\\ &< -\kappa_p < 0.\end{aligned}$$

Now, we show that $0 < \nu_0 < m$ exists such that $\mathfrak{Z}_p(\nu) < 0$, for all $0 < \nu < \nu_0$. Also, we have

$$\mathfrak{Z}_p(\mathfrak{w}, \mathrm{y}) - \mathfrak{Z}_p(0, \mathrm{y}) = \mathfrak{w} + \mathrm{o}_p\sigma_p^{-1} \sum_{q=1}^n$$
$$\left(|\vartheta_{pq}(t)|\mathcal{T}_q^{\Theta} \left(\sum_{\varsigma=0}^{\infty} \frac{(\mathfrak{w}\zeta_{pq})^{\varsigma}}{\Gamma(\varsigma\lambda+1)} - 1\right) + \mathcal{T}_q^{\Theta} |\mathrm{h}_{pq}(\mathrm{y})| \left(\int_0^{\infty} |\psi_{pq}(\mathrm{r})| \sum_{\varsigma=0}^{\infty} \frac{(\mathfrak{w}\mathrm{r})^{\varsigma}}{\Gamma(\varsigma\lambda+1)} d\mathrm{r} - \int_0^{\infty} |\psi_{pq}(\mathrm{r})|\, d\mathrm{r}\right)\right) \sigma_q.$$

If we take ϑ_p and h_p the nonnegative numbers

$$\vartheta_p = \sup_{q=1}^{n} |\vartheta_{pq}(\mathrm{y})| \mathcal{T}_q^{\Theta}, \quad \mathrm{h}_p = \sup_{q=1}^{n} |\mathrm{h}_{pq}(\mathrm{y})| \mathcal{T}_q^{\widehat{\Theta}},$$

then

$$\left|\mathfrak{Z}_p(\mathfrak{w}, \mathrm{y}) - \mathfrak{Z}_p(0, \mathrm{y})\right| \leqslant \mathfrak{w} + \mathrm{o}_p\sigma_p^{-1}\vartheta_p \sum_{q=1}^n \left(\sum_{\varsigma=0}^{\infty} \frac{(\mathfrak{w}\zeta_{pq})^{\varsigma}}{\Gamma(\varsigma\lambda+1)} - 1\right) \sigma_q$$
$$+ \mathrm{o}_q\sigma_p^{-1}\mathrm{h}_p \sum_{q=1}^n \left(\int_0^{\infty} |\psi_{pq}(\mathrm{r})| \sum_{\varsigma=0}^{\infty} \frac{(\mathfrak{w}\mathrm{r})^{\varsigma}}{\Gamma(\varsigma\lambda+1)} d\mathrm{r} - \int_0^{\infty} |\psi_{pq}(\mathrm{r})|\, d\mathrm{r}\right) \sigma_q$$

for all $\mathfrak{w} \in (0, m]$ and $\mathrm{y} \in \mathbb{R}$. Now, for every $\varrho > 0$, by continuity, there exists $0 < \delta_p^1 < m$ such that the following holds:

$$\mathfrak{w} < \delta_p^1 \Rightarrow \mathfrak{w} + \mathrm{O}_p \sigma_p^{-1} \vartheta_p \sum_{q=1}^{n} \left(\sum_{\varsigma=0}^{\infty} \frac{(\mathfrak{w} \zeta_{pq})^{\varsigma}}{\Gamma(\varsigma\lambda + 1)} - 1 \right) \sigma_q < \frac{\varrho}{2}.$$

In the sequel, assuming condition $(\mathbf{I}_3)$ and the Lebesgue dominated convergence theorem (LDCT), and due to the integrability of the function $\left|\psi_{pq}(\mathrm{r}) \sum_{\varsigma=0}^{\infty} \frac{(\varsigma \mathrm{r})^{\varsigma}}{\Gamma(\varsigma\lambda+1)}\right|$ on $(0, \infty)$ on the interval $(0, \infty)$, there exists $0 < \delta_p^2 < m$ such that $\mathfrak{w} < \delta_p^2$ implies that

$$\mathrm{O}_p \sigma_p^{-1} \mathrm{h}_p \sum_{q=1}^{n} \left(\int_0^{\infty} |\psi_{pq}(\mathrm{r})| \sum_{\varsigma=0}^{\infty} \frac{(\mathfrak{w}\mathrm{r})^{\varsigma}}{\Gamma(\varsigma\lambda + 1)} d\mathrm{r} - \int_0^{\infty} |\psi_{pq}(\mathrm{r})| \, d\mathrm{r} \right) \sigma_q < \frac{\varrho}{2}.$$

If we take $0 < \nu < \min\left(\delta_p^1, \delta_p^2\right)$ and $\varrho < \frac{\kappa_p}{2}$, we conclude that for every $\mathrm{y} \in \mathbb{R}$, we have

$$\mathfrak{Z}_p(\nu, \mathrm{y}) \leqslant \mathfrak{Z}_p(0, \mathrm{y}) + \left(\mathfrak{Z}_p(\nu, \mathrm{y}) - \mathfrak{Z}_p(0, \mathrm{y})\right) < -\kappa_p + \varrho < -\frac{\kappa}{2}.$$

□

5.2.6 Stability and Attractiveness

Theorem 5.8 *If we assume conditions* $(\mathbf{I}_1)$, $(\mathbf{I}_3)$, $(\mathbf{I}_5)$ *and* $(\mathbf{I}_6)$, *then the Mittag-Leffler stability for any solution* z^* *of Eq.* (5.2.2) *is established by the initial condition* $\{\mathrm{z}^*(\mathrm{y}) : \mathrm{y} \leqslant 0\}$. *If we add condition* $(\mathbf{I}_2)$ *to the considered conditions, then the Mittag-Leffler stability for the solution* $\mathrm{z}^* \in \mathcal{BC}(\mathbb{R}, \mathbb{R}^n)$ *of Eq.* (5.2.2) *is established. Also, if we add condition* $(\mathbf{I}_7)$ *to conditions* $(\mathbf{I}_1)$, $(\mathbf{I}_3)$, $(\mathbf{I}_5)$, $(\mathbf{I}_6)$, *and* $(\mathbf{I}_2)$, *then the Mittag-Leffler stability for the solution* $\mathrm{z}^* \in \mathcal{PAP}(\mathbb{R}, \mathbb{R}^n, \mu)$ *of Eq.* (5.2.2) *is established.*

Proof Let $\mathrm{z}(t) = \left\{\mathrm{z}_p(\mathrm{y})\right\}_{p=1}^{n}$ be a solution of Eq. (5.2.2) with initial value $\xi(\mathrm{y}) = \left\{\xi_p(\mathrm{y})\right\}_{p=1}^{n}$. Set

$$\mathrm{f}(\mathrm{y}) = \left\{\mathrm{f}_p(\mathrm{y})\right\}_{p=1}^{n} = \left\{\sigma_i^{-1}\left(\mathrm{z}_p(\mathrm{y}) - \mathrm{z}_p^*(\mathrm{y})\right)\right\}_{p=1}^{n}.$$

Then

$$\mathrm{f}_p'(\mathrm{b}) + \mathrm{e}_p(\mathrm{b})\mathrm{f}_p(\mathrm{b}) = \sigma_p^{-1} \sum_{q=1}^{n} \bar{\vartheta}_{pq}(\mathrm{b}) \left(\bar{\Theta}_q\left(\mathrm{z}_q(\mathrm{b})\right) - \bar{\Theta}_q\left(\mathrm{z}_q^*(\mathrm{b})\right) \right) + \sigma_p^{-1} \sum_{q=1}^{n} \bar{\vartheta}_{pq}(\mathrm{b}) \tag{5.2.9}$$

$$\times \left(\Theta_q \left(\mathtt{z}_q \left(\mathtt{b} - \zeta_{pq}\right)\right) - \right.$$

$$\left.\Theta_q \left(\mathtt{z}_q' \left(\mathtt{b} - \zeta_{pq}\right)\right)\right) + \sigma_p^{-1} \sum_{q=1}^{n} \mathtt{h}_{pq}(\mathtt{b})$$

$$\times \int_0^\infty \psi_{pq}(v) \left(\widehat{\Theta}_q \left(\mathtt{z}_q(\mathtt{b} - v)\right) - \widehat{\Theta}_q \left(\mathtt{z}_q^*(\mathtt{b} - v)\right)\right) dv.$$

Let

$$\left\|\xi - \mathtt{z}^*\right\|_\sigma = \sup_{\mathtt{y} \in (-\infty, 0]} \max_{p=1,2,3,\ldots,n} \sigma_p^{-1} \left|\xi_p(\mathtt{y}) - \mathtt{z}_p^*(\mathtt{y})\right|,$$

and $\aleph > 0$ be such that

$$\aleph > \sum_{p=1}^{n} \mathtt{o}_p + 1. \tag{5.2.10}$$

Then, for all $\mathtt{y} \in (-\infty, 0]$, we have

$$\|\mathtt{f}(\mathtt{y})\| \leqslant \left\|\xi - \mathtt{z}^*\right\|_\sigma. \tag{5.2.11}$$

Given ν_0 in Lemma 5.7, we assume that $0 < \nu < \min\left\{\min_{p=1,2,\ldots,n} \hat{\mathtt{e}}_p^+, \nu_0\right\}$. Then for all $\mathtt{y}, \varrho > 0$, we have

$$\|\mathtt{f}(\mathtt{y})\| < \aleph \left(\left\|\xi - \mathtt{z}^*\right\|_\sigma + \varrho\right) \sum_{\varsigma=0}^{\infty} \frac{(-\nu \mathtt{y})^\varsigma}{\Gamma(\varsigma\lambda + 1)}. \tag{5.2.12}$$

Otherwise, for $p = 1, 2, 3, \ldots, n$ and $\varkappa > 0$, given the continuity, we have

$$\begin{cases} \left|\mathtt{f}_p(\varkappa)\right| = \aleph \left(\|\xi - \mathtt{z}^*\|_\sigma + \varrho\right) \sum_{\varsigma=0}^{\infty} \frac{(-\nu\varkappa)^\varsigma}{\Gamma(\varsigma\lambda+1)}, \\ \|\mathtt{f}(\mathtt{y})\| < \aleph \left(\|\xi - \mathtt{z}^*\|_\sigma + \varrho\right) \sum_{\varsigma=0}^{\infty} \frac{(-\nu \mathtt{y})^\varsigma}{\Gamma(\varsigma\lambda+1)}, \forall \mathtt{y} \in (-\infty, \varkappa). \end{cases} \tag{5.2.13}$$

Now, first, we multiply both sides of (5.2.9) by

$$\sum_{\varsigma=0}^{\infty} \frac{\left(\int_0^{\mathtt{b}} \mathtt{e}_p(\mathtt{r}) d\mathtt{r}\right)^\varsigma}{\Gamma(\varsigma\lambda + 1)}.$$

Then we integrate the obtained equation with respect to $\mathtt{b}$ on $[0, \varkappa]$. Finally, we multiply by

$$\sum_{\varsigma=0}^{\infty} \frac{\left(-\int_0^{\mathtt{y}} \mathtt{e}_p(\mathtt{r}) d\mathtt{r}\right)^\varsigma}{\Gamma(\varsigma\lambda + 1)}.$$

Therefore, we have

$$\mathtt{f}_p(\varkappa) = \mathtt{f}_p(0)\sum_{\varsigma=0}^{\infty}\frac{(-\int_0^{\varkappa}\mathtt{e}_p(\mathtt{r})d\mathtt{r})^{\varsigma}}{\Gamma(\varsigma\lambda+1)} + \sigma_p^{-1}\int_0^{\varkappa}\sum_{\varsigma=0}^{\infty}\frac{(-\int_{\mathtt{b}}^{\varkappa}\mathtt{e}_p(\mathtt{r})d\mathtt{r})^{\varsigma}}{\Gamma(\varsigma\lambda+1)}\sum_{q=1}^{n}$$
$$\big[\bar{\vartheta}_{pq}(\mathtt{b})\left(\bar{\Theta}_q\left(\mathtt{z}_q(\mathtt{b})\right) - \bar{\Theta}_q\left(\mathtt{z}_q^*(\mathtt{b})\right)\right) + \vartheta_{pq}(\mathtt{b})\left(\Theta_q\left(\mathtt{z}_q\left(\mathtt{b}-\zeta_{pq}\right)\right)\right.$$
$$\left.\left. -\Theta_q\left(\mathtt{z}_q^*\left(\mathtt{b}-\zeta_{pq}\right)\right)\right) + \mathtt{h}_{pq}(\mathtt{b})\int_0^{\infty}\psi_{pq}(v)\left(\widehat{\Theta}_q\left(\mathtt{z}_q(\mathtt{b}-v)\right)\right.$$
$$\left. -\widehat{\Theta}_q\left(\mathtt{z}_q^*(\mathtt{b}-v)\right)\right)dv\big]\,d\mathtt{b}.$$

Next, by ($\mathbf{I}_1$) and ($\mathbf{I}_5$) we obtain that

$$|\mathtt{f}_p(\varkappa)| \leqslant \mathtt{o}_p$$
$$\Bigg(|\mathtt{f}_p(0)|\sum_{\varsigma=0}^{\infty}\frac{(-\int_0^{\varkappa}\widehat{\mathtt{e}}_p(\mathtt{r})d\mathtt{r})^{\varsigma}}{\Gamma(\varsigma\lambda+1)} + \sigma_p^{-1}\int_0^{\varkappa}\sum_{\varsigma=0}^{\infty}\frac{(-\int_{\mathtt{b}}^{n}\widehat{\mathtt{e}}_p(\mathtt{r})d\mathtt{r})^{\varsigma}}{\Gamma(\varsigma\lambda+1)}\sum_{q=1}^{n}\Big[\left|\bar{\vartheta}_{pq}(\mathtt{b})\right|\mathcal{T}_q^{\bar{\Theta}}\sigma_q\left|\mathtt{f}_q(\mathtt{b})\right|$$
$$+\left|\vartheta_{pq}(\mathtt{b})\right|\mathcal{T}_q^{\Theta}\sigma_q\left|\mathtt{f}_q\left(\mathtt{b}-\zeta_{pq}\right)\right| + \mathcal{T}_q^{\widehat{\Theta}}\left|\mathtt{h}_{pq}(\mathtt{b})\right|\int_0^{\infty}$$
$$\left|\psi_{pq}(v)\right|\sigma_q\left|\mathtt{f}_q(\mathtt{b}-v)\right|dv\Big]\,d\mathtt{b}\Bigg)$$

Now, by (5.2.11) and (5.2.13), we get

$$|\mathtt{f}_p(\varkappa)| \leqslant \mathtt{o}_p\left(\left\|\xi-\mathtt{z}^*\right\|_{\sigma}+\varrho\right)$$
$$\Bigg\{\sum_{\varsigma=0}^{\infty}\frac{(-\int_0^{\varkappa}\widehat{\mathtt{e}}_p(\mathtt{r})d\mathtt{r})^{\varsigma}}{\Gamma(\varsigma\lambda+1)} + \aleph\sigma_p^{-1}\int_0^{\varkappa}\sum_{\varsigma=0}^{\infty}\frac{(-\int_{\mathtt{b}}^{\varkappa}\widehat{\mathtt{e}}_p(\mathtt{r})d\mathtt{r})^{\varsigma}}{\Gamma(\varsigma\lambda+1)}\sum_{q=1}^{n}\sigma_q$$
$$\Bigg[\left|\bar{\vartheta}_{pq}(\mathtt{b})\right|\mathcal{T}_q^{\bar{\Theta}}\sum_{\varsigma=0}^{\infty}\frac{(-\nu\mathtt{b})^{\varsigma}}{\Gamma(\varsigma\lambda+1)} + \left|\vartheta_{pq}(\mathtt{b})\right|\mathcal{T}_q^{\Theta}\sum_{\varsigma=0}^{\infty}\frac{(-\nu\left(\mathtt{b}-\zeta_q\right))^{\varsigma}}{\Gamma(\varsigma\lambda+1)}$$
$$+\mathcal{T}_q^{\Theta}\left|\mathtt{h}_{pq}(\mathtt{b})\right|\int_0^{\infty}\left|\psi_{pq}(\mathtt{r})\right|\sum_{\varsigma=0}^{\infty}\frac{(-\nu(\mathtt{b}-\mathtt{r}))^{\varsigma}}{\Gamma(\varsigma\lambda+1)}d\mathtt{r}\Bigg]\,d\mathtt{b}\Bigg\}.$$

By virtue of Lemma 5.7, since $\mathtt{b} > 0$ and $0 < \nu < \nu_0$, we have

$$|\mathtt{f}_p(\varkappa)| \leqslant \mathtt{o}_p\left(\left\|\xi-\mathtt{z}^*\right\|_{\sigma}+\varrho\right)\Bigg\{\sum_{\varsigma=0}^{\infty}\frac{(-\int_0^{\varkappa}\widehat{\mathtt{e}}_p(\mathtt{r})d\mathtt{r})^{\varsigma}}{\Gamma(\varsigma\lambda+1)}$$
$$+\aleph\mathtt{o}_p^{-1}\int_0^{\varkappa}\sum_{\varsigma=0}^{\infty}\frac{(-\int_{\mathtt{b}}^{\varkappa}\widehat{\mathtt{e}}_p(\mathtt{r})d\mathtt{r})^{\varsigma}}{\Gamma(\varsigma\lambda+1)}\left(\widehat{\mathtt{e}}_p(\mathtt{b})-\nu\right)\sum_{\varsigma=0}^{\infty}\frac{(-\nu\mathtt{b})^{\varsigma}}{\Gamma(\varsigma\lambda+1)}\Bigg\}$$
$$= \mathtt{o}_p\left(\left\|\xi-\mathtt{z}^*\right\|_{\sigma}+\varrho\right)\Bigg\{\sum_{\varsigma=0}^{\infty}\frac{(-\nu\varkappa(-\int_0^{\varkappa}\left(\widehat{\mathtt{e}}_p(\mathtt{r})-\nu\right)d\mathtt{r}))^{\varsigma}}{\Gamma(\varsigma\lambda+1)}$$

$$+\aleph \mathrm{o}_p^{-1}(-\nu\varkappa)^{\varsigma}\int_0^{\varkappa}\sum_{\varsigma=0}^{\infty}\frac{(-\int_{\mathrm{b}}^{\varkappa}\left(\widehat{\mathrm{e}}_p(\mathrm{r})-\nu\right)d\mathrm{r})^{\varsigma}}{\Gamma(\varsigma\lambda+1)}\left(\widehat{\mathrm{e}}_p(\mathrm{b})-\nu\right)d\mathrm{b}\Bigg\}$$

$$=\mathrm{o}_p\left(\|\xi-\mathrm{z}^*\|_{\sigma}+\varrho\right)\sum_{\varsigma=0}^{\infty}\frac{(-\nu\varkappa)^{\varsigma}}{\Gamma(\varsigma\lambda+1)}\Bigg\{(-\int_0^{\varkappa}\left(\widehat{\mathrm{e}}_p(\mathrm{r})-\nu\right)d\mathrm{r})^{\varsigma}$$

$$+\aleph \mathrm{o}_p^{-1}\int_0^{\varkappa}(-\int_{\mathrm{b}}^{\varkappa}\left(\widehat{\mathrm{e}}_p(\mathrm{r})-\nu\right)d\mathrm{r})^{\varsigma}\left(\widehat{\mathrm{e}}_p(\mathrm{b})-\nu\right)d\mathrm{b}\Bigg\}$$

$$=\mathrm{o}_p\left(\|\xi-\mathrm{z}^*\|_{\sigma}+\varrho\right)\sum_{\varsigma=0}^{\infty}\frac{(-\nu\varkappa)^{\varsigma}}{\Gamma(\varsigma\lambda+1)}\Bigg\{(-\int_0^{\varkappa}\left(\widehat{\mathrm{e}}_p(\mathrm{r})-\nu\right)d\mathrm{r})^{\varsigma}$$

$$+\aleph \mathrm{o}_p^{-1}\left[(-\int_{\mathrm{b}}^{\varkappa}\left(\widehat{\mathrm{e}}_p(\mathrm{r})-\nu\right)d\mathrm{r})^{\varsigma}\right]_0^{\varkappa}\Bigg\}$$

$$=\mathrm{o}_p\left(\|\xi-\mathrm{z}^*\|_{\sigma}+\varrho\right)\sum_{\varsigma=0}^{\infty}\frac{(-\nu\varkappa)^{\varsigma}}{\Gamma(\varsigma\lambda+1)}\Bigg\{(-\int_0^{\varkappa}\left(\widehat{\mathrm{e}}_p(\mathrm{r})-\nu\right)d\mathrm{r})^{\varsigma}$$

$$+\ \aleph \mathrm{o}_p^{-1}\left(0-(-\int_0^{\varkappa}\left(\widehat{\mathrm{e}}_p(\mathrm{r})-\nu\right)d\mathrm{r})^{\varsigma}\right)\Bigg\}$$

$$=\aleph\left(\|\xi-\mathrm{z}^*\|_{\sigma}+\varrho\right)\sum_{\varsigma=0}^{\infty}\frac{(-\nu\varkappa)^{\varsigma}}{\Gamma(\varsigma\lambda+1)}\left\{-\left(1-\frac{\mathrm{o}_p}{\aleph}\right)\sum_{\varsigma=0}^{\infty}\frac{(-\int_0^{\varkappa}\left(\widehat{\mathrm{e}}_p(\mathrm{r})-\nu\right)d\mathrm{r})^{\varsigma}}{\Gamma(\varsigma\lambda+1)}\right\}.$$

We recall that $\nu<\hat{\mathrm{e}}_p^-$. This implies that $\widehat{\mathrm{e}}_p(\mathrm{r})-\nu>0$ for all $\mathrm{r}\in(0,\infty)$. Hence, by (5.2.10), we conclude that

$$\left|\mathrm{f}_p(\varkappa)\right|<\aleph\left(\|\xi-\mathrm{z}^*\|_{\sigma}+\varrho\right)\sum_{\varsigma=0}^{\infty}\frac{(-\nu\varkappa)^{\varsigma}}{\Gamma(\varsigma\lambda+1)},$$

which contradicts (5.2.13) and we conclude that what was claimed in (5.2.12) is true. We now assume that y is constant and ϱ tends to zero. Then we have

$$\|\mathrm{f}(\mathrm{y})\|\leqslant\aleph\left\|\xi-\mathrm{z}^*\right\|_{\sigma}\sum_{\varsigma=0}^{\infty}\frac{(-\nu\varkappa)^{\varsigma}}{\Gamma(\varsigma\lambda+1)}.$$

Corollary 5.1 *If we assume conditions* $(\mathbf{I}_1)$, $(\mathbf{I}_2)$, $(\mathbf{I}_3)$, $(\mathbf{I}_5)$ *and* $(\mathbf{I}_6)$, *then the Mittag-Leffler attractiveness for unique solution* $\mathrm{z}^*\in\mathcal{BC}(\mathbb{R},\mathbb{R}^n)$ *of Eq.* (5.2.2) *is established.*

Corollary 5.2 *If we assume conditions* $(\mathbf{I}_4)$ *and* $(\mathbf{I}_7)$, *then the Mittag-Leffler attractiveness for unique solution* $\mathrm{z}^*\in\mathcal{PAP}(\mathbb{R},\mathbb{R}^n)$ *of Eq.* (5.2.2) *is established.*

5.2.7 Applications

We shall provide two examples and illustrate them—see Figs. 5.1, 5.2, and 5.3.

Example 5.4 Let $n = 2$, $p, q = 1, 2$,

(I₁) We consider Lipschitz functions $\bar{\Theta}_p(\mathtt{r}) = 0$, $\Theta_p(\mathtt{r}) = \widehat{\Theta}_p(\mathtt{r}) = \frac{2}{5}\arctan \mathtt{r}$, with the Lipschitz constants $\mathcal{T}_p^{\bar{\Theta}} = 0, \mathcal{T}^{\Theta} = \mathcal{T}_p^{\Theta} = \frac{2}{5}$.

(I₂) Let

$$\psi_{pq}(\mathtt{y}) = \frac{2}{5}\sum_{\varsigma=0}^{\infty} \frac{(-8\mathtt{y})^{\varsigma}}{\Gamma(\varsigma\lambda + 1)}.$$

Then ψ_{pq} is bounded and continuous.

(I₃)

$$|\psi_{pq}|\sum_{\varsigma=0}^{\infty} \frac{(n\mathtt{y})^{\varsigma}}{\Gamma(\varsigma\lambda + 1)}$$

is integrable on $[0, +\infty)$ for $n = 1$.

(I₄) (see [15]) Let $\phi \in \mathcal{N}$, where $d\phi(\mathtt{y}) = \eta_{\psi}(\mathtt{y})d\mathtt{y}$ and

$$\eta_{\psi}(\mathtt{y}) = \sum_{\varsigma=0}^{\infty} \frac{(\psi\mathtt{y})^{\varsigma}}{\Gamma(\varsigma\lambda + 1)},\ \psi_{pq}(\mathtt{y}) = \sum_{\varsigma=0}^{\infty} \frac{(-8\mathtt{y})^{\varsigma}}{\Gamma(\varsigma\lambda + 1)},$$

with $\vartheta_0 := 2 - \psi > 0$, $-\mathtt{u} \leqslant \mathtt{b} < \mathtt{y} \leqslant \mathtt{u}$. We have for $\omega = 1$, that $\mathtt{L} = \varnothing$.

(I₅) For $\mathtt{e}_1(\mathtt{y}) = \frac{2}{5}\left(1 + \frac{3}{2}\sin \mathtt{y}\right)$, $\mathtt{e}_2(\mathtt{y}) = \frac{2}{5}\left(1 + \frac{7}{6}\cos \mathtt{y}\right)$, $\widehat{\mathtt{e}}_p(\mathtt{y}) = \frac{2}{5}$,

$$\mathtt{O}_p = \sum_{\varsigma=0}^{\infty} \frac{(\frac{4}{15})^{\varsigma}}{\Gamma(\varsigma\lambda + 1)}.$$

(I₆) For $\sigma_p = 1$, $\mathtt{R} = \pi$, $\zeta_{pq} = \frac{3}{p+q}$, $\bar{\vartheta}_{pq}(\mathtt{y}) = 0$, $\vartheta_{pq}(\mathtt{y}) = \frac{8}{10}\sin \mathtt{y}$, $\mathtt{h}_{pq}(t) = \frac{9}{10}\cos 2\mathtt{y}$, $\zeta_{\ell} = \frac{9}{100}$,

$$\sup_{\mathtt{y}\in\mathbb{R}}\left\{-\widehat{\mathtt{e}}_p(\mathtt{y}) + \mathtt{O}_p\left[\sigma^{-1}\sum_{q}^{n}\left(|\bar{\vartheta}_{pq}(\mathtt{y})|\mathcal{T}_q^{\Theta} + |\vartheta_{\sigma}(\mathtt{y})|\mathcal{T}_q^{\Theta} + \hat{\mathcal{T}}_q^{\hat{\Theta}}\,|\mathtt{h}_p(\mathtt{y})|\int_0^{\infty}|\psi_{pq}(\mathtt{r})|\,d\mathtt{r}\right)\sigma_q\right]\right\},$$
$$\le -\kappa_p \le 0.$$

(I₇) and

$$0 < \max_{p=1,2,3,n,n}\left\{\frac{\widehat{\mathtt{e}}_p^{+}}{\widehat{\mathtt{e}}_p^{-}} - \frac{\kappa_p}{\hat{\mathtt{e}}_p^{+}}\right\} < 1.$$

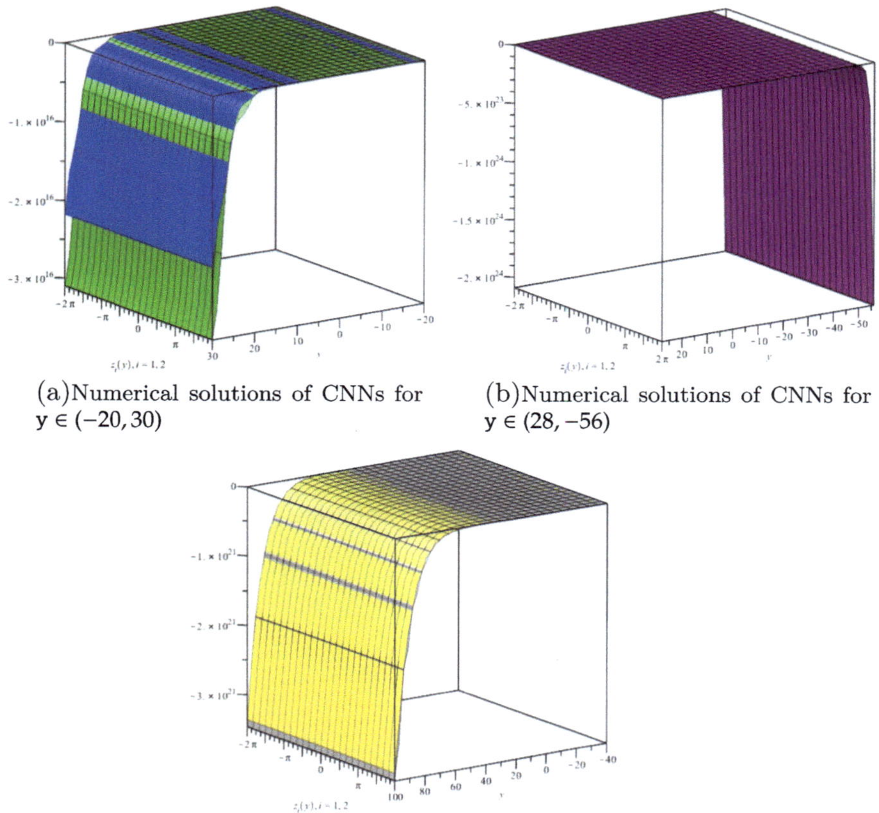

(a) Numerical solutions of CNNs for $y \in (-20, 30)$

(b) Numerical solutions of CNNs for $y \in (28, -56)$

(c) Numerical solutions of CNNs for $y \in (100, -40)$

Fig. 5.1 Graphs related to the numerical solutions of the CNNs (5.2.1) for different values

Let $L_p(y) = (20 + p)|\cos y| + U(y)$,

$$U(b) = \begin{cases} e^{-b}, & b \geqslant 0 \\ 1, & b < 0. \end{cases}$$

Therefore, all solutions of (5.2.1) are in the Mittag-Leffler form and they converge to a unique solution of Eq. (5.2.1) such as $z^*(y) \in \mathcal{PAP}\left(\mathbb{R}, \mathbb{R}^2, \phi\right)$, when $y \longrightarrow +\infty$ with convergence rate $\nu \approx 0.05 < \widehat{e}_p^-$.

Example 5.5 Assume conditions $(\mathbf{I}_1)$ to $(\mathbf{I}_7)$ and all the functions presented in Example 5.4. For $\psi > 1$, $\nu > 0$ and Dirac measure $\delta_{\frac{1}{n}}$, we also consider the following measure

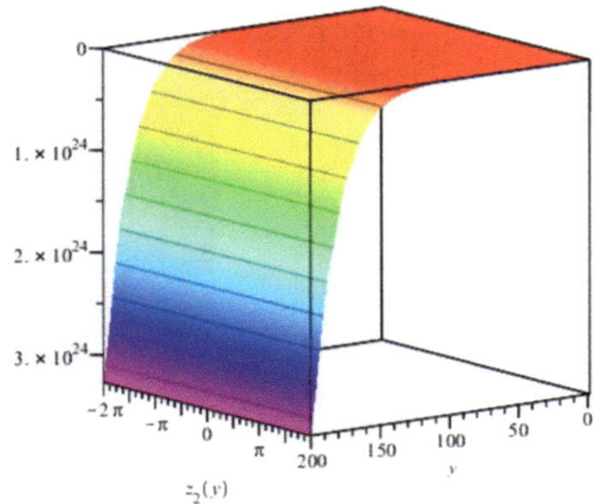

(a)Numerical solutions of CNNs for $\mathbf{y} \in (-\frac{10}{7}, 200)$

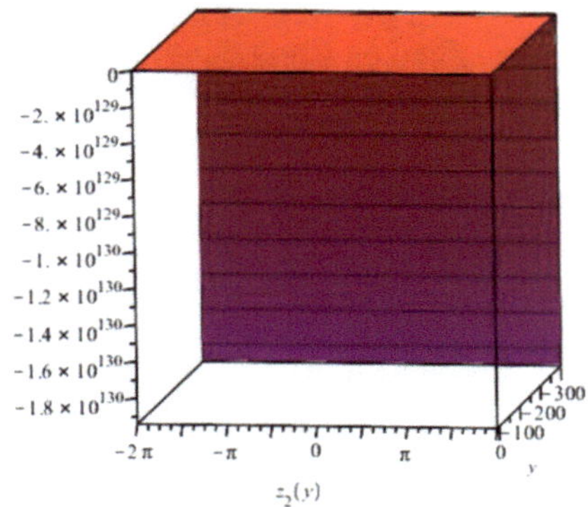

(b)Numerical solutions of CNNs for $\mathbf{y} \in (-\frac{200}{18}, -300)$

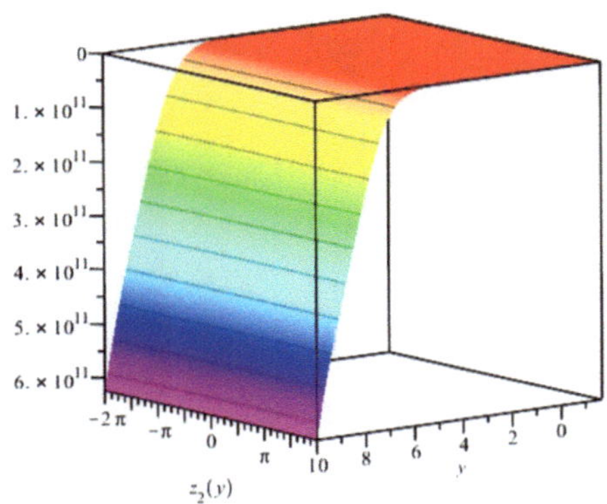

(c)Numerical solutions of CNNs for $\mathbf{y} \in (-\frac{10}{6}, 10)$

Fig. 5.2 Graphs related to the numerical solutions of the CNNs (5.2.1) for different values

$$d\phi_{\psi,\nu}(\mathrm{y}) = \sum_{\varsigma=0}^{\infty} \frac{(\psi \mathrm{y})^{\varsigma}}{\Gamma(\varsigma\lambda + 1)} d\mathrm{y} + \nu \sum_{n=1}^{\infty} \frac{1}{n^2} \delta_{\frac{1}{n}}.$$

Consider the interval $I = (-1 - |\zeta|, 1 + |\zeta|)$ For $\zeta \in \mathbb{R}$ and $\vartheta = 1$. Then, for all $\mathrm{R} \in \mathcal{A}$ and $\zeta \in \mathbb{R}$, there exist $\vartheta > 0$ and a bounded interval L such that $\varrho(\mathrm{R} + \zeta) \leqslant \vartheta \phi(\mathrm{R})$, and $\mathrm{R} \cap \mathrm{L} = \emptyset$.

Let $\mathrm{L}_p(\mathrm{y}) = (20 + p)|\cos \mathrm{y}| + \mathrm{U}(\mathrm{y})$ with

$$\mathrm{U}(\mathrm{b}) = \begin{cases} e^{-\mathrm{b}}, & \mathrm{b} \leqslant 0 \\ 0, & \mathrm{b} > 0 \end{cases}$$

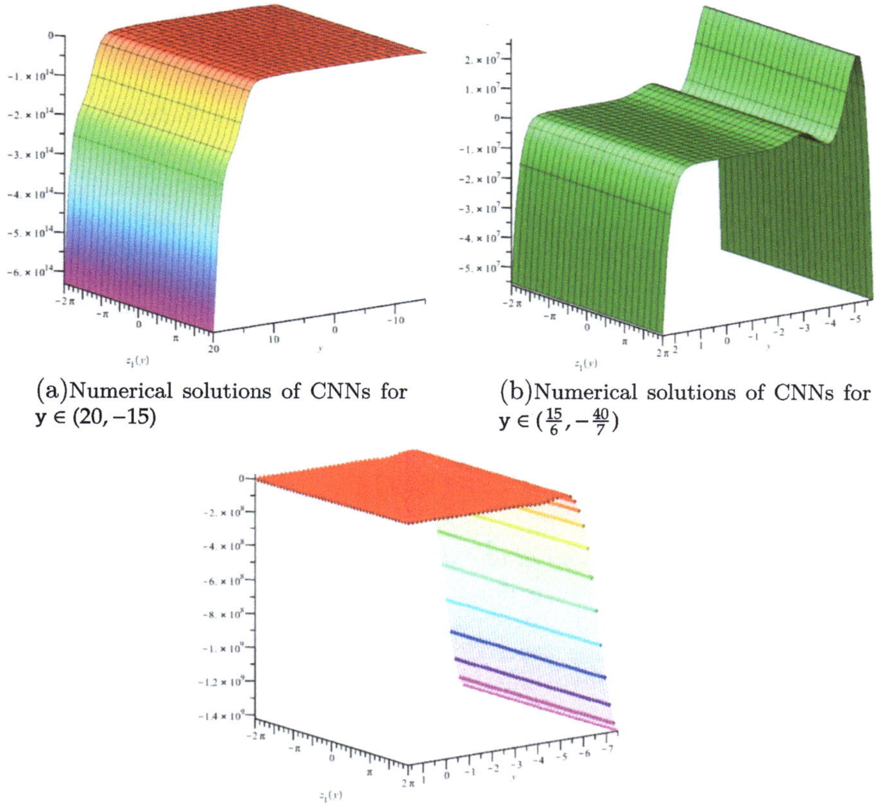

(a) Numerical solutions of CNNs for $y \in (20, -15)$

(b) Numerical solutions of CNNs for $y \in (\frac{15}{6}, -\frac{40}{7})$

(c) Numerical solutions of CNNs for $y \in (\frac{10}{6}, -\frac{60}{8})$

Fig. 5.3 Graphs related to the numerical solutions of the CNNs (5.2.1) for different values

Then $\mathtt{L}_p \in \mathcal{PAP}(\mathbb{R}, \mathbb{R}, \varrho)$. Therefore, all solutions of (5.2.1) are in the Mittag-Leffler form and they converge to a unique solution of Eq. (5.2.1) such as $\mathtt{z}^*(\mathtt{y}) \in \mathcal{PAP}\left(\mathbb{R}, \mathbb{R}^2, \phi\right)$, when $\mathtt{y} \longrightarrow +\infty$ with convergence rate $\nu \approx 0.05$.

5.3 Attractiveness of Pseudo Almost Periodic Solutions in the Context of Fuzzy Measure Theory

5.3.1 Introduction of the Delayed Cellular Neural Networks

Neural networks (NNs) are used in various fields and due to their importance, they have been widely studied. Chua and Yang [47, 48] were among those who worked on cellular neural networks (CNNs). At first, they investigated the existence of the solutions of these systems, then they studied the global exponential stability (GES) of these solutions, and in general, they played an effective role in the development of cellular neural networks (CNNs). The applications of these networks include optimization, biological systems, associative memories, face detection, image processing, etc. Considering the many applications of CNNs, Chu and Yang applied their research results to signal processing and in static image processing. Xu [25] is also among those who worked on CNNs. Using the fixed point theorem (FTP) and the method of differential inequality, he first investigated the existence of a unique solution for a cellular neural differential system. Then, he investigated the exponential stability of the solutions of this system, which are weighted pseudo almost periodic types. The term pseudo-almost periodicity, which is a natural extension of classical almost periodicity, was first introduced by Zhang [49]. After that, the authors paid a lot of attention to pseudo-almost periodic functions and were given their attention to the solutions of differential equations, partial differential equations, and functional differential equations, which are pseudo-almost periodic. They also investigated some delay differential equations and studied sufficient conditions for the existence of pseudo-almost periodic solutions for different types of delay differential equations [49, 50]. Diagana [31] introduced classical weighted pseudo almost periodic functions and ϱ-pseudo almost periodic functions were recently introduced by Blot, Ezzinbi and Cieutat [51]. Since differential equations have different solutions such as almost periodic, pseudo almost periodic, weighted pseudo almost periodic, ϱ-pseudo almost periodic, and due to the applications of these functions in various fields such as mathematical biology, physics, and control theory, investigating the existence of these solutions as well as investigating their stability in the theories of differential equations is always of interest [31, 52, 53].

We consider the following system with mixed delays (Figs. 5.4 and 5.5)

$$\begin{aligned}\mathfrak{g}_\ell'(\mathfrak{x}) = &- \mathfrak{m}_\ell(\mathfrak{x})\mathfrak{g}_\ell(\mathfrak{x}) + \sum_{\jmath=1}^{n} \bar{\omega}_{\ell\jmath}(\mathfrak{x})\bar{\mathcal{K}}_\jmath\left(\mathfrak{g}_\jmath(t)\right) + \sum_{\jmath=1}^{n} \omega_{\ell\jmath}(\mathfrak{x})\mathcal{K}_\jmath\left(\mathfrak{g}_\jmath\left(\mathfrak{x} - \kappa_{\ell\jmath}\right)\right) \\ &+ \sum_{\jmath=1}^{n} \mathfrak{p}_{\ell\jmath}(\mathfrak{x}) \int_0^\infty \varsigma_{\ell\jmath}(\mathfrak{h})\widehat{\mathcal{K}}_\jmath\left(\mathfrak{g}_\jmath(\mathfrak{x} - \mathfrak{h})\right) d\mathfrak{h} + \mathcal{H}_\ell(\mathfrak{x}), \text{ for } \mathfrak{x} \in \mathbb{R},\end{aligned} \tag{5.3.1}$$

which is called cellular neural networks (CNNs). The functions and constants of the above differential system are as follows

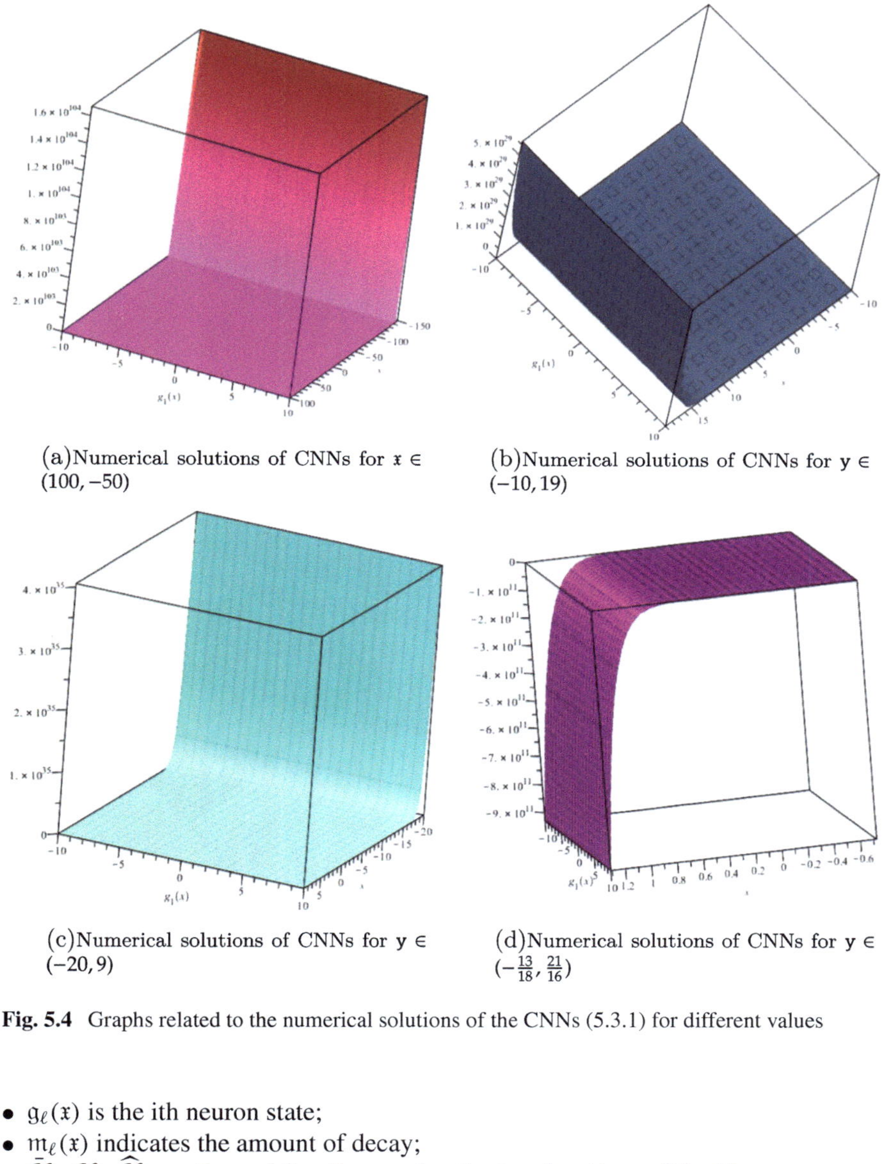

(a) Numerical solutions of CNNs for $\mathfrak{x} \in (100, -50)$

(b) Numerical solutions of CNNs for $\mathbf{y} \in (-10, 19)$

(c) Numerical solutions of CNNs for $\mathbf{y} \in (-20, 9)$

(d) Numerical solutions of CNNs for $\mathbf{y} \in (-\frac{13}{18}, \frac{21}{16})$

Fig. 5.4 Graphs related to the numerical solutions of the CNNs (5.3.1) for different values

- $\mathfrak{g}_\ell(\mathfrak{x})$ is the ith neuron state;
- $\mathfrak{m}_\ell(\mathfrak{x})$ indicates the amount of decay;
- $\bar{\mathcal{K}}_\ell, \mathcal{K}_\ell, \widehat{\mathcal{K}}_\ell$ are the real functions and activation functions of the ith neuron;
- $\bar{\omega}_{\ell_J}(\mathfrak{x}), \omega_{\ell_J}(\mathfrak{x}), \mathfrak{p}_{\ell_J}(\mathfrak{x})$ are coefficients;
- $\mathcal{H}_\ell(\mathfrak{x})$ is the input;
- In this equation the delays are κ_{ℓ_J} and $\varsigma_{\ell_J}(\mathfrak{x})$ is the transmission delay kernel.

Also, we consider the following initial value problem

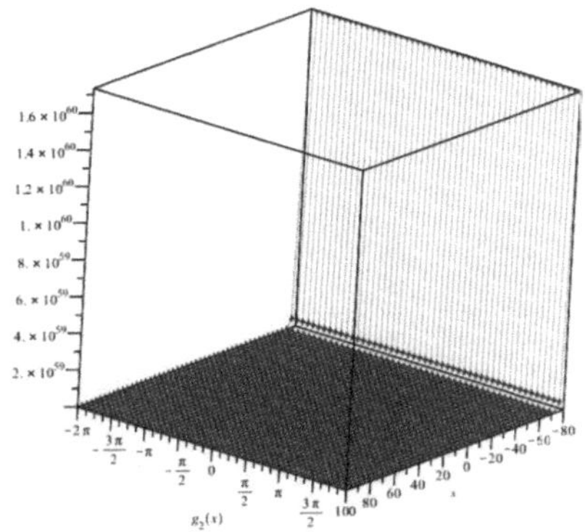

(a)Numerical solutions of CNNs for $\mathfrak{x} \in (-80, 100)$

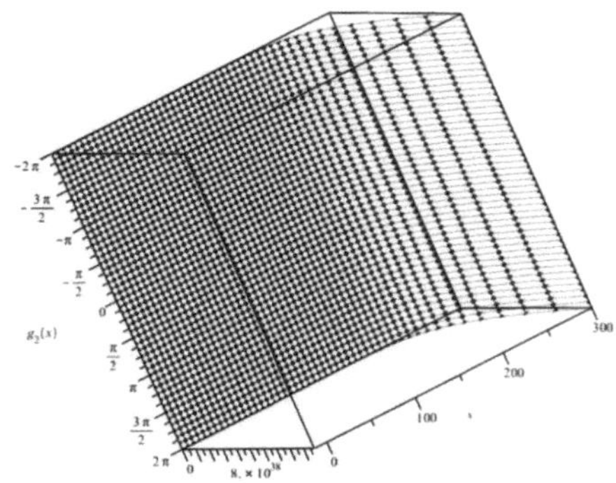

(b)Numerical solutions of CNNs for $\mathbf{y} \in (\frac{-100}{7}, 300)$

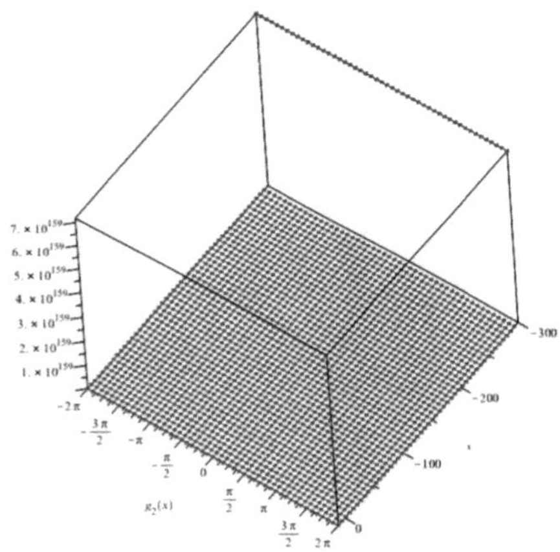

(c)Numerical solutions of CNNs for $\mathbf{y} \in (\frac{200}{18}, -300)$

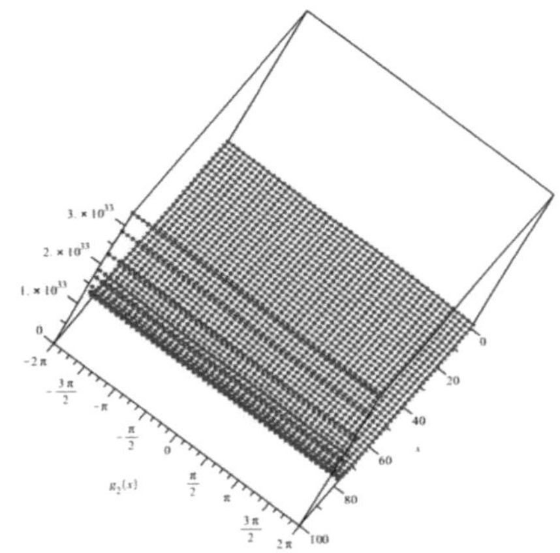

(d)Numerical solutions of CNNs for $\mathbf{y} \in (-\frac{12}{7}, 100)$

Fig. 5.5 Graphs related to the numerical solutions of the CNNs (5.3.1) for different values

$$\begin{cases} \mathfrak{g}'_\ell(\mathfrak{x}) = -\mathfrak{m}_\ell(\mathfrak{x})\mathfrak{g}_\ell(\mathfrak{x}) + \sum_{J=1}^n \bar{\omega}_{\ell J}(\mathfrak{x})\bar{\mathcal{K}}_J\left(\mathfrak{g}_J(\mathfrak{x})\right) + \sum_{J=1}^n \omega_{\ell J}(\mathfrak{x})\mathcal{K}_J\left(\mathfrak{g}_J\left(\mathfrak{x} - \kappa_{\ell J}\right)\right) \\ \quad + \sum_{J=1}^n \mathfrak{p}_{\ell J}(\mathfrak{x}) \int_0^\infty \varsigma_{\ell J}(u)\widehat{\mathcal{K}}_J\left(\mathfrak{g}_J(\mathfrak{x} - \mathfrak{h})\right) d\mathfrak{h} + \mathcal{H}_\ell(\mathfrak{x}), \text{ for } \mathfrak{x} \geqslant 0 \\ \mathfrak{g}_\ell(\mathfrak{x}) = \nu_\ell(\mathfrak{x}), \text{ for } \mathfrak{x} \leqslant 0. \end{cases} \tag{5.3.2}$$

Considering the properties and applications of fuzzy measurements and fuzzy integral, many authors have investigated these concepts. Considering the properties and applications of fuzzy measurements and fuzzy integral, many authors have investigated these concepts. For example, Wang and Klir [54] reviewed fuzzy measurement and fuzzy integral theory in general. Also, Ralescu and Adams [55], Román-Flores, and colleagues [56] have investigated some equivalent definitions of fuzzy integrals, the level-continuity of fuzzy integrals, H-continuity of fuzzy measures, and geometric inequalities for measures and fuzzy integrals, respectively. Sugeno and Choquet integrals, which are idempotent, continuous, and uniform operators, are among the most basic fuzzy integrals. Therefore, many authors have studied well-known inequalities such as general Minkowski, Chebyshev-type inequalities, Cauchy-type inequalities, etc. for this type of integrals [57, 58]. Using fuzzy measure, we get

(1) $d(\varrho(\mathfrak{x})) = \varpi(\mathfrak{x})d\mathfrak{x} + d\varrho_1(\mathfrak{x})$,
(2) $d(\varrho(\mathfrak{x})) = \varpi(\mathfrak{x})d\mathfrak{x}$, $\varrho_1 = 0$,

where $\varpi \in \mathbb{R}$ is a nonnegative measurable function, ϱ and ϱ_1 are measure with respect to the Lebesgue measure $d\mathfrak{x}$ that ϱ_1 is singular and ϱ is absolutely continuous. When we have the special form ϱ, that is, when $\varrho_1 = 0$, the ϱ-pseudo almost periodic solutions of Eq. (5.3.1) are called weighted pseudo almost periodic functions. The class of ϱ-pseudo almost periodic functions encompasses the class of weighted pseudo almost periodic functions. Investigating the existence of the solution of these systems, which are of the type functions ϱ-pseudo almost periodic, and the Mittag-Leffler stability of these solutions is one of the most important studies that have been done on these networks [33, 47, 48, 59]. Here, we first prove the existence unique solution of system (5.3.1) in matrix-valued fuzzy normed spaces. In the following, we investigate the ML stability (Mittag-Leffler stability) and ML attractiveness of the solutions of this differential system, which are of the ϱ-periodic functions using fuzzy measurement theory.

5.3.2 Preliminaries

Definition 5.16 ([61, 62]) The Mittag-Leffler function is defined by

$$E_{\lambda}(\mathrm{g}) = \sum_{\varsigma=0}^{\infty} \frac{\mathrm{g}^{\varsigma}}{\Gamma(\varsigma\lambda + 1)},$$

where λ is a real number, $\lambda \leq 0$ and g is a complex variable. The generalization of $E_{\lambda}(\mathrm{g})$ defined as

$$E_{\lambda,\mu}(\mathrm{g}) = \sum_{\varsigma=0}^{\infty} \frac{\mathrm{g}^{\varsigma}}{\Gamma(\varsigma\lambda + \mu)},$$

where $\lambda, \mu \in \mathbb{C}$, $\quad Re(\lambda) > 0, \quad Re(\mu) > 0$.

Definition 5.17 ([63]) If $0 < \lambda \leq 1$ and ν is a complex number, then

$$cosh_\lambda(\nu x^\lambda) = \sum_{\varsigma=0}^{\infty} \frac{\nu^{2\varsigma} x^{2\varsigma\lambda}}{\Gamma(2\varsigma\lambda + 1)},$$

is called λ-order fractional hyperbolic cosine function.

$$sinh_\lambda(\nu x^\lambda) = \sum_{\varsigma=0}^{\infty} \frac{\nu^{2\varsigma+1} x^{(2\varsigma+1)\lambda}}{\Gamma((2\varsigma + 1)\lambda + 1)},$$

is called λ-order fractional hyperbolic sine function.

Proposition 5.4 ([63]) *Assume that* $0 < \lambda \leq 1$ *then*

$$cosh_\lambda(\nu x^\lambda) = \frac{E_\lambda(\nu x^\lambda) + E_\lambda(-\nu x^\lambda)}{2},$$
$$sinh_\lambda(\nu x^\lambda) = \frac{E_\lambda(\nu x^\lambda) - E_\lambda(-\nu x^\lambda)}{2}.$$

5.3.3 Fuzzy Measure Theory

Definition 5.18 ([54, 55]) Let Σ be a σ-algebra of subsets of X, and let $\mu : \Sigma \rightarrow [0, \infty)$ be a non-negative extended real-valued set function. We say that μ is a fuzzy measure iff:

(1) $\mu(\emptyset) = 0$;
(2) $E, F \in \Sigma : E \subseteq F$ imply $\mu(E) \leq \mu(F)$ (monotonicity);
(3) $E_n \in \Sigma (n \in \mathbb{N})$, $E_1 \subset E_2 \subset \ldots$, imply $\lim_{n\to\infty} \mu(E_n) = \mu\left(\cup_{n=1}^{\infty} E_n\right)$ (continuity from below);
(4) $E_n \in \Sigma \quad (n \in \mathbb{N})$, $E_1 \supset E_2 \supset \ldots$, $\mu(E_1) < \infty$, imply $\lim_{n\to\infty} \mu(E_n) = \mu\left(\cap_{n=1}^{\infty} E_n\right)$ (continuity from above).

Let (X, Σ, μ) be a fuzzy measure space and f be a non-negative real-valued function on X. We denote by $\mathcal{F}_+$the set of all non-negative real valued measurable functions, and by $L_\alpha f$ the set: $L_\alpha f := \{x \in X : f(x) \geq \alpha\}$, the α-level of f for $\alpha \geq 0$.

Definition 5.19 ([54, 64]) Let (X, Σ, μ) be a fuzzy measure space. If $f \in \mathcal{F}_+$and $A \in \Sigma$, then the Sugeno integral (fuzzy integral) of f on A with respect to the fuzzy measure μ is defined by

$$(S) \int_A f d\mu := \vee_{\alpha\geq 0} (\alpha \wedge \mu (A \cap L_\alpha f)), \tag{5.3.3}$$

where $\vee$ and $\wedge$ denote the sup and inf on $[0, \infty]$, respectively.

The basic properties of Sugeno integral follow:

Theorem 5.9 ([54, 64]) *Let* (X, Σ, μ) *be a fuzzy measure space with* $A, B \in \Sigma$ *and* $f, g \in \mathcal{F}_+$. *Then*

(1) $(S) \int_A f d\mu \leq \mu(A)$;
(2) $(S) \int_A k d\mu = k \wedge \mu(A)$ *for a non-negative constant* k;
(3) *if* $f \leq g$ *on* A, *then* $(S) \int_A f d\mu \leq (S) \int_A g d\mu$;
(4) *if* $A \subset B$, *then* $(S) \int_A f d\mu \leq (S) \int_B f d\mu$;
(5) $\mu (A \cap L_\alpha f) \leq \alpha \Rightarrow (S) \int_A f d\mu \leq \alpha$;
(6) *if* $\mu(A) < \infty$, *then* $\mu (A \cap L_\alpha f) \geq \alpha \Leftrightarrow (S) \int_A f d\mu \geq \alpha$;
(7) *when* $A = X$, *then* $(S) \int_A f d\mu = \vee_{\alpha \geq 0} (\alpha \wedge \mu (L_\alpha f))$;
(8) *if* $\alpha \leq \beta$, *then* $L_\beta f \subseteq L_\alpha f$;
(9) $(S) \int_A f d\mu \geq 0$.

Theorem 5.10 ([54]) *Here,* $f \in \mathcal{F}_+$, *the class of all finite nonnegative measurable functions on* (X, Σ, μ). *Then*

(1) *if* $\mu(A) = 0$, *then* $(S) \int_A f d\mu = 0$, *for any* $f \in \mathcal{F}_+$;
(2) *if* $(S) \int_A f d\mu = 0$, *then* $\mu(A \cap \{x \mid f(x) > 0\}) = 0$;
(3) $(S) \int_A f d\mu = (S) \int_A f \cdot \chi_A d\mu$, *where* χ_A *is the characteristic function of* A;
(4) $(S) \int_A (f + a) d\mu \leq (S) \int_A f d\mu + (S) \int_A a d\mu$, *for any constant* $a \in [0, \infty)$.

Definition 5.20 ([65]) A fuzzy measure μ is subadditive iff $\mu(A \cup B) \leq \mu(A) + \mu(B)$, for all $A, B \in \Sigma$.

Theorem 5.11 ([65]) *If* μ *is subadditive, then*

$$(S) \int_X (f + g) d\mu \leq (S) \int_X f d\mu + (S) \int_X g d\mu, \tag{5.3.4}$$

for all measurable functions $f, g : X \to [0, \infty)$.

Moreover, if (5.3.4) holds for all measurable functions $f, g : X \to [0, \infty)$ and $\mu(X) < \infty$, then μ is subadditive. Notice here in (5.3.3) we have that $\alpha \in [0, \infty)$.

Corollary 5.3 *If* μ *is subadditive,* $n \in \mathbb{N}$, *and* $f : X \to [0, \infty)$ *is a measurable function, then*

$$(S) \int_X n f d\mu \leq n (S) \int_X f d\mu.$$

We consider all continuous and bounded functions (we denote the space of all these functions by $\mathcal{BC}(\mathbb{R}, \mathcal{L})$ such as $\mathfrak{w}$ from $\mathbb{R}$ to $\mathcal{L}$ in matrix-valued fuzzy norm space $(\mathcal{L}, \mathcal{N}, \circledast)$ with matrix-valued fuzzy norm $\mathcal{N}(\mathfrak{w}(\mathfrak{x}), \mathfrak{z}) = \inf_{\mathfrak{x} \in \mathbb{R}} \mathcal{K}(\mathfrak{w}(\mathfrak{x}), \mathfrak{z})$.

We also represent the space of all positive fuzzy measures ϱ with $\mathcal{H}$ that these measures can be defined in the Lebesgue ς-field. We have for these measures:

- $\varrho(\mathbb{R}) = +\infty$, for all $x, y \in \mathbb{R} (x \leqslant y)$;

- $\varrho([x, y]) < \infty$, for all $x, y \in \mathbb{R}(x \leqslant y)$.

Definition 5.21 Let $[x, y]$ be an interval of length e_ε and $\kappa \in [x, y]$. If for every $\boldsymbol{\varepsilon} > \mathbf{0}$ there exists a positive number e_ε, then a continuous function $\mathfrak{w} : \mathbb{R} \to \mathcal{L}$ is almost periodic if

$$\mathcal{N}(\mathfrak{w}(\mathfrak{x} + \kappa) - \mathfrak{w}(\mathfrak{x}), \mathfrak{z}) > \boldsymbol{\varepsilon},$$

for all $\mathfrak{x} \in \mathbb{R}$. We denote the set of all almost periodic functions with $\mathcal{AP}(\mathbb{R}, \mathcal{L})$.

Definition 5.22 For the bounded continuous function $\mathfrak{w} : \mathbb{R} \to \mathcal{L}$ and $\varrho \in \mathcal{H}$ we say that $\mathfrak{w}$ is ϱ-ergodic if

$$\mathcal{N}\left(\lim_{s \to \infty} \frac{1}{\varrho([-s, s])} (S) \int_{-s}^{s} \mathfrak{w}(\mathfrak{x}) d\varrho(\mathfrak{x}), \mathfrak{z} \right) = \mathbf{1}.$$

We represent the set of all ϱ-ergodic functions with $\mathcal{E}(\mathbb{R}, \mathcal{L}, \varrho)$.

Definition 5.23 Suppose $\mathfrak{w} : \mathbb{R} \to \mathcal{L}$ is a continuous function. For every $\varrho \in \mathcal{H}$, the function $\mathfrak{w}$ is a ϱ-pseudo almost periodic function if the following condition holds

$$\mathfrak{w} = \mathfrak{u} + \mathcal{E}.$$

We denote the set of all pseudo almost periodic functions with $\mathcal{PAP}(\mathbb{R}, \mathcal{L}, \varrho)$.

We show the space of all functions defined in 5.21, 5.22, and 5.23 along with matrix valued fuzzy norms with matrix valued fuzzy Banach spaces (MVFBS) $(\mathcal{AP}(\mathbb{R}, \mathcal{L}), \mathcal{N}, \circledast)$, $(\mathcal{E}(\mathbb{R}, \mathcal{L}, \varrho), \mathcal{N}, \circledast)$, $(\mathcal{PAP}(\mathbb{R}, \mathcal{L}, \varrho), \mathcal{N}, \circledast)$, respectively.

In the sequel, we consider two functions $\mathfrak{w}, \mathfrak{v} : \mathbb{R} \mapsto \mathbb{R}$, $\varrho \in \mathcal{H}$ and also suppose that for $\ell, \jmath = 1, 2, 3, \ldots, n$, $\bar{\omega}_{\ell\jmath}, \omega_{\ell_\jmath}, \mathfrak{p}_{\ell\jmath}, \mathcal{H}_\ell$ are ϱ-pseudo almost periodic functions and $\mathfrak{m}_\ell$ are almost periodic functions. The convolution of $\mathfrak{w}$ with $\mathfrak{v}$ is defined as follows

$$(\mathfrak{w} \star \mathfrak{v})(\mathfrak{x}) := \int_{-\infty}^{\infty} \mathfrak{w}(\mathfrak{h})\mathfrak{v}(\mathfrak{x} - \mathfrak{h}) d\mathfrak{h},$$

if it exists.

Definition 5.24 We consider two measures ϱ and $\mathfrak{r}$ in Borel space $(\mathfrak{X}, B)$. We have:

- If the disjoint sets T and U are Lebesgue ϱ-field such that $\mathfrak{X} = T \cup U$, then ϱ and $\mathfrak{r}$ are mutually singular, and $\mathfrak{r}(T) = \varrho(U) = 0$.
- Duo to the measure of ϱ, $\mathfrak{r}$ is absolutely continuous, if for each Lebesgue ϱ-field T,

$$\varrho(T) = 0, \tag{5.3.5}$$

then $\mathfrak{r}(T) = 0$.

Theorem 5.12 ([66]) *We consider two ς-finite measures ϱ, and $\mathfrak{r}$ in measurable space $(\mathcal{L}, \mathcal{Z})$. Then there are two unique measures ϱ_0 and ϱ_1 such that ϱ_0 is singular with respect to the finite measure $\mathfrak{r}$ and ϱ_1 is absolutely continuous with respect to this finite measure $\mathfrak{r}$. Also we have $\varrho = \varrho_0 + \varrho_1$.*

Theorem 5.12 is known as Lebesgue-Radon-Nikodym. What we have done is to use theorem 5.12 to consider fuzzy measure $d\varrho(\mathfrak{x}) = \varpi(\mathfrak{x})d\mathfrak{x} + d\varrho_1$ that $d\mathfrak{r} = d\mathfrak{x}$. We impose the following assumptions:

($\mathbf{M}_1$) Suppose that $\mathcal{Z}$ is the ς-field of the Lebesgue type of $\mathbb{R}$. For all $T \in \mathcal{Z}$ and $\kappa \in \mathbb{R}$, there exist $\omega > 0$ and a bounded interval $\mathcal{H}$ such that $\varrho(T + \kappa) \leqslant \omega\varrho(T)$ and $T \cap \mathcal{H} = \emptyset$.

($\mathbf{M}_2$) There exist $\kappa_0 \geqslant 0$ and $\epsilon_0 > 0$ such that if $|\kappa| \geqslant \kappa_0$ then $\varrho((x+\kappa, y+\kappa)) \geqslant \epsilon_0\varrho([\kappa, y+\kappa])$, where $x, y, s \in \mathbb{R}$ and $0 \leqslant x < y \leqslant s$.

Condition ($\mathbf{M}_1$) implies ($\mathbf{M}_2$) but the reverse is not true. It is easy to show that if ($\mathbf{M}_2$) holds then $(\mathcal{PAP}(\mathbb{R}, \mathbb{R}^n, \varrho), \mathcal{N}, \circledast)$ is a MVFBS. If condition ($\mathbf{M}_2$) is true and considering fuzzy measures ϱ and ϱ_τ, also we can easily see that ϱ and ϱ_τ are equivalent for all $\tau \in \mathbb{R}$ and for all $a > 0$ we have, $\lim_{r\to\infty} \sup \frac{\varrho([-r-a, r+a])}{\varrho([-r,r])} < \infty$. As well as, let $\varrho_1, \varrho_2 \in \mathcal{H}$. If ϱ_1 and ϱ_2 are equivalent, then $\mathcal{E}(\mathbb{R}, \mathbb{R}, \varrho_1) = \mathcal{E}(\mathbb{R}, \mathbb{R}, \varrho_2)$, therefore $\mathcal{PAP}(\mathbb{R}, \mathbb{R}^n, \varrho_1) = \mathcal{PAP}(\mathbb{R}, \mathbb{R}^n, \varrho_2)$.

Theorem 5.13 *Let ($\boldsymbol{M}_1$) holds and $\mathcal{L} = \mathbb{R}^n$. If $\mathfrak{w}$ is a ϱ- pseudo almost periodic function, then for every $\kappa \in \mathbb{R}$, $\mathfrak{w}(\mathfrak{x} - \kappa)$ is also a ϱ- pseudo almost periodic function. This means that $\mathfrak{w}(\mathfrak{x} - \kappa) \in \mathcal{PAP}(\mathbb{R}, \mathbb{R}^n, \varrho)$.*

Proof Since, $\mathcal{AP}(\mathbb{R}, \mathcal{L})$ is invariant by translation: if $\mathfrak{w} \in \mathcal{AP}(\mathbb{R}, \mathcal{L})$ implies $\mathfrak{w}_\tau \in \mathcal{AP}(\mathbb{R}, \mathcal{L})$ for all $\tau \in \mathbb{R}$. It remains to prove that if $\mathfrak{w} \in \mathcal{E}(\mathbb{R}, \mathbb{R}, \varrho)$, then $\mathfrak{w}_\tau \in \mathcal{E}(\mathbb{R}, \mathbb{R}, \varrho)$. Let $\mathfrak{w} \in \mathcal{E}(\mathbb{R}, \mathbb{R}, \varrho)$ and $\tau \in \mathbb{R}$. Since $\varrho(\mathbb{R}) = \infty$ there exists $r_0 > 0$ such that $\varrho([-r - |\tau|, r + |\tau|]) > 0$ for all $r \geq r_0$. In this proof, we always assume that $r \geq r_0$. Let us denote by

$$\boldsymbol{M}_\tau(\boldsymbol{r}) = \mathcal{N}\left(\frac{1}{\varrho_\tau([-r, r])}\ (S)\int_{-r}^{r} \mathfrak{w}(\mathfrak{x})d\varrho_\tau(\mathfrak{x}), \mathfrak{y}\right), \tag{5.3.6}$$

where ϱ_τ the positive measure defined $\varrho_\tau(A) = \varrho(\{a + \tau : a \in A\})$ for $A \in \mathcal{Z}$. Since ϱ_τ and ϱ are equivalent, then we have $\mathcal{E}(\mathbb{R}, \mathbb{R}, \varrho) = \mathcal{E}(\mathbb{R}, \mathbb{R}, \varrho_\tau)$, therefore $\mathfrak{w} \in \mathcal{E}(\mathbb{R}, \mathbb{R}, \varrho_\tau)$, that is

$$\lim_{r\to\infty} \boldsymbol{M}_\tau(\boldsymbol{r}) = \mathbf{1}, \tag{5.3.7}$$

for all $\tau \in \mathbb{R}$. For all $A \in \mathcal{Z}$ we denote by χ_A the characteristic function of A. By using definition of the measure ϱ_τ we obtain that $(S)\int_{-r}^{r} \chi_A(\mathfrak{x})d\varrho_\tau(\mathfrak{x}) = (S)\int_{-r+\tau}^{r+\tau} \chi_A(\mathfrak{x} - \tau)d\varrho(\mathfrak{x})$ for all $A \in \mathcal{Z}$ and since $\mathfrak{x} \longmapsto \mathcal{N}(\mathfrak{w}(\mathfrak{x}), \mathfrak{y})$ is the point-wise limit of an increasing sequence of linear combinations of characteristic functions [[62], Theorem 1.17, p. 15], we deduce that

$$\mathcal{N}\Big((S)\int_{-r}^{r}\mathfrak{w}(\mathfrak{x})d\varrho_\tau(\mathfrak{x}),\mathfrak{y}\Big)=\mathcal{N}\Big((S)\int_{-r+\tau}^{r+\tau}\mathfrak{w}(\mathfrak{x}-\tau)d\varrho(\mathfrak{x}),\mathfrak{y}\Big). \tag{5.3.8}$$

Using the positive measure definition and by (5.3.6) and (5.3.8), we obtain

$$\boldsymbol{M}_\tau(\boldsymbol{r})=\mathcal{N}\Big(\frac{1}{\varrho([-r+\tau,r+\tau]}(S)\int_{-r+\tau}^{r+\tau}\mathfrak{w}(\mathfrak{x}-\tau)d\varrho(\mathfrak{x}),\mathfrak{y}\Big).$$

If we denote by $\tau^+=\inf(\tau,0)$ and $\tau^-=\inf(-\tau,0)$, we have $|\tau|+\tau=2\tau^+$ and $|\tau|-\tau=2\tau^-$ and then $[-r+\tau-|\tau|,r+\tau+|\tau|]=[-r-2\tau^-,r+2\tau^+]$. Therefore we obtain

$$\boldsymbol{M}_\tau(\boldsymbol{r}+|\boldsymbol{\tau}|)=\mathcal{N}\Big(\frac{1}{\varrho([-r-2\tau^-,r+2\tau^+])}(S)\int_{-r-2\tau^-}^{r+2\tau^+}\mathfrak{w}(\mathfrak{x}-\tau)d\varrho(\mathfrak{x}),\mathfrak{y}\Big). \tag{5.3.9}$$

From (5.3.9) and the following inequality:

$$\mathcal{N}\Big(\frac{1}{\varrho([-r,r])}(S)\int_{-r}^{r}\mathfrak{w}(\mathfrak{x}-\tau)d\varrho(\mathfrak{x}),\mathfrak{y}\Big)\succeq\mathcal{N}\Big(\frac{1}{\varrho([-r,r])}(S)\int_{-r-2\tau^-}^{r+2\tau^+}\mathfrak{w}(\mathfrak{x}-\tau)d\varrho(\mathfrak{x}),\mathfrak{y}\Big),$$

we obtain

$$\mathcal{N}\Big(\frac{1}{\varrho([-r,r])}(S)\int_{-r}^{r}\mathfrak{w}(\mathfrak{x}-\tau)d\varrho(\mathfrak{x}),\mathfrak{y}\Big)\succeq\mathcal{N}\Big(\frac{\varrho([-r-2\tau^-,r+2\tau^+])}{\varrho([-r,r])}M_\tau(r+|\tau|),\mathfrak{y}\Big),$$

that implies

$$\mathcal{N}\Big(\frac{1}{\varrho([-r,r])}(S)\int_{-r}^{r}\mathfrak{w}(\mathfrak{x}-\tau)d\varrho(\mathfrak{x}),\mathfrak{y}\Big)\succeq\mathcal{N}\Big(\frac{\varrho([-r-2|\tau|,r+2|\tau|])}{\varrho([-r,r])}M_\tau(r+|\tau|),\mathfrak{y}\Big). \tag{5.3.10}$$

From (5.3.7) and (5.3.10), we deduce that

$$\mathcal{N}\Big(\frac{1}{\varrho([-r,r])}(S)\int_{-r}^{r}\mathfrak{w}(\mathfrak{x}-\tau)d\varrho(\mathfrak{x}),\mathfrak{y}\Big)=\mathbf{1},$$

that is $\mathfrak{w}_\tau\in\mathcal{E}(\mathbb{R},\mathbb{R},\varrho)$. We prove that if $\mathfrak{w}\in\mathcal{E}(\mathbb{R},\mathbb{R},\varrho)$ then $\mathfrak{w}_{-\tau}\in\mathcal{E}(\mathbb{R},\mathbb{R},\varrho)$ for all $\tau\in\mathbb{R}$ that $\mathcal{E}(\mathbb{R},\mathbb{R},\varrho)$ is invariant by translation.

Theorem 5.14 *Let $(\boldsymbol{M}_1)$ holds and $\mathfrak{K}\in\mathrm{Q}^1(\mathbb{R})$. If $\mathfrak{G}$ is a ϱ-ergodic function, then $\mathfrak{K}\star\mathfrak{G}$ is a ϱ-ergodic function, means $\mathfrak{K}\star\mathfrak{G}\in\mathcal{E}(\mathbb{R},\mathbb{R},\varrho)$.*

Proof Let $\mathfrak{G}\in\mathcal{E}(\mathbb{R},\mathbb{R},\varrho)$. Since $\mathfrak{K}\in\mathcal{BC}(\mathbb{R},\mathcal{L})$, then $\mathfrak{K}\star\mathfrak{G}\in\mathcal{BC}(\mathbb{R},\mathcal{L})$. From $\varrho(\mathbb{R})=\infty$ we deduce the existence of $r_0>0$ such that $\varrho[-r,r]>0$ for all $r>r_0$. From the following inequality:

$$\mathcal{N}\Big(\frac{1}{\varrho([-r,r])}(S)\int_{-r}^{r}\mathfrak{K}\star\mathfrak{G}(\mathfrak{x})d\varrho(\mathfrak{x}),\mathfrak{y}\Big)\succeq\mathcal{N}\Big(\frac{1}{\varrho([-r,r])}(S)\int_{-r}^{r}\int_{-\infty}^{\infty}\mathfrak{K}(\mathfrak{G}(\mathfrak{x}-s)dsd\varrho(\mathfrak{x}),\mathfrak{y}\Big),$$

and by using Fubini's theorem, we deduce

$$\mathcal{N}\Big(\frac{1}{\varrho([-r,r])}(S)\int_{-r}^{r}\mathfrak{K}\star\mathfrak{G}(\mathfrak{x})d\varrho(\mathfrak{x}),\mathfrak{y}\Big)\succeq\mathcal{N}\Big(\int_{-\infty}^{\infty}\frac{\mathfrak{K}}{\varrho([-r,r])}(S)\int_{-r}^{r}\mathfrak{G}(\mathfrak{x}-s)d\varrho(\mathfrak{x})ds,\mathfrak{y}\Big).\tag{5.3.11}$$

By using 5.13, we deduce that

$$\mathcal{N}\left(\lim_{r\to\infty}\frac{1}{\varrho([-r,r])}(S)\int_{-r}^{r}\mathfrak{G}(\mathfrak{x}-s)d\varrho(\mathfrak{x}),\mathfrak{y}\right)=\mathbf{1},$$

for all $s\in\mathbb{R}$. Since

$$\mathbf{1}\succeq\mathcal{N}\Big(\frac{\mathfrak{K}}{\varrho([-r,r])}(S)\int_{-r}^{r}\mathfrak{G}(\mathfrak{x}-s)d\varrho(\mathfrak{x}),\mathfrak{y}\Big),$$

and

$$\mathcal{N}\Big(\frac{\mathfrak{K}}{\varrho([-r,r])}(S)\int_{-r}^{r}\mathfrak{G}(\mathfrak{x}-s)d\varrho(\mathfrak{x}),\mathfrak{y}\Big)\succeq\mathcal{N}\Big(\mathfrak{K}(s)\mathfrak{G},\mathfrak{y}\Big),$$

where $\mathfrak{K}\in Q^1(\mathbb{R})$, in view of the Lebesgue dominated convergence theorem, we obtain

$$\mathcal{N}\Big(\lim_{r\to\infty}\frac{\mathfrak{K}(s)}{\varrho([-r,r])}(S)\int_{-r}^{r}\mathfrak{G}(\mathfrak{x}-s)d\varrho(\mathfrak{x})ds,\mathfrak{y}\Big)=\mathbf{1}.\tag{5.3.12}$$

From (5.3.11) and (5.3.12), we obtain that

$$\mathcal{N}\Big(\lim_{r\to\infty}\frac{1}{\varrho([-r,r])}(S)\int_{-r}^{r}\mathfrak{K}\star\mathfrak{G}(\mathfrak{x})d\varrho(\mathfrak{x}),\mathfrak{y}\Big)=\mathbf{1},$$

that is $\mathfrak{K}\star\mathfrak{G}\in\mathcal{E}(\mathbb{R},\mathbb{R},\varrho)$.

Theorem 5.15 *Suppose that* $\mathfrak{w}$ *is a almost periodic function and* $\mathfrak{v}:\mathbb{R}\longrightarrow\mathbb{R}$ *be an integrable function. Then* $\mathfrak{w}\star\mathfrak{v}$ *is a almost periodic function, means* $\mathfrak{w}\star\mathfrak{v}\in\mathcal{AP}(\mathbb{R},\mathbb{R})$.

Proof Let $\mathfrak{w}\in\mathcal{BC}(\mathbb{R},\mathcal{L})$. To state that $\mathfrak{w}\star\mathfrak{v}\in\mathcal{BC}(\mathbb{R},\mathcal{L})$, we consider the function $\Xi_n:\mathbb{R}\to\mathcal{L}$ ($n\in\mathbb{N}$ and $n\geq 1$) defined by $\Xi_n(\mathfrak{x})=\int_{-n}^{n}\mathfrak{v}\mathfrak{w}(\mathfrak{x}-s)ds$. It is clear that Ξ_n is bounded on $\mathbb{R}$ and we deduce that Ξ_n is continuous by using the uniform continuity of $\mathfrak{w}$ on all compact subsets of $\mathbb{R}$. Consequently, $\Xi_n\in\mathcal{BC}(\mathbb{R},\mathcal{L})$ and from the following inequality:

$$\mathcal{N}\Big((\mathfrak{w}\star\mathfrak{v})(\mathfrak{x})-\Xi_n(\mathfrak{x}),\mathfrak{y}\Big)\succeq\mathcal{N}\Big(\mathfrak{w}(\int_{-\infty}^{n}\mathfrak{v}(s)ds+\int_{0}^{\infty}\mathfrak{v}(s)ds),\mathfrak{y}\Big),$$

we deduce that $\lim_{n\to\infty} \Xi_n = \mathfrak{w} \star \mathfrak{v}$ uniformly on $\mathbb{R}$, therefore $\mathfrak{w} \star \mathfrak{v} \in \mathcal{BC}(\mathbb{R}, \mathcal{L})$. From the following inequality:

$$\mathcal{N}\Big(\mathfrak{w} \star \mathfrak{v}(\mathfrak{x}+\tau) - \mathfrak{w} \star \mathfrak{v}(\mathfrak{x}), \mathfrak{y}\Big) \succeq \mathcal{N}\Big(\int_{-\infty}^{\infty} \mathfrak{v}(s)(\mathfrak{w}(\mathfrak{x}+\tau-s) - \mathfrak{w}(\mathfrak{x}-s))ds, \mathfrak{y}\Big),$$

we deduce that if $\mathfrak{w}$ is a almost periodic function, then $\mathfrak{w} \star \mathfrak{v} \in \mathcal{AP}(\mathbb{R}, \mathbb{R})$.

It is also not difficult to see that $\mathcal{N}\big(\mathfrak{w}(\mathfrak{x}+\kappa) - \mathfrak{w}(\mathfrak{x}), \mathfrak{z}\big) \succeq \iota$, $\mathcal{N}\big(\mathfrak{v}(\mathfrak{x}+\kappa) - \mathfrak{v}(\mathfrak{x}), \mathfrak{z}\big) \succeq \iota$ and $\mathfrak{v}\mathfrak{w} \in \mathcal{AP}(\mathbb{R}, \mathbb{R})$ for all $\mathfrak{x} \in \mathbb{R}$, $\mathfrak{w}$, $\mathfrak{v} \in \mathcal{AP}(\mathbb{R}, \mathbb{R}))$ and $\kappa \in [x, y]$ with positive length d_ι where $\iota > 0$. Also, if $\mathcal{K}$ is a global contraction mapping on the Banach spaces $\mathcal{L}$ and $\mathcal{F}$, then $\mathcal{K} \circ \mathfrak{G} \in \mathcal{AP}(\mathbb{R}, \mathcal{F})$ for every almost periodic function $\mathfrak{G}$.

5.3.4 Applications of the Fuzzy Measures

In the sequel, we introduce three fuzzy measures that satisfy the defined conditions $(\mathbf{M}_1)$ and $(\mathbf{M}_2)$ ([7, 14]).

Example 5.6 Due to the measure of the Lebesgue type, we consider a fuzzy measure $\varrho \in \mathcal{H}$ which is not absolutely continuous and satisfying $(\mathbf{M}_2)$. This fuzzy measure is defined as $d\varrho(\mathfrak{x}) = d\mathfrak{x} + d\mathfrak{r}$ where $d\mathfrak{x}$ is a measure of the Lebesgue type. Also, $\mathfrak{r}$ is the fuzzy measure on $(\mathbb{R}, \mathcal{Z})$, which in $\mathcal{Z}$ is the ς-field of the Lebesgue type. This fuzzy measure define as follow

$$\mathfrak{r}(T) = \begin{cases} \text{card}(T \cap \mathbb{Z}) & \text{if } T \cap \mathbb{Z} \text{ is infinite,} \\ \infty & \text{if } T \cap \mathbb{Z} \text{ is infinite.} \end{cases}$$

Example 5.7 Consider the following fuzzy measure

$$d\phi_{\psi,\varkappa}(\mathrm{y}) = \sum_{\varsigma=0}^{\infty} \frac{(\psi \mathrm{y})^{\varsigma}}{\Gamma(\varsigma\lambda+1)} d\mathrm{y} + \varkappa \sum_{n=-\infty}^{\infty} \left(\sum_{\varsigma=0}^{\infty} \frac{(\psi n)^{\varsigma}}{\Gamma(\varsigma\lambda+1)} \right) \delta_n,$$

since $[\zeta, \mathrm{b}+\zeta] \subset [\zeta, [\mathrm{b}] + 1 + \mathrm{b}]$, we will show that $(\mathbf{M}_2)$ is established for $\mathrm{b} > 0$, such that $\mathrm{b} \geq \jmath$. In this measure, $\psi \geqslant 0$, $\varkappa > 0$, δ_n is Dirac measure at the integer n. $\sum_{n=-\infty}^{\infty} \left(\sum_{\varsigma=0}^{\infty} \frac{(\psi n)^{\varsigma}}{\Gamma(\varsigma\lambda+1)} \right) \delta_n$ is a generalized Dirac comb, when $\psi = 0$, it is called a Dirac comb.

$$\begin{aligned} \phi_{\psi,\varkappa}([\zeta, \mathrm{b}+\zeta]) &= (S) \int_{\zeta}^{\imath+\zeta} \sum_{\varsigma=0}^{\infty} \frac{(\psi \mathrm{y})^{\varsigma}}{\Gamma(\varsigma\lambda+1)} d\mathrm{y} + \varkappa \sum_{n \in [\zeta, \mathrm{b}+\zeta]} \left(\sum_{\varsigma=0}^{\infty} \frac{(\psi n)^{\varsigma}}{\Gamma(\varsigma\lambda+1)} \right) \\ &\leqslant \sum_{\varsigma=0}^{\infty} \frac{1}{\Gamma(\varsigma\lambda+1)} (S) \int_{\zeta}^{\imath+\zeta} (\psi \mathrm{y})^{\varsigma} d\mathrm{y} + \varkappa \sum_{n \in [\zeta, \mathrm{b}+\zeta]} \left(\sum_{\varsigma=0}^{\infty} \frac{(\psi n)^{\varsigma}}{\Gamma(\varsigma\lambda+1)} \right) \end{aligned}$$

$$\leqslant \sum_{\varsigma=0}^{\infty} \frac{1}{\Gamma(\varsigma\lambda+1)}\Big(\bigvee_{\alpha>0}\Big(\alpha \bigwedge \mathrm{y}([\tau, \mathrm{b}+\zeta] \cap \{(\psi \mathrm{y})^{\varsigma} \geq \alpha\})\Big)\Big) + \varkappa \sum_{n=[\zeta]}^{\mathrm{b}+\zeta}\Big(\sum_{\varsigma=0}^{\infty} \frac{(\psi n)^{\varsigma}}{\Gamma(\varsigma\lambda+1)}\Big)$$

$$\leqslant \sum_{\varsigma=0}^{\infty} \frac{1}{\Gamma(\varsigma\lambda+1)}\Big(\bigvee_{\alpha>0}\Big(\alpha \bigwedge \mathrm{y}([\zeta, \mathrm{b}+\zeta] \cap \{\mathrm{y} \geq \frac{\sqrt[\varsigma]{\alpha}}{\psi}\})\Big)\Big) + \varkappa \sum_{n=[\zeta]}^{\mathrm{b}+\zeta}\Big(\sum_{\varsigma=0}^{\infty} \frac{(\psi n)^{\varsigma}}{\Gamma(\varsigma\lambda+1)}\Big)$$

$$\leqslant \sum_{\varsigma=0}^{\infty} \frac{1}{\Gamma(\varsigma\lambda+1)}\Big(\bigvee_{\alpha>0}\Big(\alpha \bigwedge (\mathrm{b}-\frac{\sqrt[\varsigma]{\alpha}}{\psi})\Big)\Big) + \varkappa \sum_{n=[\zeta]}^{\mathrm{b}+\zeta}\Big(\sum_{\varsigma=0}^{\infty} \frac{(\psi n)^{\varsigma}}{\Gamma(\varsigma\lambda+1)}\Big)$$

$$\leqslant \sum_{\varsigma=0}^{\infty} \frac{1}{\Gamma(\varsigma\lambda+1)}\Big(\bigvee_{\alpha>0}\Big(\alpha \bigwedge (\mathrm{b}-\frac{\sqrt[\varsigma]{\alpha}}{\psi})\Big)\Big) + \varkappa \sum_{\varsigma=0}^{\infty} \frac{(\psi\zeta)^{\varsigma}}{\Gamma(\varsigma\lambda+1)} \sum_{z=0}^{\mathrm{b}}\Big(\sum_{\varsigma=0}^{\infty} \frac{(\psi z)^{\varsigma}}{\Gamma(\varsigma\lambda+1)}\Big)$$

$$= \sum_{\varsigma=0}^{\infty} \frac{\mathrm{b}}{\Gamma(\varsigma\lambda+1)} + \varkappa \sum_{\varsigma=0}^{\infty} \frac{(\psi\zeta)^{\varsigma}}{\Gamma(\varsigma\lambda+1)} \sum_{z=0}^{\mathrm{b}}\Big(\sum_{\varsigma=0}^{\infty} \frac{(\psi z)^{\varsigma}}{\Gamma(\varsigma\lambda+1)}\Big).$$

We also have that

$$\begin{aligned}\phi_{\psi,\varkappa}(\ell+\zeta, \jmath+\zeta) &\geqslant (S)\int_{\ell+\zeta}^{\jmath+\zeta} \sum_{\varsigma=0}^{\infty} \frac{(\psi \mathrm{y})^{\varsigma}}{\Gamma(\varsigma\lambda+1)} d\mathrm{y} \\ &= \sum_{\varsigma=0}^{\infty} \frac{1}{\Gamma(\varsigma\lambda+1)}\Big(\bigvee_{\alpha\geq 0}\Big(\alpha \bigwedge (b+a-\frac{\sqrt[k]{\alpha}}{\sigma})\Big)\Big) \\ &= \sum_{\varsigma=0}^{\infty} \frac{\ell+\jmath}{\Gamma(\varsigma\lambda+1)}.\end{aligned}$$

The conclusion follows with $\zeta_0 = 0$ and $\varrho_0 = \dfrac{\sum_{\varsigma=0}^{\infty} \frac{\ell+\jmath}{\Gamma(\varsigma\lambda+1)}}{\sum_{\varsigma=0}^{\infty} \frac{\mathrm{b}}{\Gamma(\varsigma\lambda+1)} + \varkappa \sum_{\varsigma=0}^{\infty} \frac{(\psi\zeta)^{\varsigma}}{\Gamma(\varsigma\lambda+1)} \sum_{z=0}^{\mathrm{b}}\Big(\sum_{\varsigma=0}^{\infty} \frac{(\psi z)^{\varsigma}}{\Gamma(\varsigma\lambda+1)}\Big)}$. It means that the space $\big(\mathcal{PAP}\left(\mathbb{R}, \mathbb{R}^n, \phi_{\psi,\varkappa}\right), \mathcal{N}, \circledast\big)$ is a Banach space. The measure $\phi_{\psi,\varkappa}$ does not apply to ($\mathbf{M}_1$). In the following, we will prove this issue. Let

$$\mathrm{D} = \cup_{n=-\infty}^{\infty}\left(n - \frac{1}{2} - \varrho_n, n - \frac{1}{2} + \varrho_n\right),$$

with $\varrho = \frac{1}{\psi}\sinh^{-1}\left(\frac{1}{2^{|n|+1}\sum_{\varsigma=0}^{\infty}\frac{(\psi n)^{\varsigma}}{\Gamma(\varsigma\lambda+1)}}\right)$ and $\zeta = \frac{1}{2}$. Then $\mathrm{D}+\zeta = \cup_{n=-\infty}^{\infty}(n-\varrho_n, n+\varrho_n)$ contains $\mathbb{Z}$ and

$$\phi_{\psi,\varkappa}(\mathrm{D}+\zeta) = \sum_{n=-\infty}^{\infty}(S)\int_{n-\varrho_n}^{n+\varrho_n} \sum_{\varsigma=0}^{\infty} \frac{(\psi \mathrm{y})^{\varsigma}}{\Gamma(\varsigma\lambda+1)} d\mathrm{y} + \sum_{n=-\infty}^{\infty}\sum_{\varsigma=0}^{\infty} \frac{(\psi n)^{\varsigma}}{\Gamma(\varsigma\lambda+1)}$$

$$= \sum_{n=-\infty}^{\infty}\sum_{\varsigma=0}^{\infty} \frac{\bigvee_{\alpha\geq 0}\Big(\alpha \bigwedge (2\varrho_n - \frac{\sqrt[k]{\alpha}}{\psi})\Big)}{\Gamma(\varsigma\lambda+1)} + \sum_{n=-\infty}^{\infty}\sum_{\varsigma=0}^{\infty} \frac{(\psi n)^{\varsigma}}{\Gamma(\varsigma\lambda+1)}$$

$$= \sum_{n=-\infty}^{\infty} \sum_{\varsigma=0}^{\infty} \frac{2\varrho_n}{\Gamma(\varsigma\lambda+1)} + \sum_{n=-\infty}^{\infty} \sum_{\varsigma=0}^{\infty} \frac{(\psi n)^{\varsigma}}{\Gamma(\varsigma\lambda+1)}$$

$$= \sum_{n=-\infty}^{\infty} \frac{1}{2^{|n|+1}} + \infty = \infty,$$

while $\mathtt{D} \cap \mathbb{Z} = \emptyset$. Then

$$\phi_{\psi,\varkappa}(\mathtt{D}) = \sum_{n=-\infty}^{\infty} (S) \int_{n-\frac{1}{2}-\varrho_n}^{n-\frac{1}{2}-\varrho_n} \sum_{\varsigma=0}^{\infty} \frac{(\psi \mathtt{Y})^{\varsigma}}{\Gamma(\varsigma\lambda+1)} d\mathtt{Y} + 0$$

$$= \sum_{\varsigma=0}^{\infty} \frac{\bigvee_{\alpha\geq 0}\left(\alpha \bigwedge \mathtt{Y}(n-\frac{1}{2}+\varrho_n-n+\frac{1}{2}+\varrho_n) \cap \{\mathtt{Y} \geq \frac{\sqrt[k]{\alpha}}{\psi}\}\right)}{\Gamma(\varsigma\lambda+1)}$$

$$= \sum_{n=-\infty}^{\infty} \sum_{\varsigma=0}^{\infty} \frac{\bigvee_{\alpha\geq 0}\left(\alpha \bigwedge (2\varrho_n - \frac{\sqrt[k]{\alpha}}{\psi})\right)}{\Gamma(\varsigma\lambda+1)}$$

$$= \sum_{n=-\infty}^{\infty} \frac{1}{2^{|n|+1}} < \infty.$$

Therefore, if $\mathtt{R} = \mathtt{D}/\mathtt{L}$, Where $\mathtt{L}$ is a bounded interval, we obtain

$$\phi_{\psi,\varkappa}(\mathtt{R}+\zeta) \geqslant \sum_{n\in\mathbb{Z}/\mathtt{L}} \sum_{\varsigma=0}^{\infty} \frac{(\psi n)^{\varsigma}}{\Gamma(\varsigma\lambda+1)} = \infty,$$

while $\phi_{\psi,\varkappa}(\mathtt{R}) \leqslant \phi_{\psi,\varkappa}(\mathtt{D}) < \infty$.

Example 5.8 We consider the following fuzzy measure for $\phi \in \mathcal{N}$, $\psi \geqslant 0$, $\varkappa > 0$

$$d\phi_{\psi,\varkappa}(\mathtt{Y}) = \sum_{\varsigma=0}^{\infty} \frac{(\psi \mathtt{Y})^{\varsigma}}{\Gamma(\varsigma\lambda+1)} d\mathtt{Y} + \varkappa \sum_{n=1}^{\infty} \frac{1}{n^2} \delta_{\frac{1}{n}},$$

where $\delta_{\frac{1}{n}}$ is the Dirac measure at the number $\frac{1}{n}$ and satisfying $(\mathbf{M}_1)$. For $\zeta \in \mathbb{R}$, let $\mathtt{L} = (-1 - |\zeta|, 1 + \zeta]$. We can easily see that $\mathtt{R} \cap \mathtt{L} = \emptyset$, then $\mathtt{R} \cap [0,1] = \emptyset$ and also $(\mathtt{R}+\zeta) \cap [0,1] = \emptyset$. Then,

$$\phi(\mathtt{R}+\zeta) = (S) \int_{\mathtt{R}+\zeta} \sum_{\varsigma=0}^{\infty} \frac{(\psi \mathtt{Y})^{\varsigma}}{\Gamma(\varsigma\lambda+1)} d\mathtt{Y} + 0 = \sum_{\varsigma=0}^{\infty} \frac{\bigvee \left(\alpha \bigwedge \mathtt{t}(\mathtt{R}+\zeta) \cap \{\mathtt{Y} \geq \frac{\sqrt[\varsigma]{\alpha}}{\psi}\}\right)}{\Gamma(\varsigma\lambda+1)} + 0$$

$$= \sum_{\varsigma=0}^{\infty} \frac{\bigvee \left(\alpha \bigwedge (\mathtt{R}+\zeta - \frac{\sqrt[\varsigma]{\alpha}}{\psi})\right)}{\Gamma(\varsigma\lambda+1)} + 0 = \sum_{\varsigma=0}^{\infty} \frac{\mathtt{R}+\zeta}{\Gamma(\varsigma\lambda+1)} + 0$$

$$\leq \sum_{\varsigma=0}^{\infty} \frac{\mathrm{R}}{\Gamma(\varsigma\lambda+1)} + 0 = (S)\int_{\mathrm{R}} \sum_{\varsigma=0}^{\infty} \frac{(\psi \mathrm{y})^{\varsigma}}{\Gamma(\varsigma\lambda+1)} d\mathrm{y} = \phi(\mathrm{R}).$$

The conclusion follows with $\vartheta = 1$.

5.3.5 ϱ-Pseudo Almost Periodic Solution of Eq. (5.3.1)

Theorem 5.16 *Assume that the following conditions are true*

(1) We consider the Lipschitz functions $\bar{\mathcal{K}}_{\jmath}, \mathcal{K}_{\jmath}, \widehat{\mathcal{K}}_{\jmath}$ with the Lipschitz constants $\mathrm{Q}_{\jmath}^{\bar{\mathcal{K}}}, \mathrm{Q}_{\jmath}^{\mathcal{K}}$ and $\mathrm{Q}_{\jmath}^{\widehat{\mathcal{K}}}$, respectively.
(2) The transmission delay kernel, $\varsigma_{\ell\jmath} : \mathbb{R}^{+} \longrightarrow \mathbb{R}$, is both bounded and continuous.

Now, we prove the following:

- *We consider conditions (1) and (2). For $\ell, \jmath = 1, 2, 3 \ldots, n$, if $\mathfrak{g}_{\jmath} : \mathbb{R} \to \mathbb{R}$ is a bounded continuous function, then*

$$\omega_{\ell\jmath}(\mathfrak{x})\mathcal{K}_{\jmath}\left(\mathfrak{g}_{\jmath}\left(\mathfrak{x}-\kappa_{\ell\jmath}\right)\right),\ \bar{\omega}_{\ell\jmath}(\mathfrak{x})\bar{\mathcal{K}}_{\jmath}\left(\mathfrak{g}_{\jmath}(\mathfrak{x})\right),\ d_{\ell\jmath}(\mathfrak{x})\int_0^{\infty} \varsigma_{\ell\jmath}(\mathfrak{h})\widehat{\mathcal{K}}_{\jmath}\left(\mathfrak{g}_{\jmath}(\mathfrak{x}-\mathfrak{h})\right) d\mathfrak{h}$$

  are bounded continuous functions, or

$$\omega_{\ell\jmath}(\mathfrak{x})\mathcal{K}_{\jmath}\left(\mathfrak{g}_{\jmath}\left(\mathfrak{x}-\kappa_{\ell\jmath}\right)\right),\ \bar{\omega}_{\ell\jmath}(\mathfrak{x})\bar{\mathcal{K}}_{\jmath}\left(\mathfrak{g}_{\jmath}(\mathfrak{x})\right),$$
$$d_{\ell\jmath}(\mathfrak{x})\int_0^{\infty} \varsigma_{\ell\jmath}(\mathfrak{h})\widehat{\mathcal{K}}_{\jmath}\left(\mathfrak{g}_{\jmath}(\mathfrak{x}-\mathfrak{h})\right) d\mathfrak{h} \in \mathcal{BC}(\mathbb{R}, \mathbb{R}).$$

- *We consider conditions (1), (2) and ($\boldsymbol{M}_1$). For $\ell, \jmath = 1, 2, 3 \ldots, n$, if $\mathfrak{g}_{\jmath}$ is a ϱ-pseudo almost periodic function, then $\omega_{\ell\jmath}(\mathfrak{x})\mathcal{K}_{\jmath}\left(\mathfrak{g}_{\jmath}\left(\mathfrak{x}-\kappa_{\ell\jmath}\right)\right)$, $\bar{\omega}_{\ell\jmath}(\mathfrak{x})\bar{\mathcal{K}}_{\jmath}\left(\mathfrak{g}_{\jmath}(t)\right)$ and $d_{\ell\jmath}(\mathfrak{x})\int_0^{\infty} \varsigma_{\ell\jmath}(\mathfrak{h})\widehat{\mathcal{K}}_{\jmath}\left(\mathfrak{g}_{\jmath}(\mathfrak{x}-\mathfrak{h})\right) d\mathfrak{h}$ are also ϱ-pseudo almost periodic functions, means*

$$\omega_{\ell\jmath}(\mathfrak{x})\mathcal{K}_{\jmath}\left(\mathfrak{g}_{\jmath}\left(\mathfrak{x}-\kappa_{\ell\bar{\jmath}}\right)\right),\ \bar{\omega}_{\ell\jmath}(\mathfrak{x})\bar{\mathcal{K}}_{\jmath}\left(\mathfrak{g}_{\jmath}(t)\right),$$
$$d_{\ell\jmath}(\mathfrak{x})\int_0^{\infty} \varsigma_{\ell\jmath}(\mathfrak{h})\widehat{\mathcal{K}}_{\jmath}\left(\mathfrak{g}_{\jmath}(\mathfrak{x}-\mathfrak{h})\right) d\mathfrak{h} \in \mathcal{PAP}(\mathbb{R}, \mathbb{R}, \varrho).$$

Proof In all parts $\ell, \jmath = 1, 2, 3, \ldots, n$.

Step 1. We show that if $\mathfrak{g}_j$ is a ϱ-pseudo almost periodic functions, then $\omega_{\ell\jmath}(\mathfrak{x})\mathcal{K}_{\jmath}\left(\mathfrak{g}_{\jmath}\left(\mathfrak{x}-\kappa_{\ell\jmath}\right)\right)$ is a ϱ-pseudo almost periodic function.
Let $\mathfrak{g}_{\jmath}$ is a ϱ-pseudo almost periodic function. With respect to ($\mathbf{M}_1$) and Theorem 5.13, we have $\mathfrak{g}_{\jmath}\left(\mathfrak{x}-\kappa_{\ell\jmath}\right)$ is a ϱ-pseudo almost periodic function ($\mathfrak{g}_{\jmath}\left(\mathfrak{x}-\kappa_{\ell\jmath}\right) \in$

$\mathcal{PAP}(\mathbb{R}, \mathbb{R}, \varrho))$. Let $\mathfrak{g}_J(\mathfrak{x} - \kappa_{\ell J}) = \mathfrak{g}_J^{\mathfrak{v}}(\mathfrak{x}) + \mathfrak{g}_J^{\mathcal{E}}(\mathfrak{x})$, where $\mathfrak{g}_J^{\mathfrak{v}}(\mathfrak{x})$ is a almost periodic function and $\mathfrak{g}_J^{\mathcal{E}}(\mathfrak{x})$ is a ϱ-ergodic function. Then, since $\omega_{\ell J}(\mathfrak{x})$ is a ϱ-pseudo almost periodic function, we get

$$\begin{aligned}\bar{\omega}_{\ell J}(\mathfrak{x})\mathcal{K}_J\left(\mathfrak{g}_J\left(\mathfrak{x} - \kappa_{\ell J}\right)\right) =& \left[\omega_{\ell J}^{\mathfrak{v}}(\mathfrak{x}) + \omega_{\ell J}^{\mathcal{E}}(\mathfrak{x})\right]\mathcal{K}_J\left(\mathfrak{g}_J^{\mathfrak{v}}(\mathfrak{x}) + \mathfrak{g}_J^{\mathcal{E}}(\mathfrak{x})\right)\\ =& \omega_{\ell J}^{\mathfrak{v}}(\mathfrak{x})\mathcal{K}_J\left(\mathfrak{g}_J^{\mathfrak{v}}(\mathfrak{x})\right) + \omega_{\ell J}^{\mathcal{E}}(\mathfrak{x})\mathcal{K}_J\left(\mathfrak{g}_J^{\mathfrak{v}}(\mathfrak{x})\right)\\ &+ \omega_{\ell J}(\mathfrak{x})\left[\mathcal{K}_J\left(\mathfrak{g}_J^{\mathfrak{v}}(\mathfrak{x}) + \mathfrak{g}_J^{\mathcal{E}}(\mathfrak{x})\right) - \mathcal{K}_J\left(\mathfrak{g}_J^{\mathfrak{v}}(\mathfrak{x})\right)\right],\end{aligned}$$

where $\omega_{\ell J}^{\mathfrak{v}}$ is a almost periodic function and $\omega_{\ell J}^{\mathcal{E}}$ is a ϱ-ergodic function. By (1), since $\mathfrak{g}_J^{\mathfrak{v}}(\mathfrak{x})$ is a almost periodic function then $\mathcal{K}_J\left(\mathfrak{g}_J^{\mathfrak{v}}(\mathfrak{x})\right)$ is almost periodic function. We have

$$\omega_{\ell J}^{\mathfrak{v}}(\mathfrak{x})\mathcal{K}_J\left(\mathfrak{g}_j^{\mathfrak{v}}(\mathfrak{x})\right) \in \mathcal{AP}(\mathbb{R}, \mathbb{R}), \tag{5.3.13}$$

means, $\omega_{\ell J}^{\mathfrak{v}}(\mathfrak{x})\mathcal{K}_J\left(\mathfrak{g}_J^{\mathfrak{v}}(\mathfrak{x})\right)$ is a almost periodic function. To continue proving, we let

$$A_i = \inf \mathcal{K}_J\left(\mathfrak{g}_J^{\mathfrak{v}}(\mathfrak{x})\right), \quad E_i = \inf \mathrm{Q}_J^{\mathcal{K}}\omega_{\ell J}^{\mathfrak{v}}(\mathfrak{x}),$$

then

$$\mathcal{N}\left(\lim_{s\to\infty}\frac{1}{\varrho([-s,s])}(S)\int_{-s}^{s}\omega_{\ell J}^{\mathcal{E}}(\mathfrak{x})d\varrho(\mathfrak{x}), \frac{\mathfrak{z}}{\mathcal{K}_J\left(\mathfrak{g}_J^{\mathfrak{v}}(\mathfrak{x})\right)}\right) \succeq \mathcal{N}\left(A_i\lim_{s\to\infty}\frac{1}{\varrho([-s,s))}(S)\int_{-s}^{s}\omega_{\ell J}^{\mathcal{E}}(\mathfrak{x})d\varrho(\mathfrak{x}), \mathfrak{z}\right),$$

and

$$\mathcal{N}\left(\lim_{s\to\infty}\frac{1}{\varrho[-s,s])}(S)\int_{-s}^{s}\mathfrak{g}_J^{\mathcal{E}}(\mathfrak{x})d\varrho(\mathfrak{x}), \frac{\mathfrak{t}}{\mathrm{Q}_J^{\mathcal{K}}\omega_{\ell J}^{\mathfrak{v}}(\mathfrak{x})}\right) \succeq \mathcal{N}\left(E_i\lim_{s\to\infty}\frac{1}{\varrho[-s,s])}(S)\int_{-s}^{s}\mathfrak{g}_J^{\mathcal{E}}(\mathfrak{x})d\varrho(\mathfrak{x}), \mathfrak{t}\right),$$

on the other hand

$$\mathcal{N}\left(\lim_{s\longrightarrow\infty}\frac{1}{\varrho([-s,s])}(S)\int_{-s}^{s}\omega_{\ell J}^{\mathcal{E}}(\mathfrak{x})\mathcal{K}_J\left(\mathfrak{g}_J^{\mathfrak{v}}(\mathfrak{x})\right) + \omega_{\ell J}(\mathfrak{x})\left[\mathcal{K}_J\left(\mathfrak{g}_J^{\mathfrak{v}}(\mathfrak{x}) + \mathfrak{g}_J^{\mathcal{E}}(\mathfrak{x})\right) - \mathcal{E}_J\left(\mathfrak{g}_J^{\mathfrak{v}}(\mathfrak{x})\right)\right]d\varrho(\mathfrak{x}), \mathfrak{z}+\mathfrak{t}\right) \preceq \mathbf{1}.$$

Using the fact that $\omega_{\ell J}^{\mathcal{E}}$, $\mathfrak{g}_J^{\mathcal{E}}$ are ϱ-ergodic function, then

$$\begin{aligned}&\mathcal{N}\left(\lim_{s\longrightarrow\infty}\frac{1}{\varrho([-s,s])}(S)\int_{-s}^{s}\omega_{\ell J}^{\mathcal{E}}(\mathfrak{x})\mathcal{K}_J\left(\mathfrak{g}_J^{\mathfrak{v}}(\mathfrak{x})\right) + \omega_{\ell J}(\mathfrak{x})\left[\mathcal{K}_J\left(\mathfrak{g}_J^{\mathfrak{v}}(\mathfrak{x}) + \mathfrak{g}_J^{\mathcal{E}}(\mathfrak{x})\right) - \mathcal{E}_J\left(\mathfrak{g}_J^{\mathfrak{v}}(\mathfrak{x})\right)\right]d\varrho(\mathfrak{x}), \mathfrak{z}+\mathfrak{t}\right)\\ &\succeq \mathcal{N}\left(\lim_{s\to\infty}\frac{1}{\varrho(([-s,s]))}(S)\int_{-s}^{s}\mathcal{K}_J\left(\mathfrak{g}_J^{\mathfrak{v}}(\mathfrak{x})\right)\omega_{\ell J}^{\mathcal{E}}(\mathfrak{x})d\varrho(\mathfrak{x}), \mathfrak{z}\right) \circledast \mathcal{N}\left(\lim_{s\to\infty}\frac{1}{\varrho([-s,s])}(S)\int_{-s}^{s}\mathrm{Q}_J^{\mathcal{K}}\omega_{\ell J}^{\mathfrak{v}}(\mathfrak{x})\mathfrak{g}_J^{\mathcal{E}}(\mathfrak{x})d\varrho(\mathfrak{x}), \mathfrak{t}\right)\\ &= \mathcal{N}\left(\lim_{s\to\infty}\frac{1}{\varrho([-s,s])}(S)\int_{-s}^{s}\omega_{\ell J}^{\mathcal{E}}(\mathfrak{x})d\varrho(\mathfrak{x}), \frac{\mathfrak{z}}{\mathcal{K}_J\left(\mathfrak{g}_J^{\mathfrak{v}}(\mathfrak{x})\right)}\right) \circledast \mathcal{N}\left(\lim_{s\to\infty}\frac{1}{\varrho[-s,s])}(S)\int_{-s}^{s}\mathfrak{g}_J^{\mathcal{E}}(\mathfrak{x})d\varrho(\mathfrak{x}), \frac{\mathfrak{t}}{\mathrm{Q}_J^{\mathcal{K}}\omega_{\ell J}^{\mathfrak{v}}(\mathfrak{x})}\right)\\ &\succeq \mathcal{N}\left(A_i\lim_{s\to\infty}\frac{1}{\varrho([-s,s))}(S)\int_{-s}^{s}\omega_{\ell J}^{\mathcal{E}}(\mathfrak{x})d\varrho(\mathfrak{x}), \mathfrak{z}\right) \circledast \mathcal{N}\left(E_i\lim_{s\to\infty}\frac{1}{\varrho[-s,s])}(S)\int_{-s}^{s}\mathfrak{g}_J^{\mathcal{E}}(\mathfrak{x})d\varrho(\mathfrak{x}), \mathfrak{t}\right)\\ &= \mathbf{1}.\end{aligned}$$

Therefore

$$\omega^{\mathcal{E}}_{\ell\jmath}(\mathfrak{x})\mathcal{K}_\jmath\left(\mathfrak{g}^{\mathfrak{v}}_\jmath(\mathfrak{x})\right)+\omega_{\ell\jmath}(t)\left[\mathcal{K}_\jmath\left(\mathfrak{g}^{\mathfrak{v}}_\jmath(\mathfrak{x})+\mathfrak{g}^{\mathcal{E}}_\jmath(\mathfrak{x})\right)-\mathcal{K}_\jmath\left(\mathfrak{g}^{\mathfrak{v}}_\jmath(\mathfrak{x})\right)\right]\in\mathcal{E}(\mathbb{R},\mathbb{R},\varrho).$$

Using (5.3.13), we obtain that

$$\omega_{\ell\jmath}(\mathfrak{x})\mathcal{K}_\jmath\left(\mathfrak{g}_\jmath\left(\mathfrak{x}-\kappa_{\ell\jmath}\right)\right)\in\mathcal{PAP}(\mathbb{R},\mathbb{R},\varrho),$$

and that means $\omega_{\ell\jmath}(\mathfrak{x})\mathcal{K}_\jmath\left(\mathfrak{g}_\jmath\left(\mathfrak{x}-\kappa_{\ell\jmath}\right)\right)$ is ϱ-pseudo almost periodic function.

Step 2. By **Step 1** with $\kappa_{\ell\jmath}=0$, we have that if $\mathfrak{g}_\jmath$ is a ϱ-pseudo almost periodic function, then $\bar{\omega}_{\ell\jmath}(\mathfrak{x})\bar{\mathcal{K}}_\jmath\left(\mathfrak{g}_\jmath(\mathfrak{x})\right)$ is also ϱ-pseudo almost periodic function, means $\bar{\omega}_{\ell\jmath}(\mathfrak{x})\bar{\mathcal{K}}_\jmath\left(\mathfrak{g}_\jmath(\mathfrak{x})\right)\in\mathcal{PAP}(\mathbb{R},\mathbb{R},\varrho)$.

Step 3. Here, suppose that $\mathfrak{v}_\jmath$ is a almost periodic function and $\mathcal{E}_\jmath$ is a ϱ-ergodic function such that $\mathfrak{g}_\jmath(\mathfrak{x})=\mathfrak{v}_\jmath(\mathfrak{x})+\mathcal{E}_\jmath(\mathfrak{x})$. Then

$$\begin{aligned}
&\mathfrak{p}_{\ell\jmath}(\mathfrak{x})\int_0^\infty\varsigma_{\ell\jmath}(\mathfrak{h})\widehat{\mathcal{K}}_\jmath\left(\mathfrak{g}_\jmath(\mathfrak{x}-\mathfrak{h})\right)d\mathfrak{h}\\
&=\mathfrak{p}^{\mathfrak{v}}_{\ell\jmath}(\mathfrak{x})\int_0^\infty\varsigma_{\ell\jmath}(\mathfrak{h})\widehat{\mathcal{K}}_\jmath\left(\mathfrak{v}_\jmath(\mathfrak{x}-\mathfrak{h})\right)d\mathfrak{h}\\
&\quad+\mathfrak{p}^{\mathcal{E}}_{\ell\jmath}(\mathfrak{x})\int_0^\infty\varsigma_{\ell\jmath}(\mathfrak{h})\widehat{\mathcal{K}}_\jmath\left(\mathfrak{v}_\jmath(\mathfrak{x}-\mathfrak{h})\right)d\mathfrak{h}\\
&\quad+\mathfrak{p}_{\ell\jmath}(\mathfrak{x})\int_0^\infty\varsigma_{\ell\jmath}(\mathfrak{h})\left[\widehat{\mathcal{K}}_\jmath\left(\mathcal{E}_\jmath(\mathfrak{x}-\mathfrak{h})+\mathfrak{v}_\jmath(\mathfrak{x}-\mathfrak{h})\right)\right.\\
&\quad\left.-\widehat{\mathcal{K}}_\jmath\left(\mathfrak{v}_\jmath(\mathfrak{x}-\mathfrak{h})\right)\right]d\mathfrak{h},
\end{aligned}$$

where $\mathfrak{p}_{\ell\jmath}=\mathfrak{p}^{\mathfrak{v}}_{\ell\jmath}+\mathfrak{p}^{\mathcal{E}}_{\ell\jmath}$, $\mathfrak{x}\in\mathbb{R}$, ${\mathfrak{p}_{\ell\jmath}}^{\mathfrak{v}}$ is a almost periodic function and $\mathfrak{p}^{\mathcal{E}}_{\ell\jmath}$ is a ϱ-ergodic function. with respect to (1), since $\mathfrak{v}_\jmath$ is a almost periodic function, means $\mathfrak{v}_\jmath\in\mathcal{AP}(\mathbb{R},\mathbb{R})$, then

$$\widehat{\mathcal{K}}_\jmath\left(\mathfrak{v}_\jmath(\mathfrak{x})\right)\in\mathcal{AP}(\mathbb{R},\mathbb{R}).$$

Since $\varsigma_{\ell\jmath}\in L^1(\mathbb{R})$ with respect to (2), by Theorem 5.15, we have that

$$\int_0^\infty\varsigma_{\ell\jmath}(\mathfrak{h})\widehat{\mathcal{K}}_\jmath\left(\mathfrak{v}_\jmath(\mathfrak{x}-\mathfrak{h})\right)d\mathfrak{h}\in\mathcal{AP}(\mathbb{R},\mathbb{R}). \tag{5.3.14}$$

Since $\mathfrak{p}^{\mathfrak{v}}_{\ell\jmath}$ is a almost periodic function, we have that

$$\mathfrak{p}^{\mathfrak{v}}_{\ell\jmath}(t)\int_0^\infty\varsigma_{\ell\jmath}(\mathfrak{h})\widehat{\mathcal{K}}_\jmath\left(\mathfrak{v}_\jmath(\mathfrak{x}-\mathfrak{h})\right)d\mathfrak{h}\in\mathcal{AP}(\mathbb{R},\mathbb{R}). \tag{5.3.15}$$

Using the fact that $\mathcal{E}_J$ is a ϱ-ergodic function and Theorem 5.14, we have

$$\int_0^{\infty} \left|\varsigma_{\ell_J}(\mathfrak{h})\|\mathcal{E}_J(\mathfrak{x}-\mathfrak{h})\right| d\mathfrak{h} \in \mathcal{E}(\mathbb{R}, \mathbb{R}, \varrho). \tag{5.3.16}$$

By $\inf_{\mathfrak{x}\in\mathbb{R}} \mathcal{N}\Big(\int_0^{\infty} \varsigma_{\ell_J}(\mathfrak{h})\widehat{\mathcal{K}}_J\left(\mathfrak{v}_J(\mathfrak{x}-\mathfrak{h})\right) d\mathfrak{h}, \mathfrak{z}\Big) \prec \infty$, suppose that

$$\mathcal{N}\Big(\lim_{s\to\infty} \frac{1}{\varrho[-s,s]} (S) \int_{-s}^{s} \mathfrak{p}^{\mathcal{E}}_{\ell_J}(\mathfrak{x}) \int_0^{\infty} \varsigma_{\ell_J}(\mathfrak{h})\widehat{\mathcal{K}}_J\left(\mathfrak{v}_J(\mathfrak{x}-\mathfrak{h})\right) d\mathfrak{h} d\varrho(\mathfrak{x}), \mathfrak{z}\Big) \succeq$$
$$\inf_{\mathfrak{x}\in\mathbb{R}} \mathcal{N}\Big(\int_0^{\infty} \varsigma_{\ell_J}(\mathfrak{h})\widehat{\mathcal{K}}_J\left(\mathfrak{v}_J(\mathfrak{x}-\mathfrak{h})\right) d\mathfrak{h} \lim_{s\to\infty} \frac{1}{\varrho[-s,s]} (S) \int_{-s}^{s} \mathfrak{p}^{\mathcal{E}}_{\ell_J}(\mathfrak{x}) d\varrho(\mathfrak{x}), \mathfrak{z}\Big),$$

and

$$\mathcal{N}\Big(\lim_{s\to\infty} \frac{1}{\varrho[-s,s]} (S) \int_{-s}^{s} \mathfrak{p}_{\ell_J}(\mathfrak{x}) \int_0^{\infty} \varsigma_{\ell_J}(\mathfrak{h}) \Big[\widehat{\mathcal{K}}_J\left(\mathcal{E}_J(\mathfrak{x}-\mathfrak{h})+\mathfrak{v}_J(\mathfrak{x}-\mathfrak{h})\right) - \widehat{\mathcal{K}}_J\left(\mathfrak{v}_J(\mathfrak{x}-\mathfrak{h})\right)\Big] d\mathfrak{h} d\varrho(\mathfrak{x}), \mathfrak{t}\Big) \succeq$$
$$\inf_{\mathfrak{x}\in\mathbb{R}} \mathcal{N}\Big(\varsigma_{\ell_J} Q^{\mathcal{K}}_J \lim_{s\to\infty} \frac{1}{\varrho[-s,s]} (S) \int_{-s}^{s} \mathfrak{p}_{\ell_J}(\mathfrak{x}) \int_0^{\infty} \varsigma_{\ell_J}(\mathfrak{h}) \Big[\widehat{\mathcal{K}}_J\left(\mathcal{E}_J(\mathfrak{x}-\mathfrak{h})+\mathfrak{v}_J(\mathfrak{x}-\mathfrak{h})\right) - \widehat{\mathcal{K}}_J\left(\mathfrak{v}_J(\mathfrak{x}-\mathfrak{h})\right)\Big] d\mathfrak{h} d\varrho(\mathfrak{x}), \mathfrak{t}\Big),$$

since,

$$\mathcal{N}\Big(\lim_{s\to\infty} \frac{1}{\varrho[-s,s]} (S) \int_{-s}^{s} \Big\{ \mathfrak{p}^{\mathcal{E}}_{\ell_J}(\mathfrak{x}) \int_0^{\infty} \varsigma_{\ell_J}(\mathfrak{h})\widehat{\mathcal{K}}_J\left(\mathfrak{v}_J(\mathfrak{x}-\mathfrak{h})\right) d\mathfrak{h}$$
$$+\mathfrak{p}_{\ell_J}(\mathfrak{x}) \int_0^{\infty} \varsigma_{\ell_J}(\mathfrak{h})\widehat{\mathcal{K}}_J\left(\mathcal{E}_J(\mathfrak{x}-\mathfrak{h})+\mathfrak{v}_J(\mathfrak{x}-\mathfrak{h})\right) - \widehat{\mathcal{K}}_J\left(\mathfrak{v}_J(\mathfrak{x}-\mathfrak{h})\right) d\mathfrak{h} \Big\} d\varrho(\mathfrak{x}), \mathfrak{z}+\mathfrak{t}\Big) \preceq \mathbf{1},$$

hence, we get

$$\mathcal{N}\Big(\lim_{s\to\infty} \frac{1}{\varrho[-s,s]} (S) \int_{-s}^{s} \Big\{ \mathfrak{p}^{\mathcal{E}}_{\ell_J}(\mathfrak{x}) \int_0^{\infty} \varsigma_{\ell_J}(\mathfrak{h})\widehat{\mathcal{K}}_J\left(\mathfrak{v}_J(\mathfrak{x}-\mathfrak{h})\right) d\mathfrak{h}$$
$$+\mathfrak{p}_{\ell_J}(\mathfrak{x}) \int_0^{\infty} \varsigma_{\ell_J}(\mathfrak{h})\widehat{\mathcal{K}}_J\left(\mathcal{E}_J(\mathfrak{x}-\mathfrak{h})+\mathfrak{v}_J(\mathfrak{x}-\mathfrak{h})\right) - \widehat{\mathcal{K}}_J\left(\mathfrak{v}_J(\mathfrak{x}-\mathfrak{h})\right) d\mathfrak{h} \Big\} d\varrho(\mathfrak{x}), \mathfrak{z}+\mathfrak{t}\Big)$$
$$\succeq \mathcal{N}\Big(\lim_{s\to\infty} \frac{1}{\varrho[-s,s]} (S) \int_{-s}^{s} \mathfrak{p}^{\mathcal{E}}_{\ell_J}(\mathfrak{x}) \int_0^{\infty} \varsigma_{\ell_J}(\mathfrak{h})\widehat{\mathcal{K}}_J\left(\mathfrak{v}_J(\mathfrak{x}-\mathfrak{h})\right) d\mathfrak{h} d\varrho(\mathfrak{x}), \mathfrak{z}\Big)$$
$$\circledast \mathcal{N}\Big(\lim_{s\to\infty} \frac{1}{\varrho[-s,s]} (S) \int_{-s}^{s} \mathfrak{p}_{\ell_J}(\mathfrak{x}) \int_0^{\infty} \varsigma_{\ell_J}(\mathfrak{h}) \Big[\widehat{\mathcal{K}}_J\left(\mathcal{E}_J(\mathfrak{x}-\mathfrak{h})+\mathfrak{v}_J(\mathfrak{x}-\mathfrak{h})\right) - \widehat{\mathcal{K}}_J\left(\mathfrak{v}_J(\mathfrak{x}-\mathfrak{h})\right)\Big] d\mathfrak{h} d\varrho(\mathfrak{x}), \mathfrak{t}\Big)$$
$$= \mathcal{N}\Big(\lim_{s\to\infty} \frac{1}{\varrho[-s,s]} (S) \int_{-s}^{s} \mathfrak{p}^{\mathcal{E}}_{\ell_J}(\mathfrak{x}) d\varrho(\mathfrak{x}), \frac{\mathfrak{z}}{\int_0^{\infty} \varsigma_{\ell_J}(\mathfrak{h})\widehat{\mathcal{K}}_J\left(\mathfrak{v}_J(\mathfrak{x}-\mathfrak{h})\right) d\mathfrak{h}}\Big)$$
$$\circledast \mathcal{N}\Big(Q^{\mathcal{K}}_J \lim_{s\to\infty} \frac{1}{\varrho[-s,s]} (S) \int_{-s}^{s} \mathfrak{p}_{\ell_J}(\mathfrak{x}) d\varrho(\mathfrak{x}), \frac{\mathfrak{t}}{\varsigma_{\ell_J}(\mathfrak{h}) \Big[\widehat{\mathcal{K}}_J\left(\mathcal{E}_J(\mathfrak{x}-\mathfrak{h})+\mathfrak{v}_J(\mathfrak{x}-\mathfrak{h})\right) - \widehat{\mathcal{K}}_J\left(\mathfrak{v}_J(\mathfrak{x}-\mathfrak{h})\right)\Big] d\mathfrak{h}}\Big)$$
$$\succeq \inf_{\mathfrak{x}\in\mathbb{R}} \mathcal{N}\Big(\int_0^{\infty} \varsigma_{\ell_J}(\mathfrak{h})\widehat{\mathcal{K}}_J\left(\mathfrak{v}_J(\mathfrak{x}-\mathfrak{h})\right) d\mathfrak{h} \lim_{s\to\infty} \frac{1}{\varrho[-s,s]} (S) \int_{-s}^{s} \mathfrak{p}^{\mathcal{E}}_{\ell_J}(\mathfrak{x}) d\varrho(\mathfrak{x}), \mathfrak{z}\Big)$$

$$\circledast \inf_{\mathfrak{x}\in\mathbb{R}} \mathcal{N}\Big(\varsigma_{\ell_J} Q_J^{\widehat{\mathcal{K}}} \lim_{s\to\infty} \frac{1}{\varrho[-s,s]}(S)\int_{-s}^{s} \mathfrak{p}_{\ell_J}(\mathfrak{x}) \int_0^\infty \varsigma_{\ell_J}(\mathfrak{h})$$
$$\Big[\widehat{\mathcal{K}}_J\left(\mathcal{E}_J(\mathfrak{x}-\mathfrak{h})+\mathfrak{v}_J(\mathfrak{x}-\mathfrak{h})\right) - \widehat{\mathcal{K}}_J\left(\mathfrak{v}_J(\mathfrak{x}-\mathfrak{h})\right)\Big] d\mathfrak{h}\, d\varrho(\mathfrak{x}), \mathfrak{t}\Big) = \mathbf{1}.$$

Hence, (5.3.15) and (5.3.16), imply that

$$\mathfrak{p}_{\ell_J}(\mathfrak{x}) \int_0^\infty \varsigma_{\ell_J}(\mathfrak{h})\widehat{\mathcal{K}}_J\left(\mathfrak{g}_J(\mathfrak{x}-\mathfrak{h})\right) d\mathfrak{h} \in \mathcal{PAP}(\mathbb{R}, \mathbb{R}, \varrho).$$

In the sequel, we consider $\bar{\mathfrak{g}}_\ell(\mathfrak{x}) = \aleph^{-1}\mathfrak{g}_\ell(\mathfrak{x})$. Then, for Eq. (5.3.1), we get

$$\begin{aligned}\bar{\mathfrak{g}}_\ell(\mathfrak{x}) = &- \mathfrak{m}_\ell(\mathfrak{x})\bar{\mathfrak{g}}_\ell(\mathfrak{x}) + \aleph_\ell^{-1}\sum_{J=1}^{n} \bar{\omega}_\ell(\mathfrak{x})\mathcal{K}_\ell\left(\aleph_\ell \bar{\mathfrak{g}}_J(\mathfrak{x})\right) \\ &+ \aleph_\ell^{-1}\sum_{J=1}^{n} \omega_{\ell_J}(\mathfrak{x})\mathcal{K}_J\left(\aleph_\ell \bar{\mathfrak{g}}_J\left(\mathfrak{x}-\kappa_{\ell_J}\right)\right) \\ &+ \aleph_\ell^{-1}\sum_{J=1}^{n} \mathfrak{p}_{\ell_J}(\mathfrak{x}) \int_0^\infty \varsigma_{\ell_J}(\gamma)\widehat{\mathcal{K}}_J\left(\aleph_J \bar{\mathfrak{g}}_J(\mathfrak{x}-\gamma)\right) d\gamma + \aleph_\ell^{-1}\mathcal{H}_\ell(\mathfrak{x}).\end{aligned} \tag{5.3.17}$$

Here, we show that $(\Psi\mathfrak{G})_\ell(\mathfrak{x}) : \mathcal{PAP}(\mathbb{R}, \mathbb{R}^n, \varrho) \to \mathcal{PAP}(\mathbb{R}, \mathbb{R}^n, \varrho)$ is the integral solution of Eq. (5.3.17).

Theorem 5.17 *We suppose that the following conditions are true. Also, we assume that in all parts that $\ell = 1, 2, 3, \dots, n$.*

(1) We consider the Lipschitz functions $\bar{\mathcal{K}}_J, \mathcal{K}_J, \widehat{\mathcal{K}}_J$ with the Lipschitz constants $Q_J^{\bar{\mathcal{K}}}, Q_J^{\mathcal{K}}$ and $Q_J^{\widehat{\mathcal{K}}}$, respectively.

(2) The transmission delay kernel $\varsigma_{\ell_J} : \mathbb{R}^+ \longrightarrow \mathbb{R}$ is both bounded and continuous.

(3) There are $\widehat{\mathfrak{m}}_\ell \in \mathcal{BC}(\mathbb{R}, [0, +\infty))$, $\mathcal{G}_\ell > 0$, $\zeta_\ell > 0$ and $\aleph_\ell > 0$ such that

$$\mathcal{N}\Big(\sum_{\varsigma=0}^{\infty} \frac{(-\int_{\mathfrak{i}}^{\mathfrak{x}} \mathfrak{m}_\ell(\mathfrak{h})d\mathfrak{h})^\varsigma}{\Gamma(\varsigma\lambda+1)}, \mathfrak{z}\Big) \succeq \mathcal{N}\Big(\mathcal{G}_\ell \sum_{\varsigma=0}^{\infty} \frac{(-\int_{\mathfrak{i}}^{\mathfrak{x}} \widehat{\mathfrak{m}}_\ell(\mathfrak{h})d\mathfrak{h})^\varsigma}{\Gamma(\varsigma\lambda+1)}, \mathfrak{z}\Big), \quad \forall \mathfrak{x}, \mathfrak{i} \in \mathbb{R}, \quad \mathfrak{x} \geqslant \mathfrak{i},$$

and

$$\mathcal{N}\Big(-\widehat{\mathfrak{m}}_\ell(\mathfrak{x}), \mathfrak{z}\Big) \succeq \mathcal{N}\Big(-\widehat{\mathfrak{m}}_\ell^-, \mathfrak{z}\Big), \quad \mathcal{N}\Big(\widehat{\mathfrak{m}}_\ell^+, \mathfrak{z}\Big) \succeq \mathcal{N}\Big(\widehat{\mathfrak{m}}_\ell(\mathfrak{x}), \mathfrak{z}\Big).$$

We define the nonlinear operator $\Psi : \mathcal{BC}(\mathbb{R}, \mathbb{R}^n) \longrightarrow \mathcal{BC}(\mathbb{R}, \mathbb{R}^n)$ as follows

$$(\Psi\mathfrak{G})_\ell(\mathfrak{x}) = \int_0^{\mathfrak{x}} \sum_{\varsigma=0}^{\infty} \frac{(-\int_{\mathrm{w}}^{\mathfrak{x}} \mathfrak{m}_\ell(\mathfrak{h})d\mathfrak{h})^\varsigma}{\Gamma(\varsigma\lambda+1)} \tag{5.3.18}$$

$$\Big[\aleph_\ell^{-1}\sum_{\jmath=1}^{n}\bar{\omega}_{\ell\jmath}(\mathrm{w})\bar{\mathcal{K}}_\jmath\Big(\aleph_\jmath\mathcal{E}_\jmath(\mathrm{w})\Big) + \aleph_\ell^{-1}\sum_{\jmath=1}^{n}\omega_{\ell\jmath}(\mathrm{w})\mathcal{K}_\jmath\Big(\aleph_\jmath\mathcal{E}_\jmath(\mathrm{w}-\kappa_{\ell\jmath})\Big)$$

$$+\aleph_\ell^{-1}\sum_{\ell=1}^{n}\mathfrak{p}_{\ell\jmath}(s)\int_0^\infty \varsigma_{\ell\jmath}(\gamma)\widehat{\mathcal{K}}_\jmath\Big(\aleph_\jmath\mathcal{E}_\jmath(\mathrm{w}-\gamma)\Big)d\gamma + \aleph_\ell^{-1}\mathcal{H}_\ell(\mathrm{w})\Big]ds.$$

Then, $\Psi\mathfrak{G}$ is bounded continuous functions. It means that $\Psi\mathfrak{G} \in \mathcal{BC}(\mathbb{R}, \mathbb{R}^n)$. Also assume that $(\boldsymbol{M}_1)$ holds. Therefore Ψ maps $\mathcal{PAP}(\mathbb{R}, \mathbb{R}^n, \varrho)$ into itself.

Proof We know that $\Psi\mathfrak{G} \in \mathcal{BC}(\mathbb{R}, \mathbb{R}^n)$ [45]. Also, by Theorem 5.16, there are Ω_ℓ and Υ_ℓ such that Ω_ℓ is almost periodic function and Υ_ℓ is ϱ-ergodic function. Then

$$\aleph_\ell^{-1}\sum_{\jmath=1}^{n}\bar{\omega}_{\ell\jmath}(\mathfrak{x})\bar{\mathcal{K}}_\jmath\left(\aleph_\jmath\mathfrak{G}_\jmath(\mathfrak{x})\right) + \aleph_\ell^{-1}\sum_{\jmath=1}^{n}\omega_{\ell\jmath}(\mathfrak{x})\mathcal{K}_\jmath\left(\aleph_\jmath\mathfrak{G}_\jmath\left(\mathfrak{x}-\kappa_{\ell\jmath}\right)\right) + \aleph_\ell^{-1}\sum_{\jmath=1}^{n}\mathfrak{p}_{\ell\jmath}(\mathfrak{x})$$
$$\times\int_0^\infty \varsigma_{\ell\jmath}(\gamma)\widehat{\mathcal{K}}_\jmath\left(\aleph_\jmath\mathfrak{G}_\jmath(\mathfrak{x}-\gamma)\right)d\gamma + \aleph_\ell^{-1}\mathcal{H}_\ell(\mathfrak{x})$$
$$= \Omega_\ell + \Upsilon_\ell \in \mathcal{PAP}(\mathbb{R}, \mathbb{R}, \varrho).$$

Step 1. We will show that

$$\int_0^{\mathfrak{x}} \sum_{\varsigma=0}^{\infty} \frac{(-\int_{\mathrm{w}}^{\mathfrak{x}} \mathfrak{m}_\ell(\mathfrak{h})d\mathfrak{h})^\varsigma}{\Gamma(\varsigma\lambda+1)}\Omega_\ell(\mathrm{w})d\mathrm{w} \in \mathcal{AP}(\mathbb{R}, \mathbb{R}). \tag{5.3.19}$$

Let

$$\Delta_\ell(\mathfrak{x}) =: \int_0^{\mathfrak{x}} \sum_{\varsigma=0}^{\infty} \frac{(-\int_{\mathrm{w}}^{\mathfrak{x}} \mathfrak{m}_\ell(\mathfrak{h})d\mathfrak{h})^\varsigma}{\Gamma(\varsigma\lambda+1)}\Omega_\ell(\mathrm{w})d\mathrm{w}.$$

For every $F > 0$, since $\Omega_\ell, \mathfrak{m}_\ell \in \mathcal{AP}(\mathbb{R}, \mathbb{R})$ and for every interval with positive length e_θ, there exist a number κ such that $\mathcal{N}\Big(\Omega_\ell(\mathfrak{x}+\kappa) - \Omega_\ell(\mathfrak{x}), \mathfrak{z}\Big) \succ \boldsymbol{\theta}$ and $\mathcal{N}\Big(\sum_{\varsigma=0}^{\infty}\frac{-\int_{\mathrm{w}+\kappa}^{\varkappa+\kappa}\mathfrak{m}_\ell(\mathfrak{h})d\mathfrak{h}}{\Gamma(\varsigma\lambda+1)}\mathfrak{m}_\ell(\mathfrak{x}+\kappa) - \sum_{\varsigma=0}^{\infty}\frac{-\int_{\varkappa}^{\mathfrak{x}}\mathfrak{m}_\ell(\mathfrak{h})d\mathfrak{h}}{\Gamma(\varsigma\lambda+1)}\mathfrak{m}_\ell(\mathfrak{x}), \mathfrak{z}\Big) \succ \boldsymbol{\theta}$, forall $\mathfrak{x} \in \mathbb{R}$ (see [69]). Then,

$$\mathcal{N}\Big(\Delta_\ell(\mathfrak{x}+\kappa) - \Delta_\ell(\mathfrak{x}), 2\mathfrak{z}\Big) \tag{5.3.20}$$
$$= \mathcal{N}\Big(\int_0^{\mathfrak{x}+\kappa} \sum_{\varsigma=0}^{\infty} \frac{(-\int_{\mathrm{w}}^{\mathfrak{x}+\kappa} \mathfrak{m}_\ell(\mathfrak{h})d\mathfrak{h})^\varsigma}{\Gamma(\varsigma\lambda+1)}\Omega_\ell(\mathrm{w})d\mathrm{w} - \int_0^{\mathfrak{x}} \sum_{\varsigma=0}^{\infty} \frac{(-\int_{\mathrm{w}}^{\mathfrak{x}} \mathfrak{m}_\ell(\mathfrak{h})d\mathfrak{h})^\varsigma}{\Gamma(\varsigma\lambda+1)}\Omega_\ell(\mathrm{w})d\mathrm{w}, 2\mathfrak{z}\Big)$$

$$= \mathcal{N}\Big(\int_0^{\mathfrak{x}}\sum_{\varsigma=0}^{\infty}\frac{(-\int_{\mathrm{w}+\kappa}^{\mathfrak{x}+}\mathfrak{m}_q(\mathfrak{h})d\mathfrak{h})^{\varsigma}}{\Gamma(\varsigma\lambda+1)}\Omega_\ell(\mathrm{w}+\kappa)d\mathrm{w} - \int_0^{\mathfrak{x}}\sum_{\varsigma=0}^{\infty}\frac{(-\int_{\mathrm{w}}^{\mathfrak{x}}\mathfrak{m}_\ell(\mathfrak{h})d\mathfrak{h})^{\varsigma}}{\Gamma(\varsigma\lambda+1)}\Omega_\ell(\mathrm{w})d\mathrm{w}, 2\mathfrak{z}\Big)$$

$$= \mathcal{N}\Big(\int_0^{\mathfrak{x}}\sum_{\varsigma=0}^{\infty}\frac{(-\int_{\mathrm{w}+\kappa}^{\mathfrak{x}+\kappa}\mathfrak{m}_\ell(\mathfrak{h})d\mathfrak{h})^{\varsigma}}{\Gamma(\varsigma\lambda+1)}\Omega_\ell(\mathrm{w}+\kappa)d\mathrm{w} - \int_0^{\mathfrak{x}}\sum_{\varsigma=0}^{\infty}\frac{(-\int_{\mathrm{w}}^{\mathfrak{x}}\mathfrak{m}_\ell(\mathfrak{h})d\mathfrak{h})^{\varsigma}}{\Gamma(\varsigma\lambda+1)}\Omega_\ell(\mathrm{w}+\kappa)d\mathrm{w}$$

$$+ \int_0^{\mathfrak{x}}\sum_{\varsigma=0}^{\infty}\frac{(-\int_{\mathrm{w}}^{\mathfrak{x}}\mathfrak{m}_\ell(\mathfrak{h})d\mathfrak{h})^{\varsigma}}{\Gamma(\varsigma\lambda+1)}\Omega_\ell(\mathrm{w}+\kappa)d\mathrm{w} - \int_0^{\mathfrak{x}}\sum_{\varsigma=0}^{\infty}\frac{(-\int_{\mathrm{w}}^{\mathfrak{x}}\mathfrak{m}_\ell(\mathfrak{h})d\mathfrak{h})^{\varsigma}}{\Gamma(\varsigma\lambda+1)}\Omega_\ell(\mathrm{w})d\mathrm{w}, 2\mathfrak{z}\Big)$$

$$\succeq \mathcal{N}\Big(\Big(\int_0^{\mathfrak{x}}\sum_{\varsigma=0}^{\infty}\frac{(-\int_{\mathrm{w}+\kappa}^{\mathfrak{x}+\kappa}\mathfrak{m}_\ell(\mathfrak{h})d\mathfrak{h})^{\varsigma}}{\Gamma(\varsigma\lambda+1)} - \sum_{\varsigma=0}^{\infty}\frac{(-\int_{\mathrm{w}}^{\mathfrak{x}}\mathfrak{m}_\ell(\mathfrak{h})d\mathfrak{h})^{\varsigma}}{\Gamma(\varsigma\lambda+1)}\Big)(\Omega_\ell(\mathrm{w}+\kappa))d\mathrm{w}, \mathfrak{z}\Big)$$

$$\circledast\, \mathcal{N}\Big(\int_0^{\mathfrak{x}}\Big(\sum_{\varsigma=0}^{\infty}\frac{(-\int_{\mathrm{w}}^{\mathfrak{x}}\mathfrak{m}_\ell(\mathfrak{h})d\mathfrak{h})^{\varsigma}}{\Gamma(\varsigma\lambda+1)}\Big)(\Omega_\ell(\mathrm{w}+\kappa) - \Omega_\ell(\mathrm{w}))d\mathrm{w}, \mathfrak{z}\Big).$$

Let

$$A := \mathcal{N}\left(\sum_{\varsigma=0}^{\infty}\frac{(-\int_{s+\kappa}^{\mathfrak{x}+\kappa}\mathfrak{m}_\ell(\mathfrak{h})d\mathfrak{h})^{\varsigma}}{\Gamma(\varsigma\lambda+1)} - \sum_{\varsigma=0}^{\infty}\frac{(-\int_{s}^{\mathfrak{x}}\mathfrak{m}_\ell(\mathfrak{h})d\mathfrak{h})^{\varsigma}}{\Gamma(\varsigma\lambda+1)}, \mathfrak{z}\right).$$

Using $\Big(\sum_{\varsigma=0}^{\infty}\frac{(-\int_{\mathrm{w}}^{\mathfrak{x}}\mathfrak{m}_{(\ell)}(\mathfrak{h})d\mathfrak{h})^{\varsigma}}{\Gamma(\varsigma\lambda+1)}\Big)'_{\mathfrak{x}} = \sum_{\varsigma=0}^{\infty}\frac{(\varsigma(-\int_{\mathrm{w}}^{\mathfrak{x}}\mathfrak{m}_\ell(\mathfrak{h})d\mathfrak{h})^{\varsigma-1}(-\mathfrak{m}_\ell(\mathfrak{h}))}{\Gamma(\varsigma\lambda+1)}$ and given that $\mathfrak{m}_\ell$ is almost periodic function, by (3), we obtain

$$A = \mathcal{N}\Big(\Big[\sum_{\varsigma=0}^{\infty}\frac{(-\int_{\varkappa}^{\mathfrak{x}}\mathfrak{m}_\ell(\mathfrak{h})d\mathfrak{h})^{\varsigma}}{\Gamma(\varsigma\lambda+1)}\sum_{\varsigma=0}^{\infty}\frac{(-\int_{\mathrm{w}+\kappa}^{\varkappa+\kappa}\mathfrak{m}_\ell(\mathfrak{h})d\mathfrak{h})^{\varsigma}}{\Gamma(\varsigma\lambda+1)}\Big]_{\varkappa=\mathrm{w}}^{\mathfrak{x}}, \mathfrak{z}\Big) \tag{5.3.21}$$

$$= \mathcal{N}\Big(\Big[\int_{\mathrm{w}}^{\mathfrak{x}}\sum_{\varsigma=0}^{\infty}\frac{(-\int_{\varkappa}^{\mathfrak{x}}\mathfrak{m}_\ell(\mathfrak{h})d\mathfrak{h})^{\varsigma}}{\Gamma(\varsigma\lambda+1)}\Big(\sum_{\varsigma=0}^{\infty}\frac{(-\int_{\mathrm{w}+\kappa}^{\varkappa+\kappa}\mathfrak{m}_{(\mathfrak{h})d\mathfrak{h}})^{\varsigma}}{\Gamma(\varsigma\lambda+1)}\Big)'_{\varkappa}d\varkappa$$

$$+ \int_{\mathrm{w}}^{\mathfrak{x}}\Big(\sum_{\varsigma=0}^{\infty}\frac{(-\int_{\varkappa}^{\mathfrak{x}}\mathfrak{m}_\ell(\mathfrak{h})d\mathfrak{h})^{\varsigma}}{\Gamma(\varsigma\lambda+1)}\Big)'_{\varkappa}\sum_{\varsigma=0}^{\infty}\frac{(-\int_{\mathrm{w}+\kappa}^{\varkappa+\kappa}\mathfrak{m}_\ell(\mathfrak{h})d\mathfrak{h})^{\varsigma}}{\Gamma(\varsigma\lambda+1)}d\varkappa\Big], \mathfrak{z}\Big)$$

$$= \mathcal{N}\Big(\Big[\int_{\mathrm{w}}^{\mathfrak{x}}\sum_{\varsigma=0}^{\infty}\frac{(-\int_{\varkappa}^{\mathfrak{x}}\mathfrak{m}_\ell(\mathfrak{h})d\mathfrak{h})^{\varsigma}}{\Gamma(\varsigma\lambda+1)}\Big(\sum_{\varsigma=0}^{\infty}\frac{\varsigma(-\int_{\mathrm{w}+\kappa)}^{\varkappa+\kappa}\mathfrak{m}_\ell(\mathfrak{h})d\mathfrak{h})^{\varsigma-1}(-\mathfrak{m}_\ell(\varkappa+\kappa))}{\Gamma(\varsigma\lambda+1)}d\varkappa\Big)$$

$$+ \int_{\mathrm{w}}^{\mathfrak{x}}\sum_{\varsigma=0}^{\infty}\frac{\varsigma(-\int_{\varkappa}^{\mathfrak{x}}\mathfrak{m}_\ell(\mathfrak{h})d\mathfrak{h})^{\varsigma-1}(-\mathfrak{m}_\ell(\varkappa))}{\Gamma(\varsigma\lambda+1)}\sum_{\varsigma=0}^{\infty}\frac{(-\int_{\mathrm{w}+\kappa}^{\varkappa+\kappa}\mathfrak{m}_\ell(\mathfrak{h})d\mathfrak{h})^{\varsigma}}{\Gamma(\varsigma\lambda+1)}d\varkappa\Big], \mathfrak{z}\Big)$$

$$= \mathcal{N}\Big(\int_{\mathrm{w}}^{\mathfrak{x}}\sum_{\varsigma=0}^{\infty}\frac{(-\int_{\varkappa}^{\mathfrak{x}}\mathfrak{m}_\ell(\mathfrak{h})d\mathfrak{h})^{\varsigma}}{\Gamma(\varsigma\lambda+1)}\sum_{\varsigma=0}^{\infty}\frac{(-\int_{\mathrm{w}+\kappa}^{\varkappa+\kappa}\mathfrak{m}_\ell(\mathfrak{h})d\mathfrak{h})^{\varsigma}}{\Gamma(\varsigma\lambda+1)}$$

$$\Big(\sum_{\varsigma=0}^{\infty}\frac{-\int_{\mathrm{w}+\kappa}^{\varkappa+\kappa}\mathfrak{m}_\ell(\mathfrak{h})d\mathfrak{h}}{\Gamma(\varsigma\lambda+1)}(\mathfrak{m}_\ell(\varkappa+\kappa)) - \sum_{\varsigma=0}^{\infty}\frac{-\int_{\varkappa}^{\mathfrak{x}}\mathfrak{m}_\ell(\mathfrak{h})d\mathfrak{h}}{\Gamma(\varsigma\lambda+1)}(\mathfrak{m}_\ell(\varkappa))\Big)d\varkappa, \mathfrak{z}\Big)$$

$$\succeq \mathcal{N}\Big((\mathcal{G}_\ell)^2 F \int_{\mathrm{w}}^{\mathfrak{x}} \sum_{\varsigma=0}^{\infty} \frac{(-\widehat{\mathfrak{m}}_\ell^{-}(\mathfrak{x}-\mathrm{w}))^{\varsigma}}{\Gamma(\varsigma\lambda+1)} d\varkappa, \mathfrak{z}\Big) \tag{5.3.22}$$
$$\succeq \mathcal{N}\Big(\mathcal{G}_\ell^2 F \sum_{\varsigma=0}^{\infty} \frac{(-\widehat{\mathfrak{m}}_\ell^{-}(\mathfrak{x}-\mathrm{w}))^{\varsigma}}{\Gamma(\varsigma\lambda+1)} (\mathfrak{x}-\mathrm{w}), \mathfrak{z}\Big).$$

Thus Ω_ℓ is a continuous and bounded function since Ω_ℓ is an almost periodic function. Then, by (5.3.20) and (5.3.21), we have

$$\mathcal{N}\Big(\Delta_\ell(\mathfrak{x}+\kappa) - \Delta_\ell(\mathfrak{x}), \mathfrak{z}\Big)$$
$$\succeq \mathcal{N}\Big(\mathcal{G}_\ell^2 F \int_0^{\mathfrak{x}} \sum_{\varsigma=0}^{\infty} \frac{(-\widehat{\mathfrak{m}}_\ell^{(\mathfrak{x}-\mathrm{w})})^{\varsigma}}{\Gamma(\varsigma\lambda+1)} (\mathfrak{x}-\mathrm{w})\Omega_\ell(\mathrm{w}+\kappa) d\mathrm{w}, \mathfrak{z}\Big) \circledast \mathcal{N}\Big(F \int_0^{\mathfrak{x}} \sum_{\varsigma=0}^{\infty} \frac{(-\int_{\mathrm{w}}^{\mathfrak{x}} \mathfrak{m}_\ell(\mathfrak{h}) d\mathfrak{h})^{\varsigma}}{\Gamma(\varsigma\lambda+1)} d\mathrm{w}, \mathfrak{z}\Big)$$

$$\succeq \mathcal{N}\Big(\mathcal{G}_\ell^2 F \int_0^{\mathfrak{x}} \sum_{\varsigma=0}^{\infty} \frac{(-\widehat{\mathfrak{m}}_\ell^{(\mathfrak{x}-\mathrm{w})})^{\varsigma}}{\Gamma(\varsigma\lambda+1)} (\mathfrak{x}-\mathrm{w})\Omega_\ell(\mathrm{w}+\kappa) d\mathrm{w}, \mathfrak{z}\Big) \circledast \mathcal{N}\Big(\mathcal{G}_\ell F \int_0^{\mathfrak{x}} \sum_{\varsigma=0}^{\infty} \frac{(-\int_{\mathrm{w}}^{\mathfrak{x}} \widehat{\mathfrak{m}}_\ell(\mathfrak{h}) d\mathfrak{h})^{\varsigma}}{\Gamma(\varsigma\lambda+1)} d\mathrm{w}, \mathfrak{z}\Big)$$
$$\succeq \mathcal{N}\Big(\mathcal{G}_\ell^2 F \int_0^{\mathfrak{x}} \sum_{\varsigma=0}^{\infty} \frac{(-\widehat{\mathfrak{m}}_\ell^{-}(\mathfrak{x}-\mathrm{w}))^{\varsigma}}{\Gamma(\varsigma\lambda+1)} (\mathfrak{x}-\mathrm{w}) d\mathrm{w}\Omega_\ell, \mathfrak{z}\Big) \circledast \mathcal{N}\Big(\mathcal{G}_\ell F \int_0^{\mathfrak{x}} \sum_{\varsigma=0}^{\infty} \frac{(-\widehat{\mathfrak{m}}_\ell^{-}(\mathfrak{x}-\mathrm{w})))^{\varsigma}}{\Gamma(\varsigma\lambda+1)} d\mathrm{w}, \mathfrak{z}\Big)$$
$$\succeq \mathcal{N}\Big(\frac{(\mathcal{G}_\ell)^2 F}{(\widehat{\mathfrak{m}}_\ell^{-})^2} \Omega_\ell + \frac{\mathcal{G}_\ell F}{\widehat{\mathfrak{m}}_\ell^{-}}, \mathfrak{z}\Big).$$

We obtain that

$$\int_0^{\mathfrak{x}} \sum_{\varsigma=0}^{\infty} \frac{(-\int_{\mathrm{w}}^{\mathfrak{x}} \mathfrak{m}_\ell(\mathfrak{h}) d\mathfrak{h})^{\varsigma}}{\Gamma(\varsigma\lambda+1)} \Omega_\ell(\mathrm{w}) d\mathrm{w} \in \mathcal{AP}(\mathbb{R}, \mathbb{R}).$$

Step 2. We will show that

$$\int_0^{\mathfrak{x}} \sum_{\varsigma=0}^{\infty} \frac{(-\int_{\mathrm{w}}^{\mathfrak{x}} \mathfrak{m}_\ell(\mathfrak{h}) d\mathfrak{h})^{\varsigma}}{\Gamma(\varsigma\lambda+1)} \Upsilon_\ell(\mathrm{w}) d\mathrm{w} \in \mathcal{E}(\mathbb{R}, \mathbb{R}, \varrho).$$

According to Theorem 5.14 and since $(\mathbf{M}_1)$ holds, we have for $\Upsilon_\ell \in \mathcal{E}(\mathbb{R}, \mathbb{R}, \varrho)$,

$$\mathbf{1} = \mathcal{N}\Big(\lim_{s\longrightarrow\infty} \frac{1}{\varrho([-s,s])} (S) \int_{-s}^{s} \Big(\int_0^{\infty} \sum_{\varsigma=0}^{\infty} \frac{(-\widehat{\mathfrak{m}}_\ell^{-}\mathfrak{h})^{\varsigma}}{\Gamma(\varsigma\lambda+1)} \Upsilon_\ell(\mathfrak{x}-\mathfrak{h}) d\mathfrak{h} \Big) d\varrho(\mathfrak{x}), \mathfrak{z}\Big)$$
$$= \mathcal{N}\Big(\lim_{s\longrightarrow\infty} \frac{1}{\varrho([-s,s])} (S) \int_{-s}^{s} \int_0^{\mathfrak{x}} \sum_{\varsigma=0}^{\infty} \frac{(-\widehat{\mathfrak{m}}_\ell^{-}(\mathfrak{x}-\mathrm{w}))^{\varsigma}}{\Gamma(\varsigma\lambda+1)} \Upsilon_\ell(\mathrm{w}) d\mathrm{w} \Big) d\varrho(\mathfrak{x}), \mathfrak{z}\Big).$$

Then, by (3),

$$\mathcal{N}\bigg(\lim_{s\longrightarrow\infty}\frac{1}{\varrho([-s,s])}(S)\int_{-s}^{s}\bigg(\int_{0}^{\mathfrak{x}}\sum_{\varsigma=0}^{\infty}\frac{(-\int_{s}^{\mathfrak{x}}\mathfrak{m}_\ell(\nu)d\nu)^\varsigma}{\Gamma(\varsigma\lambda+1)}\Upsilon_\ell(\mathrm{w})d\mathrm{w}\bigg)d\varrho(\mathfrak{x}),\mathfrak{z}\bigg)$$
$$\succeq\mathcal{N}\bigg(\mathcal{G}_\ell\lim_{s\to\infty}\frac{1}{\varrho([-s,s])}(S)\int_{-s}^{s}\bigg(\int_{0}^{\mathfrak{x}}\sum_{\varsigma=0}^{\infty}\frac{(-\widehat{\mathfrak{m}}_\ell^-(\mathfrak{x}-\mathrm{w}))^\varsigma}{\Gamma(\varsigma\lambda+1)}\Upsilon_\ell(\mathrm{w})d\mathrm{w}\bigg)d\varrho(\mathfrak{x}),\mathfrak{z}\bigg)$$
$$=\mathbf{1},$$

since

$$\mathcal{N}\bigg(\lim_{s\longrightarrow\infty}\frac{1}{\varrho([-s,s])}(S)\int_{-s}^{s}\bigg(\int_{0}^{\mathfrak{x}}\sum_{\varsigma=0}^{\infty}\frac{(-\int_{s}^{\mathfrak{x}}\mathfrak{m}_\ell(\nu)d\nu)^\varsigma}{\Gamma(\varsigma\lambda+1)}\Upsilon_\ell(\mathrm{w})d\mathrm{w}\bigg)d\varrho(\mathfrak{x}),\mathfrak{z}\bigg)\preceq\mathbf{1}.$$

Hence,

$$\int_{0}^{\mathfrak{x}}\sum_{\varsigma=0}^{\infty}\frac{(-\int_{\mathrm{w}}^{\mathfrak{x}}\mathfrak{m}_\ell(\mathfrak{h})d\mathfrak{h})^\varsigma}{\Gamma(\varsigma\lambda+1)}\Upsilon_\ell(\mathrm{w})d\mathrm{w}\in\mathcal{E}(\mathbb{R},\mathbb{R},\varrho),$$

using (5.3.19), we have

$$(\Psi\mathfrak{G})_\ell(\mathfrak{x})=\int_{0}^{\mathfrak{x}}\sum_{\varsigma=0}^{\infty}\frac{(-\int_{\mathrm{w}}^{\mathfrak{x}}\mathfrak{m}_\ell(\mathfrak{h})d\mathfrak{h})^\varsigma}{\Gamma(\varsigma\lambda+1)}\Omega_\ell(\mathrm{w})d\mathrm{w}+\int_{0}^{\mathfrak{x}}\sum_{\varsigma=0}^{\infty}\frac{(-\int_{\mathrm{w}}^{\mathfrak{x}}\mathfrak{m}_\ell(\mathfrak{h})d\mathfrak{h})^\varsigma}{\Gamma(\varsigma\lambda+1)}\Upsilon_\ell(\mathrm{w})d\mathrm{w}$$
$$\in\mathcal{PAP}(\mathbb{R},\mathbb{R},\varrho).$$

In the next theorem, we prove the existence of a unique solution for Eq. (5.3.1).

Theorem 5.18 *We consider the following conditions with* $\ell=1,2,3,\ldots,n$, *for all parts.*

(1) We consider the Lipschitz functions $\bar{\mathcal{K}}_J,\mathcal{K}_J,\widehat{\mathcal{K}}_J$ *with the Lipschitz constants* $\mathrm{Q}_J^{\bar{\mathcal{K}}},\mathrm{Q}_J^{\mathcal{K}}$ *and* $\mathrm{Q}_J^{\widehat{\mathcal{K}}}$, *respectively.*
(2) The transmission delay kernel $\varsigma_{\ell_J}:\mathbb{R}^+\longrightarrow\mathbb{R}$ *is both bounded and continuous.*
(3) There are $\widehat{\mathfrak{m}}_\ell\in\mathcal{BC}(\mathbb{R},[0,\infty))$, $\mathcal{G}_\ell>0$, $\zeta_\ell>0$ *and* $\aleph_\ell>0$ *such that*

$$\mathcal{N}\bigg(\sum_{\varsigma=0}^{\infty}\frac{(-\int_{\mathfrak{i}}^{\mathfrak{x}}\mathfrak{m}_\ell(\mathfrak{h})d\mathfrak{h})^\varsigma}{\Gamma(\varsigma\lambda+1)},\mathfrak{z}\bigg)\succeq\mathcal{N}\bigg(\mathcal{G}_\ell\sum_{\varsigma=0}^{\infty}\frac{(-\int_{\mathfrak{i}}^{\mathfrak{x}}\widehat{\mathfrak{m}}_\ell(\mathfrak{h})d\mathfrak{h})^\varsigma}{\Gamma(\varsigma\lambda+1)},\mathfrak{z}\bigg),\quad\forall\mathfrak{x},\mathfrak{i}\in\mathbb{R},\quad\mathfrak{x}\geqslant\mathfrak{i},$$

and

$$\mathcal{N}\Big(-\widehat{\mathfrak{m}}_\ell(\mathfrak{x}),\mathfrak{z}\Big)\succeq\mathcal{N}\Big(-\widehat{\mathfrak{m}}_\ell^-,\mathfrak{z}\Big),\quad\mathcal{N}\Big(\widehat{\mathfrak{m}}_\ell^+,\mathfrak{z}\Big)\succeq\mathcal{N}\Big(\widehat{\mathfrak{m}}_\ell(\mathfrak{x}),\mathfrak{z}\Big).$$

Then, we have

- *There is a unique solution* $\mathfrak{g}^*\in\mathcal{BC}(\mathbb{R},\mathbb{R}^n)$ *for Eq.* (5.3.1).

- *Assume that* ($\boldsymbol{M}_1$) *holds and also for* $\mathcal{G}_\ell > 0$, $\zeta_\ell > 0$ *and* $\aleph_\ell > 0$, *we have*
  (4) *There are* $\widehat{\mathfrak{m}}_\ell \in \mathcal{BC}(\mathbb{R}, [0, \infty))$, *such that*

$$\mathcal{N}\left(\sup_{\mathfrak{x}\in\mathbb{R}}\left\{-\widehat{\mathfrak{m}}_\ell(\mathfrak{x}) + \mathcal{G}_\ell\left[\aleph^{-1}\sum_{J}^{\mathfrak{m}}\left(\left|\bar{\omega}_{\ell J}(\mathfrak{x})\right| \mathrm{Q}_J^{\bar{\mathcal{K}}} + \left|\omega_{\aleph}(\mathfrak{x})\right|\mathrm{Q}_J^{\mathcal{K}} + \widehat{\mathrm{Q}}_J^{\widehat{\mathcal{K}}}\left|\mathfrak{p}_\ell(\mathfrak{x})\right|\int_0^\infty \left|\varsigma_{\ell J}(\mathfrak{h})\right|d\mathfrak{h}\right)\aleph_J\right]\right\}, \mathfrak{z}\right)$$
$$\succ \mathcal{N}\Big(-\zeta_\ell, \mathfrak{z}\Big) \succ \mathbf{0}.$$

  (5) *There are* $\widehat{\mathfrak{m}}_\ell \in \mathcal{BC}(\mathbb{R}, [0, \infty))$, *such that* $0 < \max_{i=1,2,3,n,n}\left\{\frac{\widehat{\mathfrak{m}}_\ell^+}{\widehat{\mathfrak{m}}_\ell^-} - \frac{\zeta_\ell}{\widehat{\mathfrak{m}}_\ell^+}\right\} < 1.$

Then, $\mathfrak{g}^* \in \mathcal{PAP}(\mathbb{R}, \mathbb{R}^n, \varrho)$, *where* $\mathfrak{g}^*$ *is the unique solution of Eq.* (5.3.1).

Proof Let $\mathfrak{G}, \mathfrak{D} \in BC(\mathbb{R}, \mathbb{R}^n)$ (resp. $\mathfrak{G}, \mathfrak{D} \in \mathcal{PAP}(\mathbb{R}, \mathbb{R}^n, \varrho)$). Then, with respect to Theorem 5.17, we have $\Psi\mathfrak{G}, \Psi\mathfrak{D} \in BC(\mathbb{R}, \mathbb{R}^n)$ (resp. $\Psi\mathfrak{G}, \Psi\mathfrak{D} \in \mathcal{PAP}(\mathbb{R}, \mathbb{R}^n, \mu)$). Let

$$U := \mathcal{N}\Big((\Psi\mathfrak{G})_\ell(\mathfrak{x}) - (\Psi\mathfrak{D})_\ell(\mathfrak{x}), \mathfrak{z}\Big).$$

Using (*1*) and (*4*), (*5*), we get

$$\begin{aligned}
U =&\mathcal{N}\Bigg(\int_0^{\mathfrak{x}}\sum_{\varsigma=0}^{\infty}\frac{(-\int_{\mathrm{w}}^{\mathfrak{x}}\mathfrak{m}_\ell(\mathfrak{h})d\mathfrak{h})^\varsigma}{\Gamma(\varsigma\lambda+1)}\aleph_\ell^{-1}\sum_{J=1}^{n}\Big(\bar{\omega}_{\ell J}(\mathrm{w})\Big(\bar{\mathcal{K}}_J\left(\aleph_J\mathfrak{G}_J(\mathrm{w})\right) - \bar{\mathcal{K}}_J\left(\aleph_J\mathfrak{D}_J(\mathrm{w})\right)\Big)\\
&+\aleph_\ell^{-1}\sum_{J=1}^{n}\omega_{\ell J}(\mathrm{w})\left(\mathcal{K}_J\left(\aleph_J\mathfrak{G}_J\left(\mathrm{w}-\kappa_{\ell J}\right)\right) - \mathcal{K}_J\left(\aleph_J\mathfrak{D}_J\left(-\kappa_{\ell J}\right)\right)\right)\\
&+\aleph_\ell^{-1}\sum_{J=1}^{n}\mathfrak{p}_{\ell J}(\mathrm{w})\int_0^\infty\varsigma_{\ell J}(\gamma)\Big(\widehat{\mathcal{K}}_J\left(\aleph_J\mathfrak{G}_J(\mathrm{w}-\gamma)\right) - \widehat{\mathcal{K}}_J\left(\aleph_J\mathfrak{D}_J(\mathrm{w}-\gamma)\right)\Big)d\gamma\Bigg)d\mathrm{w}, \mathfrak{z}\Bigg)\\
\succeq&\mathcal{N}\Bigg(\mathcal{G}_\ell\aleph_\ell^{-1}\int_0^{\mathfrak{x}}\sum_{\varsigma=0}^{\infty}\frac{(-\int_{\mathrm{w}}^{\mathfrak{x}}\widehat{\mathfrak{m}}_\ell(\mathfrak{h})d\mathfrak{h})^\varsigma}{\Gamma(\varsigma\lambda+1)}\sum_{J=1}^{n}\Big(\left|\bar{\omega}_{\ell J}(\mathrm{w})\right|\mathrm{Q}_J^{\bar{\mathcal{K}}} + \left|\omega_{\ell J}(\mathrm{w})\right|\mathrm{Q}_J^{\mathcal{K}}\\
&+\mathrm{Q}_J^{\widehat{\mathcal{K}}}\left|\mathfrak{p}_{\ell J}(\mathrm{w})\right|\int_0^\infty\left|\varsigma_{\ell J}(\mathfrak{h})\right|d\mathfrak{h}\Big)\aleph_J d\mathrm{w}\Bigg)(\mathfrak{G}-\mathfrak{D}), \mathfrak{z}\Bigg)\\
\succeq&\mathcal{N}\Bigg((\mathfrak{G}-\mathfrak{D})\int_0^{\mathfrak{x}}\sum_{\varsigma=0}^{\infty}\frac{(-\int_{\mathrm{w}}^{\mathfrak{x}}\widehat{\mathfrak{m}}_\ell(\mathfrak{h})d\mathfrak{h})^\varsigma}{\Gamma(\varsigma\lambda+1)}\left[\widehat{\mathfrak{m}}_\ell(\mathrm{w}) - \zeta_\ell\right]d\mathrm{w}, \mathfrak{z}\Bigg)\\
=&\mathcal{N}\Bigg((\mathfrak{G}-\mathfrak{D})\left[\widehat{\mathfrak{m}}_\ell^+\int_0^{\mathfrak{x}}\sum_{\varsigma=0}^{\infty}\frac{(-\int_{\mathrm{w}}^{\mathfrak{x}}\widehat{\mathfrak{m}}_\ell(\mathfrak{h})d\mathfrak{h})^\varsigma}{\Gamma(\varsigma\lambda+1)}d\mathrm{w} - \zeta_\ell\int_0^{\mathfrak{x}}\sum_{\varsigma=0}^{\infty}\frac{(-\int_{\mathrm{w}}^{\mathfrak{z}}\widehat{\mathfrak{m}}_\ell(\mathfrak{h})d\mathfrak{h})^\varsigma}{\Gamma(\varsigma\lambda+1)}d\mathrm{w}\right], \mathfrak{z}\Bigg)\\
\succeq&\mathcal{N}\Bigg(\max_{\ell=1,2,3,\ldots,n}\left\{\frac{\widehat{\mathfrak{m}}_\ell^+}{\widehat{\mathfrak{m}}_\ell^-} - \frac{\zeta_\ell}{\mathfrak{m}_\ell^+}\right\}(\mathfrak{G}-\mathfrak{D}), \mathfrak{z}\Bigg)\\
=&\mathcal{N}\Bigg((\mathfrak{G}-\mathfrak{D}), \frac{\mathfrak{z}}{\max_{\ell=1,2,3,\ldots,n}\left\{\frac{\widehat{\mathfrak{m}}_\ell^+}{\widehat{\mathfrak{m}}_\ell^-} - \frac{\zeta_\ell}{\widehat{\mathfrak{m}}_\ell^+}\right\}}\Bigg).
\end{aligned}$$

Then

$$\mathcal{N}\bigg((\Psi\mathfrak{G})_\ell(\mathfrak{x})-(\Psi\mathfrak{D})_\ell(\mathfrak{x}),\mathfrak{z}\bigg)\succeq\mathcal{N}\bigg(\mathfrak{G}-\mathfrak{D},\frac{\mathfrak{z}}{\max_{\ell=1,2,3,\ldots,n}\left\{\frac{\widehat{\mathfrak{m}}_\ell^+}{\widehat{\mathfrak{m}}_\ell^-}-\frac{\zeta_\ell}{\widehat{\mathfrak{m}}_\ell^+}\right\}}\bigg).$$

Using (*4*) and $0<\max_{\ell=1,2,3,\ldots,n}\left\{\frac{\widehat{\mathfrak{m}}_\ell^+}{\widehat{\mathfrak{m}}_\ell^-}-\frac{\lambda_\ell}{\widehat{\mathfrak{m}}_\ell^+}\right\}<1$, we have that G is a contraction on the MVFBS $\left(\mathcal{BC}(\mathbb{R},\mathbb{R}^n),\mathcal{N},\circledast\right)$, $\left(\text{on the MVFBS }\left(\mathcal{PAP}(\mathbb{R},\mathbb{R}^n,\varrho),\mathcal{N},\circledast\right)\right)$ and $\bar{\mathfrak{g}}^*$ is the unique fixed point G.
$(\bar{\mathfrak{g}}^*\in\mathcal{BC}(\mathbb{R},\mathbb{R}^n),\bar{\mathfrak{g}}^*\in\mathcal{PAP}(\mathbb{R},\mathbb{R}^n,\varrho))$. Moreover, (5.3.17) implies that (5.3.1) has a unique ϱ-$\mathcal{PAP}$ solution $\mathfrak{g}^*\in\mathcal{BC}(\mathbb{R},\mathbb{R}^n)\quad(\mathfrak{g}^*\in\mathcal{PAP}(\mathbb{R},\mathbb{R}^n,\varrho))$.

5.3.6 $\mathcal{PAP}(\mathbb{R},\mathbb{R}^n,\varrho)$ Solutions of Eq. (5.3.2)

Definition 5.25 We say that $\mathfrak{g}^*$ is ML stable if there exist constants $\upsilon>0$ and $\mathcal{W}_\varphi>1$ such that for every solution $\mathfrak{g}(t)=\{\mathfrak{g}_\ell(\mathfrak{x})\}_{\ell=1}^n$ of (5.3.2) with initial value $\varphi=\{\varphi_\ell(\mathfrak{x})\}$, we have

$$\mathcal{N}(\mathfrak{g}_\ell(\mathfrak{x})-\mathfrak{g}_\ell^*(\mathfrak{x}),\mathfrak{z})\succeq\mathcal{N}(\mathcal{W}_\varphi(\varphi-\mathfrak{g}^*)\sum_{\varsigma=0}^{\infty}\frac{(-\upsilon\mathfrak{x})^\varsigma}{\Gamma(\varsigma\lambda+1)},\mathfrak{z})=\mathcal{N}(\varphi-x^*,\frac{\mathfrak{z}}{\mathcal{W}_\varphi\sum_{\varsigma=0}^{\infty}\frac{(-\upsilon\mathfrak{x})^\varsigma}{\Gamma(\varsigma\lambda+1)}}),$$

where $\mathfrak{g}^*(\mathfrak{x})=\left\{\mathfrak{g}_\ell^*(\mathfrak{x})\right\}_{\ell=1}^n$ is a solution of (5.3.2), with initial value $\left\{\mathfrak{g}_\ell^*(\mathfrak{x}):\mathfrak{x}\leqslant 0\right\}$ and $\mathfrak{x}>0,\ \ell=1,2,3,\ldots,n$.

Definition 5.26 For any solution $\mathfrak{g}(\mathfrak{x})=\{\mathfrak{g}_\ell(\mathfrak{x})\}_{\ell=1}^n$ of (5.3.2), we say that $\mathfrak{g}^*$ is ML attractive if there exists $F>0$ such that

$$\lim_{\mathfrak{x}\to\infty}\mathcal{N}\bigg(\sum_{\varsigma=0}^{\infty}\frac{(F\mathfrak{x})^\varsigma}{\Gamma(\varsigma\lambda+1)}(\mathfrak{g}(\mathfrak{x})-\mathfrak{g}^*(\mathfrak{x})),\mathfrak{z}\bigg)=\mathbf{1},$$

where $\mathfrak{g}^*(\mathfrak{x})=\left\{\mathfrak{g}_\ell^*(\mathfrak{x})\right\}_{\ell=1}^n$ is a solution of (5.3.2), with initial value $\left\{\mathfrak{g}_\ell^*(\mathfrak{x}):\mathfrak{x}\leqslant 0\right\}$.

Remark 5.2 Assume that $\mathfrak{g}$ is a solution of (5.3.2). Therefore, if there is ML stability for $\mathfrak{g}$, $\mathfrak{g}$ depends uniquely on the initial value of $\mathfrak{g}$. We can also conclude from the ML stability the ML attractiveness.

Lemma 5.8 *We consider the following conditions*

(1) Assum that $\left|\varsigma_{\ell_J}(t)\right|\sum_{\varsigma=0}^{\infty}\frac{(k\mathfrak{x})^\varsigma}{\Gamma(\varsigma\lambda+1)}$ *is integrable on* $\mathbb{R}^+$ *for every* $k>0$ *that exists.*
(2) There are $\widehat{\mathfrak{m}}_\ell\in\mathcal{BC}(\mathbb{R},[0,\infty))$, *such that*

$$\mathcal{N}\left(\sup_{\mathfrak{x}\in\mathbb{R}}\left\{-\widehat{\mathfrak{m}}_\ell(\mathfrak{x})+\mathcal{G}_\ell\left[\aleph^{-1}\sum_{J}^{\mathfrak{m}}\left(|\bar{\omega}_{\ell_J}(\mathfrak{x})|\mathrm{Q}_J^{\mathcal{K}}+|\omega_\aleph(\mathfrak{x})|\mathrm{Q}_J^{\mathcal{K}}+\widehat{\mathrm{Q}}_J^{\widehat{\mathcal{K}}}|\mathfrak{p}_\ell(\mathfrak{x})|\int_0^\infty|\varsigma_{\ell_J}(\mathfrak{h})|d\mathfrak{h}\right)\aleph_J\right]\right\},\mathfrak{z}\right)$$
$$\succ\mathcal{N}\left(-\zeta_\ell,\mathfrak{z}\right)\succ\mathbf{0},$$

for every $\ell=1,2,3,\ldots,n$, $\mathcal{G}_\ell>0$, $\zeta_\ell>0$, $\aleph_\ell>0$. *We consider* $\mathfrak{Y}=\{\mathfrak{Y}_\ell\}_{\ell=1}^n:[0,k]\longrightarrow\mathbb{R}^n$ *as follows*

$$\mathfrak{Y}_\ell(\mathfrak{q})=\sup_{\mathfrak{x}\in R}\mathfrak{Y}_\ell(\mathfrak{q},\mathfrak{x}),$$

where

$$\mathfrak{Y}_\ell(\mathfrak{q},\mathfrak{x})=\mathfrak{q}-\widehat{\mathfrak{m}}_\ell(\mathfrak{x})+\mathcal{G}_\ell\aleph_\ell^{-1}\sum_{J=1}^{n}$$
$$\left(|\bar{\omega}_{\ell_J}(\mathfrak{x})|\mathrm{Q}_J^{\bar{\mathcal{K}}}+|\omega_{\ell_J}(\mathfrak{x})|\mathrm{Q}_J^{\mathcal{K}}\sum_{\varsigma=0}^{\infty}\frac{(\mathfrak{q}\kappa_{\ell_J})^\varsigma}{\Gamma(\varsigma\lambda+1)}+\widehat{\mathrm{Q}}_J^{\mathcal{K}}|\mathfrak{p}_{\ell_J}(\mathfrak{x})|\int_0^\infty|\varsigma_{\ell_J}(\mathfrak{h})|\sum_{\varsigma=0}^{\infty}\frac{(\mathfrak{q}\mathfrak{h})^\varsigma}{\Gamma(\varsigma\lambda+1)}d\mathfrak{h}\right)\aleph_J.$$

Then, there exists $0<\upsilon_0<k$ *such that for every* $0<\upsilon<\upsilon_0$, *we have* $\mathfrak{Y}_\ell(\upsilon)\succ 0$.

Proof With respect to (*1*), the function $\mathfrak{x}\mapsto\mathfrak{Y}_\ell(\mathfrak{q},\mathfrak{x})$ is defined on $[0,k]$. By (*2*), we have

$$\mathcal{N}\left(\mathfrak{Y}_\ell(0),\mathfrak{z}\right)=\mathcal{N}\left(\sup\mathfrak{Y}_\ell(0,\mathfrak{x}),\mathfrak{z}\right)$$
$$=\mathcal{N}\left(\sup\left\{-\bar{\mathfrak{m}}_\ell(\mathfrak{x})+\mathcal{G}_\ell\aleph_\ell^{-1}\sum_{J=1}^{n}\left(|\bar{\omega}_{\ell_J}(\mathfrak{x})|\mathrm{Q}_J^{\bar{\mathcal{K}}}+|\omega_{\ell_J}(t)|\mathrm{Q}_J^{\mathcal{K}}+\mathrm{Q}_J^{\bar{\mathcal{K}}}|\mathfrak{p}_{\ell_J}(t)|\int_0^\infty|\varsigma_{\ell_J}(\mathfrak{h})|d\mathfrak{h}\right)\aleph_J\right\},\mathfrak{z}\right)$$
$$\succ\mathcal{N}(-\zeta_\ell,\mathfrak{z})\succ\mathbf{0}.$$

For all $0<\upsilon<\upsilon_0$ we show that there exists $0<\upsilon_0<k$ such that $\mathfrak{Y}_\ell(\upsilon)\succ 0$. Given that

$$\mathcal{N}\left(\mathfrak{Y}(\mathfrak{q},\mathfrak{x})-\mathfrak{Y}(0,\mathfrak{x}),\mathfrak{z}\right)=\mathcal{N}\left(\mathfrak{q}+\mathcal{G}_\ell\aleph_\ell^{-1}\sum_{J=1}^{n}\right.$$
$$\left.\left(|\omega_{\ell_J}(\mathfrak{x})|\mathrm{Q}_J^J\left(\sum_{\varsigma=0}^{\infty}\frac{(\mathfrak{q}\kappa_{\ell_J})^\varsigma}{\Gamma(\varsigma\lambda+1)}-1\right)+\mathrm{Q}_J^{\widehat{\mathcal{K}}}|\mathfrak{p}_\ell(\mathfrak{q})|\left(\int_0^\infty|\varsigma_{\ell_J}(\mathfrak{h})|\sum_{\varsigma=0}^{\infty}\frac{(\mathfrak{q}\mathfrak{h})^\varsigma}{\Gamma(\varsigma\lambda+1)}d\mathfrak{h}-\int_0^\infty|\varsigma_{\ell_J}(\mathfrak{h})|d\mathfrak{h}\right)\right)\xi,\mathfrak{z}\right).$$

Then, for all $\mathfrak{q}\in(0,k)$ and $\mathfrak{x}\in\mathbb{R}$, we have

$$\mathcal{N}\Big(\mathfrak{Y}_\ell(\mathfrak{q},\mathfrak{x})-\mathfrak{Y}_\ell(0,\mathfrak{x}),\mathfrak{z}\Big)\succeq \mathcal{N}\Big(\mathfrak{q}+\mathcal{G}_\ell\aleph_\ell^{-1}\omega_{\ell_J}(\mathfrak{x})\sum_{J=1}^{n}\Big(\sum_{\varsigma=0}^{\infty}\frac{(\mathfrak{q}\kappa_{\ell_J})^{\varsigma}}{\Gamma(\varsigma\lambda+1)}-1\Big)\aleph_J$$
$$+\mathcal{G}_\ell\aleph^{-1}\mathfrak{p}_{\ell_J}(\mathfrak{x})\sum_{J=1}^{n}\Big(\int_0^\infty|\varsigma_{\ell_J}(\mathfrak{h})|\sum_{J=1}^{n}\Big(\sum_{\varsigma=0}^{\infty}\frac{(\mathfrak{q}\mathfrak{h})^{\varsigma}}{\Gamma(\varsigma\lambda+1)}d\mathfrak{h}-\int_0^\infty|\varsigma_\ell(\mathfrak{h})|d\mathfrak{h}\Big)\Big)\aleph_J,\mathfrak{z}\Big).$$

Using continuity, there exists $0<\mu_\ell^1<k$, for every $F>0$, such that

$$\mathcal{N}(\mathfrak{q},\mathfrak{z})\succ\mathcal{N}(\mu_\ell^1,\mathfrak{z})\Rightarrow\mathcal{N}\Big(\mathfrak{q}+\mathcal{G}_\ell\aleph_\ell^{-1}\omega_{\ell_J}\sum_{J=1}^{n}\Big(\sum_{\varsigma=0}^{\infty}\frac{(\mathfrak{q}\kappa_{\ell_J})^{\varsigma}}{\Gamma(\varsigma\lambda+1)}-1\Big)\aleph_J,\mathfrak{z}\Big)\succ\mathcal{N}\Big(\frac{F}{2},\mathfrak{z}\Big).$$

By the Lebesgue dominated convergence Theorem and with respect to the integrability of the function $\big|\varsigma_{\ell_J}(\mathfrak{h})\sum_{\varsigma=0}^{\infty}\frac{(k\mathfrak{h})^{\varsigma}}{\Gamma(\varsigma\lambda+1)}\big|$ on $(0,\infty)$ and utilizing *(1)*, there exists $0<\mu_\ell^2<k$ such that

$$\mathcal{N}(\mathfrak{q},\mathfrak{z})\succ\mathcal{N}(\mu_\ell^2,\mathfrak{z})\Rightarrow$$

$$\mathcal{N}\Big(\mathcal{G}_\ell\aleph_\ell^{-1}\mathfrak{p}_{\ell_J}\sum_{J=1}^{n}\Big(\int_0^\infty|\varsigma_{\ell_J}(\mathfrak{h})|\sum_{\varsigma=0}^{\infty}\frac{(\mathfrak{q}\mathfrak{h})^{\varsigma}}{\Gamma(\varsigma\lambda+1)}d\mathfrak{h}-\int_0^\infty|\varsigma_{\ell_J}(\mathfrak{h})|d\mathfrak{h}\Big)\aleph_J,\mathfrak{z}\Big)\succ\mathcal{N}\Big(\frac{F}{2},\mathfrak{z}\Big).$$

If we take $0<\upsilon<\min\left(\mu_\ell^1,\mu_\ell^2\right)$ and $F>\frac{\zeta}{2}$, we conclude that for every $\mathfrak{x}\in\mathbb{R}$, we have

$$\begin{aligned}\mathcal{N}\Big(\mathfrak{Y}_\ell(\upsilon,\mathfrak{x}),\mathfrak{z}\Big)&\succeq\mathcal{N}\Big(\mathfrak{Y}_\ell(0,\mathfrak{x})+(\mathfrak{Y}_\ell(\upsilon,\mathfrak{x})-\mathfrak{Y}_\ell(0,\mathfrak{x})),\mathfrak{z}\Big)\\&\succ\mathcal{N}\Big(-\zeta_\ell+F,\mathfrak{z}\Big)\\&\succ\mathcal{N}\Big(-\frac{\zeta}{2},\mathfrak{z}\Big).\end{aligned}$$

In the next theorem, we prove that for a solution to Eq. (5.3.2), which is of type ϱ-$\mathcal{PAP}$, there is ML stability and ML attractiveness.

Theorem 5.19 *Assume that the following conditions hold for every $\ell=1,2,3,\ldots,n$, for all parts*

(1) *We consider the Lipschitz functions $\bar{\mathcal{K}}_J,\mathcal{K}_J,\widehat{\mathcal{K}}_J$ with the Lipschitz constants $\mathrm{Q}_J^{\bar{\mathcal{K}}},\mathrm{Q}_J^{\mathcal{K}}$ and $\mathrm{Q}_J^{\widehat{\mathcal{K}}}$, respectively.*
(2) *Assum that $|\varsigma_{\ell_J}(t)|\sum_{\varsigma=0}^{\infty}\frac{(k\mathfrak{x})^{\varsigma}}{\Gamma(\varsigma\lambda+1)}$ is integrable on $\mathbb{R}^+$ for every $k>0$ that exists.*
(3) *There are $\widehat{\mathfrak{m}}_\ell\in\mathcal{BC}(\mathbb{R},[0,\infty))$, such that*

$$\mathcal{N}\Big(\sum_{\varsigma=0}^{\infty}\frac{(-\int_{\mathfrak{i}}^{\mathfrak{x}}\mathfrak{m}_\ell(\mathfrak{h})d\mathfrak{h})^{\varsigma}}{\Gamma(\varsigma\lambda+1)},\mathfrak{z}\Big)\succeq\mathcal{N}\Big(\mathcal{G}_\ell\sum_{\varsigma=0}^{\infty}\frac{(-\int_{\mathfrak{i}}^{\mathfrak{x}}\widehat{\mathfrak{m}}_\ell(\mathfrak{h})d\mathfrak{h})^{\varsigma}}{\Gamma(\varsigma\lambda+1)},\mathfrak{z}\Big),\quad\forall\mathfrak{x},\mathfrak{i}\in\mathbb{R},\quad\mathfrak{x}\geqslant\mathfrak{i},$$

and

$$\mathcal{N}\Big(-\widehat{\mathfrak{m}}_\ell(\mathfrak{x}),\mathfrak{z}\Big) \succeq \mathcal{N}\Big(-\widehat{\mathfrak{m}}_\ell^-,\mathfrak{z}\Big), \quad \mathcal{N}\Big(\widehat{\mathfrak{m}}_\ell^+,\mathfrak{z}\Big) \succeq \mathcal{N}\Big(\widehat{\mathfrak{m}}_\ell(\mathfrak{x}),\mathfrak{z}\Big),$$

$$\widehat{\mathfrak{m}}_\ell^+ = \sup_{\mathfrak{x}\in R} \widehat{\mathfrak{m}}_\ell(\mathfrak{x}), \quad \widehat{\mathfrak{m}}_\ell^- = \inf_{\mathfrak{x}\in R} \widehat{\mathfrak{m}}_\ell(\mathfrak{x}) > 0,$$

for $\mathcal{G}_\ell > 0,\ \zeta_\ell > 0$ *and* $\aleph_\ell > 0$.

(4) *There are* $\widehat{\mathfrak{m}}_\ell \in \mathcal{BC}(\mathbb{R},[0,\infty))$, *such that*

$$\mathcal{N}\Big(\sup_{\mathfrak{x}\in\mathbb{R}}\Big\{-\widehat{\mathfrak{m}}_\ell(\mathfrak{x}) + \mathcal{G}_\ell\Big[\aleph^{-1}\sum_{J}^{\mathfrak{m}}\Big(|\bar{\omega}_{\ell J}(\mathfrak{x})|Q_J^{\mathcal{K}} + |\omega_{\aleph}(\mathfrak{x})|Q_J^{\mathcal{K}} + \widehat{Q}_J^{\widehat{\mathcal{K}}}|\mathfrak{p}_\ell(\mathfrak{x})|\int_0^\infty |\varsigma_{\ell J}(\mathfrak{h})|d\mathfrak{h}\Big)\aleph_J\Big]\Big\},\mathfrak{z}\Big)$$
$$\succ \mathcal{N}\Big(-\zeta_\ell\Big) \succ \mathbf{0},$$

for $\mathcal{G}_\ell > 0,\ \zeta_\ell > 0$ *and* $\aleph_\ell > 0$. *Then,*

- *If* $\mathfrak{g}^*$ *is a solution of* (5.3.2) *with initial value* $\{\mathfrak{g}^*(\mathfrak{x}) : \mathfrak{x} \geq 0\}$, *then there is a ML stability for* $\mathfrak{g}^*$.
- *If the transmission delay kernel* $\varsigma_{\ell J} : \mathbb{R}^+ \longrightarrow \mathbb{R}$ *is both bounded and continuous, Then for* $\mathfrak{g}^* \in \mathcal{BC}(\mathbb{R},\mathbb{R}^n)$, *which is the unique solution of* (5.3.2), *there is a ML stability.*
- *Also, we have*

(a) *Consider the transmission delay kernel* $\varsigma_{\ell J} : \mathbb{R}^+ \longrightarrow \mathbb{R}$, *which is both continuous and bounded.*
(b) *Assum that* $\big|\varsigma_{\ell J}(t)\big|\sum_{\varsigma=0}^{\infty}\frac{(k\mathfrak{x})^\varsigma}{\Gamma(\varsigma\lambda+1)}$ *is integrable on* $\mathbb{R}^+$ *for every* $k > 0$ *that exists.*
(c) *There are* $\widehat{\mathfrak{m}}_\ell \in \mathcal{BC}(\mathbb{R},[0,\infty))$, *such that*

$$0 < \max_{i=1,2,3,\ldots,n}\left\{\frac{\widehat{\mathfrak{m}}_\ell^+}{\widehat{\mathfrak{m}}_\ell^-} - \frac{\zeta_\ell}{\widehat{\mathfrak{m}}_\ell^+}\right\} < 1,$$

for $\mathcal{G}_\ell > 0,\ \zeta_\ell > 0$ *and* $\aleph_\ell > 0$, *then for the unique solution, which is of type* $\mathcal{PAP}(\mathbb{R};\mathbb{R}^n;\varrho)$, *there is a ML stability.*

Proof Let $\mathfrak{g}(\mathfrak{x}) = \{\mathfrak{x}_\ell(\mathfrak{x})\}_{\ell=1}^n$ be a solution of Eq. (5.3.2) with initial value $\nu(\mathfrak{x}) = \{\nu_\ell(\mathfrak{x})\}_{\ell=1}^n$. Let $\phi(\mathfrak{x}) = \{\phi_\ell(\mathfrak{x})\}_{\ell=1}^n = \left\{\aleph_\ell^{-1}\left(\mathfrak{g}_\ell(\mathfrak{x}) - \mathfrak{g}_\ell^*(\mathfrak{x})\right)\right\}_{\ell=1}^n$. Then

$$\phi_\ell'(\mathrm{w}) + \mathfrak{m}_\ell(\mathrm{w})\phi_\ell(\mathrm{w}) = \aleph_\ell^{-1}\sum_{J=1}^{n}\bar{\omega}_{\ell J}(\mathrm{w})\Big(\bar{\mathcal{K}}_J\left(\mathfrak{g}_J(\mathrm{w})\right) - \bar{\mathcal{K}}_J\left(\mathfrak{g}_J^*(\mathrm{w})\right)\Big) + \aleph_\ell^{-1}\sum_{J=1}^{n}\omega_{\ell J} \tag{5.3.23}$$
$$\times\Big(\mathcal{K}_J\left(\mathfrak{g}_J\left(\mathrm{w}-\kappa_{\ell J}\right)\right) - \mathcal{K}_J\left(\mathfrak{g}_J^*\left(\mathrm{w}-\kappa_{\ell J}\right)\right)\Big) + \aleph_\ell^{-1}\sum_{J=1}^{n}\mathfrak{p}_{\ell J}(\mathrm{w})$$

$$\times \int_0^{\infty} \varsigma_{\ell_J}(\gamma)\left(\widehat{\mathcal{K}}_J\left(\mathfrak{g}_J(\mathrm{w}-\gamma)\right)-\widehat{\mathcal{K}}_J\left(\mathfrak{g}_J^*(\mathrm{w}-\gamma)\right)\right) d\gamma.$$

Let

$$\mathcal{N}\left(\nu-\mathfrak{g}^*, \mathfrak{z}\right)=\inf_{\mathfrak{x}\in(-\infty,0]}\ \min_{\ell=1,2,3,\ldots,n} \mathcal{N}\left(\aleph_\ell^{-1}(\nu_\ell(\mathfrak{x})-\mathfrak{g}_\ell^*(\mathfrak{x})), \mathfrak{z}\right),$$

and $\mathrm{R}>0$ be such that

$$\mathrm{R}>\sum_{\ell=1}^{n} \mathcal{G}_\ell+1. \tag{5.3.24}$$

Then, for all $\mathfrak{x} \in(-\infty, 0]$, we have

$$\mathcal{N}(\phi(\mathfrak{x}), \mathfrak{z}) \succeq \mathcal{N}(\nu-\mathfrak{g}^*, \mathfrak{z}). \tag{5.3.25}$$

Let $0<\upsilon<\min \left\{\min_{\ell=1,2,\ldots,n} \widehat{\mathfrak{m}}_\ell^{+}, \upsilon_0\right\}$, where the number υ_0 was given in Lemma 5.8. We will show that for all positive $\mathfrak{x}$ and F

$$\mathcal{N}(\phi(\mathfrak{x}), \mathfrak{z}) \succ \mathcal{N}\left(\mathrm{R}(\nu-\mathfrak{g}^{\mathfrak{x}}) \sum_{\varsigma=0}^{\infty} \frac{(-\upsilon \mathfrak{x})^\varsigma}{\Gamma(\varsigma \lambda+1)}+\mathrm{R}\, F \sum_{\varsigma=0}^{\infty} \frac{(-\upsilon \mathfrak{x})^\varsigma}{\Gamma(\varsigma \lambda+1)}, \mathfrak{z}\right). \tag{5.3.26}$$

Otherwise, there exist $\lambda>0$ such that using continuity,

$$\begin{cases}\mathcal{N}\left(\left|\phi_\ell(\lambda)\right|, \mathfrak{z}\right)=\mathcal{N}\left(\mathrm{R}(\nu-\mathfrak{g}^*) \sum_{\varsigma=0}^{\infty} \frac{(-\upsilon \lambda)^\varsigma}{\Gamma(\varsigma \lambda+1)}+\mathrm{R}F \sum_{\varsigma=0}^{\infty} \frac{(-\upsilon \lambda)^\varsigma}{\Gamma(\varsigma \lambda+1)} \mathfrak{z}\right) \\ \mathcal{N}(\phi(\mathfrak{x}), \mathfrak{z}) \succ \mathcal{N}\left(\mathrm{R}\,(\nu-\mathfrak{g}^*+F) \sum_{\varsigma=0}^{\infty} \frac{(-\upsilon \mathfrak{x})^\varsigma}{\Gamma(\varsigma \lambda+1)}, \mathfrak{z}\right), \forall \mathfrak{x} \in(-\infty, \lambda).\end{cases} \tag{5.3.27}$$

In the sequel, first, we multiply (5.3.23) by $\sum_{\varsigma=0}^{\infty} \frac{(\int_0^{\mathrm{w}} \mathfrak{m}_\ell(\mathfrak{h}) d\mathfrak{h})^\varsigma}{\Gamma(\varsigma \lambda+1)}$ then we take the integral from the equation obtained on the interval $[0, \lambda]$ relative to w, and finally, we multiply the relation by $\sum_{\varsigma=0}^{\infty} \frac{(-\int_0^{\mathfrak{x}} \mathfrak{m}_1(\mathfrak{h}) d\mathfrak{h})^\varsigma}{\Gamma(\varsigma \lambda+1)}$. Then, we obtain

$$\begin{aligned}
\phi_\ell(\lambda)= & \phi_\ell(0) \sum_{\varsigma=0}^{\infty} \frac{(-\int_0^\lambda \mathfrak{m}_\ell(\mathfrak{h}) d\mathfrak{h})^\varsigma}{\Gamma(\varsigma \lambda+1)}+\aleph_\ell^{-1} \int_0^\lambda \sum_{\varsigma=0}^{\infty} \frac{(-\int_{\mathrm{w}}^\lambda \mathfrak{m}_\ell(\mathfrak{h}) d\mathfrak{h})^\varsigma}{\Gamma(\varsigma \lambda+1)} \sum_{J=1}^{n} \\
& \left[\bar{\omega}_{\ell_J}(\mathrm{w})\left(\bar{\mathcal{K}}_J\left(\mathfrak{g}_J(\mathrm{w})\right)-\bar{\mathcal{K}}_J\left(\mathfrak{g}_J^*(\mathrm{w})\right)\right)+\omega_{\ell_J}(\mathrm{w})\left(\mathcal{K}_J\left(\mathfrak{g}_J\left(\mathrm{w}-\kappa_{\ell_J}\right)\right)\right.\right. \\
& \left.-\mathcal{K}_J\left(\mathfrak{g}_J^*\left(\mathrm{w}-\kappa_{\ell_J}\right)\right)\right)+\mathfrak{p}_{\ell_J}(\mathrm{w}) \int_0^{\infty} \varsigma_{\ell_J}(\gamma)\left(\widehat{\mathcal{K}}_J\left(\mathfrak{g}_J(\mathrm{w}-\gamma)\right)\right. \\
& \left.\left.-\widehat{\mathcal{K}}_J\left(\mathfrak{g}_J^*(\mathrm{w}-\gamma)\right)\right) d\gamma\right] d\mathrm{w}.
\end{aligned}$$

Then by *(1)* and *(3)* we obtain that

$$\mathcal{N}\Big(|\phi_\ell(\lambda)|, \mathfrak{z}\Big) \succeq \mathcal{N}\Big(\mathcal{G}_\ell \Big(|\phi_\ell(0)| \sum_{\varsigma=0}^{\infty} \frac{(-\int_0^\lambda \widehat{\mathfrak{m}}_\ell(\mathfrak{h})d\mathfrak{h})^\varsigma}{\Gamma(\varsigma\lambda+1)} + \aleph_\ell^{-1} \int_0^\lambda \sum_{\varsigma=0}^{\infty} \frac{(-\int_{\mathrm{w}}^\lambda \widehat{\mathfrak{m}}_\ell(\mathfrak{h})d\mathfrak{h})^\varsigma}{\Gamma(\varsigma\lambda+1)} \sum_{\jmath=1}^{n} \Big[|\bar{\omega}_{\ell\jmath}(\mathrm{w})| \mathrm{Q}_\jmath^{\bar{\mathcal{K}}} \aleph_\jmath |\phi_\jmath(\mathrm{w})|$$
$$+ |\omega_{\ell\jmath}(\mathrm{w})| \mathrm{Q}_\jmath^{\mathcal{K}} \aleph_\jmath |\phi_\jmath(\mathrm{w} - \kappa_{\ell\jmath})| + \mathrm{Q}_\jmath^{\mathcal{K}} |\mathfrak{p}_{\ell\jmath}(\mathrm{w})| \int_0^\infty |\varsigma_{\ell\jmath}(\gamma)| \aleph_\jmath |\phi_\jmath(\mathrm{w}-\gamma)| d\gamma\Big] d\mathrm{w}\Big), \mathfrak{z}\Big).$$

Using (5.3.25) and (5.3.26), we obtain

$$\mathcal{N}\Big(|\phi_\ell(\lambda)|, \mathfrak{z}\Big) \succeq \mathcal{N}\Big(\mathcal{G}_\ell\left((\nu - \mathfrak{g}^*) + F\right)$$
$$\Big\{\sum_{\varsigma=0}^{\infty} \frac{(-\int_0^\lambda \widehat{\mathfrak{m}}_\ell(\mathfrak{h})d\mathfrak{h})^\varsigma}{\Gamma(\varsigma\lambda+1)} + \mathrm{R}\aleph_\ell^{-1} \int_0^\lambda \sum_{\varsigma=0}^{\infty} \frac{(-\int_{\mathrm{w}}^\lambda \widehat{\mathfrak{m}}_\ell(\mathfrak{h})d\mathfrak{h})^\varsigma}{\Gamma(\varsigma\lambda+1)} \sum_{\jmath=1}^{n} \aleph_\jmath$$
$$\Big[|\bar{\omega}_{\ell\jmath}(\mathrm{w})| \mathrm{Q}_\jmath^{\bar{\mathcal{K}}} \sum_{\varsigma=0}^{\infty} \frac{(-\upsilon \mathrm{w})^\varsigma}{\Gamma(\varsigma\lambda+1)} + |\omega_{\ell\jmath}(\mathrm{w})| \mathrm{Q}_\jmath^{\mathcal{K}} \sum_{\varsigma=0}^{\infty} \frac{(-\upsilon(\mathrm{w} - \kappa_\ell))^\varsigma}{\Gamma(\varsigma\lambda+1)}$$
$$+\mathrm{Q}_\jmath^{\mathcal{K}} |\mathfrak{p}_{\ell\jmath}(\mathrm{w})| \int_0^\infty |\varsigma_{\ell\jmath}(\mathfrak{h})| \sum_{\varsigma=0}^{\infty} \frac{(-\upsilon(\mathrm{w}-\mathfrak{h}))^\varsigma}{\Gamma(\varsigma\lambda+1)} d\mathfrak{h}\Big] d\mathrm{w}\Big\}, \mathfrak{z}\Big).$$

By Lemma 5.8, since $\mathrm{w} > 0$ and $0 < \upsilon < \upsilon_0$, then

$$\begin{aligned}
\mathcal{N}(|\phi_\ell(\lambda)|, \mathfrak{z}) \succeq & \mathcal{N}\Big(\mathcal{G}_\ell\left((\nu - \mathfrak{g}^*) + F\right) \Big\{\sum_{\varsigma=0}^{\infty} \frac{(-\int_0^\lambda \widehat{\mathfrak{m}}_\ell(\mathfrak{h})d\mathfrak{h})^\varsigma}{\Gamma(\varsigma\lambda+1)} \\
& + \mathrm{R}\mathcal{G}_\ell^{-1} \int_0^\lambda \sum_{\varsigma=0}^{\infty} \frac{(-\int_{\mathrm{w}}^\lambda \widehat{\mathfrak{m}}_\ell(\mathfrak{h})d\mathfrak{h})^\varsigma}{\Gamma(\varsigma\lambda+1)} (\widehat{\mathfrak{m}}_\ell(\mathrm{w}) - \upsilon) \sum_{\varsigma=0}^{\infty} \frac{(-\upsilon \mathrm{w})^\varsigma}{\Gamma(\varsigma\lambda+1)} d\mathrm{w}\Big\}, \mathfrak{z}\Big) \\
= & \mathcal{N}\Big(\mathcal{G}_\ell\left((\nu - \mathfrak{g}^*) + F\right) \Big\{\sum_{\varsigma=0}^{\infty} \frac{(-\nu\lambda(-\int_0^\lambda (\widehat{\mathfrak{m}}_\ell(\mathfrak{h}) - \nu)\, d\mathfrak{h}))^\varsigma}{\Gamma(\varsigma\lambda+1)} \\
& + \mathrm{R}\mathcal{G}_\ell^{-1} (-\nu\lambda)^\varsigma \int_0^\lambda \sum_{\varsigma=0}^{\infty} \frac{(-\int_{\mathrm{w}}^\lambda (\widehat{\mathfrak{m}}_\ell(\mathfrak{h}) - \nu)\, d\mathfrak{h})^\varsigma}{\Gamma(\varsigma\lambda+1)} (\widehat{\mathfrak{m}}_\ell(\mathrm{w}) - \upsilon)\, d\mathrm{w}\Big\}, \mathfrak{z}\Big) \\
= & \mathcal{N}\Big(\mathcal{G}_\ell\left((\nu - \mathfrak{g}^*) + F\right) \sum_{\varsigma=0}^{\infty} \frac{(-\nu\lambda)^\varsigma}{\Gamma(\varsigma\lambda+1)} \Big\{(-\int_0^\lambda (\widehat{\mathfrak{m}}_\ell(\mathfrak{h}) - \nu)\, d\mathfrak{h})^\varsigma \\
& + \mathrm{R}\mathcal{G}_\ell^{-1} \int_0^\lambda (-\int_{\mathrm{w}}^\lambda (\widehat{\mathfrak{m}}_\ell(\mathfrak{h}) - \nu)\, d\mathfrak{h})^\varsigma (\widehat{\mathfrak{m}}_\ell(\mathrm{w}) - \upsilon)\, d\mathrm{w}\Big\}, \mathfrak{z}\Big) \\
= & \mathcal{N}\Big(\mathcal{G}_\ell\left((\nu - \mathfrak{g}^*) + F\right) \sum_{\varsigma=0}^{\infty} \frac{(-\nu\lambda)^\varsigma}{\Gamma(\varsigma\lambda+1)} \Big\{(-\int_0^\lambda (\widehat{\mathfrak{m}}_\ell(\mathfrak{h}) - \nu)\, d\mathfrak{h})^\varsigma \\
& + \mathrm{R}\mathcal{G}_\ell^{-1} \Big[(-\int_{\mathrm{w}}^\lambda (\widehat{\mathfrak{m}}_\ell(\mathfrak{h}) - \nu)\, d\mathfrak{h})^\varsigma (\widehat{\mathfrak{m}}_\ell(\mathrm{w}) - \upsilon)\Big]_0^\lambda\Big\}, \mathfrak{z}\Big)
\end{aligned}$$

$$=\mathcal{N}\Big(\mathcal{G}_\ell\left((\nu-\mathfrak{g}^*)+F\right)\sum_{\varsigma=0}^{\infty}\frac{(-\nu\lambda)^\varsigma}{\Gamma(\varsigma\lambda+1)}\Big\{(-\int_0^\lambda(\widehat{\mathfrak{m}}_\ell(\mathfrak{h})-\nu)\,d\mathfrak{h})^\varsigma$$

$$+\mathrm{R}\mathcal{G}_\ell^{-1}\Big[0-(-\int_0^\lambda(\widehat{\mathfrak{m}}_\ell(\mathfrak{h})-\nu)\,d\mathfrak{h})^\varsigma\Big]\Big\},\mathfrak{z}\Big)$$

$$=\mathcal{N}\Big(\mathrm{R}\left((\nu-\mathfrak{g}^*)+F\right)\sum_{\varsigma=0}^{\infty}\frac{(-\nu\lambda)^\varsigma}{\Gamma(\varsigma\lambda+1)}\Big\{-\Big(1-\frac{\mathcal{G}_\ell}{\mathrm{R}}\Big)\sum_{\varsigma=0}^{\infty}\frac{(-\int_0^\lambda(\widehat{\mathfrak{m}}_\ell(\mathfrak{h})-\nu)\,d\mathfrak{h})^\varsigma}{\Gamma(\varsigma\lambda+1)}\Big\},\mathfrak{z}\Big).$$

We recall that $\upsilon<\widehat{\mathfrak{m}}_\ell^-$. This implies that $\widehat{\mathfrak{m}}_\ell(\mathfrak{h})-\upsilon>0$ for all $\mathfrak{h}\in(0,\infty)$. Hence, by (5.3.24), we conclude that

$$\mathcal{N}\Big(\big|\phi_\ell(\lambda)\big|,\mathfrak{z}\Big)\succ\mathcal{N}\Bigg(\mathrm{R}\Bigg((\nu-\mathfrak{g}^*)\sum_{\varsigma=0}^{\infty}\frac{(-\upsilon\lambda)^\varsigma}{\Gamma(\varsigma\lambda+1)}+\mathrm{R}F\sum_{\varsigma=0}^{\infty}\frac{(-\upsilon\lambda)^\varsigma}{\Gamma(\varsigma\lambda+1)},\mathfrak{z}\Bigg)\Bigg).$$

This contradicts (5.3.27). Then (5.3.26) holds. Keeping $\mathfrak{x}$ fixed and letting $F\to 0$, for all positive $\mathfrak{x}$, we have

$$\mathcal{N}(\phi(\mathfrak{x}),\phi)\succeq\mathcal{N}\Big(\mathrm{R}(\nu-\mathfrak{g}^*)\sum_{\varsigma=0}^{\infty}\frac{(-\ell\lambda)^\varsigma}{\Gamma(\varsigma\lambda+1)},\mathfrak{z}\Big).$$

Therefore, by Remark 5.2, we have the following result.

Corollary 5.4 *Assume that for every $\ell=1,2,3,\ldots,n$, the following conditions are true*

(1) We consider the Lipschitz functions $\bar{\mathcal{K}}_\jmath,\mathcal{K}_\jmath,\widehat{\mathcal{K}}_\jmath$ with the Lipschitz constants $\mathrm{Q}_\jmath^{\bar{\mathcal{K}}},\mathrm{Q}_\jmath^{\mathcal{K}}$ and $\mathrm{Q}_\jmath^{\widehat{\mathcal{K}}}$, respectively.
(2) The transmission delay kernel $\varsigma_{\ell\jmath}:\mathbb{R}^+\longrightarrow\mathbb{R}$ is both bounded and continuous.
(3) Assum that $\left|\varsigma_{\ell\jmath}(t)\right|\sum_{\varsigma=0}^{\infty}\frac{(k\mathfrak{x})^\varsigma}{\Gamma(\varsigma\lambda+1)}$ is integrable on $\mathbb{R}^+$ for every $k>0$ that exists.
(4) There are $\widehat{\mathfrak{m}}_\ell\in\mathcal{BC}(\mathbb{R},[0,\infty))$, such that

$$\mathcal{N}\Big(\sum_{\varsigma=0}^{\infty}\frac{(-\int_{\mathfrak{i}}^{\mathfrak{x}}\mathfrak{m}_\ell(\mathfrak{h})d\mathfrak{h})^\varsigma}{\Gamma(\varsigma\lambda+1)},\mathfrak{z}\Big)\succeq\mathcal{N}\Big(\mathcal{G}_\ell\sum_{\varsigma=0}^{\infty}\frac{(-\int_{\mathfrak{i}}^{\mathfrak{x}}\widehat{\mathfrak{m}}_\ell(\mathfrak{h})d\mathfrak{h})^\varsigma}{\Gamma(\varsigma\lambda+1)},\mathfrak{z}\Big),\quad\forall\mathfrak{x},\mathfrak{i}\in\mathbb{R},\quad\mathfrak{x}\geqslant\mathfrak{i},$$

and

$$\mathcal{N}\Big(-\widehat{\mathfrak{m}}_\ell(\mathfrak{x}),\mathfrak{z}\Big)\succeq\mathcal{N}\Big(-\widehat{\mathfrak{m}}_\ell^-,\mathfrak{z}\Big),\quad\mathcal{N}\Big(\widehat{\mathfrak{m}}_\ell^+,\mathfrak{z}\Big)\succeq\mathcal{N}\Big(\widehat{\mathfrak{m}}_\ell(\mathfrak{x}),\mathfrak{z}\Big).$$

$$\widehat{\mathfrak{m}}_\ell^+=\sup_{\mathfrak{x}\in R}\ \widehat{\mathfrak{m}}_\ell(\mathfrak{x}),\quad\widehat{\mathfrak{m}}_\ell^-=\inf_{\mathfrak{x}\in R}\ \widehat{\mathfrak{m}}_\ell(\mathfrak{x})>0,$$

for $\mathcal{G}_\ell > 0,\ \zeta_\ell > 0$ *and* $\aleph_\ell > 0$.

(5) There are $\widehat{\mathfrak{m}}_\ell \in \mathcal{BC}(\mathbb{R}, [0, \infty))$, *such that*

$$\mathcal{N}\Bigg(\sup_{\mathfrak{x}\in\mathbb{R}}\Bigg\{-\widehat{\mathfrak{m}}_\ell(\mathfrak{x}) + \mathcal{G}_\ell\Bigg[\aleph^{-1}\sum_{J}^{\mathfrak{m}}\Bigg(\Big(|\bar{\omega}_{\ell J}(\mathfrak{x})|\mathrm{Q}_J^{\mathcal{K}} + |\omega_{\aleph}(\mathfrak{x})|\mathrm{Q}_J^{\mathcal{K}} + \widehat{\mathrm{Q}}_J^{\widehat{\mathcal{K}}}|\mathfrak{p}_\ell(\mathfrak{x})|\int_0^\infty |\varsigma_{\ell J}(\mathfrak{h})|d\mathfrak{h}\Big)\aleph_J\Bigg]\Bigg\}, \mathfrak{z}\Bigg)$$

$$\succ \mathcal{N}\Big(-\zeta_\ell, \mathfrak{z}\Big) \succ \mathbf{0},$$

for $\mathcal{G}_\ell > 0,\ \zeta_\ell > 0$ *and* $\aleph_\ell > 0$. *Then*

- *If* $\mathfrak{g}^* \in \mathcal{BC}(\mathbb{R}, \mathbb{R}^n)$ *is a solution of* (5.3.2), *then there is a ML attractive for* $\mathfrak{g}^*$.
- *If* $(\boldsymbol{M}_1)$ *holds with the following condition*

(a) There are $\widehat{\mathfrak{m}}_\ell \in \mathcal{BC}(\mathbb{R}, [0, \infty))$, *such that*

$$0 < \max_{i=1,2,3,n,n}\left\{\frac{\widehat{\mathfrak{m}}_\ell^+}{\widehat{\mathfrak{m}}_\ell^-} - \frac{\zeta_\ell}{\widehat{\mathfrak{m}}_\ell^+}\right\} < 1,$$

for $\mathcal{G}_\ell > 0, \zeta_\ell > 0$ *and* $\aleph_\ell > 0$, *then for the unique solution, which is of type* $\mathcal{PAP}(\mathbb{R}; \mathbb{R}^n; \varrho)$, *there is a ML attractive.*

5.3.7 Applications

Example 5.9 Let $n = 2$, $\ell, J = 1, 2$,

(1) we consider the Lipschitz functions $\bar{\mathcal{K}}_\ell(\mathfrak{h}) = 0$, $\mathcal{K}_\ell(\mathfrak{h}) = \widehat{\mathcal{K}}_\ell(\mathfrak{h}) = \frac{3}{4}\exp(-\frac{\mathfrak{h}}{2})$, with the Lipschitz constants $\mathrm{Q}_\ell^{\bar{\mathcal{K}}} = 0$, $\mathrm{Q}^{\mathcal{K}} = \mathrm{Q}_\ell^{\mathcal{K}} = \frac{3}{4}$.

(2) Let $\varsigma_{\ell J}(\mathfrak{x}) = \frac{3}{4}\sum_{\varsigma=0}^{\infty}\frac{(-10\mathfrak{x})^\varsigma}{\Gamma(\varsigma\lambda+1)}$. Then, $\varsigma_{\ell J}$ is bounded, continuous and $\left|\varsigma_{\ell J}\sum_{\varsigma=0}^{\infty}\frac{(k\mathfrak{x})^\varsigma}{\Gamma(\varsigma\lambda+1)}\right|$ is integrable on $[0, \infty)$ for $k = 1$.

(3) Let $\varrho \in \mathcal{M}$ where $d\varrho(\mathfrak{x}) = \varpi_\varsigma(\mathfrak{x})d\mathfrak{x}$ (see [2]) with $\varpi_\varsigma(\mathfrak{x}) = \sum_{\varsigma=0}^{\infty}\frac{(\varsigma\mathfrak{x})^\varsigma}{\Gamma(\varsigma\lambda+1)}$, $\varsigma_{\ell J}(\mathfrak{x}) = \sum_{\varsigma=0}^{\infty}\frac{(-10\mathfrak{x})^\varsigma}{\Gamma(\varsigma\lambda+1)}$, with $\omega_0 := 2 - \varsigma > 0$, $-s \leqslant \mathrm{w} < \mathfrak{x} \leqslant s$. We have for $\omega = 1$, $\kappa \in \mathbb{R}$, and $\mathcal{H} = \emptyset$. Then, the assumption $(\mathbf{M}_1)$ holds.

(4) For $\mathfrak{m}_1(\mathfrak{x}) = \frac{3}{4}\left(1 + \frac{8\mathfrak{x}^2}{9(1-\mathfrak{x})}\right)$, $\mathfrak{m}_2(\mathfrak{x}) = \frac{3}{4}\left(1 + \frac{7(1+\mathfrak{x})^{\frac{-4}{3}}}{5}\right)$, $\widehat{\mathfrak{m}}_\ell(\mathfrak{x}) = \frac{3}{4}$, $\mathcal{G}_\ell = \sum_{\varsigma=0}^{\infty}\frac{(\frac{8}{11})^\varsigma}{\Gamma(\varsigma\lambda+1)}$, we have

$$\mathcal{N}\Bigg(\sum_{\varsigma=0}^{\infty}\frac{(-\int_{\mathfrak{i}}^{\mathfrak{x}}\mathfrak{m}_\ell(\mathfrak{h})d\mathfrak{h})^\varsigma}{\Gamma(\varsigma\lambda+1)}, \mathfrak{z}\Bigg) \succeq \mathcal{N}\Bigg(\mathcal{G}_\ell\sum_{\varsigma=0}^{\infty}\frac{(-\int_{\mathfrak{i}}^{\mathfrak{x}}\widehat{\mathfrak{m}}_\ell(\mathfrak{h})d\mathfrak{h})^\varsigma}{\Gamma(\varsigma\lambda+1)}, \mathfrak{z}\Bigg), \quad \forall\mathfrak{x}, \mathfrak{i} \in \mathbb{R}, \quad \mathfrak{x} \geqslant \mathfrak{i},$$

and

$$\mathcal{N}\Big(- \widehat{\mathfrak{m}}_\ell(\mathfrak{x}), \mathfrak{z}\Big) \succeq \mathcal{N}\Big(- \widehat{\mathfrak{m}}_\ell^-, \mathfrak{z}\Big), \quad \mathcal{N}\Big(\widehat{\mathfrak{m}}_\ell^+, \mathfrak{z}\Big) \succeq \mathcal{N}\Big(\widehat{\mathfrak{m}}_\ell(\mathfrak{x}), \mathfrak{z}\Big).$$

(5) For $\aleph_\ell = 1,\ T = \pi, \kappa_{\ell_J} = \frac{10}{\ell+J},\ \bar{\omega}_{\ell_J}(\mathfrak{x}) = 0,\ \omega_{\ell_J}(\mathfrak{x}) = \frac{\mathfrak{x}^2}{19(1+\mathfrak{x})},\ \mathfrak{p}_{\ell_J}(t) = \frac{6}{11}(1 + (2\mathfrak{x}))^{\frac{-4}{3}},\ \zeta_\ell = \frac{7}{100}$, we have

$$\mathcal{N}\Big(\sup_{\mathfrak{x}\in\mathbb{R}} \Big\{ -\widehat{\mathfrak{m}}_\ell(\mathfrak{x}) + \mathcal{G}_\ell \Big[\aleph^{-1} \sum_J^{\mathfrak{m}} \Big(|\bar{\omega}_{\ell_J}(\mathfrak{x})| \mathrm{Q}_J^{\mathcal{K}} + |\omega_\aleph(\mathfrak{x})| \mathrm{Q}_J^{\mathcal{K}} + \widehat{\mathrm{Q}}_J^{\widehat{\mathcal{K}}} |\mathfrak{p}_\ell(\mathfrak{x})| \int_0^\infty |\varsigma_{\ell_J}(\mathfrak{h})| d\mathfrak{h} \Big) \aleph_J \Big] \Big\}, \mathfrak{z}\Big)$$
$$\succ \mathcal{N}\Big(- \zeta_\ell, \mathfrak{z}\Big) \succ \mathbf{0},$$

and

$$0 < \max_{i=1,2,3,n,n} \left\{ \frac{\widehat{\mathfrak{m}}_\ell^+}{\widehat{\mathfrak{m}}_\ell^-} - \frac{\zeta_\ell}{\widehat{\mathfrak{m}}_\ell^+} \right\} < 1.$$

For $\mathcal{H}_\ell(\mathfrak{x}) = (20 + \ell)|(1 + \mathfrak{x})^{\frac{-4}{3}}| + \mathfrak{G}(\mathfrak{x})$ with

$$\mathfrak{G}(\mathrm{w}) = \begin{cases} e^{-\mathrm{w}}, & \mathrm{w} \geqslant 0 \\ 1, & \mathrm{w} < 0 \end{cases}$$

therefore, all solutions of (5.3.1) with initial value (18, −16), (−10, 19), (9, −20) in the form of ML, they converge to a unique solution of Eq. (5.3.1) such as $\mathfrak{g}^*(\mathfrak{x}) \in \mathcal{PAP}\big(\mathbb{R}, \mathbb{R}^2, \varrho\big)$, when $\mathfrak{x} \longrightarrow +\infty$ with convergence rate $\upsilon \approx 0.05 < \widehat{\mathfrak{m}}_\ell^-$.

Example 5.10 Consider conditions (1) to (5) and all the functions presented in Example 5.9. For $\varsigma > 1$, $\lambda > 0$ and Dirac measure $\delta_{\frac{1}{n}}$, we also consider the following fuzzy measure

$$d\varrho_{\varsigma,\lambda}(\mathfrak{x}) = \sum_{\varsigma=0}^{\infty} \frac{(\varsigma\mathfrak{x})^\varsigma}{\Gamma(\varsigma\lambda + 1)} d\mathfrak{x} + \lambda \sum_{n=1}^{\infty} \frac{1}{n^2} \delta_{\frac{1}{n}}.$$

Let interval $I = (-1 - |\kappa|, 1 + |\kappa|)$ For $\kappa \in \mathbb{R}$ and $\omega = 1$, then for all $T \in \mathbb{Z}$ and $\kappa \in \mathbb{R}$, there exist $\omega > 0$ and a bounded interval $\mathcal{H}$ such that $\varrho(T + \kappa) \leqslant \omega\varrho(T)$, and $T \cap \mathcal{H} = \varnothing$.

For $\mathcal{H}_\ell(\mathfrak{x}) = (20 + \ell)|(1 + \mathfrak{x})^{\frac{-4}{3}}| + \mathfrak{G}(\mathfrak{x})$ with $\mathfrak{G}(\mathrm{w}) = \begin{cases} e^{-\mathrm{w}}, & \mathrm{w} \leqslant 0 \\ 0, & \mathrm{w} > 0 \end{cases}$ then $\mathcal{H}_\ell \in \mathcal{PAP}(\mathbb{R}, \mathbb{R}, \varrho)$. Therefore, all solutions of (5.3.1) with initial value in the form of ML, they converge to a unique solution of Eq. (5.3.1) such as $\mathfrak{g}^*(\mathfrak{x}) \in \mathcal{PAP}\big(\mathbb{R}, \mathbb{R}^2, \varrho\big)$, when $\mathfrak{x} \longrightarrow +\infty$ with convergence rate $\upsilon \approx 0.05$.

References

1. Bochner, S.: Continuous mappings of almost automorphic and almost periodic functions. Proc. Nat. Acad. Sci. USA **52**, 907–910 (1964)
2. N'Guerekata, G.M.: Almost Automorphic and Almost Periodic Functions in Abstract Spaces, pp. x+138. Kluwer Academic/Plenum Publishers, New York (2001). ISBN: 0-306-46686-4
3. Papageorgiou, N.S., Rădulescu, V.D., Repovš, D.D.: Periodic solutions for a class of evolution inclusions. Comput. Math. Appl. **75**(8), 3047–3065 (2018)
4. Ait Dads, E.H., Ezzinbi, K., Miraoui, M.: (μ, ν)-pseudo almost automorphic solutions for some non-autonomous differential equations. Int. J. Math. **26**(11), 1550090, 21 pp (2015)
5. Kong, F., Nieto, J.J.: Almost periodic dynamical behaviors of the hematopoiesis model with mixed discontinuous harvesting terms. Discret. Contin. Dyn. Syst. Ser. B **24**(11), 5803–5830 (2019)
6. Diagana, T.: Pseudo almost periodic solutions to some differential equations. Nonlinear Anal. **60**(7), 1277–1286 (2005)
7. Gupta, C.P.: Existence and uniqueness theorems for boundary value problems involving reflection of the argument. Nonlinear Anal. **11**(9), 1075–1083 (1987)
8. Gupta, C.P.: Two-point boundary value problems involving reflection of the argument. Int. J. Math. Math. Sci. **10**(2), 361–371 (1987)
9. Xin, N., Piao, D.: Weighted pseudo almost periodic solutions for differential equations involving reflection of the argument. Int. J. Phys. Sci. **7**(11), 1806–1810 (2012)
10. Eidinejad, Z., Saadati, R., Mesiar, R.: Optimum approximation for ς-Lie homomorphisms and jordan ς-Lie homomorphisms in ς-Lie algebras by aggregation control functions. Mathematics **10**(10), 1704 (2022)
11. Blot, J., Cieutat, P., Ezzinbi, K.: Measure theory and pseudo almost automorphic functions: new developments and applications. Nonlinear Anal. **75**(4), 2426–2447 (2012)
12. Diagana, T., Ezzinbi, K., Miraoui, M.: Pseudo-almost periodic and pseudo-almost automorphic solutions to some evolution equations involving theoretical measure theory. Cubo **16**(2), 1–31 (2014). MR3237503
13. Ait Dads, E., Fatajou, S., Khachimi, L.: Pseudo almost automorphic solutions for differential equations involving reflection of the argument. ISRN Math. Anal. Art. ID 626490, 20 (2012)
14. Miraoui, M., Repovš, D.D.: Existence results for integro-differential equations with reflection. Numer. Funct. Anal. Optim. **42**(8), 919–934 (2021). MR4284921
15. Chua, L.O., Yang, L.: Cellular neural networks: theory. IEEE Trans. Circuits Syst. **35**(10), 1257–1272 (1988)
16. Chua, L.O., Yang, L.: Cellular neural networks: applications. IEEE Trans. Circuits Syst. **35**(10), 1273–1290 (1988)
17. Liu, B., Tunç, C.: Pseudo almost periodic solutions for CNNs with leakage delays and complex deviating arguments. Neural Comput. Appl. **26**(2), 429–435 (2015)
18. Liu, B., Tunç, C.: Pseudo almost periodic solutions for a class of nonlinear Duffing system with a deviating argument. J. Appl. Math. Comput. **49**(1), 233–242 (2015)
19. Amdouni, M., Chérif, F., Tunç, C.: On the weighted piecewise pseudo almost automorphic solutions Mackey-Glass model with mixed delays and harvesting term. Iran. J. Sci. Technol. Trans. A Sci. **45**(2), 619–634 (2021)
20. Sun, Y., Zhao, J., Dimirovski, G.M.: Adaptive control for a class of state-constrained high-order switched nonlinear systems with unstable subsystems. Nonlinear Anal. Hybrid Syst. **32**, 91–105 (2019)
21. Zhang, Z., Wu, H.: Cluster synchronization in finite/fixed time for semi-Markovian switching TS fuzzy complex dynamical networks with discontinuous dynamic nodes. AIMS Math. **7**(7), 11942–11971 (2022)
22. Bai, J., Wu, H., Cao, J.: Secure synchronization and identification for fractional complex networks with multiple weight couplings under DoS attacks. Comput. Appl. Math. **41**(4), 1–18 (2022)

23. Li, R., Wu, H., Cao, J.: Impulsive exponential synchronization of fractional-order complex dynamical networks with derivative couplings via feedback control based on discrete time state observations. Acta Math. Sci. **42**(2), 737–754 (2022)
24. Blot, J., Cieutat, P., Ezzinbi, K.: New approach for weighted pseudo-almost periodic functions under the light of measure theory, basic results and applications. Appl. Anal. **92**(3), 493–526 (2013)
25. Xu, J.: Weighted pseudo almost periodic delayed cellular neural networks. Neural Comput. Appl. **30**, 2453–2458 (2018)
26. Yao, L., Wang, Z., Huang, X., Li, Y., Shen, H., Chen, G.: Aperiodic sampled data control for exponential stabilization of delayed neural networks: a refined two-sided looped-functional approach. IEEE Trans. Circuits Syst. II Express Briefs **67**(12), 3217–3221 (2020). https://doi.org/10.1109/TSMC.2020.2964605. Dec
27. Li, Y., Li, G., Meng, X.: Weighted pseudo-almost periodic solutions and global exponential synchronization for delayed QVCNNs. J. Inequalities Appl. Paper No. 231, 23 (2019)
28. Liu, B.: Pseudo almost periodic solutions for neutral type CNNs with continuously disturbed leakage delays. Neurocomputing **148**, 445–454 (2015)
29. Miraoui, M., Yaakoubi, N.: Measure pseudo almost periodic solutions of shunting inhibitory cellular neural networks with mixed delays. Numer. Funct. Anal. Optim. **40**(5), 571–585 (2019)
30. Bekolle, D., Ezzinbi, K., Fatajou, S., Danga, D.E.H., Besseme, F.M.: Attractiveness of pseudo almost periodic solutions for delayed cellular neural networks in the context of measure theory. Neurocomputing **435**, 253–263 (2021)
31. Diagana, T.: Weighted pseudo almost periodic functions and applications. C. R. Math. Acad. Sci. Paris **343**(10), 643–646 (2006)
32. Ghanmi, B., Miraoui, M.: Stability of unique pseudo almost periodic solutions with measure. Appl. Math. **65**(4), 421–445 (2020)
33. Jia, J., Huang, X., Li, Y., Cao, J., Alsaedi, A.: Global stabilization of fractional-order memristor-based neural networks with time delay. IEEE Trans. Neural Netw. Learn. Syst. **31**(3), 997–1009 (2020)
34. Yao, L., Wang, Z., Huang, X., Li, Y., Shen, H., Chen, G.: A periodic sampled-data control for exponential stabilization of delayed neural networks: a refined two-sided lopped-functional approach. IEEE Trans. Circuits Syst. II Express Briefs. https://doi.org/10.1109/TCSII.2020.2983803.
35. Yu, Y.: Exponential stability of pseudo almost periodic solutions for cellular neural networks with multi-proportional delays. Neural Process. Lett. **45**, 141–151 (2017)
36. Zhang, H.: Existence and stability of almost periodic solutions for CNNs with continuously distributed leakage delays. Neural Comput. Appl. **24**, 1135–1146 (2014)
37. Graily, E., Vaezpour, S.M., Cho, Y.J.: Generalization of fixed point theorems in ordered metric spaces concerning generalized distance. Fixed Point Theory Appl. **2011**, 30, 8 pp (2011)
38. Miraoui, M., Repovš, D.D.: Dynamics and oscillations of models for differential equations with delays. Bound. Value Probl. Paper No. 54, 17 (2020)
39. Miraoui, M., Repovšs, D.D.: Existence results for integro-differential equations with reflection. Numer. Funct. Anal. Optim. **42**(8), 919–934 (2021)
40. Zhou, Q., Shao, J.: Weighted pseudo-anti-periodic SICNNs with mixed delays. Neural Comput. Appl. **29**, 865–872 (2018)
41. Liu, B.: Global exponential convergence of non-autonomous cellular neural networks with multi-proportional delays. Neurocomputing **191**, 352–355 (2016)
42. Liu, B.: Exponential convergence of SICNNs with delays and oscillating coefficients in leakage terms. Neurocomputing **168**, 500–504 (2015)
43. Zhang, A.: Pseudo almost periodic solutions for SICNNs with oscillating leakage coefficients and complex deviating arguments. Neural Process. Lett. **45**, 183–196 (2017)
44. Royden, H.L.: Real Analysis, 3rd edn. Macmillan Publishing Company, New York (1988)
45. Fink, A.M.: Almost Periodic Differential Equations. Lecture Notes in Mathematics, vol. 377. Springer, Berlin, New York (1974)
46. N'Guérékata, G.M.: Topics in Almost Automorphy. Springer, New York (2005)

47. Chua, L.O., Yang, L.: Cellular neural networks: theory. IEEE Trans. Circuits Syst. **35**(10), 1257–1272 (1988)
48. Chua, L.O., Yang, L.: Cellular neural networks: applications. IEEE Trans. Circuits Syst. **35**(10), 1273–1290 (1988)
49. Zhang, C.Y.: Integration of vector-valued pseudo-almost periodic functions. Proc. Am. Math. Soc. **121**(1), 167–174 (1994)
50. Zhang, C.Y.: Pseudo-almost-periodic solutions of some differential equations. J. Math. Anal. Appl. **181**(1), 62–76 (1994)
51. Blot, J., Cieutat, P., Ezzinbi, K.: New approach for weighted pseudo-almost periodic functions under the light of measure theory, basic results and applications. Appl. Anal. **92**(3), 493–526 (2013)
52. Ait Dads, E., Arino, O.: Exponential dichotomy and existence of pseudo almost-periodic solutions of some differential equations. Nonlinear Anal. **27**(4), 369–386 (1996)
53. Liang, J., Xiao, T.-J., Zhang, J.: Decomposition of weighted pseudo-almost periodic functions. Nonlinear Anal. **73**(10), 3456–3461 (2010)
54. Wang, Z.Y., Klir, G.J.: Fuzzy Measure Theory, pp. x+354. Plenum Press, New York (1992). ISBN: 0-306-44260-4
55. Ralescu, D., Adams, G.: The fuzzy integral. J. Math. Anal. Appl. **75**(2), 562–570 (1980)
56. Roman-Flores, H., Chalco-Cano, Y.: Sugeno integral and geometric inequalities. Int. J. Uncertain. Fuzziness Knowl. Based Syst. **15**(1), 1–11 (2007)
57. Kaluszka, M., Okolewski, A., Boczek, M.: On Chebyshev type inequalities for generalized Sugeno integrals. Fuzzy Sets Syst. **244**, 51–62 (2014)
58. Caballero, J., Sadarangani, K.: A Cauchy-Schwarz type inequality for fuzzy integrals. Nonlinear Anal. **73**(10), 3329–3335 (2010)
59. Li, Y., Lü, G., Meng, X.: Weighted pseudo-almost periodic solutions and global exponential synchronization for delayed QVCNNs. J. Inequalities Appl. Paper No. 231, 23 (2019)
60. Klement, E.P., Mesiar, R., Pap, E.: Triangular norms-basic properties and representation theorems. In: Discovering the World with Fuzzy Logic. Studies in Fuzziness and Soft Computing, vol. 57, pp. 63–81. Physica, Heidelberg (2000)
61. Podlubny, I.: Fractional Differential Equations. An Introduction to Fractional Derivatives, Fractional Differential Equations, to Methods of their Solution and Some of their Applications. Mathematics in Science and Engineering, vol. 198, pp. xxiv+340. Academic Press, Inc., San Diego, CA (1999). ISBN: 0-12-558840-2
62. Mainardi, F.: Fractional Calculus and Waves in Linear Viscoelasticity. An Introduction to Mathematical Models, pp. xx+347. Imperial College Press, London (2010). ISBN: 978-1-84816-329-4; 1-84816-329-0
63. Yu, C.H.: A study of some fractional functions. Int. J. Math. Trends Technol. (IJMTT) **66**
64. Pap, E.: Null-Additive Set Functions. Mathematics and its Applications, vol. 337, pp. xii+315. Kluwer Academic Publishers Group, Dordrecht; Ister Science, Bratislava (1995). ISBN: 0-7923-3658-5
65. Boczek, M., Kaluszka, M.: On the Minkowski-Hölder type inequalities for generalized Sugeno integrals with an application. Kybernetika (Prague) **52**(3), 329–347 (2016)
66. Royden, H.L.: Real Analysis, 3rd edn, pp. xx+444. Macmillan Publishing Company, New York (1988). ISBN: 0-02-404151-3
67. Rudin, W.: Analiza rzeczywista i zespolona. (Polish) [[Real and complex analysis]] Translated from the English by Antoni Pierzchalski and Paweł Walczak, pp. 432. Państwowe Wydawnictwo Naukowe (PWN), Warsaw (1986). ISBN: 83-01-05124-8
68. Bekolle, D., Ezzinbi, K., Fatajou, S., Danga, D.E.H., Besseme, F.M.: Attractiveness of pseudo almost periodic solutions for delayed cellular neural networks in the context of measure theory. Neurocomputing **435**, 253–263 (2021)
69. N'Guérékata, G.M.: Topics in Almost Automorphy, pp. xii+168. Springer, New York (2005). ISBN: 0-387-22846-2
70. Yan, Z., Huang, X., Fan, Y., Xia, J., Shen, H.: Threshold-function-dependent quasisynchronization of delayed memristive neural networks via hybrid even triggered control. IEEE Trans. Syst. Man Cybern. Syst. https://doi.org/10.1109/TSMC.2020.2964605

71. Yao, L., Wang, Z., Huang, X., Li, Y., Shen, H., Chen, G.: A periodic sampled-data control for exponential stabilization of delayed neural networks: a refined two-sided lopped-functional approach. IEEE Trans. Circuits Syst. II Express Briefs. https://doi.org/10.1109/TCSII.2020.2983803

Chapter 6
Z-Numbers

To make a decision to select a suitable approximation for the solution of a functional inequality, we need reliable information. Two useful information ideas are quality and certainty, and the measure of quality and certainty approximation of the solution of a functional inequality helps us to find the optimum approximation. We define a new control function to approximate a stochastic fractional Volterra IDE using the concept of modular stability. Also, we apply Hyers–Ulam-$\mathcal{H}$-Fox (HU-$\mathcal{H}$-F) stability of Generalized Fractional Systems. To measure quality and certainty, we use the idea of the Z-number (Z-N) and we introduced the generalized Z-N (GZ-N) as a diagonal matrix of the form $diag(X, Y, X * Y)$, where X is a fuzzy set time-stamped, Y is the probability distribution function and the third part is the fuzzy-random trace of the first and the second subjects. This kind of diagonal matrix allowed us to define a new model of control functions to stabilize our problem. Using stability analysis, we obtained the most suitable approximation for functional inequalities. We are going to investigate the concept of Lebesgue–stieltjes-$*$-$\mathcal{Z}$-Number measure (L-S-$*$-$\mathcal{Z}$-N-M), $*$-$\mathcal{Z}$-Number differentiation ($*$-$\mathcal{Z}$-N-D) and prove the Lebesgue–Radon–Nikodyme theorem (LRN-T). Before this work, in [Fuzzy Sets, and Systems 430 (2022), 144–173], the authors have investigated the fuzzy measure theory and fuzzy integration considering fuzzy number-valued triangular norm-based decomposable time-stamped.

Z. Eidinejad et al., *Optimal Stability Theory and Approximate Solutions of Fractional Systems*, Studies in Systems, Decision and Control 606,
https://doi.org/10.1007/978-3-031-96704-7_6

6.1 Modular Stability Analysis of a Nonlinear Stochastic Fractional Volterra IDE

6.1.1 Introduction of the Stochastic Fractional Nonlinear Volterra-IDE and Other Preliminaries

We study the existence of solutions for the stochastic fractional nonlinear Volterra-IDE:

$$\begin{cases} {}^{H}\mathbb{D}_{0+}^{\ell,\kappa;\aleph}\hbar(\varrho,\theta) = \boldsymbol{f}(\varrho,\theta,\hbar(\varrho,\theta)) + \int_0^\theta \boldsymbol{k}(\varrho,\theta,\vartheta,\hbar(\varrho,\theta))d\vartheta, \\ \mathcal{I}_{0+}^{1-\lambda}\hbar(\varrho,0) = \vartheta, \end{cases} \tag{6.1.1}$$

with $\theta \in [0, M]$, where $\boldsymbol{f}(\varrho,\theta,\hbar)$ is a continuous random operator (in short CRO) with respect to all the variables ϱ, θ and $\hbar$ on $\Upsilon \times [0, M] \times \mathbb{R}$, $\boldsymbol{k}(\varrho,\theta,\vartheta,\hbar)$ is a CRO with respect to ϱ, θ, ϑ and $\hbar$ on $\Upsilon \times [0, M] \times \mathbb{R} \times \mathbb{R}$, ϑ is a fixed number, ${}^{H}\mathbb{D}_{0+}^{\ell,\kappa;\aleph}\hbar(.)$ is defined later in (6.1.2) where $\varrho \in \Upsilon$, $0 < \ell < 1$, $0 \le \kappa \le 1$ and $\mathcal{I}_{0+}^{1-\lambda}(.)$ is the $\aleph$-Riemann–Liouville stochastic fractional integral where $0 \le \lambda < 1$, and Υ is defined in the next section. We consider a new space called the modular space, which was first introduced in 1950 by Nakano [18]. Later, Musielak and Orlicz generalized it in [17] and we also refer the reader to [14] for more information. Let (S, ν) be a MNS. A sequence $\{\varpi_n\} \subset S$ is modular convergent to $\varpi \in S$ in MNS (S, ν), if for any $\epsilon \in \Xi_5^\circ$ and $\tau \in \Xi_2$, there exists a positive integer $N_{\epsilon,\tau} \in \Xi_2$ such that $\nu_\tau(\varpi_n - \varpi) < \epsilon$ when $n \ge N_{\epsilon,\tau}$. A sequence $\{\varpi_n\} \subset S$ is modular Cauchy in MNS (S, ν), if for any $\epsilon \in \Xi_5^\circ$ and $\tau \in \Xi_2$, there exists a positive integer $N_{\epsilon,\tau} \in \Xi_2$ such that $\nu_\tau(\varpi_n - \varpi_m) < \epsilon$ whenever $n, m \ge N_{\epsilon,\tau}$. A MNS in which every Cauchy sequence is convergent is said to be a MBS. An example of a modular norm is

$$\nu_\tau(\varpi) = \frac{\|\varpi\|}{g(\tau)},$$

in which $g : \Xi_2 \to \Xi_2$ is a nondecreasing function for all $\tau \in \Xi_2$ and ϖ is a member of a normed linear space $(W, \|.\|)$.

Consider the probability measure space (Υ, Ξ_2, Ξ) and let $(U, \boldsymbol{B}_U)$ and $(S, \boldsymbol{B}_S)$ be Borel measureable spaces, for MNS U and S. If $\{\varrho : \mathcal{F}(\varrho, \varpi) \in B\} \in \Xi_2$ for every ϖ in U and $B \in \boldsymbol{B}_S$, we say $\mathcal{F} : \Upsilon \times U \to S$ is a random operator.

To prove our main result we use an alternative fixed point theorem (AFPT) (we refer the reader to [5, 6]).

Definition 6.1 ([24]) The gamma function is defined as

$$\lambda(z) = \int_0^\infty e^{-\eth} \eth^{z-1} d\eth,$$

where $z \in \mathbb{C}$, $Re(z) > 0$.

Consider $\ell \in \mathring{\Xi}_5$ and the integrable random operator f on Ξ_1 and the nondecreasing random operator $\aleph \in C^1(\Upsilon \times \Xi_1)$ with $\aleph'(\varrho, \eth) \neq 0$, for each $\eth \in \Xi_1$. The right-sided $\aleph$-Hilfer stochastic fractional derivative, is defined by [22, 25]

$$^{H}\mathbb{D}_{0+}^{\ell,s;\aleph} f(\varrho,\theta) = \mathcal{I}_{0+}^{s(1-\ell);\aleph}\left(\frac{1}{\aleph'(\varrho,\theta)}\frac{d}{d\theta}\right)\mathcal{I}_{0+}^{(1-s)(1-\ell);\aleph} f(\varrho,\theta). \tag{6.1.2}$$

From the calculations, we consider the minimum function as the control function and define it as follows

$$\varphi_\tau(\theta) = \bigwedge\left\{Exp\left(\frac{-\|\theta\|}{\tau}\right), E_p\left(\frac{-\|\theta\|}{\tau}\right), E_p^{m_1,m_2}\left(\frac{-\|\theta\|}{\tau}\right), E_p^{m_1}\left(\frac{-\|\theta\|}{\tau}\right), \mathrm{H}_{\gamma_2,\gamma_4}^{\gamma_3,\gamma_1}\left(\frac{-\|\theta\|}{\tau}\right)\right\}.$$

Definition 6.2 Consider the continuously differentiable random operator $\hbar(\varrho,\theta)$ and let $\varphi_\tau(\theta)$ be a modular control function satisfying

$$\nu_\tau\left({}^{H}\mathbb{D}_{0+}^{\ell,\kappa;\aleph}\hbar(\varrho,\theta) - f(\varrho,\theta,\hbar(\varrho,\theta)) - \int_0^\theta k(\varrho,\theta,\vartheta,\hbar(\varrho,\theta))d\vartheta\right) \le \varphi_\tau(\theta),$$

for each $\theta \in \Xi_1$, $\tau \in \Xi_2$ and $\varrho \in \Upsilon$. If there exist a solution $\hbar_0(\varrho,\theta)$ of the VIDE (6.1.1) and a fixed number $\mathbf{C} > 0$ with

$$\nu_\tau\left(\hbar(\varrho,\theta) - \hbar_0(\varrho,\theta)\right) \le \varphi_{\frac{\tau}{\mathbf{C}}}(\theta),$$

for all $\theta \in \Xi_1$, $\tau \in \Xi_2$ and $\varrho \in \Upsilon$, in which $\mathbf{C}$ is autonomous of $\hbar(\varrho,\theta)$ and $\hbar_0(\varrho,\theta)$, then we say that (6.1.1) has Hyers–Ulam–Rassias stability.

6.1.2 *Stability Results*

We assume the following:

(H0). Let $M, L_f, L_k > 0$ be fixed numbers with $M\left(L_f + L_k\right) \in \mathring{\Xi}_5$ and let the CROs $f : \Upsilon \times \Xi_1 \times \mathbb{R} \to \mathbb{R}$ and $k : \Upsilon \times \Xi_1 \times \Xi_1 \times \mathbb{R} \to \mathbb{R}$ satisfy

$$\nu_\tau\left(f(\varrho,\theta,\hbar_1) - f(\varrho,\theta,\hbar_2)\right) \le \nu_{\frac{\tau}{L_f}}\left(\hbar_1 - \hbar_2\right), \tag{6.1.3}$$

for all $\theta \in \Xi_1, \hbar_1, \hbar_2 \in \mathbb{R}$, $\tau \in \Xi_2$ and $\varrho \in \Upsilon$, and

$$\nu_\tau\left(\boldsymbol{k}(\varrho,\theta,\vartheta,\hbar_1)-\boldsymbol{k}(\varrho,\theta,\vartheta,\hbar_2)\right)\leq \nu_{\frac{\tau}{L_k}}\left(\hbar_1-\hbar_2\right), \tag{6.1.4}$$

for all $\theta, \vartheta \in \Xi_1, \hbar_1, \hbar_2 \in \mathbb{R}$, $\tau \in \Xi_2$ and $\varrho \in \Upsilon$.

Theorem 6.1 *Assume (H0), the nondecreasing random operator $\aleph \in C(\Upsilon \times \Xi_1)$ with $\aleph'(\varrho,\theta) \neq 0$ and the continuously differentiable random operator $\hbar : \Upsilon \times \Xi_1 \to \mathbb{R}$ satisfying*

$$\nu_\tau\left({}^H\mathbb{D}_{0+}^{\ell,\kappa;\aleph}\hbar(\varrho,\theta)-\boldsymbol{f}(\varrho,\theta,\hbar(\varrho,\theta))-\int_0^\theta \boldsymbol{k}(\varrho,\theta,\vartheta,\hbar(\varrho,\vartheta))d\vartheta\right)\leq\varphi_\tau(\theta), \tag{6.1.5}$$

for all $\theta, \vartheta \in \Xi_1$, $\hbar \in \mathbb{R}$, $\tau \in \Xi_2$ and $\varrho \in \Upsilon$, where $\varphi : \Xi_1 \times \Xi_2 \to \Xi_3$ is a continuous modular set with

$$\nu_\tau\left(\frac{1}{\lambda(\ell)}\int_0^\theta \aleph'(\varrho,\varpi)(\aleph(\varrho,\theta)-\aleph(\varrho,\varpi))^{\ell-1}\varphi(\varpi,\tau)d\varpi\right)\leq\varphi_{\frac{\tau}{M}}(\theta), \tag{6.1.6}$$

for each $\theta \in \Xi_1$, $\tau \in \Xi_2$ and $\varrho \in \Upsilon$. Then, we can find a unique CRO $\hbar_0 : \Upsilon \times \Xi_1 \to \mathbb{R}$, such that

$$\begin{aligned}\hbar_0(\varrho,\theta)=&\frac{(\aleph(\varrho,\theta)-\aleph(\varrho,0))^{\lambda-1}}{\Gamma(\lambda)}\vartheta\\&+\mathcal{I}_{0+}^{\ell;\aleph}\boldsymbol{f}(\varrho,\theta,\hbar_0(\varrho,\theta))\\&+\mathcal{I}_{0+}^{\ell;\aleph}\left[\int_0^\varpi \boldsymbol{k}(\varrho,\theta,\vartheta,\hbar_0(\varrho,\vartheta))d\vartheta\right],\end{aligned} \tag{6.1.7}$$

with $\mathcal{I}_{0+}^{1-\lambda;\aleph}\hbar(\varrho,0)=\vartheta$, $0<\ell<1$, $0\leq\kappa\leq 1$ and

$$\nu_\tau\left(\hbar(\varrho,\theta)-\hbar_0(\varrho,\theta)\right)\leq\varphi_{\frac{M\tau}{1-M\left(L_f+L_k\right)}}(\theta), \tag{6.1.8}$$

for each $\theta \in \Xi_1$, $\tau \in \Xi_2$ and $\varrho \in \Upsilon$.

Proof For $\wp, \jmath \in U$, we set

$$\partial(\wp,\jmath)=\inf\left\{\mathbf{C}\in\Xi_4:\nu_\tau\left(\wp(\varrho,\theta)-\jmath(\varrho,\theta)\right)\leq\varphi_{\frac{\tau}{\mathbf{C}}}(\theta)\right\}, \tag{6.1.9}$$

for each $\theta \in \Xi_1$, $\tau \in \Xi_2$ and $\varrho \in \Upsilon$, where

$$U=\{\wp:\Upsilon\times\Xi_1\to\mathbb{R}\text{ is CRO}\}.$$

Let $\Omega : U \to U$ be given by

$$\Omega \wp(\varrho, \theta) = \frac{(\aleph(\varrho, \theta) - \aleph(\varrho, 0))^{\lambda - 1}}{\Gamma(\lambda)} \vartheta + \mathcal{I}_{0+}^{\ell;\aleph} \boldsymbol{f}(\varrho, \theta, \wp(\varrho, \theta)) + \mathcal{I}_{0+}^{\ell;\aleph}\left[\int_0^{\varpi} \boldsymbol{k}(\varrho, \theta, \vartheta, \wp(\varrho, \vartheta)) d\vartheta \right], \tag{6.1.10}$$

for all $\wp \in \Xi_1, \theta \in \Xi_1$ and $\varrho \in \Upsilon$.

First we show Ω is strictly contractive on U. Let $\partial(\wp, \jmath) \le \mathbf{C}_{\wp\jmath}$ for any $\wp, \jmath \in U$, $\mathbf{C}_{\wp\jmath} \in \Xi_4$ be a fixed number, then from (6.1.9) we have

$$\nu_\tau\left(\wp(\varrho, \theta) - \jmath(\varrho, \theta)\right) \le \varphi_{\frac{\tau}{\mathbf{C}_{\wp\jmath}}}(\theta), \tag{6.1.11}$$

for each $\theta \in \Xi_1$, $\tau \in \Xi_2$ and $\varrho \in \Upsilon$. From (6.1.3), (6.1.4), (6.1.6), (6.1.10) and (6.1.11), we have

$$\begin{aligned}
&\nu_\tau\left(\Omega\wp(\varrho, \theta) - \Omega\jmath(\varrho, \theta)\right) \\
&= \nu_\tau\Bigg(\frac{1}{\Gamma(\ell)} \int_0^\theta \aleph'(\varrho, \varpi)(\aleph(\varrho, \theta) - \aleph(\varrho, \varpi))^{\ell-1} \nu_\tau\Big(\boldsymbol{f}(\varrho, \varpi, \wp(\varrho, \varpi)) - \boldsymbol{f}(\varrho, \varpi, \jmath(\varrho, \varpi)) \\
&\quad + \int_0^{\varpi} \boldsymbol{k}(\varrho, \theta, \vartheta, \wp(\varrho, \vartheta)) - \boldsymbol{k}(\varrho, \theta, \vartheta, \jmath(\varrho, \vartheta)) d\vartheta\Big) d\varpi\Bigg) \\
&\le \nu_\tau\Bigg(\frac{1}{\Gamma(\ell)} \int_0^\theta \aleph'(\varrho, \varpi)(\aleph(\varrho, \theta) - \aleph(\varrho, \varpi))^{\ell-1} \max\Big\{\nu_\tau\left(\boldsymbol{f}(\varrho, \varpi, \wp(\varrho, \varpi)) - \boldsymbol{f}(\varrho, \varpi, \jmath(\varrho, \varpi))\right) \\
&\quad , \nu_\tau\Big(\int_0^{\varpi} \boldsymbol{k}(\varrho, \theta, \vartheta, \wp(\varrho, \vartheta)) - \boldsymbol{k}(\varrho, \theta, \vartheta, \jmath(\varrho, \vartheta)) d\vartheta\Big)\Big\} d\varpi\Bigg) \\
&\le \nu_\tau\Bigg(\frac{1}{\Gamma(\ell)} \int_0^\theta \aleph'(\varrho, \varpi)(\aleph(\varrho, \theta) - \aleph(\varrho, \varpi))^{\ell-1} \max\Big\{\nu_{\frac{\tau}{L_f}}\left(\wp(\varrho, \varpi) - \jmath(\varrho, \varpi)\right) \\
&\quad , \nu_{\frac{\tau}{L_k}}\left(\wp(\varrho, \varpi) - \jmath(\varrho, \varpi)\right)\Big\} d\varpi\Bigg) \\
&\le \nu_\tau\left(\frac{1}{\Gamma(\ell)} \int_0^\theta \aleph'(\varrho, \varpi)(\aleph(\varrho, \theta) - \aleph(\varrho, \varpi))^{\ell-1} \nu_{\frac{\tau}{L_f + L_k}}\left(\wp(\varrho, \varpi) - \jmath(\varrho, \varpi)\right) d\varpi\right) \\
&\le \nu_\tau\left(\frac{1}{\Gamma(\ell)} \int_0^\theta \aleph'(\varrho, \varpi)(\aleph(\varrho, \theta) - \aleph(\varrho, \varpi))^{\ell-1} \varphi_{\frac{\tau}{\mathbf{C}_{\wp\jmath}(L_f + L_k)}}(\varpi) d\varpi\right) \\
&\le \varphi_{\frac{\tau}{M\mathbf{C}_{\wp\jmath}(L_f + L_k)}}(\theta),
\end{aligned} \tag{6.1.12}$$

and we conclude that

$$\partial(\Omega\wp, \Omega\jmath) \le \varphi_{\frac{\tau}{M\mathbf{C}_{\wp\jmath}(L_f + L_k)}}(\theta),$$

for all $\theta \in \varXi_1$ and $\tau \in \varXi_2$. Hence, we deduce that $\partial(\Omega\wp, \Omega\jmath) \leq [M(L_f + L_k)]$ $\partial(\wp, \jmath)$ for any $\wp, \jmath \in U$, and recall $0 < M(L_f + L_k) < 1$.

Now (6.1.10), enables us to find $\mathbf{C} \in \varXi_2$, with

$$\begin{aligned}
&\nu_\tau\left(\Omega\jmath_0(\varrho,\theta) - \jmath_0(\varrho,\theta)\right) \\
&= \nu_\tau\Bigg(\frac{(\aleph(\varrho,\theta)-\aleph(\varrho,0))^{\lambda-1}}{\Gamma(\lambda)}\vartheta + \mathcal{I}_{0+}^{\ell;\aleph}\boldsymbol{f}(\varrho,\theta,\jmath_0(\varrho,\theta)) \\
&+ \mathcal{I}_{0+}^{\ell;\aleph}\left[\int_0^{\varpi}\boldsymbol{f}(\varrho,\theta,\vartheta,\jmath_0(\varrho,\vartheta))d\vartheta\right] - \jmath_0(\varrho,\theta)\Bigg) \\
&\leq \varphi_{\frac{\tau}{\mathbf{C}}}(\theta),
\end{aligned}$$

for arbitrary $\jmath_0 \in U$, for all $\theta \in \varXi_1$, $\tau \in \varXi_2$ and $\varrho \in \Upsilon$. The boundedness property of

$$\boldsymbol{f}(\varrho,\varpi,\jmath_0(\varrho,\varpi)),\ \boldsymbol{k}(\varrho,\theta,\vartheta,\jmath_0(\varrho,\vartheta)),\ \jmath_0(\varrho,\theta)$$

and (6.1.9) imply that $\partial(\Omega\jmath_0, \jmath_0) < \infty$. From the AFPT, we can find a CRO $\hbar_0 : \Upsilon \times \varXi_1 \to \mathbb{R}$ such that $\Omega^n\hbar_0 \to \hbar_0$ in (U, ∂) and $\Omega\hbar_0 = \hbar_0$.

Since $\jmath$ and $\hbar_0$ are bounded on $\varXi_1$ for each $\jmath \in U$ and $\max_{\theta\in\varXi_1}\varphi_\tau(\theta) > 0$, then we have a fixed number $\mathbf{C}_{\wp\jmath} \in \varXi_4$ with

$$\nu_\tau\left(\jmath_0(\varrho,\theta) - \jmath(\varrho,\theta)\right) \leq \varphi_{\frac{\tau}{\mathbf{C}_\jmath}}(\theta),$$

for any $\theta \in \varXi_1$, $\tau \in \varXi_2$ and $\varrho \in \Upsilon$. Thus $\partial(\jmath_0, \jmath) < \infty$ for any $\jmath \in U$.

Therefore, $U = \{\jmath \in U : \partial(\jmath_0, \jmath) < \infty\}$. Also the AFPT and (6.1.7), imply the uniqueness of $\hbar_0$.

Using (6.1.5) and [25, Theorem 5] we have

$$\begin{aligned}
&\nu_\tau\Bigg(\hbar(\varrho,\theta) - \frac{(\aleph(\varrho,\theta)-\aleph(\varrho,0))^{\lambda-1}}{\Gamma(\lambda)}\vartheta \\
&- \mathcal{I}_{0+}^{\ell;\aleph}\boldsymbol{f}(\varrho,\theta,\hbar(\varrho,\theta)) - \mathcal{I}_{0+}^{\ell;\aleph}\left[\int_0^{\varpi}\boldsymbol{k}(\varrho,\theta,\vartheta,\hbar(\varrho,\vartheta))d\vartheta\right]\Bigg) \\
&\leq \frac{1}{\Gamma(\ell)}\int_0^\theta \aleph'(\varrho,\varpi)(\aleph(\varrho,\theta)-\aleph(\varrho,\varpi))^{\ell-1}\varphi_\tau(\varpi)d\varpi.
\end{aligned}$$

Then, from (6.1.6) and (6.1.10), we obtain

$$\begin{aligned}
&\nu_\tau\left(\hbar(\varrho,\theta) - \Omega\hbar(\varrho,\theta)\right) \\
&\leq \frac{1}{\Gamma(\ell)}\int_0^\theta \aleph'(\varrho,\varpi)(\aleph(\varrho,\theta)-\aleph(\varrho,\varpi))^{\ell-1}\varphi_\tau(\varpi)d\varpi \\
&\leq \varphi_{\frac{\tau}{M}}(\theta),
\end{aligned}$$

for any $\theta \in \Xi_1$, $\tau \in \Xi_2$ and $\varrho \in \Upsilon$, which implies

$$\eth(\hbar, \Omega\hbar) \le M. \tag{6.1.13}$$

From the AFPT and (6.1.13), we deduce that

$$\eth(\hbar, \hbar_0) \le \frac{1}{1 - M(L_f + L_k)} \eth(\Omega\hbar, \hbar) \le \frac{M}{1 - M(L_f + L_k)},$$

which implies (6.1.8).

Theorem 6.2 *Consider $\ell, \kappa \in \mathring{\Xi}_5$ and the nondecreasing random operator $\aleph \in C^1(\Upsilon \times \Xi_1)$ with $\aleph'(\varrho, \theta) \neq 0$ for all $\theta \in \Xi_1$. Let $L_f, L_k \in \Xi_2$ be fixed numbers such that $\frac{(L_f+L_k)}{\Gamma(\ell+1)} \in \mathring{\Xi}_5$. Consider the CROs $\boldsymbol{f} : \Upsilon \times \Xi_1 \times \mathbb{R} \to \mathbb{R}$ and $\boldsymbol{k} : \Upsilon \times \Xi_1 \times \Xi_1 \times \mathbb{R} \to \mathbb{R}$ satisfying (6.1.3) and (6.1.4) respectively. Let $\varepsilon \in \mathring{\Xi}_5$, and consider the continuously differentiable random operator $\hbar : \Upsilon \times \Xi_1 \to \mathbb{R}$ such that*

$$\nu_\tau \left({}^H\mathbb{D}_{0+}^{\ell,\kappa;\aleph} \hbar(\varrho, \theta) - \boldsymbol{f}(\varrho, \theta, \hbar(\varrho, \theta)) - \int_0^\theta \boldsymbol{k}(\varrho, \theta, \vartheta, \hbar(\varrho, \vartheta)) d\vartheta \right) \le \varphi_\tau(\varepsilon),$$

and

$$\nu_\tau \left((\aleph(\varrho, \theta) - \aleph(\varrho, 0))^\ell \right) \le \nu_\tau(\theta),$$

for all $\theta, \vartheta \in \Xi_1$, $\hbar \in \mathbb{R}$, $\tau \in \Xi_2$ and $\varrho \in \Upsilon$. Then, we can find a unique CRO $\hbar_0 : \Upsilon \times \Xi_1 \to \mathbb{R}$ satisfying (6.1.7) and

$$\nu_\tau \left(\hbar(\varrho, \theta) - \hbar_0(\varrho, \theta) \right) \le \frac{(\aleph(\varrho, T) - \aleph(\varrho, 0))^\ell \varphi_\tau(\varepsilon)}{\Gamma(\ell+1) - (\aleph(\varrho, T) - \aleph(\varrho, 0))^\ell [T\left(L_f + L_k\right)]}, \tag{6.1.14}$$

for all $\theta \in \Xi_1$, $\hbar \in \mathbb{R}$, $\tau \in \Xi_2$ and $\varrho \in \Upsilon$.

Proof Let $U = \{\wp : \Upsilon \times \Xi_1 \to \mathbb{R} \text{ is CRO}\}$. Consider the complete Ξ_4-valued metric on U given by

$$\eth(\wp, \jmath) = \inf \left\{ \mathbf{C} \in \Xi_4 : \nu_\tau(\wp(\varrho, \theta) - \jmath(\varrho, \theta)) \le \left(\frac{\tau}{\tau + \mathbf{C}} \right) \right\}, \tag{6.1.15}$$

for each $\theta \in \Xi_1$, $\tau \in \Xi_2$ and $\varrho \in \Upsilon$ [22].

Consider $\Omega : U \to U$ in which

$$\Omega\wp(\varrho,\theta) = \frac{(\aleph(\varrho,\theta)-\aleph(\varrho,0))^{\lambda-1}}{\Gamma(\lambda)}\vartheta + \mathcal{I}_{0+}^{\ell;\aleph}\boldsymbol{f}(\varrho,\theta,\wp(\varrho,\theta)) + \mathcal{I}_{0+}^{\ell;\aleph}\left[\int_0^{\varpi}\boldsymbol{k}(\varrho,\theta,\vartheta,\wp(\varrho,\vartheta))d\vartheta\right], \tag{6.1.16}$$

for all $\theta \in \Xi_1$ and $\varrho \in \Upsilon$.

Let $\wp, \jmath \in U$ and consider a fixed number $\mathbf{C}_{\wp\jmath} \in \Xi_4$ such that $\partial(\wp,\jmath) \le \mathbf{C}_{\wp\jmath}$ and

$$\nu_\tau(\wp(\varrho,\theta)-\jmath(\varrho,\theta)) \ge \left(\frac{\tau}{\tau+\mathbf{C}_{\wp\jmath}}\right), \tag{6.1.17}$$

for each $\theta \in \Xi_1$, $\tau \in \Xi_2$ and $\varrho \in \Upsilon$. Using (6.1.3), (6.1.4), (6.1.16) and (6.1.17), we have

$$\begin{aligned}
&\nu_\tau(\Omega\wp(\varrho,\theta)-\Omega\jmath(\varrho,\theta)) \\
&= \nu_\tau\Bigg(\frac{1}{\Gamma(\ell)}\int_0^\theta \aleph'(\varrho,\varpi)(\aleph(\varrho,\theta)-\aleph(\varrho,\varpi))^{\ell-1}\nu_\tau\Big(\boldsymbol{f}(\varrho,\varpi,\wp(\varrho,\varpi))-\boldsymbol{f}(\varrho,\varpi,\jmath(\varrho,\varpi)) \\
&\quad + \int_0^{\varpi}\boldsymbol{k}(\varrho,\theta,\vartheta,\wp(\varrho,\vartheta))-\boldsymbol{k}(\varrho,\theta,\vartheta,\jmath(\varrho,\vartheta))d\vartheta\Big)d\varpi\Bigg) \\
&\le \nu_\tau\Bigg(\frac{1}{\Gamma(\ell)}\int_0^\theta \aleph'(\varrho,\varpi)(\aleph(\varrho,\theta)-\aleph(\varrho,\varpi))^{\ell-1}\max\Big\{\nu_\tau(\boldsymbol{f}(\varrho,\varpi,\wp(\varrho,\varpi))-\boldsymbol{f}(\varrho,\varpi,\jmath(\varrho,\varpi))) \\
&\quad ,\nu_\tau\Big(\int_0^{\varpi}\boldsymbol{k}(\varrho,\theta,\vartheta,\wp(\varrho,\vartheta))-\boldsymbol{k}(\varrho,\theta,\vartheta,\jmath(\varrho,\vartheta))d\vartheta\Big)\Big\}d\varpi\Bigg) \\
&\le \nu_\tau\Bigg(\frac{1}{\Gamma(\ell)}\int_0^\theta \aleph'(\varrho,\varpi)(\aleph(\varrho,\theta)-\aleph(\varrho,\varpi))^{\ell-1}\max\Big\{\nu_{\frac{\tau}{L_f}}(\wp(\varrho,\varpi)-\jmath(\varrho,\varpi)) \\
&\quad ,\nu_{\frac{\tau}{L_k}}(\wp(\varrho,\varpi)-\jmath(\varrho,\varpi))\Big\}d\varpi\Bigg) \\
&\le \nu_\tau\left(\frac{1}{\Gamma(\ell)}\int_0^\theta \aleph'(\varrho,\varpi)(\aleph(\varrho,\theta)-\aleph(\varrho,\varpi))^{\ell-1}\nu_{\frac{\tau}{L_f+L_k}}(\wp(\varrho,\varpi)-\jmath(\varrho,\varpi))\,d\varpi\right) \\
&\le \nu_\tau\left(\frac{1}{\Gamma(\ell)}\int_0^\theta \aleph'(\varrho,\varpi)(\aleph(\varrho,\theta)-\aleph(\varrho,\varpi))^{\ell-1}\left(\frac{\tau\left(L_f+L_k\right)}{\tau+\mathbf{C}_{\wp\jmath}}\right)d\varpi\right) \\
&\le \nu_\tau\left((\aleph(\varrho,\theta)-\aleph(\varrho,0))^{\ell}\left(\frac{\tau\Gamma(\ell+1)\left(L_f+L_k\right)}{\Gamma(\ell+1)\left(\tau+\mathbf{C}_{\wp\jmath}\right)}\right)\right) \\
&\le \nu_\tau\left((\aleph(\varrho,\theta)-\aleph(\varrho,0))^{\ell},\frac{\tau}{\left(\frac{\tau\left(L_f+L_k\right)}{\Gamma(\ell+1)(\tau+\mathbf{C}_{\wp\jmath})}\right)}\right) \\
&\le \nu_{\frac{\tau}{\left(\frac{\tau\left(L_f+L_k\right)}{\Gamma(\ell+1)(\tau+\mathbf{C}_{\wp\jmath})}\right)}}(\theta),
\end{aligned}$$

for each $\theta \in \Xi_1$, $\tau \in \Xi_2$ and $\varrho \in \Upsilon$. Therefore

$$\eth(\Omega\wp, \Omega\jmath) \leq \left(\frac{\tau \left(L_f + L_k \right)}{\Gamma(\ell+1) \left(\tau + \mathbf{C}_{\wp\jmath} \right)} \right) \eth(\wp, \jmath),$$

for each $\wp, \jmath \in U$ and $\varrho \in \Upsilon$. Let $\jmath_0 \in U$. We can find a fixed number $\mathbf{C} \in \Xi_2$ with

$$\begin{aligned}
&\nu_\tau \left(\Omega \jmath_0(\varrho, \theta) - \jmath_0(\varrho, \theta) \right) \\
&= \nu_\tau \Bigg(\frac{(\aleph(\varrho, \theta) - \aleph(\varrho, 0))^{\lambda-1}}{\Gamma(\lambda)} \vartheta + \mathcal{I}_{0+}^{\ell;\aleph} \boldsymbol{f}(\varrho, \theta, \jmath_0(\varrho, \theta)) \\
&+ \mathcal{I}_{0+}^{\ell;\aleph} \left[\int_0^{\varpi} \boldsymbol{k}(\varrho, \theta, \vartheta, \jmath_0(\varrho, \vartheta)) d\vartheta \right] - \jmath_0(\varrho, \theta) \Bigg) \\
&\leq \frac{\tau}{\tau + \mathbf{C}},
\end{aligned}$$

for all $\theta \in \Xi_1$ and $\varrho \in \Upsilon$.

The boundedness of

$$\boldsymbol{f}(\varrho, \varpi, \jmath_0(\varrho, \varpi)), a, \boldsymbol{k}(\varrho, \theta, \vartheta, \jmath_0(\varrho, \vartheta)), \ \jmath_0(\varrho, \theta)$$

and (6.1.15), imply that $\eth(\Omega \jmath_0, \jmath_0) < \infty$.

By the AFPT, we can find a CRO $\hbar_0 : \Upsilon \times \Xi_1 \to \mathbb{R}$ with $\Omega^n \jmath_0 \to \hbar_0$ in $(U, \eth)$ and $\Omega\hbar_0 = \hbar_0$, so, $\hbar_0$ satisfies (6.1.7). By Theorem 6.1 we get $\{\jmath \in U : \eth(\jmath_0, \jmath) < \infty\} = U$. Also the AFPT and (6.1.7) imply the uniqueness of $\hbar_0$.

Now, using (6.1.5) and [25, Theorem 5], we have

$$\begin{aligned}
&\nu_{\frac{\tau\Gamma(\ell+1)}{(\aleph(\varrho,T)-\aleph(\varrho,0))^\ell}} \Bigg(\hbar(\varrho, \theta) - \frac{(\aleph(\varrho, \theta) - \aleph(\varrho, 0))^{\lambda-1}}{\Gamma(\lambda)} \vartheta - \mathcal{I}_{0+}^{\ell;\aleph} \boldsymbol{f}(\varrho, \theta, \hbar_0(\varrho, \theta)) \\
&- \mathcal{I}_{0+}^{\ell;\aleph} \left[\int_0^{\varpi} \boldsymbol{k}(\varrho, \theta, \vartheta, \hbar_0(\varrho, \vartheta)) d\vartheta \right] \Bigg) \\
&\leq \varphi_\tau(\varepsilon),
\end{aligned}$$

for all $\theta \in \Xi_1$ and $\varrho \in \Upsilon$, which implies

$$\eth(\hbar, \Omega\hbar) \leq \varphi_\tau(\varepsilon) \frac{(\aleph(\varrho, T) - \aleph(\varrho, 0))^\ell}{\Gamma(\ell+1)}.$$

By the AFPT and (6.1.9), we deduce that

$$\nu_{\frac{\tau \left(\Gamma(\ell+1) - (\aleph(\varrho, T) - \aleph(\varrho, 0))^\ell [L_f + \frac{T}{2} L_k] \right)}{(\aleph(\varrho, T) - \aleph(\varrho, 0))^\ell}} \left(\hbar(\varrho, \theta) - \hbar_0(\varrho, \theta) \right) \leq \varphi_\tau(\varepsilon),$$

which implies (6.1.14) for all $\theta \in \Xi_1$.

6.1.3 Application

Example 6.1 We consider the stochastic fractional nonlinear Volterra-IDE:

$$\begin{cases} {}^{H}\mathbb{D}_{0+}^{\frac{1}{5},\frac{1}{10};\aleph}\hbar(\varrho,\theta) = 0.5(\sin^2(\hbar(\varrho,\theta)) + \cos^2(\hbar(\varrho,\theta))) - 0.02(\varrho+\theta) \\ + \int_0^{\theta} 0.03(\sin^2(\hbar(\varrho,\theta)) + \cos^2(\hbar(\varrho,\theta)))\cos(\varrho+\theta+\vartheta) + 4\vartheta d\vartheta, \\ \\ \mathcal{I}_{0+}^{\frac{1}{2}}\hbar(\varrho,0) = \vartheta \end{cases} \tag{6.1.18}$$

where $\ell = \frac{1}{5}$, $\kappa = \frac{1}{10}$, $\lambda = \frac{1}{2}$, $\boldsymbol{f}(\varrho,\theta,\hbar(\varrho,\theta)) = 0.5(\sin^2(\hbar(\varrho,\theta)) + \cos^2(\hbar(\varrho,\theta)))$ $- 0.02(\varrho+\theta)$, $\boldsymbol{k}(\varrho,\theta,\vartheta,\hbar(\varrho,\theta)) = 0.03(\sin^2(\hbar(\varrho,\theta)) + \cos^2(\hbar(\varrho,\theta)))\cos(\varrho+\theta+\vartheta) + 4\vartheta$. Considering $L_f = 0.5\sqrt{2}$, $L_k = 0.03\sqrt{2}$ and $M = 0.08$, for functions $\boldsymbol{f}$ and $\boldsymbol{k}$, we have

$$\begin{aligned} &\nu_\tau\Bigg(0.5(\sin^2(\hbar_1(\varrho,\theta)) + \cos^2(\hbar_1(\varrho,\theta))) - 0.02(\varrho+\theta) \\ &- 0.5(\sin^2(\hbar_2(\varrho,\theta)) - \cos^2(\hbar_2(\varrho,\theta))) + 0.02(\varrho+\theta)\Bigg) \\ &\le \nu_{\frac{\tau}{0.5\sqrt{2}}}\left(\hbar_1 - \hbar_2\right), \end{aligned} \tag{6.1.19}$$

for all $\theta \in \Xi_1, \hbar_1, \hbar_2 \in \mathbb{R}$, $\tau \in \Xi_2$ and $\varrho \in \Upsilon$, and

$$\begin{aligned} &\nu_\tau\Bigg(0.03(\sin^2(\hbar_1(\varrho,\theta)) + \cos^2(\hbar_1(\varrho,\theta)))\cos(\varrho+\theta+\vartheta) \\ &- 4\vartheta - 0.03(\sin^2(\hbar_2(\varrho,\theta)) - \cos^2(\hbar_2(\varrho,\theta)))\cos(\varrho+\theta+\vartheta) + 4\vartheta\Bigg) \\ &\le \nu_{\frac{\tau}{0.03\sqrt{2}}}\left(\hbar_1 - \hbar_2\right), \end{aligned} \tag{6.1.20}$$

for all $\theta, \vartheta \in \Xi_1, \hbar_1, \hbar_2 \in \mathbb{R}$, $\tau \in \Xi_2$ and $\varrho \in \Upsilon$.

According to the function $\aleph \in C(\Upsilon \times \Xi_1)$ with $\aleph'(\varrho,\theta) \neq 0$, if we have

$$\begin{aligned} &\nu_\tau\Bigg({}^{H}\mathbb{D}_{0+}^{\frac{1}{5},\frac{1}{10};\aleph}\hbar(\varrho,\theta) = 0.5(\sin^2(\hbar(\varrho,\theta)) + \cos^2(\hbar(\varrho,\theta))) - 0.02(\varrho+\theta) \\ &+ \int_0^{\theta} 0.03(\sin^2(\hbar(\varrho,\theta)) + \cos^2(\hbar(\varrho,\theta)))\cos(\varrho+\theta+\vartheta) + 4\vartheta d\vartheta\Bigg) \\ &\le \varphi_\tau(\theta), \end{aligned} \tag{6.1.21}$$

for all $\theta, \vartheta \in \Xi_1$, $\hbar \in \mathbb{R}$, $\tau \in \Xi_2$ and $\varrho \in \Upsilon$, where $\varphi : \Xi_1 \times \Xi_2 \to \Xi_3$ is a continuous modular set with

$$\nu_\tau\left(\frac{1}{\lambda(\ell)}\int_0^\theta \aleph'(\varrho,\varpi)(\aleph(\varrho,\theta)-\aleph(\varrho,\varpi))^{\frac{-4}{5}}\varphi(\varpi,\tau)d\varpi\right)\le\varphi_{\frac{\tau}{0.08}}(\theta), \quad (6.1.22)$$

for each $\theta\in\Xi_1$, $\tau\in\Xi_2$ and $\varrho\in\Upsilon$, then, we can find a unique CRO $\hbar_0:\Upsilon\times\Xi_1\to\mathbb{R}$, such that

$$\begin{aligned}\hbar_0(\varrho,\theta)&=\frac{(\aleph(\varrho,\theta)-\aleph(\varrho,0))^{-\frac{1}{2}}}{\Gamma(\lambda)}\vartheta \quad (6.1.23)\\&+\mathcal{I}_{0+}^{\frac{1}{5};\aleph}0.5(\sin^2(\hbar_0(\varrho,\theta))+\cos^2(\hbar_0(\varrho,\theta)))-0.02(\varrho+\theta)\\&+\mathcal{I}_{0+}^{\frac{1}{5};\aleph}\left[\int_0^\varpi 0.03(\sin^2(\hbar_0(\varrho,\theta))+\cos^2(\hbar_0(\varrho,\theta)))\cos(\varrho+\theta+\vartheta)+4\vartheta d\vartheta\right],\end{aligned}$$

with $\mathcal{I}_{0+}^{\frac{1}{2};\aleph}\hbar(\varrho,0)=\vartheta$, and

$$\nu_\tau\left(\hbar(\varrho,\theta)-\hbar_0(\varrho,\theta)\right)\le\varphi_{\frac{0.08\tau}{0.9400373450}}(\theta), \quad (6.1.24)$$

where $M\left(L_f+L_k\right)=0.05996265503$, for each $\theta\in\Xi_1$, $\tau\in\Xi_2$ and $\varrho\in\Upsilon$ (Fig. 6.1).

6.2 A New Class of Generalized-$\mathcal{H}$-FOX-$\mathcal{Z}$-Numbers

6.2.1 *Introduction of the Fractional Linear Differential Systems and Other Preliminaries*

Consider $n\times n$ constant matrix $\mathcal{Q}$, the generalized fractional derivative operators ${}_0D^{\sigma,\iota}\vec{\mathfrak{x}}(\cdot)$ and ${}_0D^{\upsilon,\iota}\vec{\mathfrak{x}}(\cdot)$ and m-dimensional continuous vector-valued mappings (VVM) $\vec{\mathfrak{w}}(\varsigma)$ and $\vec{\mathfrak{z}}(\varsigma)$. We consider the following types of fractional linear differential systems, and investigate their Hyers–Ulam-$\mathcal{H}$-Fox stability in matrix-valued fuzzy Banach (MVFB) space,

$$\begin{cases}{}_0D^{\sigma,\iota}\vec{\mathfrak{x}}(\varsigma)=\mathcal{Q}\vec{\mathfrak{x}}(\varsigma)+\vec{\mathfrak{w}}(\varsigma)+\mathcal{L}_\iota^{-1}\{\vec{\mathfrak{x}}(\gamma)\}(\varsigma),\ 0<\sigma<1,\varsigma>0\\\left({}_0I^{1-\sigma,\iota}\mathfrak{x}\right)(0+)=p, \qquad p\in\mathbb{R}^n,\end{cases} \quad (6.2.1)$$

and

$$\begin{cases}{}_0D^{\sigma,\iota}\vec{\mathfrak{x}}(\varsigma)=\mathcal{Q}_0D^{\upsilon,\iota}\vec{\mathfrak{x}}(\varsigma)+\vec{\mathfrak{s}}(\varsigma)+\mathcal{L}_\iota^{-1}\{\vec{\mathfrak{x}}(\gamma)\}(\varsigma),\ 0<\upsilon<\sigma<1,\varsigma>0\\\left({}_0I^{1-\sigma,\iota}\vec{\mathfrak{x}}\right)(0+)=p, \qquad p\in\mathbb{R}^n.\end{cases} \quad (6.2.2)$$

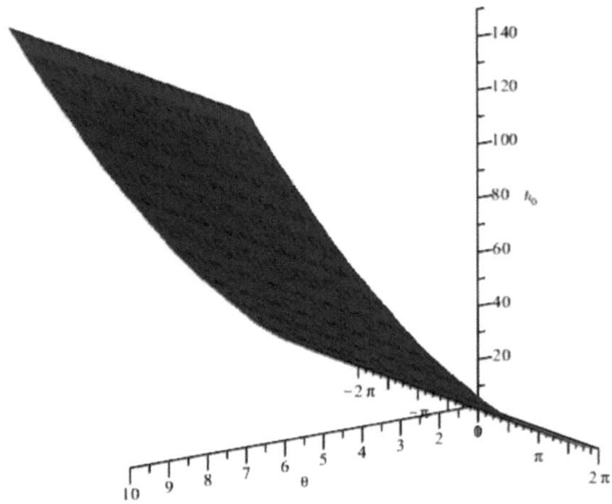

(a) The exact solution of stochastic fractional nonlinear Volterra-IDE for $\theta \in (1, 10)$

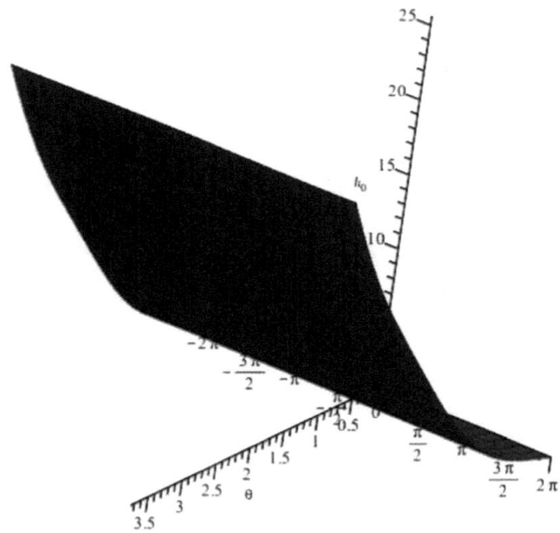

(b) The exact solution of stochastic fractional nonlinear Volterra-IDE $\theta \in (\frac{1}{10}, \frac{15}{4})$

Fig. 6.1 Graphic representation of the exact solution of Eq. (6.1.18) for different values

For $\jmath < k, \ell \in \mathbb{R}$ and $1 \leq e < \infty$, we consider the VVM-space

$$\mathcal{X}_{\ell}^{e}(\jmath, k) = \left\{ \vec{\varpi} : [\jmath, k] \to \mathbb{R}^n : \|\vec{\varpi}\|_{\mathcal{X}_{\ell}^{e}} = \left(\int_{\jmath}^{k} \left|\varsigma^{\ell}\right|^{e} \|\vec{\varpi}\|^{e} \frac{d\varsigma}{\varsigma} \right)^{\frac{1}{e}} < \infty \right\},$$

for $e = \infty$, $\|\vec{\varpi}\|_{\mathcal{X}_{\ell}^{e}} = \operatorname{ess\,sup}_{\jmath \leq \varsigma \leq k} \left\{ \varsigma^{\ell} \|\vec{\varpi}\| \right\}$.

Definition 6.3 The generalized left fractional integral of order σ of a VVM $\vec{\varpi} : [\jmath, k] \to \mathbb{R}^n$ is defined by

$$\left({}_{\jmath}I^{\sigma,\iota} \vec{\varpi} \right)(\varsigma) = \frac{1}{\Gamma(\sigma)} \int_{\jmath}^{\varsigma} \left(\frac{\varsigma^{\iota} - \gamma^{\iota}}{\iota} \right)^{\sigma - 1} \vec{\varpi}(\gamma) \frac{d\gamma}{\gamma^{1-\iota}},$$

where $\iota > 0$.

Definition 6.4 The generalized left fractional derivative of order $\sigma > 0$ of a vector-valued function $\vec{\varpi} : [\jmath, k] \to \mathbb{R}^n$ is defined by

$$\begin{aligned} \left({}_{a}D^{\sigma,\iota} \vec{\varpi} \right)(\varsigma) &= \lambda^{n} \left({}_{\jmath}I^{n-\sigma,\iota} \vec{\varpi} \right)(\varsigma) \\ &= \frac{\lambda^{n}}{\Gamma(n-\sigma)} \int_{\jmath}^{\varsigma} \left(\frac{\varsigma^{\iota} - \gamma^{\iota}}{\iota} \right)^{n-\sigma-1} \vec{\varpi}(\gamma) \frac{d\gamma}{\gamma^{1-\iota}}, \end{aligned}$$

where $\iota > 0$, $n = [R(\sigma)] + 1$ and $\lambda = t^{1-\iota} \frac{d}{d\varsigma}$.

It should mentioned that, we get Riemann–Liouville fractional integral and derivative when $\iota = 1$ in Definitions 6.3 and 6.4 also, $\iota \to 0$, in Definitions 6.3 and 6.4 implies the Hadamard fractional integral and derivative. Since the control function used in this article is a $\mathcal{H}$-Fox function, we will express the definition of the $\mathcal{H}$-Fox function.

Definition 6.5 Consider the m-dimensional VVM $\vec{\varpi}(\varsigma)$ on $[\jmath, \infty)$ and $0 < \iota \leq 1$. We define the generalized fractional Laplace transform of order ι starting from $\jmath$ of $\vec{\varpi}(\varsigma)$ as

$$\mathfrak{L}_{\iota}\{\vec{\varpi}(\varsigma)\}(s) := \int_{\jmath}^{\infty} e^{-\gamma \frac{(\varsigma - \jmath)^{\iota}}{\iota}} \vec{\varpi}(\varsigma) \frac{d\varsigma}{(\varsigma - \jmath)^{1-\iota}}, \iota > 0.$$

When $\jmath = 0$, we have

$$\mathfrak{L}_{\iota}\{\vec{\varpi}(\varsigma)\}(\gamma) := \int_{0}^{\infty} e^{-\gamma \frac{\varsigma^{\iota}}{\iota}} \vec{v}(\varsigma) \frac{d\varsigma}{\varsigma^{1-\iota}}, \iota > 0.$$

Definition 6.6 A VVM $\vec{\varpi} : [\jmath, +\infty) \to \mathbb{R}^n$ is called of ι-exponential order $e^{\ell \frac{\varsigma^{\iota}}{\iota}}$ if we can find constants $\mathcal{N}$, ℓ and $\mathcal{B}$ in $[0, +\infty)$ such that $\|\vec{\varpi}(\varsigma)\| \leq \mathcal{N} e^{\ell \frac{t^{\iota}}{\iota}}$ for $\varsigma \leq \mathcal{B}$.

Theorem 6.3 ([7, 11, 15, 23]) *If the function* $\vec{\mathfrak{x}}$ *is a solution of Eq.* (6.2.1) *then the function* $\vec{\mathfrak{x}}$ *is defined as follows*

$$\vec{\mathfrak{x}}(\varsigma) = (\gamma^{\alpha} I - Q)\mathfrak{L}_{\iota}(\vec{\mathfrak{x}}(\varsigma))(\gamma) - ({}_0 I^{1-\sigma,\iota}, \vec{\mathfrak{x}})(0_+) - \mathfrak{L}_{\iota}\{\vec{\mathfrak{w}}(\varsigma)\}(\gamma). \tag{6.2.3}$$

Proof Applying the Laplace transform to both sides of (6.2.1) and according to Laplace's definitions in the first section we derive

$$\mathfrak{L}_{\iota}\{{}_0 D^{\sigma,\iota}\vec{\mathfrak{x}}(\varsigma)\}(\gamma) = Q\mathfrak{L}_{\iota}\{\vec{\mathfrak{x}}(\varsigma)\}(\gamma) + \mathfrak{L}_{\iota}\{\vec{\mathfrak{w}}(\varsigma)\}(\gamma) + \vec{\mathfrak{x}}(\varsigma) =$$
$$\gamma^{\sigma} I\mathfrak{L}_{\iota}\{\vec{\mathfrak{x}}(\varsigma)\}(\gamma) - ({}_0 I^{1-\sigma,\iota}, \vec{\mathfrak{x}})(0_+) = Q\mathfrak{L}_{\iota}\{\vec{\mathfrak{x}}(\varsigma)\}(\gamma) + \mathfrak{L}_{\iota}\{\vec{\mathfrak{w}}(\varsigma)\}(\gamma) + \vec{\mathfrak{x}}(\varsigma).$$

We solve this equation for $\mathfrak{L}_{\iota}(\vec{\mathfrak{x}}(\varsigma))(\gamma)$, then we have

$$(\gamma^{\sigma} I - Q)\mathfrak{L}_{\iota}\{\vec{\mathfrak{x}}(\varsigma)\}(\gamma) = Q\mathfrak{L}_{\iota}\{\vec{\mathfrak{x}}(\varsigma)\}(\gamma) + \mathfrak{L}_{\iota}\{\vec{\mathfrak{w}}(\varsigma)\}(\gamma) + \vec{\mathfrak{x}}(\varsigma),$$

then

$$\vec{\mathfrak{x}}(\varsigma) = (\gamma^{\sigma} I - Q)\mathfrak{L}_{\rho}\{\vec{\mathfrak{x}}(\varsigma)\}(\gamma) - ({}_0 I^{1-\sigma,\iota}, \vec{\mathfrak{x}})(0_+) - \mathfrak{L}_{\iota}\{\vec{\mathfrak{x}}(\varsigma)\}(\gamma).$$

Theorem 6.4 *If the function* $\vec{\mathfrak{x}}$ *is a solution of Eq.* (6.2.2) *then the function* $\vec{\mathfrak{x}}$ *is defined as follows*

$$\vec{\mathfrak{x}}(\varsigma) = (\gamma^{\sigma} I - Q\gamma^{\upsilon})\,\mathfrak{L}_{\iota}\{\vec{\mathfrak{x}}(\varsigma)\}(\gamma) - \left({}_0 I^{1-\sigma,\iota}\vec{\mathfrak{x}}\right)(0+) + Q\left({}_0 I^{1-\upsilon,\iota}\vec{\mathfrak{x}}\right)(0+) - \mathfrak{L}_{\rho}\{\vec{\mathfrak{s}}(\varsigma)\}(\gamma). \tag{6.2.4}$$

Proof Applying the Laplace transform to both sides of (6.2.2) and according to Laplace's definitions in the first section we derive

$$\mathfrak{L}_{\iota}\{{}_0 D^{\sigma,\iota}\vec{\mathfrak{x}}(\varsigma)\}(\gamma) = Q_0\mathfrak{L}_{\iota}\{D^{\upsilon,\iota}\vec{\mathfrak{x}}(\varsigma)\}(\gamma) + \mathfrak{L}_{\rho}\{\vec{\mathfrak{s}}(\varsigma)\}(\gamma) + \vec{\mathfrak{x}}(\varsigma),$$

then

$$s^{\sigma} I\mathfrak{L}_{\iota}\{\vec{\mathfrak{x}}(\varsigma)\}(\gamma) - \left({}_0 I^{1-\sigma,\iota}\vec{\mathfrak{x}}\right)(0+)$$
$$= A\left\{\gamma^{\upsilon} I\mathfrak{L}_{\iota}\{\vec{\mathfrak{x}}(\varsigma)\}(\gamma) - \left({}_0 I^{1-\upsilon,\iota}\vec{\mathfrak{x}}\right)(0+)\right\} + \mathfrak{L}_{\iota}\{\vec{\mathfrak{s}}(\varsigma)\}(\gamma) + \vec{\mathfrak{x}}(\varsigma).$$

We solve this equation for $\mathfrak{L}_{\iota}(\vec{\mathfrak{x}}(\varsigma))(\gamma)$, then we have

$$(\gamma^{\sigma} I - Q\gamma^{\upsilon})\,\mathfrak{L}_{\iota}\{\vec{\mathfrak{x}}(\varsigma)\}(\gamma) = Q\left({}_0 I^{1-\sigma,\iota}\vec{\mathfrak{x}}\right)(0+) - Q\left({}_0 I^{1-\upsilon,\iota}\vec{\mathfrak{x}}\right)(0+) + \mathfrak{L}_{\iota}\{\vec{\mathfrak{s}}(\varsigma)\}(\gamma) + \vec{\mathfrak{x}}(\varsigma),$$

then

$$\vec{\mathfrak{x}}(\varsigma) = (\gamma^{\sigma} I - Q\gamma^{\upsilon})\,\mathfrak{L}_{\iota}\{\vec{\mathfrak{x}}(\varsigma)\}(\gamma) - Q\left({}_0 I^{1-\sigma,\iota}\vec{\mathfrak{x}}\right)(0+) + Q\left({}_0 I^{1-\upsilon,\iota}\vec{\mathfrak{x}}\right)(0+) - \mathfrak{L}_{\iota}\{\vec{\mathfrak{s}}(\varsigma)\}(\gamma).$$

6.2.2 Matrix-Valued Fuzzy Control Function

We will define a matrix-valued fuzzy function (in short MVFF) based on $\mathcal{H}$-Fox function. Consider the MVFF, $\mathbf{H}^{\mathbf{m,n}}_{\mathbf{p,q}} : [0, \jmath) \times (0, \infty) \to \mathrm{diag}\mathfrak{M}_n((0, 1])$ as a fuzzy control mapping which is defined as follows

$$\mathbf{H}^{\mathbf{m,n}}_{\mathbf{p,q}}\left(\frac{\|Q\|\left(\frac{-\|\varsigma\|^{\iota}}{\iota}\right)^{\sigma}}{\mathfrak{y}}\right) =$$

$$\mathrm{diag}\left[\mathcal{H}^{m,n}_{p,q}\left[\frac{\|Q\|\left(\frac{-\|\varsigma\|^{\iota}}{\iota}\right)^{\sigma}}{\mathfrak{y}}\middle| \begin{matrix}(a_j,\alpha_j)_{j=1,\cdots,p}\\(b_j,\beta_j)_{j=1,\cdots,p}\end{matrix}\right], \mathcal{H}^{m,n}_{p,q}\left[\frac{\|Q\|\left(\frac{-\|\varsigma\|^{\iota}}{\iota}\right)^{\sigma}}{\mathfrak{y}}\middle| \begin{matrix}(a_j,\alpha_j)_{j=1,\cdots,p}\\(b_j,\beta_j)_{j=1,\cdots,p}\end{matrix}\right]\right],$$

- $\mathbf{H}^{\mathbf{m,n}}_{\mathbf{p,q}}\left(\frac{\|Q\|\left(\frac{-\|\varsigma\|^{\iota}}{\iota}\right)^{\sigma}}{\mathfrak{y}}\right)$ is left continuous and increasing;
- $\lim_{\mathfrak{y}\longrightarrow\infty} \mathbf{H}^{\mathbf{m,n}}_{\mathbf{p,q}}\left(\frac{\|Q\|\left(\frac{-\|\varsigma\|^{\iota}}{\iota}\right)^{\sigma}}{\mathfrak{y}}\right) = \mathbf{1}$;
- For $\mathfrak{y} \leq 0$ we have $\mathbf{H}^{\mathbf{m,n}}_{\mathbf{p,q}}\left(\frac{\|Q\|\left(\frac{-\|\varsigma\|^{\iota}}{\iota}\right)^{\sigma}}{\mathfrak{y}}\right) = \mathbf{0}$.

In the following, we investigate the conditions $\mathfrak{A}_1, \mathfrak{A}_2$ and $\mathfrak{A}_3$ for the $\mathcal{H}$-Fox MVFF.

($\mathfrak{A}_1$) We can easily show that $\mathbf{H}^{\mathbf{m,n}}_{\mathbf{p,q}}\left(\frac{\|Q\|\left(\frac{-\|\varsigma\|^{\iota}}{\iota}\right)^{\sigma}}{\mathfrak{y}}\right) = \mathbf{1} \Longleftrightarrow \varsigma = 0$.

($\mathfrak{A}_2$)

$$\begin{aligned}\mathcal{H}^{m,n}_{p,q}\left(\frac{\|Q\|\left(\frac{-\|q\varsigma\|^{\iota}}{\iota}\right)^{\sigma}}{\mathfrak{y}}\right) &= \mathcal{H}^{m,n}_{p,q}\left[\frac{\|Q\|\left(\frac{-\|q\varsigma\|^{\iota}}{\iota}\right)^{\sigma}}{\mathfrak{y}}\middle| \begin{matrix}(a_j,\alpha_j)_{j=1,\cdots,p}\\(b_j,\beta_j)_{j=1,\cdots,p}\end{matrix}\right]\\ &= \mathcal{H}^{m,n}_{p,q}\left[\frac{\|Q\|\left(\frac{-|q|\|\varsigma\|^{\iota}}{\iota}\right)^{\sigma}}{\mathfrak{y}}\middle| \begin{matrix}(a_j,\alpha_j)_{j=1,\cdots,p}\\(b_j,\beta_j)_{j=1,\cdots,p}\end{matrix}\right]\\ &= \mathcal{H}^{m,n}_{p,q}\left[\frac{\|Q\|\left(\frac{-\|\varsigma\|^{\iota}}{\iota}\right)^{\sigma}}{\frac{\mathfrak{y}}{|q|}}\middle| \begin{matrix}(a_j,\alpha_j)_{j=1,\cdots,p}\\(b_j,\beta_j)_{j=1,\cdots,p}\end{matrix}\right] = H^{m,n}_{p,q}\left(\frac{\|Q\|\left(\frac{-\|\varsigma\|^{\iota}}{\iota}\right)^{\sigma}}{\frac{\mathfrak{y}}{|q|}}\right),\end{aligned}$$

then, $\mathbf{H}^{\mathbf{m,n}}_{\mathbf{p,q}}\left(\frac{\|Q\|\left(\frac{-\|q\varsigma\|^{\iota}}{\iota}\right)^{\sigma}}{\mathfrak{y}}\right) = \mathbf{H}^{\mathbf{m,n}}_{\mathbf{p,q}}\left(\frac{\|Q\|\left(\frac{-\|\varsigma\|^{\iota}}{\iota}\right)^{\sigma}}{\frac{\mathfrak{y}}{|q|}}\right)$.

($\mathfrak{A}_3$) We assume that $\mathcal{H}^{m,n}_{p,q}\left[\frac{\|Q\|\left(\frac{-\|\varsigma\|^{\iota}}{\iota}\right)^{\sigma}}{\mathfrak{y}}\middle| \begin{matrix}(a_j,\alpha_j)_{j=1,\cdots,p}\\(b_j,\beta_j)_{j=1,\cdots,p}\end{matrix}\right] \leq$ $\mathcal{H}^{m,n}_{p,q}\left[\frac{\|Q\|\left(\frac{-\|\varsigma'\|^{\iota}}{\iota}\right)^{\sigma}}{\mathfrak{s}}\middle| \begin{matrix}(a_j,\alpha_j)_{j=1,\cdots,p}\\(b_j,\beta_j)_{j=1,\cdots,p}\end{matrix}\right]$. Then

$$\frac{-\|\varsigma\|^{\sigma\iota}}{\iota^{\sigma}\mathfrak{y}} \leq \frac{-\|\varsigma'\|^{\sigma\iota}}{\iota^{\sigma}\mathfrak{s}} \Longrightarrow \frac{\|\varsigma\|^{\sigma\iota}}{\iota^{\sigma}\mathfrak{y}} \geq \frac{\|\varsigma'\|^{\sigma\iota}}{\iota^{\sigma}\mathfrak{s}} \Longrightarrow \frac{\mathfrak{s}\|\varsigma\|^{\sigma\iota}}{\iota^{\sigma}\mathfrak{y}} \geq \frac{\|\varsigma'\|^{\sigma\iota}}{\iota^{\sigma}} \Longrightarrow \frac{\mathfrak{s}\|\varsigma\|^{\sigma\iota}}{\iota^{\sigma}\mathfrak{y}} + \frac{\|\varsigma\|^{\sigma\iota}}{\iota^{\sigma}}$$
$$\geq \frac{\|\varsigma'\|^{\sigma\iota}}{\iota^{\sigma}} + \frac{\|\varsigma\|^{\sigma\iota}}{\iota^{\sigma}} \geq \frac{\|\varsigma+\varsigma'\|^{\sigma\iota}}{\iota^{\sigma}} \Longrightarrow \frac{\|\varsigma\|^{\sigma\iota}}{\iota^{\sigma}}(\frac{\mathfrak{s}}{\mathfrak{y}}+1) \geq \frac{\|\varsigma+\varsigma'\|^{\sigma\iota}}{\iota^{\sigma}} \Longrightarrow \frac{\mathfrak{s}+\mathfrak{y}}{\mathfrak{y}}\frac{\|\varsigma\|^{\sigma\iota}}{\iota^{\sigma}}$$
$$\geq \frac{\|\varsigma+\varsigma'\|^{\sigma\iota}}{\iota^{\sigma}} \Longrightarrow \frac{\|\varsigma\|^{\sigma\iota}}{\iota^{\sigma}\mathfrak{y}}$$
$$\geq \frac{\|\varsigma+\varsigma'\|^{\sigma\iota}}{\iota^{\sigma}(\mathfrak{y}+\mathfrak{s})} \Longrightarrow \frac{-\|\varsigma\|^{\sigma\iota}}{\iota^{\sigma}\mathfrak{y}} \leq \frac{-\|\varsigma+\varsigma'\|^{\sigma\iota}}{\iota^{\sigma}(\mathfrak{y}+\mathfrak{s})} \Longrightarrow$$
$$\mathcal{H}^{m,n}_{p,q}\left[\frac{\|Q\|\left(\frac{-\|\varsigma+\varsigma'\|^{\iota}}{\iota}\right)^{\sigma}}{\mathfrak{y}+\mathfrak{s}}\left|\begin{matrix}(a_j,\alpha_j)_{j=1,\cdots,p}\\(b_j,\beta_j)_{j=1,\cdots,p}\end{matrix}\right.\right] \geq \mathcal{H}^{m,n}_{p,q}\left[\frac{\|Q\|\left(\frac{-\|\varsigma\|^{\iota}}{\iota}\right)^{\sigma}}{\mathfrak{y}}\left|\begin{matrix}(a_j,\alpha_j)_{j=1,\cdots,p}\\(b_j,\beta_j)_{j=1,\cdots,p}\end{matrix}\right.\right],$$

then

$$\mathbf{H}^{\mathbf{m,n}}_{\mathbf{p,q}}\left(\frac{\|Q\|\left(\frac{-\|\varsigma+\varsigma'\|^{\iota}}{\iota}\right)^{\sigma}}{\mathfrak{y}+\mathfrak{s}}\right) \succeq \mathbf{H}^{\mathbf{m,n}}_{\mathbf{p,q}}\left(\frac{\|Q\|\left(\frac{-\|\varsigma\|^{\iota}}{\iota}\right)^{\sigma}}{\mathfrak{y}}\right) \circledast \mathbf{H}^{\mathbf{m,n}}_{\mathbf{p,q}}\left(\frac{\|Q\|\left(\frac{-\|\varsigma'\|^{\iota}}{\iota}\right)^{\sigma}}{\mathfrak{s}}\right).$$

Definition 6.7 We say (6.2.1) is said to be H-U-$\mathcal{H}$-Fox stable, when for a continuous mapping $\vec{\mathfrak{x}}^{\star}(\varsigma)$, satisfying

$$\mathfrak{Y}\left({}_0D^{\sigma,\iota}\vec{\mathfrak{x}}^{\star}(\varsigma) - Q\vec{\mathfrak{x}}^{\star}(\varsigma) - \vec{\mathfrak{w}}(\varsigma) - \mathcal{L}_{\iota}^{-1}\{\vec{\mathfrak{x}}^{\star}(\gamma)\}(\varsigma), \mathfrak{y}\right) \succeq \mathbf{H}^{\mathbf{m,n}}_{\mathbf{p,q}}\left(\frac{\|Q\|\left(\frac{-\|\varsigma\|^{\iota}}{\iota}\right)^{\sigma}}{\mathfrak{y}}\right),$$

$\varsigma > 0$, there is a solution $\vec{\mathfrak{x}}(\varsigma)$ of Eq. (6.2.1) such that for some $r > 0$,

$$\mathfrak{Y}\left(\vec{\mathfrak{x}}(\varsigma) - \vec{\mathfrak{x}}^{\star}(\varsigma), \mathfrak{y}\right) \succeq \mathbf{H}^{\mathbf{m,n}}_{\mathbf{p,q}}\left(\frac{\|Q\|\left(\frac{-\|\varsigma\|^{\iota}}{\iota}\right)^{\sigma}}{\frac{\mathfrak{y}}{r}}\right).$$

6.2.3 H-U-$\mathcal{H}$-F Stability of (6.2.1)

Here, we use Theorem 1.2 to investigate H-U-$\mathcal{H}$-F stability of (6.2.1) and improve recent results. for more details see [26].

Theorem 6.5 *Let* $(\mathbb{R}^n, \mathfrak{Y}, \circledast)$ *be an MVFB-space and* ϱ *and* ζ *be positive constants such that* $0 < \zeta\varrho < 1$. *Assume that the continuous mapping with MVFF,* $\mathbf{H}^{\mathbf{m,n}}_{\mathbf{p,q}}$: $[0, \jmath) \times (0, \infty) \to \text{diag}\mathfrak{M}_n((0, 1])$ *satisfying*

$$\mathbf{H}^{\mathbf{m,n}}_{\mathbf{p,q}}\left(\frac{\|Q\|\left(\frac{-\|\varsigma\|^{\iota}}{\iota}\right)^{\sigma}}{\frac{\mathfrak{y}}{(\varsigma^{\sigma}I+Q)}}\right) \succeq \mathbf{H}^{\mathbf{m,n}}_{\mathbf{p,q}}\left(\frac{\|Q\|\left(\frac{-\|\varsigma\|^{\iota}}{\iota}\right)^{\sigma}}{\frac{\mathfrak{y}}{\varrho}}\right), \tag{6.2.5}$$

and

$$\mathfrak{Y}\left(\vec{\mathfrak{x}}(\varsigma),\mathfrak{y}\right)\succeq \mathbf{H}^{\mathbf{m},\mathbf{n}}_{\mathbf{p},\mathbf{q}}\left(\frac{\|Q\|\left(\frac{-\|\varsigma\|^{\iota}}{\iota}\right)^{\sigma}}{\mathfrak{y}}\right),\ \textit{implies that}\ \mathfrak{Y}\left(\mathfrak{L}_{\iota}\{\vec{\mathfrak{x}}(\varsigma)\}(\gamma),\mathfrak{y}\right)\succeq \mathbf{H}^{\mathbf{m},\mathbf{n}}_{\mathbf{p},\mathbf{q}}\left(\frac{\|Q\|\left(\frac{-\|\varsigma\|^{\iota}}{\iota}\right)^{\sigma}}{\frac{\mathfrak{y}}{\zeta}}\right) \tag{6.2.6}$$

for every $\varsigma>0$. Let $\vec{\mathfrak{x}}^{\star}:[\jmath,\infty)\to\mathbb{R}^n$ be a continuous function satisfying

$$\mathfrak{Y}\left({}_0D^{\sigma,\iota}\vec{\mathfrak{x}}^{\star}(\varsigma)-Q\vec{\mathfrak{x}}^{\star}(\varsigma)-\vec{\mathfrak{w}}(\varsigma)-\mathfrak{L}_{\iota}^{-1}\left\{\vec{\mathfrak{x}}^{\star}(\gamma)\right\}(\varsigma),\mathfrak{y}\right)\succeq \mathbf{H}^{\mathbf{m},\mathbf{n}}_{\mathbf{p},\mathbf{q}}\left(\frac{\|Q\|\left(\frac{-\|\varsigma\|^{\iota}}{\iota}\right)^{\sigma}}{\frac{\mathfrak{y}}{\omega}}\right) \tag{6.2.7}$$

for every $\varsigma>0$ and $\mathfrak{y}>0$. Then we can find a unique continuous function $\vec{\mathfrak{x}}:[\jmath,\infty)\to\mathbb{R}^n$ such that

$${}_0D^{\sigma,\iota}\vec{\mathfrak{x}}(\varsigma)=Q\vec{\mathfrak{x}}(\varsigma)+\vec{\mathfrak{w}}(\varsigma)+\mathfrak{L}_{\iota}^{-1}\left\{\vec{\mathfrak{x}}(\gamma)\right\}(\varsigma), \tag{6.2.8}$$

and

$$\mathfrak{Y}\left(\vec{\mathfrak{x}}^{\star}(\varsigma)-\vec{\mathfrak{x}}(\varsigma),\mathfrak{y}\right)\succeq \mathbf{H}^{\mathbf{m},\mathbf{n}}_{\mathbf{p},\mathbf{q}}\left(\frac{\|Q\|\left(\frac{-\|\varsigma\|^{\iota}}{\iota}\right)^{\sigma}}{\frac{(1-\zeta\varrho)\tau}{\zeta\omega}}\right), \tag{6.2.9}$$

for every $\varsigma>0$ and $\mathfrak{y}>0$.

Proof We set

$$\mathfrak{S}:=\{\vec{\mathfrak{x}}:[\jmath,\infty)\to\mathbb{R}^n,\ \mathfrak{x}\text{ is continuous}\},$$

and introduce the $[0,\infty]$-valued metric on $\mathfrak{S}$ as,

$$\delta(\vec{\mathfrak{x}},\vec{\mathfrak{x}}^{\star})=\inf\left\{\phi>0:\mathfrak{Y}\left(\mathfrak{x}(\varsigma)-\mathfrak{x}^{\star}(\varsigma),\mathfrak{y}\right)\succeq \mathbf{H}^{\mathbf{m},\mathbf{n}}_{\mathbf{p},\mathbf{q}}\left(\frac{\|Q\|\left(\frac{-\|\varsigma\|^{\iota}}{\iota}\right)^{\sigma}}{\frac{\mathfrak{y}}{\phi}}\right)\ \forall\,\vec{\mathfrak{x}},\vec{\mathfrak{x}}^{\star}\in\mathfrak{S},\ \varsigma>0,\ \mathfrak{y}>0\right\}$$

Step 1. First we show that $(\mathfrak{S},\delta)$ is a complete $[0,1]$-valued metric space.

We prove that $\delta(\mathfrak{x},\mathfrak{x}^{\star})=0$ if and only if $\mathfrak{x}=\mathfrak{x}^{\star}$. Let $\delta(\mathfrak{x},\mathfrak{x}^{\star})=0$, we have

$$\inf\left\{\phi>0:\{\mathfrak{Y}\left(\vec{\mathfrak{x}}(\varsigma)-\vec{\mathfrak{x}}^{\star}(\varsigma),\mathfrak{y}\right)\succeq \mathbf{H}^{\mathbf{m},\mathbf{n}}_{\mathbf{p},\mathbf{q}}\left(\frac{\|Q\|\left(\frac{-\|\varsigma\|^{\iota}}{\iota}\right)^{\sigma}}{\frac{\mathfrak{y}}{\phi}}\right)\ \forall\,\vec{\mathfrak{x}},\vec{\mathfrak{x}}^{\star}\in\mathfrak{S},\ \varsigma>0,\ \mathfrak{y}>0\right\}=0,$$

and then $\mathfrak{Y}\left(\vec{\mathfrak{x}}(\varsigma)-\vec{\mathfrak{x}}^{\star}(\varsigma),\mathfrak{y}\right)\succeq \mathbf{H}^{\mathbf{m},\mathbf{n}}_{\mathbf{p},\mathbf{q}}\left(\frac{\|Q\|\left(\frac{-\|\varsigma\|^{\iota}}{\iota}\right)^{\sigma}}{\frac{\mathfrak{y}}{\phi}}\right)$, for all $\phi>0$. Tend ϕ to zero in the above inequality, we get $\mathfrak{Y}\left(\vec{\mathfrak{x}}(\varsigma)-\vec{\mathfrak{x}}^{\star}(\varsigma),\mathfrak{y}\right)=\mathbf{1}$ thus $\vec{\mathfrak{x}}(\varsigma)=\vec{\mathfrak{x}}^{\star}(\varsigma)$ for every

$\varsigma > 0$, and vice versa. Also, we have $\delta(\vec{\mathfrak{x}}, \vec{\mathfrak{x}}^\star) = \delta(\vec{\mathfrak{x}}^\star, \vec{\mathfrak{x}})$ for every $\vec{\mathfrak{x}}, \vec{\mathfrak{x}}^\star \in \mathfrak{S}$. Now, let $\delta(\vec{\mathfrak{x}}, \vec{\mathfrak{x}}^\star) = c_1 > 0$ and $\delta(\vec{\mathfrak{x}}^\star, \vec{\mathfrak{v}}) = c_2 > 0$. Then, we have $\mathfrak{Y}\left(\vec{\mathfrak{x}}(\varsigma) - \vec{\mathfrak{x}}^\star(\varsigma), \mathfrak{y}\right) \succeq \mathbf{H}^{\mathbf{m,n}}_{\mathbf{p,q}}\left(\frac{\|Q\|\left(\frac{-\|\varsigma\|^\iota}{\iota}\right)^\sigma}{\frac{\mathfrak{y}}{c_1}}\right)$, and $\mathfrak{Y}\left(\vec{\mathfrak{x}}^\star(\varsigma) - \vec{\mathfrak{v}}(\varsigma), \mathfrak{y}\right) \succeq \mathbf{H}^{\mathbf{m,n}}_{\mathbf{p,q}}\left(\frac{\|Q\|\left(\frac{-\|\varsigma\|^\iota}{\iota}\right)^\sigma}{\frac{\mathfrak{y}}{c_2}}\right)$, for every $\mathfrak{y} > 0$. Then, we have

$$\begin{aligned}
\mathfrak{Y}\left(\vec{\mathfrak{x}}(\varsigma) - \vec{\mathfrak{x}}^\star(\varsigma), (c_1 + c_2)\tau\right) \succeq & \mathfrak{Y}\left(\vec{\mathfrak{x}}(\varsigma) - \vec{\mathfrak{x}}^\star(\varsigma), c_1\tau\right) \circledast \mathfrak{Y}\left(\vec{\mathfrak{x}}^\star(\varsigma) - \vec{\mathfrak{v}}(\varsigma), c_2\tau\right) \\
\succeq & \mathbf{H}^{\mathbf{m,n}}_{\mathbf{p,q}}\left(\frac{\|Q\|\left(\frac{-\|\varsigma\|^\iota}{\iota}\right)^\sigma}{\mathfrak{y}}\right) \circledast \mathbf{H}^{\mathbf{m,n}}_{\mathbf{p,q}}\left(\frac{\|Q\|\left(\frac{-\|\varsigma\|^\iota}{\iota}\right)^\sigma}{\mathfrak{y}}\right) \\
= & \mathbf{H}^{\mathbf{m,n}}_{\mathbf{p,q}}\left(\frac{\|Q\|\left(\frac{-\|\varsigma\|^\iota}{\iota}\right)^\sigma}{\mathfrak{y}}\right),
\end{aligned}$$

and then $\delta(\vec{\mathfrak{x}}, \vec{\mathfrak{v}}) \le c_1 + c_2$. Thus, $\delta(\vec{\mathfrak{x}}, \vec{\mathfrak{v}}) \le \delta(\vec{\mathfrak{x}}, \vec{\mathfrak{x}}^\star) + \delta(\vec{\mathfrak{x}}^\star, \vec{\mathfrak{v}})$. Now, we are ready to prove $(\mathfrak{S}, \delta)$ is complete. Consider the Cauchy sequence $\{\vec{\mathfrak{x}}_k\}_k$ in $(\mathfrak{S}, \delta)$. Suppose that $\varsigma > 0$ is fixed. Assume that $\mathfrak{a} > 0$ and $\mathfrak{d} \in (0, 1)$ are arbitrary and consider $\mathfrak{y} > 0$ such that $\mathbf{H}^{\mathbf{m,n}}_{\mathbf{p,q}}\left(\frac{\|Q\|\left(\frac{-\|\varsigma\|^\iota}{\iota}\right)^\sigma}{\mathfrak{y}}\right) > 1 - \mathfrak{d}$. For $\mathfrak{e}\mathfrak{y} < \mathfrak{a}$ choose $m_0 \in \mathbb{N}$ such that $\delta(\vec{\mathfrak{x}}_m, \vec{\mathfrak{x}}_n) < \mathfrak{e}\ \forall\, m, n \ge m_0$. Then

$$\mathfrak{Y}\left(\vec{\mathfrak{x}}_m(\varsigma) - \vec{\mathfrak{x}}_n(\varsigma), \mathfrak{a}\right) \succeq \mathfrak{Y}\left(\vec{\mathfrak{x}}_m(\varsigma) - \vec{\mathfrak{x}}_n(\varsigma), \mathfrak{e}\mathfrak{y}\right) \succeq (\mathbf{H}^{\mathbf{m,n}}_{\mathbf{p,q}}\left(\frac{\|Q\|\left(\frac{-\|\varsigma\|^\iota}{\iota}\right)^\sigma}{\mathfrak{y}}\right) > 1 - \mathfrak{d}.$$

Then $\mathfrak{Y}\left(\vec{\mathfrak{x}}_m(\varsigma) - \vec{\mathfrak{x}}_n(\varsigma), \mathfrak{a}\right) > 1 - \mathfrak{d}$ and i.e., the sequence $\{\vec{\mathfrak{x}}_m(\varsigma)\}_m$ is Cauchy in complete space $(\mathbb{R}^n, \mathfrak{Y}, \circledast)$ on compact set $[\jmath, \infty)$, then it is uniformly convergent to the mapping $\vec{\mathfrak{x}} : [\jmath, \infty) \to \mathbb{R}^n$ respectively. Now, if we apply the uniform convergence we conclude that $\vec{\mathfrak{x}} \in \mathfrak{S}$ and differentiable, thus $(\mathfrak{S}, \delta)$ is complete.

Now, we define the mapping $\Psi : \mathfrak{S} \to \mathfrak{S}$ by

$$\begin{aligned}
\Psi(\mathfrak{x})(\iota)) &= \mathfrak{L}_\iota\left\{{}_0D^{\sigma,\iota}\vec{\mathfrak{x}}^\star(\varsigma) - Q\vec{\mathfrak{x}}^\star(\varsigma) - \vec{\mathfrak{w}}(\varsigma)\right\}(\gamma) \qquad (6.2.10)\\
&= \gamma^\alpha I\mathfrak{L}_\iota\{\vec{\mathfrak{x}}(\varsigma)\}(\gamma) - \left({}_0I^{1-\sigma,\iota}\vec{\mathfrak{x}}\right)(0+) - Q\mathfrak{L}_\iota\{\vec{\mathfrak{x}}(\varsigma)\}(\gamma) - \mathfrak{L}_\rho\{\vec{\mathfrak{w}}(\varsigma)\}(\gamma)\\
&= \left(\gamma^\sigma I - Q\right)\mathfrak{L}_\iota\{\vec{\mathfrak{x}}(\varsigma)\}(\gamma) - \left({}_0I^{1-\sigma,\iota}\vec{\mathfrak{x}}\right)(0+) - \mathfrak{L}_\rho\{\vec{\mathfrak{w}}(\varsigma)\}(\gamma)\\
&= \left(\gamma^\sigma I - Q\right)\mathfrak{L}_\iota\{\vec{\mathfrak{x}}(\varsigma)\}(\gamma) - p - \mathfrak{L}_\iota\{\vec{\mathfrak{w}}(\varsigma)\}(\gamma),
\end{aligned}$$

Step 2. We prove Ψ is a strictly contractive mapping.
Let $\vec{\mathfrak{x}}, \vec{\mathfrak{x}}^\star \in \mathfrak{S}$, $\phi > 0$ and $\delta(\vec{\mathfrak{x}}, \vec{\mathfrak{x}}^\star) < \mathcal{D}_{\vec{\mathfrak{x}}\vec{\mathfrak{x}}^\star}$, then we have

$$\mathfrak{Y}\left(\vec{\mathfrak{x}}(\varsigma) - \vec{\mathfrak{x}}^\star(\varsigma), \mathcal{D}_{\vec{\mathfrak{x}}\vec{\mathfrak{x}}^\star}\mathfrak{y}\right) \succeq \mathbf{H}^{\mathbf{m,n}}_{\mathbf{p,q}}\left(\frac{\|Q\|\left(\frac{-\|\varsigma\|^\iota}{\iota}\right)^\sigma}{\mathfrak{y}}\right), \quad \forall\, \vec{\mathfrak{x}}, \vec{\mathfrak{x}}^\star \in \mathfrak{S},\ \varsigma \in [\jmath, \infty],\ \mathfrak{y} > 0.$$

Using (6.2.10) and properties (MVF-1-N2) and (MVF-1-N3) of Definition 1.24, we have

$$\begin{aligned}
&\mathfrak{Y}\left(\Psi\vec{\mathfrak{x}}(\iota)-\Psi\vec{\mathfrak{x}}^{\star}(\iota),\mathcal{D}_{\vec{\mathfrak{x}}\vec{\mathfrak{x}}^{\star}}\mathfrak{y}\right) \qquad (6.2.11)\\
&=\mathfrak{Y}\left([(\gamma^{\sigma}I-Q)\,\mathfrak{L}_{\iota}\{\vec{\mathfrak{x}}(\varsigma)\}(\gamma)-p-\mathfrak{L}_{\iota}\{\vec{\mathfrak{w}}(\varsigma)\}(\gamma)-(\gamma^{\sigma}I-Q)\,\mathfrak{L}_{\iota}\left\{\vec{\mathfrak{x}}^{\star}(\varsigma)\right\}(\gamma)+p+\mathfrak{L}_{\iota}\{\vec{\mathfrak{w}}(\varsigma)\}(\gamma)],\mathcal{D}_{\vec{\mathfrak{x}}\vec{\mathfrak{x}}^{\star}\mathfrak{y}}\right)\\
&=\mathfrak{Y}\left((\gamma^{\sigma}I-Q)\,\mathfrak{L}_{\iota}\{\mathfrak{x}(\varsigma)\}(\gamma)-(\varsigma^{\sigma}I-Q)\,\mathfrak{L}_{\iota}\left\{\mathfrak{x}^{\star}(\varsigma)\right\}(\gamma),\mathcal{D}_{\vec{\mathfrak{x}}\vec{\mathfrak{x}}^{\star}}\mathfrak{y}\right)=\mathfrak{Y}\left((\gamma^{\sigma}I-Q)\,\mathfrak{L}_{\iota}\left\{\vec{\mathfrak{x}}(\varsigma)-\vec{\mathfrak{x}}^{\star}(\varsigma)\right\}(\gamma),\mathcal{D}_{\vec{\mathfrak{x}}\vec{\mathfrak{x}}^{\star}}\mathfrak{y}\right)\\
&=\mathfrak{Y}\left(\mathfrak{L}_{\iota}\left\{\vec{\mathfrak{x}}(\varsigma)-\vec{\mathfrak{x}}^{\star}(\varsigma)\right\}(\gamma),\frac{\mathcal{D}_{\vec{\mathfrak{x}}\vec{\mathfrak{x}}^{\star}}\mathfrak{y}}{(\gamma^{\sigma}I-Q)}\right).
\end{aligned}$$

On the other hand, we have

$$\mathfrak{Y}\left(\vec{\mathfrak{x}}(\varsigma)-\vec{\mathfrak{x}}^{\star}(\varsigma),\frac{\mathcal{D}_{\vec{\mathfrak{x}}\vec{\mathfrak{x}}^{\star}}\mathfrak{y}}{(\gamma^{\sigma}I-Q)}\right)\succeq\mathbf{H}^{\mathbf{m},\mathbf{n}}_{\mathbf{p},\mathbf{q}}\left(\frac{\|Q\|\left(\frac{-\|\varsigma\|^{\iota}}{\iota}\right)^{\sigma}}{\frac{\mathfrak{y}}{(\gamma^{\sigma}I-Q)}}\right). \qquad (6.2.12)$$

By (6.2.6) and (6.2.12), we get

$$\mathfrak{Y}\left(\mathfrak{L}_{\iota}\left\{\mathfrak{x}(\varsigma)-\mathfrak{x}^{\star}(\varsigma)\right\}(\gamma),\frac{\mathfrak{y}}{(\gamma^{\sigma}I-Q)}\right)\succeq\mathbf{H}^{\mathbf{m},\mathbf{n}}_{\mathbf{p},\mathbf{q}}\left(\frac{\|Q\|\left(\frac{-\|\varsigma\|^{\iota}}{\iota}\right)^{\sigma}}{\frac{\zeta\mathfrak{y}}{(\gamma^{\sigma}I-Q)}}\right). \qquad (6.2.13)$$

Form (6.2.11), (6.2.12), (6.2.13), we have

$$\mathfrak{Y}\left(\Psi\mathfrak{x}(\varsigma)-\Psi y^{\star}(\varsigma),\mathcal{D}_{\vec{\mathfrak{x}}\vec{\mathfrak{x}}^{\star}}\mathfrak{y}\right)\succeq\mathbf{H}^{\mathbf{m},\mathbf{n}}_{\mathbf{p},\mathbf{q}}\left(\frac{\|Q\|\left(\frac{-\|\varsigma\|^{\iota}}{\iota}\right)^{\sigma}}{\frac{\zeta\mathfrak{y}}{(\varsigma^{\sigma}I-Q)}}\right) \qquad (6.2.14)$$

and then

$$\mathfrak{Y}\left(\Psi\vec{\mathfrak{x}}(\varsigma)-\Psi\vec{\mathfrak{x}}^{\star}(\varsigma),\mathcal{D}_{\vec{\mathfrak{x}}\vec{\mathfrak{x}}}\mathfrak{y}\right)\succeq\mathbf{H}^{\mathbf{m},\mathbf{n}}_{\mathbf{p},\mathbf{q}}\left(\frac{\|Q\|\left(\frac{-\|\varsigma\|^{\iota}}{\iota}\right)^{\sigma}}{\frac{\mathfrak{y}}{\zeta\varrho}}\right) \qquad (6.2.15)$$

which implies that

$$\delta\left(\Psi\vec{\mathfrak{x}},\Psi\vec{\mathfrak{x}}^{\star}\right)\leq\zeta\varrho\mathcal{D}_{\vec{\mathfrak{x}}\vec{\mathfrak{x}}\mathfrak{y}},$$

and then

$$\delta\left(\Psi\vec{\mathfrak{x}},\Psi\vec{\mathfrak{x}}^{\star}\right)\leq\zeta\varrho\delta(\vec{\mathfrak{x}},\vec{\mathfrak{x}}^{\star}).$$

Then, Ψ is a strictly contractive mapping with Lipschitz constant $\zeta\varrho$.

Step 3. We show that $\delta(\Psi(\vec{\mathfrak{x}}^{\star}),\vec{\mathfrak{x}}^{\star})<\infty$, for $\vec{\mathfrak{x}}\in\mathfrak{S}$. Using (6.2.6) and (6.2.7) we get

$$\begin{aligned}
&\mathfrak{Y}\left(\Psi\left(\vec{\mathfrak{x}}^{\star}(\varsigma)\right) - \vec{\mathfrak{x}}^{\star}(\varsigma), \mathfrak{y}\right) \\
&= \mathfrak{Y}\left(\mathfrak{L}_{\varsigma}\left\{{}_0D^{\sigma,\iota}\vec{\mathfrak{x}}^{\star}(\varsigma) - Q\vec{\mathfrak{x}}^{\star}(\varsigma) - \vec{\mathfrak{w}}(\varsigma)\right\}(\gamma) - \vec{\mathfrak{x}}^{\star}(\varsigma), \mathfrak{y}\right) \\
&= \mathfrak{Y}\left(\left(\gamma^{\sigma} I - Q\right)\mathfrak{L}_{\varsigma}\{\vec{\mathfrak{x}}(\varsigma)\} - p - \mathfrak{L}_{\varsigma}\{\vec{\mathfrak{w}}(\varsigma)\} - \vec{\mathfrak{x}}^{\star}(\varsigma), \mathfrak{y}\right) \succeq \mathbf{H}^{\mathbf{m},\mathbf{n}}_{\mathbf{p},\mathbf{q}}\left(\frac{\|Q\|\left(\frac{-\|\varsigma\|^{\iota}}{\iota}\right)^{\sigma}}{\frac{\mathfrak{y}}{\zeta\omega}}\right)
\end{aligned} \tag{6.2.16}$$

for every $\mathfrak{y} > 0$. Then, we have $\delta(\Psi(\vec{\mathfrak{x}}^{\star}), \vec{\mathfrak{x}}^{\star}) < \zeta\omega < \infty$.

Now, Theorem 1.2 enables us to find an element $\vec{\mathfrak{x}}$ in $\mathfrak{S}$ satisfying the following:

(1) $\vec{\mathfrak{x}}$ is a fixed point of Ψ, i.e.,

$$\begin{aligned}
\vec{\mathfrak{x}}(\varsigma) &= \Psi(\vec{\mathfrak{x}}(\varsigma)) \\
&= \left(\gamma^{\sigma} I - Q\right)\mathfrak{L}_{\iota}(\vec{\mathfrak{x}}(\varsigma))(\gamma) - \left({}_0I^{1-\sigma,\iota}, \vec{\mathfrak{x}}\right)(0_{+}) - \mathfrak{L}_{\iota}\{\vec{\mathfrak{w}}(\varsigma)\}(\gamma),
\end{aligned} \tag{6.2.17}$$

which is unique in the set

$$\mathfrak{S}^{*} = \{\vec{\mathfrak{x}} \in \mathfrak{S}, \ \ \delta(\Psi(\vec{\mathfrak{x}}^{\star}), \vec{\mathfrak{x}}) < \infty\}.$$

Since $\vec{\mathfrak{x}}$ is a continuous function and according to the Theorem 6.3

$${}_0D^{\sigma,\iota}\vec{\mathfrak{x}}(\varsigma) = Q\vec{\mathfrak{x}}(\varsigma) + \vec{\mathfrak{w}}(\varsigma) + \mathfrak{L}_{\iota}^{-1}\{\vec{\mathfrak{x}}(\gamma)\}(\varsigma), \quad 0 < \sigma < 1, \varsigma > 0; \tag{6.2.18}$$

(2) $\delta(\Psi^{k}(\vec{\mathfrak{x}}), \vec{\mathfrak{x}}^{\star}) \to 0$ as $n \to \infty$;

(3) $\delta(y\vec{\mathfrak{x}}, y\vec{\mathfrak{x}}^{\star}) \leq \frac{1}{1-\zeta\varrho}\delta(\Psi(\vec{\mathfrak{x}}), \vec{\mathfrak{x}}) \leq \frac{\zeta\omega}{1-\zeta\varrho}$, which implies that

$$\mathfrak{Y}\left(\vec{\mathfrak{x}}(\varsigma) - \vec{\mathfrak{x}}^{\star}(\varsigma), \mathfrak{y}\right) \succeq \mathbf{H}^{\mathbf{m},\mathbf{n}}_{\mathbf{p},\mathbf{q}}\left(\frac{\|Q\|\left(\frac{-\|\varsigma\|^{\iota}}{\iota}\right)^{\sigma}}{\frac{(1-\zeta\varrho)\mathfrak{y}}{\zeta\omega}}\right), \tag{6.2.19}$$

for every $\varsigma > 0$ and $\mathfrak{y} > 0$.

Now, we prove that $\mathfrak{S}^{*} = \mathfrak{S}$. Consider $\vec{\mathfrak{z}}$ in $\mathfrak{S}$ satisfying (6.2.8) and (6.2.9), we show that $\vec{\mathfrak{z}} = \vec{\mathfrak{x}}$ and $\vec{\mathfrak{z}} \in \mathfrak{S}^{*}$. From (6.2.8) we get

$${}_0D^{\sigma,\iota}\vec{\mathfrak{z}}(\varsigma) = Q\vec{\mathfrak{z}}(\varsigma) + \vec{\mathfrak{w}}(\varsigma) + \mathfrak{L}_{\iota}^{-1}\{\vec{\mathfrak{z}}(\gamma)\}(\varsigma), \quad 0 < \sigma < 1, \varsigma > 0, \tag{6.2.20}$$

and then

$$\vec{\mathfrak{z}}(\varsigma) = \left(\gamma^{\sigma} I - Q\right)\mathfrak{L}_{\iota}(\vec{\mathfrak{z}}(\varsigma))(\gamma) - \left({}_0I^{1-\sigma,\iota}, \vec{\mathfrak{z}}\right)(0_{+}) - \mathfrak{L}_{\iota}\{\vec{\mathfrak{w}}(\varsigma)\}(\gamma). \tag{6.2.21}$$

Now, we show that

$$\vec{\mathfrak{z}} \in \{\vec{\mathfrak{x}} \in \mathfrak{S}, \ \ \delta(\Psi(\vec{\mathfrak{x}}^{\star}), \vec{\mathfrak{x}}) < \infty\},$$

i.e., $\delta(\Psi(\vec{\mathfrak{x}}^{\star}), \vec{\mathfrak{z}}) < \infty$. We set $\urcorner = \frac{(1-\zeta\varrho)\tau}{\zeta\omega}$, from (6.2.9) we get

$$\mathfrak{Y}\Big(\vec{\mathfrak{x}}(\varsigma) - \vec{\mathfrak{z}}(\varsigma), \mathfrak{y}\Big) \succeq \mathbf{H}^{\mathbf{m},\mathbf{n}}_{\mathbf{p},\mathbf{q}}\left(\frac{\|Q\| \left(\frac{-\|\varsigma\|^{\iota}}{\iota}\right)^{\sigma}}{\urcorner\mathfrak{y}}\right), \tag{6.2.22}$$

for every $\varsigma > 0$ and $\mathfrak{y} > 0$. Using (6.2.22) we get

$$\begin{aligned}
&\mathfrak{Y}\Big(\Psi\left(\vec{\mathfrak{x}}^{\star}(\varsigma)\right) - \vec{\mathfrak{z}}(\varsigma), \mathfrak{y}\Big) \qquad (6.2.23)\\
&= \mathfrak{Y}\Big(\mathfrak{L}_{\iota}\left\{{}_0D^{\sigma,\iota}\vec{\mathfrak{x}}^{\star}(\varsigma) - Q\vec{\mathfrak{x}}^{\star}(\varsigma) - \vec{\mathfrak{w}}(\varsigma)\right\}(\gamma) - \vec{\mathfrak{z}}(\varsigma), \mathfrak{y}\Big)\\
&= \mathfrak{Y}\Big(\gamma^{\sigma} I\mathfrak{L}_{\iota}\Big\{\vec{\mathfrak{x}}^{\star}(\varsigma)\Big\}(\gamma) - \Big({}_0I^{1-\sigma,\iota}\vec{\mathfrak{x}}^{\star}\Big)(0+) - Q\mathfrak{L}_{\rho}\Big\{\vec{\mathfrak{x}}^{\star}(t)\Big\}(\gamma)\\
&- \mathfrak{L}_{\rho}\{\vec{\mathfrak{w}}(\varsigma)\}(\gamma) - \gamma^{\sigma} I\mathfrak{L}_{\iota}\{\vec{\mathfrak{z}}(\varsigma)\}(\gamma) + \Big({}_0I^{1-\alpha,\rho}\mathbf{t}\Big)(0+) + Q\mathfrak{L}_{\iota}\{\vec{\mathfrak{w}}\varsigma(\iota)\}(\gamma) - \mathfrak{L}_{\rho}\{\vec{\mathfrak{w}}(\varsigma)\}(\gamma), \mathfrak{y}\Big)\\
&= \mathfrak{Y}\Big(\left(\gamma^{\sigma} I - Q\right)\mathfrak{L}_{\iota}\left\{\vec{\mathfrak{x}}^{\star}(\varsigma)\right\}(\gamma) - p - \mathfrak{L}_{\iota}\{\vec{\mathfrak{w}}(\varsigma)\}(\gamma) - \left(\gamma^{\sigma} I - Q\right)\mathfrak{L}_{\iota}\{\vec{\mathfrak{z}}(\varsigma)\}(\gamma) + p + \mathfrak{L}_{\iota}\{\vec{\mathfrak{w}}(\varsigma)\}(\gamma), \mathfrak{y}\Big)\\
&= \mathfrak{Y}\Big(\left(\gamma^{\sigma} I - Q\right)\mathfrak{L}_{\iota}\left\{\vec{\mathfrak{x}}^{\star}(\varsigma)\right\}(\gamma) - \left(\gamma^{\sigma} I - Q\right)\mathfrak{L}_{\iota}\{\vec{\mathfrak{z}}(\varsigma)\}(\gamma), \mathfrak{y}\Big)\\
&= \mathfrak{Y}\Big(\left(\gamma^{\sigma} I - Q\right)\mathfrak{L}_{\iota}\left\{\vec{\mathfrak{x}}^{\star}(\varsigma) - \vec{\mathfrak{z}}(\varsigma)\right\}(\gamma), \mathfrak{y}\Big)\\
&\succeq \mathfrak{Y}\Big(\mathfrak{L}_{\iota}\left\{\vec{\mathfrak{x}}^{\star}(\varsigma) - \vec{\mathfrak{z}}(\varsigma)\right\}(\gamma), \frac{\mathfrak{y}}{(\gamma^{\sigma} I - Q)}\Big)\\
&\succeq \mathbf{H}^{\mathbf{m},\mathbf{n}}_{\mathbf{p},\mathbf{q}}\left(\frac{\|Q\| \left(\frac{-\|\varsigma\|^{\iota}}{\iota}\right)^{\sigma}}{\urcorner\frac{\mathfrak{y}}{(\gamma^{\sigma} I - Q)}}\right)\\
&\succeq \mathbf{H}^{\mathbf{m},\mathbf{n}}_{\mathbf{p},\mathbf{q}}\left(\frac{\|Q\| \left(\frac{-\|\varsigma\|^{\iota}}{\iota}\right)^{\sigma}}{\urcorner\frac{\mathfrak{y}}{\zeta\varrho}}\right),
\end{aligned}$$

which implies that $\delta(\Psi(\vec{\mathfrak{x}}), \vec{\mathfrak{z}}) \leq \frac{\urcorner}{\zeta\varrho} < \infty$, then $\vec{\mathfrak{z}} \in \mathfrak{S}^{*}$.

6.2.4 Application

Example 6.2 Consider the generalized fractional differential system

$${}_0D^{\frac{8}{5},\frac{9}{10}}\vec{\mathfrak{x}}(\varsigma) = Q\vec{\mathfrak{x}}(\varsigma) + \vec{\mathfrak{w}}(\varsigma) + \mathfrak{L}_{\iota}^{-1}\{\vec{\mathfrak{x}}(\gamma)\}(\varsigma), \tag{6.2.24}$$

the initial value ${}_0I^{1-\frac{8}{5},\frac{9}{10}}\vec{\mathfrak{x}}(0+) = \begin{pmatrix} 0\\0\\0\\0 \end{pmatrix}$, $\vec{\mathfrak{x}}(\varsigma) = \begin{pmatrix} \vec{\mathfrak{x}}'(\varsigma)\\ \vec{\mathfrak{x}}''(\varsigma)\\ \vec{\mathfrak{x}}'''(\varsigma)\\ \vec{\mathfrak{x}}''''(\varsigma) \end{pmatrix}$, $Q = \begin{bmatrix} -2 & 3 & 0 & 1\\ 2 & 1 & 2 & 3\\ 3 & 0 & 3 & 2\\ 2 & 1 & 0 & 1 \end{bmatrix}$,

and put the forcing term

$$\vec{\mathfrak{w}}(\varsigma) = \begin{bmatrix} \frac{\Gamma(2)}{\Gamma(\frac{4}{3})}\varsigma^{\frac{1}{6}} - \varsigma^{\frac{9}{10}} - 2\varsigma^{\frac{8}{5}} + \frac{1}{10000}\sin\varsigma \\ \frac{\Gamma(\frac{4}{3})}{\Gamma(\frac{5}{3})}\varsigma^{\frac{1}{3}} + \varsigma^{\frac{9}{10}} - 3\varsigma^{\frac{3}{4}} + \frac{1}{20000}\cos\varsigma \\ \frac{\Gamma(\frac{5}{2})}{\Gamma(\frac{11}{6})}\varsigma^{\frac{5}{12}} - 2\varsigma^{\frac{9}{10}} - \varsigma^{\frac{8}{5}} + \frac{1}{30000}\sin 2\varsigma \\ \frac{\Gamma(\frac{4}{3})}{\Gamma(\frac{8}{5})}\varsigma^{\frac{1}{5}} + \varsigma^{\frac{9}{10}} - 3\varsigma^{\frac{12}{7}} + \frac{1}{40000}\cos 4\varsigma \end{bmatrix},$$

Assume that the following condition is true for the given continuous function

$$\mathbf{H}^{\mathbf{m},\mathbf{n}}_{\mathbf{p},\mathbf{q}}\left(\frac{\|Q\|\left(\frac{-\|\varsigma\|^{\iota}}{\iota}\right)^{\sigma}}{\frac{\mathfrak{y}}{(\gamma^{\sigma}I+Q)}}\right) \succeq \mathbf{H}^{\mathbf{m},\mathbf{n}}_{\mathbf{p},\mathbf{q}}\left(\frac{\|Q\|\left(\frac{-\|\varsigma\|^{\iota}}{\iota}\right)^{\sigma}}{4\mathfrak{y}}\right).$$

For $\omega = \frac{1}{10000}$, it is easy to check that the vector-valued function $\vec{\mathfrak{x}}_1(\varsigma) = \begin{bmatrix} \varsigma^{\frac{9}{10}} \\ \varsigma^{\frac{8}{5}} \\ \varsigma^{\frac{3}{4}} \\ \varsigma^{\frac{1}{5}} \end{bmatrix}$ satisfies the inequality

$$\mathfrak{Y}\left({}_{\mathbb{C}}D^{\frac{8}{5},\frac{9}{10}}\vec{\mathfrak{x}}_1(\varsigma) - Q\vec{\mathfrak{x}}_1(\varsigma) - \vec{\mathfrak{w}}(\varsigma) - \mathfrak{L}_{\iota}^{-1}\{\vec{\mathfrak{x}}_1(\varsigma)\}(\gamma), \mathfrak{y}\right) \succeq \mathbf{H}^{\mathbf{m},\mathbf{n}}_{\mathbf{p},\mathbf{q}}\left(\frac{\|Q\|\left(\frac{-\|\varsigma\|^{\iota}}{\iota}\right)^{\sigma}}{10000\mathfrak{y}}\right),$$

therefore, according to the definitions given $\vec{\mathfrak{x}}_1$ holds in the following inequality

$$\begin{aligned} &\mathfrak{Y}\left(\Psi(\vec{\mathfrak{x}}_1(\varsigma)) - \vec{\mathfrak{x}}_1(\varsigma), \mathfrak{y}\right) = \mathfrak{Y}\left(\mathfrak{L}_{\varsigma}\left\{{}_0D^{\sigma,\iota}\vec{\mathfrak{x}}_1(\varsigma) - Q\vec{\mathfrak{x}}_1(\varsigma) - \vec{\mathfrak{w}}(\varsigma)\right\}(\gamma) - \vec{\mathfrak{x}}_1(\varsigma), \mathfrak{y}\right) \\ &= \mathfrak{Y}\left(\left(\gamma^{\sigma}I - Q\right)\mathfrak{L}_{\iota}\{\vec{\mathfrak{x}}_1(\varsigma)\}(\gamma) - p - \mathfrak{L}_{\iota}\{\vec{\mathfrak{w}}(\varsigma)\}(\gamma) - \vec{\mathfrak{x}}_1(\varsigma), \mathfrak{y}\right) \succeq \mathbf{H}^{\mathbf{m},\mathbf{n}}_{\mathbf{p},\mathbf{q}}\left(\frac{\|Q\|\left(\frac{-\|\varsigma\|^{\iota}}{\iota}\right)^{\sigma}}{10000\mathfrak{y}}\right). \end{aligned} \tag{6.2.25}$$

Then, there is a unique solution $\vec{\mathfrak{x}} \in C([\jmath,\infty),\mathbb{R}^n)$ for Eq. (6.2.1), such that

(*1*) $\vec{\mathfrak{x}}$ is the fixed point of the operator Ψ, which is defined as follows

$$\Psi(\vec{\mathfrak{x}}(\varsigma)) = \left(\gamma^{\sigma}I - Q\right)\mathfrak{L}_{\iota}\{\vec{\mathfrak{x}}(\varsigma)\}(\gamma) - p - \mathfrak{L}_{\iota}\{\vec{\mathfrak{w}}(\varsigma)\}(\gamma).$$

(*2*) The fixed point $\vec{\mathfrak{x}}$ is unique.
(*3*) $\vec{\mathfrak{x}}$ satisfies in Eq. (6.2.24).

Then

$$\delta(\vec{\mathfrak{x}}, \vec{\mathfrak{x}}_1) \le \frac{\zeta}{1-4\zeta} \times \frac{1}{10000},$$

or

$$\mathfrak{Y}\left(\vec{\mathfrak{x}}(\varsigma) - \vec{\mathfrak{x}}_1(\varsigma), \mathfrak{y}\right) \succeq \mathbf{H}^{\mathbf{m,n}}_{\mathbf{p,q}}\left(\frac{\|Q\| \left(\frac{-\|\varsigma\|^{\iota}}{\iota}\right)^{\sigma}}{\frac{10000(4-\zeta)\mathfrak{y}}{4\zeta}}\right). \tag{6.2.26}$$

The table below shows the exact values for the solutions.

Solutions	$\gamma = 1$	$\gamma = 2$	$\gamma = \frac{1}{2}$	$\gamma = \frac{5}{2}$	$\gamma = 100$	$\gamma = 250$
Row 1	−188.21791 4500000	−10.46777 26134896	−4959.768587 17845	$\frac{25^{\frac{3}{5}} \times 2^{\frac{2}{5}}}{25}$ −7.356061434	$\frac{100^{\frac{3}{5}}}{50}$ −5.004252322	$\frac{250^{\frac{3}{5}}}{125}$ −5.001563276
Row 2	−382.16445 5600000	−18.976025 9204688	−11887.68977 05658	−14.24580479 $+\frac{965^{\frac{3}{5}} \times 2^{\frac{2}{5}}}{125}$	−9.001291640 $+\frac{3100^{\frac{3}{5}}}{125000}$	−9.000423743 $+\frac{3250^{\frac{3}{5}}}{1953125}$
Row 3	−267.89954 1900000	−20.69446 24702344	−7831.66369 107073	−16.44236068 $+\frac{125^{\frac{3}{5}} \times 2^{\frac{2}{5}}}{25}$	−12.00214284 $+\frac{3100^{\frac{3}{5}}}{5000}$	−12.00054493 $+\frac{3250^{\frac{3}{5}}}{31250}$
Row 4	−42.531806 9400000	−9.616741 52117202	−2990.921348 65802	−17.30036500 $+\frac{195^{\frac{3}{5}} \times 2^{\frac{2}{5}}}{125}$	−15.00242688 $+\frac{3100^{\frac{3}{5}}}{2500000}$	−15.00085242 $+\frac{3100^{\frac{3}{5}}}{97656250}$

6.2.5 H-U-$\mathcal{H}$-F Stability of (6.2.2)

Definition 6.8 We say (6.2.2) is said to be H-U-$\mathcal{H}$-F stable, when for a continuous mapping $\vec{\mathfrak{x}}^{\star}(\varsigma)$, satisfying

$$\mathfrak{Y}\left({}_0D^{\sigma,\iota}\vec{\mathfrak{x}}^{\star}(\varsigma) - Q\,{}_0D^{\upsilon,\iota}\vec{\mathfrak{x}}^{\star}(\varsigma) - \vec{\mathfrak{s}}(\varsigma) - \mathfrak{L}_{\iota}^{-1}\{\vec{\mathfrak{x}}^{\star}(\gamma)\}(\varsigma), \mathfrak{y}\right) \succeq \mathbf{H}^{\mathbf{m,n}}_{\mathbf{p,q}}\left(\frac{\|Q\| \left(\frac{-\|\varsigma\|^{\iota}}{\iota}\right)^{\sigma}}{\mathfrak{y}}\right)$$

$\varsigma \in [0, \infty]$, there is a solution $\vec{\mathfrak{x}}(\varsigma)$ of Eq. (6.2.1) such that for some $r > 0$,

$$\mathfrak{Y}\left(\vec{\mathfrak{x}}(\varsigma) - \vec{\mathfrak{x}}^{\star}(\varsigma), \mathfrak{y}\right) \succeq \mathbf{H}^{\mathbf{m,n}}_{\mathbf{p,q}}\left(\frac{\|Q\| \left(\frac{-\|\varsigma\|^{\iota}}{\iota}\right)^{\sigma}}{\frac{\mathfrak{y}}{r}}\right).$$

Theorem 6.6 *Let* $(\mathbb{R}^n, \mathfrak{Y}, \circledast)$ *be an MVFB-space and* ζ *and* ϱ *be positive constants such that* $0 < \zeta\varrho < 1$. *Assume that the continuous mapping with MVFF,* $\mathbf{H}^{\mathbf{m,n}}_{\mathbf{p,q}}$: $[0, \jmath) \times (0, \infty) \to \mathrm{diag}\mathfrak{M}_n((0, 1])$ *satisfying*

$$\mathbf{H}^{\mathbf{m,n}}_{\mathbf{p,q}}\left(\frac{\|Q\| \left(\frac{-\|\varsigma\|^{\iota}}{\iota}\right)^{\sigma}}{\frac{\mathfrak{y}}{(\gamma^{\sigma} I - Q\gamma^{\upsilon})}}\right) \succeq \mathbf{H}^{\mathbf{m,n}}_{\mathbf{p,q}}\left(\frac{\|Q\| \left(\frac{-\|\varsigma\|^{\iota}}{\iota}\right)^{\sigma}}{\frac{\mathfrak{y}}{\varrho}}\right) \tag{6.2.27}$$

and

$$\mathfrak{Y}\left(\vec{\mathfrak{x}}(t), \mathfrak{y}\right) \succeq \mathbf{H}_{\mathbf{p},\mathbf{q}}^{\mathbf{m},\mathbf{n}}\left(\frac{\|Q\| \left(\frac{-\|\varsigma\|^{\iota}}{\iota}\right)^{\sigma}}{\mathfrak{y}}\right) \tag{6.2.28}$$

$$\textit{implies that}\quad \mathfrak{Y}\left(\mathfrak{L}_{\iota}\{\vec{\mathfrak{x}}(\varsigma)\}(\gamma), \mathfrak{y}\right) \succeq \mathbf{H}_{\mathbf{p},\mathbf{q}}^{\mathbf{m},\mathbf{n}}\left(\frac{\|Q\| \left(\frac{-\|\varsigma\|^{\iota}}{\iota}\right)^{\sigma}}{\frac{\mathfrak{y}}{\zeta}}\right),$$

for every $\varsigma > 0$*. Let* $\vec{\mathfrak{x}}^{\star} : [\jmath, \infty) \to \mathbb{R}^n$ *be a continuous function satisfying*

$$\mathfrak{Y}\left({}_0D^{\sigma,\iota}\vec{\mathfrak{x}}^{\star}(\varsigma) - Q_0D^{\upsilon,\iota}\vec{\mathfrak{x}}^{\star}(\varsigma) - \vec{\mathfrak{s}}(\varsigma) - \mathfrak{L}_{\iota}^{-1}\{\vec{\mathfrak{x}}^{\star}(\gamma)\}(\varsigma), \mathfrak{y}\right) \tag{6.2.29}$$

$$\succeq \mathbf{H}_{\mathbf{p},\mathbf{q}}^{\mathbf{m},\mathbf{n}}\left(\frac{\|Q\| \left(\frac{-\|\varsigma\|^{\iota}}{\iota}\right)^{\sigma}}{\frac{\mathfrak{y}}{\omega}}\right),$$

for every $\varsigma > 0$ *and* $\mathfrak{y} > 0$*. Then we can find a unique continuous function* $\vec{\mathfrak{x}} : [\jmath, \infty) \to \mathbb{R}^n$ *such that*

$${}_0D^{\sigma,\iota}\vec{\mathfrak{x}}(\varsigma) = Q_0D^{\upsilon,\iota}\vec{\mathfrak{x}}(\varsigma) + \vec{\mathfrak{s}}(\varsigma) + \mathfrak{L}_{\iota}^{-1}\{\vec{\mathfrak{x}}(\gamma)\}(\varsigma), \tag{6.2.30}$$

and

$$\mathfrak{Y}\left(\vec{\mathfrak{x}}(\varsigma) - \vec{\mathfrak{x}}^{\star}(\varsigma), \mathfrak{y}\right) \succeq \mathbf{H}_{\mathbf{p},\mathbf{q}}^{\mathbf{m},\mathbf{n}}\left(\frac{\|Q\| \left(\frac{-\|\varsigma\|^{\iota}}{\iota}\right)^{\sigma}}{\frac{(1-\zeta\varrho)\mathfrak{y}}{\zeta\omega}}\right), \tag{6.2.31}$$

for every $\varsigma > 0$ *and* $\mathfrak{y} > 0$*.*

Proof We set

$$\mathfrak{S} := \{\vec{\mathfrak{x}} : [\jmath, \infty) \to \mathbb{R}^n, \vec{\mathfrak{x}} \text{ is continuous}\},$$

and introduce the [0, 1]-valued metric on $\mathfrak{S}$ as,

$$\delta(\vec{\mathfrak{x}}, \vec{\mathfrak{x}}^{\star}) = \inf\left\{\phi > 0 : \mathfrak{Y}\left(\vec{\mathfrak{x}}(\varsigma) - \vec{\mathfrak{x}}^{\star}(\varsigma), \mathfrak{y}\right) \succeq \mathbf{H}_{\mathbf{p},\mathbf{q}}^{\mathbf{m},\mathbf{n}}\left(\frac{\|Q\| \left(\frac{-\|\varsigma\|^{\iota}}{\iota}\right)^{\sigma}}{\frac{\mathfrak{y}}{\phi}}\right),\ \forall\, \vec{\mathfrak{x}}, \vec{\mathfrak{x}}^{\star} \in \mathfrak{S},\ \varsigma > 0,\ \mathfrak{y} > 0\right\}.$$

$(\mathfrak{S}, \delta)$ is a complete [0, 1]-valued metric space that the proof of which is similar to the **step 1** of Theorem 6.5.

Now, we define the mapping $\Psi : \mathfrak{S} \to \mathfrak{S}$ by

$$\begin{aligned}\Psi(\vec{\mathfrak{x}})(\varsigma)) &= \mathfrak{L}_\iota\left\{{}_0D^{\sigma,\iota}\vec{\mathfrak{x}}(\varsigma) - Q_0D^{\upsilon,\iota}\vec{\mathfrak{x}}(\varsigma) - \vec{\mathfrak{s}}(\varsigma)\right\}(\gamma) \\ &= \mathfrak{L}_\iota\left\{{}_0D^{\sigma,\iota}\vec{\mathfrak{x}}(\varsigma)\right\}(\gamma) - Q\mathfrak{L}_\iota\left\{{}_0D^{\upsilon,\iota}\vec{\mathfrak{x}}(\varsigma)\right\}(\gamma) - \mathfrak{L}_\iota\{\vec{\mathfrak{s}}(\varsigma)\}(\gamma) \\ &= \gamma^\sigma I\mathfrak{L}_\iota\{\vec{\mathfrak{x}}(\varsigma)\}(\gamma) - \left({}_0I^{1-\sigma,\iota}\vec{\mathfrak{x}}\right)(0+) \\ &- Q\left\{\gamma^\upsilon I\mathfrak{L}_\iota\{\vec{\mathfrak{x}}(\varsigma)\}(\gamma) - \left({}_0I^{1-\upsilon,\iota}\vec{\mathfrak{x}}\right)(0+)\right\} - \mathfrak{L}_\iota\{\vec{\mathfrak{s}}(\varsigma)\}(\gamma) \\ &= \left(\gamma^\sigma I - Q\gamma^\upsilon\right)\mathfrak{L}_\iota\{\vec{\mathfrak{x}}(\varsigma)\}(\gamma) \\ &- \left({}_0I^{1-\sigma,\iota}\vec{\mathfrak{x}}\right)(0+) + Q\left({}_0I^{1-\upsilon,\iota}\vec{\mathfrak{x}}\right)(0+) - \mathfrak{L}_\iota\{\vec{\mathfrak{s}}(\varsigma)\}(\gamma).\end{aligned} \tag{6.2.32}$$

Step 1. We prove Ψ is a strictly contractive mapping.
Let $\vec{\mathfrak{x}}, \vec{\mathfrak{x}}^\star \in \mathfrak{S}$, $\phi > 0$ and $\delta(\vec{\mathfrak{x}}, \vec{\mathfrak{x}}^\star) < \mathcal{D}_{\vec{\mathfrak{x}}\vec{\mathfrak{x}}^\star}$, then we have

$$\mathfrak{Y}\left(\vec{\mathfrak{x}}(\varsigma) - \vec{\mathfrak{x}}^\star(\varsigma), Q_{\vec{\mathfrak{x}}\vec{\mathfrak{x}}^\star}\mathfrak{y}\right) \succeq \mathbf{H}^{\mathbf{m,n}}_{\mathbf{p,q}}\left(\frac{\|Q\|\left(\frac{-\|\varsigma\|^\iota}{\iota}\right)^\sigma}{\mathfrak{y}}\right), \quad \forall\, \vec{\mathfrak{x}}, \vec{\mathfrak{x}}^\star \in \mathfrak{S}, \ \varsigma > 0, \ \mathfrak{y} > 0.$$

Using (6.2.32) and properties (MVF-1-N2) and (MVF-1-N3) of Definition 1.24, we have

$$\begin{aligned}&\mathfrak{Y}\left(\Psi\vec{\mathfrak{x}}(\varsigma) - \Psi\vec{\mathfrak{x}}^\star(\varsigma), Q_{\vec{\mathfrak{x}}\vec{\mathfrak{x}}^\star}\mathfrak{y}\right) \\ &= \mathfrak{Y}\Big(\left(\gamma^\sigma I - Q\gamma^\upsilon\right)\mathfrak{L}_\iota\{\vec{\mathfrak{x}}(\varsigma)\}(\gamma) - \left({}_0I^{1-\sigma,\iota}\vec{\mathfrak{x}}\right)(0+) + Q\left({}_0I^{1-\upsilon,\iota}\vec{\mathfrak{x}}\right)(0+) \\ &- \mathfrak{L}_\iota\{\vec{\mathfrak{s}}(\varsigma)\}(\gamma) - \left(\gamma^\sigma I - Q\gamma^\upsilon\right)\mathfrak{L}_\iota\{\vec{\mathfrak{x}}^\star(\varsigma)\}(\gamma) + \left({}_0I^{1-\sigma,\iota}\vec{\mathfrak{x}}^\star\right)(0+) - Q\left({}_0I^{1-\upsilon,\iota}\vec{\mathfrak{x}}^\star\right)(0+) + \mathfrak{L}_\iota\{\vec{\mathfrak{s}}(\varsigma)\}(\gamma), \mathcal{D}_{\vec{\mathfrak{x}}\vec{\mathfrak{x}}^\star}\mathfrak{y}\Big) \\ &= \mathfrak{Y}\left(\left(\gamma^\sigma I - Q\gamma^\upsilon\right)\mathfrak{L}_\iota\{\vec{\mathfrak{x}}(\varsigma)\}(\gamma) - \left(\gamma^\sigma I - Q\gamma^\upsilon\right)\mathfrak{L}_\iota\{\vec{\mathfrak{x}}^\star(\varsigma)\}(\gamma), \mathcal{D}_{\vec{\mathfrak{x}}\vec{\mathfrak{x}}^\star}\mathfrak{y}\right) \\ &= \mathfrak{Y}\left(\left(\gamma^\sigma I - Q\gamma^\upsilon\right)\mathfrak{L}_\iota\{\vec{\mathfrak{x}}(\varsigma) - \vec{\mathfrak{x}}^\star(\varsigma)\}(\gamma), Q_{\vec{\mathfrak{x}}\vec{\mathfrak{x}}^\star}\mathfrak{y}\right) \\ &= \mathfrak{Y}\left(\mathfrak{L}_\iota\left\{\vec{\mathfrak{x}}(\varsigma) - \vec{\mathfrak{x}}^\star(\varsigma)\right\}(\gamma), \frac{\mathcal{D}_{\vec{\mathfrak{x}}\vec{\mathfrak{x}}^\star}\mathfrak{y}}{\left(\gamma^\sigma I - Q\gamma^\upsilon\right)}\right).\end{aligned} \tag{6.2.33}$$

On the other hand, we have

$$\mathfrak{Y}\left(\vec{\mathfrak{x}}(\varsigma) - \vec{\mathfrak{x}}^\star(\varsigma), \frac{\mathcal{D}_{\vec{\mathfrak{x}}\vec{\mathfrak{x}}^\star}\mathfrak{y}}{\left(\gamma^\sigma I - Q\gamma^\upsilon\right)}\right) \succeq \mathbf{H}^{\mathbf{m,n}}_{\mathbf{p,q}}\left(\frac{\|Q\|\left(\frac{-\|\varsigma\|^\iota}{\iota}\right)^\sigma}{\frac{\mathfrak{y}}{(\gamma^\sigma I - Q\gamma^\upsilon)}}\right). \tag{6.2.34}$$

Using (6.2.28) and (6.2.34), we get

$$\mathfrak{Y}\left(\mathfrak{L}_\iota\left\{\vec{\mathfrak{x}}(\varsigma) - \vec{\mathfrak{x}}^\star(\varsigma)\right\}(\gamma), \frac{\mathfrak{y}}{\left(\gamma^\sigma I - Q\gamma^\beta\right)}\right) \succeq \mathbf{H}^{\mathbf{m,n}}_{\mathbf{p,q}}\left(\frac{\|Q\|\left(\frac{-\|\varsigma\|^\iota}{\iota}\right)^\sigma}{\frac{\zeta\mathfrak{y}}{(\gamma^\sigma I - Q\gamma^\upsilon)}}\right) \tag{6.2.35}$$

Form (6.2.33), (6.2.34), (6.2.35), we have

$$\mathfrak{Y}\left(\Psi\vec{\mathfrak{x}}(\varsigma)-\Psi\vec{\mathfrak{x}}^{\star}(\varsigma),\mathcal{D}_{\vec{\mathfrak{x}}\vec{\mathfrak{x}}^{\star}}\mathfrak{y}\right)\succeq\mathbf{H}^{\mathbf{m},\mathbf{n}}_{\mathbf{p},\mathbf{q}}\left(\frac{\|Q\|\left(\frac{-\|\varsigma\|^{\iota}}{\iota}\right)^{\sigma}}{\frac{\zeta\mathfrak{y}}{(\gamma^{\sigma}I-Q\gamma^{\upsilon})}}\right)\succeq\ \mathbf{H}^{\mathbf{m},\mathbf{n}}_{\mathbf{p},\mathbf{q}}\left(\frac{\|Q\|\left(\frac{-\|\varsigma\|^{\iota}}{\iota}\right)^{\sigma}}{\frac{\mathfrak{y}}{\zeta\varrho}}\right),\tag{6.2.36}$$

and then

$$\mathfrak{Y}\left(\Psi\vec{\mathfrak{x}}(\varsigma)-\Psi\vec{\mathfrak{x}}^{\star}(\varsigma),\mathcal{D}_{\vec{\mathfrak{x}}\vec{\mathfrak{x}}^{\star}}\mathfrak{y}\right)\succeq\mathbf{H}^{\mathbf{m},\mathbf{n}}_{\mathbf{p},\mathbf{q}}\left(\frac{\|Q\|\left(\frac{-\|\varsigma\|^{\iota}}{\iota}\right)^{\sigma}}{\frac{\mathfrak{y}}{\zeta\varrho}}\right),\tag{6.2.37}$$

which implies that

$$\delta\left(\Psi\vec{\mathfrak{x}},\Psi\vec{\mathfrak{x}}^{\star}\right)\leq\zeta\varrho\mathcal{D}_{\vec{\mathfrak{x}}\vec{\mathfrak{x}}^{\star}},$$

and then

$$\delta\left(\Psi\vec{\mathfrak{x}},\Psi\vec{\mathfrak{x}}^{\star}\right)\leq\zeta\varrho\delta(\vec{\mathfrak{x}},\vec{\mathfrak{x}}^{\star}).$$

Then, Ψ is a strictly contractive mapping with Lipschitz constant $\zeta\varrho$.

Step 2. We show that $\delta(\Psi(\vec{\mathfrak{x}}^{\star}),\vec{\mathfrak{x}}^{\star})<\infty$ for $\vec{\mathfrak{x}}\in\mathfrak{S}$.
Using (6.2.28) and (6.2.29) we get

$$\begin{aligned}&\mathfrak{Y}\left(\Psi\left(\vec{\mathfrak{x}}^{\star}(\varsigma)\right)-\vec{\mathfrak{x}}^{\star}(\varsigma),\mathfrak{y}\right)=\mathfrak{Y}\left(\mathfrak{L}_{\iota}\left\{{}_0D^{\sigma,\iota}\vec{\mathfrak{x}}^{\star}(\varsigma)-Q_0D^{\upsilon,\iota}\vec{\mathfrak{x}}^{\star}(\varsigma)-\vec{\mathfrak{s}}(\varsigma)\right\}(\gamma)-\vec{\mathfrak{x}}^{\star}(\varsigma),\mathfrak{y}\right)\\&=\mathfrak{Y}\left(\left(\gamma^{\sigma}I-Q\gamma^{\upsilon}\right)\mathfrak{L}_{\iota}\{\vec{\mathfrak{x}}^{\star}(\varsigma)\}(\gamma)-\left({}_0I^{1-\sigma,\iota}\vec{\mathfrak{x}}^{\star}\right)(0+)+Q\left({}_0I^{1-\upsilon,\iota}\vec{\mathfrak{x}}^{\star}\right)(0+)-\mathfrak{L}_{\iota}\{\vec{\mathfrak{s}}(\varsigma)\}(\gamma)-\vec{\mathfrak{x}}^{\star}(\varsigma),\mathfrak{y}\right)\\&\succeq\mathbf{H}^{\mathbf{m},\mathbf{n}}_{\mathbf{p},\mathbf{q}}\left(\frac{\|Q\|\left(\frac{-\|\varsigma\|^{\iota}}{\iota}\right)^{\sigma}}{\frac{\mathfrak{y}}{\zeta\omega}}\right),\end{aligned}$$

for every $\mathfrak{y}>0$. Then we have $\delta(\Psi(\vec{\mathfrak{x}}^{\star}),\vec{\mathfrak{x}}^{\star})<\zeta\omega<\infty$.

Now, Theorem 1.2 enables us to find an element $\vec{\mathfrak{x}}$ in $\mathfrak{S}$ satisfying the following:
(1) y is a fixed point of $\mho$, i.e.,

$$\begin{aligned}\vec{\mathfrak{x}}(\varsigma)&=\Psi(\vec{\mathfrak{x}}(\varsigma))\\&=\left(\gamma^{\sigma}I-Q\gamma^{\upsilon}\right)\mathfrak{L}_{\iota}\{\vec{\mathfrak{x}}(\varsigma)\}(\gamma)-Q\left({}_0I^{1-\sigma,\iota}\vec{\mathfrak{x}}\right)(0+)\\&\quad+Q\left({}_0I^{1-\upsilon,\iota}\vec{\mathfrak{x}}\right)(0+)-\mathfrak{L}_{\iota}\{\vec{\mathfrak{s}}(\varsigma)\}(\gamma),\end{aligned}\tag{6.2.38}$$

which is unique in the set

$$\mathfrak{S}^{*}=\{\vec{\mathfrak{x}}\in\mathfrak{S}:\delta(\Psi(\vec{\mathfrak{x}}^{\star}),\vec{\mathfrak{x}})<\infty\}.$$

Since $\vec{\mathfrak{x}}$ is a continuous function and according to the Theorem 6.4

$$ {}_0D^{\sigma,\iota}\vec{\mathfrak{x}}(\varsigma) = Q_0D^{\upsilon,\iota}\vec{\mathfrak{x}}(\varsigma) + \vec{\mathfrak{s}}(\varsigma) + \mathcal{L}_\iota^{-1}\{\vec{\mathfrak{x}}(\gamma)\}(\varsigma); \tag{6.2.39} $$

(2) $\delta(\Psi^k(\vec{\mathfrak{x}}), \vec{\mathfrak{x}}^\star) \to 0$ as $n \to \infty$;
(3) $\delta(\vec{\mathfrak{x}}, \vec{\mathfrak{x}}^\star) \le \frac{1}{1-\zeta\varrho}\delta(\Psi(\vec{\mathfrak{x}}), \vec{\mathfrak{x}}) \le \frac{\zeta\omega}{1-\zeta\varrho}$, which implies that

$$ \mathfrak{Y}\left(\vec{\mathfrak{x}}(\varsigma) - \vec{\mathfrak{x}}^\star(\varsigma), \mathfrak{y}\right) \succeq \mathbf{H}^{\mathbf{m,n}}_{\mathbf{p,q}}\left(\frac{\|Q\|\left(\frac{-\|\varsigma\|^\iota}{\iota}\right)^\sigma}{\frac{(1-\zeta\varrho)\mathfrak{y}}{\zeta\omega}}\right), \tag{6.2.40} $$

for every $\varsigma > 0$ and $\mathfrak{y} > 0$.

Now, we prove that $\mathfrak{S}^* = \mathfrak{S}$. Consider $\vec{\mathfrak{z}}$ in $\mathfrak{S}$ satisfying (6.2.8) and (6.2.9), we show that $\vec{\mathfrak{z}} = \vec{\mathfrak{x}}$ and $\vec{\mathfrak{z}} \in \mathfrak{S}^*$. From (6.2.30) we get

$$ {}_0D^{\sigma,\iota}\vec{\mathfrak{z}}(\varsigma) = Q_0D^{\upsilon,\iota}\vec{\mathfrak{z}}(\varsigma) + \vec{\mathfrak{s}}(\varsigma) + \mathcal{L}_\iota^{-1}\{\vec{\mathfrak{z}}(\gamma)\}(\varsigma), \quad 0 < \upsilon < \sigma < 1, \varsigma > 0, \tag{6.2.41} $$

and then

$$ \vec{\mathfrak{z}}(\varsigma) = \left(\gamma^\sigma I - Q\gamma^\upsilon\right)\mathfrak{L}_\iota\{\vec{\mathfrak{x}}(\varsigma)\}(\gamma) - Q\left({}_0I^{1-\sigma,\iota}\vec{\mathfrak{x}}\right)(0+) + Q\left({}_0I^{1-\upsilon,\iota}\vec{\mathfrak{x}}\right)(0+) - \mathfrak{L}_\iota\{\vec{\mathfrak{s}}(\varsigma)\}(\gamma). \tag{6.2.42} $$

Now, we show that

$$ \vec{\mathfrak{z}} \in \{\vec{\mathfrak{x}} \in \mathfrak{S}, \ \ \delta(\Psi(\vec{\mathfrak{x}}^\star), \vec{\mathfrak{x}}) < \infty\}, $$

i.e., $\delta(\Psi(\vec{\mathfrak{x}}^\star), \vec{\mathfrak{z}}) < \infty$. We set $\daleth = \frac{(1-\zeta\varrho)\mathfrak{y}}{\zeta\omega}$, from (6.2.9) we get

$$ \mathfrak{Y}\left(\vec{\mathfrak{x}}(\varsigma) - \vec{\mathfrak{z}}(\varsigma)_{\mathfrak{y}} \succeq (\mathbf{W}_{\sigma-\upsilon,\sigma+1})^\varsigma, \daleth\mathfrak{y}\right), \tag{6.2.43} $$

for every $\varsigma > 0$ and $\tau > 0$. Using (6.2.43), we get

$$ \begin{aligned} &\mathfrak{Y}_{\mathfrak{y}}^{\Psi(\vec{\mathfrak{x}}(\varsigma))-\vec{\mathfrak{z}}(\varsigma)} \qquad (6.2.44) \\ &= \mathfrak{Y}\left(\mathfrak{L}_\varsigma\left\{{}_0D^{\sigma,\iota}\vec{\mathfrak{x}}(\varsigma) - Q_0D^{\upsilon,\iota}\vec{\mathfrak{x}}(\varsigma) - \vec{\mathfrak{s}}(\varsigma)\right\}(\gamma) - \vec{\mathfrak{z}}(\varsigma), \mathfrak{y}\right) \\ &= \mathfrak{Y}\Big(\left(\gamma^\sigma I - Q\gamma^\upsilon\right)\mathfrak{L}_\iota\{\vec{\mathfrak{x}}^\star(\varsigma)\}(\gamma) - \left({}_0I^{1-\sigma,\iota}\vec{\mathfrak{x}}^\star\right)(0+) + Q\left({}_0I^{1-\upsilon,\iota}\vec{\mathfrak{x}}^\star\right)(0+) \\ &\quad - \mathfrak{L}_\iota\{\vec{\mathfrak{s}}(\varsigma)\}(\gamma) - \left(\gamma^\sigma I - Q\gamma^\upsilon\right)\mathfrak{L}_\iota\{\vec{\mathfrak{z}}(\varsigma)\}(\gamma) + \left({}_0I^{1-\sigma,\iota}\vec{\mathfrak{z}}\right)(0+) \\ &\quad - Q\left({}_0I^{1-\upsilon,\iota}\vec{\mathfrak{z}}\right)(0+) + \mathfrak{L}_\iota\{\vec{\mathfrak{s}}(\varsigma)\}(\gamma), \mathfrak{y}\Big) \\ &= \mathfrak{Y}\left(\left(\gamma^\sigma I - Q\gamma^\upsilon\right)\mathfrak{L}_\iota\{\vec{\mathfrak{x}}^\star(\varsigma)\}(\gamma) - \left(\gamma^\sigma I - Q\gamma^\upsilon\right)\mathfrak{L}_\iota\{\vec{\mathfrak{z}}(\varsigma)\}(\gamma), \mathfrak{y}\right) \end{aligned} $$

$$= \mathfrak{Y}\left(\left(\gamma^{\sigma} I - Q\gamma^{\upsilon}\right) \mathfrak{L}_{\iota}\{\vec{\mathfrak{x}}^{\star}(\varsigma) - \vec{\mathfrak{z}}(\varsigma)\}(\gamma), \mathfrak{y}\right)$$

$$= \mathfrak{Y}\left(\left(\gamma^{\sigma} I - Q\gamma^{\upsilon}\right) \mathfrak{L}_{\iota}\left\{\vec{\mathfrak{x}}^{\star}(\varsigma) - \vec{\mathfrak{z}}(\varsigma)\right\}(\gamma), \mathfrak{y}\right)$$

$$\succeq \mathfrak{Y}\left(\mathfrak{L}_{\iota}\left\{\vec{\mathfrak{x}}^{\star}(\varsigma) - \vec{\mathfrak{z}}(\varsigma)\right\}(\gamma), \frac{\mathfrak{y}}{\left(\gamma^{\sigma} I - Q\gamma^{\upsilon}\right)}\right)$$

$$\succeq \mathbf{H}^{\mathbf{m},\mathbf{n}}_{\mathbf{p},\mathbf{q}}\left(\frac{\|Q\|\left(\frac{-\|\varsigma\|^{\iota}}{\iota}\right)^{\sigma}}{\daleth\frac{\mathfrak{y}}{(\gamma^{\sigma} I - Q\gamma^{\upsilon})}}\right)$$

$$\succeq \mathbf{H}^{\mathbf{m},\mathbf{n}}_{\mathbf{p},\mathbf{q}}\left(\frac{\|Q\|\left(\frac{-\|\varsigma\|^{\iota}}{\iota}\right)^{\sigma}}{\daleth\frac{\mathfrak{y}}{\zeta\varrho}}\right),$$

which implies that $\delta(\Psi(\vec{\mathfrak{x}}), \vec{\mathfrak{z}}) \leq \frac{\daleth}{\zeta\varrho} < \infty$, then $\vec{\mathfrak{z}} \in \mathfrak{S}^{*}$.

6.2.6 Application

Example 6.3 Consider the following generalized fractional differential system

$$_{0}D^{\frac{4}{7},\frac{3}{8}}\vec{\mathfrak{x}}(\varsigma) = Q\ {}_{0}D^{\frac{1}{25},\frac{3}{8}}\vec{\mathfrak{x}}(\varsigma) + \vec{\mathfrak{s}}(\varsigma) + \mathfrak{L}_{\iota}^{-1}\{\vec{\mathfrak{x}}(\gamma)\}(\varsigma), \tag{6.2.45}$$

where the initial value $\left({}_{0}I^{1-\frac{4}{7},\frac{3}{8}}\vec{\mathfrak{x}}\right)(0+) = \begin{bmatrix} 0 \\ 0 \\ 0 \end{bmatrix}$, $\vec{\mathfrak{x}}(\varsigma) = \begin{bmatrix} \vec{\mathfrak{x}}'(\varsigma) \\ \vec{\mathfrak{x}}''(\varsigma) \\ \vec{\mathfrak{x}}'''(\varsigma) \end{bmatrix}$, $Q = \begin{bmatrix} 3 & 1 & 4 \\ 2 & -4 & 1 \\ 1 & 3 & 2 \end{bmatrix}$, and

$$\vec{\mathfrak{s}}(\varsigma) = \begin{bmatrix} \frac{\Gamma(\frac{5}{2})}{\Gamma(\frac{19}{10})}\varsigma^{\frac{7}{12}} - \frac{\Gamma(\frac{5}{2})}{\Gamma(\frac{5}{4})}\varsigma^{\frac{7}{15}} - \frac{2\Gamma(2)}{\Gamma(\frac{7}{4})}\varsigma^{\frac{1}{25}} + \tau\cos 2\beta\varsigma \\ \frac{\Gamma(2)}{\Gamma(\frac{7}{5})}\varsigma^{\frac{1}{19}} + \frac{3\Gamma(2)}{\Gamma(\frac{7}{4})}\varsigma^{\frac{1}{25}} + \frac{1}{2}\tau\sin 3\beta\varsigma \\ \frac{\Gamma(\frac{10}{3})}{\Gamma(\frac{17}{10})}\varsigma^{\frac{14}{12}} - \frac{\Gamma(\frac{5}{2})}{\Gamma(\frac{5}{4})}\varsigma^{\frac{7}{12}} - \frac{2\Gamma(\frac{18}{3})}{\Gamma(\frac{1}{16})}\varsigma^{\frac{1}{25}} + \varsigma\cos 4\beta\varsigma \end{bmatrix},$$

Assume that the following condition is true for the given continuous function

$$\mathbf{H}^{\mathbf{m},\mathbf{n}}_{\mathbf{p},\mathbf{q}}\left(\frac{\|Q\|\left(\frac{-\|\varsigma\|^{\iota}}{\iota}\right)^{\sigma}}{\frac{\mathfrak{y}}{(\gamma^{\sigma} I + Q\gamma^{\upsilon})}}\right) \succeq \mathbf{H}^{\mathbf{m},\mathbf{n}}_{\mathbf{p},\mathbf{q}}\left(\frac{\|Q\|\left(\frac{-\|\varsigma\|^{\iota}}{\iota}\right)^{\sigma}}{8\mathfrak{y}}\right).$$

For $\omega = \frac{1}{10000}$, the vector-valued function $\vec{\mathfrak{x}}_2(\varsigma) = \begin{bmatrix} \varsigma^{\frac{1}{2}} \\ \varsigma^{\frac{3}{8}} \\ \varsigma^{\frac{7}{12}} \end{bmatrix}$ satisfies

$$\mathfrak{Y}\left({}_0D^{\frac{4}{7},\frac{3}{8}}\vec{\mathfrak{x}}_2(\varsigma) - Q\,{}_0D^{\frac{1}{25},\frac{3}{8}}\vec{\mathfrak{x}}_2(\varsigma) - \vec{\mathfrak{s}}(\varsigma) - \mathfrak{L}_\iota^{-1}\{\vec{\mathfrak{x}}_1(\varsigma)\}, \mathfrak{y}\right) \succeq \mathbf{H}^{\mathbf{m},\mathbf{n}}_{\mathbf{p},\mathbf{q}}\left(\frac{\|Q\|\left(\frac{-\|\varsigma\|^\iota}{\iota}\right)^\sigma}{10000\mathfrak{y}}\right),$$

therefore, according to the definitions given $\vec{\mathfrak{x}}_2$ holds in the following inequality

$$\begin{aligned} &\mathfrak{Y}\left(\Psi\left(\vec{\mathfrak{x}}_2(\varsigma)\right) - \vec{\mathfrak{x}}_2(\varsigma), \mathfrak{y}\right) \qquad (6.2.46)\\ &= \mathfrak{Y}\left(\mathfrak{L}_\iota\{{}_0D^{\frac{4}{7},\frac{3}{8}}\vec{\mathfrak{x}}_2(\varsigma) - Q_0D^{\frac{1}{25},\frac{3}{8}}\vec{\mathfrak{x}}_2(\varsigma) - \vec{\mathfrak{s}}(\varsigma)\} - \vec{\mathfrak{x}}_2(\varsigma), \mathfrak{y}\right)\\ &= \mathfrak{Y}\Big(\left(\gamma^\sigma I - Q\gamma^\upsilon\right)\mathfrak{L}_\iota\{\vec{\mathfrak{x}}_2(\varsigma)\}(\gamma) - Q\left({}_0I^{1-\sigma,\iota}\vec{\mathfrak{x}}_2\right)(0+)\\ &\quad + Q\left({}_0I^{1-\upsilon,\iota}\vec{\mathfrak{x}}_2\right)(0+) - \mathfrak{L}_\iota\{\vec{\mathfrak{s}}(\varsigma)\}(\gamma) - \vec{\mathfrak{x}}_2(\varsigma), \mathfrak{y}\Big)\\ &\succeq \mathbf{H}^{\mathbf{m},\mathbf{n}}_{\mathbf{p},\mathbf{q}}\left(\frac{\|Q\|\left(\frac{-\|\varsigma\|^\iota}{\iota}\right)^\sigma}{\frac{10000\mathfrak{y}}{\zeta}}\right). \end{aligned}$$

Then, there is a unique solution $\vec{\mathfrak{x}} \in C([\jmath, \infty), \mathbb{R}^n)$ for Eq. (6.2.2), such that

(*1*) $\vec{\mathfrak{x}}$ is the fixed point of the operator Ψ, which is defined as follows

$$\Psi(\vec{\mathfrak{x}}(\varsigma)) = \left(\gamma^\sigma I - Q\gamma^\upsilon\right)\mathfrak{L}_\iota\{\vec{\mathfrak{x}}(\varsigma)\}(\gamma) - Q\left({}_0I^{1-\sigma,\iota}\vec{\mathfrak{x}}\right)(0+) + Q\left({}_0I^{1-\upsilon,\iota}\vec{\mathfrak{x}}\right)(0+) - \mathfrak{L}_\iota\{\vec{\mathfrak{s}}(\varsigma)\}(\gamma).$$

(*2*) The fixed point $\vec{\mathfrak{x}}$ is unique.
(*3*) $\vec{\mathfrak{x}}$ satisfies in Eq. (6.2.45).

Then

$$\delta(\vec{\mathfrak{x}}\vec{\mathfrak{x}}_2) \leq \frac{\zeta}{1-8\zeta} \times \frac{1}{10000},$$

or

$$\mathfrak{Y}\left(\vec{\mathfrak{x}}(\varsigma) - \vec{\mathfrak{x}}_2(\varsigma), \mathfrak{y}\right) \succeq \mathbf{H}^{\mathbf{m},\mathbf{n}}_{\mathbf{p},\mathbf{q}}\left(\frac{\|Q\|\left(\frac{-\|\varsigma\|^\iota}{\iota}\right)^\sigma}{\frac{10000(8-\zeta)\tau}{8\zeta}}\right). \qquad (6.2.47)$$

The table below shows the exact values for the solutions.

Solutions	$\gamma = 10$	$\gamma = \frac{125}{4}$	$\gamma = 300$
Row 1	$\frac{6110^{\frac{4}{7}}}{500} - \frac{251310^{\frac{1}{25}}}{2500} + 0.06368\,625452$	$\frac{16657125^{\frac{4}{7}} \times 4^{\frac{3}{7}}}{1953125} - \frac{7028759125^{\frac{1}{25}} \times 4^{\frac{24}{25}}}{244140625} + 0.01657711334$	$\frac{45301300^{\frac{1}{25}}}{13500000} - \frac{4590367300^{\frac{4}{7}}}{450000000} + 0.001301757873$
Row 2	$-\frac{15110^{\frac{1}{25}}}{625} + \frac{4310^{\frac{4}{7}}}{5000} + 0.1831715502$	$-\frac{4290214125^{\frac{1}{25}} \times 4^{\frac{24}{25}}}{244140625} + \frac{16384125^{\frac{4}{7}} \times 4^{\frac{3}{7}}}{244140625} + 0.05831820269$	$-\frac{568981300^{\frac{1}{25}}}{84375000} + \frac{401300^{\frac{4}{7}}}{13500000} + 0.005752054996$
Row 3	$-\frac{105910^{\frac{1}{25}}}{5000} + \frac{410^{\frac{4}{7}}}{125} + 1.389006508$	$-\frac{2514277125^{\frac{1}{25}} \times 4^{\frac{24}{25}}}{244140625} + \frac{1532125^{\frac{4}{7}} \times 4^{\frac{3}{7}}}{1953125} + 0.4262023240$	$-\frac{1540501300^{\frac{1}{25}}}{450000000} + \frac{451300^{\frac{4}{7}}}{13500000} + 0.04038844526$

6.3 Measure of Quality and Certainty Approximation of Functional Inequalities

6.3.1 *Preliminaries*

If $\mathcal{M}_1$ is a vector space, $\mathcal{M}_2$ is a Banach space and $\Delta : \mathcal{M}_1 \to \mathcal{M}_2$; is an approximately-additive mapping (A-M) such that for any positive constant κ and for all $\alpha, \beta \in \mathcal{M}$, we have

$$\|\Delta(\alpha + \beta) - \Delta(\alpha) - \Delta(\beta)\| \leq \kappa, \tag{6.3.1}$$

then; there exists a unique (A-M) $\Upsilon : \mathcal{M}_1 \to \mathcal{M}_2$, such that for all $\alpha \in \mathcal{M}_1$ defined as $\Upsilon(\alpha) = \lim_{n\to\infty} 2^{-n}\Delta(2^n\alpha)$, we get $\|\Delta(\alpha) - \Upsilon(\alpha)\| \leq \kappa$. The linearity and continuity of the mapping $\Upsilon : \mathcal{M}_1 \to \mathcal{M}_2$ is easily obtained from Δ. Our effort is to measure the quality and certainty of an approximation so that we can achieve the optimal approximation. In [8, 19, 29], the authors presented a model to increase information about a subject by studying the measure of quality and certainty via Z-numbers (Z-Ns) [1, 2, 12, 27]. We apply the generalized Z-N (GZ-N) theory to get more information about certainty and reliability of the approximation for the functional inequality (FI)

$$\|\Delta(\alpha + \beta) - \Delta(\alpha) - \Delta(\beta)\| \leq \sigma||\alpha||^m||\beta||^n, \tag{6.3.2}$$

which is known as the Cauchy–Rassias inequality; where $\sigma \geq 0$ and $c = m + n \neq 1$ and

$$\begin{aligned} &\mathcal{Z}_{\Omega,\Lambda}\Big(\Delta(\alpha) + \Delta(\beta) + \Delta(\gamma) + \Delta(\zeta), \mathcal{T}\Big) \\ &\succeq \mathcal{Z}_{\Omega,\Lambda}\Big(\Delta(\alpha + \beta) + \Delta(\gamma + \zeta), \mathcal{T}\Big). \end{aligned} \tag{6.3.3}$$

Definition 6.9 ([16]) On the set $\mathcal{W}_1$, which is a nonempty set, we consider the function $\delta_{\mathcal{W}_1} : \mathcal{W}_1 \times \mathcal{W}_1 \to [0, +\infty]$. The pair $(W_1, \delta_{\mathcal{W}_1})$ is a generalized metric space (GMS) if: For each $\alpha, \beta, \gamma \in \mathcal{W}_1$, $\delta_{\mathcal{W}_1}(\alpha, \beta) = 0$ iff $\alpha = \beta$, $\delta_{\mathcal{W}_1}(\alpha, \beta) = \delta_{\mathcal{W}_1}(\beta, \alpha)$ and $\delta_{\mathcal{W}_1}(\alpha, \gamma) \leq \delta_{\mathcal{W}_1}(\alpha, \beta) + \delta_{\mathcal{W}_1}(\beta, \gamma)$.

In [16, pp. 542, 543] the author presented an example of a GMS. For example, let $W = \{z \in C[0, 1] \text{ with } z(0) = 0\}$ and put

$$d(x, y) = \sup_{t \in [0,1]} t^{-\frac{1}{2}} |x(t) - y(t)|,$$

for $x, y \in W$. Now, we present another example. The function $\delta_{\mathbb{R}} : \mathbb{R} \times \mathbb{R} \to [0, +\infty]$ with

$$\delta_{\mathbb{R}}(\alpha, \beta) = \begin{cases} \left|\frac{1}{\alpha} - \frac{1}{\beta}\right| & \alpha, \beta \neq 0 \\ 0 & \alpha, \beta = 0 \\ +\infty & \text{otherwise,} \end{cases}$$

is a GMS. We denote a complete GMS by CGMS.

Definition 6.10 ([21]) A mapping $\mathcal{E}_{\mathcal{T}}$ from $[-\infty, +\infty]$ to $[0, 1]$ is a distribution-distance mapping (D-DM) if it has the following properties

- It is continuous from the left and increasing on $\mathbb{R}$;
- $\mathcal{E}_0 = 0$ and $\mathcal{E}_{+\infty} = 1$.

We denote the set of all D-DMs by $\aleph^+$. The subset $\mathcal{L}^+$ of $\aleph^+$ contains those D-DMs whose left limit at $+\infty$ is one; that is, $\ell^- \mathcal{E}_{+\infty} = 1$. Also, we define

$$\Pi^0_\tau = \begin{cases} 0, \text{ if } \mathcal{T} \in \mathbb{R} - (0, +\infty), \\ 1, \quad \text{if } \mathcal{T} \in (0, +\infty). \end{cases}$$

Definition 6.11 ([21]) We consider the vector space $\mathcal{W}_1$ and define a D-DM $\Lambda_{\mathcal{W}_1} : \mathcal{W}_1 \to \mathcal{L}^+$ on this space, and we consider the ternary space $(\mathcal{W}_1, \Lambda_{\mathcal{W}_1}, \circledast)$. We call this space a Menger norm space (M-N-S) if:

(1) For every $\mathcal{T} \in (0, +\infty)$ $\Lambda^{\varkappa}_{\mathcal{W}_1\mathcal{T}} = \Pi^0{}_{\mathcal{T}}$ if, and only if, $\varkappa = 0$;
(2) For every $\varkappa \in \mathcal{W}_1$ and $\mathrm{a} \in \mathbb{C}$ with $\mathrm{a} \neq 0$, $\Lambda^{\mathrm{a}\varkappa}_{\mathcal{W}_1\mathcal{T}} = \Lambda^{\varkappa}_{\mathcal{W}_1 \frac{\mathcal{T}}{|\mathrm{a}|}}$;
(3) For every $\varkappa_1, \varkappa_2 \in \mathcal{W}_1$ and $\mathcal{T}, \varsigma \in (0, +\infty)$, $\Lambda^{\varkappa_1+\varkappa_2}_{\mathcal{W}_1\ \mathcal{T}+\varsigma} \geq \Lambda^{\varkappa}_{\mathcal{W}_1\mathcal{T}} \circledast \Lambda^{\varkappa_2}_{\mathcal{W}_1\varsigma}$.

A complete M-N-S is called a Menger Banach space (M-B-S). Note $\Lambda^{\varkappa}_{\mathcal{W}_1\mathcal{T}}$ is an M-N-S where we define

$$\Lambda^{\varkappa}_{\mathcal{W}_1\mathcal{T}} = \begin{cases} 0, & \text{if } \tau \in \mathbb{R} - (0, +\infty), \\ \sum_{\mathrm{p}=0}^{\infty} \frac{(-\frac{\|\varkappa\|}{\mathcal{T}})^{\mathrm{p}}}{\Gamma(\mathrm{pb}+\mathrm{c})}, & \text{if } \tau \in (0, +\infty). \end{cases}$$

Now, we apply the concept of Z-N [2, 12] to introduce GZ-N. We consider a GZ-N

$$\mathcal{Z}_{\Omega,\Lambda} : \mathcal{W}_1 \times \mathbb{R} \to \mathcal{Y}$$

as follows

$$\mathcal{Z}_{\Omega,\Lambda}(\alpha, \mathcal{T}) = \operatorname{diag}\left[\Omega_{\mathcal{W}_1}(\alpha, \mathcal{T}), \Lambda_{\mathcal{W}_{1\alpha}}(\mathcal{T}), \Omega_{\mathcal{W}_1}(\alpha, \mathcal{T}) \circledast \Lambda_{\mathcal{W}_{1\alpha}}(\mathcal{T})\right], \tag{6.3.4}$$

where, $\mathcal{W}_1$ is a vector space. This GZ-N is a member of

$$\mathcal{Z}_{\Omega,\Lambda} = \operatorname{diag}\left[\mathcal{K}(\alpha), \mathcal{H}(\alpha), \mathcal{K}(\alpha) \circledast \mathcal{H}(\alpha)\right],$$

where $\alpha \in \mathcal{W}_1$ and $\mathcal{K}(\alpha)$ is a fuzzy norm, $\mathcal{H}(\alpha)$ is a Menger norm and $\mathcal{K}(\alpha) \circledast \mathcal{H}(\alpha)$ is a t-norm (in general, $\mathcal{K}(\alpha)$ is a fuzzy set and $\mathcal{H}(\alpha)$ is a probability distribution function; also, $\mathcal{K}(\alpha) \circledast \mathcal{H}(\alpha)$ is the degree of reliability of $\mathcal{K}(\alpha)$). Note all the properties mentioned in Definitions 1.1 and 6.11 are valid for $\mathcal{Z}_{\Omega,\Lambda}(\alpha, \mathcal{T})$. In the proofs, for the fuzzy set ϕ and the probability distribution function ψ_α, the control function $\Xi_{\phi,\psi} : \mathcal{W}_1^2 \to \mathcal{Y}$ is a matrix function given by

$$\Xi_{\phi,\psi}(\alpha, \mathcal{T}) = \operatorname{diag}\left[\phi(\alpha, \mathcal{T}), \psi_\alpha(\mathcal{T}), \phi(\alpha, \mathcal{T}) \circledast \psi_\alpha(\mathcal{T})\right], \tag{6.3.5}$$

where the fuzzy set ϕ has properties such as increasing, continuity and $\lim_{\zeta\to+\infty} \phi(\varkappa, \zeta) = 1$ for every $\zeta \in (0, +\infty)$.

6.3.2 Cauchy–Rassias Inequality of Type (6.3.2)

Theorem 6.7 *For each $\alpha, \beta, \gamma, \zeta, \vartheta \in \mathcal{W}_1$, suppose that $\Delta : \mathcal{W}_1 \to \mathcal{W}_2$ is a mapping such that*

$$\begin{aligned} &\mathcal{Z}_{\Omega,\Lambda,\mathcal{W}_2}\Big(\vartheta\Delta(\alpha) + \Delta(\beta) + \Delta(\gamma) + \Delta(\zeta), \mathcal{T}\Big) \\ &\succeq \mathcal{Z}_{\Omega,\Lambda,\mathcal{W}_2}\Big(\Delta(\vartheta\alpha + \beta) + \Delta(\gamma + \zeta)\Big), \end{aligned} \tag{6.3.6}$$

then (see [4]); Δ is linear.

Theorem 6.8 *Consider the approximately additive odd mapping (AOM) $\Delta : \mathcal{W}_1 \to \mathcal{W}_2$ on the vector space $\mathcal{W}_1$ and the complete space $\mathcal{W}_2$ with the GZ-N norm, such that*

$$\mathcal{Z}_{\Omega,\Lambda}\Big(\Delta(\alpha)+\Delta(\beta)+\Delta(\gamma)+\Delta(\zeta),\mathcal{T}\Big) \tag{6.3.7}$$
$$\succeq \mathcal{Z}_{\Omega,\Lambda}\Big(\Delta(\alpha+\beta+\gamma+\zeta),\mathcal{T}\Big)\circledast \Xi_{\phi,\psi}\Big(\alpha,\beta,\gamma,\zeta,\mathcal{T}\Big),$$

where, $\Xi_{\phi,\psi}:\mathcal{W}_1^5\to\mathcal{W}$ *is a given function and* $\alpha,\beta,\gamma,\zeta\in\mathcal{W}_1$. *If there exists a* $0<\Upsilon<1$ *such that*

$$\Xi_{\phi,\psi}\Big(\alpha,\beta,\gamma,\zeta,\mathcal{T}\Big)\succeq\Xi_{\phi,\psi}\Big(3\alpha,3\beta,3\gamma,3\zeta,\frac{3\mathcal{T}}{\Upsilon}\Big), \tag{6.3.8}$$

then, for all $\alpha\in\mathcal{W}_1$, *there exists a unique-AM* $\mathcal{J}:\mathcal{W}_1\to\mathcal{W}_2$ *satisfying*

$$\mathcal{Z}_{\Omega,\Lambda}\Big(\Delta(\alpha)-\mathcal{J}(\alpha),\mathcal{T}\Big)\succeq\Xi_{\phi,\psi}\Big(\alpha,\alpha,\alpha,-3\alpha,\frac{3\mathcal{T}-3\mathcal{T}\Upsilon}{\Upsilon}\Big). \tag{6.3.9}$$

If for every $\mathcal{T}\in(0,+\infty)$ *and* $\alpha\in\mathcal{W}_1$, $\Delta(\mathcal{T}\alpha)$ *is continuous, then* $\mathcal{J}$ *is* $\mathbb{R}$*-linear.*

Proof Considering the characteristic of the odd mapping Δ, for all $\alpha\in\mathcal{W}_1$; we have

$$\Delta(0)=0,\qquad \Delta(-\alpha)=-\Delta(\alpha).$$

Now, we define the generalized metric by considering the function $\Xi_{\phi,\psi}$ and introducing the set $\mathcal{I}:=\{\Phi:\mathcal{W}_1\to\mathcal{W}_2\}$ as

$$\delta_{\mathcal{W}_1}(\Phi,\Psi)=\inf\Big\{\varrho\in\mathbb{R}_+:\mathcal{Z}_{\Omega,\Lambda}\Big(\Phi(\alpha)-\Psi(\alpha),\mathcal{T}\Big)\succeq\Xi_{\phi,\psi}\Big(\alpha,\alpha,\alpha,-3\alpha,\frac{\mathcal{T}}{\varrho}\Big),\ \forall\alpha\in\mathcal{W}_1\Big\}.$$

First, we prove that $(\mathcal{I},\delta_{\mathcal{W}_1})$ is a complete space. If $\{\Psi_n\}$ is a Cauchy sequence, then; for every $\upsilon>0$ there exists an integer $N_\upsilon>0$ such that for all $m,n\geq\mathcal{O}_\upsilon$, we have $\delta_{\mathcal{W}_1}(\Psi_m,\Psi_n)<\upsilon$. Next, for $\alpha\in\mathcal{W}_1$, there exists $\varrho\in(0,\upsilon)$ such that

$$\mathcal{Z}_{\Omega,\Lambda}\Big(\Psi_m(\alpha)-\Psi_n(\alpha),\mathcal{T}\Big)\succeq\Xi_{\phi,\psi}\Big(\alpha,\alpha,\alpha,-3\alpha,\frac{\mathcal{T}}{\varrho}\Big) \tag{6.3.10}$$
$$\succeq\Xi_{\phi,\psi}\Big(\alpha,\alpha,\alpha,-3\alpha,\frac{\mathcal{T}}{\upsilon}\Big).$$

Due to the completeness of $\mathcal{W}_2$, the sequence $\{\Psi_n(\alpha)\}$ converges to every $\alpha\in\mathcal{W}_1$, then, for each $\alpha\in\mathcal{W}_1$ the mapping $\Psi:\mathcal{W}_1\to\mathcal{W}_2$ is defined as $\Psi(\alpha):=\lim_{n\to\infty}\Psi_n(\alpha)$. Considering (6.3.10) when $n\to\infty$, for all $\alpha\in\mathcal{W}_1$ and $m\geq\mathcal{O}_\upsilon$, we have

$$\mathcal{Z}_{\Omega,\Lambda}\Big(\Psi_m(\alpha)-\Psi_n(\alpha),\mathcal{T}\Big)\succeq\Xi_{\phi,\psi}\Big(\alpha,\alpha,\alpha,-3\alpha,b\frac{\mathcal{T}}{\upsilon}\Big),$$

and

$$\upsilon \in \left\{ \varrho \in \mathbb{R}_+ : \mathcal{Z}_{\Omega,\Lambda}\Big(\Phi(\alpha) - \Psi(\alpha), \mathcal{T}\Big) \succeq \Xi_{\phi,\psi}\Big(\alpha, \alpha, \alpha, -3\alpha, \frac{\mathcal{T}}{\varrho}\Big), \ \forall \alpha \in \mathcal{W}_1 \right\},$$

and, as a result, $\delta_{\mathcal{W}_1}(\Psi_m, \Psi) \leq \upsilon$. That is, $\{\Psi_n\}$ in $(\mathcal{I}, \delta_{\mathcal{W}_1})$ converges to Ψ and the completeness of $(\mathcal{I}, \delta_{\mathcal{W}_1})$ is proved. Next, consider $\mathcal{X} : \mathcal{I} \to \mathcal{I}$ such that $\Lambda\Phi(\alpha) := 3\Phi\left(\frac{\alpha}{3}\right)$ for all $\alpha \in \mathcal{W}_1$. We show that $\mathcal{X}$ on $\mathcal{I}$ is strictly contractive. For this purpose, we consider $\varrho_{\Phi,\Psi} \geq 0$ for every $\Phi, \Psi \in \mathcal{I}$, such that $\delta_{\mathcal{W}_1}(\Phi, \Psi) \leq \varrho_{\Phi,\Psi}$, then for all $\alpha \in \mathcal{W}_1$, we get

$$\begin{aligned}
\delta_{\mathcal{W}_1}(\Phi, \Psi) \leq \varrho_{\Phi,\Psi} &\Longrightarrow \mathcal{Z}_{\Omega,\Lambda}\Big(\Phi(\alpha) - \Psi(\alpha), \mathcal{T}\Big) \succeq \Xi_{\phi,\psi}\Big(\alpha, \alpha, \alpha, -3\alpha, \frac{\mathcal{T}}{C_{\Phi,\Psi}}\Big) \\
&\Longrightarrow \mathcal{Z}_{\Omega,\Lambda}\Big(3\Phi\left(\frac{\alpha}{3}\right) - 3\Psi\left(\frac{\alpha}{3}\right), \mathcal{T}\Big) \succeq \Xi_{\phi,\psi}\left(\frac{\alpha}{3}, \frac{\alpha}{3}, \frac{\alpha}{3}, -\alpha, \frac{\mathcal{T}}{3\varrho_{\Phi,\Psi}}\right) \\
&\Longrightarrow \mathcal{Z}_{\Omega,\Lambda}\Big(3\Phi\left(\frac{\alpha}{3}\right) - 3\Psi\left(\frac{\alpha}{3}\right), \mathcal{T}\Big) \succeq \Xi_{\phi,\psi}\Big(\alpha, \alpha, \alpha, -3\alpha, \frac{\mathcal{T}}{\Upsilon\varrho_{\Phi,\Psi}}\Big),
\end{aligned}$$

or, $\delta_{\mathcal{W}_1}(\mathcal{X}\Phi, \mathcal{X}\Psi) \leq \Upsilon\varrho_{\Phi,\Psi}$. On the other hand, for every $\Phi, \Psi \in \mathcal{I}$, we have $\delta_{\mathcal{W}_1}(\mathcal{X}\Phi, \mathcal{X}\Psi) \leq \Upsilon\delta_{\mathcal{W}_1}(\Phi, \Psi)$. Therefore, Λ on $\mathcal{I}$ with the Lipschitz constant $0 < \Upsilon < 1$ is strictly contractive. Now, considering (6.3.7) and setting $\beta = \gamma = \alpha$ and $\zeta = -3\alpha$; for all $\alpha \in \mathcal{W}_1$, we have

$$\mathcal{Z}_{\Omega,\Lambda}\Big(3\Delta(\alpha) - \Delta(3\alpha), \mathcal{T}\Big) \succeq \Xi_{\phi,\psi}\Big(\alpha, \alpha, \alpha, -3\alpha, \mathcal{T}\Big). \tag{6.3.11}$$

For all $\alpha \in \mathcal{W}_1$,

$$\mathcal{Z}_{\Omega,\Lambda}\left(\Delta(\alpha) - 3\Delta\left(\frac{\alpha}{3}\right), \mathcal{T}\right) \succeq \Xi_{\phi,\psi}\left(\frac{\alpha}{3}, \frac{\alpha}{3}, \frac{\alpha}{3}, -\alpha, \mathcal{T}\right) \succeq \Xi_{\phi,\psi}\Big(\alpha, \alpha, \alpha, -3\alpha, \frac{3\mathcal{T}}{\Upsilon}\Big).$$

Thus, $\delta_{\mathcal{W}_1}(\Delta, \mathcal{X}\Delta) \leq \frac{\Upsilon}{3}$ and, according to Theorem 1.2, $\{\mathcal{X}^s\Delta\}$ converges to a unique fixed point $\mathcal{J}$ of $\mathcal{X}$ or, equivalently,

$$\mathcal{J}(\alpha) = (\mathcal{X}\Delta)(\alpha) = \lim_{s\to\infty} 3^s\Delta\left(\frac{\alpha}{3^s}\right),$$

and for all $\alpha \in \mathcal{W}_1$, we get $\mathcal{J}(3\alpha) = 3\mathcal{J}(\alpha)$. Also,

$$\delta_{\mathcal{W}_1}(\mathcal{J}, \Delta) \leq \frac{1}{1-\Upsilon}\delta_{\mathcal{W}_1}(\mathcal{X}\Delta, \Delta) \leq \frac{\Upsilon}{3-3\Upsilon},$$

so (6.3.9) holds. If for every $\mathcal{T} \in (0, +\infty)$ and $\alpha \in \mathcal{W}_1$, $\Delta(\mathcal{T}\alpha)$ is continuous, then Theorem 6.7 guarantees that $\mathcal{J}$ is an $\mathbb{R}$-linear mapping.

6.3.3 Applications

First, we consider the function $\Xi_{\phi,\psi}$ as

$$\Xi_{\phi,\psi}(\alpha,\mathcal{T}) = \tag{6.3.12}$$
$$\mathrm{diag}\Bigg[\frac{\Gamma(\gamma)}{\Gamma(\alpha)\Gamma(\beta)}\sum_{p=0}^{\infty}\frac{\Gamma(\alpha+p)\Gamma(\beta+p)}{\Gamma(\gamma+p)}\frac{(\frac{-\|\alpha\|}{\mathcal{T}})^{p}}{p!},$$
$$\begin{cases} 0, & \text{if } \mathcal{T}\in\mathbb{R}-(0,+\infty),\\ \sum_{p=0}^{\infty}\frac{(-\frac{\|\alpha\|}{\mathcal{T}})^{p}}{\Gamma(pb+c)}, & \text{if } \mathcal{T}\in(0,+\infty)\end{cases}\Bigg],$$

and for all examples, $\mathcal{W}_1$ is a vector space and $\mathcal{W}_2$ is a complete space with the GZ-N norm.

Example 6.4 For every $\alpha,\beta,\gamma,\zeta\in\mathcal{W}_1$ and all $g\in\mathbb{T}:=\{d\in\mathbb{C}\mid |d|=1\}$, suppose that $\Delta:\mathcal{W}_1\to\mathcal{W}_2$ is a multiplicative mapping such that

$$\mathcal{Z}_{\Omega,\Lambda}\Big(g\Delta(\alpha)+\Delta(\beta)+\Delta(\gamma)+\Delta(\zeta),\mathcal{T}\Big) \tag{6.3.13}$$
$$\succeq \mathcal{Z}_{\Omega,\Lambda}\Big(\Delta(g\alpha+\beta)+\Delta(\gamma+\zeta),\mathcal{T}\Big).$$

Thus, the mapping Δ is an A-homomorphism.

Proof Using Theorem 6.7, we conclude that the multiplicative mapping Δ is linear and it follows that Δ is an A-homomorphism.

Example 6.5 Consider the approximately-AOM $\Delta:\mathcal{W}_1\to\mathcal{W}_2$, such that for every $m\in\mathbb{R}$ and $4m\neq 1$, we have

$$\mathcal{Z}_{\Omega,\Lambda}\Big(\Delta(\alpha)+\Delta(\beta)+\Delta(\gamma)+\Delta(\zeta),\mathcal{T}\Big) \tag{6.3.14}$$
$$\succeq \mathcal{Z}_{\Omega,\Lambda}\Big(\Delta(\alpha+\beta+\gamma+\zeta),\mathcal{T}\Big)\circledast$$
$$\mathcal{Z}_{\Omega,\Lambda}\Big(||\alpha||^{m}||\beta||^{m}||\gamma||^{m}||\zeta||^{m},\frac{\mathcal{T}}{\sigma}\Big),$$

in which $\sigma\geq 0$. For all $\alpha\in\mathcal{W}_1$, we can find a unique AM $\mathcal{J}:\mathcal{W}_1\to\mathcal{W}_2$ satisfying

$$\mathcal{Z}_{\Omega,\Lambda}\Big(\Delta(\alpha)-\Upsilon(\alpha),\mathcal{T}\Big)\succeq\mathcal{Z}_{\Omega,\Lambda}\Big(||\alpha||^{4m},\frac{|81^m-3|\mathcal{T}}{3^{m-1}\sigma}\Big). \tag{6.3.15}$$

If for every $\mathcal{T}\in(0,+\infty)$ and $\alpha\in\mathcal{W}_1$, $\Delta(\mathcal{T}\alpha)$ is continuous, then $\mathcal{J}$ is $\mathbb{R}$-linear.

Proof We consider Theorem 6.8 and define the function $\Xi_{\phi,\psi}:\mathcal{W}_1^5\to\mathcal{W}$ as

$$\Xi_{\phi,\psi}\Big(\alpha,\beta,\gamma,\zeta,\mathcal{T}\Big)= \tag{6.3.16}$$

$$\mathrm{diag}\left[\frac{\Gamma(\gamma)}{\Gamma(\alpha)\Gamma(\beta)}\sum_{\mathrm{p}=0}^{\infty}\frac{\Gamma(\alpha+\mathrm{p})\Gamma(\beta+\mathrm{p})}{\Gamma(\gamma+\mathrm{p})}\frac{(\frac{\sigma||\alpha||^m||\beta||^m||\gamma||^m||\zeta||^m}{\mathcal{T}})^{\mathrm{p}}}{\mathrm{p}!},\right.$$

$$\left.\begin{cases}0, & \text{if }\mathcal{T}\in\mathbb{R}-(0,+\infty),\\ \sum_{\mathrm{p}=0}^{\infty}\frac{(-\frac{\sigma||\alpha||^m||\beta||^m||\gamma||^m||\zeta||^m}{\mathcal{T}})^{\mathrm{p}}}{\Gamma(\mathrm{pb+c})}, & \text{if }\mathcal{T}\in(0,+\infty)\end{cases}\right].$$

Now, by considering the function $\Xi_{\phi,\psi}\Big(\alpha,\beta,\gamma,\zeta,\mathcal{T}\Big)$ defined above in Theorem 6.8; and choosing $\Upsilon=81^{1-m}$; for all $\alpha,\beta,\gamma,\zeta\in\mathcal{W}_1$, the result follows.

Example 6.6 We consider the approximately-AOM $\Delta:\mathcal{W}_1\to\mathcal{W}_2$, such that for every $m\in\mathbb{R}$ and $4m\neq1$, we have

$$\begin{aligned}&\mathcal{Z}_{\Omega,\Lambda}\Big(\Delta(\alpha)+\Delta(\beta)+\Delta(\gamma)+\Delta(\zeta),\mathcal{T}\Big)\\&\succeq\mathcal{Z}_{\Omega,\Lambda}\Big(\Delta(\alpha+\beta+\gamma+\zeta),\mathcal{T}\Big)\\&\circledast\mathcal{Z}_{\Omega,\Lambda}\Big((||\alpha||^m+||\beta||^m+||\gamma||^m+||\zeta||^m),\frac{\mathcal{T}}{\sigma}\Big),\end{aligned} \tag{6.3.17}$$

in which $\sigma\geq0$. For all $\alpha\in\mathcal{W}_1$, there exists a unique A-M $\mathcal{J}:\mathcal{W}_1\to\mathcal{W}_2$ satisfying

$$\mathcal{Z}_{\Omega,\Lambda}\Big(\Delta(\alpha)-\mathcal{J}(\alpha),\mathcal{T}\Big)\succeq\mathcal{Z}_{\Omega,\Lambda}\Big(||\alpha||^m,\frac{|3^m-3|\mathcal{T}}{(3^m+3)\sigma}\Big). \tag{6.3.18}$$

If for every $\mathcal{T}\in(0,+\infty)$ and $\alpha\in\mathcal{W}_1$, $\Delta(\mathcal{T}\alpha)$ is continuous, then $\mathcal{J}$ is an $\mathbb{R}$-linear mapping.

Proof We consider Theorem 6.8 and define the function $\Xi_{\phi,\psi} : \mathcal{W}_1^5 \to \mathcal{Y}$ as

$$\Xi_{\phi,\psi}\Bigg(\alpha, \beta, \gamma, \zeta, \mathcal{T}\Bigg) = \tag{6.3.19}$$

$$\mathrm{diag}\Bigg[\frac{\Gamma(\gamma)}{\Gamma(\alpha)\Gamma(\beta)}\sum_{\mathrm{p}=0}^{\infty}\frac{\Gamma(\alpha+\mathrm{p})\Gamma(\beta+\mathrm{p})}{\Gamma(\gamma+\mathrm{p})}\frac{(\frac{\sigma\|\alpha\|^m+\|\beta\|^m+\|\gamma\|^m+\|\zeta\|^m}{\mathcal{T}})^{\mathrm{p}}}{\mathrm{p}!},$$

$$\begin{cases} 0, & \text{if } \mathcal{T} \in \mathbb{R} - (0, +\infty), \\ \sum_{\mathrm{p}=0}^{\infty} \frac{(-\frac{\sigma\|\alpha\|^m+\|\beta\|^m+\|\gamma\|^m+\|\zeta\|^m}{\mathcal{T}})^{\mathrm{p}}}{\Gamma(\mathrm{pb+c})}, & \text{if } \mathcal{T} \in (0, +\infty) \end{cases}\Bigg].$$

Now, by considering the function $\Xi_{\phi,\psi}\Bigg(\alpha, \beta, \gamma, \zeta, \mathcal{T}\Bigg)$ defined above in Theorem 6.8; and choosing $\Upsilon = 3^{1-m}$; for all $\alpha, \beta, \gamma, \zeta \in \mathcal{W}_1$, the result follows.

6.4 ∗-Generalized-$\mathcal{Z}$-Number-Bounded-Borel Measures

There are many problems in the real world where there is a lot of uncertainty and the information is somewhat reliable. We can address this aspect of real problems by mathematically formalizing many scientific and engineering problems. Until now, many theories such as fuzzy functions, functions of random variables, and functions with distance values have been worked to reduce uncertainty and investigate these kinds of problems. But these theories examine only one type of uncertainty, and this is what made researchers seek to expand this topic to address real-world issues. This issue started with the official introduction of $\mathcal{G}$-$\mathcal{Z}$-$\mathcal{N}$. Researchers have done a lot of research on these numbers so far. In [9, 20, 30], the parametric form of these numbers and linear regression based on $\mathcal{Z}$-$\mathcal{N}$s have been examined. References [3, 13, 28] provide more information about $\mathcal{Z}$-$\mathcal{N}$s.

We consider a ∗-$\mathcal{G}$-$\mathcal{Z}$-$\mathcal{N}$-Bounded Borel measure (∗-$\mathcal{G}$-$\mathcal{Z}$-$\mathcal{N}$-B-B-M) $\mathcal{Z}_\mu$ on $\mathbb{R}$ and for each $z \in \mathbb{R}$, we define

$$\mathcal{Z}_{\mathcal{F}}(z, t) = \mathcal{Z}_\mu((-\infty, z], t),$$

that the following characteristics are valid for $\mathcal{Z}_{\mathcal{F}}$:

- Using definition of $\mathcal{Z}_\mu$ shows that $\mathcal{Z}_{\mathcal{F}}$ is decreasing.
- $\mathcal{Z}_{\mathcal{F}}$ is right continuous, because $(-\infty, z] = \bigcap_{n=1}^{+\infty}(-\infty, z_n]$ whenever $z_n \downarrow z$.
- If $b > a$ and $(-\infty, b] = (-\infty, a] \cup (a, b]$ we have $\mathcal{Z}_\mu((-\infty, b], t) = \mathcal{Z}_\mu((-\infty, a] \cup (a, b], t) = \mathcal{Z}_\mu((-\infty, a], t) \times \mathcal{Z}_\mu((a, b], t)$, then $\mathcal{Z}_{\mathcal{F}}(b, t) = \mathcal{Z}_{\mathcal{F}}(a, t) \times \mathcal{Z}_\mu((a, b], t)$, hence $0 \leq \mathcal{Z}_\mu((a, b], t) = \dfrac{\mathcal{Z}_{\mathcal{F}}(b, t)}{\mathcal{Z}_{\mathcal{F}}(a, t)} \leq 1$.

We will be to construct a $*$-$\mathcal{G}$-$\mathcal{Z}$-$\mathcal{N}$ $\mathcal{Z}_\mu$ starting from a decreasing, right-continuous function $\mathcal{Z}_\mathcal{F}$. The structural components for our theory will be the left-open, right-close intervals in $\mathbb{R}$, we show that, sets of the form $(a, b]$ or $(a, +\infty)$ or $\emptyset$, where $-\infty \leq a < b < +\infty$. Clearly if we consider $(a, b]$ and $(c, d]$ then $(a, b] \cap (c, d]$ is an h-interval, since $(a, b]^c$ is an h-interval or the disjoint union of two h-intervals. In the classic measure theory the collection $\mathcal{A}$ of finite disjoint unions of h-intervals is an algebra and the σ-algebra generated by $\mathcal{A}$ is $\mathcal{B}_\mathbb{R}$.

Theorem 6.9 *We consider $\mathcal{Z}_\mathcal{F} : \mathcal{W}_1 \times (0, +\infty) \to \mathcal{Y}$ such that it is descending and continuous from the right. We choose intervals $(a_\ell, b_\ell]$, for $\ell = 1, ..., k$ and define*

$$\mathcal{Z}_{\mu_0}\left(\bigcup_{\ell=1}^{k}(a_\ell, b_\ell], t\right) = \prod_{\ell=1}^{k} \frac{\mathcal{Z}_\mathcal{F}(b_\ell, t)}{\mathcal{Z}_\mathcal{F}(a_\ell, t)}$$

and let $\mathcal{Z}_{\mu_0}(\phi, t) = 1$, then $\mathcal{Z}_{\mu_0}$ is a $*$-fuzzy premeasure on the algebra $\mathcal{A}$. What is meant by the product of the above multiplication is the multiplication of the first components, the second components, the third components, etc., up to the Kth components.

Proof Given that $E \in \mathcal{A}$ can be represented in various ways as disjoint unions of h intervals, the proof of this theorem begins by showing that $\mathcal{Z}_{\mu_0}$ is well-defined. If $\{(a_\ell, b_\ell]\}_{\ell=1}^{k}$ are disjoint and $(a, b] = \bigcup_{\ell=1}^{k}(a_\ell, b_\ell]$ then, after perhaps relabeling the index ℓ, we must have $a = a_1 < b_1 = a_2 < b_2 = ... < b_k = b$, then

$$\begin{aligned}
\prod_{\ell=1}^{k} \frac{\mathcal{Z}_\mathcal{F}(b_\ell, t)}{\mathcal{Z}_\mathcal{F}(a_\ell, t)} &= \frac{\mathcal{Z}_\mathcal{F}(b_1, t)}{\mathcal{Z}_\mathcal{F}(a_1, t)} \times \frac{\mathcal{Z}_\mathcal{F}(b_2, t)}{\mathcal{Z}_\mathcal{F}(a_2, t)} \times \cdots \times \frac{\mathcal{Z}_\mathcal{F}(b_k, t)}{\mathcal{Z}_\mathcal{F}(a_k, t)} \\
&= \frac{\mathcal{Z}_\mathcal{F}(b_1, t)}{\mathcal{Z}_\mathcal{F}(a, t)} \times \frac{\mathcal{Z}_\mathcal{F}(b_2, t)}{\mathcal{Z}_\mathcal{F}(b_1, t)} \times \cdots \times \frac{\mathcal{Z}_\mathcal{F}(b, t)}{\mathcal{Z}_\mathcal{F}(b_{k-1}, t)} \\
&= \frac{\mathcal{Z}_\mathcal{F}(b, t)}{\mathcal{Z}_\mathcal{F}(a, t)} < 1.
\end{aligned}$$

We also consider two finite sequences of disjoint h-intervals and denote them by $\{I_\ell\}_{\ell=1}^{n}$ and $\{J_\jmath\}_{\jmath=1}^{m}$. In these two sequences, we have $\bigcup_{\ell=1}^{n} I_\ell = \bigcup_{\jmath=1}^{m} J_\jmath$. Therefore

$$\begin{aligned}
\mathcal{Z}_{\mu_0}\left(\bigcup_{\ell=1}^{n} I_\ell, t\right) &= \prod_{\ell=1}^{n} \mathcal{Z}_{\mu_0}(I_\ell, t) = \prod_{\ell=1}^{n} \mathcal{Z}_{\mu_0}\left(\bigcup_{\jmath=1}^{m}(I_\ell \cap I_\jmath), t\right) = \prod_{\ell=1}^{n}\prod_{\jmath=1}^{m} \mathcal{Z}_{\mu_0}(I_\ell \cap I_\jmath, t) \\
&= \prod_{\jmath=1}^{m}\prod_{\ell=1}^{n} \mathcal{Z}_{\mu_0}(I_\ell \cap I_\jmath, t) = \prod_{\jmath=1}^{m} \mathcal{Z}_{\mu_0}\left(\bigcup_{\ell=1}^{n}(I_\ell \cap I_\jmath), t\right) = \prod_{\jmath=1}^{m} \mathcal{Z}_{\mu_0}(I_\jmath, t) = \mathcal{Z}_{\mu_0}\left(\bigcup_{\jmath=1}^{m} I_\jmath, t\right).
\end{aligned}$$

Thus $\mathcal{Z}_{\mu_0}$ is well defined, clearly, it is finitely additive. To continue the proof, we consider $\{I_\jmath\}_{\jmath=1}^{+\infty}$ with the following properties

- $I_\jmath \cap I_\ell = \emptyset$, $\jmath \neq \ell$;
- I_ℓ's are h-intervals with $\bigcup_{\jmath=1}^{+\infty} I_\jmath \in \mathcal{A}$.

Then, $\mathcal{Z}_{\mu_0}\left(\bigcup_{\jmath=1}^{+\infty} I_\jmath, t\right) = \prod_{\jmath=1}^{+\infty} \mathcal{Z}_{\mu_0}(I_\jmath, t)$. On the other hand, given that $\bigcup_{\jmath=1}^{+\infty} I_\jmath \in \mathcal{A}$, then $\bigcup_{\jmath=1}^{+\infty} I_\jmath = \bigcup_{\kappa=1}^{\rho} J_\kappa$ where J_κ's be h-intervals. By considering $\{J_\kappa\}_{\kappa=1}^{\rho}$ and using the finite additivity of $\mathcal{Z}_{\mu_0}$, we have that $\bigcup_{\jmath=1}^{+\infty} I_\jmath$ is an h-interval $I = (a, b]$, then

$$\begin{aligned}
\mathcal{Z}_{\mu_0}(I, t) = \mathcal{Z}_{\mu_0}((a, b], t) &= \mathcal{Z}_{\mu_0}\left(\left(\bigcup_{\jmath=1}^{n} I_\jmath\right) \bigcup \left(I - \bigcup_{\jmath=1}^{n} I_\jmath\right), t\right) \\
&= \mathcal{Z}_{\mu_0}\left(\bigcup_{\jmath=1}^{n} I_\jmath, t\right) \times \mathcal{Z}_{\mu_0}\left(\left(I - \bigcup_{\jmath=1}^{n} I_\jmath\right), t\right) \\
&\leq \mathcal{Z}_{\mu_0}\left(\bigcup_{\jmath=1}^{n} I_\jmath, t\right) = \prod_{\jmath=1}^{n} \mathcal{Z}_{\mu_0}(I_\jmath, t).
\end{aligned}$$

If n tends to $+\infty$, then

$$\mathcal{Z}_{\mu_0}(I, t) \leq \prod_{\jmath=1}^{+\infty} \mathcal{Z}_{\mu_0}(I_\jmath, t). \tag{6.4.1}$$

Reciprocally, it is initially assumed that $0 < \varepsilon < 1$ and a and b are finite. Considering that $\mathcal{Z}_{\mathcal{F}}$ is right continuous, then for $0 < \varepsilon < 1$ there exists $\delta > 0$ such that $\mathcal{Z}_{\mathcal{F}}(a, t) - \mathcal{Z}_{\mathcal{F}}(a + \delta, t) < \varepsilon$. Since for every $I_\jmath(a_\jmath, b_\jmath]$ there exist $\delta_\jmath > 0$ such that $\mathcal{Z}_{\mathcal{F}}(b_\jmath, t) - \mathcal{Z}_{\mathcal{F}}(b_\jmath + \delta_\jmath, t) < \varepsilon$. The sequence $\{(a_\jmath, b_\jmath + \delta_\jmath]\}_{\jmath=1}^{+\infty}$ cover the set $[a + \delta, b]$, according to property of compact sets, there is $\{(a_\jmath, b_\jmath + \delta_\jmath]\}_{\jmath=1}^{\kappa}$ such that $[a + \delta, b] \subseteq \bigcup_{\jmath=1}^{\kappa} (a_\jmath, b_\jmath + \delta_\jmath]$, then $b_\jmath + \delta_\jmath \in (a_{\jmath+1}, b_{\jmath+1} + \delta_{\jmath+1})$, $\jmath = 1, \ldots, \kappa - 1$ and $a_1 < a_2 < \cdots < a_\kappa$. But, then

$$\begin{aligned}
\mathcal{Z}_{\mu_0}(I, t) = \mathcal{Z}_{\mu_0}((a, b], t) &= \frac{\mathcal{Z}_{\mathcal{F}}(b, t)}{\mathcal{Z}_{\mathcal{F}}(a, t)} \\
&\geq \frac{\mathcal{Z}_{\mathcal{F}}(b, t)}{\mathcal{Z}_{\mathcal{F}}(a + \delta, t) \circledast \mathcal{Z}(\varepsilon, t)} \\
&\geq \frac{\mathcal{Z}_{\mathcal{F}}(b_\kappa + \delta_\kappa, t)}{\mathcal{Z}_{\mathcal{F}}(a_1, t) \circledast \mathcal{Z}(\varepsilon, t)}
\end{aligned}$$

$$
\begin{aligned}
&= \frac{\mathcal{Z}_{\mathcal{F}}(b_\kappa + \delta_\kappa, t)}{\mathcal{Z}_{\mathcal{F}}(a_\kappa, t)} \times \frac{\mathcal{Z}_{\mathcal{F}}(a_\kappa, t)}{\mathcal{Z}_{\mathcal{F}}(a_{\kappa-1}, t)} \times \cdots \times \frac{\mathcal{Z}_{\mathcal{F}}(a_2, t)}{\mathcal{Z}_{\mathcal{F}}(a_1, t) \circledast \mathcal{Z}(\varepsilon, t)} \\
&\geq \frac{\mathcal{Z}_{\mathcal{F}}(b_\kappa + \delta_\kappa, t)}{\mathcal{Z}_{\mathcal{F}}(a_\kappa, t)} \times \frac{\mathcal{Z}_{\mathcal{F}}(b_{\kappa-1} + \delta_{\kappa-1}, t)}{\mathcal{Z}_{\mathcal{F}}(a_{\kappa-1}, t)} \times \cdots \times \frac{\mathcal{Z}_{\mathcal{F}}(b_1 + \delta_1, t)}{\mathcal{Z}_{\mathcal{F}}(a_1, t) \circledast \mathcal{Z}(\varepsilon, t)} \\
&\geq \frac{\mathcal{Z}_{\mathcal{F}}(b_\kappa + \delta_\kappa, t)}{\mathcal{Z}_{\mathcal{F}}(b_\kappa, t)} \times \frac{\mathcal{Z}_{\mathcal{F}}(b_\kappa, t)}{\mathcal{Z}_{\mathcal{F}}(a_\kappa, t)} \times \cdots \times \frac{\mathcal{Z}_{\mathcal{F}}(b_1 + \delta_1, t)}{\mathcal{Z}_{\mathcal{F}}(b_1, t)} \times \frac{\mathcal{Z}_{\mathcal{F}}(b_1, t)}{\mathcal{Z}_{\mathcal{F}}(a_1, t) \circledast \mathcal{Z}(\varepsilon, t)} \\
&\geq \frac{\mathcal{Z}_{\mathcal{F}}(b_\kappa + \delta_\kappa, t)}{\mathcal{Z}_{\mathcal{F}}(b_\kappa + \delta_\kappa, t) \circledast \mathcal{Z}(\varepsilon, t)} \times \mathcal{Z}_{\mu_0}((a_\kappa, b_\kappa], t) \times \cdots \times \frac{\mathcal{Z}_{\mathcal{F}}(b_1 + \delta_1, t)}{\mathcal{Z}_{\mathcal{F}}(b_1 + \delta_1, t) + \varepsilon} \\
&\times \frac{\mathcal{Z}_{\mathcal{F}}(b_1, t)}{\mathcal{Z}_{\mathcal{F}}(a_1, t) \circledast \mathcal{Z}(\varepsilon, t)} \to 1 \times \mathcal{Z}_{\mu_0}((a_\kappa, b_\kappa], t) \times \cdots \times \mathcal{Z}_{\mu_0}((a_1, b_1], t) \\
&= \prod_{\ell=1}^{\kappa} \mathcal{Z}_{\mu_0}((a_\ell, b_\ell], t) \\
&\geq \prod_{\ell=1}^{+\infty} \mathcal{Z}_{\mu_0}((a_\ell, b_\ell], t),
\end{aligned}
$$

we obtain

$$
\mathcal{Z}_{\mu_0}(I, t) \geq \prod_{\ell=1}^{+\infty} \mathcal{Z}_{\mu_0}((a_\ell, b_\ell], t). \tag{6.4.2}
$$

Considering (6.4.1), (6.4.2), we get $\mathcal{Z}_{\mu_0}(I, t) = \mathcal{Z}_{\mu_0}\left(\bigcup_{\ell=1}^{+\infty} I_\ell, t\right) = \prod_{\ell=1}^{+\infty} \mathcal{Z}_{u_0}((a_\ell, b_\ell], t)$. For each $\mathcal{M} > 0$, we have the following:

- If $a = -\infty$, $\{(a_\jmath, b_\jmath + \delta_\jmath)\}_{\jmath=1}^{+\infty}$ cover $[-\mathcal{M}, b]$, then we obtain

$$
\frac{\mathcal{Z}_{\mathcal{F}}(b, t)}{\mathcal{Z}_{\mathcal{F}}(-\mathcal{M}, t)} \geq \prod_{\ell=1}^{+\infty} \mathcal{Z}_{\mu_0}((a_\ell, b_\ell], t).
$$

- If $b = +\infty$, we get

$$
\frac{\mathcal{Z}_{\mathcal{F}}(\mathcal{M}, t)}{\mathcal{Z}_{\mathcal{F}}(a, t)} \geq \prod_{\ell=1}^{+\infty} \mathcal{Z}_{\mu_0}((a_\ell, b_\ell], t).
$$

When $\mathcal{M} \to +\infty$, the desired result is obtained.

Theorem 6.10 *If we consider the decreasing and right continuous function* $\mathcal{Z}_{\mathcal{F}} : \mathcal{W}_1 \times (0, +\infty) \to \mathcal{Y}$, *then on* $\mathcal{B}_{\mathbb{R}}$, *there is a* $*$-$\mathcal{G}$-$\mathcal{Z}$-$\mathcal{N}$-$\mathcal{B}$-$\mathcal{M}$ *as follows*

$$
\mathcal{Z}_{\mu_{\mathcal{F}}}((a, b], t) = \frac{\mathcal{Z}_{\mathcal{F}}(b, t)}{\mathcal{Z}_{\mathcal{F}}(a, t)},
$$

for every $a, b \in \mathbb{R}$. If $\mathcal{G}$ is another such function, we have $\mathcal{Z}_{\mu_{\mathcal{F}}} = \mathcal{Z}_{\mu_{\mathcal{G}}}$ iff $\frac{\mathcal{G}}{\mathcal{F}}$ is constant. Conversely, if $\mathcal{Z}_\mu$ is a $*$-$\mathcal{G}$-$\mathcal{Z}$-$\mathcal{N}$-B-M on $\mathbb{R}$ such that $\mathcal{Z}_\mu$ is finite on every bounded Borel sets, since

$$\mathcal{Z}_{\mathcal{F}}(z,t) = \begin{cases} \mathcal{Z}_{\mu((0,z],t)}, & \text{if } z > 0, \\ 1, & \text{if } z = 0, \\ \mathcal{Z}_{\mu((z,0],t)}, & \text{if } z < 0, \end{cases}$$

then $\mathcal{Z}_{\mathcal{F}}$ is decreasing and right continuous, and $\mathcal{Z}_{\mu_{\mathcal{F}}} = \mathcal{Z}_\mu$.

Proof Considering Theorem 6.9 $\mathcal{Z}_{\mathcal{F}}$ is a $*$-$\mathcal{G}$-$\mathcal{Z}$-$\mathcal{N}$-premeasure on $\mathcal{A}$. If $\frac{\mathcal{Z}_{\mathcal{G}}}{\mathcal{Z}_{\mathcal{F}}}$ is constant, it is clear that $\mathcal{Z}_{\mathcal{F}}$ and $\mathcal{Z}_{\mathcal{G}}$ are the same $*$-$\mathcal{G}$-$\mathcal{Z}$-$\mathcal{N}$-premeasure. In this regard, if $\frac{\mathcal{Z}_{\mathcal{G}}(z,t)}{\mathcal{Z}_{\mathcal{F}}(z,t)} = c$, we have

$$\mathcal{Z}_{\mu_{\mathcal{G}}}((a,b],t) = \frac{\mathcal{Z}_{\mathcal{G}}(b,t)}{\mathcal{Z}_{\mathcal{G}}(a,t)} = \frac{c \cdot \mathcal{Z}_{\mathcal{F}}(b,t)}{c \cdot \mathcal{Z}_{\mathcal{F}}(a,t)} = \frac{\mathcal{Z}_{\mathcal{F}}(b,t)}{\mathcal{Z}_{\mathcal{F}(a,t)}} = \mathcal{Z}_{\mu_{\mathcal{F}}}((a,b],t),$$

then, $\mathcal{Z}_{\mu_{\mathcal{F}}} = \mathcal{Z}_{\mu_G}$ on $\mathcal{A}$. Conversely, if $\mathcal{Z}_{\mu_{\mathcal{G}}} = \mathcal{Z}_{\mu_{\mathcal{F}}}$ on $\mathcal{A}$. Then

$$\mathcal{Z}_{\mu_{\mathcal{G}}}((0,z],t) = \mathcal{Z}_{\mu_{\mathcal{F}}}((0,z],t) \text{ for every } z \geq 0,$$

hence $\dfrac{\mathcal{Z}_{\mathcal{G}}(z,t)}{\mathcal{Z}_{\mathcal{G}}(0,t)} = \dfrac{\mathcal{Z}_{\mathcal{F}}(z,t)}{\mathcal{Z}_{\mathcal{F}}(0,t)}$, therefore $\dfrac{\mathcal{Z}_{\mathcal{F}}(z,t)}{\mathcal{Z}_{\mathcal{G}}(z,t)} = \dfrac{\mathcal{Z}_{\mathcal{F}}(0,t)}{\mathcal{Z}_{\mathcal{G}}(0,t)} = \text{constant}$, and we have $\mathcal{Z}_{\mu_{\mathcal{G}}}((z,0],t) = \mathcal{Z}_{\mu_{\mathcal{F}}}((z,0],t)$ for every $z < 0$, hence $\dfrac{\mathcal{Z}_{\mathcal{G}}(0,t)}{\mathcal{Z}_{\mathcal{G}}(z,t)} = \dfrac{\mathcal{Z}_{\mathcal{F}}(0,t)}{\mathcal{Z}_{\mathcal{F}}(z,t)}$, therefore $\dfrac{\mathcal{Z}_{\mathcal{F}}(z,t)}{\mathcal{Z}_{\mathcal{G}}(z,t)} = \dfrac{\mathcal{Z}_{\mathcal{F}}(0,t)}{\mathcal{Z}_{\mathcal{G}}(0,t)} = \text{constant}$. Therefore, $*$-$\mathcal{G}$-$\mathcal{Z}$-$\mathcal{N}$-premeasures are σ-bounded because $\mathbb{R} = \bigcup_{j=-\infty}^{+\infty} (j, j+1]$. Then, there exists an unique extension ($\mathcal{Z}_\mu$) of the $*$-$\mathcal{G}$-$\mathcal{Z}$-$\mathcal{N}$-premeasure $\mathcal{Z}_{\mu_{\mathcal{F}}}$ to a $*$-$\mathcal{G}$-$\mathcal{Z}$-$\mathcal{N}$-M on $\mathcal{B}_{\mathbb{R}}$ such that $\mu \mid_{\mathcal{A}} = \mu_{\mathcal{F}}$. Therefore $\mathcal{Z}_{\mu_{\mathcal{F}}} = \mathcal{Z}_{\mu_{\mathcal{G}}}$ on $\mathcal{B}_{\mathbb{R}}$.

We conclude that $\mathcal{Z}_{\mathcal{F}}$ is monotonicity from the monotonicity of $\mathcal{Z}_\mu$. Also, for $z \geq 0$, and $z < 0$, the right continuity of $\mathcal{Z}_{\mathcal{F}}$ is obtained according to the continuity of $\mathcal{Z}_\mu$ from above and below. If $z \geq 0$, we set $F = (0, z]$, $F_n = (0, z_n]$ such that $z_n \uparrow z$ then, $F_1 \subset F_2 \cdots \subset F_n \subset \cdots$ and $\bigcup_{n=1}^{+\infty} F_n = (0, z]$. Now, we have

$$\mathcal{Z}_{\mathcal{F}}(z,t) = \mathcal{Z}_\mu((0,z],t) = \lim_{n \to +\infty} \mathcal{Z}_\mu(F_n, t) = \lim_{n \to +\infty} \mathcal{Z}_{\mathcal{F}}(z_n),$$

if $z < 0$, is as same as above.

Description 1: All our results are valid considering the interval $[a, b)$ and left continuous functions $\mathcal{F}$.

Description 2: Each decreasing and right continuous $\mathcal{F}$, not only the $*$-$\mathcal{G}$-$\mathcal{Z}$-$\mathcal{N}$-B-M $\mathcal{Z}_{\mu_{\mathcal{F}}}$ but also a complete-$*$-$\mathcal{G}$-$\mathcal{Z}$-$\mathcal{N}$-M (C-$*$-$\mathcal{G}$-$\mathcal{Z}$-$\mathcal{N}$-M) $\mathcal{Z}_{\bar{\mu}_{\mathcal{F}}}$ with domain includes $\mathcal{B}_{\mathbb{R}}$. Actually, $\mathcal{Z}_{\bar{\mu}_{\mathcal{F}}}$ is only the completion of $\mathcal{Z}_{\mu_{\mathcal{F}}}$, and can be shown that it's domain is always strictly larger than $\mathcal{B}_{\mathbb{R}}$. We show this C-$*$-$\mathcal{G}$-$\mathcal{Z}$-$\mathcal{N}$-M with $\mathcal{Z}_{\mu_{\mathcal{F}}}$; it is named the Lebesgue–Stieltjes-$*$-$\mathcal{G}$-$\mathcal{Z}$-$\mathcal{N}$-M (L-S-$*$-$\mathcal{G}$-$\mathcal{Z}$-$\mathcal{N}$-M) associated to $\mathcal{F}$.

There are several regular properties for L-S-$*$-$\mathcal{G}$-$\mathcal{Z}$-$\mathcal{N}$-M that will be investigated. In this discussion we consider a complete L-S-$*$-$\mathcal{G}$-$\mathcal{Z}$-$\mathcal{N}$-M $\mathcal{Z}_\mu$ on $\mathbb{R}$ associated to the decreasing, right continuous function $\mathcal{F}$, and we show the domain of $\mathcal{Z}_\mu$ with M_μ. Then, for each $F \in M_\mu$

$$\begin{aligned}\mathcal{Z}_\mu(F,t) &= \sup\left\{ *_{j=1}^{+\infty} \frac{\mathcal{Z}_{\mathcal{F}}(b_j,t)}{\mathcal{Z}_{\mathcal{F}}(a_j,t)} : F \subseteq \bigcup_{j=1}^{+\infty}(a_j,b_j] \right\}\\ &= \sup\left\{ *_{j=1}^{+\infty} \mathcal{Z}_\mu((a_j,b_j],t) : F \subseteq \bigcup_{j=1}^{+\infty}(a_j,b_j] \right\}.\end{aligned}$$

Now, we show that, we can replace h-intervals by open-intervals for $\mathcal{Z}_\mu(F,t)$.

Lemma 6.1 *Let $F \in M_\mu$ then,*

$$\mathcal{Z}_\mu(F,t) = \sup\left\{ *_{j=1}^{+\infty} \mathcal{Z}_\mu((a_j,b_j),t) : F \subseteq \bigcup_{j=1}^{+\infty}(a_j,b_j) \right\}.$$

Proof Suppose $\mathcal{Z}_\nu(F,t) = \sup\left\{ *_{j=1}^{+\infty} \mathcal{Z}_\mu((a_j,b_j),t) : F \subseteq \bigcup_{j=1}^{+\infty}(a_j,b_j) \right\}$. We know each (a_j,b_j) is countable disjoint union of h-intervals $I_j^k = (c_j^k, c_j^{k+1}]$ such that $c_j^1 = a_j$ and $c_j^k \uparrow b_j$. Thus $F \subseteq \bigcup_{k,j=1}^{+\infty} (c_j^k, c_j^{k+1}]$, then

$$\begin{aligned}*_{j=1}^{+\infty} \mathcal{Z}_\mu((a_j,b_j),t) &= *_{k,j=1}^{+\infty} \mathcal{Z}_\mu((c_j^k, c_j^{k+1}],t)\\ &\le \mathcal{Z}_\mu\left(\bigcup_{k,j=1}^{+\infty} (c_j^k, c_j^{k+1}],t \right)\\ &\le \mathcal{Z}_\mu(F,t),\end{aligned}$$

and hence

$$\begin{aligned}\mathcal{Z}_\nu(F,t) &= \sup\left\{ *_{j=1}^{+\infty} \mu((a_j,b_j),t) : F \subseteq \bigcup_{j=1}^{+\infty}(a_j,b_j) \right\} \qquad (6.4.3)\\ &\le \mathcal{Z}_\mu(F,t),\end{aligned}$$

on the other hand $\mathcal{Z}_\mu(F,t) = \sup\left\{ *_{j=1}^{+\infty} \mathcal{Z}_\mu((a_j,b_j],t) : F \subseteq \bigcup_{j=1}^{+\infty}(a_j,b_j)\right\}$, therefore for $0<\varepsilon<1$, there exists $\{(a_j,b_j]\}_{j=1}^{+\infty}$ with $F \subseteq \bigcup_{j=1}^{+\infty}(a_j,b_j]$ and $*_{j=1}^{+\infty}\mathcal{Z}_\mu((a_j,b_j],t) > \mathcal{Z}_\mu(F,t) \circledast \mathcal{Z}(\varepsilon,t)$.

Since for every j there exists $\delta_j > 0$ such that $\mathcal{Z}_\mathcal{F}(b_j,t) \circledast \mathcal{Z}_\mathcal{F}(b_j+\delta_j,t) < \varepsilon$, then $\frac{\mathcal{Z}_\mathcal{F}(b_j+\delta_j,t)}{\mathcal{Z}_\mathcal{F}(b_j,t)} \to 1$, and

$$\begin{aligned} *_{j=1}^{+\infty} \mathcal{Z}_\mu((a_j,b_j+\delta_j),t) &= *_{j=1}^{+\infty}\mathcal{Z}_\mu((a_j,b_j] \cup (b_j,b_j+\delta_j),t) \\ &\geq *_{j=1}^{+\infty}[\mathcal{Z}_\mu((a_j,b_j],t) * \mathcal{Z}_\mu((b_j,b_j+\delta_j),t)] \\ &\geq *_{j=1}^{+\infty}\left[(\mathcal{Z}_\mu(F,t)-\varepsilon) * \frac{\mathcal{Z}_\mathcal{F}(b_j+\delta_j,t)}{\mathcal{Z}_\mathcal{F}(b_j,t)}\right] \\ &\to \mathcal{Z}_\mu(F,t), \end{aligned}$$

and hence

$$\begin{aligned} \mathcal{Z}_\nu(F,t) &= \sup\left\{ *_{j=1}^{+\infty} \mathcal{Z}_\mu((a_j,b_j),t) : F \subseteq \bigcup_{j=1}^{+\infty}(a_j,b_j)\right\} \\ &\geq \mathcal{Z}_\mu(F,t). \end{aligned} \tag{6.4.4}$$

From (6.4.3) and (6.4.4) we have $\mathcal{Z}_\nu(F,t) = \mathcal{Z}_\mu(F,t)$.

Theorem 6.11 *If $F \in M_\mu$, then*

$$\begin{aligned} \mathcal{Z}_\mu(F,t) &= \sup\{\mathcal{Z}_\mu(u,t) : F \subseteq u \text{ and } u \text{ is open}\} \\ &= \inf\{\mathcal{Z}_\mu(k,t) : k \subseteq F \text{ and } k \text{ is compact}\}. \end{aligned}$$

Proof It is clear that if $\mathcal{Z}_\mu(F,t) = 0$, the result is proof. Now, we consider $\mathcal{Z}_\mu(F,t) > 0$ and according to the properties of the sup, for every $0<\varepsilon<1$ there exist intervals (a_j,b_j) such that $F \subseteq \bigcup_{j=1}^{+\infty}(a_j,b_j)$ and $\mathcal{Z}_\mu(F,t) - \mathcal{Z}(\varepsilon,t) \leq *_{j=1}^{+\infty}\mathcal{Z}_\mu((a_j,b_j),t))$. If $u_0 = \bigcup_{j=1}^{+\infty}(a_j,b_j)$ then u_0 is open and $\mathcal{Z}_\mu(u_0,t) = \mathcal{Z}_\mu\left(\bigcup_{j=1}^{+\infty}(a_j,b_j),t\right) \geq *_{j=1}^{+\infty}\mathcal{Z}_\mu((a_j,b_j),t) \geq \mathcal{Z}_\mu(F,t) \circledast \mathcal{Z}(\varepsilon,t)$, therefore

$$\sup\{\mathcal{Z}_\mu(u,t) : F \subseteq u,\, u \text{ is open}\} \geq \mathcal{Z}_\mu(u_0,t) \geq \mathcal{Z}_\mu(F,t) \circledast \mathcal{Z}(\varepsilon,t) \tag{6.4.5}$$

Considering that $F \subseteq u$, according to the monotonicity property, $\mathcal{Z}_\mu(F,t) \geq \mathcal{Z}_\mu(u,t)$ hence

$$\mathcal{Z}_\mu(F,t) \geq \sup\{\mathcal{Z}_\mu(u,t) : F \subseteq u \text{ and } u \text{ is open}\}. \tag{6.4.6}$$

From (6.4.5) and (6.4.6) we have $\mathcal{Z}_\mu(F,t) = \sup\{\mathcal{Z}_\mu(u,t) : F \subseteq u$ and u is open$\}$. For the second one, at first, we suppose that F is bounded. If F is closed, then F is compact and the equality is clear. Otherwise, for $0 < \varepsilon < 1$ we have an open $u \supset \bar{F} \setminus F$ such that $\mathcal{Z}_\mu(u,t) \geq \mathcal{Z}_\mu(\bar{F} \setminus F, t) - \mathcal{Z}(\varepsilon, t)$. We set $k_0 = \bar{F} - u$. Then k_0 is compact, $k_0 \subset F$, and $k_0 = \bar{F} - u = \bar{F} \cap u^c \subset \bar{F} \cap (\bar{F} - F)^c = F$, and $F - k_0 = F \cap k_0^c = F \cap (\bar{F}^c \cup u) = F \cap u$, therefore $\mathcal{Z}_\mu(F,t) = \mathcal{Z}_\mu((k_0 \cup (F - k_0)), t) = \mathcal{Z}_\mu(k_0, t) * \mathcal{Z}_\mu((F - k_0), t)$, hence $\mathcal{Z}_\mu(k_0, t) = \dfrac{\mathcal{Z}_\mu(F,t)}{\mathcal{Z}_\mu((F - k_0), t)} = \dfrac{\mathcal{Z}_\mu(F,t)}{\mathcal{Z}_\mu(F \cap u, t)}$, but $u = (u \cap F) \cup (u - F)$, then, $\mathcal{Z}_\mu(u,t) = \mathcal{Z}_\mu(u \cap F, t) * \mathcal{Z}_\mu(u - F, t)$, then,

$$\mathcal{Z}_\mu(k_0,t) = \frac{\mathcal{Z}_\mu(F,t)}{\mathcal{Z}_\mu(F \cap u, t)} = \frac{\mathcal{Z}_\mu(F,t) \times \mathcal{Z}_\mu(u - F, t)}{\mathcal{Z}_\mu(u,t)} \leq \frac{\mathcal{Z}_\mu(F,t) \times \mathcal{Z}_\mu(\bar{F} - F, t)}{\mathcal{Z}_\mu(u,t)}$$
$$\leq \frac{\mathcal{Z}_\mu(F,t) \times (\mathcal{Z}_\mu(u,t) \circledast \mathcal{Z}(\varepsilon, t))}{\mathcal{Z}_\mu(u,t)} \to \mathcal{Z}_\mu(F,t),$$

then, $\mu(k_0, t) \leq \mu(F, t)$, therefore

$$\inf\{\mathcal{Z}_\mu(k,t) : k \subseteq F,\ k \text{ is compact}\} \leq \mathcal{Z}_\mu(k_0,t) \leq \mathcal{Z}_\mu(F,t), \tag{6.4.7}$$

also, the monotonicity of $\mathcal{Z}_\mu$ implies that, $k \subseteq F \Rightarrow \mathcal{Z}_\mu(F,t) \leq \mathcal{Z}_\mu(k,t)$, hence

$$\mu(F,t) \leq \inf\{\mu(k,t) : k \subseteq F\} \text{ for compact set } k, \tag{6.4.8}$$

considering (6.4.7), (6.4.8), we have $\mathcal{Z}_\mu(F,t) = \inf\{\mathcal{Z}_\mu(k,t) : k \subseteq F$, k is compact$\}$. Let F is unbounded, and $F_\jmath = F \cap (\jmath, \jmath + 1]$. By the preceding argument, for any $0 < \varepsilon < 1$, there exist compact $k_\jmath \subset F_\jmath$ such that $\mathcal{Z}_\mu(k_\jmath, t) \leq \mathcal{Z}_\mu(F_\jmath, t) \circledast \mathcal{Z}(\varepsilon, t)$. We set $H_n = \bigcup_{\jmath=-n}^{n} k_\jmath$, then H_n is compact, $H_n \subseteq F$ and $\mathcal{Z}_\mu(H_n, t) = \mathcal{Z}_\mu\left(\bigcup_{\jmath=-n}^{n} k_\jmath, t\right) = *_{\jmath=-n}^{n} \mathcal{Z}_\mu(k_\jmath, t) \leq *_{\jmath=-n}^{n} \mathcal{Z}_\mu(F_\jmath, t) \circledast \mathcal{Z}(\varepsilon, t) \to *_{\jmath=-n}^{n} \mathcal{Z}_\mu(F_\jmath, t)$, then

$$\inf\{\mu(k,t) : k \subseteq F,\ k \text{ is compact}\} \leq \mu(H_n, t) \leq *_{\jmath=-n}^{n} \mu(F_\jmath, t) \leq \mu\left(\bigcup_{\jmath=-n}^{n} F_\jmath, t\right),$$

then $\inf\{\mathcal{Z}_\mu(k,t) : k \subseteq F,\ k \text{ is compact}\} \leq \lim_{n \to +\infty} \mathcal{Z}_\mu\left(\bigcup_{\jmath=-n}^{n} F_\jmath, t\right) = \mathcal{Z}_\mu(F,t)$, we conclude that $\inf\{\mathcal{Z}_\mu(k,t) : k \subseteq F,\ k \text{ is compact}\} \leq \mathcal{Z}_\mu(F,t)$.

Theorem 6.12 *Suppose $F \subseteq \mathbb{R}$, then the following are equivalent:*

(i) $F \in M_\mu$.
(ii) $F = W \setminus N_1$ *where W is a G_δ set and* $\mathcal{Z}_\mu(N_1, t) = \mathbf{1}$.
(iii) $F = H \cup N_2$ *where H is a F_σ set and* $\mathcal{Z}_\mu(N_2, t) = \mathbf{1}$.

Proof Considering the completeness of $\mathcal{Z}_\mu$ on M_μ, clearly (ii), (iii) result in (i). Now, we prove that (ii) results in (i). Let $F \in M_\mu$ and $\mathcal{Z}_\mu(F, t) > 0$. By Theorem 6.11, for every $\jmath \in \mathbb{N}$ we can choose an open $u_\jmath \supset F$ such that $\mathcal{Z}_\mu(F, t) \circledast \mathcal{Z}_\mu(u_\jmath, t) < \mathcal{Z}(\frac{1}{2^\jmath}, t)$. We assume that $W = \bigcap_{\jmath=1}^{+\infty} u_\jmath$, then $F \subseteq W$ and $\mathcal{Z}_\mu(W, t) \geq \mathcal{Z}_\mu(u_\jmath, t) \geq \mathcal{Z}_\mu(F, t) \circledast \mathcal{Z}(\frac{1}{2^\jmath}, t)$, then, for any $\jmath \to +\infty$, we have

$$\mathcal{Z}_\mu(W, t) \geq \mathcal{Z}_\mu(F, t), \tag{6.4.9}$$

also, $F \subseteq W$, by the property of monotonicity we get

$$\mathcal{Z}_\mu(W, t) \leq \mathcal{Z}_\mu(F, t). \tag{6.4.10}$$

From (6.4.9), (6.4.10) we get

$$\mathcal{Z}_\mu(W, t) = \mathcal{Z}_\mu(F, t). \tag{6.4.11}$$

Due to the $\mathcal{Z}_\mu(W, t) = \mathcal{Z}_\mu(F \cup (W \setminus F), t) = \mathcal{Z}_\mu(F, t) \times \mathcal{Z}_\mu(W \setminus F), t)$ and from (6.4.11) we have $\mathcal{Z}_\mu(W \setminus F, t) = \mathbf{1}$. If we set $W \setminus F = N_1$ we have $W \setminus N_1 = F$. Now, if $\mathcal{Z}_\mu(F, t) = \mathbf{0}$, we set $F_n = F \cap ((-n, -n+1] \cup (n, n+1])$, therefore $F = \bigcup_{n \in \mathbb{Z}^+} F_n$ and $\mathcal{Z}_\mu(F_n, t) > 0$ for every $n \in \mathbb{Z}^+$. According to Theorem 6.11, there exist $F_n \subseteq u_n^\jmath$, $\jmath \in \mathbb{N}$ such that $\mathcal{Z}_\mu(F_n, t) \circledast \mathcal{Z}(\frac{1}{2^{n+\jmath}, t}) < \mathcal{Z}_\mu(u_n^\jmath, t)$ then, $\mathcal{Z}_\mu(u_n^\jmath, t) = \mathcal{Z}_\mu(F_n \cup (u_n^\jmath - F_n), t) = \mathcal{Z}_\mu(F_n, t) * \mathcal{Z}_\mu(u_n^\jmath - F_n, t)$ hence $\mathcal{Z}_\mu(u_n^\jmath - F_n, t) = \frac{\mathcal{Z}_\mu(u_n^\jmath, t)}{\mathcal{Z}_\mu(F_n, t)}$. We set $u^\jmath = \bigcup_{n \in \mathbb{Z}^+} u_n^\jmath$ and $W = \bigcap_{\jmath=1}^{+\infty} u^\jmath$, thus, we have $u^\jmath = \bigcup_{n \in \mathbb{Z}^+} u_n^\jmath \supseteq \bigcup_{n \in \mathbb{Z}^+} F_n = F$, and $W = \bigcap_{\jmath=1}^{+\infty} u^\jmath \supseteq F$, hence

$$W - F = \bigcap_{\jmath=1}^{+\infty} (u^\jmath) - F = \bigcap_{\jmath=1}^{+\infty} (u^\jmath - F), \tag{6.4.12}$$

but $u^\jmath - F = \bigcup_{n \in \mathbb{Z}^+} (u_n^\jmath - F) \subseteq \bigcup_{n \in \mathbb{Z}^+} (u_n^\jmath - F_n)$, therefore, $W - F = \bigcap_{\jmath=1}^{+\infty} (u^\jmath - F) \subseteq \bigcap_{\jmath=1}^{+\infty} \bigcup_{n \in \mathbb{Z}^+} (u_n^\jmath - F_n)$, then,

$$\mathcal{Z}_\mu(W - F, t) \geq \mathcal{Z}_\mu\left(\bigcap_{j=1}^{+\infty} \bigcup_{n\in\mathbb{Z}^+} (u_n^j - F_n), t\right) \geq \mathcal{Z}_\mu\left(\bigcup_{n\in\mathbb{Z}^+} (u_n^j - F_n), t\right)$$

$$= *_{n\in\mathbb{Z}^+}\mathcal{Z}_\mu((u_n^j - F_n), t) \geq *_{n\in\mathbb{Z}^+}\frac{\mathcal{Z}_\mu(u_n^j, t)}{\mathcal{Z}_\mu(F_n, t)} \geq \frac{\mathcal{Z}_\mu(F_n, t) \circledast \mathcal{Z}(\frac{1}{2^{n+j}}, t)}{\mathcal{Z}_\mu(F_n, t)}$$

$$\to 1,$$

hence $\mathcal{Z}_\mu(W - F, t) = 1$, we set $N_1 = W - F$, hence $F = W - N_1$. In the following, we show that (iii) results in (i). Let $F \in M_\mu$ and $\mathcal{Z}_\mu(F, t) > 0$. Considering Theorem 6.11, for every $j \in \mathbb{N}$ we can choose a compact set $k_j \subset F$ such that $\mathcal{Z}_\mu(F, t) \circledast \mathcal{Z}(\frac{1}{2^j}, t) > \mathcal{Z}_\mu(k_j, t)$.

Let $H = \bigcup_{j=1}^{+\infty} k_j$, then $H \subseteq F$ and $\mathcal{Z}_\mu(H, t) \leq \mathcal{Z}_\mu(k_j, t) \leq \mathcal{Z}_\mu(F, t) \circledast \mathcal{Z}(\frac{1}{2^j}, t)$, then, for every $j \to +\infty$

$$\mathcal{Z}_\mu(H, t) \leq \mathcal{Z}_\mu(F, t). \tag{6.4.13}$$

Because $H \subseteq F$ and according to the monotonicity property, we obtain

$$\mathcal{Z}_\mu(F, t) \leq \mathcal{Z}_\mu(H, t), \tag{6.4.14}$$

considering (6.4.13), (6.4.14), we have

$$\mathcal{Z}_\mu(F, t) = \mathcal{Z}_\mu(H, t), \tag{6.4.15}$$

since $\mathcal{Z}_\mu(F, t) = \mathcal{Z}_\mu(H \cup (F - H), t) = \mathcal{Z}_\mu(H, t) \times \mathcal{Z}_\mu(F - H, t)$, from (6.4.15) we have $\mathcal{Z}_\mu(F - H, t) = \mathbf{1}$, now, set $F - H = N_2$, thus $H \cup N_2 = F$.

In the sequel, if $\mathcal{Z}_\mu(F, t) = 0$, we set $F_n = F \cap ((-n, -n+1] \cup (n, n+1])$, therefore, for every $n \in \mathbb{Z}^+$, $F = \bigcup_{n\in\mathbb{Z}^+} F_n$ and $\mathcal{Z}_\mu(F_n, t) > 0$. Similarly, there is a F_σ, a set H_n, and a null set N_n for every set F_n such that $F_n = H_n \cup N_n$. We set $H = \bigcup_{n\in\mathbb{Z}^+} H_n$, then $H \subseteq F$ is a F_σ set and $F - H = \bigcup_{n\in\mathbb{Z}^+} F_n - \bigcup_{n\in\mathbb{Z}^+} H_n \subseteq \bigcup_{n\in\mathbb{Z}^+} (F_n - H_n) = \bigcup_{n\in\mathbb{Z}^+} N_n$, therefore,

$$\mathcal{Z}_\mu(F - H, t) \geq \mathcal{Z}_\mu\left(\bigcup_{n\in\mathbb{Z}^+} (F_n - H_n), t\right) \geq *_{n\in\mathbb{Z}^+}\mathcal{Z}_\mu(F_n - H_n, t)$$

$$\geq *_{n\in\mathbb{Z}^+}\mathcal{Z}_\mu(N_n, t) = *_{n\in\mathbb{Z}^+} 1 = \mathbf{1}.$$

It means that $\mathcal{Z}_\mu(F - H, t) = \mathbf{1}$, if we set $N = F - H$, we get $F = H \cup N$.

All Borel sets are reasonably simple modular sets of measure 1 =. This is all that is stated in Theorem 6.12. In the next theorem, we prove that measurable sets are approximated by "simple" sets.

Theorem 6.13 *If A is a finite union of open intervals, then for every $F \in M_\mu$ and $\mathcal{Z}_\mu(F,t) > 0$ and every $0 < \varepsilon < 1$ there is a set A such that $\mathcal{Z}_\mu(F \triangle A, t) > \mathcal{Z}(1-\varepsilon, t)$.*

Proof To start the proof, we consider Theorem 6.11. Then, for every $0 < \varepsilon < 1$ there is an open set $u \supset F$ such that

$$\mathcal{Z}_\mu(F,t) \circledast \mathcal{Z}(\varepsilon, t) < \mathcal{Z}_\mu(u,t) \tag{6.4.16}$$

For every open interval I_n where $I_\ell \cap I_\jmath = \emptyset$, $\ell \neq \jmath$, we know that $u = \bigcup_{n=1}^{+\infty} I_n$. If we consider $A_n = \bigcup_{\ell=1}^{n} I_\ell$, we have $u = \bigcup_{n=1}^{+\infty} A_n$, and $\mathcal{Z}_\mu(u,t) = \inf_{n\in\mathbb{N}} \mathcal{Z}_\mu(A_n)$, then

$$\mathcal{Z}_\mu(A_n, t) < \mathcal{Z}_\mu(u,t) \circledast \mathcal{Z}(\varepsilon, t), \tag{6.4.17}$$

but $A_n \triangle F = (A_n - F) \cup (F - A_n) \subseteq (u - F) \cup (u - A_n)$, therefore

$$\mathcal{Z}_\mu(A_n \triangle F, t) \geq \mathcal{Z}_\mu((u-F) \cup (u - A_n), t) \geq \mathcal{Z}_\mu(u - F, t) \times \mathcal{Z}_\mu(u - A_n, t).$$

From (6.4.16), (6.4.17) we get

$$\mathcal{Z}_\mu(A_n \triangle F, t) \geq \mathcal{Z}_\mu(u - F, t) \times \mathcal{Z}_\mu(u - A_n, t) \geq \frac{\mathcal{Z}_\mu(u,t)}{\mathcal{Z}_\mu(F,t)} \times \frac{\mathcal{Z}_\mu(u,t)}{\mathcal{Z}_\mu(A_n,t)}$$
$$\geq \left(\mathbf{1} \circledast \frac{\mathcal{Z}(\varepsilon,t)}{\mathcal{Z}_\mu(F,t)}\right) \times \left(\mathbf{1} \circledast \frac{\mathcal{Z}(\varepsilon,t)}{\mathcal{Z}_\mu(A_n,t)}\right) \to \mathbf{1} \times \mathbf{1} = \mathbf{1}.$$

6.4.1 Lebesgue-$*$-Generalized-$\mathcal{Z}$-Number-Fuzzy Measure

Next, we will discuss the introduction of the Lebesgue-$*$-$\mathcal{G}$-$\mathcal{Z}$-$\mathcal{N}$-fuzzy measure (L-$*$-$\mathcal{G}$-$\mathcal{Z}$-$\mathcal{N}$-F-M), which is one of the most important $*$-$\mathcal{G}$-$\mathcal{Z}$-$\mathcal{N}$-fuzzy measures ($*$-$\mathcal{G}$-$\mathcal{Z}$-$\mathcal{N}$-F-M) on $\mathbb{R}$. Considering the function $\mathcal{F}(z,t) = e^{\frac{-|z|}{t}}$, $\mathcal{Z}_{\mu_\mathcal{F}}$ is the C-$*$-$\mathcal{G}$-$\mathcal{Z}$-$\mathcal{N}$-M. We represent this C-$*$-$\mathcal{G}$-$\mathcal{Z}$-$\mathcal{N}$-M with $\mathcal{Z}_m$, whose domain is called the class of Lebesgue $*$-$\mathcal{G}$-$\mathcal{Z}$-$\mathcal{N}$-F-M sets, and we denote these sets with $\mathcal{L}$ for convenience. We shall also refer to the restriction of $\mathcal{Z}_m$ to $\mathcal{B}_\mathbb{R}$ as L-$*$-$\mathcal{G}$-$\mathcal{Z}$-$\mathcal{N}$-F-M. If $F \subseteq \mathbb{R}$ and $\alpha, \beta \in \mathbb{R}$, we define $F + \beta = \{z + \beta : z \in F\}$, $\alpha F = \{\alpha z : z \in F\}$.

Theorem 6.14 *If we assume that* $F \in \mathcal{L}$, *then*

- $F + \beta \in \mathcal{L}$, *for all* $\alpha, \beta \in \mathbb{R}$;
- $\alpha F \in \mathcal{L}$ *for all* $\alpha, \beta \in \mathbb{R}$;
- $\mathcal{Z}_m(F + \beta, t) = \mathcal{Z}_m(F, t)$;
- $\mathcal{Z}_m(\alpha F, t) = (\mathcal{Z}_m(F, t))^{|\alpha|}$.

Proof For each $a, b \in \mathbb{R}$, we assume $F = (a, b)$. Then, $\mathcal{Z}_m((a, b), t) = e^{\frac{-(b-a)}{t}}$, since

$$\mathcal{Z}_m(\alpha F, t) = \mathcal{Z}_\mu((\alpha a, \alpha b), t) = e^{\frac{-\alpha(b-a)}{t}} = [\mathcal{Z}_m((a, b), t)]^{|\alpha|}, \tag{6.4.18}$$

and

$$\mathcal{Z}_m(F + \beta, t) = \mathcal{Z}_m((a + \beta, b + \beta), t) = e^{\frac{-(b-a)}{t}} = \mathcal{Z}_m(F, t), \tag{6.4.19}$$

since, we have

$$\mathcal{Z}_m\left(\bigcup_{\ell=1}^{+\infty}(a_\ell, b_\ell), t\right) = \prod_{\ell=1}^{+\infty}\mathcal{Z}_m((a_\ell, b_\ell), t),$$

therefore

$$\mathcal{Z}_m\left(\alpha\left(\bigcup_{\ell=1}^{+\infty}(a_\ell, b_\ell)\right), t\right) = \mathcal{Z}_m\left(\bigcup_{\ell=1}^{+\infty}(\alpha a_\ell, \alpha b_\ell), t\right) = \prod_{\ell=1}^{+\infty}\mathcal{Z}_m((\alpha a_\ell, \alpha b_\ell), t),$$

with respect to (6.4.18) we have

$$\mathcal{Z}_m\left(\alpha\left(\bigcup_{\ell=1}^{+\infty}(a_\ell, b_\ell)\right), t\right) = \prod_{\ell=1}^{+\infty}\mathcal{Z}_m((\alpha a_\ell, \alpha b_\ell), t) = \left(\prod_{\ell=1}^{+\infty}\mathcal{Z}_m((a_\ell, b_\ell), t)\right)^{|\alpha|}$$
$$= \left(\mathcal{Z}_m\left(\bigcup_{\ell=1}^{+\infty}(a_\ell, b_\ell), t\right)\right)^{|\alpha|},$$

and

$$\mathcal{Z}_m\left(\bigcup_{\ell=1}^{+\infty}(a_\ell, b_\ell) + \beta, t\right) = \mathcal{Z}_m\left(\bigcup_{\ell=1}^{+\infty}(a_\ell + \beta, b_\ell + \beta), t\right) = \prod_{\ell=1}^{+\infty}\mathcal{Z}_m((a_\ell + \beta, b_\ell + \beta), t),$$

with respect to (6.4.19) we have

$$\mathcal{Z}_m\left(\bigcup_{\ell=1}^{+\infty}(a_\ell, b_\ell) + \beta, t\right) = \prod_{\ell=1}^{+\infty}\mathcal{Z}_m((a_\ell + \beta, b_\ell + \beta), t) = \prod_{\ell=1}^{+\infty}\mathcal{Z}_m((a_\ell, b_\ell), t)$$
$$= \mathcal{Z}_m\left(\bigcup_{\ell=1}^{+\infty}(a_\ell, b_\ell), t)\right).$$

According to Theorem 6.12 we have $\mathcal{Z}_m(\alpha F, t) = (\mathcal{Z}_m(F, t))^{|\alpha|}$, similarly $\mathcal{Z}_m((F + \beta), t) = \mathcal{Z}_m(F, t)$.

6.4.2 Lebesgue–Radon–Nikodym Theorem

Definition 6.12 We consider $(\chi, C, \mu, *)$ and $(\chi, C, \upsilon, *)$ which are two ∗-$\mathcal{G}$-$\mathcal{Z}$-$\mathcal{N}$-F-M-Ss. $\mathcal{Z}_\upsilon$ is absolutely continuous with respect to $\mathcal{Z}_\mu$ and write $\mathcal{Z}_\upsilon \ll \mathcal{Z}_\mu$, if we have $\mathcal{Z}_\upsilon(H, t) = \mathbf{1}$ and $\mathcal{Z}_\mu(H, t) = \mathbf{1}$ for every $H \in C$.

Definition 6.13 We consider$(\chi, C, \mu, *)$ and $(\chi, C, \upsilon, *)$, which are both ∗-$\mathcal{G}$-$\mathcal{Z}$-$\mathcal{N}$-F-M-Ss. $\mathcal{Z}_\mu, \mathcal{Z}_\upsilon$ are mutually singular and write $\mathcal{Z}_\upsilon \perp \mathcal{Z}_\mu$, if there exist $H, K \in C$ such that $H \cap K = \emptyset$, $H \cup K = \chi$, $\mathcal{Z}_\mu(H, t) = \mathbf{1}$ and $\mathcal{Z}_\upsilon(K, t) = \mathbf{1}$. If $\mathcal{Z}_\upsilon \perp \mathcal{Z}_\mu$ and $\mathcal{Z}_\upsilon \ll \mathcal{Z}_\mu$ then $\mathcal{Z}_\upsilon = \mathbf{1}$, in fact Absolute continuity is in a sense the antithesis of mutual singularity.

Lemma 6.2 *We consider $\mathcal{Z}_\upsilon$ and $\mathcal{Z}_\mu$, which are two ∗-$\mathcal{G}$-$\mathcal{Z}$-$\mathcal{N}$-B-B-M, on $(\chi, C, *)$. There exist $\mathcal{Z}_\upsilon \perp \mathcal{Z}_\mu$ or $0 < \varepsilon < 1$ and $H \in M$ such that $\mathcal{Z}_\mu(H, t) < \mathbf{1}$ and* $\dfrac{\mathcal{Z}_\upsilon(H, t)}{\mathcal{Z}_\mu(H, \frac{t}{\varepsilon})} \leq \mathbf{1}$.

Proof We define $\mathcal{Z}_{\lambda_n}$ as follows and show that $\mathcal{Z}_{\lambda_n}$ is a ∗-$\mathcal{G}$-$\mathcal{Z}$-$\mathcal{N}$-F-M:

$$\mathcal{Z}_{\lambda_n}(H, t) = \begin{cases} \frac{\mathcal{Z}_\upsilon(H,t)}{\mathcal{Z}_\mu(H,nt)}, & \text{if, } \frac{\mathcal{Z}_\upsilon(H,t)}{\mathcal{Z}_\mu(H,nt)} < \mathbf{1}, \\ \mathbf{1}, & \text{if, } \frac{\mathcal{Z}_\upsilon(H,t)}{\mathcal{Z}_\mu(H,nt)} \geq \mathbf{1}, \end{cases} \quad \text{for every } n \in \mathbb{N}.$$

Step 1. $\mathcal{Z}_{\lambda_n}(\phi, t) = \frac{\mathcal{Z}_\upsilon(\phi,t)}{\mathcal{Z}_\mu(\phi,nt)} = \frac{\mathbf{1}}{\mathbf{1}}$.
Step 2. for $\ell = 1, 2, \cdots$, if $H_\ell \in C$ such that $H_\ell \cap H_\jmath = \emptyset$, $\ell \neq \jmath$, therefore

$$\mathcal{Z}_{\lambda_n}\left(\bigcup_{\ell=1}^{+\infty} H_\ell, t\right) = \frac{\mathcal{Z}_\upsilon\left(\bigcup_{\ell=1}^{+\infty} H_\ell, t\right)}{\mathcal{Z}_\mu\left(\bigcup_{\ell=1}^{+\infty} H_\ell, nt\right)} = \frac{*_{\ell=1}^{+\infty}\mathcal{Z}_\upsilon(H_\ell, t)}{*_{\ell=1}^{+\infty}\mathcal{Z}_\mu(H_\ell, nt)} = \frac{\prod_{\ell=1}^{+\infty}\mathcal{Z}_\upsilon(H_\ell, t)}{\prod_{\ell=1}^{+\infty}\mathcal{Z}_\mu(H_\ell, nt)}$$
$$= \prod_{\ell=1}^{+\infty}\mathcal{Z}_{\lambda_n}(H_\ell, t).$$

Let $\chi = P_n \cup N_n$ be a decomposition for λ_n, and we assume that $P = \bigcup_{n=1}^{+\infty} P_n$ and $N = \bigcap_{n=1}^{+\infty} N_n$. We have

$$\mathbf{1} \geq \frac{\mathcal{Z}_\upsilon(N, t)}{\mathcal{Z}_\mu(N, nt)} = \mathcal{Z}_{\lambda_n}(N, t) \geq \mathcal{Z}_{\lambda_n}(N_n, t) \geq \frac{\mathcal{Z}_\upsilon(N_n, t)}{\mathcal{Z}_\lambda(N_n, nt)} = \mathbf{1}.$$

Then, for every $n \in \mathbb{N}$, $\mathcal{Z}_{\lambda_n}(N, t) = \frac{\mathcal{Z}_\upsilon(N,t)}{\mathcal{Z}_\mu(N,nt)} = \mathbf{1} \Rightarrow \mathcal{Z}_\upsilon(N, t) = \mathcal{Z}_\mu(N, nt)$.

- If $n \to +\infty$ we have $\mathcal{Z}_\mu(N_n, t) \to \mathbf{1}$, then $\mathcal{Z}_\upsilon(N, t) = \mathbf{1}$.
- If $\mathcal{Z}_\mu(P, t) = \mathbf{1}$, then $\mathcal{Z}_\upsilon \perp \mathcal{Z}_\mu$.
- If $\mathcal{Z}_\mu(P, t) < \mathbf{1}$, then $\mathcal{Z}_\mu(P, t) \leq \mathcal{Z}_\mu(P_{n_0}, t)$ for some $n_0 \in \mathbb{N}$ and P_{n_0} is a positive set for λ_n, then $\mathcal{Z}_{\lambda_n}(P, t) \leq \mathcal{Z}_{\lambda_n}(P_{n_0}, t) \leq \mathbf{1}$, hence $\frac{\mathcal{Z}_\upsilon(P,t)}{\mathcal{Z}_\mu(P,nt)} < \frac{\mathcal{Z}_\upsilon(P_{n_0},t)}{\mathcal{Z}_\mu(P_{n_0},nt)} \leq \mathbf{1}$, and P_{n_0} is a positive set for λ_n.

In the following, we will discuss the Lebesgue–Radon–Nikodym Theorem.

Theorem 6.15 *We consider $\mathcal{Z}_\upsilon$ and $\mathcal{Z}_\mu$, which are two σ-bounded-$*$-$\mathcal{G}$-$\mathcal{Z}$-$\mathcal{N}$-F-M (σ-B-$*$-$\mathcal{G}$-$\mathcal{Z}$-$\mathcal{N}$-F-M) on $(\chi, C, *)$. Then, $\mathcal{Z}_\lambda$ and $\mathcal{Z}_\rho$ exist on $(\chi, C, *)$, which are unique σ-B-$*$-$\mathcal{G}$-$\mathcal{Z}$-$\mathcal{N}$-F-Ms and we have $\mathcal{Z}_\rho \ll \mathcal{Z}_\mu$, $\mathcal{Z}_\lambda \perp \mathcal{Z}_\mu$ and $\mathcal{Z}_\upsilon = \mathcal{Z}_\lambda * \mathcal{Z}_\rho$.*

Proof By referring to Theorem 5.6 in [10], we find that there exists a unique σ-B-$*$-$\mathcal{G}$-$\mathcal{Z}$-$\mathcal{N}$-F-M, $\mathcal{Z}_\rho$ such that $\mathcal{Z}_\rho \ll \mathcal{Z}_\mu$. On the other hand, there exist a μ-integrable function $f : \chi \to \mathbb{R}^+$ such that $d\mathcal{Z}_\rho = f d\mathcal{Z}_\mu$. The continuation of the proof is checked in two cases:

(*I*): Again, considering Theorem 5.6 in [10], we consider two σ-B-$*$-$\mathcal{G}$-$\mathcal{Z}$-$\mathcal{N}$-F-MS such as $\mathcal{Z}_\upsilon$ and $\mathcal{Z}_\mu$ and we define $\mathcal{F} = \{f : X \to [0, +\infty) : \int_H f d\mathcal{Z}_\mu(z, t) \geq \mathcal{Z}_\upsilon(H, t)\}$. We choose $\{f_n\} \subseteq \mathcal{F}$ such that $f = \sup f_n$ and $\int f_n d\mathcal{Z}_\mu(z, t) \to a$, then, $\int f d\mathcal{Z}_\mu(z, t) = a$. Also, we define $\mathcal{Z}_\lambda(H, t) = \frac{\mathcal{Z}_\upsilon(H,t)}{\int_H f d\mathcal{Z}_\mu(z,t)}$. Its well-defined is clear and $\mathcal{Z}_\lambda$ is a σ-$*$-$\mathcal{G}$-$\mathcal{Z}$-$\mathcal{N}$-F-MS.

(1) : $\mathcal{Z}_\lambda(\phi, t) = \frac{\mathcal{Z}_\upsilon(\phi,t)}{\int_\phi f(z) d\mathcal{Z}_\mu(z,t)} = \mathbf{1}$.

(2) :

$$\mathcal{Z}_\lambda\left(\bigcup_{\ell=1}^{+\infty} H_\ell, t\right) = \frac{\mathcal{Z}_\upsilon\left(\bigcup_{\ell=1}^{+\infty} H_\ell, t\right)}{\int_{\bigcup_{\ell=1}^{+\infty} H_\ell} f(z) d\mathcal{Z}_\mu(z,t)} = \frac{*_{\ell=1}^{+\infty} \mathcal{Z}_\upsilon(H_\ell, t)}{*_{\ell=1}^{+\infty} \int_{H_\ell} f(z) d\mathcal{Z}_\mu(z,t)}$$

$$= \frac{\prod_{\ell=1}^{+\infty} \mathcal{Z}_\upsilon(H_\ell, t)}{\prod_{\ell=1}^{+\infty} \int_{H_\ell} f(z) d\mathcal{Z}_\mu(z,t)} = \prod_{\ell=1}^{+\infty} \frac{\mathcal{Z}_\upsilon(H_\ell, t)}{\int_{H_\ell} f(z) d\mathcal{Z}_\mu(z,t)} = \prod_{\ell=1}^{+\infty} \mathcal{Z}_\lambda(H_\ell, t).$$

We know that $\mathcal{Z}_\lambda \perp \mathcal{Z}_\mu$, that is, $\mathcal{Z}_\lambda$ is singular with respect to $\mathcal{Z}_\mu$. Because if it is not like this, according to Lemma 6.2, there exist $H \in M$ and $0 < \varepsilon < 1$ such that $\mathcal{Z}_\mu(H,t) < \mathbf{1}$ and $\frac{\mathcal{Z}_\lambda(H,t)}{\mathcal{Z}_\mu(H,\frac{t}{\varepsilon})} \leq \mathbf{1}$ on H. Then, $\mathcal{Z}_\lambda(H,t) \leq \mathcal{Z}_\mu(H, \frac{t}{\varepsilon}) \frac{\mathcal{Z}_\upsilon(H,t)}{\int_H f(z) d\mathcal{Z}_\mu(z,t)} \leq \mathcal{Z}_\mu(H, \frac{t}{\varepsilon})$, hence $\mathcal{Z}_\upsilon(H,t) \leq \mathcal{Z}_\mu(H, \frac{t}{\varepsilon}). \int_H f(z) d\mathcal{Z}_\mu(z,t)$, therefore, $\mathcal{Z}_\upsilon(H,t) \leq \int (f + \varepsilon\chi_H) d\mathcal{Z}_\mu(z,t)$, then, $f + \varepsilon\chi_H \in \mathcal{F}$. On the other hand $\int (f + \varepsilon\chi_H) d\mathcal{Z}_\mu(z,t) < \int_H f(z) d\mathcal{Z}_\mu(z,t) = a$, hence $f + \varepsilon\chi_H \notin \mathcal{F}$, it is contradicting the definition of a.

For uniqueness, if there exist $\mathcal{Z}_{\rho'}, \mathcal{Z}_{\lambda'}$ such that $\mathcal{Z}_{\lambda'}(H,t) = \frac{\mathcal{Z}_\upsilon(H,t)}{\int_H f'(z) d\mathcal{Z}_\mu(z,t)}$, we have $\mathcal{Z}_{\lambda'}(H,t). \int_H f'(z) d\mathcal{Z}_\mu(z,t) = \mathcal{Z}_\lambda(H,t). \int_H f(z) d\mathcal{Z}_\mu(z,t)$, then

$$\frac{\mathcal{Z}_{\lambda'}(H,t)}{\mathcal{Z}_\lambda(H,t)} = \frac{\int_H f(z) d\mathcal{Z}_\mu(z,t)}{\int_H f'(z) d\mathcal{Z}_\mu(z,t)} \leq 1 \text{ or } \frac{\mathcal{Z}_\lambda}{\mathcal{Z}_{\lambda'}} = \frac{\int_H f}{\int_H f'} \leq \mathbf{1}. \quad (6.4.20)$$

If we assume that $\frac{\mathcal{Z}_{\lambda'}}{\mathcal{Z}_\lambda} \leq \mathbf{1}$, we will show that $\frac{\mathcal{Z}_{\lambda'}}{\mathcal{Z}_\lambda} \perp \mathcal{Z}_\mu$, $\frac{\int_H f}{\int_H f'} \ll \mathcal{Z}_\mu$.

Since $\mathcal{Z}_\lambda \perp \mathcal{Z}_\mu$, thus, there exist $H, K \in C$ such that $H \cap K = \emptyset$, $H \cup K = \chi$, $\mathcal{Z}_\mu(H,t) = \mathbf{1}$ and $\mathcal{Z}_\lambda(K,t) = \mathbf{1}$. Also, $\mathcal{Z}_{\lambda'} \perp \mathcal{Z}_\mu$, then, there exist $H', K' \in C$ such that $H' \cap K' = \emptyset$, $H' \cup K' = \chi$, $\mathcal{Z}_\mu(H',t) = \mathbf{1}$ and $\mathcal{Z}_{\lambda'}(K',t) = \mathbf{1}$. We set $H'' = H \cup H'$, $K'' = K \cap K'$, therefore

$$\begin{aligned} H'' \cap K'' &= (H \cup H') \cap (K \cap K') \\ &= [H \cap (K \cap K')] \cup [H' \cap (K \cap K')] \\ &= [(H \cap K) \cap K'] \cup [(H' \cap K') \cap K] \\ &= [\emptyset \cap K'] \cup [\emptyset \cap K] \\ &= \emptyset \cup \emptyset \\ &= \emptyset, \end{aligned}$$

and

$$\begin{aligned} H'' \cup K'' &= (H \cup H') \cup (K \cap K') \\ &= [H \cup (K \cap K')] \cup [H' \cup (K \cap K')] \\ &= [(H \cup K) \cap (H \cup K')] \cup [(H' \cup K) \cap (H' \cup K')] \\ &= (H \cup K') \cup (H' \cup K) \\ &= (H \cup H') \cup (K \cup K') \\ &= \chi \cup \chi \\ &= \chi, \end{aligned}$$

then, $\mathbf{1} \geq \frac{\mathcal{Z}_{\lambda'}(K'',t)}{\mathcal{Z}_{\lambda}(K'',t)} \geq \frac{\mathcal{Z}_{\lambda'}(K',t)}{\mathcal{Z}_{\lambda}(K'',t)} \geq \frac{\mathcal{Z}_{\lambda'}(K',t)}{\mathbf{1}} = \mathbf{1}$, hence

$$\frac{\mathcal{Z}_{\lambda'}(K'',t)}{\mathcal{Z}_{\lambda}(K'',t)} = \mathbf{1}, \tag{6.4.21}$$

since $\mathbf{1} \geq \mathcal{Z}_{\mu}(H'',t) \geq \mathcal{Z}_{\mu}(H,t) = \mathbf{1}$, then, $\mathcal{Z}_{\mu}(H'',t) = \mathbf{1}$, we conclude that

$$\frac{\mathcal{Z}_{\lambda'}}{\mathcal{Z}_{\lambda}} \perp \mathcal{Z}_{\mu}. \tag{6.4.22}$$

We claim $\frac{\int_H f(z)d\mathcal{Z}_{\mu}(z,t)}{\int_H f'(z)d\mathcal{Z}_{\mu}(z,t)} \ll \mathcal{Z}_{\mu}$, because, if $\mathcal{Z}_{\mu}(H,t) = \mathbf{1}$, then $\mathcal{Z}_{\rho}(H,t) = \int_H f(z)d\mathcal{Z}_{\mu}(z,t) = \mathbf{1}$ and $\mathcal{Z}_{\rho'}(H,t) = \int_H f'(z)d\mathcal{Z}_{\mu}(z,t) = \mathbf{1}$, then, $\frac{\int_H f(z)d\mathcal{Z}_{\mu}(z,t)}{\int_H f'(z)d\mathcal{Z}_{\mu}(z,t)} = \frac{\mathbf{1}}{\mathbf{1}} = \mathbf{1}$, hence

$$\frac{\int_H f(z)d\mathcal{Z}_{\mu}(z,t)}{\int_H f'(z)d\mathcal{Z}_{\mu}(z,t)} \ll \mathcal{Z}_{\mu}. \tag{6.4.23}$$

From (6.4.20), (6.4.22), (6.4.23) we have $\frac{\mathcal{Z}_{\lambda'}}{\mathcal{Z}_{\lambda}} = \frac{\int_H f}{\int_H f'} = \mathbf{1}$, therefore $\mathcal{Z}_{\lambda'} = \mathcal{Z}_{\lambda}$ and $\int_H f = \int_H f'$ and as a result $\mathcal{Z}_{\rho} = \mathcal{Z}_{\rho'}$.

(***II***): If we assume that $\mathcal{Z}_{\mu}$ and $\mathcal{Z}_{\upsilon}$ are two σ-B-$*$-$\mathcal{G}$-$\mathcal{Z}$-$\mathcal{N}$-F-M, then χ is a unique countable disjoint of the μ-bounded and υ-bounded sets. By taking intersections of these, for all $\jmath$ and $\chi = \bigcup_{\jmath=1}^{+\infty} A_{\jmath}$ we obtain a disjoint sequence $\{A_{\jmath}\}_{\jmath} \subseteq C$ such that $\mathcal{Z}_{\mu}(A_{\jmath},t)$ and $\mathcal{Z}_{\upsilon}(A_{\jmath},t)$ are bounded. Define $\mathcal{Z}_{\mu_{\jmath}}(H,t) = \mathcal{Z}_{\mu}(H \cap A_{\jmath},t)$ and $\mathcal{Z}_{\upsilon_{\jmath}}(H,t) = \mathcal{Z}_{\upsilon}((H \cap A_{\jmath},t)$. By the reasoning above, for each $\jmath$ we have $\mathcal{Z}_{\upsilon_{\jmath}} = \mathcal{Z}_{\lambda_{\jmath}} \times \int f_{\jmath} d\mathcal{Z}_{\mu_{\jmath}}(z,t)$, where $\mathcal{Z}_{\lambda_{\jmath}} \perp \mathcal{Z}_{\mu_{\jmath}}$ and $\mathcal{Z}_{\upsilon_{\jmath}} = \mathcal{Z}_{\lambda_{\jmath}} \times \int f_{\jmath} d\mathcal{Z}_{\mu_{\jmath}}(z,t)$. Since $\mathcal{Z}_{\mu_{\jmath}}(A_{\jmath}^c,t) = \mathcal{Z}_{\mu}(A_{\jmath}^c \cap A_{\jmath},t) = \mathcal{Z}_{\mu}(\phi,t) = \mathbf{1}$, and $\mathcal{Z}_{\upsilon_{\jmath}}(A_{\jmath}^c,t) = \mathcal{Z}_{\upsilon}(A_{\jmath}^c \cap A_{\jmath},t) = \mathcal{Z}_{\upsilon}(\phi,t) = \mathbf{1}$, then, $\mathcal{Z}_{\lambda_{\jmath}}(A_{\jmath}^c,t) = \frac{\mathcal{Z}_{\upsilon_{\jmath}}(A_{\jmath}^c,t)}{\int_{A_{\jmath}^c} f_{\jmath} d\mathcal{Z}_{\mu_{\jmath}}(z,t)} = \frac{\mathbf{1}}{\int f_{\jmath}\chi_{A_{\jmath}^c} d\mathcal{Z}_{\mu_{\jmath}}(z,t)}$, hence

$\mathcal{Z}_{\lambda_j}(A_j^c, t) \times \int_{A_j^c} f_j d\mathcal{Z}_{\mu_j}(z,t) = \mathbf{1}$, and we may assume that $f_j = 0$ on A_j^c. If $\mathcal{Z}_\lambda = *_{j=1}^{+\infty}\mathcal{Z}_{\lambda_j}$ and $f = \sum_{j=1}^{+\infty} f_j$ then $\mathcal{Z}_\upsilon = \mathcal{Z}_\lambda \times \int f d\mathcal{Z}_\mu(z,t)$; $\mathcal{Z}_\lambda \perp \mathcal{Z}_\mu$, and $\mathcal{Z}_\lambda$, $\int f d\mathcal{Z}_\mu(z,t)$ are σ-B, as desired. Uniqueness follows as before.

$*$-$\mathcal{G}$-$\mathcal{Z}$-$\mathcal{N}$-F-M-S-Lebesgue-decomposition of $\mathcal{Z}_\upsilon$ is the decomposition $\mathcal{Z}_\upsilon = \mathcal{Z}_\lambda \times \mathcal{Z}_\rho$ with properties $\mathcal{Z}_\lambda \perp \mathcal{Z}_\mu$ and $\mathcal{Z}_\rho \ll \mathcal{Z}_\mu$. Due to the Theorem 6.15, if $\mathcal{Z}_\upsilon \ll \mathcal{Z}_\mu$ we have $\mathcal{Z}_\upsilon = \mathcal{Z}_\rho$. This is the Radon–Nikodym theorem, f is called the Radon–Nikodym derivative with respect to μ and we denote it by $\frac{d\mathcal{Z}_\upsilon}{d\mathcal{Z}_\mu}$, and $d\mathcal{Z}_\upsilon = \frac{d\mathcal{Z}_\upsilon}{d\mathcal{Z}_\mu} d\mathcal{Z}_\mu = f d\mathcal{Z}_\mu$.

Theorem 6.16 *We consider $\mathcal{Z}_{\upsilon_1}, \mathcal{Z}_{\upsilon_2}$ and $\mathcal{Z}_\mu$ which are $*$-$\mathcal{G}$-$\mathcal{Z}$-$\mathcal{N}$-F-M-S on $(\chi, C, *)$, then* $\frac{d(\mathcal{Z}_{\upsilon_1}\times\mathcal{Z}_{\upsilon_2})(z,t)}{d\mathcal{Z}_\mu(z,t)} = \frac{d\mathcal{Z}_{\upsilon_1}(z,t)}{d\mathcal{Z}_\mu(z,t)} \times \frac{d\mathcal{Z}_{\upsilon_2}(z,t)}{d\mathcal{Z}_\mu(z,t)}$.

Proof If $f = \frac{d\mathcal{Z}_{\upsilon_1}(z,t)}{d\mathcal{Z}_\mu}$, $g = \frac{d\mathcal{Z}_{\upsilon_2}(z,t)}{d\mathcal{Z}_\mu(z,t)}$ and $h = \frac{d(\mathcal{Z}_{\upsilon_1}\times\mathcal{Z}_{\upsilon_2})(z,t)}{d\mathcal{Z}_\mu(z,t)}$ then, according to define $\int_H h d\mathcal{Z}_\mu(z,t) = \inf\left\{ \int_H h' d\mathcal{Z}_\mu(z,t) : \int_H h' d\mathcal{Z}_\mu(z,t) \geq (\mathcal{Z}_{\nu_1} \times \mathcal{Z}_{\nu_2})(H,t), \forall H \in C \right\}$, we have $\int_H (f' + g') d\mathcal{Z}_\mu(z,t) = \int_H f' d\mathcal{Z}_\mu(z,t) \times \int_H g' d\mathcal{Z}_\mu(z,t) \geq \mathcal{Z}_{\upsilon_1}(H,t) \times \mathcal{Z}_{\upsilon_2}(H,t)$, then

$$\int_H h d\mathcal{Z}_\mu(z,t) < \int_H f d\mathcal{Z}_\mu(z,t) \times \int_H g d\mathcal{Z}_\mu(z,t) = \int_H (f+g) d\mathcal{Z}_\mu(z,t). \tag{6.4.24}$$

On the other hand $\int_H h' d\mathcal{Z}_\mu(z,t) \geq (\mathcal{Z}_{\upsilon_1} \times \mathcal{Z}_{\nu_2})(H,t) = \mathcal{Z}_{\upsilon_1}(H,t) \times \mathcal{Z}_{\nu_2}(H,t) = \int_H f d\mathcal{Z}_\mu(z,t) \times \int_H g d\mathcal{Z}_\mu(z,t) = \int_H (f+g) d\mathcal{Z}_\mu(z,t)$, hence

$$\begin{aligned} \int_H h d\mathcal{Z}_\mu(z,t) &= \inf\left\{ \int_H h' d\mathcal{Z}_\mu(z,t) : \int_H h' d\mathcal{Z}_\mu(z,t) \geq (\mathcal{Z}_{\upsilon_1} \times \mathcal{Z}_{\upsilon_2})(H,t), \forall H \in C \right\} \\ &\geq \int_H (f+g) d\mathcal{Z}_\mu(z,t). \end{aligned} \tag{6.4.25}$$

From (6.4.24), (6.4.25) we have $\int_H h d\mathcal{Z}_\mu(z,t) = \int_H (f+g) d\mathcal{Z}_\mu(z,t) = \int_H f d\mathcal{Z}_\mu(z,t) \times \int_H g d\mathcal{Z}_\mu(z,t)$, therefore, $\frac{d(\mathcal{Z}_{\upsilon_1}\times\mathcal{Z}_{\upsilon_2})(z,t)}{d\mathcal{Z}_\mu(z,t)} = \frac{d\mathcal{Z}_{\upsilon_1}(z,t)}{d\mathcal{Z}_\mu(z,t)} \times \frac{d\mathcal{Z}_{\upsilon_2}(z,t)}{d\mathcal{Z}_\mu(z,t)}$.

Lemma 6.3 *We assume that $f, g \in L^+$, then $\int_H f d\mathcal{Z}_\mu(z,t) = \int_H g d\mathcal{Z}_\mu(z,t)$ for all $H \in C$ iff $f = g$.*

Proof We consider the simple functions $\phi_n = \sum_{\ell_n=1}^{\kappa_n} a_{\ell_n} \chi_{H_{\ell_n}}$ and $\psi_m = \sum_{J_m=1}^{\iota_m} b_{J_m} \chi_{K_{J_m}}$ and assume that $f = g$ or equivalently $\phi_n \uparrow f$, $\psi_n \uparrow g$. Then

$$\lim_{n\to+\infty} \phi_n = \lim_{m\to+\infty} \psi_m$$

$$\lim_{n\to+\infty} \sum_{\ell_n=1}^{\kappa_n} a_{\ell_n} \chi_{H_{\ell_n}} = \lim_{m\to+\infty} \sum_{J_m=1}^{\iota_m} b_{J_m} \chi_{K_{J_m}}$$

$$\lim_{m,n\to+\infty} \sum_{\ell_n, J_m} a_{\ell_n} \chi_{H_{\ell_n}\cap K_{J_m}} = \lim_{m,n\to+\infty} \sum_{\ell_n, J_m} b_{J_m} \chi_{H_{\ell_n}\cap K_{J_m}}.$$

Therefore $\lim_{m,n\to+\infty} \sum_{\ell_n, J_m} (a_{\ell_n} - b_{J_m}) \chi_{H_{\ell_n}\cap K_{J_m}} = 0$, hence $a_{\ell_n} - b_{J_m} = 0$, then $a_{\ell_n} = b_{J_m}$, since

$$\begin{aligned}
\int_H g d\mathcal{Z}_\mu(z,t) &= \int_H \lim_{m\to+\infty} \psi_m d\mathcal{Z}_\mu(z,t) \\
&= \int_H \lim_{m\to+\infty} \sum_{J_m=1}^{\iota_m} b_{J_m} \chi_{K_{J_m}} d\mathcal{Z}_\mu(z,t) \\
&= \int_H \lim_{m,n\to+\infty} \sum_{\ell_n, J_m} b_{J_m} \chi_{K_{J_m}\cap H_{\ell_n}} d\mathcal{Z}_\mu(z,t) \\
&= \lim_{m,n\to+\infty} \int_H \sum_{\ell_n, J_m} b_{J_m} \chi_{K_{J_m}\cap H_{\ell_n}} d\mathcal{Z}_\mu(z,t) \\
&= \lim_{m,n\to+\infty} \int_H \sum_{\ell_n, J_m} a_{\ell_n} \chi_{K_{J_m}\cap H_{\ell_n}} d\mathcal{Z}_\mu(z,t) \\
&= \int_H \lim_{n\to+\infty} \sum_{\ell_n=1}^{\kappa_n} a_{\ell_n} \chi_{H_{\ell_n}} d\mathcal{Z}_\mu(z,t) \\
&= \int_H \lim_{n\to+\infty} \phi_n d\mathcal{Z}_\mu(z,t) \\
&= \int_H f d\mathcal{Z}_\mu(z,t).
\end{aligned}$$

For every $H \in C$, $\int_H g d\mathcal{Z}_\mu(z,t) = \int_H f d\mathcal{Z}_\mu(z,t)$ and let $g \neq f$. We consider $E = \{z : (f-g)(z) \neq 0\}$ and $F = \{z : (f-g)(z) = 0\}$, then $\chi_E \neq 0$. Since

$$\int_E f d\mathcal{Z}_\mu(z,t) = \int_E g d\mathcal{Z}_\mu(z,t)$$

$$\int_E \lim_{n\to+\infty} \phi_n d\mathcal{Z}_\mu(z,t) = \int_E \lim_{m\to+\infty} \psi_m d\mathcal{Z}_\mu(z,t)$$

$$\lim_{n\to+\infty} \int_E \phi_n d\mathcal{Z}_\mu(z,t) = \lim_{m\to+\infty} \int_E \psi_m d\mathcal{Z}_\mu(z,t)$$

$$\lim_{n\to+\infty}\int_E\sum_{\ell=1}^{\kappa}a_{n,\ell}\chi_{H_{n,\ell}}d\mathcal{Z}_\mu(x,t)=\lim_{m\to+\infty}\int_E\sum_{\jmath=1}^{\iota}b_{m,\jmath}\chi_{K_{m,\jmath}}d\mathcal{Z}_\mu(z,t)$$

$$\lim_{m,n\to+\infty}\int\sum_{\ell,\jmath}a_{n,\ell}\chi_{H_{n,\ell}\cap K_{m,\jmath}\cap E}d\mathcal{Z}_\mu(z,t)=\lim_{m,n\to+\infty}\int\sum_{\ell,\jmath}b_{m,\jmath}\chi_{H_{n,\ell}\cap K_{m,\jmath}\cap E}d\mathcal{Z}_\mu(z,t),$$

then,

$$\frac{\lim_{m,n\to+\infty}*_{\ell,\jmath}\mathcal{Z}_\mu(H_{n,\ell}\cap K_{m,\jmath}\cap E,\frac{t}{a_{n,\ell}})}{\lim_{m,n\to+\infty}*_{\ell,\jmath}\mathcal{Z}_\mu(H_{n,\ell}\cap K_{m,\jmath}\cap E,\frac{t}{b_{m,\jmath}})}=\mathbf{1}$$

$$\lim_{m,n\to+\infty}\frac{*_{\ell,\jmath}\mathcal{Z}_\mu(H_{n,\ell}\cap K_{m,\jmath}\cap E,\frac{t}{a_{n,\ell}})}{*_{\ell,\jmath}\mathcal{Z}_\mu(H_{n,\ell}\cap K_{m,\jmath}\cap E,\frac{t}{b_{m,\jmath}})}=\mathbf{1}$$

$$\lim_{m,n\to+\infty}*_{\ell,\jmath}\frac{\mathcal{Z}_\mu(H_{n,\ell}\cap K_{m,\jmath}\cap E,\frac{t}{a_{n,\ell}})}{\mathcal{Z}_\mu(H_{n,\ell}\cap K_{m,\jmath}\cap E,\frac{t}{b_{m,\jmath}})}=\mathbf{1}$$

$$*_{\ell,\jmath}\lim_{m,n\to+\infty}\frac{\mathcal{Z}_\mu(H_{n,\ell}\cap K_{m,\jmath}\cap E,\frac{t}{a_{n,\ell}})}{\mathcal{Z}_\mu(H_{n,\ell}\cap K_{m,\jmath}\cap E,\frac{t}{b_{m,\jmath}})}=\mathbf{1},$$

hence $a_{n,\ell}=b_{m,\jmath}$ or $f=g$. It is contradiction.

Next, we have the chain rule.

Theorem 6.17 *We consider $\mathcal{Z}_\upsilon,\mathcal{Z}_\mu$ and $\mathcal{Z}_\lambda$ which are $*$-B-$\mathcal{G}$-$\mathcal{Z}$-$\mathcal{N}$-F-M on $(\chi,C,*)$ such that $\upsilon\ll\mu$ and $\mu\ll\lambda$. Then*

(1) If $g\in L^+(\upsilon)$, then $g(\frac{d\mathcal{Z}_\upsilon(z,t)}{d\mathcal{Z}_\mu(z,t)})\in L^+(\mu)$ and $\int gd\mathcal{Z}_\upsilon(z,t)=\int g(\frac{d\mathcal{Z}_\upsilon(z,t)}{d\mathcal{Z}_\mu(z,t)})d\mathcal{Z}_\mu(z,t)$.

(2) We have $\mathcal{Z}_\upsilon\ll\mathcal{Z}_\lambda$, and $\frac{d\mathcal{Z}_\upsilon}{d\mathcal{Z}_\lambda}=\frac{d\mathcal{Z}_\upsilon(z,t)}{d\mathcal{Z}_\mu(z,t)}\cdot\frac{d\mathcal{Z}_\mu(z,t)}{d\mathcal{Z}_\lambda(z,t)}$.

Proof (1) If $g=\chi_H$, then

$$\int gd\mathcal{Z}_\upsilon(z,t)=\int\chi_Hd\mathcal{Z}_\upsilon(z,t)=\int_Hd\mathcal{Z}_\upsilon(z,t)=\mathcal{Z}_\upsilon(H,t),\quad(6.4.26)$$

on the other hand

$$\begin{aligned}\int g\frac{d\mathcal{Z}_\upsilon(z,t)}{d\mathcal{Z}_\mu(z,t)}d\mathcal{Z}_\mu(z,t)&=\int\chi_H\frac{d\mathcal{Z}_\upsilon(z,t)}{d\mathcal{Z}_\mu(z,t)}d\mathcal{Z}_\mu(z,t)\\&=\int_H\frac{d\mathcal{Z}_\upsilon(z,t)}{d\mathcal{Z}_\mu(z,t)}d\mathcal{Z}_\mu(z,t)\\&=\int_Hd\mathcal{Z}_\upsilon(z,t)=\mathcal{Z}_\upsilon(H,t).\end{aligned}\quad(6.4.27)$$

From (6.4.26), (6.4.27) we have $\int \chi_H d\mathcal{Z}_\upsilon(z,t) = \int \chi_H \frac{d\mathcal{Z}_\upsilon(z,t)}{d\mathcal{Z}_\mu(z,t)} d\mathcal{Z}_\mu(z,t)$, let $g = \sum_{\ell=1}^{n} a_\ell \chi_{H_\ell}$, we have $\frac{d\mathcal{Z}_\upsilon(z,t)}{d\mathcal{Z}_\mu(z,t)} = f(z)$ and $\frac{d\mathcal{Z}_\upsilon(z,\frac{t}{a})}{d\mathcal{Z}_\mu(z,\frac{t}{a})} = f(z)$, then $\frac{d\mathcal{Z}_\upsilon(z,t)}{d\mathcal{Z}_\mu(z,t)} = \frac{d\mathcal{Z}_\upsilon(z,\frac{t}{a})}{d\mathcal{Z}_\mu(z,\frac{t}{a})} = f(z)$, hence

$$\begin{aligned}
\int g d\upsilon(z,t) &= \int \sum_{\ell=1}^{n} a_\ell \chi_{H_\ell} d\mathcal{Z}_\upsilon(z,t) \\
&= *_{\ell=1}^{n} \int \chi_{H_\ell} d\mathcal{Z}_\upsilon\left(z, \frac{t}{a_\ell}\right) \\
&= *_{\ell=1}^{n} \int \chi_{H_\ell} \frac{d\mathcal{Z}_\upsilon\left(z, \frac{t}{a_\ell}\right)}{d\mathcal{Z}_\mu\left(z, \frac{t}{a_\ell}\right)} d\mathcal{Z}_\mu\left(z, \frac{t}{a_\ell}\right) \\
&= *_{\ell=1}^{n} \int_{H_\ell} \frac{d\mathcal{Z}_\upsilon\left(z, \frac{t}{a_\ell}\right)}{d\mathcal{Z}_\mu\left(z, \frac{t}{a_\ell}\right)} d\mathcal{Z}_\mu\left(z, \frac{t}{a_\ell}\right) \\
&= *_{\ell=1}^{n} \int_{H_\ell} d\mathcal{Z}_\upsilon\left(z, \frac{t}{a_\ell}\right) \\
&= *_{\ell=1}^{n} \mathcal{Z}_\upsilon(H_\ell, \frac{t}{a_\ell}),
\end{aligned} \tag{6.4.28}$$

on the other hand

$$\begin{aligned}
\int g \frac{d\mathcal{Z}_\upsilon(z,t)}{d\mathcal{Z}_\mu(z,t)} d\mathcal{Z}_\mu(z,t) &= \int \sum_{\ell=1}^{n} a_\ell \chi_{H_\ell} \frac{d\mathcal{Z}_\upsilon(z,t)}{d\mathcal{Z}_\mu(z,t)} d\mathcal{Z}_\mu(z,t) \\
&= *_{\ell=1}^{n} \int \chi_{H_\ell} \frac{d\mathcal{Z}_\upsilon\left(z, \frac{t}{a_\ell}\right)}{d\mathcal{Z}_\mu(z, \frac{t}{a_\ell})} d\mathcal{Z}_\mu(z, \frac{t}{a_\ell}) \\
&= *_{\ell=1}^{n} \int \chi_{H_\ell} d\mathcal{Z}_\upsilon(z, \frac{t}{a_\ell}) \\
&= *_{\ell=1}^{n} \int_{H_\ell} d\mathcal{Z}_\upsilon(z, \frac{t}{a_\ell}) \\
&= *_{\ell=1}^{n} \mathcal{Z}_\upsilon(H_\ell, \frac{t}{a_\ell}).
\end{aligned} \tag{6.4.29}$$

From (6.4.28), (6.4.29) we have $\int g d\mathcal{Z}_\upsilon(z,t) = \int g \frac{d\mathcal{Z}_\upsilon(z,t)}{d\mathcal{Z}_\mu(z,t)} d\mathcal{Z}_\mu(z,t)$. If $g \in L^+(\upsilon)$, according to the monotone convergence theorem, there exist $\phi_n \uparrow g$ and

$$\begin{aligned}
\int g d\mathcal{Z}_{\upsilon}(z,t) &= \int \lim_{n\to+\infty} \phi_n d\mathcal{Z}_{\upsilon}(z,t) \\
&= \lim_{n\to+\infty} \int \phi_n d\mathcal{Z}_{\upsilon}(z,t) \\
&= \lim_{n\to+\infty} \int \phi_n \frac{d\mathcal{Z}_{\upsilon}(z,t)}{d\mathcal{Z}_{\mu}(z,t)} d\mathcal{Z}_{\mu}(z,t) \\
&= \int \lim_{n\to+\infty} \phi_n \frac{d\mathcal{Z}_{\upsilon}(z,t)}{d\mathcal{Z}_{\mu}(z,t)} d\mathcal{Z}_{\mu}(z,t) \\
&= \int g \frac{d\mathcal{Z}_{\upsilon}(z,t)}{d\mathcal{Z}_{\mu}(z,t)} d\mathcal{Z}_{\mu}(z,t).
\end{aligned}$$

(2) From (1) we have $\int g d\mathcal{Z}_{\upsilon}(z,t) = \int g \frac{d\mathcal{Z}_{\upsilon}(z,t)}{d\mathcal{Z}_{\mu}(z,t)} d\mathcal{Z}_{\mu}(z,t)$, replacing $\mathcal{Z}_{\upsilon}, \mathcal{Z}_{\mu}$ by $\mathcal{Z}_{\mu}, \mathcal{Z}_{\lambda}$ we have $\int g d\mathcal{Z}_{\mu}(z,t) = \int g \frac{d\mathcal{Z}_{\mu}(z,t)}{d\mathcal{Z}_{\lambda}(z,t)} d\mathcal{Z}_{\lambda}(z,t)$, Now, we set $g = \chi_H(\frac{d\mathcal{Z}_{\upsilon}(z,t)}{d\mathcal{Z}_{\mu}(z,t)})$, then $\int \chi_H(\frac{d\mathcal{Z}_{\upsilon}(z,t)}{d\mathcal{Z}_{\mu}(z,t)}) d\mathcal{Z}_{\mu}(z,t) = \int \chi_H(\frac{d\mathcal{Z}_{\upsilon}(z,t)}{d\mathcal{Z}_{\mu}(z,t)}).(\frac{d\mathcal{Z}_{\mu}(z,t)}{d\mathcal{Z}_{\lambda}(z,t)}) d\mathcal{Z}_{\lambda}$, hence

$$\begin{aligned}
\int \chi_H d\mathcal{Z}_{\upsilon}(z,t) &= \int \chi_H\left(\frac{d\mathcal{Z}_{\upsilon}(z,t)}{d\mathcal{Z}_{\mu}(z,t)}\right).\left(\frac{d\mathcal{Z}_{\mu}(z,t)}{d\mathcal{Z}_{\lambda}(z,t)}\right) d\mathcal{Z}_{\lambda} \\
\int \chi_H\left(\frac{d\mathcal{Z}_{\upsilon}(z,t)}{d\mathcal{Z}_{\lambda}(z,t)}\right) d\mathcal{Z}_{\lambda} &= \int \chi_H\left(\frac{d\mathcal{Z}_{\upsilon}(z,t)}{d\mathcal{Z}_{\mu}(z,t)}\right).\left(\frac{d\mathcal{Z}_{\mu}(z,t)}{d\mathcal{Z}_{\lambda}(z,t)}\right) d\mathcal{Z}_{\lambda} \\
\int_H \left(\frac{d\mathcal{Z}_{\upsilon}(z,t)}{d\mathcal{Z}_{\lambda}(z,t)}\right) d\mathcal{Z}_{\lambda} &= \int_H \left(\frac{d\mathcal{Z}_{\upsilon}(z,t)}{d\mathcal{Z}_{\mu}(z,t)}\right).\left(\frac{d\mathcal{Z}_{\mu}(z,t)}{d\mathcal{Z}_{\lambda}(z,t)}\right) d\mathcal{Z}_{\lambda},
\end{aligned}$$

from Lemma 6.3 we have $\frac{d\mathcal{Z}_{\upsilon}(z,t)}{d\mathcal{Z}_{\lambda}(z,t)} = \left(\frac{d\mathcal{Z}_{\upsilon}(z,t)}{d\mathcal{Z}_{\mu}(z,t)}\right) \cdot \left(\frac{d\mathcal{Z}_{\mu}(z,t)}{d\mathcal{Z}_{\lambda}(z,t)}\right)$.

6.4.3 *Differentiation in Euclidean Space*

Now, by considering the Radon–Nikodym theorem, the abstract concept of the "derivative" of $*$-$\mathcal{G}$-$\mathcal{Z}$-$\mathcal{N}$-F-M with respect to a $*$-$\mathcal{G}$-$\mathcal{Z}$-$\mathcal{N}$-F-M $\mathcal{Z}_{\mu}$ was presented. Now, we investigate the $*$-B-$\mathcal{G}$-$\mathcal{Z}$-$\mathcal{N}$-F special case that $(\chi, C, *) = (\mathbb{R}^{\kappa}, \mathcal{B}_{\mathbb{R}^{\kappa}}, *)$ and $\mathcal{Z}_{\mu} = \mathcal{Z}_m$ is Lebesgue $*$-$\mathcal{G}$-$\mathcal{Z}$-$\mathcal{N}$-F-M. Here, we define a pointwise derivative of $\mathcal{Z}_{\upsilon}$ with respect to $\mathcal{Z}_m$. We consider a $\mathcal{G}$-$\mathcal{Z}$-$\mathcal{N}$ as $\mathcal{Z}_N(z,t) = \exp(\frac{-|z|}{t})$. If $\mathcal{B}(r,z,t) = \{y \in \chi : \mathcal{Z}_N(z-y,t) > \mathbf{1-r}\}$ $0 \le r \le 1$, is the open ball of radius r about z in $\mathbb{R}^{\kappa}$, then we define

$$\mathcal{F}(z) = \lim_{r\to 0} \frac{\nu(\mathcal{B}(r,z,t),t)}{m(\mathcal{B}(r,z,t),t)},$$

If $\mathcal{Z}_\upsilon \ll \mathcal{Z}_m$, such that $d\mathcal{Z}_\upsilon = f d\mathcal{Z}_m$, then $\frac{\mathcal{Z}_\upsilon(\mathcal{B}(r,z,t),t)}{\mathcal{Z}_m(\mathcal{B}(r,z,t),t)}$ is simply the average value of f on $\mathcal{B}(r, z, t)$, therefore we have that $\mathcal{F} = f$. This is the desired state provided that $\mathcal{Z}_\upsilon(\mathcal{B}(r, z, t), t)$ is finite for every r, z. f is the derivative of the indefinite integral of υ. From here on in the article, expressions like "$*$-$\mathcal{G}$-$\mathcal{Z}$-$\mathcal{N}$-F-integrable" and "almost everywhere" refer to $*$-$\mathcal{G}$-$\mathcal{Z}$-$\mathcal{N}$-F-M.

Definition 6.14 Let $f : \mathbb{R}^n \to \mathbb{R}^+$ be a $*$-$\mathcal{G}$-$\mathcal{Z}$-$\mathcal{N}$-F-measurable-function. f is locally-$*$-$\mathcal{G}$-$\mathcal{Z}$-$\mathcal{N}$-F-integrable with respect to L-$*$-$\mathcal{G}$-$\mathcal{Z}$-$\mathcal{N}$-M, if for every bounded measurable set $H \subseteq \mathbb{R}^n$, $\int_H f(z)dm(z, t) > 0$ and denote it $(L^+)_{loc}$.

Theorem 6.18 *If $f \in (L^+)_{loc}$, then for $z \in \mathbb{R}^n$,* $\lim_{r\to 0} \frac{\int_{\mathcal{B}(r,z,t)} \mathcal{Z}_N(f(z)-f(y),t)d\mathcal{Z}_m(y,t)}{\mathcal{Z}_m(\mathcal{B}(r,z,t),t)} = \mathbf{1}$.

Proof For $n_0 \in \mathbb{N}$ and $z \in \mathbb{R}^n$, it is enough to show that $\frac{\int_{\mathcal{B}} \mathcal{Z}_N((f(z)-f(y)),t)d\mathcal{Z}_m(y,t)}{\mathcal{Z}_m(\mathcal{B}(r,z,t),t)} \to \mathbf{1}$ with $|z| \le n_0$. But the value of $\mathcal{Z}_N((f(z) - f(y)), t) = \exp(\frac{-|f(z)-f(y)|}{t})$ for $|z| \le n_0$, $r \le 1$ and $|y| \le n_0 + 1$ depends only on the value of $f(y)$. Now, by placing $f\chi_{\mathcal{B}(\frac{1}{n_0+1},0,t)}$ instead of f, we have $f \in L^+$. If $0 < \varepsilon < 1$, there exist a continuous integrable function g such that $|f - g| < \varepsilon$, since $\mathcal{Z}_N(f(z) - f(y), t) \le \mathbf{1}$, then $\int_{\mathcal{B}(r,z,t)} \mathcal{Z}_N(f(z) - f(y), t)d\mathcal{Z}_m(y, t) \ge \int_{\mathcal{B}(r,z,t)} d\mathcal{Z}_m(y, t)$, hence

$$\begin{aligned}
\mathbf{1} &\le \frac{\int_{\mathcal{B}(r,z,t)} \mathcal{Z}_N((f(z) - f(y)), t)d\mathcal{Z}_m(y, t)}{\mathcal{Z}_m(\mathcal{B}(r, z, t), t)} \\
&= \frac{\int_{\mathcal{B}(r,z,t)} \mathcal{Z}_N((f(z) - g(z)) + (g(z) - g(y)) + (g(y) - f(y)), t)d\mathcal{Z}_m(y, t)}{\mathcal{Z}_m(\mathcal{B}(r, z, t), t)} \\
&\le \frac{\int_{\mathcal{B}(r,z,t)} \mathcal{Z}_N((f(z) - g(z)), \frac{t}{3}) \times \mathcal{Z}_N((g(z) - g(y)), \frac{t}{3}) \times \mathcal{Z}_N((g(y) - f(y)), \frac{t}{3})d\mathcal{Z}_m(y, t)}{\mathcal{Z}_m(\mathcal{B}(r, z, t), t)} \\
&\le \frac{\int_{\mathcal{B}(r,z,t)} \exp(\frac{-3|f(z)-g(z)|}{t}) \times \exp(\frac{-3|g(z)-g(y)|}{t}) \times \exp(\frac{-3|g(y)-f(y)|}{t})d\mathcal{Z}_m(y, t)}{\mathcal{Z}_m(\mathcal{B}(r, z, t), t)} \\
&\le \frac{\int_{\mathcal{B}(r,z,t)} \exp(\frac{-3(\varepsilon+\varepsilon+\varepsilon)}{t})d\mathcal{Z}_m(y, t)}{\mathcal{Z}_m(\mathcal{B}(r, z, t), t)} \\
&\to \frac{\int_{\mathcal{B}(r,z,t)} d\mathcal{Z}_m(y, t)}{\mathcal{Z}_m(\mathcal{B}(r, z, t), t)} \quad \text{whenever } \varepsilon \to 0 \\
&= \frac{\mathcal{Z}_m(\mathcal{B}(r, z, t), t)}{\mathcal{Z}_m(\mathcal{B}(r, z, t), t)} \\
&= \mathbf{1}.
\end{aligned}$$

Definition 6.15 If $f \in (L^+)_{loc}$, we define the Lebesgue set of f as follows

$$L_f = \left\{ z : \lim_{r\to 0} \frac{\int_{\mathcal{B}(r,z,t)} \mathcal{Z}_N(f(z) - f(y), t)d\mathcal{Z}_m(y, t)}{\mathcal{Z}_m(\mathcal{B}(r, z, t), t)} = \mathbf{1} \right\}.$$

Theorem 6.19 *If $f \in (L^+)_{loc}$ then $\mathcal{Z}_m((L_f)^c, t) = \mathbf{1}$.*

Proof For every $a \in \mathbb{R}$, we consider $g_a(z) = |z - a|$. According to Theorem 6.18, expect on Lebesgue null set H_a, we have

$$\lim_{r\to 0} \frac{\int_{\mathcal{B}(r,z,t)} \mathcal{Z}_N(g_a(z) - g_a(y), t)d\mathcal{Z}_m(y,t)}{\mathcal{Z}_m(\mathcal{B}(r,z,t),t)} = \mathbf{1}. \tag{6.4.30}$$

If $D \subseteq \mathbb{R}$ be a countable dense, we set $H = \bigcup_{a\in D} H_a$, then, $\mathcal{Z}_m(H,t) = \mathcal{Z}_m\left(\bigcup_{a\in D} H_a\right) = *_{a\in D}\mathcal{Z}_m(H_a,t) = *_{a\in D}\mathbf{1} = \mathbf{1}$, and if $z \notin H$, for any $0 < \varepsilon < 1$, there exist $a \in D$ such that $|f(z) - a| < \varepsilon$, then $|f(z) - f(y)| \le |(f(z) - a) + (a - f(y))| \le |f(z) - a| + |f(y) - a| \le \varepsilon + |f(y) - a|$, then, for $z \in D$ the relation (6.4.30) convert to

$$\begin{aligned}
&\lim_{r\to 0} \frac{\int_{\mathcal{B}(r,z,t)} \mathcal{Z}_N(g_a(z) - g_a(y), t)d\mathcal{Z}_m(y,t)}{\mathcal{Z}_m(\mathcal{B}(r,z,t),t)} \\
&= \frac{\lim_{r,\varepsilon\to 0} \int_{\mathcal{B}(r,z,t)} \exp(\frac{-|g_a(z)-g_a(y)|}{t})d\mathcal{Z}_m(y,t)}{\lim_{r\to 0} \mathcal{Z}_m(\mathcal{B}(r,z,t),t)} \\
&= \frac{\int_{\mathcal{B}(r,z,t)} \lim_{r,\varepsilon\to 0} \exp(\frac{-||f(z)-a|-|f(y)-a||}{t})d\mathcal{Z}_m(y,t)}{\lim_{r\to 0} \mathcal{Z}_m(\mathcal{B}(r,z,t),t)} \\
&= \frac{\int_{\mathcal{B}(r,z,t)} \exp(\frac{-|f(y)-a|}{t})d\mathcal{Z}_m(y,t)}{\lim_{r\to 0} \mathcal{Z}_m(\mathcal{B}(r,z,t),t)} \\
&= \mathbf{1}.
\end{aligned} \tag{6.4.31}$$

On the other hand

$$\begin{aligned}
1 &\le \lim_{r\to 0} \frac{\int_{\mathcal{B}(r,z,t)} \mathcal{Z}_N(f(z) - f(y), t)d\mathcal{Z}_m(y,t)}{\mathcal{Z}_m(\mathcal{B}(r,z,t),t)} \\
&= \frac{\lim_{r,\varepsilon\to 0} \int_{\mathcal{B}(r,z,t)} \exp(\frac{-|f(z)-f(y)|}{t})d\mathcal{Z}_m(y,t)}{\lim_{r\to 0} \mathcal{Z}_m(\mathcal{B}(r,z,t),t)} \\
&\le \frac{\lim_{r,\varepsilon\to 0} \int_{\mathcal{B}(r,z,t)} \exp(\frac{-||f(z)-a|+|f(y)-a||}{t})d\mathcal{Z}_m(y,t)}{\lim_{r\to 0} \mathcal{Z}_m(\mathcal{B}(r,z,t),t)} \\
&\le \frac{\lim_{r,\varepsilon\to 0} \int_{\mathcal{B}(r,z,t)} \exp(\frac{-\varepsilon-|f(y)-a|}{t})d\mathcal{Z}_m(y,t)}{\lim_{r\to 0} \mathcal{Z}_m(\mathcal{B}(r,z,t),t)}
\end{aligned}$$

$$= \frac{\int_{\mathcal{B}(r,z,t)} \lim_{r,\varepsilon \to 0} \exp(\frac{-\varepsilon - |f(y)-a|}{t}) d\mathcal{Z}_m(y,t)}{\lim_{r \to 0} \mathcal{Z}_m(\mathcal{B}(r,z,t),t)}$$

$$= \frac{\int_{\mathcal{B}(r,z,t)} \exp(\frac{-|f(y)-a|}{t}) d\mathcal{Z}_m(y,t)}{\lim_{r \to 0} \mathcal{Z}_m(\mathcal{B}(r,z,t),t)}. \tag{6.4.32}$$

From (6.4.31), (6.4.32), we have

$$\mathbf{1} \le \lim_{r \to 0} \frac{\int_{\mathcal{B}(r,z,t)} \mathcal{Z}_N(f(z)-f(y),t) d\mathcal{Z}_m(y,t)}{\mathcal{Z}_m(\mathcal{B}(r,z,t),t)} \le \frac{\int_{\mathcal{B}(r,z,t)} \exp(\frac{-|f(y)-a|}{t}) d\mathcal{Z}_m(y,t)}{\lim_{r \to 0} \mathcal{Z}_m(\mathcal{B}(r,z,t),t)} = \mathbf{1},$$

hence

$$\lim_{r \to 0} \frac{\int_{\mathcal{B}(r,z,t)} \mathcal{Z}_N(f(z)-f(y),t) d\mathcal{Z}_m(y,t)}{\mathcal{Z}_m(\mathcal{B}(r,z,t),t)} = \mathbf{1}.$$

References

1. Ahadi, A., Saadati, R.: Generalized Z-number approximation for the fractional two-point iterative equation with a boundary condition. Inform. Sci. **649**, 119673 (2023)
2. Aliev, R.A., Alizadeh, A.V., Huseynov, O.H.: The arithmetic of discrete Z-numbers. Inform. Sci. **290**, 134–155 (2015)
3. Aliev, R.A., Alizadeh, A.V., Huseynov, O.H.: The arithmetic of discrete Z-numbers. Inform. Sci. **290**, 134–155 (2015)
4. An, J.S.: On an additive functional inequality in normed modules over a C^*-algebra. J. Korea Soc. Math. Educ. Ser. B Pure Appl. Math. **15**, 393–400 (2008)
5. Cădariu, Liviu, Radu, Viorel. Fixed points and the stability of Jensen's functional equation. *JIPAM. J. Inequal. Pure Appl. Math* 4 **2003**, no. 1, Article 4, 7 pp
6. Diaz, J., Margolis, B.: A fixed point theorem of the alternative for contractions on a generalized complete metric space. Bull. Amer. Math. Soc. **74**, 305–309 (1968)
7. Eidinejad, Z; Saadati, R.: Hyers-Ulam-Rassias-Wright Stability for Fractional Oscillation Equation. Discrete Dynamics in Nature and Society, no. 1–7 (2022)
8. Ezadi, S., Allahviranloo, T.: Numerical solution of linear regression based on Z-numbers by improved neural network. Intell. Autom. Soft Comput. 1–11 (2017)
9. Ezadi, S., Allahviranloo, T.: Numerical solution of linear regression based on Z-numbers by improved neural network. Intell. Autom. Soft Comput. 1–11 (2017)
10. Ghaffari, A.; Saadati, R. $*$-fuzzy measure model for COVID-19 disease. Adv. Difference Equ. Paper No. 202, 18 pp (2021)
11. Jarad, F., Abdeljawad, T.: A modified Laplace transform for certain generalized fractional operators. Results Nonlinear Anal. **1**, 88–98 (2018)
12. Kang, B., Wei, D., Li, Y., Deng, Y.: Decision making using Z-numbers under uncertain environment. J. Comput. Inf. Syst. **8**(7), 2807–2814 (2012)
13. Kang, B., Wei, D., Li, Y., Deng, Y.: Decision making using Z-numbers under uncertain environment. J. Comput. Inf. Syst. **8**(7), 2807–2814 (2012)
14. Krasnoselskii, M.A., Rutickii, YB.: Convex. Functions and Orlicz Spaces. Russian, Fizmatgiz, Moscow. Noordhoff, Groningen, English translation (1961)

15. Li, K., Peng, J.: Laplace transform and fractional differential equations. Appl. Math. Lett. **24**(12), 2019–2023 (2011)
16. Luxemburg, W.A.J.: On the convergence of successive approximations in the theory of ordinary differential equations, II, Koninkl, Nederl. Akademie van Wetenschappen, Amsterdam, Proc. Ser. A (5) **61**, and Indag. Math. (5) **20**, 540–546 (1958)
17. Musielak, J., Orlicz, W.: On modular space. Stud. Math. **18**, 49–65 (1959)
18. Nakano, H.: Modular Semi-Ordered Spaces. Maruzen Co., Ltd., Tokyo, Japan (1950)
19. Pirmuhammadi, S., Allahviranloo, T., Keshavarz, M.: The parametric form of Z-number and its application in Z-number initial value problem. Int. J. Intell. Syst. **32**(10), 1030–1061 (2017)
20. Pirmuhammadi, S., Allahviranloo, T., Keshavarz, M.: The parametric form of Z-number and its application in Z-number initial value problem. Int. J. Intell. Syst. **32**(10), 1030–1061 (2017)
21. Saadati, Reza: Nonlinear contraction and fuzzy compact operator in fuzzy Banach algebras. Fixed Point Theory **20**(1), 289–297 (2019)
22. Sevgin, S., Sevli, H.: Stability of a nonlinear Volterra integro-differential equation via a fixed point approach. J. Nonlinear Sci. Appl. **9**(1), 200–207 (2016)
23. Shen, Y., Chen, W.: Laplace transform method for the Ulam stability of linear fractional differential equations with constant coefficients. Mediterr. J. Math. **14**(1), Paper No. 25, 17 pp (2017)
24. Sousa, J.V.daC., de Oliveira, E.C.: On a new operator in fractional calculus and applications. J. Fixed Point Theory Appl. **20**(3), Paper No. 96, 21 pp (2018)
25. Sousa, J.V.daC., de Oliveira, E.C.: On the Φ-Hilfer fractional derivative. Commun. Nonlinear Sci. Numer. Simul. **60**, 72–91 (2018)
26. Sousa, J.V.daC., Fabio G., Capelas de O.E.: Stability of the fractional Volterra integro-differential equation by means of ψ-Hilfer operator. Math. Methods Appl. Sci. **42**(9), 3033–3043 (2019)
27. Yager, R.R.: On Z-valuations using Zadeh's Z-numbers. Int. J. Intell. Syst. **27**(3), 259–278 (2012)
28. Yager, R.R.: On Z-valuations using Zadeh's Z-numbers. Int. J. Intell. Syst. **27**(3), 259-278 (2012)
29. Zadeh, L.A.: A note on Z-numbers. Inform. Sci. **181**(14), 2923–2932 (2011)
30. Zadeh, L.A.: A note on Z-numbers. Inform. Sci. **181**(14), 2923–2932 (2011)

Chapter 7
Analysis of Numerical Methods

Considering the importance of using nonlinear evolution equations in the investigation of many natural phenomena, we consider the $(2+1)$-dimensional Date–Jimbo–Kashiwara–Miwa ($(2+1)$-dimensional DJKM) equation. Using the multiple exp functions method, we obtain analytical solutions for this equation, which are one-soliton, two-soliton, and three-soliton solutions and these solutions include three categories of soliton wave solutions, i.e., one-wave solutions, two-wave solutions, and three wave solutions. We have performed all calculations with a computer algebra system such as Maple and have also provided a graphical representation of the obtained solutions. In the following, we use the new clique polynomial method to obtain a numerical solution to the two-dimensional stochastic time-fractional Sine-Gordon equation (2D ST-FS-G). In this technique, the clique polynomial is considered a basic function for operational matrices. By converting the 2D ST-FS-G equation to algebraic equations, a solution is obtained for the desired equation, which clearly shows that this approach is suitable and accurate for dealing with the equation. We present the error boundary for the obtained approximation of the desired three-variable function based on the clique polynomial. Finally, we compare some numerical results obtained with the exact ones by a few practical examples. Another equation that is investigated with the new clique polynomial method (CP-M) is a mathematical model of a new virus called Coronavirus, which causes an infectious disease called COVID-19. By solving this system of equations, we check the transmissibility of the Coronavirus by the CPs method. In particular, we turn the system of differential equations into an algebraic system to obtain solutions. Here, we compare the numerical results obtained by the CP method with the numerical results of other methods. After checking the stability of the fractional hybrid integrodifferential equations with Dirichlet boundary conditions, we present the reproducing kernel Hilbert space numerical method to obtain approximate solutions to the equation. After the complete explanation of the presented method, we obtain the numerical

Z. Eidinejad et al., *Optimal Stability Theory and Approximate Solutions of Fractional Systems*, Studies in Systems, Decision and Control 606,
https://doi.org/10.1007/978-3-031-96704-7_7

results by considering the numerical examples of the equation. For Navier–Stokes equations with dimensions two and three with initial and boundary conditions, both in constructing the formula and in solving numerical examples of the N-SE, the integer order, and fractional order are considered and investigated.

7.1 The Multiple Exp-Function Method

7.1.1 Introduction of the Conformable Date–Jimbo–Kashiwara–Miwa Equation

In nonlinear sciences, nonlinear partial equations are widely used in fields such as nonlinear optics, quantum mechanics, molecular biology, surface water waves, and other cases. On the other hand, many physical phenomenon that exist in nature are described by exact solutions of nonlinear partial differential equations (PDEs). Nonlinear wave solutions play a fundamental role in various scientific fields such as geochemistry, optical fibers, chemical physics, fluid mechanics, plasma physics, solid-state physics, chemical kinematics, etc. Therefore, for nonlinear models in nature and industry, analytical and semi-analytical solutions have been investigated with many methods including the tanh-coth method, the sine-cosine method, the tanh-function method, the homogeneous balance method, inverse scattering transformation, Darboux transformation, Hirota bilinear method and so on [22, 23, 32, 72, 83, 87, 126, 131]. For example, the generalized bilinear equation of Kadomtsev Petviashvili–Benjamin–Bona–Mahony was investigated by Hirota's efficient bilinear method and exact solutions were obtained from this equation. By converting nonlinear equations to ordinary differential equations (ODEs) with variable coefficients, we can use these techniques to obtain logical solutions. Among the methods for accurate solutions of nonlinear equations, there is now a single method that determines the relationship between ordinary solvable differential equations and nonlinear partial differential equations. Nowadays, various solutions such as periodic waves, single waves and so on, have been proposed for scattered and nonlinear scattered equations. The authors also found in their studies that by using the substitution of high-order polynomial functions in bilinear equations, the exact solutions of some nonlinear equations that are of the type of rogue-type multiple lumps waves can be constructed [13, 48, 49, 107, 135]. There are various solutions to nonlinear equations, and these solutions are not limited to traveling wave solutions. Then, there are direct methods to obtain these solutions for nonlinear equations. For example, for important equations such as the Korteweg–deVries (KdV) equations, the the Toda lattice equations, and the Hirota bilinear equations, there are multi-soliton multiple wave solutions and multiple periodic wave solutions, respectively [53, 82, 84, 143]. Lump-type solutions and interaction phenomenon as well as lump wave solutions have been obtained for the (2+1) dimensional Breaking Soliton equation, variable-coefficient Kadomtsev–Petviashvili equation, and Kadomtsev–Petviashvili–Benjamin–Bona–Mahony equation, respectively. Describing the propagation of nonlinear waves with non-uniform velocities is another task provided by the nonlinear evolution equations and their solutions. Many characteristics of natural phenomena with continuous and variable fields are shown by converting a nonlinear evolution equation into a non-linear evolution equation of a integrated. One of the important long water wave equations that describe the propagation of nonlinear and weakly scattered waves in heterogeneous media is Date–Jimbo–Kashiwara Miwa equation [119]. We investigate the conformable Date–Jimbo–Kashiwara Miwa (CDJKM) equation as follows [58]:

$$\Phi_{uuuuv} + 4\Phi_{uuv}\Phi_u + 2\Phi_{uuu}\Phi_v + 6\Phi_{uv}\Phi_{uu} - \lambda\Phi_{vvv} - 2\mu\frac{\partial}{\partial u}\left(\frac{\partial}{\partial u}\left(\frac{\partial^\zeta \Phi}{\partial s^\zeta}\right)\right) = 0, \tag{7.1.1}$$

where $\Phi = \Phi(u, v, s)$ and μ is a constant and $0 < \zeta \leq 1$. For $0 < \zeta < 1$, the desired derivative is the conformable fractional derivative which is defined as follows for the function $\mathfrak{u} : [a, \infty) \to \mathbb{R}$ and starting from a of order ζ:

$$\mathfrak{D}^a_\zeta \mathfrak{u}(\mathfrak{d}) = \lim_{\iota \to 0} \frac{\mathfrak{u}\left(\mathfrak{d}+\iota(\mathfrak{d}-a)^{1-\zeta}\right)-\mathfrak{u}(\mathfrak{d})}{\iota}, \quad \mathfrak{d} > a, \ 0 < \zeta < 1.$$

If on (a, b), $\mathfrak{D}^a_\zeta \mathfrak{u}(\mathfrak{d})$ exists, then $\mathfrak{D}^a_\zeta \mathfrak{u}(a) = \lim_{\mathfrak{d} \to a^+} \mathfrak{D}^a_\zeta \mathfrak{u}(\mathfrak{d})$. The CDJKM equation turns out to be $(2+1)$-dimensional Date–Jimbo–Kashiwara–Miwa ($(2+1)$-dimensional DJKM) equation when the fractional derivative $\zeta = 1$. For the first time, Guo and Lin introduced the CDJKM equation and Kumar et al. successfully investigated the solutions for this equation via the sine-Gordon expansion and the improved Bernoulli sub equation methods [43, 69]. Many researchers have investigated the $(2+1)$-dimensional Date–Jimbo–Kashiwara–Miwa ($(2+1)$-dimensional DJKM) equation [117, 125, 139], but no one used our desired method to solve the CDJKM equation. The multiple exp function method is a direct and systematic solution method for nonlinear partial differential equations in mathematical physics that generalizes Hirota's perturbation scheme and obtains exact multiple wavelet solutions. This method is a suitable method for obtaining construct solitary and soliton solutions of nonlinear evolution equations because it has a wider application for use in nonlinear wave equations. This method was first presented by Hi and Wu and it can be used in all kinds of nonlinear evolution equations. Considering the optimality of the multiple exp function method in dealing with nonlinear equations, we choose this method to obtain the exact traveling wave solutions of the desired equation. Therefore, one of our reasons for choosing this method is to emphasize its power because it can be used for various nonlinear equations. This method is a new tool in mathematics to obtain solitary and periodic solutions and also is a coherent framework in which other methods can be placed.

We propose a multiple exp function method and investigate precise wave one-soliton, wave two-soliton and wave three-soliton solutions to the $(2+1)$-dimensional Date–Jimbo–Kashiwara–Miwa ($(2+1)$-dimensional DJKM) equation.

7.1.2 Description of Multiple Exp-Function Method

We consider the $(1+1)$-dimensional PDE

$$\mathcal{H}(u, s, \Phi_u, \Phi_s, \ldots) = 0.$$

We present the proposed method for this equation, which is a type of differential polynomials. This method is also valid for nonlinear equations and higher dimensions of equations.

- In the first step, we introduce a sequence of new variables that includes partial differential equations with a solution. $\Theta_\ell = \Theta_\ell(u, s)$, $1 \leqslant \ell \leqslant \mathrm{m}$, by solvable PDEs, where, $\Theta_{\ell,u} = \gamma_\ell \Theta_\ell, \quad \Theta_{\ell,s} = -\tau_\ell \Theta_\ell, \quad 1 \leqslant \ell \leqslant \mathrm{m}$ and $\gamma_\ell, \quad 1 \leqslant \ell \leqslant \mathrm{m}$, are the angular wave numbers and τ_ℓ, $1 \leqslant \ell \leqslant \mathrm{m}$, are the wave frequencies. Since nonlinear differential equations are not directly solvable, by going through the first step we can make accurate solutions of nonlinear equations. By solving linear equations as follows,

  $$\Theta_\ell = \kappa_\ell \exp(\varkappa_\ell), \quad \varkappa_\ell = \gamma_\ell u - \tau_\ell s, \quad 1 \leqslant \ell \leqslant \mathrm{m}$$

  where κ_ℓ, $1 \leqslant \ell \leqslant \mathrm{m}$, are any constants, positive or negative, we can obtain exponential function solutions for these equations.
  By selecting the desired κ_ℓ, $1 \leqslant \ell \leqslant \mathrm{m}$ constants, there will be more choices for the solutions obtained, see [68] for more details. The function Θ_ℓ, $1 \leqslant \ell \leqslant \mathrm{m}$ is a combination of all single waves. Each function Θ_ℓ also describes a single wave and a multiple wave solution.
  According to the obtained solutions, the reason for naming the introduced method is determined Solutions obtained above offer reasons as to why the approach is called the multiple exp-function method.
  The introduction of linear differential equations and their application to other systems is also used, in which case the solutions are varied [81]. The use of linear differential equations is used in various methods to accurately solve partial differential equations (PDEs). For example, in T-F-type method, the mapping method, the F-E-type methods, the JEF method, and G'/G-E method uses this idea [57, 58, 73, 99, 140].
- In this step, we consider solutions using the new variables Θ_ℓ, $1 \leqslant \ell \leqslant \mathrm{m}$

$$\Phi(u,s) = \frac{\Psi\,(\Theta_1, \Theta_2, \ldots, \Theta_{\rm m})}{\Upsilon\,(\Theta_1, \Theta_2, \ldots, \Theta_{\rm m})}, \tag{7.1.2}$$

$$\Psi = \sum_{p,q=1}^{\rm m} \sum_{\ell,J=0}^{\rm I} \Psi_{pq,\ell J}\,\Theta_p^{\ell}\Theta_q^{J},$$

$$\Upsilon = \sum_{p,q=1}^{\rm m} \sum_{\ell,J=0}^{\rm J} \Upsilon_{pq,\ell J}\,\Theta_p^{\ell}\Theta_q^{J}$$

where $\Psi_{pq,\ell J}$ and $\Upsilon_{pq,\ell J}$ are constants. These constants are determined by the main $(1+1)$-dimensional partial differential equation ($(1+1)$-dimensional PDE). Consider the differential equations $\Theta_{\ell,u}$ and $\Theta_{\ell,s}$. Using these equations, we can express the partial derivatives of Φ in terms of Θ_ℓ, $1 \leqslant \ell \leqslant {\rm m}$ with u and s. If Ψ_{Θ_ℓ} is a partial derivative of Ψ in terms of Θ_ℓ and Υ_{Θ_ℓ} is a partial derivative of Υ with respect to Θ_ℓ, then we can have:

$$\begin{aligned}\Phi_s &= \frac{\Upsilon \sum_{\ell=1}^{\rm m} \Psi_{\Theta_\ell}\Theta_{\ell,s} - \Psi \sum_{\ell=1}^{\rm m} \Upsilon_{\Theta_\ell}\Theta_{\ell,s}}{\Upsilon^2}\\ &= \frac{-\Upsilon \sum_{\ell=1}^{\rm m} \tau_\ell \Psi_{\Theta_\ell}\Theta_\ell + \Psi \sum_{\ell=1}^{\rm m} \tau_\ell \Upsilon_{\Theta_\ell}\Theta_\ell}{\Upsilon^2}\end{aligned}$$

and

$$\begin{aligned}\Phi_u &= \frac{\Upsilon \sum_{\ell=1}^{\rm m} \Psi_{\Theta_\ell}\Theta_{\ell,u} - \Psi \sum_{\ell=1}^{\rm m} \Upsilon_{\Theta_\ell}\Theta_{\ell,u}}{\Upsilon^2}\\ &= \frac{\Upsilon \sum_{\ell=1}^{\rm m} \gamma_\ell \Psi_{\Theta_\ell}\Theta_\ell - \Psi \sum_{\ell=1}^{\rm m} \gamma_\ell \Upsilon_{\Theta_\ell}\Theta_\ell}{\Upsilon^2}\end{aligned}$$

On this account, all the partial derivatives, including Φ_s and Φ_u, represented by the new variables Θ_ℓ, $1 \leqslant \ell \leqslant {\rm m}$ are rational functions. Next, we place the newly obtained partial derivatives in the principal $(1+1)$-dimensional partial differential equation ($(1+1)$-dimensional PDE). Therefore, with the new variables Θ_ℓ, $1 \leqslant \ell \leqslant {\rm m}$, a logical function equation is created as follows

$$\mathcal{M}(u, s, \Theta_1, \Theta_2, \ldots, \Theta_{\rm m}) = 0.$$

This is called the transformed equation of the original equation $(1+1)$-dimensional PDE.

- In this step we can obtain solutions of differential equations using computer algebraic systems. Here, we come to a system of algebraic equations of all variables $\gamma_\ell, \tau_\ell, \Psi_{pq,\ell J}, \Upsilon_{pq,\ell J}$, which we arrive at by equating the numerator of the logical function $\mathcal{M}(u, s, \Theta_1, \Theta_2, \ldots, \Theta_{\rm m})$ with zero. We solve the system obtained from algebraic equations to determine the polynomials of Ψ and Υ as well as the power of the wave $\varkappa_\ell$, $1 \leqslant \ell \leqslant {\rm m}$. Since both the algorithm presented in this work and the other algorithms obtained from these algebraic systems may be complex, we can use computer programs to obtain solutions for these systems. Here, we use the

Maple algebraic system to obtain the solutions of the desired system. Thus, the multiple wave solitone solution Φ is obtained, which is shown as follows

$$\Phi(u, s) = \frac{\Psi\ (\kappa_1 \exp(\gamma_1 u - \tau_1 s), \ldots, \kappa_{\mathrm{m}} \exp(\gamma_{\mathrm{m}} u - \tau_{\mathrm{m}} s))}{\Upsilon\ (\kappa_1 \exp(\gamma_1 u - \tau_1 s), \ldots, \kappa_{\mathrm{m}} \exp(\gamma_{\mathrm{m}} u - \tau_{\mathrm{m}} s))}. \tag{7.1.3}$$

According to the solutions of linear equations obtained at the beginning of the algorithm, which are of exponential function type, we call this algorithm the multiple exp-function (MEF) method. By considering different linear equations, we can obtain different solutions for nonlinear equations. For example, by considering linear sine-cosine equations, we can obtain multiple periodic wave solutions for nonlinear equations by multiple sine-cosine method. If $\mathrm{m} = 1$ is considered in the proposed multiple exp-function method, this method becomes the so-called exp-function method provided by He and Wu. See [50] for more information. The algorithm presented in this paper is a direct and systematic method that is used to obtain multi-soliton wave solutions. To perform complex calculations of this algorithm, we can use computer algebraic systems such as Maple, Mathematica, MuPAD and Matlab. This method also offers a generalization of the Hirotas perturbation scheme to create multi-soliton solution [53]. In this work, we define different Ψ and Υ polynomials in each step to obtain the multiple soliton wave solutions of the $(2 + 1)$-dimensional Date–Jimbo–Kashiwara–Miwa equation ($(2 + 1)$-dimensional DJKM).

7.1.3 Solutions of (2 + 1)*-Dimensional Date–Jimbo–Kashiwara–Miwa Equation via the Multiple Exp-Function Method*

We consider the $(2 + 1)$-dimensional Date–Jimbo–Kashiwara–Miwa equation ($(2 + 1)$-dimensional DJKM), which is defined as follows

$$\Phi_{uuuuv} + 4\Phi_{uuv}\Phi_u + 2\Phi_{uuu}\Phi_v + 6\Phi_{uv}\Phi_{uu} - \Phi_{vvv} - 2\frac{\partial}{\partial u}\left(\frac{\partial}{\partial u}\left(\frac{\partial \Phi}{\partial s}\right)\right) = 0, \tag{7.1.4}$$

where $\lambda = 1$, $\mu = 1$ and $\zeta = 1$. To obtain the one-soliton wave, two-soliton wave, and three-soliton wave solutions of the above equation, we use the proposed multiple exp-function method (MEF method). To obtain these solutions, we introduce different polynomials of Ψ and Υ at each staps. These steps are as follows:

7.1.3.1 One-Wave Solutions

We consider the function $\Phi = \Phi(u, v, s)$ as follows

$$\Phi(u, v, s) = \frac{\Psi(u, v, s)}{\Upsilon(u, v, s)}; \tag{7.1.5}$$

Where Ψ and Υ are one degree polynomials defined as follows

$$\Psi(u, v, s) = f_0 + f_1\Theta_1,\ \Upsilon(u, v, s) = g_0 + g_1\Theta_1,\ \Theta_1 = \exp(\gamma_1 u + \varsigma_1 v - \tau_1 s), \tag{7.1.6}$$

where f_0, f_1, g_0 and g_1 are constants to be determined. Then, due to the (7.1.5), we obtain

$$\Phi(u, v, s) = \frac{f_0 + f_1 \exp(\gamma_1 u + \varsigma_1 v - \tau_1 s)}{g_0 + g_1 \exp(\gamma_1 u + \varsigma_1 v - \tau_1 s)}, \tag{7.1.7}$$

Now substituting Eq. (7.1.7) into the (7.1.4), we have

$$\varsigma_1^4\gamma_1^4\Phi^{(5)} + 6\varsigma_1\gamma_1^3\Phi'''\Phi' + 6\varsigma_1\gamma_1^3(\Phi'')^2 - \varsigma_1^3\Phi''' + 2\tau_1\gamma_1^2\Phi''' = 0 \tag{7.1.8}$$

By the multiple exp-function method and and simplifying, we have

$$\begin{aligned}\frac{1}{\mathcal{Y}}\Big(&\mathcal{G}_5 \exp(5(\gamma_1 u + \varsigma_1 v - \tau_1 s)) + \mathcal{G}_4 \exp(4(\gamma_1 u + \varsigma_1 v - \tau_1 s)) \\ &+ \mathcal{G}_3 \exp(3(\gamma_1 u + \varsigma_1 v - \tau_1 s)) \\ &+ \mathcal{G}_2 \exp(2(\gamma_1 u + \varsigma_1 v - \tau_1 s)) + \mathcal{G}_1 \exp(1(\gamma_1 u + \varsigma_1 v - \tau_1 s))\Big) = 0,\end{aligned} \tag{7.1.9}$$

where $\mathcal{Y} = (g_0 + g_1(\exp(\gamma_1 u + \varsigma_1 v - \tau_1 s)))^6$.

Setting each coefficient of $\exp(\mathrm{m}(\gamma_1 u + \varsigma_1 v - \tau_1 s))$, $\mathrm{m} = 1, 2, 3, 4, 5$, we obtain a set of algebraic equations for f_0, f_1, g_0, g_1. Then,

$$\begin{aligned}
\mathcal{G}_5 &= -f_0 g_1^5\gamma_1^4\varsigma_1 + f_1 g_0 g_1^4\gamma_1^4\varsigma_1 - 2f_0 g_1^5\gamma_1^2\tau_1 + f_0 g_1^5\varsigma_1^3 + 2f_1 g_0 g_1^4\gamma_1^2\tau_1 - f_1 g_0 g_1^4\varsigma_1^3 = 0, \qquad (7.1.10)\\
\mathcal{G}_4 &= 26f_0 g_0 g_1^4\gamma_1^4\varsigma_1 - 26f_1 g_0^2 g_1^3\gamma_1^4\varsigma_1 + 12\varsigma_1\gamma_1^3 g_1^4 f_0^2 - 24f_0 f_1 g_0 g_1^3\gamma_1^3\varsigma_1 + 12f_1^2 g_0^2 g_1^2\gamma_1^3\varsigma_1 \\
&\quad + 4f_0 g_0 g_1^4\gamma_1^2\tau_1 - 2f_0 g_0 g_1^4\varsigma_1^3 - 4f_1 g_0^2 g_1^3\gamma_1^2\tau_1 + 2f_1 g_0^2 g_1^3\varsigma_1^3 = 0,\\
\mathcal{G}_3 &= -66f_0 g_0^2 g_1^3\gamma_1^4\varsigma_1 + 66f_1 g_0^3 g_1^2\gamma_1^4\varsigma_1 - 36f_0^2 g_0 g_1^3\gamma_1^3\varsigma_1 + 72f_0 f_1 g_0^2 g_1^2\gamma_1^3\varsigma_1 - 36f_1^2 g_0^3 g_1\gamma_1^3\varsigma_1 \\
&\quad + 12f_0 g_0^2 g_1^3\gamma_1^2\tau_1 - 6f_0 g_0^2 g_1^3\varsigma_1^3 - 12f_1 g_0^3 g_1^2\gamma_1^2\tau_1 + 6f_1 g_0^3 g_1^2\varsigma_1^3 = 0,\\
\mathcal{G}_2 &= 26f_0 g_0^3 g_1^2\gamma_1^4\varsigma_1 - 26f_1 g_0^4 g_1\gamma_1^4\varsigma_1 + 12f_0^2 g_0^2 g_1^2\gamma_1^3\varsigma_1 - 24f_0 f_1 g_0^3 g_1\gamma_1^3\varsigma_1 + 12f_1^2 g_0^4\gamma_1^3\varsigma_1 \\
&\quad + 4f_0 g_0^3 g_1^2\gamma_1^2\tau_1 - 2f_0 g_0^3 g_1^2\varsigma_1^3 - 4f_1 g_0^4 g_1\gamma_1^2\tau_1 + 2f_1 g_0^4 g_1\varsigma_1^3 = 0,\\
\mathcal{G}_1 &= -f_0 g_0^4 g_1\gamma_1^4\varsigma_1 + f_1 g_0^5\gamma_1^4\varsigma_1 - 2f_0 g_0^4 g_1\gamma_1^2\tau_1 + f_0 g_0^4 g_1\varsigma_1^3 + 2f_1 g_0^5\gamma_1^2\tau_1 - f_1 g_0^5\varsigma_1^3 = 0.
\end{aligned}$$

Solving the system of the aforementioned algebraic equations with the aid of Maple, we obtain

$$f_0 = f_0,\; f_1 = \frac{g_1(2g_0\gamma_1 + f_0)}{g_0},\; g_0 = g_0,\; g_1 = g_1, \tag{7.1.11}$$
$$\gamma_1 = \gamma_1,\; \varsigma_1 = \varsigma_1,\; \tau_1 = -\frac{\varsigma_1(\gamma_1^4 - \varsigma_1^2)}{2\gamma_1^2}$$

By placing these items in Eq. (7.1.7), we have

$$\Phi(u, v, s) = \tag{7.1.12}$$
$$-\frac{\begin{array}{c}2\exp(\gamma_1 u + \varsigma_1 v - \tau_1 s)g_0 g_1^2\gamma_1 + \exp(\gamma_1 u + \varsigma_1 v - \tau_1 s)f_0 g_1^2 \\ -2\exp(\gamma_1 u + \varsigma_1 v - \tau_1 s)g_0 g_1\gamma_1 + 2g_0^2 g_1\gamma_1 - \exp(\gamma_1 u + \varsigma_1 v - \tau_1 s)f_0 g_1\end{array}}{g_0(g_0 + g_1\exp(\gamma_1 u + \varsigma_1 v - \tau_1 s))}.$$

The solution diagram, for some values, is drawn as follows.

7.1.3.2 Two-Wave Solutions

Consider the polynomials Ψ and Υ of degree two as follows

$$\Psi(\Theta_1, \Theta_2) = 2[\gamma_1\Theta_1 + \gamma_2\Theta_2 + f_{12}(\gamma_1 + \gamma_2)\Theta_1\Theta_2] \tag{7.1.13}$$
$$\Upsilon(\Theta_1, \Theta_2) = 1 + \Theta_1 + \Theta_2 + f_{12}\Theta_1\Theta_2,$$

where $\Theta_\ell = \kappa_\ell\exp(\gamma_1 u + \varsigma_1 v - \tau_1 s)$, $\quad 1 \le \ell \le 2$. Now, using the above polynomials and according to the Eq. (7.1.5), we define the Φ function, which gives us two-wave solutions, as follows:

$$\Phi(u, v, s) = \frac{2[\gamma_1\Theta_1 + \gamma_2\Theta_2 + f_{12}(\gamma_1 + \gamma_2)\Theta_1\Theta_2]}{1 + \Theta_1 + \Theta_2 + f_{12}\Theta_1\Theta_2} \tag{7.1.14}$$

By placing Eq. (7.1.14) in Eq. (7.1.4) and using the multiple exponential method and solving the algebraic system as in the first case, we have

$$f_{12} = \frac{\gamma_1^4\gamma_2^2 - 2\gamma_1^3\gamma_2^3 + \gamma_1^2\gamma_2^4 + \gamma_1^2\varsigma_2^2 - 2\gamma_1\gamma_2\varsigma_1\varsigma_2 + \gamma_2^2\varsigma_1^2}{\gamma_1^4\gamma_2^2 + 2\gamma_1^3\gamma_2^3 + \gamma_1^2\gamma_2^4 + \gamma_1^2\varsigma_2^2 - 2\gamma_1\gamma_2\varsigma_1\varsigma_2 + \gamma_2^2\varsigma_1^2}, \tag{7.1.15}$$
$$\kappa_1 = \kappa_1, \kappa_2 = \kappa_2, \gamma_1 = \gamma_1, \gamma_2 = \gamma_2, \varsigma_1 = \varsigma_1, \varsigma_2 = \varsigma_2,$$
$$\tau_1 = \frac{\varsigma_1(\gamma_1^4 - \varsigma_1^2)}{2\gamma_1^2}, \tau_2 = \frac{\varsigma_2(\gamma_2^4 - \varsigma_2^2)}{2\gamma_2^2}$$

By placing these items in Eq. (7.1.14), two-waves solutions are obtained.

7.1.3.3 Three-Wave Solutions

Similar to the second case, for $\Theta_\ell = \exp(\gamma_\ell u + \varsigma_\ell v - \tau_\ell s),\ \ 1 \le \ell \le 3$, we consider the repentance of Ψ and Υ as follows

$$\begin{aligned}
&\Psi(\Theta_1, \Theta_1, \Theta_1) = 2[\gamma_1\Theta_1 + \gamma_2\Theta_2 + \gamma_3\Theta_3 + f_{12}(\gamma_1 + \gamma_2)\Theta_1\Theta_2 \\
&+ f_{13}(\gamma_1 + \gamma_3)\Theta_1\Theta_3 + f_{23}(\gamma_2 + \gamma_3)\Theta_2\Theta_3 \\
&+ f_{12}f_{13}f_{23}(\gamma_1 + \gamma_2 + \gamma_3)\Theta_1\Theta_2\Theta_3] \\
&\Upsilon(\Theta_1, \Theta_2, \Theta_3) = 1 + \Theta_1 + \Theta_2 + \Theta_3 \\
&+ f_{12}\Theta_1\Theta_2 + f_{13}\Theta_1\Theta_3 + f_{23}\Theta_2\Theta_3 + f_{12}f_{13}f_{23}\Theta_1\Theta_2\Theta_3,
\end{aligned} \tag{7.1.16}$$

and using the above polynomials and according to the Eq. (7.1.5), we define the u function, which gives us three-wave solutions, as follows:

$$\Phi(u, v, s) = \frac{\begin{array}{r}2[\gamma_1\Theta_1 + \gamma_2\Theta_2 + \gamma_3\Theta_3 + f_{12}(\gamma_1 + \gamma_2)\Theta_1\Theta_2 + f_{13}(\gamma_1 + \gamma_3)\Theta_1\Theta_3 + f_{23}(\gamma_2 + \gamma_3)\Theta_2\Theta_3 \\ + f_{12}f_{13}f_{23}(\gamma_1 + \gamma_2 + \gamma_3)\Theta_1\Theta_2\Theta_3]\end{array}}{1 + \Theta_1 + \Theta_2 + \Theta_3 + f_{12}\Theta_1\Theta_2 + f_{13}\Theta_1\Theta_3 + f_{23}\Theta_2\Theta_3 + f_{12}f_{13}f_{23}\Theta_1\Theta_2\Theta_3} \tag{7.1.17}$$

By placing Eq. (7.1.17) in Eq. (7.1.4) and using the multiple exponential method and solving the algebraic system as in the first case, we have

$$\begin{aligned}
&f_{12} = \frac{\gamma_1^4\gamma_2^2 - 2\gamma_1^3\gamma_2^3 + \gamma_1^2\gamma_2^4 + \gamma_1^2\varsigma_2^2 - 2\gamma_1\gamma_2\varsigma_1\varsigma_2 + \gamma_2^2\varsigma_1^2}{\gamma_1^4\gamma_2^2 + 2\gamma_1^3\gamma_2^3 + \gamma_1^2\gamma_2^4 + \gamma_1^2\varsigma_2^2 - 2\gamma_1\gamma_2\varsigma_1\varsigma_2 + \gamma_2^2\varsigma_1^2}, \\
&f_{13} = 0, f_{23} = \frac{\gamma_2^4\gamma_3^2 - 2\gamma_2^3\gamma_3^3 + \gamma_2^2\gamma_3^4 + \gamma_2^2\varsigma_3^2 - 2\gamma_2\gamma_3\varsigma_2\varsigma_3 + \gamma_3^2\varsigma_2^2}{\gamma_2^4\gamma_3^2 + 2\gamma_2^3\gamma_3^3 + \gamma_2^2\gamma_3^4 + \gamma_2^2\varsigma_3^2 - 2\gamma_2\gamma_3\varsigma_2\varsigma_3 + \gamma_3^2\varsigma_2^2}, \\
&\kappa_1 = \kappa_1, \kappa_2 = \kappa_2, \kappa_3 = \kappa_3, \gamma_1 = \gamma_1, \gamma_2 = \gamma_2, \gamma_3 = \gamma_3, \varsigma_1 = \varsigma_1, \varsigma_2 = \varsigma_2, \varsigma_3 = \varsigma_3, \\
&\tau_1 = \frac{\varsigma_1(\gamma_1^4 - \varsigma_1^2)}{2\gamma_1^2}, \tau_2 = \frac{\varsigma_2(\gamma_2^4 - \varsigma_2^2)}{2\gamma_2^2}, \\
&\tau_3 = \frac{\begin{array}{r}\gamma_1^6\varsigma_3 + 2\gamma_1^5\gamma_3\varsigma_1 - 4\gamma_1^5\gamma_3\varsigma_3 - 5\gamma_1^4\gamma_3^2\varsigma_1 + 6\gamma_1^4\gamma_3^2\varsigma_3 + 4\gamma_1^3\gamma_3^3\varsigma_1 - 4\gamma_1^3\gamma_3^3\varsigma_3 - \gamma_1^2\gamma_3^4\varsigma_1 + \gamma_1^2\gamma_3^4\varsigma_3 \\ - 3\gamma_1^2\varsigma_1^2\varsigma_3 + 3\gamma_1^2\varsigma_1\varsigma_3^2 - \gamma_1^2\varsigma_3^3 + 2\gamma_1\gamma_3\varsigma_1^3 - \gamma_3^2\varsigma_1^3\end{array}}{2(\gamma_1 - \gamma_3)^2\gamma_1^2}.
\end{aligned} \tag{7.1.18}$$

By placing these items in Eq. (7.1.17), three-waves solutions are obtained. The following diagrams show the solution of three waves for different values.

7.1.4 Remark

Remark 7.1 Figure 7.1 illustrates the one-wave solution for the values of :

(a) $g_0 = -1, g_1 = -2, f_0 = -1, \gamma_1 = -2, s = 0, \varsigma_1 = -50, u = [-2, 2], v = [-2, 2]$,
(b) $g_0 = 5, g_1 = 8, f_0 = 5, \gamma_1 = 8, \varsigma_1 = 50, v = 1, \tau_1 = -3, u = [-2, 2], s = [-2, 2]$,

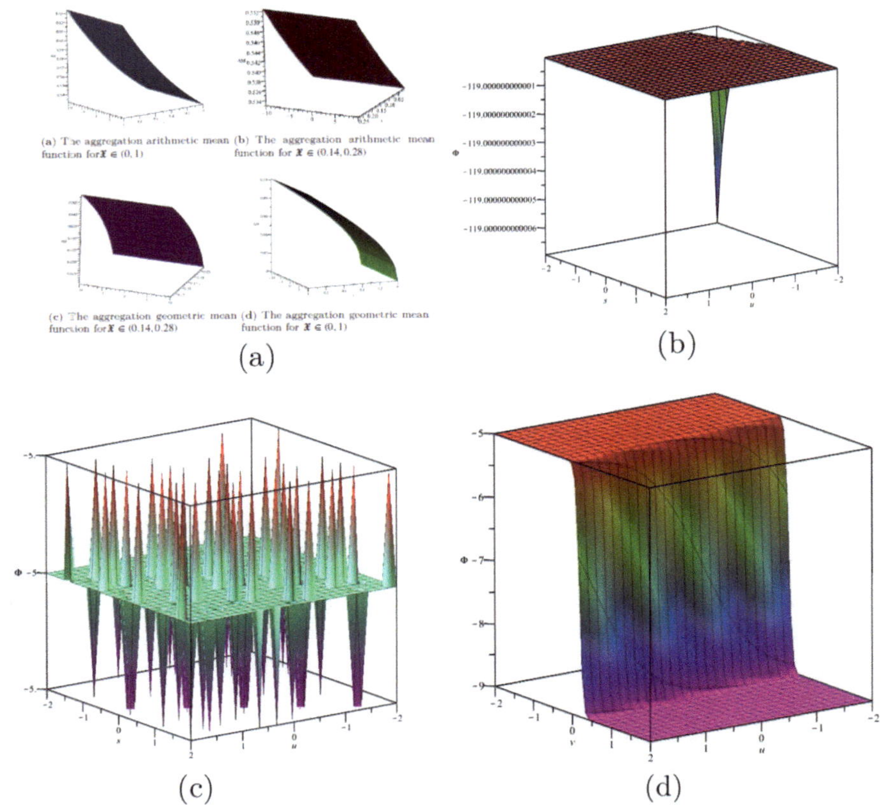

Fig. 7.1 The figure of the one-wave solutions

(c) $g_0 = 1, g_1 = 2, f_0 = 1, \gamma_1 = 2, \varsigma_1 = 50, v = 1, \tau[1] = -3, u = [-2, 2], s = [-2, 2]$,
(d) $g_0 = 1, g_1 = 2, f_0 = 1, \gamma_1 = 2, s = 0, \varsigma_1 = -25, u = [-2, 2], v = [-2, 2]$.

Figure 7.2 illustrates the two-wave solution for the values of :

(a) $\gamma_1 = 1, \gamma_2 = -2, \kappa_1 = 1, \kappa_2 = 2, s = \frac{1}{4}, \varsigma_1 = -50, \varsigma_2 = -20, v = 1, \tau_1 = -3, \tau_2 = -2, u = [-60, 60]$,
(b) $\gamma_1 = 1, \gamma_2 = -2, \kappa_1 = 1, \kappa_2 = 2, u = 1, \varsigma_1 = -50, \varsigma_2 = -20, v = 1, \tau_1 = -3, \tau_2 = -2, s = [-60, 60]$,
(c) $\gamma_1 = 1, \gamma_2 = -2, \kappa_1 = 1, \kappa_2 = 2, u = 1, s = 1, \varsigma_1 = -50, \varsigma_2 = -20, \tau_1 = -3, \tau_2 = -2, v = [-60, 60]$,
(d) $\gamma_1 = 1, \gamma_2 = -2, \kappa_1 = 1, \kappa_2 = 2, s = \frac{1}{4}, \varsigma_1 = -50, \varsigma_2 = -20, v = 1, \tau_1 = -3, \tau_2 = -2, u = [-2, 2], v = [-2, 2]$.

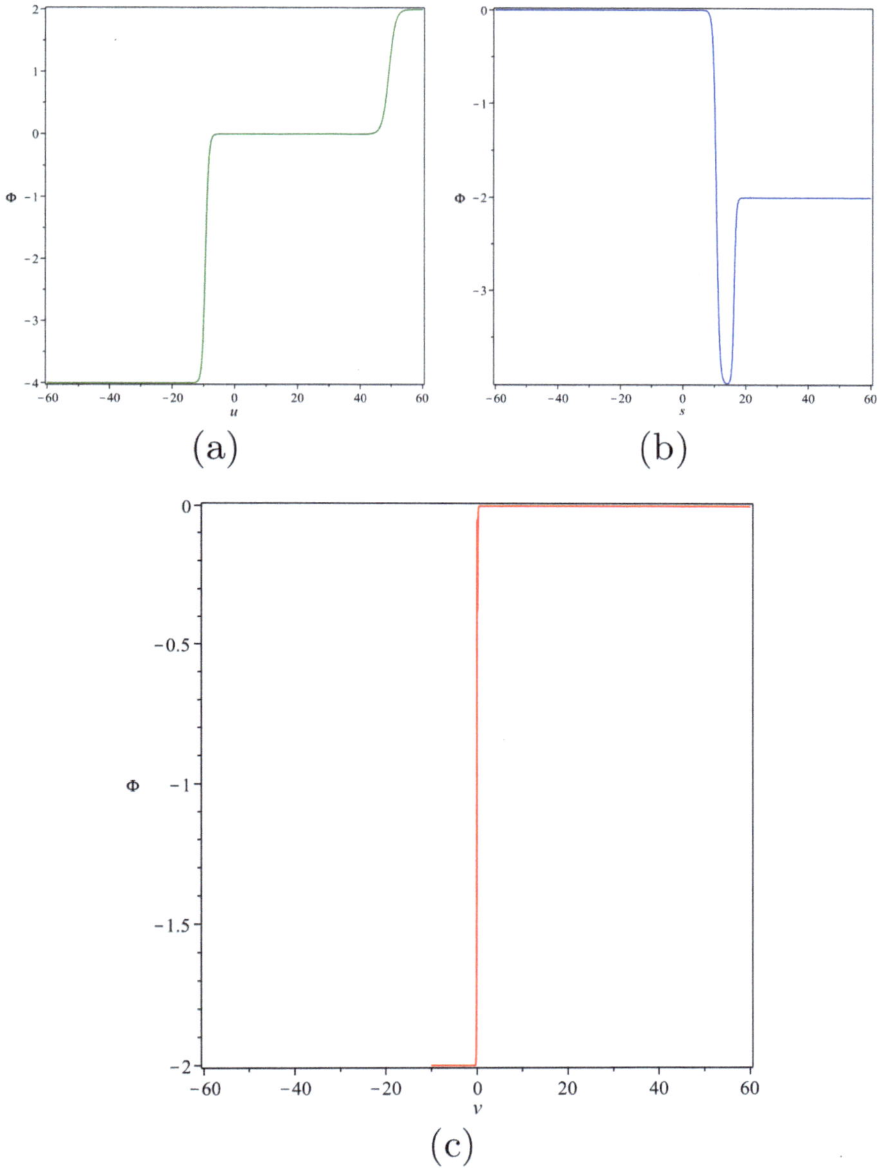

Fig. 7.2 The figure of the two-wave solutions

Fig. 7.2 (continued)

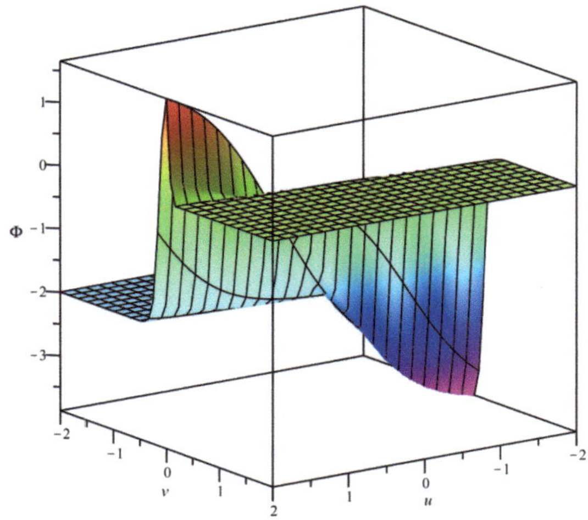

(d)

Figure 7.3 illustrates the two-wave solution for the values of :

(a) $\gamma_1 = 1, \gamma_2 = -2, \kappa_1 = 1, \kappa_2 = 2, s = \frac{1}{4}, \varsigma_1 = -50, \varsigma_2 = -20, v = 1, \tau_1 = -3, \tau_2 = -2, u = [-60, 60], v = [-60, 60]$,
(b) $\gamma_1 = 1, \gamma_2 = -2, \kappa_1 = 1, \kappa_2 = 2, u = 1, \varsigma_1 = -50, \varsigma_2 = -20, v = 1, \tau_1 = -3, \tau_2 = -2, v = [-60, 60], s = [-60, 60]$,
(c) $\gamma_1 = 1, \gamma_2 = -2, \kappa_1 = 1, \kappa_2 = 2, \varsigma_1 = -50, \varsigma_2 = -20, v = 1, \tau_1 = -3, \tau_2 = -2, u = [-60, 60], s = [-60, 60]$.

Figure 7.4 illustrates the three-wave solution for the values of:

(a) $\gamma_1 = 1, \gamma_2 = -2, \gamma_3 = 3, \kappa_1 = 1, \kappa_2 = 2, \kappa_3 = -3, s = \frac{1}{4}, \varsigma_1 = -50, \varsigma_2 = -20, \varsigma_3 = -20, v = 1, \tau_1 = -3, \tau_2 = -2, \tau_3 = -4, u = [-60, 60]$,
(b) $\gamma_1 = 1, \gamma_2 = -2, \gamma_3 = 3, \kappa_1 = 1, \kappa_2 = 2, \kappa_3 = -3, u = 1, \varsigma_1 = -50, \varsigma_2 = -20, \varsigma_3 = -20, v = 1, \tau_1 = -3, \tau_2 = -2, \tau_3 = -4, s = [-2, 2]$,
(c) $\gamma_1 = 1, \gamma_2 = -2, \gamma_3 = 3, \kappa_1 = 1, \kappa_2 = 2, \kappa_3 = -3, u = 1, \varsigma_1 = -50, \varsigma_2 = -20, \varsigma_3 = -20, v = 1, \tau_1 = -3, \tau_2 = -2, \tau_3 = -4, s = [-60, 60]$,
(d) $\gamma_1 = 1, \gamma_2 = -2, \gamma_3 = 3, \kappa_1 = 1, \kappa_2 = 2, \kappa_3 = -3, u = 1, s = 1, \varsigma_1 = -50, \varsigma_2 = -20, \varsigma_3 = -20, \tau_1 = -3, \tau_2 = -2, \tau_3 = -4, v = [-60, 60]$.

And then Fig. 7.5 is drawn for the following values:

(a) $\gamma_1 = 1, \gamma_2 = -2, \gamma_3 = 3, \kappa_1 = 1, \kappa_2 = 2, \kappa_3 = -3, s = \frac{1}{4}, \varsigma_1 = -50, \varsigma_2 = -20, \varsigma_3 = -20, \tau_1 = -3, \tau_2 = -2, \tau_3 = -4, u = [-60, 60], v = [-60, 60]$,
(b) $\gamma_1 = 1, \gamma_2 = -2, \gamma_3 = 3, \kappa_1 = 1, \kappa_2 = 2, \kappa_3 = -3, u = 1, \varsigma_1 = -50, \varsigma_2 = -20, \varsigma_3 = -20, \tau_1 = -3, \tau_2 = -2, \tau_3 = -4, s = [-60, 60], v = [-60, 60]$,
(c) $\gamma_1 = 1, \gamma_2 = -2, \gamma_3 = 3, \kappa_1 = 1, \kappa_2 = 2, \kappa_3 = -3, v = \frac{1}{4}, \varsigma_1 = -50, \varsigma_2 = -20, \varsigma_3 = -20, v = 1, \tau_1 = -3, \tau_2 = -2, \tau_3 = -4, s = [-60, 60], v = [-60, 60]$.

Fig. 7.3 The figure of the two-wave solutions

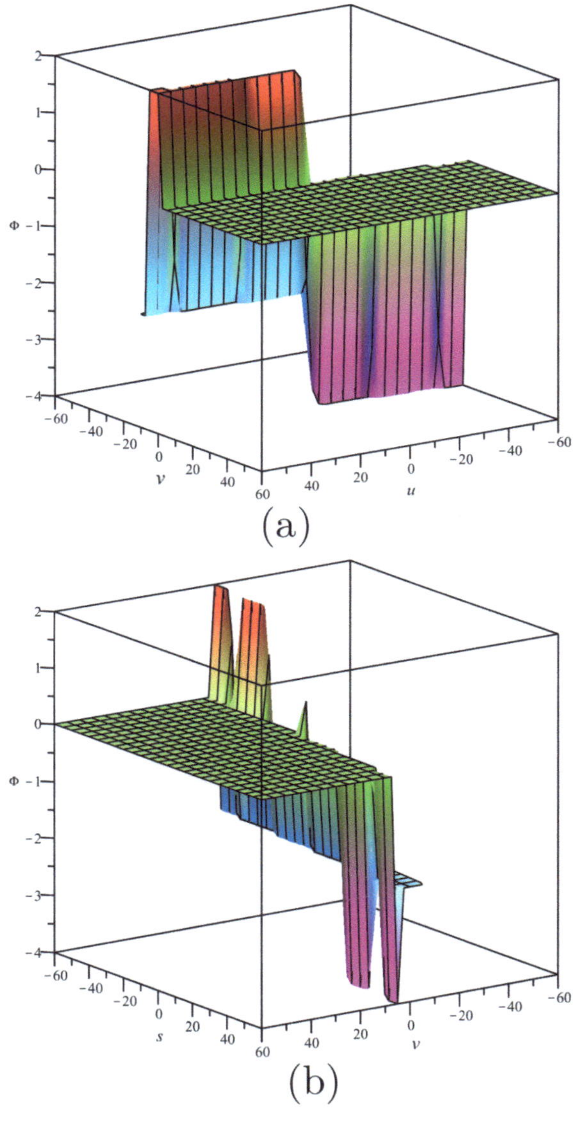

Fig. 7.3 (continued)

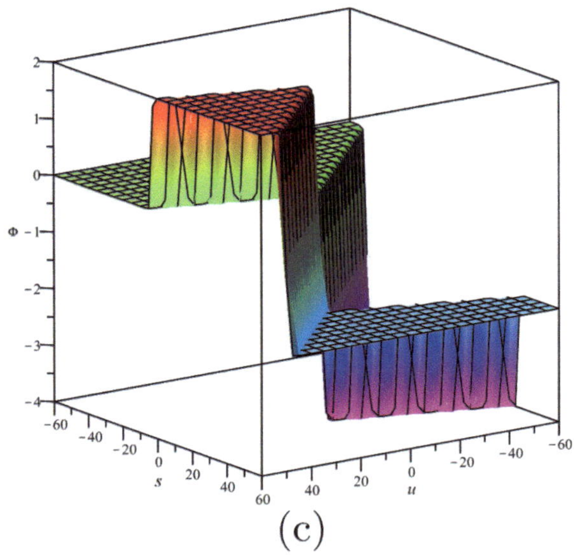

7.2 The Clique Polynomial Method Based on the Operational Matrix

7.2.1 Introduction of the Stochastic Time-Fractional Sine-Gordon Equation in the Caputo Sense

The Sine-Gordon nonlinear partial differential equation (S-GN PDE) first introduced by Edmond Bour (1862) is one of the most important equations in the modern nonlinear wave hypothesis, which was initially proposed in the study of the differential geometry of surfaces with Gaussian constant curvature [21]. The S-GN PDE is also used in other subjects such as nonlinear optics solid state physics, metal displacement and so on. For this reason, researchers have conducted extensive studies to investigate these equations to find accurate solutions [61, 78, 90, 127, 129], [?], [3, 9]. For the nonlinear sine-Gordon equation and generalized sine-Gordon equation, elements Galerkin method is used by Li et al. [75]. The spectral collocation method based on Legendre Wavelets is presented by Yin et al. [138] to solve the nonlinear Klein-sine-Gordon equations. Shan et al. [116] investigated the two-dimensional generalized space-time Legendre-Gauss-Lobatto collocation method. In [17], a finite difference scheme and a meshless method were used in time and place variables, respectively. It has been applied in [16], meshless procedures with the Neumann boundary conditions. In [2], a combination of the local discontinuous Galerkin method and the finite difference has been used to obtain solutions of the time-space fractional sine-Gordon equation. Soliton structured methods represent special wave-like solutions for non-

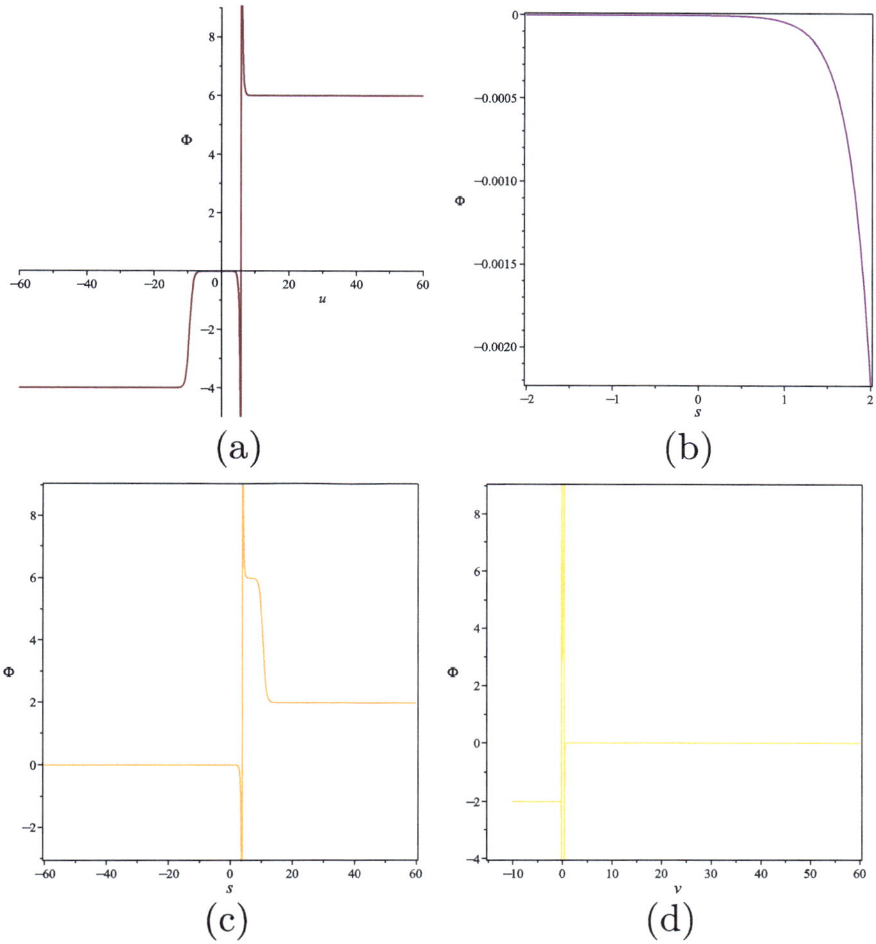

Fig. 7.4 The figure of the three-wave solutions

linear dynamic equations, which have attracted the attention of mathematicians. Various nonlinear differential equations such as S-GN PDE and KVE (Kortewegde Vries equation) have been investigated using methods with soliton structures. Many analytical methods have been proposed to numerically solve the S-GN PDE at higher dimensions. For example, Zagrodzinsky, Balcklund, and Painleve developed solutions for the S-GN PDE using the Hirota method. Bratsos, Guo et al. have proposed a numerical solution for the third-order S-GN PDE and the two-dimensional S-GN PDE respectively [10, 33, 59, 108]. Since the occurrence of many real phenomena depends on random factors and there is no certainty about them, stochastic partial differential equations (SPDEs) are used to investigate and explain such problems. SPDEs are applied in various scientific fields, such as nanotechnology, physics, chemistry, biology, economics, etc., which have made researchers to pay attention to solving these equations. One of the most important partial differential equations (SPDEs) is the S-GN PDE. Since there is no exact solution to this equation in most cases, numerical solutions to these equations have always been of interest to researchers. Various methods have been proposed for the numerical solutions. For example, there are the boundary elements method, Taylor expansion method, finite

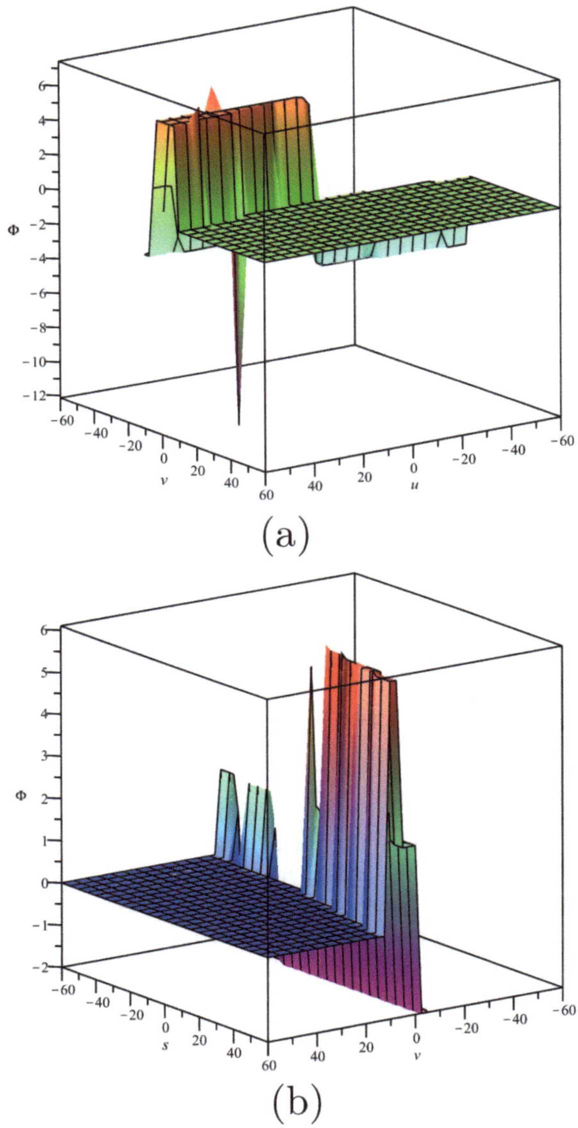

Fig. 7.5 The figure of the three-wave solutions

Fig. 7.5 (continued)

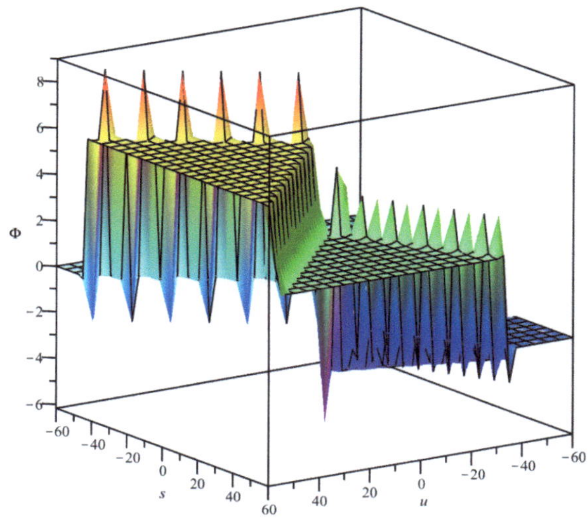

(c)

element methods, meshless method, etc. [14, 52, 141]. Among the meshless methods for solving high-order equations, we can name the element free Galerkin methods, the moving least squares, reproducing kernel particle, and so on [111]. Another well-known method of solving time-independent and time-dependent SPDEs is the meshless method based on the radial basis functions (RBFs). We study the generalized form of two-dimensional stochastic time-fractional Sine-Gordon equation (2D ST-FS-G) as follows:

$$\frac{\partial^{\kappa}\Psi(u,w,\tau)}{\partial\tau^{\kappa}} + \mu_1\frac{\partial^{\kappa-1}\Psi(u,w,\tau)}{\partial\tau^{\kappa-1}} = \mu_2\frac{\partial^2\Psi(u,w,\tau)}{\partial u^2} + \mu_3\frac{\partial^2\Psi(u,w,\tau)}{\partial w^2} \tag{7.2.1}$$

$$+ s(u,w)\sin(\Psi(u,w,\tau)) + \phi(u,w,\tau) + \mu_4\frac{d\mathrm{M}(\tau)}{d\tau}, \quad (u,w)\in\mathcal{Y}, \tau\in[0,\mathrm{R}],$$

with the initial and boundary conditions:

$$\begin{cases} \Psi(u,w,0) = \Psi_0(u,w), & (u,w)\in\mathcal{Y}, \\ \left.\frac{\partial\Psi(u,w,\tau)}{\partial\tau}\right|_{\tau=0} = z(u,w), & (u,w)\in\mathcal{Y}, \end{cases}$$

and

$$\Psi(u,w,\tau) = k(\tau), \quad 0\le\tau\le\mathrm{R}, (u,w)\in\partial\mathcal{Y},$$

where $\mu_\jmath$, $\jmath = 1, 2, 3, 4$, are known positive constants, the functions $\phi(u, w, \tau)$, $s(u, w)$, $\Psi_0(u, w)$, $z(u, w)$ and $k(\tau)$ are known functions and $\Psi(u, w, \tau)$ is an unknown function. Also, $\mathrm{M}(\tau)$ is the standard Brownian motion process defined on the probability space $(\Xi, \mathcal{H}, \mathcal{J})$, which is given below, and $\frac{\partial^\kappa \Psi(u,w,\tau)}{\partial \tau^\kappa}$ is the Caputo fractional derivative of order κ, and $1 < \kappa < 2$.

Definition 7.1 ([89]) The Brownian motion process $\mathrm{M}(\tau)$ satisfies the following properties:

(1) $\mathrm{M}(0) = 0$ (with probability 1).
(2) For $0 \le \sigma < \tau \le \mathrm{R}$, the increment $\mathrm{M}(\tau) - \mathrm{M}(\sigma)$ is a random variable with mean zero and variance $\tau - \sigma$. Consequently, $\mathrm{M}(\tau) - \mathrm{M}(\sigma) \sim \sqrt{\tau - \sigma}\mathcal{N}(0, 1)$, where $\mathcal{N}(0, 1)$ denotes the normal distribution with zero mean and unit variance.
(3) For $0 \le \sigma < \tau < \lambda < \zeta \le T$, the increments $\mathrm{M}(\tau) - \mathrm{M}(\sigma)$ and $B(\zeta) - B(\lambda)$ are independent.

Our goal is to obtain the numerical solution to Eq. (7.2.1) using the method based on clique polynomials. In mathematics, there is a branch called graph theory that deals with the subject of vertices and edges. Considering a set of $\mathfrak{e}$ vertices called $V(\mathcal{Q})$ and a set of edges of $\mathcal{Q}$ called $E(\mathcal{Q})$, we have a graph $\mathcal{Q} = \mathcal{Q}(V(\mathcal{Q}), E(\mathcal{Q}))$ which is pairwise ordered. Our goal is to obtain the numerical solutions of Eq. (7.2.1) using the method based on clique polynomials. The clique polynomial related to the maximal clique problem was first introduced by Hoede and Li which is defined as $\Omega(\mathcal{Q}, \tau) = \sum_{\iota=0}^{J} \mathfrak{m}_\mathfrak{e}(\mathcal{Q})\tau^\mathfrak{e}$ where $\mathfrak{m}_0(\mathcal{Q}) = 1$ and $\mathfrak{m}_\mathfrak{e}(\mathcal{Q})$ is the number of $\mathfrak{e}$-cliques of $\mathcal{E}$ [55, 91]. The maximal clique problem, which is a hybrid optimization problem, is used in various fields such as economics, information retrieval and biomedical engineering, graph coloring, classification theory, set packing, and optimal winner determination [134]. Polynomials are used to solve many equations i.e. they solve these equations by expanding the solution by polynomials [37, 122]. There are different types of graph polynomials such as characteristic polynomials, Tutte polynomials, chromatic polynomials, matching polynomials [79, 115] and clique polynomials. Using the technique of the operational method, which is a new method, we obtain the operational matrix with respect to the polynomial of the complete category of the graph. This work is actually a link between graph theory and numerical solutions of differential equations.

7.2.2 *Preliminaries*

We will introduce the basic concepts and the main technique of the article using clique polynomials.

Definition 7.2 ([103]) The κ-order ($\kappa > 0$) Riemann–Liouville fractional integral (RL-integral) is defined by

$$ {}^{RL}I_{\rho}^{\kappa}\Psi(\rho)=\frac{1}{\Gamma(\kappa)}\int_0^{\rho}(\rho-r)^{\kappa-1}\Psi(r)dr. $$

Definition 7.3 ([103]) The κ-order Caputo derivative for $\Psi\in C[p,q]$ is given by

$$ {}^{C}D_{\rho}^{\kappa}\Psi(\rho)=\begin{cases}\frac{1}{\Gamma(\ell-\kappa)}\int_p^{\rho}(\rho-r)^{\ell-\kappa-1}\Psi^{\ell}(r)dr, & \ell-1<\kappa<\ell\\ \frac{d^{\ell}\Psi(\rho)}{d\rho^{\ell}}, & \kappa=\ell\in\mathbb{N}.\end{cases} $$

The clique polynomial of a graph $\mathcal{Q}$, which denotes by $\Omega(\mathcal{Q},\tau)$, is given by [54]

$$ \Omega(\mathcal{Q},\tau)=\sum_{\mathfrak{e}=0}^{J}\mathfrak{m}_{\mathfrak{e}}(\mathcal{Q})\tau^{\mathfrak{e}},\quad \mathfrak{m}_0(\mathcal{Q})=1 $$

where $\mathfrak{m}_{\mathfrak{e}}(\mathcal{Q})$ is the number of $\mathfrak{e}$-cliques of $\mathcal{Q}$. The CP of a complete graph $\mathcal{D}_{J}$ with J vertices is given by

$$ \Omega\left(\mathcal{D}_{J},\tau\right)=(1+\tau)^{J}. $$

We can approximate any function $\Psi(\tau)$ in $L^2(0,\mathrm{R})$ in terms of the CP as

$$ \Psi(\tau)\simeq\tilde{\Psi}(\tau)=\sum_{J=0}^{E-1}\Psi_{J}\Omega\left(\mathcal{D}_{J},\tau\right)=\Delta^{T}\mathfrak{G}(\tau), \tag{7.2.2} $$

where $\Delta=\left[\Psi_0,\Psi_1,\ldots,\Psi_{E-1}\right]^{T}$, and

$$ \mathfrak{G}(\tau)=\left[\Omega\left(\mathcal{D}_0,\tau\right),\Omega\left(\mathcal{D}_1,\tau\right),\ldots,\Omega\left(\mathcal{D}_{E-1},\tau\right)\right]^{T}. \tag{7.2.3} $$

From (7.2.2), we have $\Delta=\mathrm{A}^{-1}\langle\Psi(\tau),\mathfrak{G}(\tau)\rangle$, where $\mathrm{A}=\langle\mathfrak{G}(\tau),\mathfrak{G}(\tau)\rangle$ and $\langle\cdot,\cdot\rangle$ indicates the inner product of two arbitrary functions. Also, an arbitrary function $\Psi(u,\tau)\in L^2([p,q]\times[0,\mathrm{R}])$ can be expanded in terms of the CP by the following infinite series

$$ \Psi(u,\tau)=\sum_{J=0}^{\infty}\sum_{\iota=0}^{\infty}\Psi_{J\iota}\Omega\left(\mathcal{D}_{J},u\right)\Omega\left(\mathcal{D}_{j},\tau\right). \tag{7.2.4} $$

We consider the first E terms in (7.2.4), as an approximation of the function $\Psi(u,\tau)$. It is obtained as

$$ \Psi(u,\tau)\simeq\sum_{J=0}^{E-1}\sum_{\iota=0}^{E-1}\Psi_{J\iota}\Omega\left(\mathcal{D}_{J},u\right)\Omega\left(\mathcal{D}_{\iota},\tau\right)=\mathfrak{G}^{T}(u)\Delta\mathfrak{G}(\tau), $$

where $\Delta = \mathrm{A}^{-1}\langle \mathfrak{G}(p), \langle \Psi(u,\tau), \mathfrak{G}(\tau)\rangle\rangle \mathrm{A}^{-1}$. Also, for function $\Psi(u, w, \tau) \in L^2([p,q] \times [p,q] \times [0,\mathrm{R}])$, we have

$$\Psi(u, w, \tau) = \sum_{j=0}^{\infty}\sum_{\iota=0}^{\infty}\sum_{\epsilon=0}^{\infty} \Psi_{j\iota\epsilon}\Omega\left(\mathcal{D}_j, u\right)\Omega\left(\mathcal{D}_\epsilon, y\right)\Omega\left(\mathcal{D}_\iota, \tau\right).$$

Since in the process of solving differential equations using the clique polynomials technique, assume that $\mathfrak{G}(\tau)$ is the matrix defined in (7.2.3). This vector is written based on Taylor functions as follows

$$\mathfrak{G}(\tau) = AT_M(\tau), \tag{7.2.5}$$

where A is an invertible matrix defined as follows

$$A = [a_{ij}], \qquad i, j = 0, 1, \ldots, M-1, \qquad a_{ij} = \begin{cases} \binom{i}{j}, & i \geq j \\ 0, & i < j. \end{cases}$$

Also, $T_M(\tau)$ is the vector of Taylor's basic functions as follows

$$T_M(\tau) = \begin{bmatrix} 1 \\ \tau \\ \tau^2 \\ \vdots \\ \tau^{M-1} \end{bmatrix}. \tag{7.2.6}$$

By taking the integral from Eq. (7.2.5), we have

$$\begin{aligned} \int_0^{\tau} \mathfrak{G}(t)dt &= \int_0^{\tau} AT_M(t)dt \\ &= A\int_0^{u} T_M(t)dt \\ &= B\bar{T}(\tau), \end{aligned} \tag{7.2.7}$$

where

$$B = A\begin{bmatrix} 1 & 0 & \cdots & 0 \\ 0 & \frac{1}{2} & \cdots & 0 \\ \vdots & \vdots & \ddots & \vdots \\ 0 & 0 & \cdots & 0 \end{bmatrix} \tag{7.2.8}$$

and

$$\bar{T}(\tau) = \begin{bmatrix} \tau \\ \tau^2 \\ \tau^3 \\ \vdots \\ \tau^M \end{bmatrix}. \tag{7.2.9}$$

Now we approximate the τ^k functions based on the clique base and using Eq. (7.2.5) by

$$\tau^k \simeq A^{-1}_{[k+1]}\mathfrak{G}(\tau), k = 1, 2, \ldots, M-1,$$

where $A^{-1}_{[k+1]}$ is the $(k+1)$th row of the matrix A^{-1}. Furthermore, we estimate the function t^M based on the clique base as follows

$$\tau^M \simeq c_M^T \mathfrak{G}(\tau),$$

where c_M is defined as

$$c_M = Q^{-1} \int_0^1 \tau^M \mathfrak{G}(\tau) d\tau$$

and $Q =< \mathfrak{G}(\tau), \mathfrak{G}(\tau) >$. We suppose that

$$E = \begin{bmatrix} A^{-1}_{[2]} \\ A^{-1}_{[3]} \\ A^{-1}_{[4]} \\ \vdots \\ A^{-1}_{[M]} \\ c_M^T \end{bmatrix}, \tag{7.2.10}$$

then $\bar{T}(\tau) \simeq E\mathfrak{G}(\tau)$. Therefore, the integral operating matrix based on the clique base is derived as

$$\int_0^\tau \mathfrak{G}(\tau) d\tau \simeq P\mathfrak{G}(\tau), \tag{7.2.11}$$

and $P = BE$ the operational matrix is called the clique base integral.

By taking twice the integral of Eq. (7.2.3), we have

$$\int_0^{\mathcal{T}}\int_0^{\mathcal{T}} \mathfrak{G}(t)dtdt = \int_0^{\mathcal{T}} P\mathfrak{G}(t)dt = P\int_0^{\mathcal{T}} \mathfrak{G}(t)dt = P^2\mathfrak{G}(\mathcal{T}). \tag{7.2.12}$$

Consider the following set

$$\mathfrak{X} = span\{\Omega(\mathcal{D}_0, u)\Omega(\mathcal{D}_0, w)\Omega(\mathcal{D}_0, \tau), \ldots, \Omega(\mathcal{D}_0, u)\Omega(\mathcal{D}_0, w)\Omega(\mathcal{D}_{E-1}, \tau), \ldots, \Omega(\mathcal{D}_{E-1}, u)\Omega(\mathcal{D}_{E-1}, w)\Omega(\mathcal{D}_{E-1}, \tau)\}, \tag{7.2.13}$$

which is the space of all three-variable polynomials. If F is the degree of polynomials on u, w, τ, then $F \leq E$. The best approximation for the function $\Psi(u, w, \tau) \in \mathrm{L}^*$ to $\mathfrak{X}$ is the function $\bar{\Psi}(u, w, \tau)$.

In the next theorem, we obtain the error of approximation of the three-variable function using clique polynomials.

Theorem 7.1 *Suppose $\Psi(u, w, \tau)$ is a sufficiently smooth function on $\mathrm{L} = [p, q] \times [p, q] \times [0, \mathrm{R}]$ and $\tilde{\Psi}(u, w, \tau)$ is its approximation using clique polynomials. Then we have*

$$\left\|\Psi(u, w, \tau) - \tilde{\Psi}(u, w, \tau)\right\|_2 \leq \left(\frac{\aleph_1(q-p)^E}{E!2^{2E-1}} + \frac{\aleph_2(q-p)^E}{E!2^{2E-1}} + \frac{\aleph_3(\mathrm{R})^E}{E!2^{2E-1}} + \frac{\aleph_4(q-p)^{2E}\mathrm{R}^E}{(E!)^3 2^{6E-3}}\right)(q-p)\mathrm{R}. \tag{7.2.14}$$

Proof Let $u_{\jmath}, w_{\iota}, \jmath, \iota = 0, 1, \ldots, E-1$ and $\tau_{\epsilon}, \epsilon = 0, 1, \ldots, E-1$ be the roots of E-degree shifted Chebyshev polynomials on [0, 1]. By considering the interpolating polynomial Υ_E to $\Psi(u, w, \tau)$ at points $(u_{\jmath}, w_{\iota}, \tau_{\epsilon})$, we have

$$\begin{aligned}\Psi(u, w, \tau) - \Upsilon_E(u, w, \tau) = {} & \frac{\partial^E\Psi(\alpha, w, \tau)}{\partial u^E E!}\prod_{\jmath=0}^{E-1}\left(u - u_{\jmath}\right) + \frac{\partial^E\Psi(u, \beta, \tau)}{\partial w^E E!}\prod_{\iota=0}^{E-1}(w - w_{\iota}) \\ & + \frac{\partial^E\Psi(u, w, \nu)}{\partial \tau^E E!}\prod_{\epsilon=0}^{E-1}(\tau - \tau_{\epsilon}) \\ & - \frac{\partial^{3E}\Psi(\alpha', \beta', \nu')}{\partial u^E\partial w^E\partial \tau^E(E!)^3}\prod_{\jmath=0}^{E-1}\left(u - u_{\jmath}\right)\prod_{\iota=0}^{E-1}(w - w_{\iota})\prod_{\epsilon=0}^{E-1}(\tau - \tau_{\epsilon}),\end{aligned} \tag{7.2.15}$$

where $\alpha, \alpha' \in [p, q]$, $\beta, \beta' \in [p, q]$ and $\nu, \nu' \in [0, \mathrm{R}]$. Then we further imply

$$\begin{aligned}|\Psi(u, w, \tau) - \Upsilon_E(u, w, \tau)| \leqslant {} & \max_{(u,w,\tau)\in\mathrm{L}}\left|\frac{\partial^E\Psi(\alpha, w, \tau)}{\partial u^E}\right|\frac{\prod_{\jmath=0}^{E-1}|u - u_{\jmath}|}{E!} \\ & + \max_{(u,w,\tau)\in\mathrm{L}}\left|\frac{\partial^E\Psi(u, \beta, \tau)}{\partial w^E}\right|\frac{\prod_{\iota=0}^{E-1}|w - w_{\iota}|}{E!} + \max_{(u,w,\nu)\in\mathrm{L}}\left|\frac{\partial\Psi(u, w, \nu)^E}{\partial \tau^E}\right|\frac{\prod_{\epsilon=0}^{E-1}|\tau - \tau_{\epsilon}|}{E!} \\ & + \max_{(u,w,\tau)\in\mathrm{L}}\left|\frac{\partial^{3E}\Psi(\alpha', \beta', \nu')}{\partial u^E\partial w^E\partial \tau^E}\right|\frac{\prod_{\jmath=0}^{E-1}|u - u_{\jmath}|\prod_{\jmath=0}^{E-1}|w - w_{\iota}|\prod_{\tau_E}^{E-1}|\tau - \tau_{\epsilon}|}{(E!)^3}.\end{aligned} \tag{7.2.16}$$

According to the smooth function $\Psi(u, w, \tau)$ on L^*, there are constants $\aleph_1, \aleph_2, \aleph_3, \aleph_4$ such that

$$\max_{(u,w,\tau)\in L}\left|\frac{\partial^{L}\Psi(u, w, \tau)}{\partial u^E}\right| \leq \aleph_1, \max_{(u,w,\tau)\in L}\left|\frac{\partial^{E}\Psi(u, w, \tau)}{\partial w^E}\right| \leq \aleph_2, \max_{(u,w,\tau)\in L}\left|\frac{\partial\Psi(u, w, \nu)^{E}}{\partial \tau^E}\right| \leq \aleph_3, \tag{7.2.17}$$

$$\max_{(u,w,\tau)\in L}\left|\frac{\partial^{3E}\Psi(\aleph', \beta', \nu')}{\partial u^E \partial w^E \partial \tau^E}\right| \leq \aleph_4.$$

By substituting (7.2.17) into (7.2.16) and employing the estimates for Chebyshev interpolation nodes, we come to

$$|\Psi(u, w, \tau) - \Upsilon_E(u, w, \tau)| \leqslant \frac{\aleph_1(q-p)^E}{E!2^{2E-1}} + \frac{\aleph_2(q-p)^E}{E!2^{2E-1}} + \frac{\aleph_3(\mathrm{R})^E}{E!2^{2E-1}} + \frac{\aleph_4(q-p)^{2E}\mathrm{R}^E}{(E!)^3 2^{6E-3}}. \tag{7.2.18}$$

Since $\tilde{\Psi}(u, w, \tau)$ is the best approximation of $\Psi(u, w, \tau)$ in $\mathfrak{X}$ and using (7.2.18), we can write

$$\begin{aligned}
\left\|\Psi(u, w, \tau) - \tilde{\Psi}(u, w, \tau)\right\|_2^2 &= \int_0^{\mathrm{R}}\int_p^q\int_p^q |\Psi(u, w, \tau) - \tilde{\Psi}(u, w, \tau)|^2 du dw d\tau \\
&\leqslant \int_0^{\mathrm{R}}\int_p^q\int_p^q |\Psi(u, w, \tau) - \Upsilon_E(u, w, \tau)|^2 du dw d\tau \\
&= \int_0^{\mathrm{R}}\int_p^q\int_p^q \left(\frac{\aleph_1(q-p)^E}{E!2^{2E-1}} + \frac{\aleph_2(q-p)^E}{E!2^{2E-1}} + \frac{\aleph_3(\mathrm{R})^E}{E!2^{2E-1}} + \frac{\aleph_4(q-p)^{2E}\mathrm{R}^E}{(E!)^3 2^{6E-3}}\right)^2 du dw d\tau \\
&= \left(\frac{\aleph_1(q-p)^E}{E!2^{2E-1}} + \frac{\aleph_2(q-p)^E}{E!2^{2E-1}} + \frac{\aleph_3(\mathrm{R})^E}{E!2^{2E-1}} + \frac{\aleph_4(q-p)^{2E}\mathrm{R}^E}{(E!)^3 2^{6E-3}}\right)^2 (q-p)^2\mathrm{R},
\end{aligned} \tag{7.2.19}$$

which gives the desired result. Note that for the special case, when $p = 0, q = 1$ and $\mathrm{R} = 1$, we have

$$\left\|\Psi(u, w, \tau) - \tilde{\Psi}(u, w, \tau)\right\|_2 \leqslant \frac{\aleph_1}{E!2^{2E-1}} + \frac{\aleph_2}{E!2^{2E-1}} + \frac{\aleph_3}{E!2^{2E-1}} + \frac{\aleph_4}{(E!)^3 2^{6E-3}}, \tag{7.2.20}$$

therefore

$$\left\|\Psi(u, w, \tau) - \tilde{\psi}(u, w, \tau)\right\|_2 = O\left(\frac{1}{E!2^{2E-1}}\right).$$

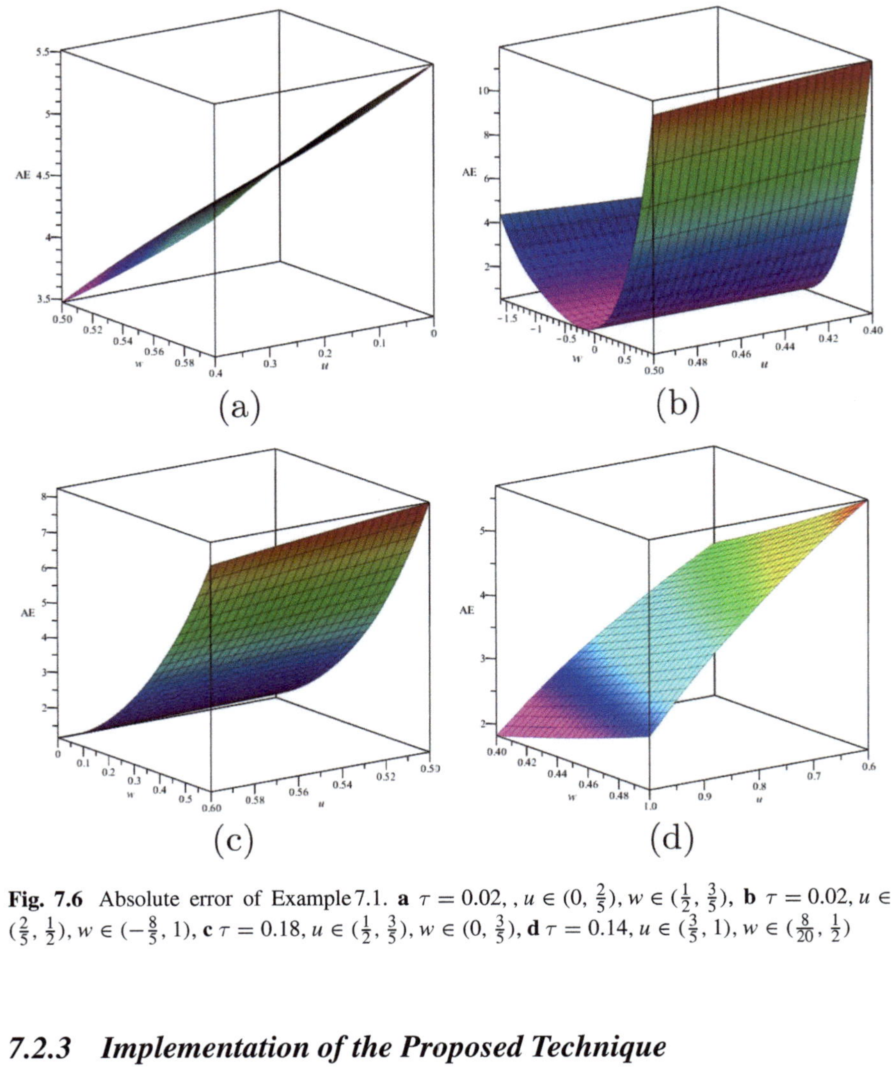

Fig. 7.6 Absolute error of Example 7.1. **a** $\tau = 0.02,$, $u \in (0, \frac{2}{5})$, $w \in (\frac{1}{2}, \frac{3}{5})$, **b** $\tau = 0.02$, $u \in (\frac{2}{5}, \frac{1}{2})$, $w \in (-\frac{8}{5}, 1)$, **c** $\tau = 0.18$, $u \in (\frac{1}{2}, \frac{3}{5})$, $w \in (0, \frac{3}{5})$, **d** $\tau = 0.14$, $u \in (\frac{3}{5}, 1)$, $w \in (\frac{8}{20}, \frac{1}{2})$

7.2.3 Implementation of the Proposed Technique

We use the clique polynomials technique to obtain the solution of an equation of type (7.2.1). For this purpose, we consider the following 2D ST-FS-G equation:

$$\frac{\partial^2 \Psi(u,w,\tau)}{\partial \tau^2} + \mu_1 \frac{\partial \Psi(u,w,\tau)}{\partial \tau} = \mu_2 \frac{\partial^2 \Psi(u,w,\tau)}{\partial u^2} + \mu_3 \frac{\partial^2 \Psi(u,w,\tau)}{\partial w^2} \tag{7.2.21}$$
$$+ s(u,w)\sin(\Psi(u,w,\tau)) + \phi(u,w,\tau) + \mu_4 \frac{d\mathrm{M}(\tau)}{d\tau}, \quad (u,w) \in \mathcal{Y}, \tau \in [0, \mathrm{R}],$$

with the initial and boundary conditions in Eq. (7.2.1). Assume that,

$$\frac{\partial^5 \Psi(u,w,\tau)}{\partial u^2 \partial w^2 \partial \tau} = \sum_{J=0}^{\infty}\sum_{\iota=0}^{\infty}\sum_{\epsilon=0}^{\infty} \Psi_{J\iota\epsilon}\Omega\left(\mathcal{D}_J,u\right)\Omega\left(\mathcal{D}_\epsilon,w\right)\Omega\left(\mathcal{D}_\iota,\tau\right) = \mathfrak{G}(\tau)^T \Delta\mathfrak{G}(u)\mathfrak{G}(w), \tag{7.2.22}$$

and

$$\begin{aligned}
\mathfrak{G}(\tau)^T &= \left[\Omega\left(\mathcal{D}_0,\tau\right), \Omega\left(\mathcal{D}_1,\tau\right), \Omega\left(\mathcal{D}_2,\tau\right), \Omega\left(\mathcal{D}_3,\tau\right), \ldots, \Omega\left(\mathcal{D}_\ell,\tau\right)\right] \\
\mathfrak{G}(u) &= \left[\Omega\left(\mathcal{D}_0,u\right), \Omega\left(\mathcal{D}_1,u\right), \Omega\left(\mathcal{D}_2,u\right), \Omega\left(\mathcal{D}_3,u\right), \ldots, \Omega\left(\mathcal{D}_\ell,u\right)\right]^T \\
\mathfrak{G}(w) &= \left[\Omega\left(\mathcal{D}_0,u\right), \Omega\left(\mathcal{D}_1,u\right), \Omega\left(\mathcal{D}_2,u\right), \Omega\left(\mathcal{D}_3,u\right), \ldots, \Omega\left(\mathcal{D}_\ell,u\right)\right]^T.
\end{aligned}$$

Integrate Eq (7.2.22) regarding τ from 0 to τ,

$$\begin{aligned}
\frac{\partial^4 \Psi(u,w,\tau)}{\partial u^2 \partial w^2} &= \frac{\partial^4 \Psi(u,w,0)}{\partial u^2 \partial w^2} + \int_0^\tau \mathfrak{G}(\tau)^T \Delta\mathfrak{G}(u)\mathfrak{G}(w)d\tau, \qquad (7.2.23)\\
\frac{\partial^4 \Psi(u,w,\tau)}{\partial u^2 \partial w^2} &= \frac{\partial^4 \Psi(u,w,0)}{\partial u^2 \partial w^2} + \Delta\mathfrak{G}(u)\mathfrak{G}(w)\int_0^\tau \mathfrak{G}(\tau)^T d\tau,\\
\frac{\partial^4 \Psi(u,w,\tau)}{\partial u^2 \partial w^2} &= \frac{\partial^4 \Psi(u,w,0)}{\partial u^2 \partial w^2} + [P\mathfrak{G}(\tau)]^T \Delta\mathfrak{G}(u)\mathfrak{G}(w).
\end{aligned}$$

Integrating Eq (7.2.23) regarding u from 0 to u,

$$\frac{\partial^3 \Psi(u,w,\tau)}{\partial u \partial w^2} = \frac{\partial^3 \Psi(0,w,\tau)}{\partial u \partial w^2} - \frac{\partial^3 \Psi(0,w,0)}{\partial u \partial w^2} + \frac{\partial^3 \Psi(u,w,0)}{\partial u \partial w^2} + \int_0^u [P\mathfrak{G}(\tau)]^T \Delta\mathfrak{G}(u)\mathfrak{G}(w)du, \tag{7.2.24}$$

$$\frac{\partial^3 \Psi(u,w,\tau)}{\partial u \partial w^2} = \frac{\partial^3 \Psi(0,w,\tau)}{\partial u \partial w^2} - \frac{\partial^3 \Psi(0,w,0)}{\partial u \partial w^2} + \frac{\partial^3 \Psi(u,w,0)}{\partial u \partial w^2} + \Delta\mathfrak{G}(w)[P\mathfrak{G}(\tau)]^T \int_0^u \mathfrak{G}(u)du,$$

$$\frac{\partial^3 \Psi(u,w,\tau)}{\partial u \partial w^2} = \frac{\partial^3 \Psi(0,w,\tau)}{\partial u \partial w^2} - \frac{\partial^3 \Psi(0,w,0)}{\partial u \partial w^2} + \frac{\partial^3 \Psi(u,w,0)}{\partial u \partial w^2} + \Delta\mathfrak{G}(w)[P\mathfrak{G}(\tau)]^T [P\mathfrak{G}(u)].$$

Again, by taking the integral from 0 to u relative to u

$$\begin{aligned}
\frac{\partial^2 \Psi(u,w,\tau)}{\partial w^2} &= \frac{\partial^2 \Psi(0,w,\tau)}{\partial w^2} + u\left[-\frac{\partial^2 \Psi(0,w,0)}{\partial w^2} + \frac{\partial^2 \Psi(0,w,\tau)}{\partial w^2}\right] - \frac{\partial^2 \Psi(0,w,0)}{\partial w^2} + \frac{\partial^2 \Psi(u,w,0)}{\partial w^2} \qquad (7.2.25)\\
&+ \int_0^u \Delta\mathfrak{G}(w)[P\mathfrak{G}(\tau)]^T [P\mathfrak{G}(u)]du,\\
\frac{\partial^2 \psi(u,w,\tau)}{\partial w^2} &= \frac{\partial^2 \Psi(0,w,\tau)}{\partial w^2} + u\left[-\frac{\partial^2 \Psi(0,w,0)}{\partial w^2} + \frac{\partial^2 \Psi(0,w,\tau)}{\partial w^2}\right] - \frac{\partial^2 \Psi(0,w,0)}{\partial w^2} + \frac{\partial^2 \Psi(u,w,0)}{\partial w^2}\\
&+ \Delta\mathfrak{G}(w)[P\mathfrak{G}(\tau)]^T [p^2\mathfrak{G}(u)].
\end{aligned}$$

Now we take an integral from Eq. (7.2.23) regarding w from 0 to w

$$\frac{\partial^3\Psi(u,w,\tau)}{\partial u^2\partial w}=\frac{\partial^3\Psi(u,0,\tau)}{\partial u^2\partial w}-\frac{\partial^3\Psi(u,0,0)}{\partial u^2\partial w}+\frac{\partial^3\Psi(u,w,0)}{\partial u^2\partial w}+\Delta\mathfrak{G}(u)[P\mathfrak{G}(\tau)]^T[P\mathfrak{G}(w)], \tag{7.2.26}$$

and again, by taking the integral from 0 to w relative to w

$$\frac{\partial^2\Psi(u,w,\tau)}{\partial u^2}=\frac{\partial^2\Psi(u,0,\tau)}{\partial u^2}+w\left[-\frac{\partial^2\Psi(u,0,0)}{\partial u^2}+\frac{\partial^2\Psi(u,0,\tau)}{\partial u^2}\right]-\frac{\partial^2\Psi(u,0,0)}{\partial u^2}+\frac{\partial^2\Psi(u,w,0)}{\partial u^2}+\Delta\mathfrak{G}(u)[P\mathfrak{G}(\tau)]^T[p^2\mathfrak{G}(w)]. \tag{7.2.27}$$

Integrating twice from 0 to u relative to u from the above equation, we have

$$\begin{aligned}\Psi(u,w,\tau)&=\Psi(0,w,\tau)+\Psi(u,0,\tau)-\Psi(0,0,\tau)\\&+w\left[\Psi(u,0,\tau)-\Psi(0,0,\tau)-\Psi(u,0,0)+\Psi(0,0,0)+u\left[-\frac{\partial\Psi(0,0,\tau)}{\partial u}+\frac{\partial\Psi(0,0,0)}{\partial u}\right]\right]\\&-\Psi(u,0,0)+\Psi(0,0,0)+\Psi(u,w,0)-\Psi(0,w,0)\\&+\Delta[P\mathfrak{G}(\tau)]^T[p^2\mathfrak{G}(w)][P^2\mathfrak{G}(u)],\end{aligned} \tag{7.2.28}$$

by putting $u=c$, $w=d$ in the equation, we have obtained

$$\begin{aligned}\Psi(c,d,\tau)&=\Psi(0,d,\tau)+\Psi(c,0,\tau)-\Psi(0,0,\tau)\\&+d\left[\Psi(c,0,\tau)-\Psi(0,0,\tau)-\Psi(c,0,0)+\Psi(0,0,0)+c\left[-\frac{\partial\Psi(0,0,\tau)}{\partial u}+\frac{\partial\Psi(0,0,0)}{\partial u}\right]\right]\\&-\Psi(c,0,0)+\Psi(0,0,0)+\Psi(c,d,0)-\Psi(0,d,0)\\&+\Delta[P\mathfrak{G}(\tau)]^T[p^2\mathfrak{G}(d)][P^2\mathfrak{G}(c)].\end{aligned} \tag{7.2.29}$$

Then

$$\begin{aligned}&\left[-\frac{\partial\Psi(0,0,\tau)}{\partial u}+\frac{\partial\Psi(0,0,0)}{\partial u}\right]=\\&\frac{1}{dc}\Big[\Psi(c,d,\tau)-\Psi(0,d,\tau)-\Psi(c,0,\tau)+\Psi(0,0,\tau)-d\Psi(c,0,\tau)+d\Psi(0,0,\tau)+d\Psi(c,0,0)-d\Psi(0,0,0)\\&+\Psi(c,0,0)-\Psi(0,0,0)-\Psi(c,d,0)+\Psi(0,d,0)\\&-\Delta[P\mathfrak{G}(\tau)]^T[p^2\mathfrak{G}(d)][P^2\mathfrak{G}(c)]\Big].\end{aligned} \tag{7.2.30}$$

Using Eq. (7.2.28), we get

$$\begin{aligned}\Psi(u,w,\tau)&=\Psi(0,w,\tau)+\Psi(u,0,\tau)-\Psi(0,0,\tau)\\&+w\Big[\Psi(u,0,\tau)-\Psi(0,0,\tau)-\Psi(u,0,0)+\Psi(0,0,0)+u\Big[\frac{1}{dc}\Big[\Psi(c,d,\tau)-\Psi(0,d,\tau)-\Psi(c,0,\tau)\\&+\Psi(0,0,\tau)-d\Psi(c,0,\tau)+d\Psi(0,0,\tau)+d\Psi(c,0,0)-d\Psi(0,0,0)\\&+\Psi(c,0,0)-\Psi(0,0,0)-\Psi(c,d,0)+\Psi(0,d,0)-\Delta[P\mathfrak{G}(\tau)]^T[p^2\mathfrak{G}(d)][P^2\mathfrak{G}(c)]\Big]\Big]\Big]\\&-\Psi(u,0,0)+\Psi(0,0,0)+\Psi(u,w,0)-\Psi(0,w,0)+\Delta[P\mathfrak{G}(\tau)]^T[p^2\mathfrak{G}(w)][P^2\mathfrak{G}(u)].\end{aligned} \tag{7.2.31}$$

Continuing the process we derive that

$$\begin{aligned}
\frac{\partial \Psi(u,w,\tau)}{\partial \tau} &= \frac{\partial \Psi(0,w,\tau)}{\partial \tau} + \frac{\partial \Psi(u,0,\tau)}{\partial \tau} - \frac{\partial \Psi(0,0,\tau)}{\partial \tau} \\
&+ w\frac{d}{d\tau}\Bigg[\Psi(u,0,\tau) - \Psi(0,0,\tau) - \Psi(u,0,0) + \Psi(0,0,0) + u\Bigg[\frac{1}{dc}\Bigg[\Psi(c,d,\tau) - \Psi(0,d,\tau) - \Psi(c,0,\tau) \\
&+ \Psi(0,0,\tau) - d\Psi(c,0,\tau) + d\Psi(0,0,\tau) + d\Psi(c,0,0) - d\Psi(0,0,0) \\
&+ \Psi(c,0,0) - \Psi(0,0,0) - \Psi(c,d,0) + \Psi(0,d,0) - \Delta[P\mathfrak{G}(\tau)]^T[p^2\mathfrak{G}(d)][P^2\mathfrak{G}(c)]\Bigg]\Bigg]\Bigg] \\
&+ \frac{d}{d\tau}\Delta[P\mathfrak{G}(\tau)]^T[P^2\mathfrak{G}(w)][P^2\mathfrak{G}(u)],
\end{aligned} \tag{7.2.32}$$

and

$$\begin{aligned}
\frac{\partial^2 \Psi(u,w,\tau)}{\partial \tau^2} &= \frac{\partial^2 \Psi(0,w,\tau)}{\partial \tau^2} + \frac{\partial^2 \Psi(u,0,\tau)}{\partial \tau^2} - \frac{\partial^2 \Psi(0,0,\tau)}{\partial \tau^2} \\
&+ w\frac{d^2}{d\tau^2}\Bigg[\Psi(u,0,\tau) - \Psi(0,0,\tau) - \Psi(u,0,0) + \Psi(0,0,0) + u\Bigg[\frac{1}{dc}\Bigg[\Psi(c,d,\tau) - \Psi(0,d,\tau) - \Psi(c,0,\tau) \\
&+ \Psi(0,0,\tau) - d\Psi(c,0,\tau) + d\Psi(0,0,\tau) + d\Psi(c,0,0) - d\Psi(0,0,0) \\
&+ \Psi(c,0,0) - \Psi(0,0,0) - \Psi(c,d,0) + \Psi(0,d,0) - \Delta[P\mathfrak{G}(\tau)]^T[p^2\mathfrak{G}(d)][P^2\mathfrak{G}(c)]\Bigg]\Bigg]\Bigg] \\
&+ \frac{d^2}{d\tau^2}\Delta[P\mathfrak{G}(\tau)]^T[P^2\mathfrak{G}(w)][P^2\mathfrak{G}(u)].
\end{aligned} \tag{7.2.33}$$

To complete the algorithm, we put Eqs. (7.2.33), (7.2.32), (7.2.31), (7.2.27), and (7.2.25) in Eq. (7.2.1)

$$\begin{aligned}
&\frac{\partial^2 \Psi(0,w,\tau)}{\partial \tau^2} + \frac{\partial^2 \Psi(u,0,\tau)}{\partial \tau^2} - \frac{\partial^2 \Psi(0,0,\tau)}{\partial \tau^2} \\
&+ w\frac{d^2}{d\tau^2}\Bigg[\Psi(u,0,\tau) - \Psi(0,0,\tau) - \Psi(u,0,0) + \Psi(0,0,0) + u\Bigg[\frac{1}{dc}\Bigg[\Psi(c,d,\tau) - \Psi(0,d,\tau) - \Psi(c,0,\tau) \\
&+ \Psi(0,0,\tau) - d\Psi(c,0,\tau) + d\Psi(0,0,\tau) + d\Psi(c,0,0) - d\Psi(0,0,0) \\
&+ \Psi(c,0,0) - \Psi(0,0,0) - \Psi(c,d,0) + \Psi(0,d,0) - \Delta[P\mathfrak{G}(\tau)]^T[p^2\mathfrak{G}(d)][P^2\mathfrak{G}(c)]\Bigg]\Bigg]\Bigg] \\
&+ \frac{d^2}{d\tau^2}\Delta[P\mathfrak{G}(\tau)]^T[P^2\mathfrak{G}(w)][P^2\mathfrak{G}(u)], \\
&+ \mu_1\Bigg(\frac{\partial \Psi(0,w,\tau)}{\partial \tau} + \frac{\partial \Psi(u,0,\tau)}{\partial \tau} - \frac{\partial \Psi(0,0,\tau)}{\partial \tau} \\
&+ w\frac{d}{d\tau}\Bigg[\Psi(u,0,\tau) - \Psi(0,0,\tau) - \Psi(u,0,0) + \Psi(0,0,0) + u\Bigg[\frac{1}{dc}\Bigg[\Psi(c,d,\tau) - \Psi(0,d,\tau) - \Psi(c,0,\tau) \\
&+ \Psi(0,0,\tau) - d\Psi(c,0,\tau) + d\Psi(0,0,\tau) + d\Psi(c,0,0) - d\Psi(0,0,0) \\
&+ \Psi(c,0,0) - \Psi(0,0,0) - \Psi(c,d,0) + \Psi(0,d,0) - \Delta[P\mathfrak{G}(\tau)]^T[p^2\mathfrak{G}(d)][P^2\mathfrak{G}(c)]\Bigg]\Bigg]\Bigg] \\
&+ \frac{d}{d\tau}\Delta[P\mathfrak{G}(\tau)]^T[P^2\mathfrak{G}(w)][P^2\mathfrak{G}(u)]\Bigg) \\
&= \mu_2\Bigg(\frac{\partial^2 \Psi(u,0,\tau)}{\partial u^2} + w\Bigg[-\frac{\partial^2 \Psi(u,0,0)}{\partial u^2} + \frac{\partial^2 \Psi(u,0,\tau)}{\partial u^2}\Bigg] - \frac{\partial^2 \Psi(u,0,0)}{\partial u^2} + \frac{\partial^2 \Psi(u,w,0)}{\partial u^2} \\
&+ \Delta\mathfrak{G}(u)[P\mathfrak{G}(\tau)]^T[p^2\mathfrak{G}(w)]\Bigg) \\
&+ \mu_3\Bigg(\frac{\partial^2 \Psi(0,w,\tau)}{\partial w^2} + u\Bigg[-\frac{\partial^2 \Psi(0,w,0)}{\partial w^2} + \frac{\partial^2 \Psi(0,w,\tau)}{\partial w^2}\Bigg] - \frac{\partial^2 \Psi(0,w,0)}{\partial w^2} + \frac{\partial^2 \Psi(u,w,0)}{\partial w^2} \\
&+ \Delta\mathfrak{G}(w)[P\mathfrak{G}(\tau)]^T[p^2\mathfrak{G}(u)]\Bigg)
\end{aligned} \tag{7.2.34}$$

$$+ s(u, w) \sin \Bigg(\Psi(0, w, \tau) + \Psi(u, 0, \tau) - \Psi(0, 0, \tau)$$
$$+ w \Bigg[\Psi(u, 0, \tau) - \Psi(0, 0, \tau) - \Psi(u, 0, 0) + \Psi(0, 0, 0) + u \Bigg[\frac{1}{dc} \Bigg[\Psi(c, d, \tau) - \Psi(0, d, \tau) - \Psi(c, 0, \tau)$$
$$+ \Psi(0, 0, \tau) - d\Psi(c, 0, \tau) + d\Psi(0, 0, \tau) + d\Psi(c, 0, 0) - d\Psi(0, 0, 0)$$
$$+ \Psi(c, 0, 0) - \Psi(0, 0, 0) - \Psi(c, d, 0) + \Psi(0, d, 0) - \Delta[P\mathfrak{G}(\tau)]^T [p^2 \mathfrak{G}(d)][P^2 \mathfrak{G}(c)] \Bigg]\Bigg]\Bigg]$$
$$- \Psi(u, 0, 0) + \Psi(0, 0, 0) + \Psi(u, w, 0) - \Psi(0, w, 0) + \Delta[P\mathfrak{G}(\tau)]^T [p^2 \mathfrak{G}(w)][P^2 \mathfrak{G}(u)] \Bigg) + \phi(u, w, \tau) + \mu_4 \frac{d\mathrm{M}(\tau)}{d\tau}.$$

By solving the above system at different points (Local points $u_J, w_J, \tau_J = \frac{2J-1}{2E^2}$), the values of polynomial coefficients of the category are derived. Then by placing these coefficients in (7.2.31), we achieve the desired solution.

7.2.4 Numerical Examples

A performance of the clique polynomial method for a type of Eq. (7.2.1) is demonstrated. We consider the domain Ξ and these results are calculated in this domain. We shall computer the errors, $L_2 - error$, $L_\infty - error$ and $RMS - error$. Therefore, in the following, we will first introduce these errors and then present them in the examples.

$$\begin{aligned} L^2 - error &= \sqrt{\sum_{J=1}^{\ell} e_J^2} \\ L^\infty - error &= \max(e_J) \quad 1 \le J \le \ell - 1 \\ RMS - error &= \sqrt{\sum_{J=1}^{\ell} \frac{e_J^2}{\ell}}, \end{aligned} \tag{7.2.35}$$

where e_J is an absolute error.

Example 7.1 Consider the following nonlinear 2D stochastic timefractional Sine-Gordon equation (2D ST-FS-G)

$$\begin{aligned} \frac{\partial^\kappa \Psi(u, w, \tau)}{\partial \tau^\kappa} + 3\frac{\partial^{\kappa-1} \Psi(u, w, \tau)}{\partial \tau^{\kappa-1}} &= 2\frac{\partial^2 \Psi(u, w, \tau)}{\partial u^2} + \frac{\partial^2 \Psi(u, w, \tau)}{\partial \tau^2} \\ &+ \frac{\pi}{3} \sin(\psi(u, w, \tau)) + \phi(u, w, \tau) + 4\frac{d\mathrm{M}(\tau)}{d\tau}, \end{aligned} \tag{7.2.36}$$

by choosing $\kappa = 2$ and with the following initial and boundary conditions

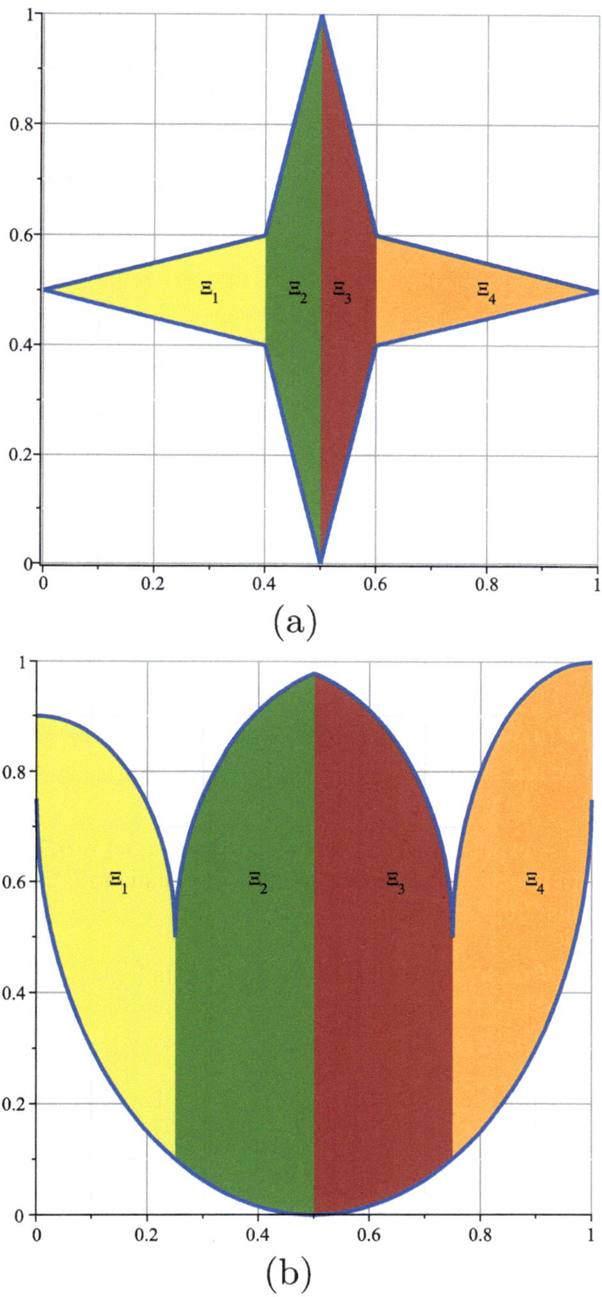

Fig. 7.7 The intended domain of Examples 7.1 and 7.2

Table 7.1 Compare the errors of our proposed method and other methods for Example 7.1

R	$L_2 - error$	$RMS - error$	$L_\infty - error$
Numerical results using Gaussian RBF			
1	–	$4.2234e-3$	$5.1057e-3$
2	–	$2.6358e-3$	$3.3081e-3$
3	–	$8.9071e-4$	$9.0937e-4$
4	–	$5.9450e-4$	$5.9999e-4$
5	–	–	–
Numerical results using multi-quadrics RBF			
1	–	$9.3207e-4$	$1.2396e-3$
2	–	$5.5302e-4$	$6.2397e-4$
3	–	$3.4501e-4$	$4.2139e-4$
4	–	$1.4540e-4$	$1.4999e-4$
5	–	–	–
Numerical results using our method			
1	$3.2e-5$	$8.1e-6$	$1.6e-5$
2	$4.6e-5$	$1.1e-5$	$2.3e-5$
3	$5.6e-5$	$1.4e-5$	$2.8e-5$
4	$6.5e-5$	$1.6e-5$	$3.2e-5$
5	$7.2e-5$	$1.8e-5$	$3.6e-5$

$$\begin{cases} \Psi(u, w, 0) = -\sin\left(\frac{\pi}{2}u\right)\sin\left(\frac{\pi}{2}w\right), & (u, w) \in \Xi, \\ \left.\frac{\partial\Psi(u,w,\tau)}{\partial\tau}\right|_{\tau=0} = -2\sin\left(\frac{\pi}{2}u\right)\sin\left(\frac{\pi}{2}w\right), & (u, w) \in \Xi, \end{cases}$$

and $\Psi(u, w, \tau) = -\exp(2\tau)\sin\left(\frac{\pi}{2}u\right)\sin\left(\frac{\pi}{2}w\right)$, where $(u, w) \in \partial\Xi, 0 \le \tau \le T$. We consider $\phi(u, w, \tau)$ such that the exact solution of equation (7.2.36) is as follows (Table 7.1 and Figs. 7.6, 7.7, 7.8)

$$\Psi(u, w, \tau) = -\exp(2\tau)\sin\left(\frac{\pi}{2}u\right)\sin\left(\frac{\pi}{2}w\right),$$

where $\Xi = \Xi_1 \cup \Xi_2 \cup \Xi_3 \cup \Xi_4$

$$\Xi_1 = \left\{(u, w) \in \mathbb{R}^2 : \quad 0 < u < \frac{2}{5}, \quad -\frac{1}{4}u + \frac{1}{2} < w < \frac{1}{4}u + \frac{1}{2}\right\}$$

$$\Xi_2 = \left\{(u, w) \in \mathbb{R}^2 : \quad \frac{2}{5} < u < \frac{1}{2}, \quad -4u + 2 < w < 4u - 1\right\}$$

$$\Xi_3 = \left\{(u, w) \in \mathbb{R}^2 : \quad \frac{1}{2} < u < \frac{3}{5}, \quad 4u - 2 < w < -4u + 3\right\}$$

$$\Xi_4 = \left\{(u, w) \in \mathbb{R}^2 : \quad \frac{3}{5} < u < 1, \quad \frac{1}{4}u + \frac{1}{4} < w < -\frac{1}{4}u + \frac{3}{4}\right\}.$$

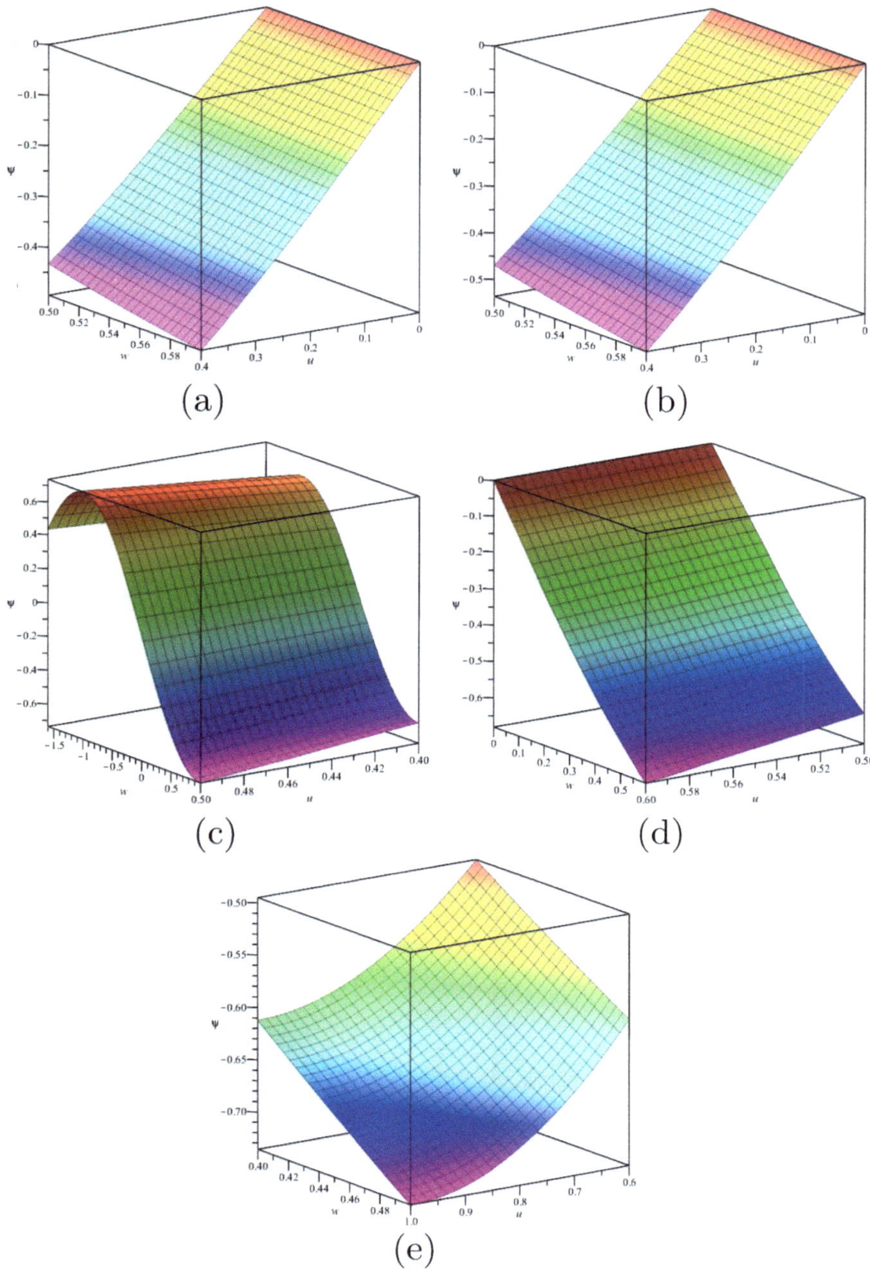

Fig. 7.8 Exact solution for Example 7.1. **a** $\tau = 0.02, u \in (0, \frac{2}{5}), w \in (\frac{1}{2}, \frac{3}{5})$, **b** $\tau = 0.06, u \in (0, \frac{2}{5}), w \in (\frac{1}{2}, \frac{3}{5})$, **c** $\tau = 0.02, u \in (\frac{2}{5}, \frac{1}{2}), w \in (-\frac{8}{5}, 1)$, **d** $\tau = 0.02, u \in (\frac{1}{2}, \frac{3}{5}), w \in (0, \frac{3}{5})$, **e** $\tau = 0.02, u \in (\frac{3}{5}, 1), w \in (\frac{8}{20}, \frac{1}{2})$

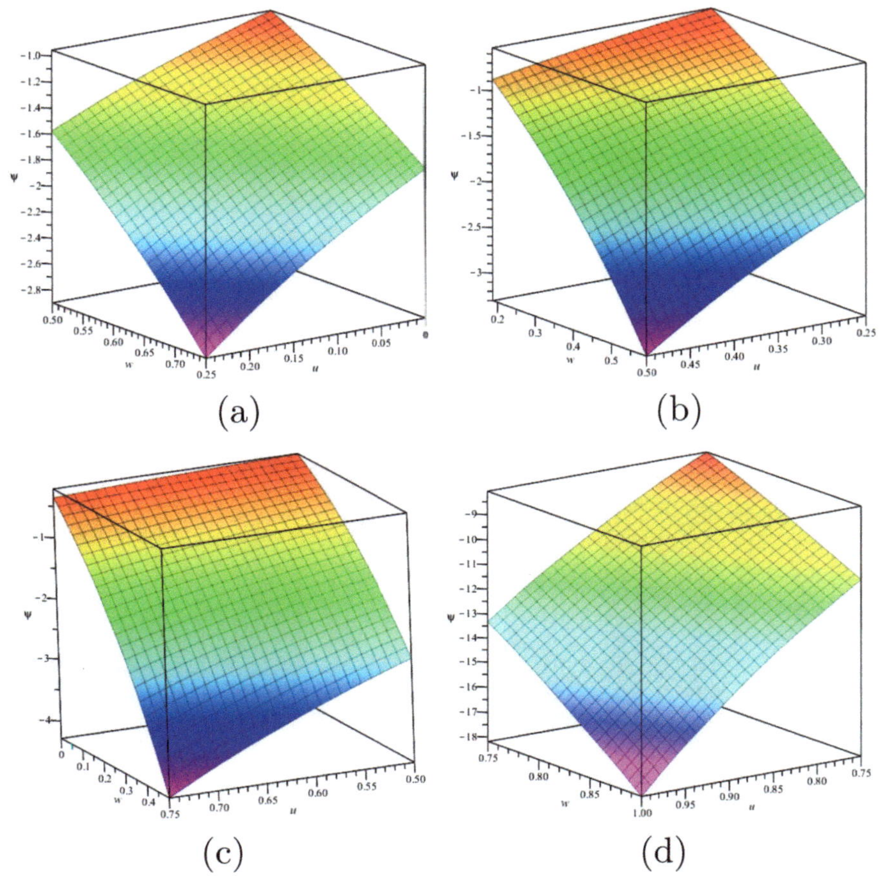

Fig. 7.9 Exact solution for Example 7.2. **a** $\tau = 0.1$, $u \in (0, 0.25)$, $w \in (0.75, 0.5)$, **b** $\tau = 0.1$, $u \in (0.25, 0.5)$, $w \in (-0.1875, 0.5918)$, **c** $\tau = 0.1$, $u \in (0.5, 0.75)$, $w \in (0, 0.5)$, **d** $\tau = 0.14$, $u \in (0.75, 1)$, $w \in (0.75, 0.9)$

Example 7.2 Consider the following nonlinear 2D stochastic timefractional Sine-Gordon equation (2D ST-FS-G) (Figs. 7.9, 7.10 and Tables 7.2, 7.3)

$$\frac{\partial^{\alpha}\Psi(u,w,\tau)}{\partial\tau^{\kappa}} + \frac{\partial^{\kappa-1}\Psi(u,w,\tau)}{\partial\tau^{\kappa-1}} = \pi\frac{\partial^{2}\Psi(u,w,\tau)}{\partial u^{2}} + \frac{\pi}{2}\frac{\partial^{2}\Psi(u,w,\tau)}{\partial w^{2}} + \sin(\Psi(u,w,\tau)) + \phi(u,w,\tau) + \frac{\pi}{3}\frac{d\mathrm{M}(\tau)}{d\tau}, \tag{7.2.37}$$

for $\kappa = 2$ and with the following initial and boundary conditions:

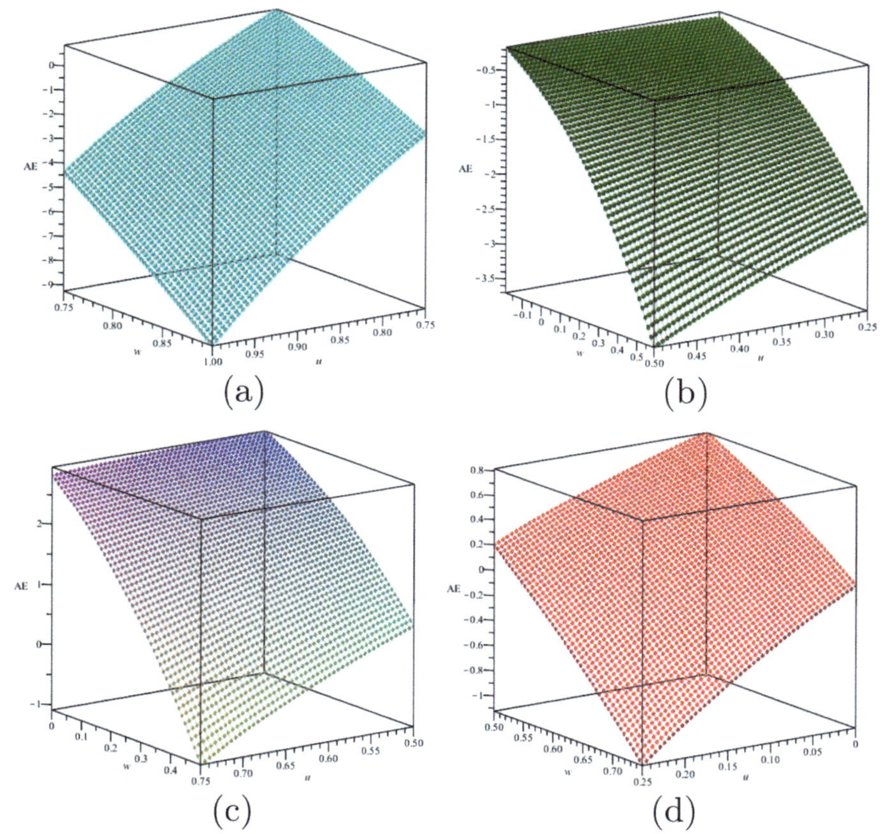

Fig. 7.10 Absolute error for Example 7.2. **a** $\tau = 0.1$, $u \in (0, 0.25)$, $w \in (0.75, 0.5)$, **b** $\tau = 0.1$, $u \in (0.25, 0.5)$, $w \in (-0.1875, 0.5918)$, **c** $\tau = 0.1$, $u \in (0.5, 0.75)$, $w \in (0, 0.5)$, **d** $\tau = 0.14$, $u \in (0.75, 1)$, $w \in (0.75, 0.9)$

$$\begin{cases} \Psi(u, w, 0) = -w \exp(2u + w), & (u, w) \in \Xi, \\ \left.\frac{\partial \Psi(u,w,\tau)}{\partial \tau}\right|_{\tau=0} = -\exp(2u + w), & (u, w) \in \Xi, \end{cases}$$

and $\Psi(u, w, \tau) = \left(2\tau^2 - \tau - w\right) \exp(2u + w)$, where $(u, w) \in \partial\Xi$, $0 \leq \tau \leq T$. We consider $\phi(u, w, \tau)$ such that the exact solution of Eq. (7.2.37) is as follows

$$\Psi(u, w, \tau) = \left(2\tau^2 - \tau - w\right) \exp(2u + w),$$

where $\Xi = \Xi_1 \cup \Xi_2 \cup \Xi_3 \cup \Xi_4$ and

$$\Xi_1 = \left\{ (u, w) \in \mathbb{R}^2 : \quad 0 < u < 0.25, \quad 0.75 - 0.75\sqrt{1 - \frac{(u - 0.5)^2}{0.25}} < w < 0.5 + 0.4\sqrt{1 - \frac{u^2}{0.0625}} \right\}$$

Table 7.2 Compare the errors of our proposed method and other methods for Example 7.2

R	$L_2 - error$	$RMS - error$	$L_\infty - error$
Numerical results using Gaussian RBF			
6	–	$5.0091e-4$	$5.8107e-4$
7	–	$2.5407e-4$	$3.1279e-4$
8	–	$9.8531e-5$	$1.0125e-4$
9	–	$7.5442e-5$	$7.5978e-5$
10	–	–	–
Numerical results using multi-quadrics RBF			
6	–	$9.3728e-5$	$1.1215e-4$
7	–	$8.5703e-5$	$9.2271e-5$
8	–	$5.7301e-5$	$6.5317e-5$
9	–	$3.2362e-5$	$3.2832e-5$
10	–	–	–
Numerical results using our method			
6	$7.9e-5$	$1.9e-5$	$3.9e-5$
7	$8.6e-5$	$2.1e-5$	$4.3e-5$
8	$9.6e-5$	$2.3e-5$	$4.6e-5$
9	$9.5e-5$	$2.4e-5$	$4.8e-5$
10	–	$2.5e-5$	$5.1e-5$

$$\Xi_2 = \left\{(u, w) \in \mathbb{R}^2 : \quad 0.25 < u < 0.5, \quad 0.75 - 0.75\sqrt{1 - \frac{(u-0.5)^2}{0.25}} < w \quad < 0.5 + 0.5\sqrt{1 - \frac{(u-0.6)^2}{0.1225}}\right\}$$

$$\Xi_3 = \left\{(u, w) \in \mathbb{R}^2 : \quad 0.5 < u < 0.75, \quad 0.75 - 0.75\sqrt{1 - \frac{(u-0.5)^2}{0.25}} < w \quad < 0.5 + 0.5\sqrt{1 - \frac{(u-0.4)^2}{0.1225}}\right\}$$

$$\Xi_4 = \left\{(u, w) \in \mathbb{R}^2 : \quad 0.75 < u < 1, \quad 0.75 - 0.75\sqrt{1 - \frac{(u-0.5)^2}{0.25}} < w \quad < 0.5 + 0.4\sqrt{1 - \frac{(u-1)^2}{0.0625}}\right\}.$$

Example 7.3 Consider the following nonlinear 2D ST-FS-G

$$\frac{\partial^\alpha \Psi(u, w, \tau)}{\partial \tau^\kappa} + \frac{\partial^{\kappa-1} \Psi(u, w, \tau)}{\partial \tau^{\kappa-1}} = \frac{\partial^2 \Psi(u, w, \tau)}{\partial u^2} + \frac{\partial^2 \Psi(u, w, \tau)}{\partial w^2} \tag{7.2.38}$$
$$- \sin(\Psi(u, w, \tau)) + \phi(u, w, \tau) + \frac{d\mathrm{M}(\tau)}{d\tau},$$

where $\kappa = 2$. We consider the function $\phi(u, w, \tau)$ such that $\Psi(u, w, \tau) = -u^2 - w^2 + \tau^2 + (u - w - \tau)\cos(\tau)$, is the exact solution of the equation. We have the initial and boundary conditions as follows

$$\begin{cases} \Psi(u, w, 0) = -u^2 - w^2 + u - w, & (u, w) \in \Xi, \\ \left.\frac{\partial \Psi(u,w,\tau)}{\partial \tau}\right|_{\tau=0} = -1, & (u, w) \in \Xi, \end{cases}$$

Table 7.3 Exact and approximate solutions of Examples 7.1 and 7.2

R	*ApproximateSolutions*	*ExactSolutions*
Numerical results for example (7.1)		
1	−0.9797428447	−0.2445680150
2	1.210786774	−0.04126143009
3	3.288801672	0.6776125542
4	3.233305841	−0.4651960087
5	4.211685926	−0.5741286545
Numerical results for example (7.2)		
6	4.120551377	−1.480747753
7	6.731720887	0.04310902677
8	6.362513321	−1.141583089
9	−6.548391730	−14.59614450
10	−5.155084770	−14.01832209

and where $(u, w) \in \partial \Xi,\ 0 \leq \tau \leq T,\ \Xi = \Xi_1 \cup \Xi_2 \cup \Xi_3 \cup \Xi_4$ and

$$\Xi_1 = \left\{(u, w) \in \mathbb{R}^2 :\ \ 0 < u < \frac{1}{6},\ \ -\frac{1}{3}u + \frac{1}{6} < w < \frac{1}{3}u + \frac{1}{6}\right\}$$
$$\Xi_2 = \left\{(u, w) \in \mathbb{R}^2 :\ \ \frac{1}{6} < u < \frac{1}{3},\ \ -3u + 6 < w < 3u + 2\right\}$$
$$\Xi_3 = \left\{(u, w) \in \mathbb{R}^2 :\ \ \frac{1}{3} < u < \frac{1}{2},\ \ 3u - 6 < w < -3u + 4\right\}$$
$$\Xi_4 = \left\{(u, w) \in \mathbb{R}^2 :\ \ \frac{1}{2} < u < 1,\ \ \frac{1}{6}u + \frac{1}{6} < w < -\frac{1}{6}u + \frac{2}{3}\right\}.$$

In the following, we present the numerical results obtained for Eq. (7.2.38) in Table 7.4. In this table the exact solutions and approximate solutions obtained from the presented method are given. Also, the comparison of these solutions is shown graphically in Fig. 7.11.

Example 7.4 Consider the following nonlinear 2D ST-FS-G

$$\frac{\partial^{\alpha}\Psi(u, w, \tau)}{\partial \tau^{\kappa}} + \frac{\partial^{\kappa-1}\Psi(u, w, \tau)}{\partial \tau^{\kappa-1}} = \frac{\partial^2\Psi(u, w, \tau)}{\partial u^2} - \frac{\partial^2\Psi(u, w, \tau)}{\partial w^2}$$
$$- \sin(\Psi(u, w, \tau)) + \phi(u, w, \tau) + \pi\frac{d\mathrm{M}(\tau)}{d\tau}, \tag{7.2.39}$$

Table 7.4 Numerical results for Example 7.3

R	*Approximate Solutions*	*Exact Solutions*
1	−0.6056681223	−0.6136571156
2	−0.8800356855	−0.9700367267
3	−0.2913271607	−0.3110146471
4	−0.4623267413	−0.2627466525
5	−0.3354488719	−0.2312488719
6	−0.3927865112	−0.3475555741
7	−0.9455632147	−1.444063416
8	−2.874625172	−4.576032249
9	−1.126851245	−1.376880030
10	−1.864202633	−2.065604301

where $\kappa = 2$. We consider the function $\phi(u, w, \tau)$ such that $\Psi(u, w, \tau) = \frac{1}{4}(\sin(\frac{\pi}{2}(u + w + \tau)) + \sin(\frac{\pi}{2}(u + w - \tau)))$, is the exact solution of the equation. We have the initial and boundary conditions as follows

$$\begin{cases} \Psi(u, w, 0) = \frac{1}{2}(\sin(\frac{\pi}{2}(u + w)), & (u, w) \in \Xi, \\ \left.\frac{\partial \Psi(u,w,\tau)}{\partial \tau}\right|_{\tau=0} = 0, & (u, w) \in \Xi, \end{cases}$$

and where $(u, w) \in \partial\Xi, 0 \le \tau \le T$, $\Xi = \Xi_1 \cup \Xi_2 \cup \Xi_3 \cup \Xi_4$ and

$$\Xi_1 = \Big\{(u, w) \in \mathbb{R}^2 : \quad 0 < u < 0.15, \quad 0.35 - 0.35(0.15 - \sqrt{1-u^2}) < w < 0.75 + 0.55(1 - \sqrt{1-u^2})\Big\}$$

$$\Xi_2 = \Big\{(u, w) \in \mathbb{R}^2 : \quad 0.15 < u < 0.35, \quad 0.35 - 0.35(0.15 - \sqrt{1-u^2}) < w < 0.75 + 0.75(1 - \sqrt{u^2})\Big\}$$

$$\Xi_3 = \Big\{(u, w) \in \mathbb{R}^2 : \quad 0.35 < u < 0.55, \quad 0.35 - 0.35(0.15 - \sqrt{1-u^2}) < w < 0.75 + 0.75(1 - \sqrt{u^2})\Big\}$$

$$\Xi_4 = \Big\{(u, w) \in \mathbb{R}^2 : \quad 0.55 < u < 1, \quad 0.35 - 0.35(0.15 - \sqrt{1-u^2}) < w < 0.75 + 0.55(1 - \sqrt{1-u^2})\Big\}.$$

In the following, we present the numerical results obtained for Eq. (7.2.39) in Table 7.5. In this table the exact solutions and approximate solutions obtained from the presented method are given. Also, the comparison of these solutions is shown graphically in Fig. 7.11.

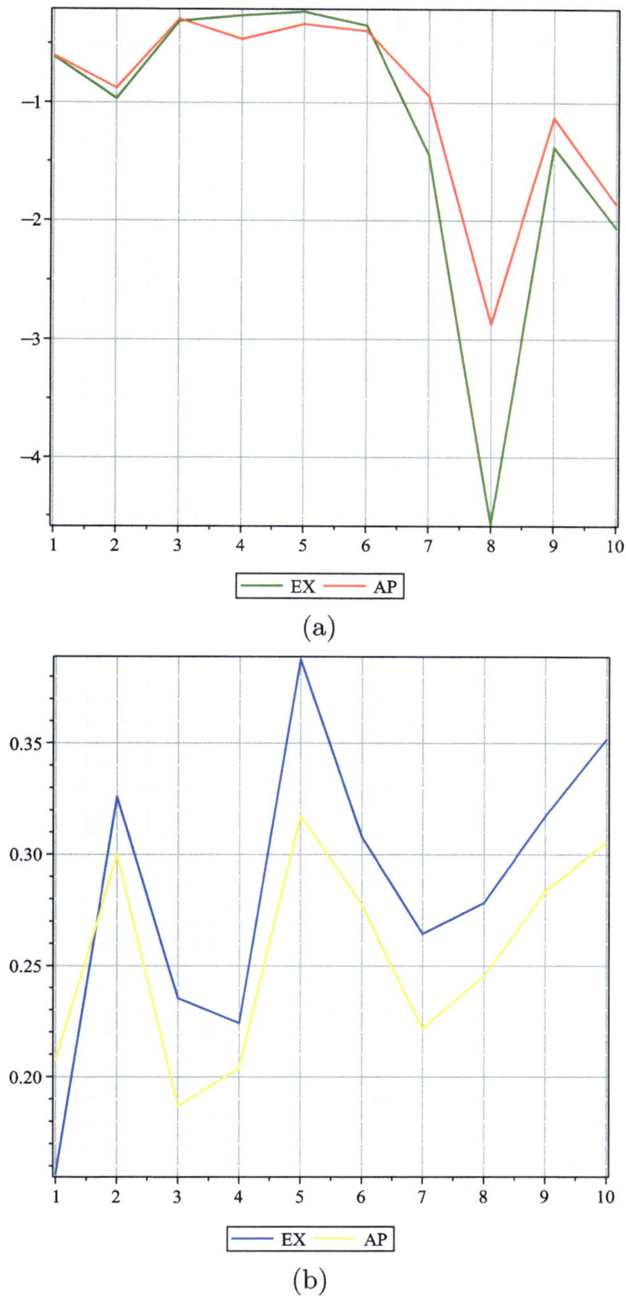

Fig. 7.11 Comparison of exact solutions and approximate solutions for Examples 7.3, 7.4 for different values of u, w in the given domains and for $\tau = 0.2,\ 0.3$

Table 7.5 Numerical results for Example 7.4

R	*Approximate Solutions*	*Exact Solutions*
1	0.15734821368	0.2461101360
2	0.3000472826	0.3260483918
3	0.1867232491	0.2351866150
4	0.2040013769	0.2240029381
5	0.3175440995	0.3880140310
6	0.2773576314	0.3078128528
7	0.2213675422	0.2643731790
8	0.2455000038	0.2783018098
9	0.2833147637	0.3173109120
10	0.3057133399	0.3519093811

7.3 Coronavirus (COVID 19) Mathematical Model

7.3.1 *Introduction and Preliminaries*

In 2019, in Wuhan, China, a new infectious disease called Corona (the new Coronavirus) spread rapidly throughout the country and the rest of the world. As more than 28 million people worldwide are infected with the disease (COVID-19), the fight against the new Coronavirus is still ongoing in many countries [96, 120, 132]. Considering that most natural phenomena are modeled through differential equations, it is necessary to provide numerical and analytical solutions for them, since there is no exact solution for all of these equations. These numerical methods provide suitable approximations of these equations (see [7, 26, 27, 65, 77, 112]).

We consider the COVID-19 differential system (C-19-DS) as follows (see [12, 38]),

$$\begin{cases} \frac{d\mathrm{S}_{\sigma 1}}{d\tau} = \Delta_\sigma - \varsigma_\sigma \mathrm{S}_{\sigma 1} - \mathcal{L}_\sigma \mathrm{S}_{\sigma 1}\left(\mathcal{I}_{\sigma 1} + \mathcal{M}A_{\sigma 1}\right) - \mathcal{L}_{\mathrm{W}} \mathrm{S}_{\sigma 1}\mathrm{W} \\ \frac{d\mathrm{E}_{\sigma 11}}{d\tau} = \mathcal{L}_\sigma \mathrm{S}_{\sigma 1}\left(\mathcal{I}_{\sigma 1} + \mathcal{M}A_{\sigma 1}\right) + \mathcal{L}_{\mathrm{W}} \mathrm{S}_{\sigma 1}\mathrm{W}_1 - (1-\nu_\sigma)\,\zeta_\sigma \mathrm{E}_{\sigma 1} - \nu_\sigma \zeta'_\sigma \mathrm{E}_{\sigma 1} - \varsigma_\sigma \mathrm{E}_{\sigma 1} \\ \frac{d\mathcal{I}_{\sigma 1}}{d\tau} = (1-\nu_\sigma)\,\zeta_\sigma \mathrm{E}_{\sigma 1} - (\kappa_\sigma + \varsigma_\sigma)\,\mathcal{I}_{\sigma 1} \\ \frac{dA_{\sigma 1}}{d\tau} = \nu_\sigma \zeta'_\sigma \mathrm{E}_{\sigma 1} - \left(\kappa'_\sigma + \varsigma_\sigma\right) A_{\sigma 1} \\ \frac{dR_{\sigma 1}}{d\tau} = \kappa_\sigma \mathcal{I}_{\sigma 1} + \kappa'_\sigma A_{\sigma 1} - \varsigma_\sigma R_{\sigma 1} \\ \frac{d\mathrm{W}_1}{d\tau} = \eta_\sigma \mathcal{I}_{\sigma 1} + \eta'_\sigma A_{\sigma 1} - \varkappa \mathrm{W}_1 \end{cases} \tag{7.3.1}$$

where λ_σ is the birth rate of people, ς_σ the death rate of people, $\frac{1}{\zeta_\sigma}$ the incubation period of people, $\frac{1}{\zeta'_\sigma}$ the latent period of people, $\frac{1}{\kappa_\sigma}$ the infectious period of symptomatic infection in people, $\frac{1}{\kappa'_\sigma}$ the infectious period of asymptomatic infection in people, η_σ the shedding coefficients from $\mathcal{I}_{\sigma 1}$ to W_1, η'_σ the shedding coefficients from A_σ to W, ν_σ the proportion of asymptomatic infection rate of people, $\mathcal{L}_\sigma$ the transmission rate from $\mathcal{I}_{\sigma 1}$ to $\mathrm{S}_{\sigma 1}$, $\mathcal{L}_\mathrm{W}$ the transmission rate from W_1 to $\mathrm{S}_{\sigma 1}$, $\mathcal{M}$ the multiple of the transmissibility of $A_{\sigma 1}$ to that of $\mathcal{I}_{\sigma 1}$, $\frac{1}{\varkappa}$ the lifetime of the virus in W_1, $\mathcal{R}_{\sigma 1}$ the relative shedding rate of $A_{\sigma 1}$ compared to $\mathcal{I}_{\sigma 1}$, $\Delta_\sigma = \lambda_\sigma \times \mathcal{A}_\sigma$ where $\mathcal{A}_\sigma$ refers to the total number of people and

- $\mathrm{S}_{\sigma 1}(\tau)$ is susceptible people,
- $\mathrm{E}_{\sigma 1}(\tau)$ is exposed people,
- $\mathcal{I}_{\sigma 1}(\tau)$ is symptomatic infected people,
- $A_{\sigma 1}(\tau)$ is asymptomatic infected people,
- $R_{\sigma 1}(\tau)$ is recovered and dead people,
- $\mathrm{W}_1(\tau)$ is COVID-19 in reservoir in time τ.

By obtaining the unknowns $\mathrm{S}_{\sigma 1}(\tau)$, $\mathrm{E}_{\sigma 1}(\tau)$, $\mathcal{I}_{\sigma 1}(\tau)$, $A_{\sigma 1}(\tau)$, $R_{\sigma 1}(\tau)$ and $\mathrm{W}_1(\tau)$, we investigate the transmittance of COVID-19. Since the definitive cure for this disease has not yet been found, variables obtained in the medical sciences are used to obtain a better treatment and measure the progression of COVID-19. For more research in this field, we refer the reader to [4, 5, 64, 123]. Using the operational technique, which is a new method, we obtain the operational matrix with respect to the polynomial of the complete category of the graph.

The clique polynomial of a graph $\mathcal{E}$, denoted by $C(\mathcal{E}, \tau)$, is given by [54, 55]

$$C(\mathcal{E}, \tau) = \sum_{\iota=0}^{\mathrm{m}} \mathrm{g}_\iota(\mathcal{E})\tau^\iota, \quad \mathrm{g}_0(\mathcal{E}) = 1$$

where $\mathrm{g}_\iota(\mathcal{E})$ is the number of ι-cliques of $\mathcal{E}$. The CP of a complete graph $\mathcal{E}_\mathrm{m}$ with m vertices is given by [70]

$$C\left(\mathcal{E}_\mathrm{m}, \tau\right) = (1 + \tau)^\mathrm{m}.$$

We can approximate any function $\mathrm{y}(\tau)$ in $L^2(0, \mathrm{R})$ in terms of the CP as

$$\mathrm{y}(\tau) \simeq \tilde{\mathrm{y}}(\tau) = \sum_{\mathrm{m}=0}^{\mathcal{J}-1} \mathrm{y}_\mathrm{m} C\left(\mathcal{E}_\mathrm{m}, \tau\right) = \mathcal{Y}^T \mathcal{G}_\mathcal{J}(\tau), \tag{7.3.2}$$

where $\mathcal{Y} = \left[\mathrm{y}_0, \mathrm{y}_1, \ldots, \mathrm{y}_{\mathcal{J}-1}\right]^T$, and

$$\mathcal{G}_\mathcal{J}(\tau) = \left[C\left(\mathcal{E}_0, \tau\right), C\left(\mathcal{E}_1, \tau\right), \ldots, C\left(\mathcal{E}_{\mathcal{J}-1}, \tau\right)\right]^T. \tag{7.3.3}$$

From (7.3.2), we have $\mathcal{Y} = \mathcal{D}^{-1}\langle \mathrm{y}(\tau), \mathcal{G}_{\mathcal{J}}(\tau)\rangle$, where $\mathcal{D} = \langle \mathcal{G}_{\mathcal{J}}(\tau), \mathcal{G}_{\mathcal{J}}(\tau)\rangle$ and $\langle \cdot, \cdot\rangle$ indicates the inner product of two arbitrary functions. Also, an arbitrary function $\mathcal{Y}(\mathrm{x}, \tau) \in L^2([i, j] \times [0, \mathrm{A}])$ can be expanded in terms of the CP by the following infinite series

$$\mathcal{Y}(\mathrm{x}, \tau) = \sum_{\mathrm{m}=0}^{\infty}\sum_{\mathrm{n}=0}^{\infty} \mathcal{Y}_{\mathrm{mn}} C\left(\mathcal{E}_{\mathrm{m}}, \mathrm{x}\right) C\left(\mathcal{E}_{\mathrm{n}}, \tau\right). \tag{7.3.4}$$

We consider the first $\mathcal{J}$ terms in (7.3.4), as an approximation of the function $\mathcal{Y}(\mathrm{x}, \tau)$. It is obtained as

$$\mathcal{Y}(\mathrm{x}, \tau) \simeq \sum_{\mathrm{m}=0}^{\mathcal{J}-1}\sum_{\mathrm{n}=0}^{\mathcal{J}-1} \mathcal{Y}_{\mathrm{mn}} C\left(\mathcal{E}_{\mathrm{m}}, \mathrm{x}\right) C\left(\mathcal{E}_{\mathrm{n}}, \tau\right) = \mathcal{G}_{\mathcal{J}}^{T}(\mathrm{x}) \mathcal{Y} \mathcal{G}_{\mathcal{J}}(\tau),$$

where $\mathcal{Y} = \mathcal{D}^{-1}\langle \mathcal{G}_{\mathcal{J}}(\mathrm{x}), \langle \mathrm{y}(\mathrm{x}, \tau), \mathcal{G}_{\mathcal{J}}(\tau)\rangle\rangle \mathcal{D}^{-1}$.

In the process of solving differential equations using the clique polynomials technique, assume that $\mathcal{G}_{\mathcal{J}}(\tau)$ is the matrix defined in (7.3.3). This vector is written based on Taylor functions as follows

$$\mathcal{G}_{\mathcal{J}}(\tau) = \mathcal{W} \mathcal{Z}_{\mathcal{J}}(\tau), \tag{7.3.5}$$

where $\mathcal{W}$ is an invertible matrix defined as follows

$$\mathcal{W} = [w_{\imath\jmath}], \qquad \imath, \jmath = 0, 1, \ldots, \mathcal{J} - 1, \qquad w_{\imath\jmath} = \begin{cases} \binom{\imath}{\jmath}, & \imath \geq \jmath \\ 0, & \imath < \jmath \end{cases}$$

Also, $\mathcal{Z}_{\mathcal{J}}(\tau)$ is the vector of Taylor's basic functions as follows

$$\mathcal{Z}_{\mathcal{J}}(\tau) = \begin{bmatrix} 1 \\ \tau \\ \tau^2 \\ \vdots \\ \tau^{\mathcal{J}-1} \end{bmatrix}.$$

By taking the integral in Eq. (7.3.5), we have

$$\begin{aligned} \int_0^{\tau} \mathcal{G}_{\mathcal{J}}(\tau) d\tau &= \int_0^{\tau} \mathcal{W} \mathcal{Z}_{\mathcal{J}}(t) dt \\ &= \mathcal{W} \int_0^{\tau} \mathcal{Y}_{\mathcal{J}}(t) dt \\ &= G \bar{\mathcal{Z}}(\tau) \end{aligned}$$

where

$$G = \mathcal{W}\begin{bmatrix} 1 & 0 & \cdots & 0 \\ 0 & \frac{1}{2} & \cdots & 0 \\ \vdots & \vdots & \ddots & \vdots \\ 0 & 0 & \cdots & 0 \end{bmatrix}$$

and

$$\bar{\mathcal{Z}}(\tau) = \begin{bmatrix} \tau \\ \tau^2 \\ \tau^3 \\ \vdots \\ \tau^{\mathcal{J}} \end{bmatrix}.$$

Now, we approximate the τ^{p} functions based on the clique base and using Eq. (7.3.5) as follows

$$\tau^{\mathrm{p}} \simeq W^{-1}_{[\mathrm{p}+1]}\mathcal{G}_{\mathcal{J}}(\tau), \mathrm{p} = 1, 2, \ldots, \mathcal{J} - 1;$$

here $\mathcal{W}^{-1}_{[\mathrm{p}+1]}$ is the $(\mathrm{p}+1)$th row of the matrix $\mathcal{W}^{-1}$. Now we approximate the function $\tau^{\mathcal{J}}$ based on the clique base as follows

$$\tau^{\mathcal{J}} \simeq \mathrm{q}^T_{\mathcal{J}}\mathcal{G}_{\mathcal{J}}(\tau),$$

where $\mathrm{q}_{\mathcal{J}}$ is calculated as follows

$$\mathrm{q}_{\mathcal{J}} = \mathcal{D}^{-1}\int_0^1 \tau^{\mathcal{J}}\mathcal{G}_{\mathcal{J}}(\tau)d\tau$$

where $\mathcal{D} =< \mathcal{G}_{\mathcal{J}}(\tau), \mathcal{G}_{\mathcal{J}}(\tau) >$. We suppose that

$$\mathrm{K} = \begin{bmatrix} \mathcal{W}^{-1}_{[2]} \\ \mathcal{W}^{-1}_{[3]} \\ \mathcal{W}^{-1}_{[4]} \\ \vdots \\ \mathcal{W}^{-1}_{[\mathcal{J}]} \\ \mathrm{q}^T_{\mathcal{J}} \end{bmatrix},$$

then, $\bar{\mathcal{Z}}(\tau) \simeq \mathrm{K}\mathcal{G}_{\mathcal{J}}(\tau)$. Therefore, the integral operating matrix based on the clique base is obtained as follows

$$\int_0^{\tau} \mathcal{G}_{\mathcal{J}}(t)dt \simeq \mathcal{S}\mathcal{G}_{\mathcal{J}}(\tau),$$

and that $\mathcal{S} = G\mathrm{K}$ the operational matrix is called the clique base integral.

By taking twice the integral of Eq. (7.3.3), we have

$$\begin{aligned}\int_0^{\tau}\int_0^{\tau} \mathcal{G}_{\mathcal{J}}(t)dtdt &= \int_0^{\tau} \mathcal{S}\mathcal{G}_{\mathcal{J}}(t)dt\\ &= \mathcal{S}\int_0^{\tau} \mathcal{G}_{\mathcal{J}}(t)dt\\ &= \mathcal{S}^2\mathcal{G}_{\mathcal{J}}(\tau),\end{aligned}$$

Consider the set $\mathrm{T} = span\{C(\mathcal{E}_0, \tau), \ldots, C(\mathcal{E}_{\mathcal{J}-1}, \tau)\}$. The set T is considered to be the space of all one-variable polynomials. If ϱ is the degree of polynomials on τ, then $\varrho \leq \mathcal{J}$. The best approximation for the function $\mathrm{y}(\tau) \in L^2([0, \mathrm{A}])$ to T is the function $\tilde{\mathrm{y}}(\tau)$.

In the next theorem, we obtain the error of approximation of the one-variable function obtained using clique polynomials. It should be noted that $\|.\|_2$ used in the following theorem is the L^2-norm which on $X \subseteq \mathbb{R}^n$ is defined as follows

$$\|\mathrm{y}\|_2 = \left(\int_X |\mathrm{y}(\tau)|^2 d\tau\right)^{\frac{1}{2}}.$$

Theorem 7.2 *Suppose* $\mathrm{y}(\tau)$ *be a sufficiently smooth function on* $\mathrm{I} = [0, \mathrm{A}]$ *and* $\tilde{\mathrm{y}}(\tau)$ *be its approximation using clique polynomials. Then, we have*

$$\|\mathrm{y}(\tau) - \tilde{\mathrm{y}}(\tau)\|_2 \leq \left(\frac{\varpi_1(\mathrm{A})^{\mathcal{J}}}{\mathcal{J}!2^{2\mathcal{J}-1}} + + \frac{\varpi_3(\mathrm{A})^{\mathcal{J}}}{\mathcal{J}!2^{2\mathcal{J}-1}}\right)\mathrm{A}.$$

Proof Let $\tau_{\mathrm{m}}, \mathrm{m} = 0, 1, \ldots, \mathcal{J} - 1$ be the root of $\mathcal{J}$-degree shifted Chebyshev polynomials on [0, 1]. By considering the interpolating polynomial $\mathrm{H}_{\mathcal{J}}$ to $\mathrm{y}(\tau)$ at point (τ_{m}), we have

$$\mathrm{y}(\tau) - \mathrm{H}_{\mathcal{J}}(\tau) = \frac{\partial^{\mathcal{J}}\mathrm{y(a)}}{\partial\tau^{\mathcal{J}}\mathcal{J}!}\prod_{\mathrm{m}=0}^{\mathcal{J}-1}(\tau - \tau_{\mathrm{m}}) - \frac{\partial^{\mathcal{J}}\mathrm{y(a')}}{\partial\tau^{\mathcal{J}}\mathcal{J}!}\prod_{\mathrm{m}=0}^{\mathcal{J}-1}(\tau - \tau_{\mathrm{m}}),$$

where $\mathrm{a}, \mathrm{a}' \in [0, \mathrm{A}]$. Then we obtain

$$\left|\mathrm{y}(\tau) - \mathrm{H}_{\mathcal{J}}(\tau)\right| \leqslant \max_{\tau\in\mathrm{I}}\left|\frac{\partial^{\mathcal{J}}\mathrm{y(a)}}{\partial\tau^{\mathcal{J}}}\right|\frac{\prod_{\mathrm{m}=0}^{\mathcal{J}-1}|\tau - \tau_{\mathrm{m}}|}{\mathcal{J}!} + \max_{\tau\in\mathrm{I}}\left|\frac{\partial^{\mathcal{J}}\mathrm{y(a')}}{\partial\tau^{\mathcal{J}}}\right|\frac{\prod_{\mathrm{m}=0}^{\mathcal{J}-1}|\tau - \tau_{\mathrm{m}}|}{\mathcal{J}!}. \tag{7.3.6}$$

According to the smooth function $\mathrm{y}(\tau)$ on $L^2([0, \mathrm{A}])$, there are constants ϖ_1, ϖ_2 such that

$$\max_{\tau\in\mathrm{I}}\left|\frac{\partial^{\mathcal{J}}\mathrm{y(a)}}{\partial\tau^{\mathcal{J}}}\right| \leq \varpi_1, \max_{\tau\in\mathrm{I}}\left|\frac{\partial^{\mathcal{J}}\mathrm{y(a')}}{\partial\tau^{\mathcal{J}}}\right| \leq \varpi_2. \tag{7.3.7}$$

By substituting (7.3.7) into (7.3.6) and employing the estimates for Chebyshev interpolation nodes we have

$$\left|y(\tau) - H_{\mathcal{J}}(\tau)\right| \leqslant \frac{\varpi_1 (A)^{\mathcal{J}}}{\mathcal{J}! 2^{2\mathcal{J}-1}} + \frac{\varpi_2 (A)^{\mathcal{J}}}{\mathcal{J}! 2^{2\mathcal{J}-1}}. \tag{7.3.8}$$

Since $\tilde{y}(\tau)$ is the best approximation of $y(\tau)$ in T and using (7.3.8), we can write

$$\begin{aligned}
\|y(\tau) - \tilde{y}(\tau)\|_2^2 &= \int_0^A |y(\tau) - \tilde{y}(\tau)|^2 d\tau \\
&\leqslant \int_0^A \left|y(\tau) - H_{\mathcal{J}}(\tau)\right|^2 d\tau \\
&= \int_0^A \left(\frac{\varpi_1 (A)^{\mathcal{J}}}{\mathcal{J}! 2^{2\mathcal{J}-1}} + \frac{\varpi_2 (A)^{\mathcal{J}}}{\mathcal{J}! 2^{2\mathcal{J}-1}}\right)^2 d\tau \\
&= \left(\frac{\varpi_1 (A)^{\mathcal{J}}}{\mathcal{J}! 2^{2\mathcal{J}-1}} + \frac{\varpi_2 (A)^{\mathcal{J}}}{\mathcal{J}! 2^{2\mathcal{J}-1}}\right)^2 A,
\end{aligned}$$

which gives the desired result. Note that for the special case, when $A = 1$, we have

$$\|y(\tau) - \tilde{y}(\tau)\|_2 \leqslant \frac{\varpi_1}{\mathcal{J}! 2^{2\mathcal{J}-1}} + \frac{\varpi_2}{\mathcal{J}! 2^{2\mathcal{J}-1}},$$

therefore

$$\|y(\tau) - \tilde{y}(\tau)\|_2 = O\left(\frac{1}{\mathcal{J}! 2^{2\mathcal{J}-1}}\right).$$

7.3.2 The Proposed Numerical Technique of the COVID-19 Mathematical Model (C-19-DS)

We will present a numerical method for solving the COVID-19 differential system having the initial conditions using a clique-based operating matrix. Before solving numerically, we first normalize system (7.3.1). Therefore, for this purpose, we consider the sets $S_\sigma = \frac{S_{\sigma 1}}{\mathcal{A}_\sigma}$, $E_\sigma = \frac{E_{\sigma 1}}{\mathcal{A}_\sigma}$, $\mathcal{I}_\sigma = \frac{\mathcal{I}_{\sigma 1}}{\mathcal{A}_\sigma}$, $A_\sigma = \frac{A_{\sigma 1}}{\mathcal{A}_\sigma}$, $R_\sigma = \frac{R_\sigma}{\mathcal{A}_\sigma}$, $W_\sigma = \frac{\varkappa W_{\sigma 1}}{\eta_\sigma \mathcal{A}_\sigma}$, $\eta'_\sigma = \mathcal{R}\eta_\sigma$, $f_\sigma = \mathcal{L}_\sigma \mathcal{A}_\sigma$, $f_W = \frac{\eta_\sigma \mathcal{L}_W \mathcal{A}_\sigma}{\varkappa}$. Then, the normalized system is as follows

$$\begin{cases}
\frac{dS_\sigma}{d\tau} = \lambda_\sigma - \varsigma_\sigma S_\sigma - f_\sigma S_\sigma \left(\mathcal{I}_\sigma + \mathcal{M} A_\sigma\right) - f_W S_\sigma W_\sigma \\
\frac{dE_\sigma}{d\tau} = f_\sigma S_\sigma \left(\mathcal{I}_\sigma + \mathcal{M} A_\sigma\right) + f_W S_\sigma W_\sigma - (1 - \nu_\sigma) \zeta_\sigma E_\sigma - \nu_\sigma \zeta'_\sigma E_\sigma - \varsigma_\sigma E_\sigma \\
\frac{d\mathcal{I}_\sigma}{d\tau} = (1 - \nu_\sigma) \zeta_\sigma E_\sigma - (\kappa_\sigma + \varsigma_\sigma) \mathcal{I}_\sigma \\
\frac{dA_\sigma}{d\tau} = \nu_\sigma \zeta'_\sigma E_\sigma - \left(\kappa'_\sigma + \varsigma_\sigma\right) A_\sigma \\
\frac{dR_\sigma}{d\tau} = \kappa_\sigma \mathcal{I}_\sigma + \kappa'_\sigma A_\sigma - \varsigma_\sigma R_\sigma \\
\frac{dW_\sigma}{d\tau} = \eta_\sigma \mathcal{I}_\sigma + \eta'_\sigma A_\sigma - \varkappa W_\sigma
\end{cases} \tag{7.3.9}$$

To start the numerical solution of the system, we consider

$$\frac{\partial^2 \mathrm{S}_\sigma(\tau)}{\partial \tau^2} = \sum_{\mathrm{m}=0}^{\infty} \xi_{\mathrm{m}} C\left(\mathcal{E}_{\mathrm{m}}, \tau\right) = \boldsymbol{\xi}^T \mathcal{G}_{\mathcal{J}}(\tau), \tag{7.3.10}$$

and

$$\mathcal{G}_{\mathcal{J}}(\tau) = \left[C\left(\mathcal{E}_0, \tau\right), C\left(\mathcal{E}_1, \tau\right), C\left(\mathcal{E}_2, \tau\right), C\left(\mathcal{E}_3, \tau\right), \ldots, C\left(\mathcal{E}_{\mathcal{J}-1}, \tau\right)\right]^T$$

Integrate Eq (7.3.10) regarding τ from 0 to τ to obtain

$$\begin{aligned}
\frac{\partial \mathrm{S}_\sigma(\tau)}{\partial \tau} &= \frac{\partial \mathrm{S}_\sigma(0)}{\partial \tau} + \int_0^\tau \boldsymbol{\xi}^T \mathcal{G}_{\mathcal{J}}(t)dt \\
&= \frac{\partial \mathrm{S}_\sigma(0)}{\partial \tau} + \boldsymbol{\xi}^T \int_0^\tau \mathcal{G}_{\mathcal{J}}(t)dt \\
&= \frac{\partial \mathrm{S}_\sigma(0)}{\partial \tau} + \boldsymbol{\xi}^T [\mathcal{S}\mathcal{G}_{\mathcal{J}}(\tau)].
\end{aligned} \tag{7.3.11}$$

Integrating Eq (7.3.11) regarding τ from 0 to τ gives

$$\begin{aligned}
\mathrm{S}_\sigma(\tau) &= \mathrm{S}_\sigma(0) + \tau\left(\frac{\partial \mathrm{S}_\sigma(0)}{\partial \tau}\right) + \int_0^\tau \boldsymbol{\xi}^T [\mathcal{S}\mathcal{G}_{\mathcal{J}}(t)]dt. \\
&= \mathrm{S}_\sigma(0) + \tau\left(\frac{\partial \mathrm{S}_\sigma(0)}{\partial \tau}\right) + \boldsymbol{\xi}^T \mathcal{S} \int_0^\tau \mathcal{G}_{\mathcal{J}}(t)dt \\
&= \mathrm{S}_\sigma(0) + \tau\left(\frac{\partial \mathrm{S}_\sigma(0)}{\partial \tau}\right) + \boldsymbol{\xi}^T \mathcal{S}^2 \mathcal{G}_{\mathcal{J}}(\tau).
\end{aligned} \tag{7.3.12}$$

By putting $\tau = \beta$ in Eq. (7.3.12), we have

$$\mathrm{S}_\sigma(\beta) = \mathrm{S}_\sigma(0) + \beta\left(\frac{\partial \mathrm{S}_\sigma(0)}{\partial \tau}\right) + \boldsymbol{\xi}^T \mathcal{S}^2 \mathcal{G}_{\mathcal{J}}(\tau)|_{\tau=\beta},$$

then

$$\frac{\partial \mathrm{S}_\sigma(0)}{\partial \tau} = \frac{1}{\beta}\left(\mathrm{S}_\sigma(\beta) - \mathrm{S}_\sigma(0) - \boldsymbol{\xi}^T \mathcal{S}^2 \mathcal{G}_{\mathcal{J}}(\tau)|_{\tau=\beta}\right).$$

By placing it in (7.3.12), we have

$$\mathrm{S}_\sigma(\tau) = \mathrm{S}_\sigma(0) + \tau\left(\frac{1}{\beta}\left(\mathrm{S}_\sigma(\beta) - \mathrm{S}_\sigma(0) - \boldsymbol{\xi}^T \mathcal{S}^2 \mathcal{G}_{\mathcal{J}}(\tau)|_{\tau=\beta}\right)\right) + \boldsymbol{\xi}^T \mathcal{S}^2 \mathcal{G}_{\mathcal{J}}(\tau) \tag{7.3.13}$$

We have exactly the same situation as above

$$\frac{\partial^2 \mathrm{E}_\sigma(\tau)}{\partial \tau^2} = \sum_{\mathrm{m}=0}^{\infty} \theta_{\mathrm{m}} C\left(\mathcal{E}_{\mathrm{m}}, \tau\right) = \boldsymbol{\theta}^T \mathcal{G}_{\mathcal{J}}(\tau), \tag{7.3.14}$$

and by taking the integral of Eq. (7.3.14) with respect to τ from 0 to τ, we have

$$\frac{\partial \mathrm{E}_\sigma(\tau)}{\partial \tau} = \frac{\partial \mathrm{E}_\sigma(0)}{\partial \tau} + \boldsymbol{\theta}^T [\mathcal{S} \mathcal{G}_{\mathcal{J}}(\tau)]. \tag{7.3.15}$$

Again, by taking the integral in Eq. (7.3.15) from 0 to τ, with regard to τ, we have

$$\mathrm{E}_\sigma(\tau) = \mathrm{E}_\sigma(0) + \tau\left(\frac{\partial \mathrm{E}_\sigma(0)}{\partial \tau}\right) + \boldsymbol{\theta}^T \mathcal{S}^2 \mathcal{G}_{\mathcal{J}}(\tau). \tag{7.3.16}$$

By placing $\tau = \beta$ in relation (7.3.16), we have the same situation as above

$$\mathrm{E}_\sigma(\tau) = \mathrm{E}_\sigma(0) + \tau\left(\frac{1}{\beta}\left(\mathrm{E}_\sigma(\beta) - \mathrm{E}_\sigma(0) - \boldsymbol{\theta}^T \mathcal{S}^2 \mathcal{G}_{\mathcal{J}}(\tau)|_{\tau=\beta}\right)\right) + \boldsymbol{\theta}^T \mathcal{S}^2 \mathcal{G}_{\mathcal{J}}(\tau). \tag{7.3.17}$$

Repeat the same process for the function $\mathcal{I}_\sigma(\tau)$. Then we consider

$$\frac{\partial^2 \mathcal{I}_\sigma(\tau)}{\partial \tau^2} = \sum_{\mathrm{m}=0}^{\infty} \upsilon_{\mathrm{m}} C\left(\mathcal{E}_{\mathrm{m}}, \tau\right) = \boldsymbol{\upsilon}^T \mathcal{G}_{\mathcal{J}}(\tau), \tag{7.3.18}$$

and by taking the integral in Eq. (7.3.18) from 0 to τ, with regard to τ, we have

$$\frac{\partial \mathcal{I}_\sigma(\tau)}{\partial \tau} = \frac{\partial \mathcal{I}_\sigma(0)}{\partial \tau} + \boldsymbol{\upsilon}^T [\mathcal{S} \mathcal{G}_{\mathcal{J}}(\tau)]. \tag{7.3.19}$$

Then

$$\mathcal{I}_\sigma(\tau) = \mathcal{I}_\sigma(0) + \tau\left(\frac{\partial \mathcal{I}_\sigma(0)}{\partial \tau}\right) + \boldsymbol{\upsilon}^T \mathcal{S}^2 \mathcal{G}_{\mathcal{J}}(\tau), \tag{7.3.20}$$

and

$$\mathcal{I}_\sigma(\tau) = \mathcal{I}_\sigma(0) + \tau\bigg(\frac{1}{\beta}\Big(\mathcal{I}_\sigma(\beta) - \mathcal{I}_\sigma(0) - \boldsymbol{\upsilon}^T \mathcal{S}^2 \mathcal{G}_{\mathcal{J}}(\tau)|_{\tau=\beta}\Big)\bigg) + \boldsymbol{\upsilon}^T \mathcal{S}^2 \mathcal{G}_{\mathcal{J}}(\tau). \tag{7.3.21}$$

For the function $A_\sigma(\tau)$, we have

$$\frac{\partial^2 A_\sigma(\tau)}{\partial \tau^2} = \sum_{\mathrm{m}=0}^{\infty} \phi_{\mathrm{m}} \mathcal{C}\left(\mathcal{E}_{\mathrm{m}}, \tau\right) = \boldsymbol{\phi}^T \mathcal{G}_{\mathcal{J}}(\tau), \tag{7.3.22}$$

and by taking the integral of Eq. (7.3.22) with respect to τ from 0 to τ, we have

$$\frac{\partial A_\sigma(\tau)}{\partial \tau} = \frac{\partial A_\sigma(0)}{\partial \tau} + \boldsymbol{\phi}^T [\mathcal{S}\mathcal{G}_{\mathcal{J}}(\tau)], \tag{7.3.23}$$

and by the integral with respect to τ, we have

$$A_\sigma(\tau) = A_\sigma(0) + \tau\left(\frac{\partial A_\sigma(0)}{\partial \tau}\right) + \boldsymbol{\phi}^T \mathcal{S}^2 \mathcal{G}_{\mathcal{J}}(\tau) \tag{7.3.24}$$

and

$$A_\sigma(\tau) = A_\sigma(0) + \tau\bigg(\frac{1}{\beta}\Big(A_\sigma(\beta) - A_\sigma(0) - \boldsymbol{\phi}^T \mathcal{S}^2 \mathcal{G}_{\mathcal{J}}(\tau)|_{t=b}\Big)\bigg) + \boldsymbol{\phi}^T \mathcal{S}^2 \mathcal{G}_{\mathcal{J}}(\tau) \tag{7.3.25}$$

Consider the function $R_\sigma(t)$, W as follows

$$\begin{aligned} \frac{\partial^2 R_\sigma(\tau)}{\partial \tau^2} &= \sum_{\mathrm{m}=0}^{\infty} \psi_{\mathrm{m}} \mathcal{C}\left(\mathcal{E}_{\mathrm{m}}, \tau\right) = \boldsymbol{\psi}^T \mathcal{G}_{\mathcal{J}}(\tau), \\ \frac{\partial^2 \mathrm{W}(\tau)}{\partial \tau^2} &= \sum_{\mathrm{m}=0}^{\infty} \omega_{\mathrm{m}} \mathcal{C}\left(\mathcal{E}_{\mathrm{m}}, \tau\right) = \boldsymbol{\omega}^T \mathcal{G}_{\mathcal{J}}(\tau). \end{aligned} \tag{7.3.26}$$

Now, we take an integral on each line of Eq. (7.3.26), then

$$\frac{\partial R_\sigma(\tau)}{\partial\tau} = \frac{\partial R_\sigma(0)}{\partial\tau} + \psi^T[\mathcal{S}\mathcal{G}_{\mathcal{J}}(\tau)] \tag{7.3.27}$$
$$\frac{\partial \mathrm{W}(\tau)}{\partial\tau} = \frac{\partial \mathrm{W}(0)}{\partial\tau} + \omega^T[\mathcal{S}\mathcal{G}_{\mathcal{J}}(\tau)].$$

We take integrals from $\frac{\partial R_\sigma(\tau)}{\partial\tau}$ and $\frac{\partial \mathrm{W}(\tau)}{\partial\tau}$ with respect to τ from 0 to τ and we obtain

$$R_\sigma(\tau) = R_\sigma(0) + \tau\left(\frac{\partial R_\sigma(0)}{\partial\tau}\right) + \psi^T\mathcal{S}^2\mathcal{G}_{\mathcal{J}}(\tau) \tag{7.3.28}$$
$$\mathrm{W}(\tau) = \mathrm{W}(0) + \tau\left(\frac{\partial \mathrm{W}(0)}{\partial\tau}\right) + \omega^T\mathcal{S}^2\mathcal{G}_{\mathcal{J}}(\tau)$$

and

$$R_\sigma(\tau) = R_\sigma(0) + \tau\left(\frac{1}{\beta}\left(R_\sigma(\beta) - R_\sigma(0) - \omega^T\mathcal{S}^2\mathcal{G}_{\mathcal{J}}(\tau)|_{\tau=\beta}\right)\right) + \psi^T\mathcal{S}^2\mathcal{G}_{\mathcal{J}}(\tau) \tag{7.3.29}$$
$$\mathrm{W}(\tau) = \mathrm{W}(0) + \tau\left(\frac{1}{\beta}\left(\mathrm{W}(\beta) - \mathrm{W}(0) - \omega^T\mathcal{S}^2\mathcal{G}_{\mathcal{J}}(\tau)|_{\tau=\beta}\right)\right) + \omega^T\mathcal{S}^2\mathcal{G}_{\mathcal{J}}(\tau).$$

By placing (7.3.29), (7.3.27), (7.3.25), (7.3.23), (7.3.21), (7.3.19), (7.3.17), (7.3.15), (7.3.13) and (7.3.11) in Eq. (7.3.1), we obtain:

$$\begin{aligned}
\frac{\partial \mathrm{S}_\sigma(0)}{\partial\tau} + \xi^T[\mathcal{S}\mathcal{G}_{\mathcal{J}}(\tau)] = {}& \Delta_\sigma - \varsigma_\sigma\left(\mathrm{S}_\sigma(0) + \tau\left(\frac{1}{\beta}\left(\mathrm{S}_\sigma(\beta) - \mathrm{S}_\sigma(0) - \xi^T\mathcal{S}^2\mathcal{G}_{\mathcal{J}}(\tau)|_{\tau=\beta}\right)\right) + \xi^T\mathcal{S}^2\mathcal{G}_{\mathcal{J}}(\tau)\right) \\
& - \mathcal{L}_\sigma\left(\mathrm{S}_\sigma(0) + \tau\left(\frac{1}{\beta}\left(\mathrm{S}_\sigma(\beta) - \mathrm{S}_\sigma(0) - \xi^T\mathcal{S}^2\mathcal{G}_{\mathcal{J}}(\tau)|_{\tau=\beta}\right)\right) + \xi^T\mathcal{S}^2\mathcal{G}_{\mathcal{J}}(\tau)\right) \\
& \left(\mathcal{I}_\sigma + \mathcal{M}(\mathrm{A}_\sigma(0) + \tau\left(\frac{1}{\beta}\left(\mathrm{A}_\sigma(\beta) - \mathrm{A}_\sigma(0) - \phi^T\mathcal{S}^2\mathcal{G}_{\mathcal{J}}(\tau)|_{\tau=\beta}\right)\right) + \phi^T\mathcal{S}^2\mathcal{G}_{\mathcal{J}}(\tau))\right) \\
& - \mathcal{L}_{\mathrm{W}}\left(\mathrm{S}_\sigma(0) + \tau\left(\frac{1}{\beta}\left(\mathrm{S}_\sigma(\beta) - \mathrm{S}_\sigma(0) - \xi^T\mathcal{S}^2\mathcal{G}_{\mathcal{J}}(\tau)|_{\tau=\beta}\right)\right) + \xi^T\mathcal{S}^2\mathcal{G}_{\mathcal{J}}(\tau)\right) \\
& \left(\mathrm{W}(0) + \tau\left(\frac{1}{\beta}\left(\mathrm{W}(\beta) - \mathrm{W}(0) - \omega^T\mathcal{S}^2\mathcal{G}_{\mathcal{J}}(\tau)|_{\tau=\beta}\right)\right) + \omega^T\mathcal{S}^2\mathcal{G}_{\mathcal{J}}(\tau)\right) \\
\frac{\partial \mathrm{E}_\sigma(0)}{\partial\tau} + \theta^T[\mathcal{S}\mathcal{G}_{\mathcal{J}}(\tau)] = {}& \mathcal{L}_\sigma\left(\mathrm{S}_\sigma(0) + \tau\left(\frac{1}{\beta}\left(\mathrm{S}_\sigma(\beta) - \mathrm{S}_\sigma(0) - \xi^T\mathcal{S}^2\mathcal{G}_{\mathcal{J}}(\tau)|_{\tau=\beta}\right)\right) + \xi^T\mathcal{S}^2\mathcal{G}_{\mathcal{J}}(\tau)\right) \\
& \left(\mathcal{I}_\sigma + \mathcal{M}\left(\mathrm{A}_\sigma(0) + \tau\left(\frac{1}{\beta}\left(\mathrm{A}_\sigma(\beta) - \mathrm{A}_\sigma(0) - \phi^T\mathcal{S}^2\mathcal{G}_{\mathcal{J}}(\tau)|_{\tau=\beta}\right)\right) + \phi^T\mathcal{S}^2\mathcal{G}_{\mathcal{J}}(\tau)\right)\right) \\
& + \mathcal{L}_{\mathrm{W}}\left(\mathrm{S}_\sigma(0) + \tau\left(\frac{1}{\beta}\left(\mathrm{S}_\sigma(b) - \mathrm{S}_\sigma(0) - \xi^T\mathcal{S}^2\mathcal{G}_{\mathcal{J}}(\tau)|_{\tau=\beta}\right)\right) + \xi^T\mathcal{S}^2\mathcal{G}_{\mathcal{J}}(\tau)\right) \\
& \left(\mathrm{W}(0) + \tau\left(\frac{1}{\beta}\left(\mathrm{W}(\beta) - \mathrm{W}(0) - \omega^T\mathcal{S}^2\mathcal{G}_{\mathcal{J}}(\tau)|_{\tau=\beta}\right)\right) + \omega^T\mathcal{S}^2\mathcal{G}_{\mathcal{J}}(\tau)\right) \\
& - (1-\nu_\sigma)\,\zeta_\sigma\left(\mathrm{E}_\sigma(0) + \tau\left(\frac{1}{\beta}\left(\mathrm{E}_\sigma(\beta) - \mathrm{E}_\sigma(0) - \theta^T\mathcal{S}^2\mathcal{G}_{\mathcal{J}}(\tau)|_{\tau=\beta}\right)\right) + \theta^T\mathcal{S}^2\mathcal{G}_{\mathcal{J}}(\tau)\right) \\
& - \nu_\sigma\zeta'_\sigma\left(\mathrm{E}_\sigma(0) + \tau\left(\frac{1}{\beta}\left(\mathrm{E}_\sigma(\beta) - \mathrm{E}_\sigma(0) - \theta^T\mathcal{S}^2\mathcal{G}_{\mathcal{J}}(\tau)|_{\tau=\beta}\right)\right) + \theta^T\mathcal{S}^2\mathcal{G}_{\mathcal{J}}(\tau)\right)
\end{aligned} \tag{7.3.30}$$

$$
\begin{aligned}
&-\varsigma_\sigma\Big(\mathrm{E}_\sigma(0)+\tau\Big(\frac{1}{\beta}\Big(\mathrm{E}_\sigma(\beta)-\mathrm{E}_\sigma(0)-\theta^T\mathcal{S}^2\mathcal{G}_{\mathcal{J}}(\tau)|_{\tau=\beta}\Big)\Big)+\theta^T\mathcal{S}^2\mathcal{G}_{\mathcal{J}}(\tau)\Big)\\
\frac{\partial \mathcal{I}_\sigma(0)}{\partial\tau}+\upsilon^T[\mathcal{S}\mathcal{G}_{\mathcal{J}}(\tau)]&=(1-\nu_\sigma)\zeta_\sigma\Big(\mathrm{E}_\sigma(0)+\tau\Big(\frac{1}{\beta}\Big(\mathrm{E}_\sigma(\beta)-\mathrm{E}_\sigma(0)-\theta^T\mathcal{S}^2\mathcal{G}_{\mathcal{J}}(\tau)|_{\tau=\beta}\Big)\Big)+\theta^T\mathcal{S}^2\mathcal{G}_{\mathcal{J}}(\tau)\Big)\\
&-(\kappa_\sigma+\varsigma_\sigma)\Big(\mathcal{I}_\sigma(0)+\tau\Big(\frac{\partial \mathcal{I}_\sigma(0)}{\partial\tau}\Big)+\upsilon^T\mathcal{S}^2\mathcal{G}_{\mathcal{J}}(\tau)\Big)\\
\frac{\partial A_\sigma(0)}{\partial\tau}+\phi^T[\mathcal{S}\mathcal{G}_{\mathcal{J}}(\tau)]&=\nu_\sigma\zeta'_\sigma\Big(\mathrm{E}_\sigma(0)+\tau\Big(\frac{1}{\beta}\Big(\mathrm{E}_\sigma(\beta)-\mathrm{E}_\sigma(0)-\theta^T\mathcal{S}^2\mathcal{G}_{\mathcal{J}}(\tau)|_{\tau=\beta}\Big)\Big)+\theta^T\mathcal{S}^2\mathcal{G}_{\mathcal{J}}(\tau)\Big)\\
&-(\kappa'_\sigma+\varsigma_\sigma)\Big(A_\sigma(0)+\tau\Big(\frac{1}{\beta}\Big(A_\sigma(\beta)-\mathrm{E}_\sigma(0)-\phi^T\mathcal{S}^2\mathcal{G}_{\mathcal{J}}(\tau)|_{\tau=\beta}\Big)\Big)+\phi^T\mathcal{S}^2\mathcal{G}_{\mathcal{J}}(\tau)\Big)\\
\frac{\partial R_\sigma(0)}{\partial\tau}+\psi^T[\mathcal{S}\mathcal{G}_{\mathcal{J}}(\tau)]&=\kappa_\sigma\Big(\mathcal{I}_\sigma(0)+\tau\Big(\frac{1}{\beta}\Big(\mathcal{I}_\sigma(\beta)-\mathcal{I}_\sigma(0)-\upsilon^T\mathcal{S}^2\mathcal{G}_{\mathcal{J}}(\tau)|_{\tau=\beta}\Big)\Big)+\upsilon^T\mathcal{S}^2\mathcal{G}_{\mathcal{J}}(\tau)\Big)\\
&+\kappa'_\sigma\Big(A_\sigma(0)+\tau\Big(\frac{1}{\beta}\Big(A_\sigma(\beta)-A_\sigma(0)-\phi^T\mathcal{S}^2\mathcal{G}_{\mathcal{J}}(\tau)|_{\tau=\beta}\Big)\Big)+\phi^T\mathcal{S}^2\mathcal{G}_{\mathcal{J}}(\tau)\Big)\\
&-\varsigma_\sigma\Big(R_\sigma(0)+\tau\Big(\frac{1}{\beta}\Big(R_\sigma(\beta)-R_\sigma(0)-\psi^T\mathcal{S}^2\mathcal{G}_{\mathcal{J}}(\tau)|_{\tau=\beta}\Big)\Big)+\psi^T\mathcal{S}^2\mathcal{G}_{\mathcal{J}}(\tau)\Big).\\
\frac{\partial \mathrm{W}(0)}{\partial\tau}+\omega^T[\mathcal{S}\mathcal{G}_{\mathcal{J}}(\tau)]&=\eta_\sigma\Big(\mathcal{I}_\sigma(0)+\tau\Big(\frac{\partial \mathcal{I}_\sigma(0)}{\partial\tau}\Big)+\upsilon^T\mathcal{S}^2\mathcal{G}_{\mathcal{J}}(\tau)\Big)\\
&+\eta'_\sigma\Big(A_\sigma(0)+\tau\Big(\frac{1}{\beta}\Big(A_\sigma(\beta)-A_\sigma(0)-\phi^T\mathcal{S}^2\mathcal{G}_{\mathcal{J}}(\tau)|_{\tau=\beta}\Big)\Big)+\phi^T\mathcal{S}^2\mathcal{G}_{\mathcal{J}}(\tau)\Big)\\
&-\varepsilon\Big(\mathrm{W}(0)+\tau\Big(\frac{1}{\beta}\Big(\mathrm{W}(\beta)-\mathrm{W}(0)-\omega^T\mathcal{S}^2\mathcal{G}_{\mathcal{J}}(\tau)|_{\tau=\beta}\Big)\Big)+\omega^T\mathcal{S}^2\mathcal{G}_{\mathcal{J}}(\tau)\Big).
\end{aligned}
$$

By considering the points $\tau_{\mathrm{m}}=\frac{2\mathrm{m}-1}{2\mathcal{J}^2}$ for $\mathrm{m}=1,2,\ldots,\mathcal{J}$ in Eq. (7.3.30) and by solving this system, the equations of the unknown coefficients are obtained, and finally, by placing S, E, $\mathcal{I}$, A, R, W in (7.3.29), (7.3.25), (7.3.21), (7.3.17), (7.3.13), the numerical solution of Eq. (7.3.1) is obtained. Also, Fig. 7.12a–f show the solutions obtained from the RK4, ALP, and CP methods.

7.3.3 Numerical Examples

In this section, we evaluate the proposed method by presenting a numerical problem.

Example 7.5 Considering the differential system (7.3.1) and the following values, we solve the system using the Clique polynomials method (CPM) based on the operational matrix. The values of the initial and boundary conditions and parameters are given as

$$
\begin{aligned}
&\mathcal{A}_\sigma=1000000000,\quad \nu_\sigma=\mathcal{M}=\mathcal{L}_\sigma=\mathcal{L}_{\mathrm{W}}=\eta_\sigma=\kappa'_\sigma=\mathcal{R}=0.5,\quad \lambda_\sigma=\varsigma_\sigma=0.0018,\\
&\varkappa=0/1,\quad \zeta_\sigma=\zeta'_\sigma=0.1923,\quad \kappa_\sigma=0.1724
\end{aligned}
$$

with the initial and boundary conditions of unknown parameters as follows

$$
\begin{aligned}
&\mathrm{S}_\sigma(0)=2,\quad \mathrm{E}_\sigma(0)=4,\quad \mathcal{I}_\sigma(0)=3,\quad A_\sigma(0)=4,\quad R_\sigma(0)=2,\quad \mathrm{W}_\sigma(0)=3.5,\\
&\mathrm{S}_\sigma(1)=3\quad \mathrm{E}_\sigma(1)=5,\quad \mathcal{I}_\sigma(1)=4,\quad A_\sigma(1)=5,\quad R_\sigma(1)=3,\quad \mathrm{W}_\sigma(1)=4.5
\end{aligned}
$$

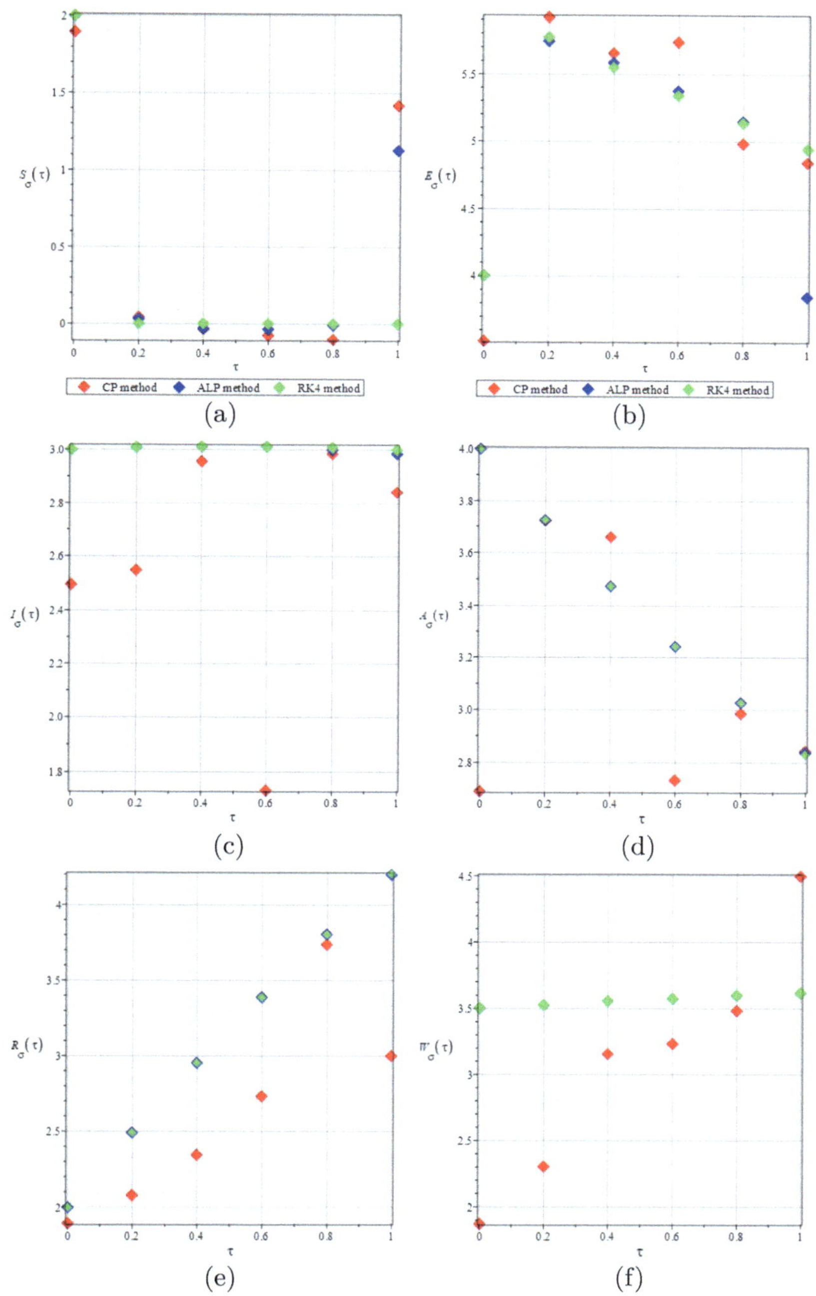

Fig. 7.12 Diagrams related to the numerical solution obtained from the RK4 and ALP methods and the CP method

Table 7.6 Comparison of the results obtained from our proposed method and other methods

τ	Numerical results using RK4 method	Numerical results using ALP method	CPM
0.0	2.000000000	2.00000000	1.89520000
0.2	$1.59905E-13$	0.03002862	−1.07872737
0.4	$1.59871E-13$	−0.02923894	−0.034354705
0.6	$1.59898E-13$	−0.03456393	−0.073501708
0.8	$1.59982E-13$	−0.01139075	−0.098500956
1	$1.60124E-13$	1.12482514	1.42108547

Table 7.7 Comparison of the results obtained from our proposed method and other methods

τ	Numerical results using RK4 method	Numerical results using ALP method	CPM method
0.0	4	4	3.520930000000000
0.2	5.77189614477	5.74277771538	5.921272622080000
0.4	5.55247760052	5.58266126424	5.65645294656000
0.6	5.34141367556	5.37685723359	5.73501708540000
0.8	5.13838625205	5.15034353745	4.98500956990000
1	4.94308933537	3.80689507404	4.84217094304040

Table 7.8 Comparison of the results obtained from our proposed method and other methods

τ	Numerical results using RK4 method	Numerical results using ALP method	CPM method
0.0	3	3	2.49520000000000
0.2	3.00849107875	3.00849108363	2.5487273779200002
0.4	3.01246360734	3.01246360563	2.956452946560000
0.6	3.01223326389	3.01223326126	1.73501708540000
0.8	3.00809881420	3.00809880689	2.98500956990000
1	3.00034288194	3.00034287358	2.84217094302879

In the following, we compare the results obtained from the discussed method with the results obtained from other methods used to solve this system. This information is presented in Tables 7.6, 7.7, 7.8, 7.10, 7.9 and 7.11. In all cases, we have considered the interval [0, 1] and obtained our result on this interval.

Considering and different values for τ in the range of [0, 1], we have presented the results for $S_\sigma(\tau)$, $E_\sigma(\tau)$, $\mathcal{I}_\sigma(\tau)$, $A_\sigma(\tau)$, $R_\sigma(\tau)$ and $W_\sigma(\tau)$ respectively in Tables 7.6, 7.7, 7.8, 7.10, 7.9 and 7.11.

The absolute errors of $S_\sigma(\tau)$, $E_\sigma(\tau)$, $\mathcal{I}_\sigma(\tau)$, $A_\sigma(\tau)$, $R_\sigma(\tau)$ and $W_\sigma(\tau)$ are presented in Fig. 7.13a, b, c, d, e and f respectively.

Table 7.9 Comparison of the results obtained from our proposed method and other methods

τ	Numerical results using RK4 method	Numerical results using ALP method	CRM method
0.0	4	4	2.69521000000000
0.2	3.72569099766	3.725275447302	3.721272622080000
0.4	3.47348293584	3.473083082484	3.65745294656000
0.6	3.24142145708	3.241082166196	2.73501708540000
0.8	3.02773254616	3.027574786909	2.98500956990000
1	2.83080613397	2.836696723317	2.84217094304040

Table 7.10 Comparison of the results obtained from our proposed method and other methods

τ	Numerical results using RK4 method	Numerical results using ALP method	CRM method
0.0	2	2	1.89526
0.2	2.48888268589	2.48882899953	1.078788578
0.4	2.95149948421	2.95140029238	0.343669453
0.6	3.38981976531	3.38968102039	0.734833485
0.8	3.80563689580	3.80546221628	0.734833485
1	4.20058431510	4.20029857270	2.999694

Table 7.11 Comparison of the results obtained from our proposed method and other methods

t	Numerical results using RK4 method	Numerical results using ALP method	CRM method
0.0	3.5	3.5	1.87520000000000
0.2	3.52841252047	3.52840049359	2.301272622080000
0.4	3.55378078473	3.55375851466	3.15645294656000
0.6	3.57628769500	3.57625646333	3.37501708540000
0.8	3.59610020735	3.59606081175	3.48500956990000
1	3.61337096445	3.61330663649	4.50000000000000

7.4 Reproducing Kernel Hilbert Space Method

7.4.1 Introduction and Preliminaries

A lot of research has been done on hybrid differential equations and hybrid integrodifferential equations. Dhage and Lakshmikantham considered a classical Cauchy problem of hybrid differential equations and evaluated the existence of a unique solution for this problem. Herzallah and Baleanu have investigated the Cauchy problems

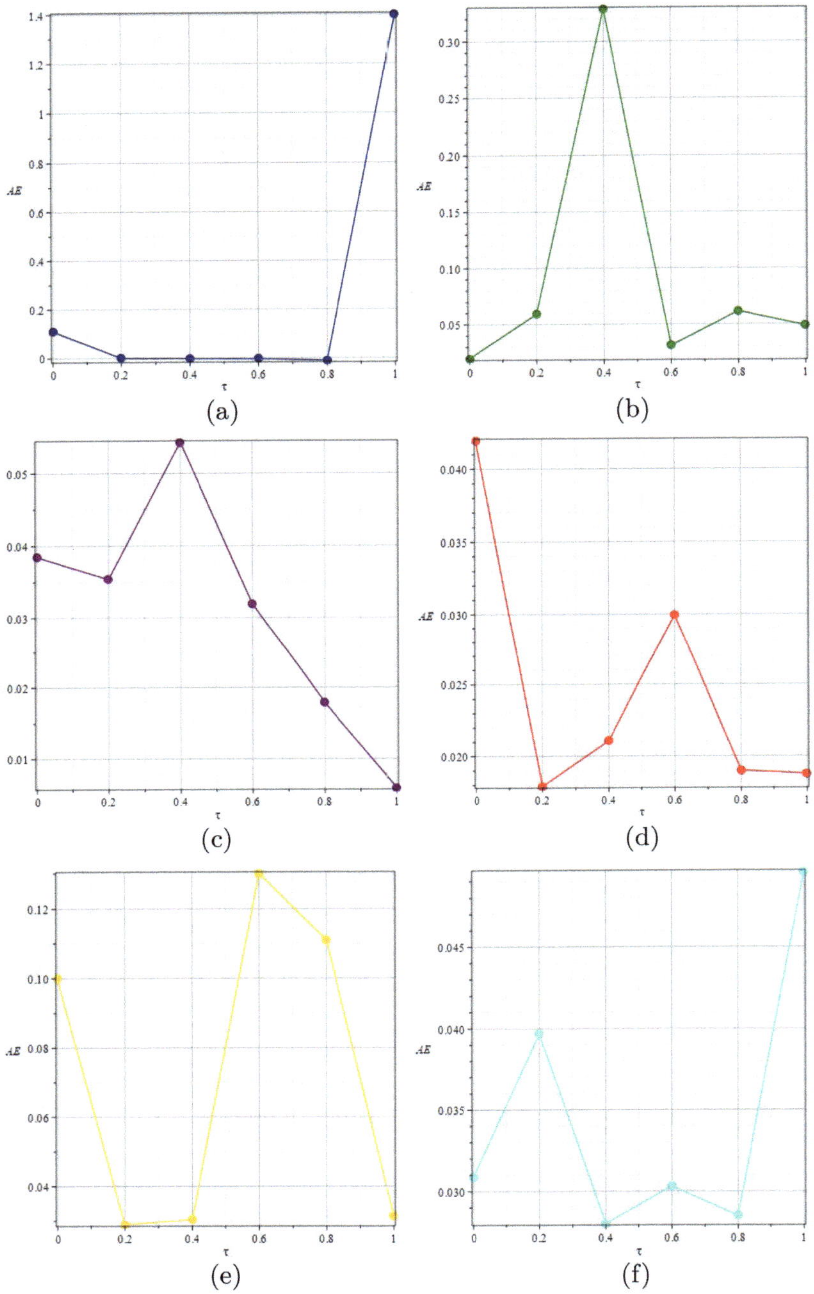

Fig. 7.13 Diagrams related to the absolute error for $S_\sigma(\tau)$, $E_\sigma(\tau)$, $\mathcal{I}_\sigma(\tau)$, $A_\sigma(\tau)$, $R_\sigma(\tau)$ and $W_\sigma(\tau)$ respectively

of hybrid differential equations with the Caputo–Liouville derivative. Bashir et al. considered a nonlocal boundary value problem of hybrid fractional integrodifferential inclusions and proved the existence of solutions for this problem. Considering an initial value problem of nonlinear hybrid fractional integrodifferential equations, Doug has done his research on the existence of solutions to this problem. See [1, 18, 24, 51, 121] for more details.

We consider the following hybrid proportional fractional integro-differential equations with Dirichlet boundary conditions (HPFIDE)

$$\begin{cases} {}^{C}_{m}(\Xi_D)^{\varsigma_1,\varsigma_2}_{0^+}\left(\frac{\phi(\zeta)-\sum_{r=1}^{r=m}\Pi_{I}{}^{\varsigma_{3r}\varsigma_2}_{0^+}\mathcal{W}_r(\zeta,\phi(\zeta))}{\mathcal{Y}(\zeta,\phi(\zeta))}\right) = \mathcal{K}(\zeta,\phi(\zeta)), & \zeta \in [0,\mathrm{F}], \\ \phi(0) = \phi_0, \phi(\mathrm{F}) = \phi_{\mathrm{F}}, \end{cases} \tag{7.4.1}$$

where ${}^{C}_{m}(\Xi_D)^{\varsigma_1,\varsigma_2}_{0^+}$ represents the Caputo–Liouville proportional fractional derivative (C-LPFD) of order ς_1 and $(\Pi_I)^{\varsigma_{3r},\varsigma_2}_{0^+}$ represents the Reimann–Liouville proportional fractional integral (R-LPFD) of order ς_{3r} for $\varsigma_2 \in (0,1]$, $1 < \varsigma_1 \leq 2$, $\varsigma_{3r} > 0$, $(r = 1,\ldots,m)$, $m \in \mathbb{N}^*$ and $\phi_0, \phi_{\mathrm{F}} \in \mathbb{R}$ and also, for functions $\mathcal{W}_r : [0,\mathrm{F}] \times \mathbb{R} \longrightarrow \mathbb{R}$, $\mathcal{Y} : [0,\mathrm{F}] \times \mathbb{R} \longrightarrow \mathbb{R}\backslash\{0\}$, $\mathcal{K} : [0,\mathrm{F}] \times \mathbb{R} \longrightarrow \mathbb{R}$.

The different modes of HPFIDE (7.4.1) are as follows:

- In the case that $\varsigma_2 = 1$, HPFIDE (7.4.1) becomes a combined integrodifferential equation with Caputo–Liouville fractional derivative and Reimann–Liouville fractional integral.
- For all $(\zeta, \phi(\zeta)) \in [0,\mathrm{F}] \times \mathbb{R}$ and $r = 1,\ldots,m$ in the case that $\mathcal{Y}(\zeta,\phi(\zeta)) = 1$ and $\mathcal{W}_r(\zeta,\phi(\zeta)) = 0$, HPFIDE (7.4.1) becomes the Caputo–Liouville proportional fractional boundary value problem (C-LPFBVP), which is defined as follows

$$\begin{cases} {}^{C}_{m}(\Xi_D)^{\varsigma_1,\varsigma_2}_{0^+}(\phi(\zeta)) = \mathcal{K}(\zeta,\phi(\zeta)), & \zeta \in [0,\mathrm{F}], \\ \phi(0) = \phi_0, \phi(\mathrm{F}) = \phi_{\mathrm{F}}. \end{cases} \tag{7.4.2}$$

- For all $(\zeta, \phi(\zeta)) \in [0,\mathrm{F}] \times \mathbb{R}$ and $r = 1,\ldots,m$ in the case that $\mathcal{Y}(\zeta,\phi(\zeta)) = 1$ and $\varsigma_2 = 1$, HPFIDE (7.4.1) becomes the C-LPFBVP, which is defined as follows

$$\begin{cases} {}^{C}(\Xi_D)^{\varsigma_1,\varsigma_2}_{0^+}(\phi(\zeta)) = \mathcal{K}(\zeta,\phi(\zeta)), & \zeta \in [0,\mathrm{F}], \\ \phi(0) = \phi_0, \phi(\mathrm{F}) = \phi_{\mathrm{F}}. \end{cases} \tag{7.4.3}$$

In addition to the qualitative research that has been done on differential equations, especially in the fractional system, many authors have so far provided many numerical solutions for boundary value problems or initial value problems. The method of variational iteration, Adomian decomposition method, Laplace transform, finite difference method, Fourier transform, collocation method, etc. can be considered among the methods of solving fractional differential equations [6, 25, 34, 42, 47, 98, 104, 109, 124, 136, 137]. Belmekki et al., Javidi and Saedshoar Heris and other authors have investigated some fractional differential equations with periodic conditions. They investigated unique minimum and maximum solutions, method of fractional backward differentiation formulas and monotone iterative techniques on fractional differential equations, respectively [8, 110, 128, 130]. In the early 20th century, the reproducing kernel method was proposed and used for the first time in the authors' research on boundary value problems. In 1907 and 1980, Zarmba and Cui used methods based on the reproducing kernel space. Zarmba introduced the kernel of some functions of their reproducing properties, and Cui presented the reproducing kernel functions of Hilbert space in the form of very simple polynomials [39, 60, 146]. The reproducing kernel method is used in creating approximate solutions to various problems, as well as along with neural networks in machine learning [45, 56, 145]. Recently, this method has also been considered for fractional differential equations [15, 74, 95].

We start with definitions of integral and derivative of fractional order and explanations of gamma function used in fractional calculations [71].

Definition 7.4 For $Re(\varsigma_1) > 0$ and $\varsigma_1 \in Re(\varsigma_1)$, the upper incomplete gamma function, the lower incomplete gamma function, the upper regularized incomplete gamma function and lower regularized incomplete gamma function are defined as follows, respectively

- $\Gamma_1(\varsigma_1, \zeta) = \int_{\zeta}^{+\infty} h^{\varsigma_1 - 1} \exp(-h) dh, \quad \zeta \geq 0;$
- $\Gamma_2(\varsigma_1, \zeta) = \int_0^{\zeta} h^{\varsigma_1 - 1} \exp(-h) dh, \quad \zeta \geq 0;$
- $\Gamma_3(\varsigma_1, \zeta) = \frac{\Gamma_1(\varsigma_1, \zeta)}{\Gamma_1(\varsigma_1)};$
- $\Gamma_4(\varsigma_1, \zeta) = 1 - \Gamma_3(\varsigma_1, \zeta) = \frac{\Gamma_2(\varsigma_1, \zeta)}{\Gamma(\varsigma_1)}.$

Definition 7.5 For each function $\mathcal{Z} \in L^1([0, \mathrm{F}], \mathbb{R})$, The left fractional Reimann–Liouville integral (L-FRLI) and the left generalized proportional Reimann–Liouville fractional integral (L-GPRLFI) are defined as follows, respectively

- $(\Pi_I)_{p^+}^{0} \mathcal{Z}(\zeta) = \mathcal{Z}(\zeta);$
- $(\Pi_I)_{p^+}^{\varsigma_1} \mathcal{Z}(\zeta) = \frac{1}{\Gamma(\varsigma_1)} \int_p^{\zeta} (\zeta - \mu)^{\varsigma_1 - 1} \mathcal{Z}(\mu) d\mu;$
- $(\Pi_I)_{p^+}^{0,\varsigma_2} \mathcal{Z}(\zeta) = \mathcal{Z}(\zeta);$
- $(\Pi_I)_{p^+}^{\varsigma_1,\varsigma_2} \mathcal{Z}(\zeta) = \frac{1}{\varsigma_2^{\varsigma_1} \Gamma(\varsigma_1)} \int_p^{\zeta} (\zeta - \mu)^{\varsigma_1 - 1} \exp(\frac{\varsigma_2 - 1}{\varsigma_2}(\zeta - \mu)) \mathcal{Z}(\mu) d\mu;$

where $\zeta \in [p, q], \quad \varsigma_2 \in (0, 1]$ and $\varsigma_1 \geq 0$.

Definition 7.6 For each function $\mathcal{Z} \in C^e([0, \mathrm{F}], \mathbb{R})$, The left fractional Caputo–Liouville derivative (L-FCLD) and the left generalized proportional Caputo–Liouville fractional derivative (L-GPCLFD) are defined as follows, respectively

- ${}_m^C(\Xi_D)_{p^+}^{0} \mathcal{Z}(\zeta) = \mathcal{Z}(\zeta);$
- ${}^C(\Xi_D)_{p^+}^{\varsigma_1} \mathcal{Z}(\zeta) = (\Pi_I)_{p^+}^{e-\varsigma_1} \left(D^e \mathcal{Z} \right)(\zeta) = \frac{1}{\Gamma(n-\varsigma_1)} \int_p^{\zeta} (\zeta - \mu)^{n-\varsigma_1-1} D^e \mathcal{Z}(\mu);$
- ${}_m^C(\Xi_D)_{p^+}^{0,\varsigma_2} \mathcal{Z}(\zeta) = \mathcal{Z}(\zeta);$

•

$$\begin{aligned}
{}_{m}^{C}(\Xi_D)_{p^+}^{\varsigma_1,\varsigma_2}\mathcal{Z}(\zeta) &= (\Pi_I)_{p^+}^{e-\varsigma_1,\varsigma_2}\left((\Xi_D)^{n,\varsigma_2\varsigma_2}\mathcal{Z}\right)(\zeta)\\
&= \frac{1}{\varsigma_2^{\varsigma_1}\Gamma(e-\varsigma_1)}\int_p^{\zeta}(\zeta-\mu)^{e-\varsigma_1-1}\exp(\frac{e-1}{\varsigma_2}(\zeta-\mu))\left((\Xi_D)^{e,\varsigma_2}\mathcal{Z}\right)(\mu)d\mu;
\end{aligned}$$

where $\zeta \in [p,q]$, $e-1<\varsigma_1 \le e, e \in \mathbb{N}$, $\varsigma_2 \in (0,1]$ and $\varsigma_1 \ge 0$. Also, we have for the last item $\left((\Xi_D)^{1,\varsigma_2}\mathcal{Z}\right)(\zeta) = \left((\Xi_D)^{\varsigma_2}\mathcal{Z}\right)(\zeta) = (1-\varsigma_2)\mathcal{Z}(\zeta)+\varsigma_2\mathcal{Z}'(\zeta)$, $\left((\Xi_D)^{e,\varsigma_2}\mathcal{Z}\right)(\zeta) = \mathcal{Z}(\zeta)$ for $e=0$ and $\left((\Xi_D)^{e,\varsigma_2}\mathcal{Z}\right)(\zeta) = \underbrace{((\Xi_D)^{\varsigma_2}D^{\varsigma_2}\cdots(\Xi_D)^{\varsigma_2}}_{e\text{-times}}\mathcal{Z})(\zeta)$ for $e \ge 1$.

For L-GPRLFI and L-GPCLFD, the following properties always hold:

- $(\Pi_I)_{p^+}^{\varsigma_1,\varsigma_2}\left((\Pi_I)_{p^+}^{\varsigma_3,\varsigma_2}\mathcal{Z}\right)(\zeta) = (\Pi_I)_{p^+}^{\varsigma_3,\varsigma_2}\left((\Pi_I)_{p^+}^{\varsigma_1\varsigma_2\varsigma_2}\mathcal{Z}\right)(\zeta) = (\Pi_I)_{p^+}^{\varsigma_1+\varsigma_3,\varsigma_2}\mathcal{Z}(\zeta)$;
- ${}_{m}^{C}(\Xi_D)_{p^+}^{\varsigma_1,\varsigma_2}\left((\Pi_I)_{p^+}^{\varsigma_1\varsigma_2\varsigma_2}\mathcal{Z}\right)(\zeta) = \mathcal{Z}(\zeta)$;
- $(\Pi_I)_{p^+}^{\varsigma_1,\tau}\left({}_{m}^{C}(\Xi_D)_{p^+}^{\varsigma_1,\varsigma_2}\mathcal{Z}\right)(\zeta) = \mathcal{Z}(\zeta) - \sum_{i=0}^{p-1} d_i(\zeta-p)^i\exp(\frac{\varsigma_2-1}{\varsigma_2}(\zeta-p))$,

where, $\varsigma_2 \in (0,1], \varsigma_3 > 0$, $\varsigma_1 > 0$ with $e-1 < \varsigma_1 \le e$, $\mathcal{Z} \in L^1([0,\mathrm{F}],\mathbb{R})$, $\mathcal{Z} \in C^e([0,\mathrm{F}],\mathbb{R}))$ and $d_i = \frac{((\Xi_D)^{i,\varsigma_2}\cdot\mathcal{Z})(p)}{\varsigma_2^i i!}$. Also, for $1 < \varsigma_1 \le 2$ and $p=0$

- $(\Pi_I)_{0^+}^{\varsigma_1,\varsigma_2}\left({}_{m}^{C}(\Xi_D)_{0^+}^{\varsigma_1,\varsigma_2}\mathcal{Z}\right)(\zeta) = \mathcal{Z}(\zeta) - d_0 e^{\frac{\varsigma_2-1}{\varsigma_2}\zeta} - d_1\zeta\exp(\frac{\varsigma_2-1}{\varsigma_2}\zeta)$.

By considering the lower regularized incomplete gamma function Γ_4, for $\varsigma \in \mathbb{R}^+$, $\varsigma_1 \in \mathbb{C}$, $\mathrm{Re}(\varsigma_1) > 0$, $\varsigma_2 \in (0,1]$ and $g > 0$, Always, we have

- For $\zeta \in [p,q]$, $\Gamma_4(\varsigma_1, x(\zeta-a))$ is a non-decreasing function;
- For all $\zeta \ge p$, $\Gamma_4(\varsigma_1, x(\zeta-p)) \in [0,1]$;
- $\max_{\zeta\in[p,q]}\Gamma_4(\varsigma_1,x(\zeta-p)) = \Gamma_4(\varsigma_1,x(\zeta-p))|_{\zeta=q} = \Gamma_4(\varsigma_1,x(q-p))$;
- $\min_{\zeta\in[p,q]}\Gamma_4(\varsigma_1,x(\zeta-p)) = \Gamma_4(\varsigma_1,x(\zeta-p))|_{\zeta=p} = 0$;
- For $\varsigma_2 \in (0,1)$, $\left((\Pi_I)_{p^+}^{g,\varsigma_2}1\right)(\zeta) = \frac{\Gamma_4\left(g,\frac{1-\varsigma_2}{\varsigma_2}(\zeta-p)\right)}{(1-\varsigma_2)^g}$;
- For $\varsigma_2 = 1$, $\left((\Pi_I)_{p^+}^{g,\varsigma_2}1\right)(\zeta) = \left((\Pi_I)_{p^+}^{g}\right)(\zeta) = \frac{(\zeta-p)^{\mu}}{\Gamma_1(g+1)}$;
- $\lim_{\varsigma_2\to 1}\frac{\Gamma_4\left(g,\frac{1-\varsigma_2}{\varsigma_2}(\zeta-p)\right)}{(1-\varsigma_2)^g} = \left((\Pi_I)_{p^+}^{g}1\right)(\zeta) = \frac{(\zeta-p)^g}{\Gamma_1(g+1)}$;
- $\max_{g\in[p,q]}\left(\lim_{\varsigma_2\to 1}\frac{\Gamma_4\left(g,\frac{1-\varsigma_2}{\varsigma_2}(\zeta-p)\right)}{(1-\varsigma_2)^g}\right) = \frac{(q-p)^g}{\Gamma_1(g+1)}$;

Proposition 7.1 *If Γ_4 is the lower regularized incomplete gamma function, we have*

(1) For $\varsigma_2 \in (0,1)$, $\zeta_1, \zeta_2 \in [p,q]$ $(\zeta_1 \le \zeta_2)$ and $g > 0$

$$\int_{\zeta_1}^{\zeta_2} (q-\mu)^{g-1} \exp(\frac{\varsigma_2 - 1}{\varsigma_2}(q-\mu))d\mu =$$
$$\frac{\varsigma_2^g \Gamma(g)}{(1-\varsigma_2)^g}\left[\Gamma_4\left(g, \frac{1-\varsigma_2}{\varsigma_2}(q-\zeta_1)\right) - \Gamma_4\left(g, \frac{1-\varsigma_2}{\varsigma_2}(q-\zeta_2)\right)\right];$$

(2) For $\varsigma_2 \in (0,1], \varsigma_1 > 0$ *and* $p \leq \mu \leq \zeta_1 < \zeta_2 \leq q$

$$\lim_{\zeta_2 \to \zeta_1} \int_p^{\zeta_1} \left|(\zeta_2-\mu)^{\varsigma_1 - 1} \exp(\frac{\varsigma_2 - 1}{\varsigma_2}(\zeta_2-\mu)) - (\zeta_1-\mu)^{\varsigma_1 - 1} \exp(\frac{-1}{\varsigma_2}(\zeta_1-\mu))\right| d\mu$$
$$= 0.$$

Definition 7.7 ([28–30]) We say that Eq. (7.4.1) is stable if for the differentiable function $\phi(\zeta)$ and the controller $\Omega(\zeta)$

$$\left|{}_m^C\Xi D_{0^+}^{\varsigma_1,\varsigma_2}\left(\frac{\phi(\zeta) - \sum_{r=1}^{r=m} \Pi_{I_{0^+}}^{\varsigma_{3r}\varsigma_2} \mathcal{W}_r(\zeta, \phi(\zeta))}{\mathcal{Y}(\zeta, \phi(\zeta))}\right) - \mathcal{K}(\zeta, \phi(\zeta))\right| \leq \delta\, \Omega(\zeta), \tag{7.4.4}$$

for $\delta > 0, \quad \zeta \in \mathrm{F}$. Then, exists a solution $\psi(\zeta)$ of Eq. (7.4.1) such that

$$|\psi(\zeta) - \phi(\zeta)| \leq \omega\, \delta\, \Omega(\zeta),$$

for $\zeta \in \mathrm{F}, \quad \omega > 0$.

Definition 7.8 We consider the space $\mathcal{A}$ and the Hilbert space H and assume that $\Theta_{[rk]}(\zeta, \eta)$ is a function in the Hilbert space H. Then, we say that the function $\Theta_{[rk]}(\zeta, \eta)$ is a reproduction kernel function and the Hilbert space H is a reproducing kernel space, if we have

(RKF-1) $\Theta_{[rk]}(\zeta, \eta) \in H$ for $\eta \in \mathcal{A}$;
(RKF-2) $\langle \mathcal{Y}(\zeta), \Theta_{[rk]}(\zeta, \eta)\rangle_H = \mathcal{Y}(\eta)$,

where $\langle ., . \rangle_H$ is the inner product.

Definition 7.9 By considering the set of square Lebesgue integrable functions on $[0, \mathrm{F}]$, means $L^2[0, \mathrm{F}]$, a Hilbert space function is a space of functions defined as follows

$$\mathrm{U}_2^1[0, \mathrm{F}] = \{\psi(\zeta) \mid \psi \text{ is absolutely continuous}, \psi' \in L^2[0, \mathrm{F}]\},$$

in which inner product and norm, are considered as follows, respectively

(IP-U_2^1) $\langle \psi(\zeta), z(\zeta)\rangle_{\mathrm{U}_2^1} = \psi(0)z(0) + \int_0^{\mathrm{F}} \psi'(\zeta)z'(\zeta)d\zeta$;
(N-U_2^1) $\|\psi(\zeta)\|_{\mathrm{U}_2^1} = \langle \psi(\zeta), \psi(\zeta)\rangle_{\mathrm{U}_2^1}^{\frac{1}{2}}$.

Definition 7.10 A Hilbert space function is a space of functions defined as follows

$$\mathrm{U}_2^2[0,\mathrm{F}] = \left\{\psi(\zeta) \mid \psi(\zeta)' \text{ is absolutely continuous, } \psi'' \in L^2[0,\mathrm{F}], \psi(0) = \psi(\mathrm{F})\right\},$$

in which inner product and norm, are considered as follows, respectively

(IP-U_2^2) $\langle \psi(\zeta), z(\zeta)\rangle_{\mathrm{U}_2^2} = \sum_{r=0}^{1} \psi^{(r)}(0) z^{(r)}(0) + \int_0^{\mathrm{F}} \psi^{(2)}(\zeta) z^{(2)}(\zeta) \mathrm{d}\zeta;$
(N-U_2^2) $\|\psi(\zeta)\|_{\mathrm{U}_2^2} = \langle \psi(\zeta), \psi(\zeta)\rangle_{\mathrm{U}_2^2}^{1/2}.$

Lemma 7.1 *The reproducing kernel function $\Theta_{[rk]}(\zeta,\eta)$ of the space $\mathrm{U}_2^1[0,\mathrm{F}]$ is as follows*

$$\Theta_{[rk]}(\zeta,\eta) = \begin{cases} \zeta + 1, & \zeta \le \eta \\ \eta + 1, & \zeta > \eta \end{cases} \tag{7.4.5}$$

Proof Considering (IP-U_2^1), we have

$$\langle \psi(\zeta), \Theta_{[rk]}(\zeta,\eta)\rangle_{\mathrm{U}_2^2} = \psi(0)\Theta_{[rk]\eta}(0) + \int_0^{\mathrm{F}} \psi'(\zeta)\Theta_{[rk]\eta}{}'(\zeta) d\zeta, \tag{7.4.6}$$

By taking several integrals, we have

$$\langle \psi(\zeta), \Theta_{[rk]}(\zeta,\eta)\rangle_{\mathrm{U}_2^2} = \psi(0)\Theta_{[rk]\eta}(0) + \psi(\mathrm{F})\Theta_{[rk]\eta}{}'(\mathrm{F}) - \psi(0)\Theta'_{[rk]\eta}(0) - \int_0^{\mathrm{F}} \psi(\zeta)\Theta''_{[rk]\eta}(\zeta) d\zeta. \tag{7.4.7}$$

Now, we let $\Theta_{[rk]\eta}(0) - \Theta'_{[rk]\eta}(0) = 0$ and $\Theta'_{[rk]\eta}(\mathrm{F}) = 0$. Therefore, $-\Theta''_{[rk]\eta}(\zeta) = \delta(\zeta - \eta)$, where δ denotes the Dirac delta function. When, $\zeta \ne \eta$, we have $\Theta''_{[rk]\eta}(\zeta) = 0$. Then

$$\Theta_{[rk]}(\zeta,\eta) = \begin{cases} \tau_1(\eta) + \tau_2(\eta)\zeta, & \zeta \le \eta \\ \tau_3(\eta) + \tau_4(\eta)\zeta, & \zeta > \eta. \end{cases} \tag{7.4.8}$$

Now, using the properties of δ, we have $\Theta_{[rk]\eta^+}(\zeta) = \Theta_{[rk]\eta^-}(\zeta)$ and $\Theta'_{[rk]\eta^+}(\zeta) = \Theta'_{[rk]\eta^-}(\zeta) = -1$. Then, considering (7.4.8) and the obtained properties, the coefficients of τ_{1r}, τ_{2r}, τ_{1r} and τ_{4r} are obtained for $r = 1, 2$, that's means $\tau_1(\zeta) = 1$, $\tau_2(\zeta) = 1$, $\tau_3(\zeta) = 1 + \zeta$, $\tau_4(\zeta) = 0$.

Lemma 7.2 *The reproducing kernel function $\Lambda_{[rk]}(\zeta,\eta)$ of the space $\mathrm{U}_2^2[0,\mathrm{F}]$ is as follows*

$$\Lambda_{[rk]}(\zeta,\eta) = \begin{cases} \zeta\eta + \frac{1}{2}\zeta^2\eta - \frac{1}{6}\zeta^3, & \zeta \le \eta, \\ \eta\zeta + \frac{1}{2}\zeta\eta^2 - \frac{1}{6}\eta^3, & \zeta > \eta. \end{cases}$$

Proof Similar to what was proved in the previous theorem, considering (IP-U_2^2) and taking several integrals, we obtain

$$\begin{aligned} \langle \psi(\zeta), \Lambda_{[rk]}(\zeta,\eta)\rangle_{\mathrm{U}_2^2} &= \psi(0)\Lambda_{[rk]\eta}(0) + \psi'(0)\Lambda'_{[rk]\eta}(0) + \psi'(\mathrm{F})\Lambda''_{[rk]\eta}(\mathrm{F}) - \psi'(0)\Lambda''_{[rk]\eta}(0) \\ &\quad - \psi(\mathrm{F})\Lambda^3_{[rk]\eta}(\mathrm{F}) + \psi(0)\Lambda^3_{[rk]\eta}(0) + \int_0^{\mathrm{F}} \psi(\zeta)\Lambda^4_{[rk]\eta}(\mathrm{F}) d\zeta. \end{aligned} \tag{7.4.9}$$

Now, we let $\Lambda'_{[rk]\eta}(0) - \Lambda''_{[rk]\eta}(0) = 0,\ \Lambda_{[rk]\eta}(0) = 0,\ \Lambda''_{[rk]\eta}(\mathrm{F}) = 0,\ \Lambda^{3}_{[rk]\eta}(\mathrm{F}) = 0$. Therefore, $\Lambda^{4}_{[rk]\eta}(\zeta) = \delta(\zeta - \eta)$, where δ denotes the Dirac delta function. When, $\zeta \neq \eta$, we have $\Lambda^{4}_{[rk]\eta}(\zeta) = 0$. Then

$$\Lambda_{[rk]}(\zeta, \eta) = \begin{cases} \tau_1(\eta) + \tau_2(\eta)\zeta + \tau_5(\eta)\zeta^2 + \tau_6(\eta)\zeta^3, & \zeta \leq \eta \\ \tau_3(\eta) + \tau_4(\eta)\zeta + \tau_7(\eta)\zeta^2 + \tau_8(\eta)\zeta^3, & \zeta > \eta. \end{cases} \tag{7.4.10}$$

Now, using the properties of δ, we have $\Lambda_{[rk]\eta^+}(\zeta) = \Lambda_{[rk]\eta^-}(\zeta),\ \Lambda'_{[rk]\eta^+}(\zeta) = \Lambda'_{[rk]\eta^-}(\zeta),\ \Lambda''_{[rk]\eta^+}(\zeta) = \Lambda''_{[rk]\eta^-}(\zeta),\ \Lambda^{3}_{[rk]\eta^+}(\zeta) - \Lambda^{3}_{[rk]\eta^-}(\zeta) = 1$. Then, considering (7.4.10) and the obtained properties, the coefficients of τ_r and τ_r are obtained for $r = 1, 2, 3, 4$, that's means $\tau_1(\zeta) = 0,\ \tau_2(\zeta) = \zeta,\ \tau_5(\zeta) = \frac{\zeta}{2},\ \tau_6(\zeta) = \frac{-1}{6},\ \tau_3(\zeta) = \frac{-\zeta^3}{6},\ \tau_4(\zeta) = \frac{1}{2}\zeta(\zeta + 2),\ \tau_7(\zeta) = 0,\ \tau_8(\zeta) = 0$.

7.4.2 Stability Results for HPFIDE (7.4.1)

In this section, in order to obtain a suitable approximation and stability of Eq. (7.4.1), we state and prove the main theorem of this subject. We show that Eq. (7.4.1) has a unique solution. Before starting the proofs, we introduce a mapping that will be used throughout the stability results. Also, the value of the boundary conditions of the equation is considered the same throughout the calculations and proofs.

Definition 7.11 Considering the set $\mathcal{N} = \{\phi : \mathrm{F} \to \mathbb{R},\ \phi \text{ is differentiable}\}$, the operator Δ_{CON} on this set i.e. $\Delta_{CON} : \mathcal{N} \times \mathcal{N} \to [0, \infty]$ is defined as follows

$$\Delta_{CON}(\phi, \psi) = \inf\{\mathcal{E} \in [0, \infty] : |\phi(\zeta) - \psi(\zeta)| \leq \mathcal{E}\, \Omega(\zeta),\ \zeta \in \mathrm{F}\}. \tag{7.4.11}$$

Lemma 7.3 *The set $\mathcal{N}$ along with the operator defined on it in Definition 7.11 is a generalized complete metric space (GCMS).*

Proof We assume $\Delta_{CON}(\phi, \psi) > \Delta_{CON}(\phi, \upsilon) + \Delta_{CON}(\upsilon, \psi)$ for $\phi, \psi \in \mathcal{N}$. Then, we have $|\phi(\zeta_0) - \psi(\zeta_0)| > \Delta_{CON}(\phi, \upsilon) + \Delta_{CON}(\upsilon, \psi)\ \Omega(\zeta)$. Now, with respect to Δ_{CON}, we come to a contradiction. That is, we have $|\phi(\zeta_0) - \psi(\zeta_0)| > |\phi(\zeta_0) - \upsilon(\zeta_0)| + |\upsilon(\zeta_0) - \psi(\zeta_0)|$. We consider a Cauchy sequence like $\{\phi_n\}$ on $(\mathcal{N}, \Delta_{CON})$. For $\delta > 0$ we can find $N_\delta \in \mathbb{N}$, such that $|\phi_n(\zeta) - \phi_m(\zeta)| < \delta\ \Omega(\zeta)$, for each $m, n \geq N_\delta$ and $\zeta \in \mathrm{F}$. Now, considering the recent disparity and taking into account the continuity property of the control function $\Omega(\zeta)$ on $[0, \mathrm{F}]$ (which is a compact set), we conclude that the sequence $\{\text{Œ}_n\}$ is uniformly convergent to Œ where Œ $\in \mathcal{N}$. Then, for $\delta > 0$, we can find $N_\delta \in \mathbb{N}$, such that $|\phi_n(\zeta) - \phi(\zeta)| < \delta\ \Omega(\zeta)$, for $n \geq N_\delta$ and $\zeta \in \mathcal{N}$. We assume that $m \to \infty$, then $\Delta_{CON}(\phi_n, \phi) \leq \delta$. Therefore, we conclude that $(\mathcal{N}, \Delta_{CON})$ is a GCMS.

Theorem 7.3 *We consider positive constants* $\mathcal{M}_1, \mathcal{M}_2, \mathcal{M}_3, \mathcal{M}_4, \mathcal{M}_5, \varrho$ *such that we have*

$$0 < \mathcal{M}_1 \frac{\psi_0}{\mathcal{Y}(0, \psi_0)} \left(1 - \frac{\zeta}{\mathrm{F}}\right) \exp(\frac{\varsigma_2 - 1}{\varsigma_2}\zeta) + \mathcal{M}_1 \frac{\psi_{\mathrm{F}} \frac{\zeta}{\mathrm{F}} \exp(\frac{\varsigma_3 - 1}{\varsigma_2}\zeta)}{\mathcal{Y}(\mathrm{F}, \psi_{\mathrm{F}}) \exp(\frac{\varsigma_3 - 1}{\varsigma_2}\mathrm{F})} \tag{7.4.12}$$

$$+ \frac{\zeta \exp(\frac{\varsigma_2 - 1}{\varsigma_2}\zeta)}{\mathcal{Y}(\mathrm{F}, \psi_{\mathrm{F}}) \mathrm{F} \exp(\frac{\varsigma_3 - 1}{\varsigma_2}\mathrm{F})} \sum_{r=1}^{r=m} \frac{1}{\varsigma_2^{\varsigma_{3r}} \Gamma(\varsigma_{3r})} \mathcal{M}_2 + \mathcal{M}_3 \frac{\zeta \exp(\frac{\varsigma_3 - 1}{\varsigma_2}\zeta)}{\mathrm{F} \exp(\frac{\varsigma_2 - 1}{\varsigma_2}\mathrm{F})} \frac{1}{\varsigma_2^{\varsigma_1} \Gamma(\varsigma_1)}$$

$$+ \mathcal{M}_4 \frac{1}{\varsigma_2^{\varsigma_1} \Gamma(\varsigma_1)} + \sum_{r=1}^{m} \frac{1}{\varsigma_2^{\varsigma_{3r}} \Gamma(\varsigma_{3r})} \int_0^{\zeta} (\zeta - \mu)^{\varsigma_{3r} - 1} \exp(\frac{\varsigma_2 - 1}{\varsigma_2}(\zeta - \mu)) \mathcal{M}_5 \varrho < 1.$$

For the continuous functions $\mathcal{Y} : [0, \mathrm{F}] \times \mathbb{R} \to \mathbb{R} - \{0\}$, $\mathcal{W}_r : [0, \mathrm{F}] \times \mathbb{R} \to \mathbb{R}$, $\mathcal{K} : [0, \mathrm{F}] \times \mathbb{R} \to \mathbb{R}$ *and* $\Omega : \mathbb{R} \longrightarrow \mathbb{R}$, *the following inequalities are established*

$$|\mathcal{Y}(\zeta, \phi(\zeta)) - \mathcal{Y}(\zeta, \psi(\zeta))| \leq \mathcal{M}_1 |\phi(\zeta) - \psi(\zeta)|, \tag{7.4.13}$$

$$\begin{aligned} &\Big|\mathcal{Y}(\zeta, \psi(\zeta)) \int_0^{\mathrm{F}} (\mathrm{F} - \mu)^{\varsigma_{3r} - 1} \exp(\frac{\varsigma_3 - 1}{\varsigma_2}(\mathrm{F} - \mu)) \mathcal{W}_r(\mu, \psi(\mu)) d\mu \\ &- \mathcal{Y}(\zeta, \phi(\zeta)) \int_0^{\mathrm{F}} (\mathrm{F} - \mu)^{\varsigma_{3r} - 1} \exp(\frac{\varsigma_3 - 1}{\varsigma_2}(\mathrm{F} - \mu)) \mathcal{W}_r(\mu, \phi(\mu)) d\mu\Big| \\ &\leq \mathcal{M}_2 |\phi(\zeta) - \psi(\zeta)|, \end{aligned} \tag{7.4.14}$$

$$\begin{aligned} &\Big|\mathcal{Y}(\zeta, \psi(\zeta)) \int_0^{\mathrm{F}} (\mathrm{F} - \mu)^{\varsigma_1 - 1} \exp(\frac{\varsigma_2 - 1}{\varsigma_2}(\mathrm{F} - \mu)) \mathcal{K}(\mu, \psi(\mu)) d\mu - \\ &- \mathcal{Y}(\zeta, \phi(\zeta)) \int_0^{\mathrm{F}} (\mathrm{F} - \mu)^{\varsigma_1 - 1} \exp(\frac{\varsigma_2 - 1}{\varsigma_2}(\mathrm{F} - \mu)) \mathcal{K}(\mu, \phi(\mu)) d\mu\Big| \\ &\leq \mathcal{M}_3 |\phi(\zeta) - \psi(\zeta)|, \end{aligned} \tag{7.4.15}$$

$$\begin{aligned} &\Big|\mathcal{Y}(\zeta, \phi(\zeta)) \int_0^{\zeta} (\zeta - \mu)^{\varsigma_1 - 1} \exp(\frac{\varsigma_3 - 1}{\varsigma_2}(\zeta - \mu)) \mathcal{K}(\mu, \phi(\mu)) d\mu \\ &- \mathcal{Y}(\zeta, \psi(\zeta)) \int_0^{\zeta} (\zeta - \mu)^{\varsigma_1 - 1} \exp(\frac{\varsigma_3 - 1}{\varsigma_2}(\zeta - \mu)) \mathcal{K}(\mu, \psi(\mu)) d\mu\Big| \\ &\leq \mathcal{M}_4 |\phi(\zeta) - \psi(\zeta)|, \end{aligned} \tag{7.4.16}$$

$$\frac{1}{\varsigma_2{}^{\varsigma_1}\Gamma(\varsigma_1)}\int_0^{\zeta}(\zeta-\mu)^{\varsigma_1-1}\exp(\varsigma_2-1(\zeta-\mu))\Omega(\zeta)d\mu \le \varrho\Omega(\zeta) \tag{7.4.17}$$

We assume that $\phi:[0,\mathrm{F}]\to\mathbb{R}$ is a function such that for $\delta>0$ and $\zeta\in[0,\mathrm{F}]$, applies in the following conditions

$$\left|{}^{C}_{m}\Xi D_{0^+}^{\varsigma_1,\varsigma_2}\left(\frac{\phi(\zeta)-\sum_{r=1}^{r=m}\Pi_{I_{0^+}}^{\varsigma_{3r},\varsigma_2}\mathcal{W}_r(\zeta,\phi(\zeta))}{\mathcal{Y}(\zeta,\phi(\zeta))}\right)-\mathcal{K}(\zeta,\phi(\zeta))\right|\le\delta\,\Omega(\zeta), \tag{7.4.18}$$

$$\begin{aligned}&\Bigg|\mathcal{K}(\zeta,\phi(\zeta))-\Bigg(\mathcal{Y}(\zeta,\phi(\zeta))\Bigg[\frac{\phi_0}{\mathcal{Y}(0,\phi_0)}\left(1-\frac{\zeta}{\mathrm{F}}\right)\exp(\frac{\varsigma_2-1}{\varsigma_2}\zeta)+\frac{\phi_{\mathrm{F}}\frac{\zeta}{\mathrm{F}}\exp(\frac{\varsigma_3-1}{\varsigma_2}\zeta)}{\mathcal{Y}(\mathrm{F},\phi_{\mathrm{F}})\exp(\frac{\varsigma_3-1}{\varsigma_2}\mathrm{F})}\\
&-\frac{\zeta\exp(\frac{\varsigma_2-1}{\varsigma_2}\zeta)}{\mathcal{Y}(\mathrm{F},\phi_{\mathrm{F}})\,\mathrm{F}\exp(\frac{\varsigma_3-1}{\varsigma_2}\mathrm{F})}\sum_{r=1}^{r=m}\frac{1}{\varsigma_2^{\varsigma_{3r}}\Gamma(\varsigma_{3r})}\int_0^{\mathrm{F}}(\mathrm{F}-\mu)^{\varsigma_{3r}-1}\exp(\frac{\varsigma_3-1}{\varsigma_2}(\mathrm{F}-\mu))\mathcal{W}_r(\mu,\phi(\mu))d\mu\\
&-\frac{\zeta\exp(\frac{\varsigma_3-1}{\varsigma_2}\zeta)}{\mathrm{F}\exp(\frac{\varsigma_2-1}{\varsigma_2}\mathrm{F})}\frac{1}{\varsigma_2^{\varsigma_1}\Gamma(\varsigma_1)}\int_0^{\mathrm{F}}(\mathrm{F}-\mu)^{\varsigma_1-1}\exp(\frac{\varsigma_2-1}{\varsigma_2}(\mathrm{F}-\mu)\mathcal{K}(\zeta,\phi(\zeta))d\mu\\
&+\frac{1}{\varsigma_2^{\varsigma_1}\Gamma(\varsigma_1)}\int_0^{\zeta}(\zeta-\mu)^{\varsigma_1-1}\exp(\frac{\varsigma_3-1}{\varsigma_2}(\zeta-\mu))\mathcal{K}(\zeta,\phi(\zeta))d\mu\Bigg]\\
&+\sum_{r=1}^{m}\frac{1}{\varsigma_2^{\varsigma_{3r}}\Gamma(\varsigma_{3r})}\int_0^{\zeta}(\zeta-\mu)^{\varsigma_{3r}-1}\exp(\frac{\varsigma_2-1}{\varsigma_2}(\zeta-\mu))\mathcal{W}_r(\mu,\phi(\mu))d\mu\Bigg)\Bigg|\\
&\le\delta\frac{1}{\varsigma_2^{\varsigma_1}\Gamma(\varsigma_1)}\int_a^{\zeta}(\zeta-\mu)^{\varsigma_1-1}\exp(\varsigma_2-1(\zeta-\mu))\Omega(\zeta)d\mu,\end{aligned} \tag{7.4.19}$$

then, HPFIDE (7.4.1) has a unique solution $\psi:[0,\mathrm{F}]\to\mathbb{R}$ such that $|\psi(\zeta)-\phi(\zeta)|\le\frac{\delta\varrho}{1-\upsilon}\,\Omega(\zeta)$, where υ, is Eq. (7.4.12).

Proof The first step to prove is as follows: We define the mapping $\Psi:\mathcal{N}\to\mathcal{N}$ as follows

$$\begin{aligned}\Psi(\phi(\zeta))=&\,\mathcal{Y}(\zeta,\phi(\zeta))\Bigg[\frac{\phi_0}{\mathcal{Y}(0,\phi_0)}\left(1-\frac{\zeta}{\mathrm{F}}\right)\exp(\frac{\varsigma_2-1}{\varsigma_2}\zeta)+\frac{\phi_{\mathrm{F}}\frac{\zeta}{\mathrm{F}}\exp(\frac{\varsigma_3-1}{\varsigma_2}\zeta)}{\mathcal{Y}(\mathrm{F},\phi_{\mathrm{F}})\exp(\frac{\varsigma_3-1}{\varsigma_2}\mathrm{F})}\\
&-\frac{\zeta\exp(\frac{\varsigma_2-1}{\varsigma_2}\zeta)}{\mathcal{Y}(\mathrm{F},\phi_{\mathrm{F}})\,\mathrm{F}\exp(\frac{\varsigma_3-1}{\varsigma_2}\mathrm{F})}\sum_{r=1}^{r=m}\frac{1}{\varsigma_2^{\varsigma_{3r}}\Gamma(\varsigma_{3r})}\int_0^{\mathrm{F}}(\mathrm{F}-\mu)^{\varsigma_{3r}-1}\exp(\frac{\varsigma_3-1}{\varsigma_2}(\mathrm{F}-\mu))\mathcal{W}_r(\mu,\phi(\mu))d\mu\\
&-\frac{\zeta\exp(\frac{\varsigma_3-1}{\varsigma_2}\zeta)}{\mathrm{F}\exp(\frac{\varsigma_2-1}{\varsigma_2}\mathrm{F})}\frac{1}{\varsigma_2^{\varsigma_1}\Gamma(\varsigma_1)}\int_0^{\mathrm{F}}(\mathrm{F}-\mu)^{\varsigma_1-1}\exp(\frac{\varsigma_2-1}{\varsigma_2}(\mathrm{F}-\mu))\mathcal{K}(\mu,\phi(\mu))d\mu\\
&+\frac{1}{\varsigma_2^{\varsigma_1}\Gamma(\varsigma_1)}\int_0^{\zeta}(\zeta-\mu)^{\varsigma_1-1}\exp(\frac{\varsigma_3-1}{\varsigma_2}(\zeta-\mu))\mathcal{K}(\mu,\phi(\mu))d\mu\Bigg]\\
&+\sum_{r=1}^{m}\frac{1}{\varsigma_2^{\varsigma_{3r}}\Gamma(\varsigma_{3r})}\int_0^{\zeta}(\zeta-\mu)^{\varsigma_{3r}-1}\exp(\frac{\varsigma_2-1}{\varsigma_2}(\zeta-\mu))\mathcal{W}_r(\mu,\phi(\mu))d\mu.\end{aligned} \tag{7.4.20}$$

Now, we show that this mapping is contractional. For this purpose, we assume that $\Delta_{CON}(\phi, \psi) \leq \mathcal{E}_{\phi\psi}$, for $\phi, \psi \in \mathcal{N}$ and the coefficient $\mathcal{E}_{\phi\psi} \in [0, \infty]$, then, $|\phi(\zeta) - \psi(\zeta)| < \mathcal{E}_{\phi\psi}\ \Omega(\zeta)$, for $\zeta \in [0, \mathrm{F}]$. Using (7.4.13), (7.4.14), (7.4.15), (7.4.16) and (7.4.17), we get

$$
\begin{aligned}
|\Psi\phi(\zeta) - \Psi\psi(\zeta)| &= \Bigg|\mathcal{Y}(\zeta, \phi(\zeta)) \Bigg[\frac{\phi_0}{\mathcal{Y}(0, \phi_0)} \left(1 - \frac{\zeta}{\mathrm{F}}\right) \exp(\frac{\varsigma_2 - 1}{\varsigma_2}\zeta) + \frac{\phi_{\mathrm{F}} \frac{\zeta}{\mathrm{F}} \exp(\frac{\varsigma_3 - 1}{\varsigma_2}\zeta)}{\mathcal{Y}(\mathrm{F}, \phi_{\mathrm{F}}) \exp(\frac{\varsigma_3 - 1}{\varsigma_2}\mathrm{F})} \\
&- \frac{\zeta \exp(\frac{\varsigma_2 - 1}{\varsigma_2}\zeta)}{\mathcal{Y}(\mathrm{F}, \phi_{\mathrm{F}})\, \mathrm{F} \exp(\frac{\varsigma_3 - 1}{\varsigma_2}\mathrm{F})} \sum_{r=1}^{r=m} \frac{1}{\varsigma_2^{\varsigma_{3r}} \Gamma(\varsigma_{3r})} \int_0^{\mathrm{F}} (\mathrm{F} - \mu)^{\varsigma_{3r} - 1} \exp(\frac{\varsigma_3 - 1}{\varsigma_2}(\mathrm{F} - \mu)) \mathcal{W}_r(\mu, \phi(\mu)) d\mu \\
&- \frac{\zeta \exp(\frac{\varsigma_3 - 1}{\varsigma_2}\zeta)}{\mathrm{F} e^{\frac{\varsigma_2 - 1}{\varsigma_2}\mathrm{F}}} \frac{1}{\varsigma_2^{\varsigma_1} \Gamma(\varsigma_1)} \int_0^{\mathrm{F}} (\mathrm{F} - \mu)^{\varsigma_1 - 1} \exp(\frac{\varsigma_2 - 1}{\varsigma_2}(\mathrm{F} - \mu)) \mathcal{K}(\mu, \phi(\mu)) d\mu \\
&+ \frac{1}{\varsigma_2^{\varsigma_1} \Gamma(\varsigma_1)} \int_0^{\zeta} (\zeta - \mu)^{\varsigma_1 - 1} \exp(\frac{\varsigma_3 - 1}{\varsigma_2}(\zeta - \mu)) \mathcal{K}(\mu, \phi(\mu)) d\mu \Bigg] \\
&+ \sum_{r=1}^{m} \frac{1}{\varsigma_2^{\varsigma_{3r}} \Gamma(\varsigma_{3r})} \int_0^{\zeta} (\zeta - \mu)^{\varsigma_{3r} - 1} \exp(\frac{\varsigma_2 - 1}{\varsigma_2}(\zeta - \mu)) \mathcal{W}_r(\mu, \phi(\mu)) d\mu \\
&- \Bigg(\mathcal{Y}(\zeta, \psi(\zeta)) \Bigg[\frac{\psi_0}{\mathcal{Y}(0, \psi_0)} \left(1 - \frac{\zeta}{\mathrm{F}}\right) \exp(\frac{\varsigma_2 - 1}{\varsigma_2}\zeta) + \frac{\psi_{\mathrm{F}} \frac{\zeta}{\mathrm{F}} \exp(\frac{\varsigma_3 - 1}{\varsigma_2}\zeta)}{\mathcal{Y}(\mathrm{F}, \psi_{\mathrm{F}}) \exp(\frac{\varsigma_3 - 1}{\varsigma_2}\mathrm{F})} \\
&- \frac{\zeta \exp(\frac{\varsigma_2 - 1}{\varsigma_2}\zeta)}{\mathcal{Y}(\mathrm{F}, \psi_{\mathrm{F}})\, \mathrm{F} \exp(\frac{\varsigma_3 - 1}{\varsigma_2}\mathrm{F})} \sum_{r=1}^{r=m} \frac{1}{\varsigma_2^{\varsigma_{3r}} \Gamma(\varsigma_{3r})} \int_0^{\mathrm{F}} (\mathrm{F} - \mu)^{\varsigma_{3r} - 1} \exp(\frac{\varsigma_3 - 1}{\varsigma_2}(\mathrm{F} - \mu)) \mathcal{W}_r(\mu, \psi(\mu)) d\mu \\
&- \frac{\zeta \exp(\frac{\varsigma_3 - 1}{\varsigma_2}\zeta)}{\mathrm{F} \exp(\frac{\varsigma_2 - 1}{\varsigma_2}\mathrm{F})} \frac{1}{\varsigma_2^{\varsigma_1} \Gamma(\varsigma_1)} \int_0^{\mathrm{F}} (\mathrm{F} - \mu)^{\varsigma_1 - 1} \exp(\frac{\varsigma_2 - 1}{\varsigma_2}(\mathrm{F} - \mu)) \mathcal{K}(\mu, \psi(\mu)) d\mu \\
&+ \frac{1}{\varsigma_2^{\varsigma_1} \Gamma(\varsigma_1)} \int_0^{\zeta} (\zeta - \mu)^{\varsigma_1 - 1} \exp(\frac{\varsigma_3 - 1}{\varsigma_2}(\zeta - \mu)) \mathcal{K}(\mu, \psi(\mu)) d\mu \Bigg] \\
&+ \sum_{r=1}^{m} \frac{1}{\varsigma_2^{\varsigma_{3r}} \Gamma(\varsigma_{3r})} \int_0^{\zeta} (\zeta - \mu)^{\varsigma_{3r} - 1} \exp(\frac{\varsigma_2 - 1}{\varsigma_2}(\zeta - \mu)) \mathcal{W}_r(\mu, \psi(\mu)) d\mu \Bigg)\Bigg| \\
&\leq \frac{\psi_0}{\mathcal{Y}(0, \psi_0)} \left(1 - \frac{\zeta}{\mathrm{F}}\right) \exp(\frac{\varsigma_2 - 1}{\varsigma_2}\zeta) \left(\Big|\mathcal{Y}(\zeta, \phi(\zeta)) - \mathcal{Y}(\zeta, \psi(\zeta))\Big|\right) \\
&+ \frac{\psi_{\mathrm{F}} \frac{\zeta}{\mathrm{F}} \exp(\frac{\varsigma_3 - 1}{\varsigma_2}\zeta)}{\mathcal{Y}(\mathrm{F}, \psi_{\mathrm{F}}) \exp(\frac{\varsigma_3 - 1}{\varsigma_2}\mathrm{F})} \left(\Big|\mathcal{Y}(\zeta, \phi(\zeta)) - \mathcal{Y}(\zeta, \psi(\zeta))\Big|\right) \\
&+ \frac{\zeta \exp(\frac{\varsigma_2 - 1}{\varsigma_2}\zeta)}{\mathcal{Y}(\mathrm{F}, \psi_{\mathrm{F}})\, \mathrm{F} \exp(\frac{\varsigma_3 - 1}{\varsigma_2}\mathrm{F})} \sum_{r=1}^{r=m} \frac{1}{\varsigma_2^{\varsigma_{3r}} \Gamma(\varsigma_{3r})} \Bigg(\Bigg|\mathcal{Y}(\zeta, \psi(\zeta)) \int_0^{\mathrm{F}} (\mathrm{F} - \mu)^{\varsigma_{3r} - 1} \exp(\frac{\varsigma_3 - 1}{\varsigma_2}(\mathrm{F} - \mu)) \mathcal{W}_r(\mu, \psi(\mu)) d\mu \\
&- \mathcal{Y}(\zeta, \phi(\zeta)) \int_0^{\mathrm{F}} (\mathrm{F} - \mu)^{\varsigma_{3r} - 1} \exp(\frac{\varsigma_3 - 1}{\varsigma_2}(\mathrm{F} - \mu)) \mathcal{W}_r(\mu, \phi(\mu)) d\mu\Bigg|\Bigg) \\
&+ \frac{\zeta \exp(\frac{\varsigma_3 - 1}{\varsigma_2}\zeta)}{\mathrm{F} \exp(\frac{\varsigma_2 - 1}{\varsigma_2}\mathrm{F})} \frac{1}{\varsigma_2^{\varsigma_1} \Gamma(\varsigma_1)} \Bigg(\mathcal{Y}(\mu, \psi(\mu))\Bigg| \int_0^{\mathrm{F}} (\mathrm{F} - \mu)^{\varsigma_1 - 1} \exp(\frac{\varsigma_2 - 1}{\varsigma_2}(\mathrm{F} - \mu)) \mathcal{K}(\mu, \psi(\mu)) d\mu \\
&- \mathcal{Y}(\zeta, \phi(\zeta)) \int_0^{\mathrm{F}} (\mathrm{F} - \mu)^{\varsigma_1 - 1} \exp(\frac{\varsigma_2 - 1}{\varsigma_2}(\mathrm{F} - \mu)) \mathcal{K}(\mu, \phi(\mu)) d\mu\Bigg|\Bigg) \\
&+ \frac{1}{\varsigma_2^{\varsigma_1} \Gamma(\varsigma_1)} \Bigg(\Bigg|\mathcal{Y}(\zeta, \phi(\zeta)) \int_0^{\zeta} (\zeta - \mu)^{\varsigma_1 - 1} \exp(\frac{\varsigma_3 - 1}{\varsigma_2}(\zeta - \mu)) \mathcal{K}(\mu, \phi(\mu)) d\mu \\
&- \mathcal{Y}(\zeta, \psi(\zeta)) \int_0^{\zeta} (\zeta - \mu)^{\varsigma_1 - 1} \exp(\frac{\varsigma_3 - 1}{\varsigma_2}(\zeta - \mu)) \mathcal{K}(\mu, \psi(\mu)) d\mu\Bigg|\Bigg)
\end{aligned}
$$

$$
\begin{aligned}
&+\sum_{r=1}^{m}\frac{1}{\varsigma_2^{\varsigma_{3r}}\Gamma(\varsigma_{3r})}\int_0^{\zeta}(\zeta-\mu)^{\varsigma_{3r}-1}\exp(\frac{\varsigma_2-1}{\varsigma_2}(\zeta-\mu))\Big|\mathcal{W}_r(\mu,\phi(\mu))-\mathcal{W}_r(\mu,\psi(\mu))\Big|d\mu\\
&\le\frac{\psi_0}{\mathcal{Y}(0,\psi_0)}\left(1-\frac{\zeta}{\mathrm{F}}\right)\exp(\frac{\varsigma_2-1}{\varsigma_2}\zeta)\Big(\mathcal{M}_1\Big|\phi(\zeta)-\psi(\zeta)\Big|\Big)+\frac{\psi_{\mathrm{F}}\frac{\zeta}{\mathrm{F}}\exp(\frac{\varsigma_3-1}{\varsigma_2}\zeta)}{\mathcal{Y}(\mathrm{F},\psi_{\mathrm{F}})\exp(\frac{\varsigma_3-1}{\varsigma_2}\mathrm{F})}\Big(\mathcal{M}_1\Big|\phi(\zeta)-\psi(\zeta)\Big|\Big)\\
&+\frac{\zeta\exp(\frac{\varsigma_2-1}{\varsigma_2}\zeta)}{\mathcal{Y}(\mathrm{F},\psi_{\mathrm{F}})\mathrm{F}\exp(\frac{\varsigma_3-1}{\varsigma_2}\mathrm{F})}\sum_{r=1}^{r=m}\frac{1}{\varsigma_2^{\varsigma_{3r}}\Gamma(\varsigma_{3r})}\Big(\mathcal{M}_2\Big|\phi(\zeta)-\psi(\zeta)\Big|\Big)\\
&+\frac{\zeta\exp(\frac{\varsigma_3-1}{\varsigma_2}\zeta)}{\mathrm{F}\exp(\frac{\varsigma_2-1}{\varsigma_2}\mathrm{F})}\frac{1}{\varsigma_2^{\varsigma_1}\Gamma(\varsigma_1)}\Big(\mathcal{M}_3\Big|\phi(\zeta)-\psi(\zeta)\Big|\Big)\\
&+\frac{1}{\varsigma_2^{\varsigma_1}\Gamma(\varsigma_1)}\Big(\mathcal{M}_4\Big|\phi(\zeta)-\psi(\zeta)\Big|\Big)\\
&+\sum_{r=1}^{m}\frac{1}{\varsigma_2^{\varsigma_{3r}}\Gamma(\varsigma_{3r})}\int_0^{\zeta}(\zeta-\mu)^{\varsigma_{3r}-1}\exp(\frac{\varsigma_2-1}{\varsigma_2}(\zeta-\mu))\Big(\mathcal{M}_5\Big|\phi(\zeta)-\psi(\zeta)\Big|\Big)d\mu\\
&\le\frac{\psi_0}{\mathcal{Y}(0,\psi_0)}\left(1-\frac{\zeta}{\mathrm{F}}\right)\exp(\frac{\varsigma_2-1}{\varsigma_2}\zeta)\Big(\mathcal{M}_1\mathcal{E}_{\phi\psi}\,\Omega(\zeta)\Big)+\frac{\psi_{\mathrm{F}}\frac{\zeta}{\mathrm{F}}\exp(\frac{\varsigma_3-1}{\varsigma_2}\zeta)}{\mathcal{Y}(\mathrm{F},\psi_{\mathrm{F}})\exp(\frac{\varsigma_3-1}{\varsigma_2}\mathrm{F})}\Big(\mathcal{M}_1\mathcal{E}_{\phi\psi}\,\Omega(\zeta)\Big)\\
&+\frac{\zeta\exp(\frac{\varsigma_2-1}{\varsigma_2}\zeta)}{\mathcal{Y}(\mathrm{F},\psi_{\mathrm{F}})\mathrm{F}\exp(\frac{\varsigma_3-1}{\varsigma_2}\mathrm{F})}\sum_{r=1}^{r=m}\frac{1}{\varsigma_2^{\varsigma_{3r}}\Gamma(\varsigma_{3r})}\Big(\mathcal{M}_2\mathcal{E}_{\phi\psi}\,\Omega(\zeta)\Big)\\
&+\frac{\zeta\exp(\frac{\varsigma_3-1}{\varsigma_2}\zeta)}{\mathrm{F}\exp(\frac{\varsigma_2-1}{\varsigma_2}\mathrm{F})}\frac{1}{\varsigma_2^{\varsigma_1}\Gamma(\varsigma_1)}\Big(\mathcal{M}_3\mathcal{E}_{\phi\psi}\,\Omega(\zeta)\Big)\\
&+\frac{1}{\varsigma_2^{\varsigma_1}\Gamma(\varsigma_1)}\Big(\mathcal{M}_4\mathcal{E}_{\phi\psi}\,\Omega(\zeta)\Big)\\
&+\sum_{r=1}^{m}\frac{1}{\varsigma_2^{\varsigma_{3r}}\Gamma(\varsigma_{3r})}\int_0^{\zeta}(\zeta-\mu)^{\varsigma_{3r}-1}\exp(\frac{\varsigma_2-1}{\varsigma_2}(\zeta-\mu))\Big(\mathcal{M}_5\mathcal{E}_{\phi\psi}\,\Omega(\zeta)\Big)d\mu\\
&\le\Big(\mathcal{M}_1\frac{\psi_0}{\mathcal{Y}(0,\psi_0)}\left(1-\frac{\zeta}{\mathrm{F}}\right)\exp(\frac{\varsigma_2-1}{\varsigma_2}\zeta)+\mathcal{M}_1\frac{\psi_{\mathrm{F}}\frac{\zeta}{\mathrm{F}}\exp(\frac{\varsigma_3-1}{\varsigma_2}\zeta)}{\mathcal{Y}(\mathrm{F},\psi_{\mathrm{F}})\exp(\frac{\varsigma_3-1}{\varsigma_2}\mathrm{F})}\\
&+\frac{\zeta\exp(\frac{\varsigma_2-1}{\varsigma_2}\zeta)}{\mathcal{Y}(\mathrm{F},\psi_{\mathrm{F}})\mathrm{F}\exp(\frac{\varsigma_3-1}{\varsigma_2}\mathrm{F})}\sum_{r=1}^{r=m}\frac{1}{\varsigma_2^{\varsigma_{3r}}\Gamma(\varsigma_{3r})}\mathcal{M}_2+\mathcal{M}_3\frac{\zeta\exp(\frac{\varsigma_3-1}{\varsigma_2}\zeta)}{\mathrm{F}\exp(\frac{\varsigma_2-1}{\varsigma_2}\mathrm{F})}\frac{1}{\varsigma_2^{\varsigma_1}\Gamma(\varsigma_1)}\\
&+\mathcal{M}_4\frac{1}{\varsigma_2^{\varsigma_1}\Gamma(\varsigma_1)}+\sum_{r=1}^{m}\frac{1}{\varsigma_2^{\varsigma_{3r}}\Gamma(\varsigma_{3r})}\int_0^{\zeta}(\zeta-\mu)^{\varsigma_{3r}-1}\exp(\frac{\varsigma_2-1}{\varsigma_2}(\zeta-\mu))\mathcal{M}_5\varrho\Big)\mathcal{E}_{\phi\psi}\,\Omega(\zeta).
\end{aligned}
$$

Thus, we have

$$
\begin{aligned}
\Delta_{CON}(\Psi\phi,\Psi\psi)\le&\Big[\mathcal{M}_1\frac{\psi_0}{\mathcal{Y}(0,\psi_0)}\left(1-\frac{\zeta}{\mathrm{F}}\right)\exp(\frac{\varsigma_2-1}{\varsigma_2}\zeta)+\mathcal{M}_1\frac{\psi_{\mathrm{F}}\frac{\zeta}{\mathrm{F}}\exp(\frac{\varsigma_3-1}{\varsigma_2}\zeta)}{\mathcal{Y}(\mathrm{F},\psi_{\mathrm{F}})\exp(\frac{\varsigma_3-1}{\varsigma_2}\mathrm{F})}\\
&+\frac{\zeta\exp(\frac{\varsigma_2-1}{\varsigma_2}\zeta)}{\mathcal{Y}(\mathrm{F},\psi_{\mathrm{F}})\mathrm{F}\exp(\frac{\varsigma_3-1}{\varsigma_2}\mathrm{F})}\sum_{r=1}^{r=m}\frac{1}{\varsigma_2^{\varsigma_{3r}}\Gamma(\varsigma_{3r})}\mathcal{M}_2+\mathcal{M}_3\frac{\zeta\exp(\frac{\varsigma_3-1}{\varsigma_2}\zeta)}{\mathrm{F}\exp(\frac{\varsigma_2-1}{\varsigma_2}\mathrm{F})}\frac{1}{\varsigma_2^{\varsigma_1}\Gamma(\varsigma_1)}\\
&+\mathcal{M}_4\frac{1}{\varsigma_2^{\varsigma_1}\Gamma(\varsigma_1)}+\sum_{r=1}^{m}\frac{1}{\varsigma_2^{\varsigma_{3r}}\Gamma(\varsigma_{3r})}\int_0^{\zeta}(\zeta-\mu)^{\varsigma_{3r}-1}\exp(\frac{\varsigma_2-1}{\varsigma_2}(\zeta-\mu))\mathcal{M}_5\varrho\Big]\Delta_{CON}(\phi,\psi).
\end{aligned}
\tag{7.4.21}
$$

Since,

$$0 < \mathcal{M}_1 \frac{\psi_0}{\mathcal{Y}(0, \psi_0)} \left(1 - \frac{\zeta}{\mathrm{F}}\right) \exp(\frac{\varsigma_2 - 1}{\varsigma_2}\zeta) + \mathcal{M}_1 \frac{\psi_{\mathrm{F}} \frac{\zeta}{\mathrm{F}} \exp(\frac{\varsigma_3 - 1}{\varsigma_2}\zeta)}{\mathcal{Y}(\mathrm{F}, \psi_{\mathrm{F}}) \exp(\frac{\varsigma_3 - 1}{\varsigma_2}\mathrm{F})} \tag{7.4.22}$$
$$+ \frac{\zeta \exp(\frac{\varsigma_2 - 1}{\varsigma_2}\zeta)}{\mathcal{Y}(\mathrm{F}, \psi_{\mathrm{F}}) \mathrm{F} \exp(\frac{\varsigma_3 - 1}{\varsigma_2}\mathrm{F})} \sum_{r=1}^{r=m} \frac{1}{\varsigma_2^{\varsigma_{3r}} \Gamma(\varsigma_{3r})} \mathcal{M}_2 + \mathcal{M}_3 \frac{\zeta \exp(\frac{\varsigma_3 - 1}{\varsigma_2}\zeta)}{\mathrm{F} \exp(\frac{\varsigma_2 - 1}{\varsigma_2}\mathrm{F})} \frac{1}{\varsigma_2^{\varsigma_1} \Gamma(\varsigma_1)}$$
$$+ \mathcal{M}_4 \frac{1}{\varsigma_2^{\varsigma_1} \Gamma(\varsigma_1)} + \sum_{r=1}^{m} \frac{1}{\varsigma_2^{\varsigma_{3r}} \Gamma(\varsigma_{3r})} \int_0^{\zeta} (\zeta - \mu)^{\varsigma_{3r} - 1} \exp(\frac{\varsigma_2 - 1}{\varsigma_2}(\zeta - \mu)) \mathcal{M}_5 \varrho < 1,$$

therefore Λ is a contraction mapping. The next step of the proof is to show that $\Delta_{CON}(\Psi\psi, \psi) < \infty$. Then, for the function $\psi \in \mathcal{N}$, we get

$$|\Psi(\psi(\zeta)) - \psi(\zeta)| = \Bigg| \mathcal{Y}(\zeta, \phi(\zeta)) \Bigg[\frac{\phi_0}{\mathcal{Y}(0, \phi_0)} \left(1 - \frac{\zeta}{\mathrm{F}}\right) \exp(\frac{\varsigma_2 - 1}{\varsigma_2}\zeta) + \frac{\phi_{\mathrm{F}} \frac{\zeta}{\mathrm{F}} \exp(\frac{\varsigma_3 - 1}{\varsigma_2}\zeta)}{\mathcal{Y}(\mathrm{F}, \phi_{\mathrm{F}}) \exp(\frac{\varsigma_3 - 1}{\varsigma_2}\mathrm{F})}$$
$$- \frac{\zeta \exp(\frac{\varsigma_2 - 1}{\varsigma_2}\zeta)}{\mathcal{Y}(\mathrm{F}, \phi_{\mathrm{F}}) \mathrm{F} \exp(\frac{\varsigma_3 - 1}{\varsigma_2}\mathrm{F})} \sum_{r=1}^{r=m} \frac{1}{\varsigma_2^{\varsigma_{3r}} \Gamma(\varsigma_{3r})} \int_0^{\mathrm{F}} (\mathrm{F} - \mu)^{\varsigma_{3r} - 1} \exp(\frac{\varsigma_3 - 1}{\varsigma_2}(\mathrm{F} - \mu)) \mathcal{W}_r(\mu, \phi(\mu)) d\mu$$
$$- \frac{\zeta \exp(\frac{\varsigma_3 - 1}{\varsigma_2}\zeta)}{\mathrm{F} \exp(\frac{\varsigma_2 - 1}{\varsigma_2}\mathrm{F})} \frac{1}{\varsigma_2^{\varsigma_1} \Gamma(\varsigma_1)} \int_0^{\mathrm{F}} (\mathrm{F} - \mu)^{\varsigma_1 - 1} \exp(\frac{\varsigma_2 - 1}{\varsigma_2}(\mathrm{F} - \mu)) \mathcal{K}(\zeta, \phi(\zeta)) d\mu$$
$$+ \frac{1}{\varsigma_2^{\varsigma_1} \Gamma(\varsigma_1)} \int_0^{\zeta} (\zeta - \mu)^{\varsigma_1 - 1} \exp(\frac{\varsigma_3 - 1}{\varsigma_2}(\zeta - \mu)) \mathcal{K}(\zeta, \phi(\zeta)) d\mu \Bigg]$$
$$+ \sum_{r=1}^{m} \frac{1}{\varsigma_2^{\varsigma_{3r}} \Gamma(\varsigma_{3r})} \int_0^{\zeta} (\zeta - \mu)^{\varsigma_{3r} - 1} \exp(\frac{\varsigma_2 - 1}{\varsigma_2}(\zeta - \mu)) \mathcal{W}_r(\mu, \phi(\mu)) d\mu - \phi(\zeta) \Bigg|$$
$$\leq \delta \frac{1}{\varsigma_2^{\varsigma_1} \Gamma(\varsigma_1)} \int_p^{\zeta} (\zeta - \mu)^{\varsigma_1 - 1} \exp(\varsigma_2 - 1(\zeta - \mu)) \Omega(\zeta) d\mu$$
$$\leq \delta \varrho\, \Omega(\zeta).$$

Consequently, $\Delta_{CON}(\Psi\psi, \psi) \leq \delta\varrho < \infty,\ \varrho < 1$. Therefore, in this situation, by referring to Theorem 1.2, we find that all the cases of the theorem are true, that is, the sequence $\{\Psi^{\mathbf{b}}\phi\}$ converges to a fixed point ϕ of Ψ, where ϕ is in the set $\mathcal{N}^* = \{\psi \in \mathcal{N} : \Delta_{CON}(\Psi\psi, \phi) < \infty\}$. In other words $\Psi\phi = \phi$ or

$$\Psi(\phi(\zeta)) = \mathcal{Y}(\zeta, \phi(\zeta)) \Bigg[\frac{\phi_0}{\mathcal{Y}(0, \phi_0)} \left(1 - \frac{\zeta}{\mathrm{F}}\right) \exp(\frac{\varsigma_2 - 1}{\varsigma_2}\zeta) + \frac{\phi_{\mathrm{F}} \frac{\zeta}{\mathrm{F}} \exp(\frac{\varsigma_3 - 1}{\varsigma_2}\zeta)}{\mathcal{Y}(\mathrm{F}, \phi_{\mathrm{F}}) \exp(\frac{\varsigma_3 - 1}{\varsigma_2}\mathrm{F})}$$
$$- \frac{\zeta \exp(\frac{\varsigma_2 - 1}{\varsigma_2}\zeta)}{\mathcal{Y}(\mathrm{F}, \phi_{\mathrm{F}}) \mathrm{F} \exp(\frac{\varsigma_3 - 1}{\varsigma_2}\mathrm{F})} \sum_{r=1}^{r=m} \frac{1}{\varsigma_2^{\varsigma_{3r}} \Gamma(\varsigma_{3r})} \int_0^{\mathrm{F}} (\mathrm{F} - \mu)^{\varsigma_{3r} - 1} \exp(\frac{\varsigma_3 - 1}{\varsigma_2}(\mathrm{F} - \mu)) \mathcal{W}_r(\mu, \phi(\mu)) d\mu$$
$$- \frac{\zeta \exp(\frac{\varsigma_3 - 1}{\varsigma_2}\zeta)}{\mathrm{F} \exp(\frac{\varsigma_2 - 1}{\varsigma_2}\mathrm{F})} \frac{1}{\varsigma_2^{\varsigma_1} \Gamma(\varsigma_1)} \int_0^{\mathrm{F}} (\mathrm{F} - \mu)^{\varsigma_1 - 1} \exp(\frac{\varsigma_2 - 1}{\varsigma_2}(\mathrm{F} - \mu)) \mathcal{K}(\zeta, \phi(\zeta)) d\mu$$
$$+ \frac{1}{\varsigma_2^{\varsigma_1} \Gamma(\varsigma_1)} \int_0^{\zeta} (\zeta - \mu)^{\varsigma_1 - 1} \exp(\frac{\varsigma_3 - 1}{\varsigma_2}(\zeta - \mu)) \mathcal{K}(\zeta, \phi(\zeta)) d\mu \Bigg]$$

$$+\sum_{r=1}^{m}\frac{1}{\varsigma_2^{\varsigma_{3r}}\Gamma(\varsigma_{3r})}\int_0^{\zeta}(\zeta-\mu)^{\varsigma_{3r}-1}\exp(\frac{\varsigma_2-1}{\varsigma_2}(\zeta-\mu))\mathcal{W}_r(\mu,\phi(\mu))d\mu. \tag{7.4.23}$$

Since ϕ is a differentiable function, we have

$$ {}_{m}^{C}\Xi D_{0^+}^{\varsigma_1,\varsigma_2}\left(\frac{\phi(\zeta)-\sum_{r=1}^{r=m}\Pi_{I_{0^+}}^{\varsigma_{3r}\varsigma_2}\mathcal{W}_r(\zeta,\phi(\zeta))}{\mathcal{Y}(\zeta,\phi(\zeta))}\right)=\mathcal{K}(\zeta,\phi(\zeta)) \tag{7.4.24}$$

and as a result

$$\Delta_{CON}(\psi,\phi)\leq\frac{1}{1-\upsilon}\Delta_{CON}(\Psi\psi,\psi)\leq\frac{\delta\varrho}{1-\upsilon}\,\Omega(\zeta).$$

In the following, to prove the uniqueness of the solution, we consider the differentiable function $\mathcal{L}$ and assume that it applies to Eq. (7.4.24). Then

$$ {}_{m}^{C}\Xi D_{0^+}^{\varsigma_1,\varsigma_2}\left(\frac{\mathcal{L}(\zeta)-\sum_{r=1}^{r=m}\Pi_{I_{0^+}}^{\varsigma_{3r}\varsigma_2}\mathcal{W}_r(\zeta,\mathcal{L}(\zeta))}{\mathcal{Y}(\zeta,\mathcal{L}(\zeta))}\right)=\mathcal{K}(\zeta,\mathcal{L}(\zeta)), \tag{7.4.25}$$

To finish proving the uniqueness of the solution, we show that $\mathcal{L}\in\mathcal{N}^*$ is a fixed point for Ψ. According to (7.4.25), $\Psi\mathcal{L}=\mathcal{L}$. we show that $\Delta_{CON}(\Psi\psi,\mathcal{L})<\infty$. Considering (7.4.25) and assuming $\psi\in\mathcal{N}$, $\Delta_{CON}(\psi,\mathcal{L})<\frac{\delta\varrho}{1-\upsilon}$, we have

$$\begin{aligned}
|\Psi(\psi(\zeta))-\mathcal{L}(\zeta)| &= \Bigg|\mathcal{Y}(\zeta,\phi(\zeta))\Bigg[\frac{\phi_0}{\mathcal{Y}(0,\phi_0)}\left(1-\frac{\zeta}{\mathrm{F}}\right)\exp(\frac{\varsigma_2-1}{\varsigma_2}\zeta)+\frac{\phi_{\mathrm{F}}\frac{\zeta}{\mathrm{F}}\exp(\frac{\varsigma_3-1}{\varsigma_2}\zeta)}{\mathcal{Y}(\mathrm{F},\phi_{\mathrm{F}})\exp(\frac{\varsigma_3-1}{\varsigma_2}\mathrm{F})}\\
&-\frac{\zeta\exp(\frac{\varsigma_2-1}{\varsigma_2}\zeta)}{\mathcal{Y}(\mathrm{F},\phi_{\mathrm{F}})\mathrm{F}\exp(\frac{\varsigma_3-1}{\varsigma_2}\mathrm{F})}\sum_{r=1}^{r=m}\frac{1}{\varsigma_2^{\varsigma_{3r}}\Gamma(\varsigma_{3r})}\int_0^{\mathrm{F}}(\mathrm{F}-\mu)^{\varsigma_{3r}-1}\exp(\frac{\varsigma_3-1}{\varsigma_2}(\mathrm{F}-\mu))\mathcal{W}_r(\mu,\phi(\mu))d\mu\\
&-\frac{\zeta\exp(\frac{\varsigma_3-1}{\varsigma_2}\zeta)}{\mathrm{F}\exp(\frac{\varsigma_2-1}{\varsigma_2}\mathrm{F})}\frac{1}{\varsigma_2^{\varsigma_1}\Gamma(\varsigma_1)}\int_0^{\mathrm{F}}(\mathrm{F}-\mu)^{\varsigma_1-1}\exp(\frac{\varsigma_2-1}{\varsigma_2}(\mathrm{F}-\mu))\mathcal{K}(\zeta,\phi(\zeta))d\mu\\
&+\frac{1}{\varsigma_2^{\varsigma_1}\Gamma(\varsigma_1)}\int_0^{\zeta}(\zeta-\mu)^{\varsigma_1-1}\exp(\frac{\varsigma_3-1}{\varsigma_2}(\zeta-\mu))\mathcal{K}(\zeta,\phi(\zeta))d\mu\Bigg]\\
&+\sum_{r=1}^{m}\frac{1}{\varsigma_2^{\varsigma_{3r}}\Gamma(\varsigma_{3r})}\int_0^{\zeta}(\zeta-\mu)^{\varsigma_{3r}-1}\exp(\frac{\varsigma_2-1}{\varsigma_2}(\zeta-\mu))\mathcal{W}_r(\mu,\phi(\mu))d\mu\\
&-\Bigg(\mathcal{Y}(\zeta,\mathcal{L}(\zeta))\Bigg[\frac{\phi_0}{\mathcal{Y}(0,\mathcal{L}_0)}\left(1-\frac{\zeta}{\mathrm{F}}\right)\exp(\frac{\varsigma_2-1}{\varsigma_2}\zeta)+\frac{\mathcal{L}_T\frac{\zeta}{\mathrm{F}}\exp(\frac{\varsigma_3-1}{\varsigma_2}\zeta)}{\mathcal{Y}(\mathrm{F},\mathcal{L}_T)\exp(\frac{\varsigma_3-1}{\varsigma_2}\mathrm{F})}\\
&-\frac{\zeta\exp(\frac{\varsigma_2-1}{\varsigma_2}\zeta)}{\mathcal{Y}(\mathrm{F},\mathcal{L}_T)\mathrm{F}\exp(\frac{\varsigma_3-1}{\varsigma_2}\mathrm{F})}\sum_{r=1}^{r=m}\frac{1}{\varsigma_2^{\varsigma_{3r}}\Gamma(\varsigma_{3r})}\int_0^{\mathrm{F}}(\mathrm{F}-\mu)^{\varsigma_{3r}-1}\exp(\frac{\varsigma_3-1}{\varsigma_2}(\mathrm{F}-\mu))\mathcal{W}_r(\mu,\mathcal{L}(\mu))d\mu\\
&-\frac{\zeta\exp(\frac{\varsigma_3-1}{\varsigma_2}\zeta)}{\mathrm{F}\exp(\frac{\varsigma_2-1}{\varsigma_2}\mathrm{F})}\frac{1}{\varsigma_2^{\varsigma_1}\Gamma(\varsigma_1)}\int_0^{\mathrm{F}}(\mathrm{F}-\mu)^{\varsigma_1-1}\exp(\frac{\varsigma_2-1}{\varsigma_2}(\mathrm{F}-\mu))\mathcal{K}(\zeta,\mathcal{L}(\zeta))d\mu
\end{aligned}$$

$$+\frac{1}{\varsigma_2^{\varsigma_1}\Gamma(\varsigma_1)}\int_0^{\zeta}(\zeta-\mu)^{\varsigma_1-1}\exp(\frac{\varsigma_3-1}{\varsigma_2}(\zeta-\mu))\mathcal{K}(\zeta,\mathcal{L}(\zeta))d\mu\Big]$$

$$+\sum_{r=1}^{m}\frac{1}{\varsigma_2^{\varsigma_{3r}}\Gamma(\varsigma_{3r})}\int_0^{\zeta}(\zeta-\mu)^{\varsigma_{3r}-1}\exp(\frac{\varsigma_2-1}{\varsigma_2}(\zeta-\mu))\mathcal{W}_r(\mu,\mathcal{L}(\mu))d\mu\Big)\Big|$$

$$\leq\frac{\psi_0}{\mathcal{Y}(0,\psi_0)}\left(1-\frac{\zeta}{\mathrm{F}}\right)\exp(\frac{\varsigma_2-1}{\varsigma_2}\zeta)\Big(\Big|\mathcal{Y}(\zeta,\phi(\zeta))-\mathcal{Y}(\zeta,\mathcal{L}(\zeta))\Big|\Big)$$

$$+\frac{\psi_{\mathrm{F}}\frac{\zeta}{\mathrm{F}}\exp(\frac{\varsigma_3-1}{\varsigma_2}\zeta)}{\mathcal{Y}(\mathrm{F},\psi_{\mathrm{F}})\exp(\frac{\varsigma_3-1}{\varsigma_2}\mathrm{F})}\Big(\Big|\mathcal{Y}(\zeta,\phi(\zeta))-\mathcal{Y}(\zeta,\mathcal{L}(\zeta))\Big|\Big)$$

$$+\frac{\zeta\exp(\frac{\varsigma_2-1}{\varsigma_2}\zeta)}{\mathcal{Y}(\mathrm{F},\psi_{\mathrm{F}})\,\mathrm{F}\exp(\frac{\varsigma_3-1}{\varsigma_2}\mathrm{F})}\sum_{r=1}^{r=m}\frac{1}{\varsigma_2^{\varsigma_{3r}}\Gamma(\varsigma_{3r})}\Big(\Big|\mathcal{Y}(\zeta,\mathcal{L}(\zeta))\int_0^{\mathrm{F}}(\mathrm{F}-\mu)^{\varsigma_{3r}-1}\exp(\frac{\varsigma_3-1}{\varsigma_2}(\mathrm{F}-\mu))\mathcal{W}_r(\mu,\mathcal{L}(\mu))d\mu$$

$$-\mathcal{Y}(\zeta,\phi(\zeta))\int_0^{\mathrm{F}}(\mathrm{F}-\mu)^{\varsigma_{3r}-1}\exp(\frac{\varsigma_3-1}{\varsigma_2}(\mathrm{F}-\mu))\mathcal{W}_r(\mu,\phi(\mu))d\mu\Big|\Big)$$

$$+\frac{\zeta\exp(\frac{\varsigma_3-1}{\varsigma_2}\zeta)}{\mathrm{F}\exp(\frac{\varsigma_2-1}{\varsigma_2}\mathrm{F})}\frac{1}{\varsigma_2^{\varsigma_1}\Gamma(\varsigma_1)}\Big(\mathcal{Y}(\mu,\mathcal{L}(\mu))\Big|\int_0^{\mathrm{F}}(\mathrm{F}-\mu)^{\varsigma_1-1}\exp(\frac{\varsigma_2-1}{\varsigma_2}(\mathrm{F}-\mu))\mathcal{K}(\mu,\mathcal{L}(\mu))d\mu$$

$$-\mathcal{Y}(\zeta,\phi(\zeta))\int_0^{\mathrm{F}}(\mathrm{F}-\mu)^{\varsigma_1-1}\exp(\frac{\varsigma_2-1}{\varsigma_2}(\mathrm{F}-\mu))\mathcal{K}(\mu,\phi(\mu))d\mu\Big|\Big)$$

$$+\frac{1}{\varsigma_2^{\varsigma_1}\Gamma(\varsigma_1)}\Big(\Big|\mathcal{Y}(\zeta,\phi(\zeta))\int_0^{\zeta}(\zeta-\mu)^{\varsigma_1-1}\exp(\frac{\varsigma_3-1}{\varsigma_2}(\zeta-\mu))\mathcal{K}(\mu,\phi(\mu))d\mu$$

$$-\mathcal{Y}(\zeta,\mathcal{L}(\zeta))\int_0^{\zeta}(\zeta-\mu)^{\varsigma_1-1}\exp(\frac{\varsigma_3-1}{\varsigma_2}(\zeta-\mu))\mathcal{K}(\mu,\mathcal{L}(\mu))d\mu\Big|\Big)$$

$$+\sum_{r=1}^{m}\frac{1}{\varsigma_2^{\varsigma_{3r}}\Gamma(\varsigma_{3r})}\int_0^{\zeta}(\zeta-\mu)^{\varsigma_{3r}-1}\exp(\frac{\varsigma_2-1}{\varsigma_2}(\zeta-\mu))\Big|\mathcal{W}_r(\mu,\phi(\mu))-\mathcal{W}_r(\mu,\mathcal{L}(\mu))\Big|d\mu$$

$$\leq\frac{\psi_0}{\mathcal{Y}(0,\psi_0)}\left(1-\frac{\zeta}{\mathrm{F}}\right)\exp(\frac{\varsigma_2-1}{\varsigma_2}\zeta)\Big(\mathcal{M}_1\Big|\phi(\zeta)-\mathcal{L}(\zeta)\Big|\Big)+\frac{\psi_{\mathrm{F}}\frac{\zeta}{\mathrm{F}}\exp(\frac{\varsigma_3-1}{\varsigma_2}\zeta)}{\mathcal{Y}(\mathrm{F},\psi_{\mathrm{F}})\exp(\frac{\varsigma_3-1}{\varsigma_2}\mathrm{F})}\Big(\mathcal{M}_1\Big|\phi(\zeta)-\mathcal{L}(\zeta)\Big|\Big)$$

$$+\frac{\zeta\exp(\frac{\varsigma_2-1}{\varsigma_2}\zeta)}{\mathcal{Y}(\mathrm{F},\psi_{\mathrm{F}})\,\mathrm{F}\exp(\frac{\varsigma_3-1}{\varsigma_2}\mathrm{F})}\sum_{r=1}^{r=m}\frac{1}{\varsigma_2^{\varsigma_{3r}}\Gamma(\varsigma_{3r})}\Big(\mathcal{M}_2\Big|\phi(\zeta)-\mathcal{L}(\zeta)\Big|\Big)$$

$$+\frac{\zeta\exp(\frac{\varsigma_3-1}{\varsigma_2}\zeta)}{\mathrm{F}\exp(\frac{\varsigma_2-1}{\varsigma_2}\mathrm{F})}\frac{1}{\varsigma_2^{\varsigma_1}\Gamma(\varsigma_1)}\Big(\mathcal{M}_3\Big|\phi(\zeta)-\mathcal{L}(\zeta)\Big|\Big)$$

$$+\frac{1}{\varsigma_2^{\varsigma_1}\Gamma(\varsigma_1)}\Big(\mathcal{M}_4\Big|\phi(\zeta)-\mathcal{L}(\zeta)\Big|\Big)$$

$$+\sum_{r=1}^{m}\frac{1}{\varsigma_2^{\varsigma_{3r}}\Gamma(\varsigma_{3r})}\int_0^{\zeta}(\zeta-\mu)^{\varsigma_{3r}-1}\exp(\frac{\varsigma_2-1}{\varsigma_2}(\zeta-\mu))\Big(\mathcal{M}_5\Big|\phi(\zeta)-\mathcal{L}(\zeta)\Big|\Big)d\mu$$

$$\leq\frac{\psi_0}{\mathcal{Y}(0,\psi_0)}\left(1-\frac{\zeta}{\mathrm{F}}\right)\exp(\frac{\varsigma_2-1}{\varsigma_2}\zeta)\Big(\mathcal{M}_1\frac{\delta\varrho}{1-\upsilon}\,\Omega(\zeta)\Big)+\frac{\psi_{\mathrm{F}}\frac{\zeta}{\mathrm{F}}\exp(\frac{\varsigma_3-1}{\varsigma_2}\zeta)}{\mathcal{Y}(\mathrm{F},\psi_{\mathrm{F}})\exp(\frac{\varsigma_3-1}{\varsigma_2}\mathrm{F})}\Big(\mathcal{M}_1\frac{\delta\varrho}{1-\upsilon}\,\Omega(\zeta)\Big)$$

$$+\frac{\zeta\exp(\frac{\varsigma_2-1}{\varsigma_2}\zeta)}{\mathcal{Y}(\mathrm{F},\psi_{\mathrm{F}})\,\mathrm{F}\exp(\frac{\varsigma_3-1}{\varsigma_2}\mathrm{F})}\sum_{r=1}^{r=m}\frac{1}{\varsigma_2^{\varsigma_{3r}}\Gamma(\varsigma_{3r})}\Big(\mathcal{M}_2\frac{\delta\varrho}{1-\upsilon}\,\Omega(\zeta)\Big)$$

$$+\frac{\zeta\exp(\frac{\varsigma_3-1}{\varsigma_2}\zeta)}{\mathrm{F}\exp(\frac{\varsigma_2-1}{\varsigma_2}\mathrm{F})}\frac{1}{\varsigma_2^{\varsigma_1}\Gamma(\varsigma_1)}\Big(\mathcal{M}_3\frac{\delta\varrho}{1-\upsilon}\,\Omega(\zeta)\Big)$$

$$+\frac{1}{\varsigma_2^{\varsigma_1}\Gamma(\varsigma_1)}\Big(\mathcal{M}_4\frac{\delta\varrho}{1-\upsilon}\,\Omega(\zeta)\Big)$$

$$+\sum_{r=1}^{m}\frac{1}{\varsigma_2^{\varsigma_{3r}}\Gamma(\varsigma_{3r})}\int_0^{\zeta}(\zeta-\mu)^{\varsigma_{3r}-1}\exp(\frac{\varsigma_2-1}{\varsigma_2}(\zeta-\mu))\Big(\mathcal{M}_5\frac{\delta\varrho}{1-\upsilon}\,\Omega(\zeta)\Big)d\mu$$

$$\leq\Big(\mathcal{M}_1\frac{\psi_0}{\mathcal{Y}(0,\psi_0)}\Big(1-\frac{\zeta}{\mathrm{F}}\Big)\exp(\frac{\varsigma_2-1}{\varsigma_2}\zeta)+\mathcal{M}_1\frac{\psi_{\mathrm{F}}\frac{\zeta}{\mathrm{F}}\exp(\frac{\varsigma_3-1}{\varsigma_2}\zeta)}{\mathcal{Y}(\mathrm{F},\psi_{\mathrm{F}})\exp(\frac{\varsigma_3-1}{\varsigma_2}\mathrm{F})}$$

$$+\frac{\zeta\exp(\frac{\varsigma_2-1}{\varsigma_2}\zeta)}{\mathcal{Y}(\mathrm{F},\psi_{\mathrm{F}})\,\mathrm{F}\exp(\frac{\varsigma_3-1}{\varsigma_2}\mathrm{F})}\sum_{r=1}^{r=m}\frac{1}{\varsigma_2^{\varsigma_{3r}}\Gamma(\varsigma_{3r})}\mathcal{M}_2+\mathcal{M}_3\frac{\zeta\exp(\frac{\varsigma_3-1}{\varsigma_2}\zeta)}{\mathrm{F}\exp(\frac{\varsigma_2-1}{\varsigma_2}\mathrm{F})}\frac{1}{\varsigma_2^{\varsigma_1}\Gamma(\varsigma_1)}$$

$$+\mathcal{M}_4\frac{1}{\varsigma_2^{\varsigma_1}\Gamma(\varsigma_1)}+\sum_{r=1}^{m}\frac{1}{\varsigma_2^{\varsigma_{3r}}\Gamma(\varsigma_{3r})}\int_0^{\zeta}(\zeta-\mu)^{\varsigma_{3r}-1}\exp(\frac{\varsigma_2-1}{\varsigma_2}(\zeta-\mu))\mathcal{M}_5\varrho\Big)\frac{\delta\varrho}{1-\upsilon}\,\Omega(\zeta)$$

that means

$$\Delta_{CON}(\Psi\psi,\mathcal{L})\leq\Big(\mathcal{M}_1\frac{\psi_0}{\mathcal{Y}(0,\psi_0)}\Big(1-\frac{\zeta}{\mathrm{F}}\Big)\exp(\frac{\varsigma_2-1}{\varsigma_2}\zeta)+\mathcal{M}_1\frac{\psi_{\mathrm{F}}\frac{\zeta}{\mathrm{F}}\exp(\frac{\varsigma_3-1}{\varsigma_2}\zeta)}{\mathcal{Y}(\mathrm{F},\psi_{\mathrm{F}})\exp(\frac{\varsigma_3-1}{\varsigma_2}\mathrm{F})}$$

$$+\frac{\zeta\exp(\frac{\varsigma_2-1}{\varsigma_2}\zeta)}{\mathcal{Y}(\mathrm{F},\psi_{\mathrm{F}})\,\mathrm{F}\exp(\frac{\varsigma_3-1}{\varsigma_2}\mathrm{F})}\sum_{r=1}^{r=m}\frac{1}{\varsigma_2^{\varsigma_{3r}}\Gamma(\varsigma_{3r})}\mathcal{M}_2+\mathcal{M}_3\frac{\zeta\exp(\frac{\varsigma_3-1}{\varsigma_2}\zeta)}{\mathrm{F}\exp(\frac{\varsigma_2-1}{\varsigma_2}\mathrm{F})}\frac{1}{\varsigma_2^{\varsigma_1}\Gamma(\varsigma_1)}$$

$$+\mathcal{M}_4\frac{1}{\varsigma_2^{\varsigma_1}\Gamma(\varsigma_1)}+\sum_{r=1}^{m}\frac{1}{\varsigma_2^{\varsigma_{3r}}\Gamma(\varsigma_{3r})}\int_0^{\zeta}(\zeta-\mu)^{\varsigma_{3r}-1}\exp(\frac{\varsigma_2-1}{\varsigma_2}(\zeta-\mu))\mathcal{M}_5\varrho\Big)\frac{\delta\varrho}{1-\upsilon}<\infty.$$

7.4.3 *Implementation of Reproducing Kernel Method for HPFIDE* (7.4.1)

To implement the reproducing kernel method of Eq. (7.4.1) and obtain the solution, we first define a linear differential operator. Therefore, considering $\mathcal{J}:\mathrm{U}_2^2[0,\mathrm{F}]\to\mathrm{U}_2^1[0,\mathrm{F}]$, we define this operator as follows

$$\mathcal{J}\phi(\zeta)={}_m^C\Xi D_{0^+}^{\varsigma_1,\varsigma_2}\left(\frac{\phi(\zeta)-\sum_{r=1}^{r=m}\Pi I_{0^+}^{\varsigma_{3r}\varsigma_2}\mathcal{W}_r(\zeta,\phi(\zeta))}{\mathcal{Y}(\zeta,\phi(\zeta))}\right).\tag{7.4.26}$$

Then, according to the above differential operator, we have the following system

$$\begin{cases}\mathcal{J}\phi(\zeta)=\mathcal{K}(\zeta,\phi(\zeta)), & \zeta\in[0,\mathrm{F}],\\ \phi(0)=\phi_0,\ \phi(\mathrm{F})=\phi_{\mathrm{F}},\end{cases}\tag{7.4.27}$$

Theorem 7.4 *The operator* $\mathcal{J}:\mathrm{U}_2^2[0,\mathrm{F}]\to\mathrm{U}_2^1[0,\mathrm{F}]$ *defined in* (7.4.26) *is a linear and bounded operator.*

Proof Step **1.** First, we show that $\mathcal{J}$ is a linear operator. For this purpose, we assume that $\psi(\zeta),\phi(\zeta)\in\mathrm{U}_2^2[0,\mathrm{F}]$, therefore, we have

$$\mathcal{J}(\psi+\phi)(\zeta)={}_m^C\Xi D_{0^+}^{\varsigma_1,\varsigma_2}\left(\frac{(\psi(\zeta)+\phi(\zeta))-\sum_{r=1}^{r=m}\Pi I_{0^+}^{\varsigma_{3r}\varsigma_2}\mathcal{W}_r(\zeta,(\psi(\zeta)+\phi(\zeta)))}{\mathcal{Y}(\zeta,(\psi(\zeta)+\phi(\zeta)))}\right)$$

$$= \frac{1}{\varsigma_2^{\varsigma_1}\Gamma(n-\varsigma_1)}\int_0^{\zeta}(\zeta-\mu)^{n-\varsigma_1-1}\exp(\frac{n-1}{\varsigma_2}(\zeta-\mu))$$
$$\left(\Xi_D{}^{n,\varsigma_2}\left(\frac{(\psi(\mu)+\phi(\mu))-\sum_{r=1}^{r=m}\Pi_{I_{0+}}^{\varsigma_3 r\varsigma_2}\mathcal{W}_r(\mu,(\psi(\mu)+\phi(\mu)))}{\mathcal{Y}(\mu,(\psi(\mu)+\phi(\mu)))}\right)\right)(\mu)d\mu$$
$$= {}_m^C\Xi_{D_{0+}}^{\varsigma_1,\varsigma_2}\left(\frac{(\psi(\zeta))-\sum_{r=1}^{r=m}\Pi_{I_{0+}}^{\varsigma_3 r\varsigma_2}\mathcal{W}_r(\zeta,\psi(\zeta))}{\mathcal{Y}(\zeta,\psi(\zeta))}\right)+{}_m^C\Xi_{D_{0+}}^{\varsigma_1,\varsigma_2}\left(\frac{\phi(\zeta)-\sum_{r=1}^{r=m}\Pi_{I_{0+}}^{\varsigma_3 r\varsigma_2}\mathcal{W}_r(\zeta,\phi(\zeta))}{\mathcal{Y}(\zeta,\phi(\zeta))}\right)$$
$$= \mathcal{J}\psi(\zeta)+\mathcal{J}\phi(\zeta).$$

To complete the proof of linearity, we assume $\psi(\zeta)\in \mathrm{U}_2^2[0,\mathrm{F}]$ and $s\in\mathbb{R}$, then, we have

$$\mathcal{J}(s\psi)(\zeta) = {}_m^C\Xi_{D_{0+}}^{\varsigma_1,\varsigma_2}\left(\frac{(s\psi(\zeta))-\sum_{r=1}^{r=m}\Pi_{I_{0+}}^{\varsigma_3 r\varsigma_2}\mathcal{W}_r(\zeta,(s\psi(\zeta)))}{\mathcal{Y}(\zeta,(s\psi(\zeta)))}\right)$$
$$= \frac{1}{\varsigma_2^{\varsigma_1}\Gamma(n-\varsigma_1)}\int_0^{\zeta}(\zeta-\mu)^{n-\varsigma_1-1}\exp(\frac{n-1}{\varsigma_2}(\zeta-\mu))\left(\Xi_D{}^{n,\varsigma_2}\left(\frac{(s\psi(\mu))-\sum_{r=1}^{r=m}\Pi_{I_{0+}}^{\varsigma_3 r\varsigma_2}\mathcal{W}_r(\mu,(s\psi(\mu)))}{\mathcal{Y}(\mu,(s\psi(\mu)))}\right)\right)(\mu)d\mu$$
$$= s\ {}_m^C\Xi_{D_{0+}}^{\varsigma_1,\varsigma_2}\left(\frac{(\psi(\zeta))-\sum_{r=1}^{r=m}\Pi_{I_{0+}}^{\varsigma_3 r\varsigma_2}\mathcal{W}_r(\zeta,(\psi(\zeta)))}{\mathcal{Y}(\zeta,(\psi(\zeta)))}\right) = s\ \mathcal{J}(\psi)(\zeta).$$

***Step* 2.** To start the proof of boundedness, we consider (IP-U_2^1) and (N-U_2^1)

$$\|\mathcal{J}\psi(\zeta)\|_{\mathrm{U}_2^1}^2 = \langle\mathcal{J}\psi(\zeta),\mathcal{J}\psi(\zeta)\rangle_{\mathrm{U}_2^1} = [(\mathcal{J}\psi)(0)]^2+\int_0^{\mathrm{F}}\left[(\mathcal{J}\psi)'(\zeta)\right]^2\,d\zeta. \tag{7.4.28}$$

On the other hand, according to (RKF-2), we have

- $\psi(\zeta) = \langle\psi(\eta),\Lambda_{[rk]}(\eta,\zeta)\rangle_{\mathrm{U}_2^2[0,1]}$;
- $(\mathcal{J}\psi)(\zeta) = \langle\psi(\eta),\mathcal{J}\Lambda_{[rk]}(\eta,\zeta)\rangle_{\mathrm{U}_2^2[0,1]}$;
- $(\mathcal{J}\psi)'(\zeta) = \langle\psi(\eta),(\mathcal{J}\Lambda'_{[rk]}(\eta,\zeta))\rangle_{\mathrm{U}_2^2[0,1]}$.

Using Schwartz's inequality and considering positive constants $\mathcal{B}_1,\mathcal{B}_2>0$ and the continuity $\Lambda_{[rk]}(\zeta,\eta)$, we have

$$|(\mathcal{J}\psi)(\zeta)| = \left|\langle\psi(\eta),\mathcal{J}\Lambda_{[rk]\zeta}(\eta)\rangle_{\mathrm{U}_2^2}\right| \le \left\|\mathcal{J}\Lambda_{[rk]\zeta}\right\|_{\mathrm{U}_2^1}\|\psi(\zeta)\|_{\mathrm{U}_2^2} = \mathcal{B}_1\|\psi(\zeta)\|_{\mathrm{U}_2^2},$$

with the same process $\left|(\mathcal{J}\psi)'(\zeta)\right| \le \mathcal{B}_2\|\psi\|_{\mathrm{U}_2^2}$. Therefore,

$$\|(\mathcal{J}\psi(\zeta))\|_{\mathrm{U}_2^1}^2 \le \mathcal{B}_1^2\|\psi(\zeta)\|_{\mathrm{U}_2^2}^2+\int_0^{\mathrm{F}}\mathcal{B}_2^2\|\psi(\zeta)\|_{\mathrm{U}_2^2}^2d\zeta = \left(\mathcal{B}_1^2+\mathrm{F}\mathcal{B}_2^2\right)\|\psi(\zeta)\|_{\mathrm{U}_2^2}^2.$$

Now, to complete the reproducing kernel Hilbert space method, we make an orthogonal system of $\mathrm{U}_2^2[0,\mathrm{F}]$. For this purpose, we consider the following

(O-1) $\mathcal{Q}_{1r}(\zeta) = \Lambda_{[rk]\zeta_r}(\zeta)$;
(O-2) $\mathcal{Q}_{2r}(\zeta) = \mathcal{J}^*\mathcal{Q}_{1r}(\zeta)$;

where $\{\zeta_r\}_{r=1}^{\infty}$ is a dense sequence in $[0,\mathrm{F}]$ and $\mathcal{J}^\star : \mathrm{U}_2^1[0,\mathrm{F}] \longrightarrow \mathrm{U}_2^2[0,\mathrm{F}]$ is conjugate operator of $\mathcal{J}$ and $\left\{\mathcal{Q}_{2r}(.)\right\}_{r=1}^{\infty}$ is a complete system of $\mathrm{U}_2^2[0,\mathrm{F}]$. Then, we have

$$\begin{aligned}\mathcal{Q}_{2r}(\zeta) = \mathcal{J}^\star\mathcal{Q}_{1r}(\zeta) &= \left\langle \mathcal{J}^\star\mathcal{Q}_{1r}(\zeta), \Lambda_{[rk]\zeta}(\eta)\right\rangle_{\mathrm{U}_2^2} = \left\langle \mathcal{Q}_{1r}(\eta), \mathcal{J}\Lambda_{[rk]\zeta}(\eta)\right\rangle_{\mathrm{U}_2^1} \\ &= \left\langle \Theta_{[rk]\zeta_r}(\eta), \mathcal{J}\Lambda_{[rk]\zeta}(\eta)\right\rangle_{\mathrm{U}_2^2} = L_y\Lambda_{[rk]\zeta}(\eta)\big|_{\eta=\zeta_r}.\end{aligned}$$

Considering the Gram-Schmidt process on $\mathrm{U}_2^2[0,\mathrm{F}]$, an orthogonal basis is obtained from $\left\{\mathcal{Q}_{2r}(\zeta)\right\}_{r=1}^{\infty}$, which is shown as follows

$$\bar{\mathcal{Q}}_{2r}(\zeta) = \sum_{i=1}^{r} \vartheta_{ji}\mathcal{Q}_{2i}(\zeta), \tag{7.4.29}$$

where the orthogonalization coefficients of ϑ_{jk} are considered as follows

(OC-1) $\vartheta_{11} = \frac{1}{\|\mathcal{Q}_{21}\|}$;
(OC-2) $\vartheta_{rr} = \frac{1}{\sqrt{\|\mathcal{Q}_{2r}'\|^2 - \sum_{i=1}^{r-1} f_{rk}^2}}$;
(OC-3) $\vartheta_{rj} = \frac{-\sum_{i=j}^{r-1} f_{ri}\vartheta_{ij}}{\sqrt{\|\mathcal{Q}_{2r}\|^2 - \sum_{i=1}^{r-1} f_{ri}^2}}$,

such that $f_{ri} = \left\langle \mathcal{Q}_{2r}, \bar{\mathcal{Q}}_{2i}\right\rangle_{\mathrm{U}_2^2}$.

In the following, we present several lemmas regarding exact and approximate solutions obtained by reproducing kernel Hilbert space method.

Lemma 7.4 *We assume that $\psi(\zeta)$ is an exact solution of* (7.4.27). *Then, for any dense set such as $\{\zeta_r\}_{r=1}^{\infty}$ in $\mathrm{U}_2^2[0,\mathrm{F}]$, $\psi(\zeta)$ is as follows*

$$\psi(\zeta) = \sum_{r=1}^{\infty}\sum_{i=1}^{r} \vartheta_{ri}\mathcal{K}\left(\zeta_k, \psi\left(\zeta_k\right)\right)\bar{\mathcal{Q}}_{2r}(\zeta) \tag{7.4.30}$$

If we substitute a finite sum in (7.4.30) *instead of an infinite sum, an approximate solution is obtained, which is written as*

$$\psi_n(\zeta) = \sum_{r=1}^{n}\sum_{i=1}^{r} \vartheta_{ri}\mathcal{K}\left(\zeta_k, \psi\left(\zeta_k\right)\right)\bar{\mathcal{Q}}_{2r}(\zeta). \tag{7.4.31}$$

Lemma 7.5 *If* $\psi \in \mathrm{U}_2^2[0, \mathrm{F}]$, *then there exists constant* $C > 0$ *such that*

$$\left|\psi^{(r)}(\zeta)\right| \leq C\|\psi\|_{\mathrm{U}_2^2}, \quad r = 0, 1$$

Lemma 7.6 *The approximate solutions* $\psi_n(\zeta)$ *and* $\psi_n'(\zeta)$ *uniformly converge to the exact solutions* $\psi(\zeta)$ *and* $\psi'(\zeta)$, *respectively.*

7.4.4 Numerical Example

We will analyze qualitatively and numerically an HPFID equation. First, we investigate the optimal stability of the equation using specific controller functions. After optimal stability and obtaining the best approximation, we obtain the numerical solutions of the equation by considering the reproducing kernel Hilbert space method (Fig. 7.14, 7.15).

Example 7.6 We consider an HPFIDE system, which is as follows

$$\begin{cases} {}_{m}^{C}\Xi_{D_{0^+}}^{\frac{5}{6},\frac{5}{8}}\left(\dfrac{\phi(\zeta)-\sum_{r=1}^{r=4}\Pi_{I_{0^+}}^{\frac{3r-2}{8},\frac{5}{8}}\frac{\cos\phi(\zeta)\zeta}{\zeta+150r}}{\frac{1}{20}\sqrt{4+|\phi(\zeta)|^2}}\right) = \dfrac{\epsilon(2+\zeta)\sin|\phi(\zeta)|}{4+|\phi(\zeta)|}, \\ \phi(0) = \frac{1}{100}, \phi(1) = \frac{1}{200}, \end{cases} \tag{7.4.32}$$

where in, $\zeta \in [0, \mathrm{F}], \mathrm{F} = 1,\quad \varsigma_1 = \frac{5}{6},\quad \varsigma_2 = \frac{5}{8},\quad \varsigma_{3r} = \frac{3r-2}{5},\quad \phi_0 = \frac{1}{100},\quad \phi_1 = \frac{1}{200},\ \varrho = \frac{1}{4}$, $\mathcal{W}_r(\zeta, \phi(\zeta)) = \frac{\cos|\phi(\zeta)|}{\zeta+150r}$, $(r = 1, \ldots, 4)$, $\mathcal{K}(\zeta, \phi(\zeta)) = \frac{\epsilon(2+\zeta)\sin|\phi(\zeta)|}{4+|\phi(\zeta)|}$, and $\mathcal{Y}(\zeta, \phi(\zeta)) = \frac{1}{20}\sqrt{4+|\phi(\zeta)|^2}$. For the continuous functions $\mathcal{Y} : [0, \mathrm{F}] \times \mathbb{R} \to \mathbb{R} - \{0\}$, $\mathcal{W}_r : [0, 1] \times \mathbb{R} \to \mathbb{R}$, $\mathcal{K} : [0, 1] \times \mathbb{R} \to \mathbb{R}$ and $\Omega : \mathbb{R} \longrightarrow \mathbb{R}$, the following inequalities are

$$|\mathcal{Y}(\zeta, \phi(\zeta)) - \mathcal{Y}(\zeta, \psi(\zeta))| \leq \frac{1}{20}|\phi(\zeta) - \psi(\zeta)|, \tag{7.4.33}$$

$$\begin{aligned} &\left|\frac{1}{20}\sqrt{4+|\psi(\zeta)|^2}\int_0^1 (1-\mu)^{\frac{3r-2}{5}-1}\exp\Big(\frac{-3}{5}(1-\mu)\Big)\frac{\cos|\psi(\mu)|}{\mu+150r}d\mu\right. \\ &\left.-\frac{1}{20}\sqrt{4+|\phi(\zeta)|^2}\int_0^1 (1-\mu)^{\frac{3r-2}{5}-1}\exp\Big(\frac{-3}{5}(1-\mu)\Big)\frac{\cos|\phi(\mu)|}{\mu+150r}d\mu\right| \\ &\leq \frac{1}{40}|\phi(\zeta) - \psi(\zeta)|, \end{aligned} \tag{7.4.34}$$

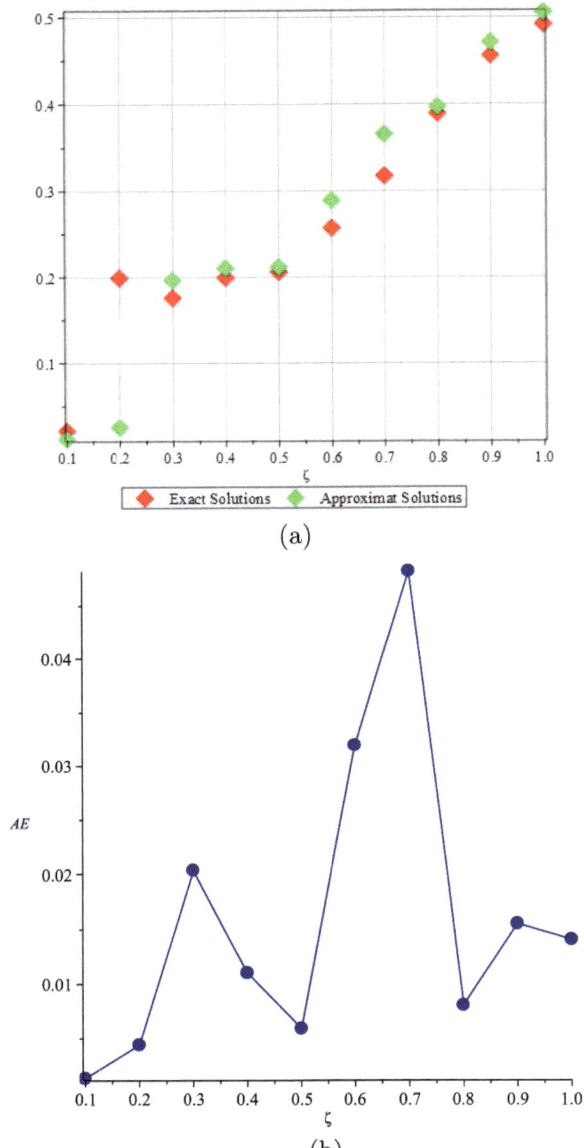

Fig. 7.14 Diagrams of approximate and exact solutions and absolute error of example 7.6 by selecting 15 points

$$\begin{aligned}
&\Big|\frac{1}{20}\sqrt{4+|\phi(\zeta)|^2}\int_0^1 (1-\mu)^{-\frac{1}{6}}\exp(\frac{-3}{5}(1-\mu))\frac{\epsilon(2+\mu)\sin|\psi(\mu)|}{4+|\psi(\mu)|}d\mu \\
&-\frac{1}{20}\sqrt{4+|\phi(\zeta)|^2}\int_0^1 (1-\mu)^{\varsigma_1-1}\exp(\frac{-3}{5}(1-\mu))\frac{\epsilon(2+\mu)\sin|\phi(\mu)|}{4+|\phi(\mu)|}d\mu\Big| \\
&\leq \frac{1}{30}|\phi(\zeta)-\psi(\zeta)|,
\end{aligned} \tag{7.4.35}$$

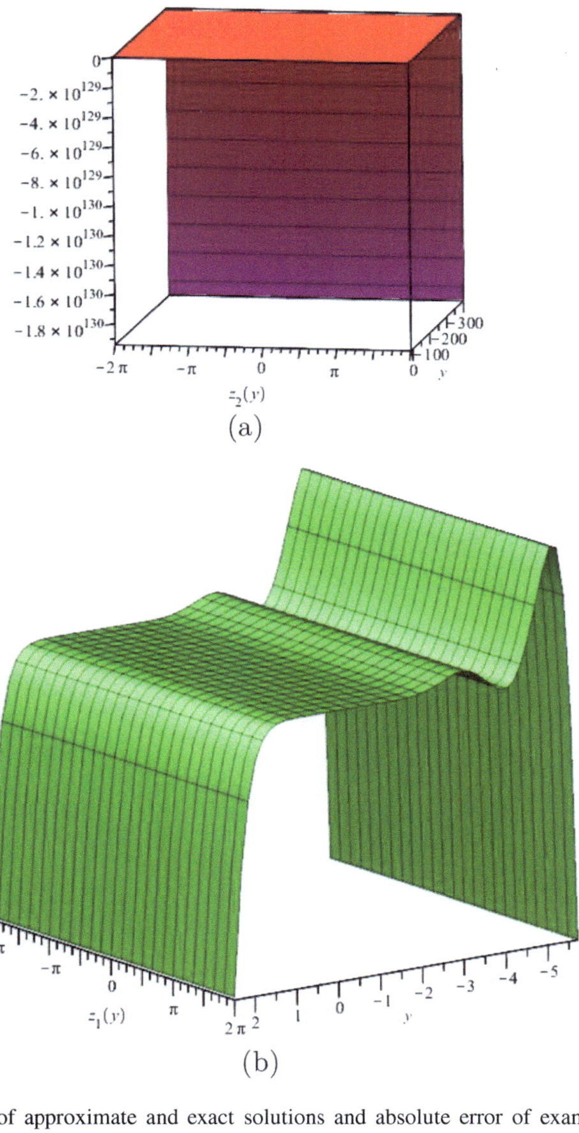

Fig. 7.15 Diagrams of approximate and exact solutions and absolute error of example 7.6 by selecting 60 points

$$\left|\frac{1}{20}\sqrt{4+|\phi(\zeta)|^2}\int_0^{\zeta}(\zeta-\mu)^{-\frac{-1}{6}}\exp(\frac{-3}{5}(\zeta-\mu))\frac{\epsilon(2+\mu)\sin|\phi(\mu)|}{4+|\phi(\mu)|}d\mu\right. \tag{7.4.36}$$

$$\left.-\frac{1}{20}\sqrt{4+|\psi(\zeta)|^2}\int_0^{\zeta}(\zeta-\mu)^{-\frac{-1}{6}}\exp(\frac{-3}{5}(\zeta-\mu))\frac{\epsilon(2+\mu)\sin|\psi(\mu)|}{4+|\psi(\mu)|}d\mu\right|$$

$$\leq\frac{1}{50}|\phi(\zeta)-\psi(\zeta)|,$$

$$\frac{1}{\frac{5}{8}^{\frac{5}{6}}\Gamma(\frac{5}{6})}\int_0^{\zeta}(\zeta-\mu)^{\frac{-1}{6}}\exp(\varsigma_2-1(\zeta-\mu))\left(MIN_{OPT}(\mathcal{R})\right)d\mu\leq\frac{1}{4}\left(MIN_{OPT}(\mathcal{R})\right), \tag{7.4.37}$$

where, $\mathcal{M}_1=\frac{1}{20},\ \mathcal{M}_2=\frac{1}{40},\ \mathcal{M}_3=\frac{1}{30},\ \mathcal{M}_4=\frac{1}{50},\ \mathcal{M}_5=\frac{1}{60}$. We assume that $\phi:[0,1]\to\mathbb{R}$ is a function such that for $\delta>0$ and $\zeta\in[0,1]$, applies in the following conditions

$$\left|{}^{C}_{m}\Xi D_{0^+}^{\frac{5}{6},\frac{5}{8}}\left(\frac{\phi(\zeta)-\sum_{r=1}^{r=4}\Pi I_{0^+}^{\frac{3r-2}{8},\frac{5}{8}}\frac{\cos\phi(\zeta)\zeta}{\zeta+150r}}{\frac{1}{20}\sqrt{4+|\phi(\zeta)|^2}}\right)-\frac{\epsilon(2+\zeta)\sin|\phi(\zeta)|}{4+|\phi(\zeta)|}\right|\leq\frac{1}{8}\left(MIN_{OPT}(\mathcal{R})\right), \tag{7.4.38}$$

$$\left|\mathcal{K}(\zeta,\phi(\zeta))-\left(\frac{1}{20}\sqrt{4+|\phi(\zeta)|^2}\left[\frac{\phi_0}{\mathcal{Y}(0,\phi_0)}(1-\zeta)\exp(\frac{-3}{5}\zeta)+\frac{\phi_F\zeta\exp(\frac{-3}{5}\zeta)}{\mathcal{Y}(F,\phi_F)\exp(\frac{-3}{5})}\right.\right.\right. \tag{7.4.39}$$

$$-\frac{\zeta\exp(\frac{-3}{5}\zeta)}{\mathcal{Y}(F,\phi_F)\exp(\frac{-3}{5})}\sum_{r=1}^{r=m}\frac{1}{\frac{5}{8}^{\frac{3i-2}{8}}\Gamma\left(\frac{3i-2}{8}\right)}\int_0^1(1-\mu)^{\frac{3i-2}{8}-1}\exp(\frac{-3}{5}(1-\mu))\frac{\cos\phi(\mu)\mu)}{\mu+150r}d\mu$$

$$-\frac{\zeta\exp(\frac{-3}{5}\zeta)}{\exp(\frac{-3}{5})}\frac{1}{\frac{5}{8}^{\frac{5}{6}}\Gamma(\frac{5}{6})}\int_0^1(1-\mu)^{-\frac{1}{6}}\exp(\frac{-3}{5}(1-\mu))\frac{\epsilon(2+\mu)\sin|\phi(\mu)|}{4+|\phi(\mu)|}d\mu$$

$$\left.+\frac{1}{\frac{5}{8}^{\frac{5}{6}}\Gamma(\frac{5}{6})}\int_0^{\zeta}(\zeta-\mu)^{-\frac{1}{6}}\exp(\frac{-3}{5}(\zeta-\mu))\frac{\epsilon(2+\zeta)\sin|\phi(\zeta)|}{4+|\phi(\zeta)|}d\mu\right]$$

$$\left.\left.+\sum_{r=1}^{m}\frac{1}{\varsigma_2^{\frac{3i-2}{8}}\Gamma\left(\frac{3i-2}{8}\right)}\int_0^{\zeta}(\zeta-\mu)^{\frac{3i-2}{8}-1}\exp(\frac{\varsigma_2-1}{\varsigma_2}(\zeta-\mu))\mathcal{W}_r(\mu,\phi(\mu))d\mu\right)\right|$$

$$\leq\frac{1}{8}\frac{1}{\frac{5}{8}^{\frac{5}{6}}\Gamma(\frac{5}{6})}\int_0^{\zeta}(\zeta-\mu)^{-\frac{1}{6}}\exp(-3(\zeta-\mu))\left(MIN_{OPT}(\mathcal{R})\right)d\mu$$

then, HPFIDE (7.4.32) has a unique solution $\psi:[0,1]\to\mathbb{R}$ such that

$$|\psi(\zeta)-\phi(\zeta)|\leq\frac{1}{8-8\upsilon}\left(MIN_{OPT}(\mathcal{R})\right),$$

where

$$\upsilon = \mathcal{M}_1 \frac{\psi_0}{\mathcal{Y}(0, \psi_0)} \left(1 - \frac{\zeta}{\mathrm{F}}\right) \exp(\frac{\varsigma_2 - 1}{\varsigma_2}\zeta) + \mathcal{M}_1 \frac{\psi_{\mathrm{F}} \frac{\zeta}{\mathrm{F}} \exp(\frac{\varsigma_3 - 1}{\varsigma_2}\zeta)}{\mathcal{Y}(\mathrm{F}, \psi_{\mathrm{F}}) \exp(\frac{\varsigma_3 - 1}{\varsigma_2}\mathrm{F})}$$
$$+ \frac{\zeta \exp(\frac{\varsigma_2 - 1}{\varsigma_2}\zeta)}{\mathcal{Y}(\mathrm{F}, \psi_{\mathrm{F}}) \mathrm{F} \exp(\frac{\varsigma_3 - 1}{\varsigma_2}\mathrm{F})} \sum_{r=1}^{r=m} \frac{1}{\varsigma_2^{\varsigma_{3r}} \Gamma(\varsigma_{3r})} \mathcal{M}_2 + \mathcal{M}_3 \frac{\zeta \exp(\frac{\varsigma_3 - 1}{\varsigma_2}\zeta)}{\mathrm{F} \exp(\frac{\varsigma_2 - 1}{\varsigma_2}\mathrm{F})} \frac{1}{\varsigma_2^{\varsigma_1} \Gamma(\varsigma_1)}$$
$$+ \mathcal{M}_4 \frac{1}{\varsigma_2^{\varsigma_1} \Gamma(\varsigma_1)} + \sum_{r=1}^{m} \frac{1}{\varsigma_2^{\varsigma_{3r}} \Gamma(\varsigma_{3r})} \int_0^\zeta (\zeta - \mu)^{\varsigma_{3r} - 1} \exp(\frac{\varsigma_2 - 1}{\varsigma_2}(\zeta - \mu)) \mathcal{M}_5 \varrho \, d\mu,$$

and $\upsilon = 0.657124821 < 1$. Now, we will implement the reproducing kernel Hilbert space method for (7.4.32). Therefore, we have

$$\mathcal{J} = {}_m^C \Xi_{D_{0^+}}^{\frac{5}{6}, \frac{5}{8}} \left(\frac{\phi(\zeta) - \sum_{r=1}^{r=4} \Pi_{I_{0^+}}^{\frac{3r-2}{8}, \frac{5}{8}} \frac{\cos\phi(\zeta)\zeta}{\zeta + 150r}}{\frac{1}{20}\sqrt{4 + |\phi(\zeta)|^2}} \right), \tag{7.4.40}$$

and

$$\mathcal{K}(\zeta, \phi(\zeta)) = \frac{\epsilon(2 + \zeta)\sin|\phi(\zeta)|}{4 + |\phi(\zeta)|}, \tag{7.4.41}$$

and

$$\mathcal{Q}_{2r}(\zeta) = \mathcal{J} \Lambda_{[rk]\eta}(\zeta)|_{\eta = \zeta_r} = {}_m^C \Xi_{D_{0^+}}^{\frac{5}{6}, \frac{5}{8}} \left(\frac{\Lambda_{\zeta_r}(\zeta) - \sum_{r=1}^{r=4} \Pi_{I_{0^+}}^{\frac{3r-2}{8}, \frac{5}{8}} \frac{\cos(\Lambda_{\zeta_r}(\zeta))\zeta}{\zeta + 150r}}{\frac{1}{20}\sqrt{4 + |\Lambda_{\zeta_r}(\zeta)|^2}} \right). \tag{7.4.42}$$

Here, in the interval [0, 1], we consider 15, 60, and 95 points, respectively, and using the introduced method, we obtain approximate numerical solutions. For the solutions obtained in each of these points, we calculate the absolute errors. Tables 7.12, 7.13 and 7.14 show all numerical results obtained for exact solutions, approximate solutions, and absolute errors (Fig. 7.16).

7.5 Navier–Stokes Equations

7.5.1 Introduction and Preliminaries

One of the simple and useful tools in mathematics to obtain solutions to many differential equations is fractional calculus. There is a class of mathematical modeling

Table 7.12 Exact solution, approximate solution and absolute error for Example 7.6 using 15 points

ζ	Exact solution	Approximate solution	Approximate solution
0.1	0.01024838364	0.01164548364	0.00139710000
0.2	0.02132564272	0.0257943685	0.00446872578
0.3	0.17599125413	0.19635721478	0.0203659607
0.4	0.19921237451	0.21025681527	0.0110444408
0.5	0.205598631164	0.2115635268	0.0059648956
0.6	0.257000186351	0.289000369573	0.0320001832
0.7	0.317221400000	0.365437500000	0.0482161000
0.8	0.388924536100	0.397001573410	0.0080770373
0.9	0.455245160370	0.470763894721	0.0155187343
1	0.491100000275	0.505178569316	0.0140785690

Table 7.13 Exact solution, approximate solution and absolute error for Example 7.6 using 60 points

ζ	Exact solution	Approximate solution	Approximate solution
0.1	0.03247084555	0.04658193455	0.01411108900
0.2	0.05452367100	0.06614758213	0.01162391113
0.3	0.06971289364	0.075223741587	0.00551084795
0.4	0.07141258369	0.079051478216	0.00763889453
0.5	0.08030479612	0.087475111357	0.00717031524
0.6	0.08568713549	0.089581746920	0.00389461143
0.7	0.08442100030	0.092458697125	0.00803769682
0.8	0.08976395721	0.099478963581	0.00971500637
0.9	0.10214007362	0.104596327100	0.0024562535
1	0.109971463827	0.127854126903	0.0178826631

Table 7.14 Exact solution, approximate solution and absolute error for Example 7.6 using 95 points

ζ	Exact solution	Approximate solution	Approximate solution
0.1	0.06657142301	0.07375472970	0.00718330669
0.2	0.06914523781	0.07991753624	0.01077229843
0.3	0.10156789413	0.11157368910	0.0100057950
0.4	0.11943692817	0.15067813281	0.0312412046
0.5	0.13652147369	0.18302016870	0.0464986950
0.6	0.15111200748	0.21314285364	0.0620308461
0.7	0.19423147569	0.24856931407	0.0543378384
0.8	0.22110036710	0.28136874501	0.0602683779
0.9	0.280005320144	0.30100268234	0.0209973622
1	0.310781267382	0.35615732815	0.0453760608

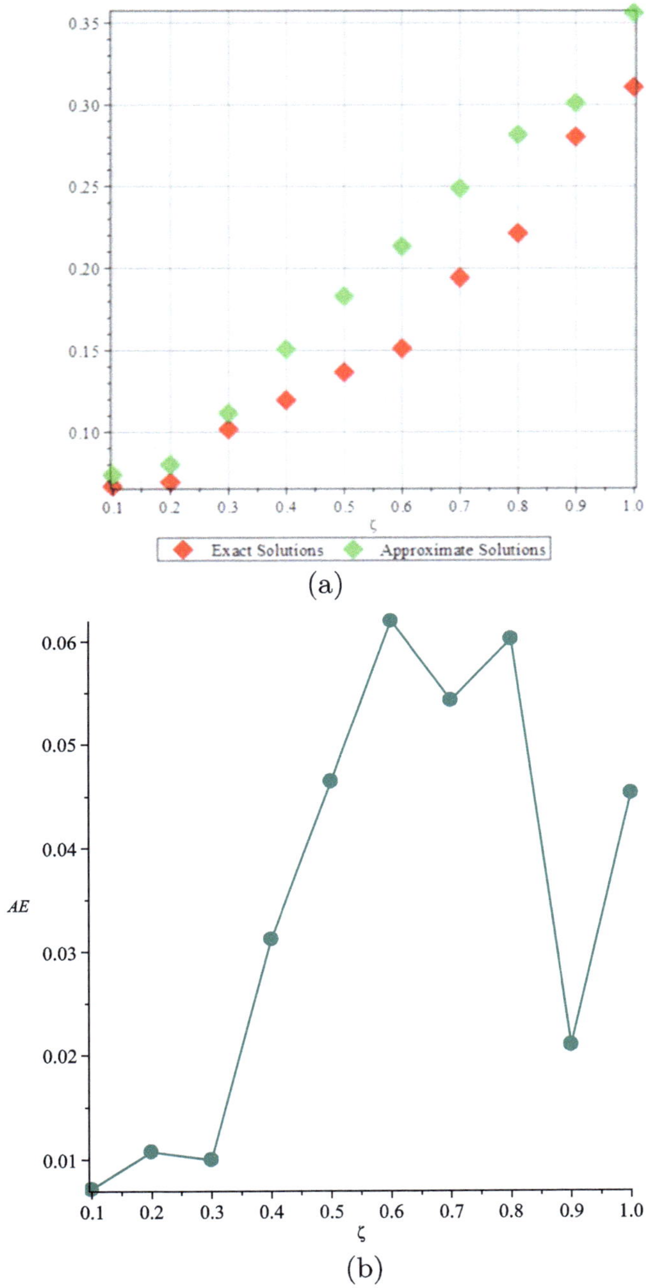

Fig. 7.16 Diagrams of approximate and exact solutions and absolute error of Example 7.6 by selecting 90 points

that includes partial differential equations (FPDEs) and fractional differential equations (FODEs), which are the most important equations. This class of equations is obtained by generalizing the integer-order to the fractional order [11, 41, 101]. In fractional calculations, the incorrect powers of differential equations are studied. Therefore, this field is one of the important scientific fields that has many applications. Examples include Hepatitis B disease model (HBDM), F- COVID-19 Model (F-C19M), viscoelastic materials (VM), electrical networks (ENs), control theory of dynamic systems (CTDS), control theory (CT), thermodynamics (TH), electro-magnetics (E-M), fluid flow (FF), probability and statistics (PS), etc. [19, 20, 62, 67, 76, 85, 86, 88, 97, 100, 142]. Researchers have done a lot of research to solve partial differential equations. Since solving these equations is a priority for researchers, and given the weaknesses in solving these equations, they have developed many analytical and numerical methods. Among the methods studied by scientists, we can name the methods of the exp-function method (E-FM), Laplace Adomian Decomposition Method (LADM), the functional variable (FV), Natural Transform Method (NTM), finite Hankel transform procedure (FHTP), Variational iteration method (VIM), modified auxiliary equation method (MAEM), the $(\frac{G}{G})$-expansion method ($(\frac{G}{G})$-EM), RBF collocation method (RBF-CM), generalized Kudryashov method (GKM) and so on [7, 40, 44, 63, 66, 77, 80, 94, 113, 114, 133, 144].

Navier–Stokes equations (N-SE), which are a combination of equations of motion and energy, are also referred to as Newton's second law of motion. These equations, first calculated in 1822, are the basic equations of motion [93]. Many physical phenomena are described using these equations. These phenomena include ocean currents, blood flow, liquid flow in pipes, airflow around the wings of aircraft, and weather changes. N-SEs are also used in fluid mechanics, which is a field of physics. The field of study in fluid mechanics, and physical mechanics of fluids includes plasma, gases, and liquids and the forces acting on them. They are also widely used in fields such as biology, geophysics, oceanography, astrophysics, and meteorology. Fluid statics and fluid mechanics are two classes of fluid mechanics, and this field can be divided into these two classes. Investigating fluids at rest introduces static fluids, and analyzing the effect of forces on moving fluids shows fluid dynamics. In the field of fluid mechanics, the flow of viscous fluids is expressed by Navier–Stokes equations. At first, Leonard Euler presented equations to explain the flow of frictionless and incompressible fluids, and the Navier–Stokes equations can be considered a generalization of these equations. These equations are also used in the study and modeling of magnetohydrodynamics, analysis of dense liquids, pollution investigation, and power design. As it was said, N-SEs are very important in various problems of physics, mathematical, chemistry, and pure mathematics. Therefore, researchers have always been looking for solutions to solve these equations. For example, these equations were solved in 2055 [31]. Ragab et al., Chaurasia and Kumar have solved these equations using the method of homotopic analysis and the combining Laplace and Finite Hankel transforms, respectively [35, 46, 105]. N-SEs of nonlinear type have also been studied by Ganji et al. [118]. Researchers have also used other methods to solve these equations, see [92, 106] for details. Here, we have presented a method to obtain approximate solutions to this class of equations using the clique polynomial technique. The solutions obtained from the presented method provide the best approximation compared to other methods of solving these equations. We have performed calculations on various errors for this method and other methods and according to the results presented in the table, we see that the clique polynomial technique has less error than other methods, and this shows the effectiveness of the proposed method. We study a generalized form of Navier–Stokes equation (N-SE) as follows:

$$D_{\lambda}^{\mathrm{m}}\phi(\mu,\lambda)=\mathcal{K}(\phi)+\mathcal{Z}(\phi), \tag{7.5.1}$$

with the initial condition:

$$\phi(\mu,0)=\mathrm{g}(\mu)$$

where, $\mathcal{K}(\phi)$ and $\mathcal{Z}(\phi)$ are linear and nonlinear terms respectively.

Our goal is to obtain the numerical solutions of Eq. (7.5.1) using the exact method based on clique polynomials.

Definition 7.12 ([102]) The m-order Riemann–Liouville fractional integral (R-LFI) is defined by

$$^{RL}\mathcal{I}_{\alpha}^{\mathrm{m}}\phi(\alpha)=\frac{1}{\Gamma(\mathrm{m})}\int_{0}^{\alpha}(\alpha-\kappa)^{\mathrm{m}-1}\phi(\kappa)d\kappa.$$

Definition 7.13 ([102]) The m-order Caputo derivative (CD) for $\phi\in C[i,j]$ is given by

$$^{C}D_{\alpha}^{\mathrm{m}}\phi(\alpha)=\begin{cases}\frac{1}{\Gamma(\mathrm{n}-\mathrm{m})}\int_{i}^{\alpha}(\alpha-\kappa)^{\mathrm{n}-\mathrm{m}-1}\phi^{\ell}(\kappa)d\kappa, & \mathrm{n}-1<\mathrm{m}<\mathrm{n}\\ \frac{d^{\mathrm{n}}\phi(\alpha)}{d\alpha^{\mathrm{n}}}, & \mathrm{m}=\mathrm{n}\in\mathbb{N}\end{cases}$$

The clique polynomial of a graph $\mathcal{N}$, which denotes by $\Xi(\mathcal{N},\lambda)$, is given by [54]

$$\mathcal{W}(\mathcal{N},\lambda)=\sum_{\mathfrak{e}=0}^{\jmath}\mathfrak{p}_{\mathfrak{e}}(\mathcal{N})\lambda^{\mathfrak{e}},\quad \mathfrak{p}_{0}(\mathcal{N})=1$$

where $\mathfrak{p}_{\mathfrak{e}}(\mathcal{N})$ is the number of $\mathfrak{e}$-cliques of $\mathcal{N}$. The CP of a complete graph $\mathcal{S}_{\jmath}$ with $\jmath$ vertices is given by [70]

$$\mathcal{W}\left(\mathcal{S}_{\jmath},\lambda\right)=(1+\lambda)^{\jmath}.$$

We can approximate any function $\phi(\lambda)$ in $L^2(0,\mathrm{R})$ in terms of the CP as

$$\phi(\lambda) \simeq \tilde{\phi}(\lambda) = \sum_{\jmath=0}^{\mathcal{L}-1} \phi_\jmath \mathcal{W}\left(\mathcal{S}_\jmath, \lambda\right) = \Phi^T \Xi_{\mathcal{L}}(\lambda), \tag{7.5.2}$$

where $\Phi = \left[\phi_0, \phi_1, \ldots, \phi_{\mathcal{L}-1}\right]^T$, and

$$\Xi_{\mathcal{L}}(\lambda) = \left[\mathcal{W}\left(\mathcal{S}_0, \lambda\right), \mathcal{W}\left(\mathcal{S}_1, \lambda\right), \ldots, \mathcal{W}\left(\mathcal{S}_{\mathcal{L}-1}, \lambda\right)\right]^T. \tag{7.5.3}$$

From (7.5.2), we have $\Phi = \mathcal{G}^{-1}\langle\Phi(\lambda), \Xi(\lambda)\rangle$, where $\mathcal{G} = \langle\Xi(\lambda), \Xi(\lambda)\rangle$ and $\langle\cdot,\cdot\rangle$ indicates the inner product of two arbitrary functions. Also, an arbitrary function $\Phi(\mu, \lambda) \in L^2([i, j] \times [0, \mathrm{R}])$ can be expanded in terms of the CP by the following infinite series

$$\Phi(\mu, \lambda) = \sum_{\jmath=0}^{\infty}\sum_{\iota=0}^{\infty} \Phi_{\jmath\iota} \mathcal{W}\left(\mathcal{S}_\jmath, \mu\right) \mathcal{W}\left(\mathcal{S}_j, \lambda\right). \tag{7.5.4}$$

We consider the first $\mathcal{L}$ terms in (7.5.4), as an approximation of the function $\Phi(\mu, \lambda)$. It is obtained as

$$\Phi(\mu, \lambda) \simeq \sum_{\jmath=0}^{\mathcal{L}-1}\sum_{\iota=0}^{\mathcal{L}-1} \Phi_{\jmath\iota} \mathcal{W}\left(\mathcal{S}_\jmath, \mu\right) \mathcal{W}\left(\mathcal{S}_\iota, \lambda\right) = \Xi^T(\mu)\Phi\Xi(\lambda),$$

where $\Phi = \mathcal{G}^{-1}\langle\Xi(i), \langle\phi(\mu, \lambda), \Xi(\lambda)\rangle\rangle\mathcal{G}^{-1}$. Also, for function $\phi(\mu, \sigma, \lambda) \in L^2$ $([i, j] \times [i, j] \times [0, \mathrm{R}])$, we have

$$\phi(\mu, \sigma, \lambda) = \sum_{\jmath=0}^{\infty}\sum_{\iota=0}^{\infty}\sum_{\epsilon=0}^{\infty} \Psi_{\jmath\iota\epsilon} \mathcal{W}\left(\mathcal{S}_\jmath, \mu\right) \mathcal{W}\left(\mathcal{S}_\epsilon, \sigma\right) \mathcal{W}\left(\mathcal{S}_\iota, \lambda\right).$$

Since in the process of solving differential equations using the clique polynomials technique, assume that $\Xi(\lambda)$ is the matrix defined in (7.5.3). This vector is written based on Taylor functions as follows

$$\Xi(\lambda) = \mathcal{G}\mathcal{Y}_{\mathcal{L}}(\lambda), \tag{7.5.5}$$

where $\mathcal{G}$ is an invertible matrix defined as follows

$$\mathcal{G} = [g_{ij}], \qquad i, j = 0, 1, \ldots, \mathcal{L} - 1, \qquad g_{ij} = \begin{cases} \binom{i}{j}, & i \geq j \\ 0, & i < j \end{cases}$$

Also, $\mathcal{Y}_{\mathcal{L}}(\lambda)$ is the vector of Taylor's basic functions as follows

$$\mathcal{Y}_{\mathcal{L}}(\lambda) = \begin{bmatrix} 1 \\ \lambda \\ \lambda^2 \\ \vdots \\ \lambda^{\mathcal{L}-1} \end{bmatrix}. \tag{7.5.6}$$

By taking the integral from Eq. (7.5.5), we have

$$\begin{aligned} \int_0^{\lambda} \varXi(t)dt &= \int_0^{\lambda} \mathcal{G}\mathcal{Y}_{\mathcal{L}}(t)dt \\ &= \mathcal{G}\int_0^{\lambda} \mathcal{Y}_{\mathcal{L}}(t)dt \\ &= \varXi_{\mathcal{L}}\bar{\mathcal{Y}}(\lambda) \end{aligned} \tag{7.5.7}$$

where

$$\varXi = \mathcal{G}\begin{bmatrix} 1 & 0 & \cdots & 0 \\ 0 & \frac{1}{2} & \cdots & 0 \\ \vdots & \vdots & \ddots & \vdots \\ 0 & 0 & \cdots & 0 \end{bmatrix} \tag{7.5.8}$$

and

$$\bar{\mathcal{Y}}(\lambda) = \begin{bmatrix} \lambda \\ \lambda^2 \\ \lambda^3 \\ \vdots \\ \lambda^{\mathcal{L}} \end{bmatrix}. \tag{7.5.9}$$

Now, we approximate the λ^x functions based on the clique base and using Eq. (7.5.5) as follows

$$\lambda^x \simeq G^{-1}_{[x+1]}\varXi(\lambda), x = 1, 2, \ldots, \mathcal{L} - 1,$$

Where $\mathcal{G}^{-1}_{[x+1]}$ is the $(x+1)$th row of the matrix $\mathcal{G}^{-1}$. Now, we approximate the function $\lambda^{\mathcal{L}}$ based on the clique base as follows

$$\lambda^{\mathcal{L}} \simeq v_{\mathcal{L}}^T\varXi(\lambda),$$

where $v_{\mathcal{L}}$ is calculated as follows

$$v_{\mathcal{L}} = Q^{-1} \int_0^1 \lambda^{\mathcal{L}} \Xi(\lambda) d\lambda$$

that $Q =< \Xi(\lambda), \Xi(\lambda) >$. We suppose that

$$\mathrm{H} = \begin{bmatrix} \mathcal{G}_{[2]}^{-1} \\ \mathcal{G}_{[3]}^{-1} \\ \mathcal{G}_{[4]}^{-1} \\ \vdots \\ \mathcal{G}_{[\mathcal{L}]}^{-1} \\ v_{\mathcal{L}}^{T} \end{bmatrix}, \tag{7.5.10}$$

then, $\bar{\mathrm{Y}}(\lambda) \simeq \mathrm{H}\Xi(\lambda)$. Therefore, the integral operating matrix based on the clique base is obtained as follows

$$\int_0^{\lambda} \Xi_{\mathcal{L}}(t)dt \simeq \Upsilon \Xi_{\mathcal{L}}(\lambda), \tag{7.5.11}$$

that $\Upsilon = \Xi\mathrm{H}$ the operational matrix is called the clique base integral.

By taking twice the integral of Eq. (7.5.3), we have

$$\begin{aligned} \int_0^{\lambda} \int_0^{\lambda} \Xi_{\mathcal{L}}(t)dtdt &= \int_0^{\lambda} P \Xi_{\mathcal{L}}(t)dt \\ &= \Upsilon \int_0^{\lambda} \Xi_{\mathcal{L}}(t)dt \\ &= \Upsilon^2 \Xi_{\mathcal{L}}(\lambda). \end{aligned} \tag{7.5.12}$$

Theorem 7.5 *The* m*-order RL-fractional integral (R-LFI) of the given vector* $\Xi_{\mathcal{L}}(\lambda)$ *in* (7.5.3) *can be approximated by*

$$^{RL}I_{\lambda}^{\mathrm{m}} \approx \mathcal{R}^{\mathrm{m}} \Xi_{\mathcal{L}}(\lambda)$$

where $\mathcal{R}^{\mathrm{m}}$ *is called the operational matrix (OM) based on the CP, which is given by*

$$\mathcal{R}^{\mathrm{m}} = \begin{bmatrix} \mathrm{r}_{0,0,0} & \mathrm{r}_{0,1,0} & \cdots & \mathrm{r}_{0,\mathcal{L},0} \\ \sum_{v=0}^{1} \mathrm{r}_{1,0,v} & \sum_{v=0}^{1} \mathrm{r}_{1,1,v} & \cdots & \sum_{v=0}^{1} \mathrm{r}_{1,\mathcal{L}-1,v} \\ \vdots & \vdots & \ddots & \vdots \\ \sum_{v=0}^{\mathcal{L}-1} \mathrm{r}_{\mathcal{L}-1,0,v} & \sum_{v=0}^{\mathcal{L}-1} \mathrm{r}_{\mathcal{L}-1,1,v} & \cdots & \sum_{v=0}^{\mathcal{L}-1} \mathrm{r}_{\mathcal{L}-1,\mathcal{L}-1,v} \end{bmatrix},$$

with

$$\mathrm{r}_{i,j,\mathrm{v}} = \frac{\Gamma(i+1)e_{\mathrm{v},j}}{\Gamma(i-\mathrm{v}+1)\Gamma(\mathrm{v}+\mathrm{m}+1)}.$$

Proof See [36].

Consider the following set

$$\mathrm{U} = span\{\mathcal{W}(\mathcal{S}_0, \mu)\mathcal{W}(\mathcal{S}_0, \sigma)\mathcal{W}(\mathcal{S}_0, \lambda), \cdots, \mathcal{W}(\mathcal{S}_0, \mu)\mathcal{W}(\mathcal{S}_0, \sigma)\mathcal{W}(\mathcal{S}_{\mathcal{L}-1}, \lambda), \cdots, \mathcal{W}(\mathcal{S}_{\mathcal{L}-1}, \mu)\mathcal{W}(\mathcal{S}_{\mathcal{L}-1}, \sigma)\mathcal{W}(\mathcal{S}_{\mathcal{L}-1}, \lambda)\}. \tag{7.5.13}$$

The set U is considered to be the space of all Three-variable polynomials. If ϱ is the degree of polynomials on μ, σ, λ, then $\varrho \leq \mathcal{L}$. The best approximation for the function $\phi(\mu, \sigma, \lambda) \in \mathrm{J}^*$ to U is the function $\tilde{\phi}(\mu, \sigma, \lambda)$.

In the next theorem, we obtain the error of approximation of the three-variable function obtained using clique polynomials.

Theorem 7.6 *Suppose $\phi(\mu, \sigma, \lambda)$ be a sufficiently smooth function on $\mathrm{J} = [i, j] \times [i, j] \times [0, \mathrm{R}]$ and $\tilde{\phi}(\mu, \sigma, \lambda)$ be its approximation using clique polynomials. Then, we have*

$$\|\phi(\mu, \sigma, \lambda) - \tilde{\phi}(\mu, \sigma, \lambda)\|_2 \leq \left(\frac{\Delta_1 (j-i)^{\mathcal{L}}}{\mathcal{L}! 2^{2\mathcal{L}-1}} + \frac{\Delta_2 (j-i)^{\mathcal{L}}}{\mathcal{L}! 2^{2\mathcal{L}-1}} + \frac{\Delta_3 (\mathcal{F})^{\mathcal{L}}}{\mathcal{L}! 2^{2\mathcal{L}-1}} + \frac{\Delta_4 (j-i)^{2\mathcal{L}} \mathcal{F}^{\mathcal{L}}}{(\mathcal{L}!)^3 2^{6\mathcal{L}-3}}\right) (j-i)\mathcal{F}. \tag{7.5.14}$$

Proof Let $\mu_\jmath, \sigma_\iota, \jmath, \iota = 0, 1, \ldots, \mathcal{L}-1$ and $\lambda_\epsilon, \epsilon = 0, 1, \ldots, \mathcal{L}-1$ are the roots of $\mathcal{L}$-degree shifted Chebyshev polynomials on [0, 1]. By considering the interpolating polynomial $\psi_{\mathcal{L}}$ to $\phi(\mu, \sigma, \lambda)$ at points $(\mu_\jmath, \sigma_\iota, \lambda_\epsilon)$, we have

$$\begin{aligned}\phi(\mu, \sigma, \lambda) - \psi_{\mathcal{L}}(\mu, \sigma, \lambda) = {} & \frac{\partial^{\mathcal{L}} \phi(\rho, \sigma, \lambda)}{\partial \mu^{\mathcal{L}} \mathcal{L}!} \prod_{\jmath=0}^{\mathcal{L}-1} (\mu - \mu_\jmath) + \frac{\partial^{\mathcal{L}} \Psi(\mu, \varsigma, \lambda)}{\partial \sigma^{\mathcal{L}} \mathcal{L}!} \prod_{\iota=0}^{\mathcal{L}-1} (\sigma - \sigma_\iota) + \frac{\partial^{\mathcal{L}} \Psi(\mu, \sigma, \gamma)}{\partial \lambda^{\mathcal{L}} \mathcal{L}!} \prod_{\epsilon=0}^{\mathcal{L}-1} (\lambda - \lambda_\epsilon) \\ & - \frac{\partial^{3\mathcal{L}} \Psi(\rho', \varsigma', \gamma')}{\partial \mu^{\mathcal{L}} \partial \sigma^{\mathcal{L}} \partial \lambda^{\mathcal{L}} (\mathcal{L}!)^3} \prod_{\jmath=0}^{\mathcal{L}-1} (\mu - \mu_\jmath) \prod_{\iota=0}^{\mathcal{L}-1} (\sigma - \sigma_\iota) \prod_{\epsilon=0}^{\mathcal{L}-1} (\lambda - \lambda_\epsilon),\end{aligned} \tag{7.5.15}$$

where $\rho, \rho' \in [i, j]$, $\varsigma, \varsigma' \in [i, j]$ and $\gamma, \gamma' \in [0, \mathcal{F}]$. Then we obtain

$$\begin{aligned}|\phi(\mu, \sigma, \lambda) - \psi_{\mathcal{L}}(\mu, \sigma, \lambda)| \leqslant {} & \max_{(\mu,\sigma,\lambda)\in\mathcal{J}} \left|\frac{\partial^{\mathcal{L}} \phi(\rho, \sigma, \lambda)}{\partial \mu^{\mathcal{L}}}\right| \frac{\prod_{\jmath=0}^{\mathcal{L}-1} |\mu - \mu_\jmath|}{\mathcal{L}!} \\ & + \max_{(\mu,\sigma,\lambda)\in\mathcal{J}} \left|\frac{\partial^{\mathcal{L}} \phi(\mu, \sigma, \lambda)}{\partial \sigma^{\mathcal{L}}}\right| \frac{\prod_{\iota=0}^{\mathcal{L}-1} |\sigma - \sigma_\iota|}{\mathcal{L}!} + \max_{(\mu,\sigma,\gamma)\in\mathcal{F}} \left|\frac{\partial \phi(\mu, \sigma, \gamma)^{\mathcal{L}}}{\partial \lambda^{\mathcal{L}}}\right| \frac{\prod_{\epsilon=0}^{\mathcal{L}-1} |\lambda - \lambda_\epsilon|}{\mathcal{L}!} \\ & + \max_{(\mu,\sigma,\lambda)\in\mathcal{F}} \left|\frac{\partial^{3\mathcal{L}} \phi(\rho', \varsigma', \gamma')}{\partial \mu^{\mathcal{L}} \partial \sigma^{\mathcal{L}} \partial \lambda^{\mathcal{L}}}\right| \frac{\prod_{\jmath=0}^{\mathcal{L}-1} |\mu - \mu_\jmath| \prod_{\jmath=0}^{\mathcal{L}-1} |\sigma - \sigma_\iota| \prod_{\lambda\mathcal{L}}^{\mathcal{L}-1} |\lambda - \lambda_\epsilon|}{(\mathcal{L}!)^3}.\end{aligned} \tag{7.5.16}$$

According to the smooth function $\phi(\mu, \sigma, \lambda)$ on J^*, there are constants $\Delta_1, \Delta_2, \Delta_3, \Delta_4$ such that

$$\max_{(\mu,\sigma,\lambda)\in\mathrm{J}} \left|\frac{\partial^{\mathrm{J}} \phi(\mu, \sigma, \lambda)}{\partial \mu^{\mathcal{L}}}\right| \leq \Delta_1, \quad \max_{(\mu,\sigma,\lambda)\in\mathrm{J}} \left|\frac{\partial^{\mathcal{L}} \phi(\mu, \sigma, \lambda)}{\partial \sigma^{\mathcal{L}}}\right| \leq \Delta_2, \quad \max_{(\mu,\sigma,\lambda)\in\mathrm{J}} \left|\frac{\partial \phi(\mu, \sigma, \gamma)^{\mathcal{L}}}{\partial \lambda^{\mathcal{L}}}\right| \leq \Delta_3, \tag{7.5.17}$$

$$\max_{(\mu,\sigma,\lambda)\in\mathrm{J}} \left|\frac{\partial^{3\mathcal{L}} \phi(\Delta', \varsigma', \gamma')}{\partial \mu^{\mathcal{L}} \partial \sigma^{\mathcal{L}} \partial \lambda^{\mathcal{L}}}\right| \leq \Delta_4.$$

By substituting (7.5.17) into (7.5.16) and employing the estimates for Chebyshev interpolation nodes gives

$$|\phi(\mu,\sigma,\lambda)-\psi_{\mathcal{L}}(\mu,\sigma,\lambda)|\leqslant\frac{\Delta_1(j-i)^{\mathcal{L}}}{\mathcal{L}!2^{2\mathcal{L}-1}}+\frac{\Delta_2(j-i)^{\mathcal{L}}}{\mathcal{L}!2^{2\mathcal{L}-1}}+\frac{\Delta_3(\mathcal{F})^{\mathcal{L}}}{\mathcal{L}!2^{2\mathcal{L}-1}}+\frac{\Delta_4(j-i)^{2\mathcal{L}}\mathcal{F}^{\mathcal{L}}}{(\mathcal{L}!)^3 2^{6\mathcal{L}-3}}. \tag{7.5.18}$$

Since $\tilde{\phi}(\mu,\sigma,\lambda)$ is the best approximation of $\phi(\mu,\sigma,\lambda)$ in U and using (7.5.18), we can write

$$\begin{aligned}
\left\|\phi(\mu,\sigma,\lambda)-\tilde{\phi}(\mu,\sigma,\lambda)\right\|_2^2 &= \int_0^{\mathcal{F}}\int_i^j\int_i^j |\phi(\mu,\sigma,\lambda)-\tilde{\phi}(\mu,\sigma,\lambda)|^2 d\mu d\sigma d\lambda \\
&\leqslant \int_0^{\mathcal{F}}\int_i^j\int_i^j \left|\phi(\mu,\sigma,\lambda)-\psi_{\mathcal{L}}(\mu,\sigma,\lambda)\right|^2 d\mu d\sigma d\lambda \\
&= \int_0^{\mathcal{F}}\int_i^j\int_i^j \left(\frac{\Delta_1(j-i)^{\mathcal{L}}}{\mathcal{L}!2^{2\mathcal{L}-1}}+\frac{\Delta_2(j-i)^{\mathcal{L}}}{\mathcal{L}!2^{2\mathcal{L}-1}}+\frac{\Delta_3(\mathcal{F})^{\mathcal{L}}}{\mathcal{L}!2^{2\mathcal{L}-1}}+\frac{\Delta_4(j-i)^{2\mathcal{L}}\mathcal{F}^{\mathcal{L}}}{(\mathcal{L}!)^3 2^{6\mathcal{L}-3}}\right)^2 d\mu d\sigma d\lambda \\
&= \left(\frac{\Delta_1(j-i)^{\mathcal{L}}}{\mathcal{L}!2^{2\mathcal{L}-1}}+\frac{\Delta_2(j-i)^{\mathcal{L}}}{\mathcal{L}!2^{2\mathcal{L}-1}}+\frac{\Delta_3(\mathcal{F})^{\mathcal{L}}}{\mathcal{L}!2^{2\mathcal{L}-1}}+\frac{\Delta_4(j-i)^{2\mathcal{L}}\mathcal{F}^{\mathcal{L}}}{(\mathcal{L}!)^3 2^{6\mathcal{L}-3}}\right)^2 (j-i)^2\mathcal{F},
\end{aligned} \tag{7.5.19}$$

which gives the desired result. Note that for the special case, when $i=0, j=1$ and $\mathcal{F}=1$, we have

$$\left\|\phi(\mu,\sigma,\lambda)-\tilde{\phi}(\mu,\sigma,\lambda)\right\|_2\leqslant\frac{\Delta_1}{\mathcal{L}!2^{2\mathcal{L}-1}}+\frac{\Delta_2}{\mathcal{L}!2^{2\mathcal{L}-1}}+\frac{\Delta_3}{\mathcal{L}!2^{2\mathcal{L}-1}}+\frac{\Delta_4}{(\mathcal{L}!)^3 2^{6\mathcal{L}-3}}, \tag{7.5.20}$$

therefore

$$\left\|\phi(\mu,\sigma,\lambda)-\tilde{\phi}(\mu,\sigma,\lambda)\right\|_2 = O\left(\frac{1}{\mathcal{L}!2^{2\mathcal{L}-1}}\right).$$

7.5.2 The Clique Polynomial Method for the Navier–Stokes Equation (N-SE) of Integer Order

We will present a numerical method for solving the Navier–Stokes equations with two and three dimensions having the initial conditions using a clique-based operating matrix. For this purpose, we consider the following NS equation with the initial condition

$$D_\lambda^{\mathrm{m}}\phi(\mu,\lambda)=\mathcal{K}(\phi)+\mathcal{Z}(\phi), \tag{7.5.21}$$

such that m = 3. Assume that,

$$\frac{\partial^5 \phi(\mu, \lambda)}{\partial \mu^2 \partial \lambda^3} = \sum_{\jmath=0}^{\infty} \sum_{\jmath=0}^{\infty} \phi_{\jmath} \iota \mathcal{W}\left(\mathcal{S}_{\jmath}, \mu\right) \mathcal{W}\left(\mathcal{S}_{\iota}, \lambda\right) = \Xi_{\mathcal{L}}(\lambda)^T \Phi \Xi_{\mathcal{L}}(\mu), \tag{7.5.22}$$

that

$$\Xi_{\mathcal{L}}(\lambda)^T = \left[\mathcal{W}\left(\mathcal{S}_0, \lambda\right), \mathcal{W}\left(\mathcal{S}_1, \lambda\right), \mathcal{W}\left(\mathcal{S}_2, \lambda\right), \mathcal{W}\left(\mathcal{S}_3, \lambda\right), \ldots, \mathcal{W}\left(\mathcal{S}_{\mathcal{L}-1}, \lambda\right)\right]$$
$$\Xi_{\mathcal{L}}(\mu) = \left[\mathcal{W}\left(\mathcal{S}_0, \mu\right), \mathcal{W}\left(\mathcal{S}_1, \mu\right), \mathcal{W}\left(\mathcal{S}_2, \mu\right), \mathcal{W}\left(\mathcal{S}_3, \mu\right), \ldots, \mathcal{W}\left(\mathcal{S}_{\mathcal{L}-1}, \mu\right)\right]^T.$$

Integrate Eq (7.5.22) regarding μ from 0 to μ,

$$\begin{aligned} \frac{\partial^4 \phi(\mu, \lambda)}{\partial \mu \partial \lambda^3} &= \frac{\partial^4 \phi(0, \lambda)}{\partial \mu \partial \lambda^3} + \int_0^{\mu} \Xi_{\mathcal{L}}(\lambda)^T \Phi \Xi_{\mathcal{L}}(\mu) d\mu \\ &= \frac{\partial^4 \phi(0, \lambda)}{\partial \mu \partial \lambda^3} + \Xi_{\mathcal{L}}(\lambda)^T \Phi \int_0^{\mu} \Xi_{\mathcal{L}}(\mu) d\mu \\ &= \frac{\partial^4 \phi(0, \lambda)}{\partial \mu \partial \lambda^3} + \Xi_{\mathcal{L}}(\lambda)^T \Phi [\Upsilon \Xi_{\mathcal{L}}(\mu)]. \end{aligned} \tag{7.5.23}$$

Integrating Eq (7.5.23) regarding μ from 0 to μ,

$$\begin{aligned} \frac{\partial^3 \phi(\mu, \lambda)}{\partial \lambda^3} &= \frac{\partial^3 \phi(0, \lambda)}{\partial \lambda^3} + \mu[\frac{\partial^4 \phi(0, \lambda)}{\partial \mu \partial \lambda^3}] + \int_0^{\mu} \Xi_{\mathcal{L}}(\lambda)^T \Phi [\Upsilon \Xi_{\mathcal{L}}(\mu)] \\ &= \frac{\partial^3 \phi(0, \lambda)}{\partial \lambda^3} + \mu[\frac{\partial^4 \phi(0, \lambda)}{\partial \mu \partial \lambda^3}] + \Xi_{\mathcal{L}}(\lambda)^T \Phi \int_0^{\mu} [\Upsilon \Xi_{\mathcal{L}}(\mu)] \\ &= \frac{\partial^3 \phi(0, \lambda)}{\partial \lambda^3} + \mu[\frac{\partial^4 \phi(0, \lambda)}{\partial \mu \partial \lambda^3}] + \Xi_{\mathcal{L}}(\lambda)^T \Phi [\Upsilon^2 \Xi_{\mathcal{L}}(\mu)]. \end{aligned} \tag{7.5.24}$$

Now, we get an integral from Eq. (7.5.24) regarding λ from 0 to λ,

$$\frac{\partial^2 \phi(\mu, \lambda)}{\partial \lambda^2} = \frac{\partial^2 \phi(\mu, 0)}{\partial \lambda^2} + \frac{\partial^2 \phi(0, \lambda)}{\partial \lambda^2} - \frac{\partial^2 \phi(0, 0)}{\partial \lambda^2} + \mu\left[\frac{\partial^3 \phi(0, \lambda)}{\partial \mu \partial \lambda^2} - \frac{\partial^3 \phi(0, 0)}{\partial \mu \partial \lambda^2}\right] \tag{7.5.25}$$

$$\begin{aligned} &+ \int_0^{\lambda} \Xi_{\mathcal{L}}(\lambda)^T \Phi [\Upsilon^2 \Xi_{\mathcal{L}}(\mu)] d\lambda \\ &= \frac{\partial^2 \phi(\mu, 0)}{\partial \lambda^2} + \frac{\partial^2 \phi(0, \lambda)}{\partial \lambda^2} - \frac{\partial^2 \phi(0, 0)}{\partial \lambda^2} \\ &+ \mu\left[\frac{\partial^3 \phi(0, \lambda)}{\partial \mu \partial \lambda^2} - \frac{\partial^3 \phi(0, 0)}{\partial \mu \partial \lambda^2}\right] + \Phi [\Upsilon^2 \Xi_{\mathcal{L}}(\mu)] \int_0^{\lambda} \Xi_{\mathcal{L}}(\lambda)^T d\lambda \\ &= \frac{\partial^2 \phi(\mu, 0)}{\partial \lambda^2} \end{aligned}$$

$$+\frac{\partial^2\phi(0,\lambda)}{\partial\lambda^2}-\frac{\partial^2\phi(0,0)}{\partial\lambda^2}+\mu\Big[\frac{\partial^3\phi(0,\lambda)}{\partial\mu\partial\lambda^2}-\frac{\partial^3\phi(0,0)}{\partial\mu\partial\lambda^2}\Big]$$
$$+\Phi[\Upsilon^2\Xi_{\mathcal{L}}(\mu)][\Upsilon\Xi_{\mathcal{L}}(\lambda)]^T.$$

Again, by taking the integral from Eq. (7.5.25) from 0 to λ, with regard to λ, we have

$$\begin{aligned}\frac{\partial\phi(\mu,\lambda)}{\partial\lambda}&=\frac{\partial\phi(\mu,0)}{\partial\lambda}+\frac{\partial\phi(0,\lambda)}{\partial\lambda}-\frac{\partial\phi(0,0)}{\partial\lambda}+\lambda\Big[\frac{\partial^2\phi(\mu,0)}{\partial\lambda^2}-\frac{\partial^2\phi(0,0)}{\partial\lambda^2}\Big]\\&+\mu\Big[\frac{\partial^2\phi(0,t)}{\partial\mu\partial\lambda}-\frac{\partial^2\phi(0,0)}{\partial\mu\partial\lambda}-\lambda\frac{\partial^3\phi(0,0)}{\partial\mu\partial\lambda}]\Big]+\int_0^\lambda\Phi[\Upsilon^2\Xi_{\mathcal{L}}(\mu)][\Upsilon\Xi_{\mathcal{L}}(\lambda)]^T d\lambda\\&=\frac{\partial\phi(\mu,0)}{\partial\lambda}+\frac{\partial\phi(0,\lambda)}{\partial\lambda}-\frac{\partial\phi(0,0)}{\partial\lambda}+\lambda\Big[\frac{\partial^2\phi(\mu,0)}{\partial\lambda^2}-\frac{\partial^2\phi(0,0)}{\partial\lambda^2}\Big]\\&+\mu\Big[\frac{\partial^2\phi(0,\lambda)}{\partial\mu\partial\lambda}-\frac{\partial^2\phi(0,0)}{\partial\mu\partial\lambda}-\lambda\frac{\partial^3\phi(0,0)}{\partial\mu\partial\lambda}]\Big]+\Phi[\Upsilon^2\Xi_{\mathcal{L}}(\xi)]\int_0^\lambda[\Upsilon\Xi_{\mathcal{L}}(\lambda)]^T d\lambda\\&=\frac{\partial\phi(\mu,0)}{\partial\lambda}+\frac{\partial\phi(0,\lambda)}{\partial\lambda}-\frac{\partial\phi(0,0)}{\partial\lambda}+\lambda\Big[\frac{\partial^2\phi(\mu,0)}{\partial\lambda^2}-\frac{\partial^2\phi(0,0)}{\partial\lambda^2}\Big]\\&+\mu\Big[\frac{\partial^2\phi(0,\lambda)}{\partial\mu\partial\lambda}-\frac{\partial^2\phi(0,0)}{\partial\mu\partial\lambda}-\lambda\frac{\partial^3\phi(0,0)}{\partial\mu\partial\lambda}]\Big]+\Phi[\Upsilon^2\Xi_{\mathcal{L}}(\mu)][\Upsilon^2\Xi_{\mathcal{L}}(\lambda)]^T.\end{aligned}\tag{7.5.26}$$

Now, we take the integral with regard to λ from 0 to λ again from the above equation. Then we obtain

$$\begin{aligned}\phi(\mu,\lambda)&=\phi(\mu,0)+\phi(0,\lambda)-\phi(0,0)+\lambda\Big[\frac{\partial\phi(\mu,0)}{\partial\lambda}-\frac{\partial\phi(0,0)}{\partial\lambda}\Big]+\lambda^2\Big[\frac{\partial^2\phi(\mu,0)}{\partial\lambda^2}-\frac{\partial^2\phi(0,0)}{\partial\lambda^2}\Big]\\&+\mu\lambda\Big[\frac{\partial^2\phi(0,\lambda)}{\partial\mu\partial\lambda}-\frac{\partial^2\phi(0,0)}{\partial\mu\partial\lambda}-\lambda\frac{\partial^3\phi(0,0)}{\partial\mu\partial\lambda}]\Big]+\int_0^\lambda\Phi[\Upsilon^2\Xi_{\mathcal{L}}(\mu)][\Upsilon^2\Xi_{\mathcal{L}}(\lambda)]^T d\lambda\\&=\phi(\mu,0)+\phi(0,\lambda)-\phi(0,0)+\lambda\Big[\frac{\partial\phi(\mu,0)}{\partial\lambda}-\frac{\partial\phi(0,0)}{\partial\lambda}\Big]+\lambda^2\Big[\frac{\partial^2\phi(\mu,0)}{\partial\lambda^2}-\frac{\partial^2\phi(0,0)}{\partial\lambda^2}\Big]\\&+\mu\lambda\Big[\frac{\partial^2\phi(0,\lambda)}{\partial\mu\partial\lambda}-\frac{\partial^2\phi(0,0)}{\partial\mu\partial\lambda}-\lambda\frac{\partial^3\phi(0,0)}{\partial\mu\partial\lambda}]\Big]+\Phi[\Upsilon^2\Xi_{\mathcal{L}}(\mu)]\int_0^\lambda[\Upsilon^2\Xi_{\mathcal{L}}(\lambda)]^T d\lambda\\&=\phi(\mu,0)+\phi(0,\lambda)-\phi(0,0)+\lambda\Big[\frac{\partial\phi(\mu,0)}{\partial\lambda}-\frac{\partial\phi(0,0)}{\partial\lambda}\Big]+\lambda^2\Big[\frac{\partial^2\phi(\mu,0)}{\partial\lambda^2}-\frac{\partial^2\phi(0,0)}{\partial\lambda^2}\Big]\\&+\mu\lambda\Big[\frac{\partial^2\phi(0,\lambda)}{\partial\mu\partial\lambda}-\frac{\partial^2\phi(0,0)}{\partial\mu\partial\lambda}-\lambda\frac{\partial^3\phi(0,0)}{\partial\mu\partial\lambda}]\Big]+\Phi[\Upsilon^2\Xi_{\mathcal{L}}(\mu)][\Upsilon^3\Xi_{\mathcal{L}}(\lambda)]^T.\end{aligned}\tag{7.5.27}$$

In the obtained equation we put $\mu=j$. Then

$$\begin{aligned}\phi(j,\lambda)&=\phi(j,0)+\phi(0,\lambda)-\phi(0,0)+\lambda\Big[\frac{\partial\phi(\mu,0)}{\partial\lambda}-\frac{\partial\phi(0,0)}{\partial\lambda}\Big]+\lambda^2\Big[\frac{\partial^2\phi(\mu,0)}{\partial\lambda^2}-\frac{\partial^2\phi(0,0)}{\partial\lambda^2}\Big]\\&+j\lambda\Big[\frac{\partial^2\phi(0,\lambda)}{\partial\mu\partial\lambda}-\frac{\partial^2\phi(0,0)}{\partial\mu\partial\lambda}-\lambda\frac{\partial^3\phi(0,0)}{\partial\mu\partial\lambda}]\Big]+\Phi[\Upsilon^2\Xi_{\mathcal{L}}(\mu)][\Upsilon^3\Xi_{\mathcal{L}}(\lambda)]^T|_{\mu=j},\end{aligned}\tag{7.5.28}$$

and

$$\left[\frac{\partial\phi(\mu,0)}{\partial\lambda}-\frac{\partial\phi(0,0)}{\partial\lambda}+\lambda\frac{\partial^2\phi(\mu,0)}{\partial\lambda^2}-\lambda\frac{\partial^2\phi(0,0)}{\partial\lambda^2}+j\frac{\partial^2\phi(0,\lambda)}{\partial\mu\partial\lambda}-j\frac{\partial^2\phi(0,0)}{\partial\mu\partial\lambda}-j\lambda[\frac{\partial^3\phi(0,0)}{\partial\mu\partial\lambda}]\right] \tag{7.5.29}$$
$$=\frac{1}{\lambda}\left[\phi(j,\lambda)-\phi(j,0)-\phi(0,\lambda)+\phi(0,0)-\Phi[\Upsilon^2\Xi_{\mathcal{L}}(\mu)][\Upsilon^3\Xi_{\mathcal{L}}(\lambda)]^T|_{\mu=j}\right].$$

By putting (7.5.29) in (7.5.27) we have

$$\phi(\mu,\lambda)=\phi(\mu,0)+\phi(0,\lambda)-\phi(0,0) \tag{7.5.30}$$
$$+\left[\phi(j,\lambda)-\phi(j,0)-\phi(0,\lambda)+\phi(0,0)-\Phi[\Upsilon^2\Xi_{\mathcal{L}}(\mu)][\Upsilon^3\Xi_{\mathcal{L}}(\lambda)]^T|_{\mu=j}\right]$$
$$+\Phi[\Upsilon^2\Xi_{\mathcal{L}}(\mu)][\Upsilon^3\Xi_{\mathcal{L}}(\lambda)]^T.$$

By putting Eqs. (7.5.24) and (7.5.30) in Eq. (7.5.21), we have

$$\frac{\partial^3\phi(0,\lambda)}{\partial\lambda^3}+\mu[\frac{\partial^4\phi(0,\lambda)}{\partial\mu\partial\lambda^3}]+\Xi_{\mathcal{L}}(\lambda)^T\Phi[\Upsilon^2\Xi_{\mathcal{L}}(\mu)]= \tag{7.5.31}$$
$$\mathcal{K}\Big(\phi(\mu,0)+\phi(0,\lambda)-\phi(0,0)$$
$$+\left[\phi(j,\lambda)-\phi(j,0)-\phi(0,\lambda)+\phi(0,0)-\Phi[\Upsilon^2\Xi_{\mathcal{L}}(\mu)][\Upsilon^3\Xi_{\mathcal{L}}(\lambda)]^T|_{\mu=j}\right]$$
$$+\Phi[\Upsilon^2\Xi_{\mathcal{L}}(\mu)][\Upsilon^3\Xi_{\mathcal{L}}(\lambda)]^T\Big)$$
$$+\mathcal{Z}\Big(\phi(\mu,0)+\phi(0,\lambda)-\phi(0,0)$$
$$+\left[\phi(j,\lambda)-\phi(j,0)-\phi(0,\lambda)+\phi(0,0)-\Phi[\Upsilon^2\Xi_{\mathcal{L}}(\mu)][\Upsilon^3\Xi_{\mathcal{L}}(\lambda)]^T|_{\mu=j}\right]$$
$$+\Phi[\Upsilon^2\Xi_{\mathcal{L}}(\mu)][\Upsilon^3\Xi_{\mathcal{L}}(\lambda)]^T\Big).$$

By considering the points $\mu_\jmath, \lambda_\jmath = \frac{2\jmath-1}{2\mathcal{L}^2}$ for $\jmath, \iota = 1, 2, \ldots, \mathcal{L}$ in Eq. (7.5.31) and by solving this system, the equations of unknown coefficients are obtained, and finally, by placing the coefficients obtained in Eq. (7.5.30), the numerical solution of Eq. (7.5.21) is obtained.

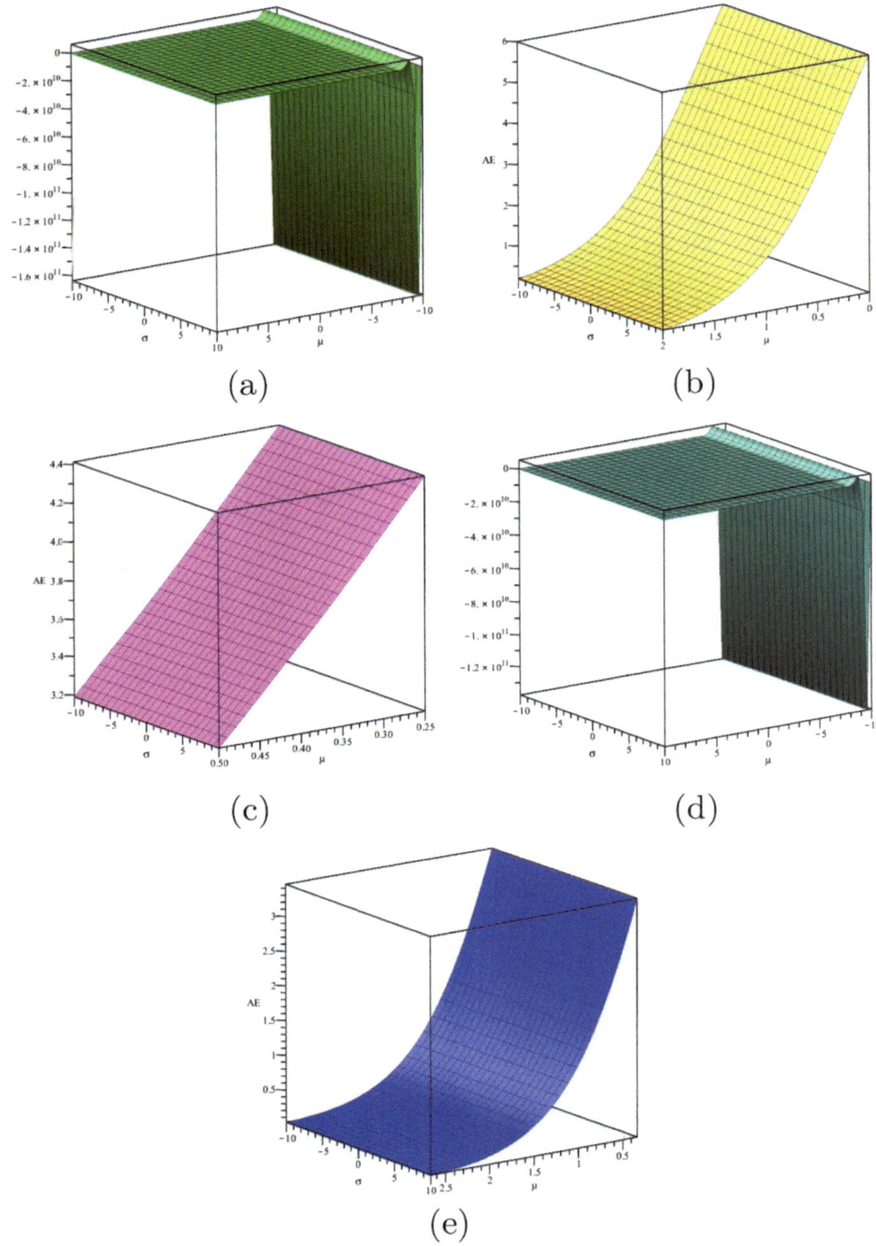

Fig. 7.17 Absolute error of Example 7.7 for different values of $\lambda, \sigma,\ \mu \in (0, 2),\ \mu \in (\frac{1}{4}, \frac{1}{2}), \mu \in (\frac{1}{3}, \frac{8}{3})$

7.5.3 The Clique Polynomial Method for the Navier–Stokes Equation (N-SE) of Fractional Order

We will generalize the ordinary Navier–Stokes equation to the Navier–Stokes equation (N-SE) of the fractional order, and with the help of the integral operations matrix resulting from the clique polynomial, a method for solving the Navier–Stokes equation (N-SE) of the fractional order is presented. We consider the N-SE of the fraction order as follows

$$ {}^{C}D_{\lambda}^{\mathrm{m}}\phi(\mu,\lambda) - \mathcal{K}(\phi(\mu,\lambda)) - \mathcal{Z}(\phi(\mu,\lambda)) = 0, \tag{7.5.32}$$

with initial condition

$$\begin{aligned} \phi(\mu,0) &= \varpi_1(\xi), \qquad & D_\lambda\phi(\mu,0) &= \varpi_2(\mu) \\ \theta(\mu,0) &= \varpi_1'(\mu) =, \qquad & D_\lambda\theta(\mu,0) &= \varpi_2'(\mu) \end{aligned} \tag{7.5.33}$$

and boundary conditions

$$\phi(i,\lambda) = \upsilon_1(\lambda), \quad \phi(j,\lambda) = \upsilon_2(\lambda), \quad \theta(i,\lambda) = \upsilon_3'(\lambda), \quad \theta(j,\lambda) = \upsilon_4'(\lambda), \tag{7.5.34}$$

where $(\mu,\lambda) \in (i,j) \times [0,\mathcal{F}]$, $1 < \mathrm{m} \le 2$ and ${}^{C}D_{\lambda}^{\mathrm{m}}$ is a Caputo fractional derivative. First we approximate $D_{\mu\mu}\phi(\mu,\lambda)$

$$D_{\mu\mu}\phi(\mu,\lambda) \simeq \sum_{\jmath=1}^{\mathcal{L}-1}\sum_{\epsilon=1}^{\mathcal{L}-1} \mathcal{W}(\mathcal{S}_\jmath,\mu)\mathcal{W}(\mathcal{S}_\epsilon,\lambda) = \varXi_{\mathcal{L}}(\mu)\varPhi[\varXi_{\mathcal{L}}(\lambda)]^T. \tag{7.5.35}$$

We take the integral of Eq. (7.5.35) with respect to μ twice. Then

$$\phi(\mu,\lambda) \simeq [\varUpsilon^2\varXi_{\mathcal{L}}(\mu)]\varPhi\varXi_{\mathcal{L}}(\lambda)^T + \mu D_\mu\phi(0,\lambda) + \upsilon_1(\lambda). \tag{7.5.36}$$

We put $\mu = j$ in Eq. (7.5.36). Then we have

$$D_\mu\phi(0,\lambda) = \frac{1}{j}\Big(\upsilon_2(\lambda) - \upsilon_1(\lambda) - [\varUpsilon^2\varXi_{\mathcal{L}}(\mu)]\varPhi\varXi_{\mathcal{L}}(\lambda)^T\Big). \tag{7.5.37}$$

By placing (7.5.37) in (7.5.36), we get

$$\begin{aligned}\phi(\mu,\lambda) \simeq{}& [\varUpsilon^2\varXi_{\mathcal{L}}(\mu)]\varPhi\varXi_{\mathcal{L}}(\lambda)^T + \frac{\mu}{j}\Big(\upsilon_2(\lambda) - \upsilon_1(\lambda) - [\varUpsilon^2\varXi_{\mathcal{L}}(\mu)]\varPhi\varXi_{\mathcal{L}}(\lambda)^T\Big) \\ &+ \upsilon_1(\lambda).\end{aligned} \tag{7.5.38}$$

Now, we approximate the $\mathcal{K}$ and $\mathcal{Z}$ functions based on clique polynomials. We consider

$$\mathcal{K}(\phi(\mu,\lambda)) \simeq \Xi_{\mathcal{L}}(\mu)\mathcal{K}[\Xi_{\mathcal{L}}(\lambda)]^T, \tag{7.5.39}$$

and

$$\mathcal{Z}(\phi(\mu,\lambda)) \simeq \Xi_{\mathcal{L}}(\mu)\mathcal{Z}[\Xi_{\mathcal{L}}(\lambda)]^T. \tag{7.5.40}$$

By placing Eqs. (7.5.40), (7.5.39), and (7.5.38) in Eq. (7.5.32) and taking the Riemann–Liouville fractional integral (R-LFI) of order m, we have

$$\begin{aligned}&[\Upsilon^2\Xi_{\mathcal{L}}(\mu)]\Phi\Xi_{\mathcal{L}}(\lambda)^T + \frac{\mu}{j}\Big(v_2(\lambda) - v_1(\lambda) - [\Upsilon^2\Xi_{\mathcal{L}}(\mu)]\Phi\Xi_{\mathcal{L}}(\lambda)^T\Big) + v_1(\lambda) - \varpi_1(\lambda) - \lambda\varpi_2(\lambda)\\ &+ \Xi_{\mathcal{L}}(\mu)\mathcal{K}\mathcal{R}^{\mathrm{m}}[\Xi_{\mathcal{L}}(\lambda)]^T + \Xi_{\mathcal{L}}(\mu)\mathcal{Z}\mathcal{R}^{\mathrm{m}}[\Xi_{\mathcal{L}}(\lambda)]^T = 0.\end{aligned} \tag{7.5.41}$$

Finally, by placing (7.5.36) in (7.5.39) and (7.5.40) we have

$$\begin{aligned}&\mathcal{K}\Big([\Upsilon^2\Xi_{\mathcal{L}}(\mu)]\Phi\Xi_{\mathcal{L}}(\lambda)^T + \frac{\mu}{j}\Big(v_2(\lambda) - v_1(\lambda) - [\Upsilon^2\Xi_{\mathcal{L}}(\mu)]\Phi\Xi_{\mathcal{L}}(\lambda)^T\Big) + v_1(\lambda)\Big)\\ &- \Xi_{\mathcal{L}}(\mu)\mathcal{K}[\Xi_{\mathcal{L}}(\lambda)]^T = 0,\end{aligned} \tag{7.5.42}$$

and

$$\begin{aligned}&\mathcal{Z}\Big([\Upsilon^2\Xi_{\mathcal{L}}(\mu)]\Phi\Xi_{\mathcal{L}}(\lambda)^T + \frac{\mu}{j}\Big(v_2(\lambda) - v_1(\lambda) - [\Upsilon^2\Xi_{\mathcal{L}}(\mu)]\Phi\Xi_{\mathcal{L}}(\lambda)^T\Big) + v_1(\lambda))\Big)\\ &- \Xi_{\mathcal{L}}(\mu)\mathcal{Z}[\Xi_{\mathcal{L}}(\lambda)]^T = 0.\end{aligned} \tag{7.5.43}$$

By placing the local points $\mu_j = \frac{j(j-i)}{\mathcal{L}+2}$, $\lambda_\epsilon = \frac{\epsilon}{\mathcal{L}+2}T$ in Eqs. (7.5.43), (7.5.42), and (7.5.41) and solving the system, the coefficients $\phi_j\epsilon$ are obtained for $j, \epsilon = 1, 2, \ldots, \mathcal{L}$. Next, by placing the matrix Φ in (7.5.38) approximate solutions are obtained (Fig. 7.17).

7.5.4 Numerical Examples

Example 7.7 Given the method proposed to obtain the approximate solutions, we consider the following Navier–Stokes equation (N-SE) of the integer orders equation (Fig. 7.18)

$$D_{\lambda}^{\mathrm{m}}\phi(\mu,\sigma,\lambda)+\phi(\mu,\sigma,\lambda)\frac{\partial\phi(\mu,\sigma,\lambda)}{\partial\mu}+\phi(\mu,\sigma,\lambda)\frac{\partial\theta(\mu,\sigma,\lambda)}{\partial\sigma}=\delta_0\left(\frac{\partial^2\phi(\mu,\sigma,\lambda)}{\partial\mu^2}+\frac{\partial^2\phi(\mu,\sigma,\lambda)}{\partial\sigma^2}\right)+\mathrm{f},$$

$$D_{\lambda}^{\mathrm{m}}\theta(\mu,\sigma,\lambda)+\phi(\mu,\sigma,\lambda)\frac{\partial\theta(\mu,\sigma,\lambda)}{\partial\mu}+\theta(\mu,\sigma,\lambda)\frac{\partial\theta(\mu,\sigma,\lambda)}{\partial\sigma}=\delta_0\left(\frac{\partial^2\theta(\mu,\sigma,\lambda)}{\partial\mu^2}+\frac{\partial^2\theta(\mu,\sigma,\lambda)}{\partial\theta^2}\right)-\mathrm{f}, \tag{7.5.44}$$

where m = 3 and with the initial conditions

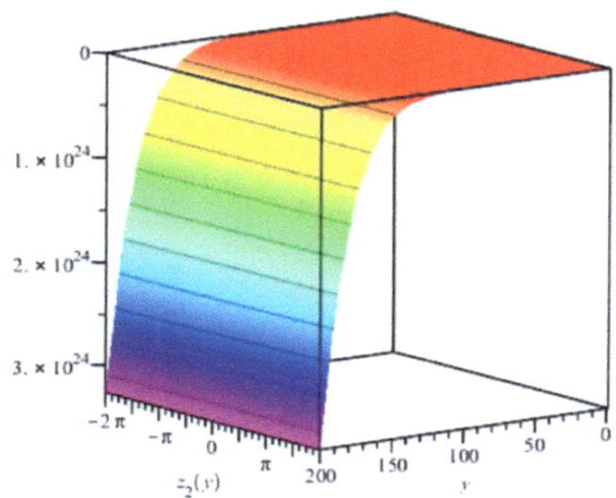

(a)Exact solution for example (7.7) for $\lambda = 0.2, 0.3, 0.4, 0.5, 0.6, 0.7, 0.8, 0.9, \mu \in (0, 2\pi), \sigma \in (0, 2\pi)$.

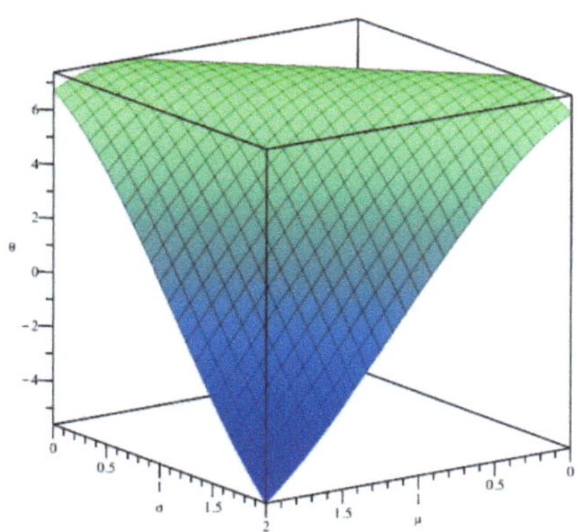

(b)

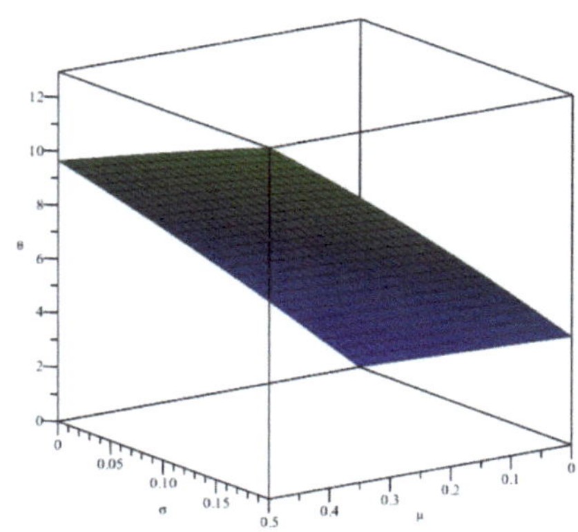

(c)

Fig. 7.18 Exact solution for Example 7.7. **a** $\lambda = 2$, $\mu \in (0, 2)$, $\sigma \in (0, 2)$, **b** $\lambda = 3$, $\mu \in (0, \frac{1}{2})$, $w \in (0, \frac{1}{5})$, **c** $\lambda = 5$, $u \in (0, \frac{1}{8})$, $w \in (0, \frac{9}{5})$, **d** $\lambda = 4$, $u \in (0, \frac{2}{7})$, $w \in (0, \frac{6}{7})$, **e** $\lambda = 4$, $\mu \in (0, \frac{1}{7})$, $w \in (0, \frac{3}{5})$, **f** $\lambda = 3$, $\mu \in (0, \frac{7}{10})$, $w \in (0, \frac{8}{15})$

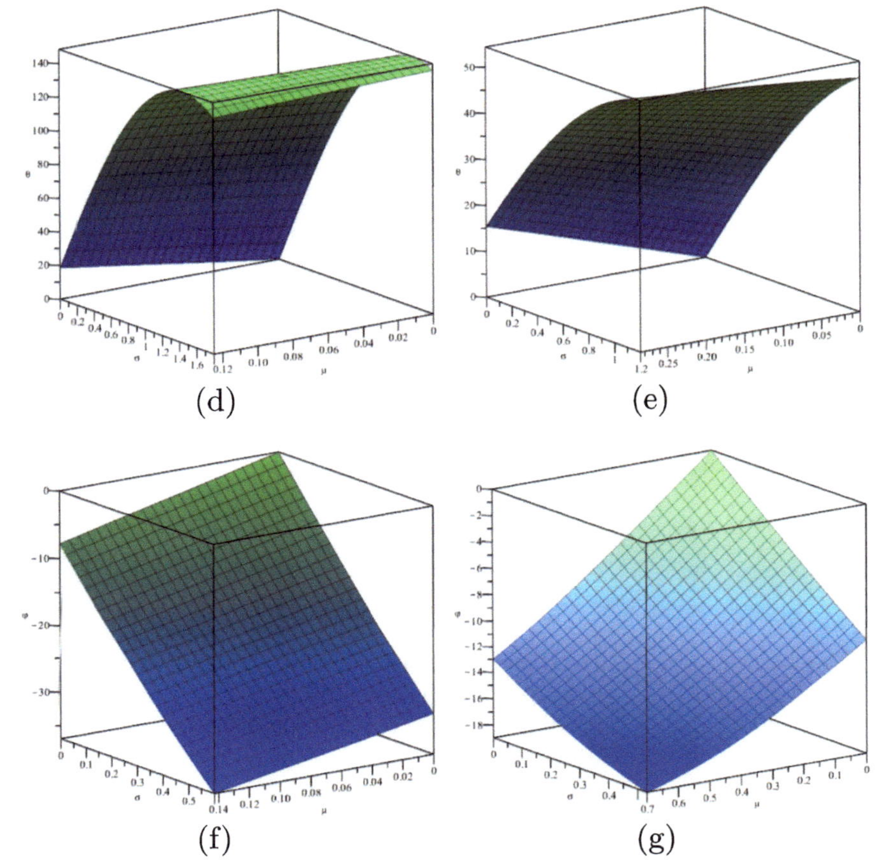

Fig. 7.18 (continued)

$$
\begin{aligned}
\phi(\mu, \sigma, 0) &= -\sin(\mu + \sigma) \\
\theta(\mu, \sigma, 0) &= \sin(\mu + \sigma).
\end{aligned} \tag{7.5.45}
$$

Assume that

$$
\frac{\partial^5 \phi(\mu, \sigma, \lambda)}{\partial \mu^2 \partial \sigma^2 \partial \lambda} = \sum_{\jmath=1}^{\mathcal{L}-1} \sum_{\iota=1}^{\mathcal{L}-1} \sum_{\epsilon=1}^{\mathcal{L}-1} \mathcal{W}(\mathcal{S}_\jmath, \mu) \mathcal{W}(\mathcal{S}_\iota, \sigma) \mathcal{W}(\mathcal{S}_\epsilon, \lambda) = \Xi_{\mathcal{L}}(\lambda)^T \Phi \, \Xi_{\mathcal{L}}(\mu) \Xi_{\mathcal{L}}(\sigma). \tag{7.5.46}
$$

By taking the integral from 0 to λ with respect to λ

$$
\begin{aligned}
\frac{\partial^4\phi(\mu,\sigma,\lambda)}{\partial\mu^2\partial\sigma^2} &= \frac{\partial^4\phi(\mu,\sigma,0)}{\partial\mu^2\partial\sigma^2} + \int_0^{\lambda} \varXi_{\mathcal{L}}(\lambda)^T \varPhi\, \varXi_{\mathcal{L}}(\mu)\varXi_{\mathcal{L}}(\sigma)d\lambda \\
&= \frac{\partial^4\phi(\mu,\sigma,0)}{\partial\mu^2\partial\sigma^2} + \varPhi\, \varXi_{\mathcal{L}}(\mu)\varXi_{\mathcal{L}}(\sigma)\int_0^{\lambda} \varXi_{\mathcal{L}}(\lambda)^T d\lambda \\
&= \frac{\partial^4\phi(\mu,\sigma,0)}{\partial\mu^2\partial\sigma^2} + [\varUpsilon\, \varXi_{\mathcal{L}}(\lambda)]^T \varPhi\, \varXi_{\mathcal{L}}(\mu)\varXi_{\mathcal{L}}(\sigma).
\end{aligned}
\tag{7.5.47}
$$

Next, we have an integral from 0 to μ with respect to μ

$$
\begin{aligned}
\frac{\partial^3\phi(\mu,\sigma,\lambda)}{\partial\nu\partial\sigma^2} &= \frac{\partial^3\phi(0,\sigma,\lambda)}{\partial\mu\partial\sigma^2} + \frac{\partial^3\phi(\mu,\sigma,0)}{\partial\mu\partial\sigma^2} - \frac{\partial^3\phi(0,\sigma,0)}{\partial\mu\partial\sigma^2} + \int_0^{\mu} \varXi_{\mathcal{L}}(\lambda)^T \varPhi\, \varXi_{\mathcal{L}}(\mu)\varXi_{\mathcal{L}}(\sigma)d\mu \\
&= \frac{\partial^3\phi(0,\sigma,\lambda)}{\partial\mu\partial\sigma^2} + \frac{\partial^3\phi(\mu,\sigma,0)}{\partial\mu\partial\sigma^2} - \frac{\partial^3\phi(0,\sigma,0)}{\partial\mu\partial\sigma^2} + [\varUpsilon\, \varXi_{\mathcal{L}}(\lambda)]^T \varPhi[\varUpsilon\, \varXi_{\mathcal{L}}(\mu)]\varXi_{\mathcal{L}}(\sigma),
\end{aligned}
\tag{7.5.48}
$$

again, by taking the integral from 0 to μ with respect to μ from Eq. (7.5.48) , we have

$$
\begin{aligned}
\frac{\partial^2\phi(\mu,\sigma,\lambda}{\partial\sigma^2} &= \frac{\partial^2\phi(0,\sigma,\lambda)}{\partial\sigma^2} + \frac{\partial^2\phi(\mu,\sigma,0)}{\partial\sigma^2} - \frac{\partial^2\phi(0,\sigma,0)}{\partial\sigma^2} + \mu\left[\frac{\partial^3\phi(0,\sigma,\lambda)}{\partial\mu\partial\sigma^2} - \frac{\partial^3\phi(0,\sigma,0)}{\partial\mu\partial\sigma^2}\right] \\
&+ \int_0^{\mu} [\varUpsilon\, \varXi_{\mathcal{L}}(\lambda)]^T \varPhi[\varUpsilon\, \varXi_{\mathcal{L}}(\mu)]\varXi_{\mathcal{L}}(\sigma)d\sigma, \\
&= \frac{\partial^2\phi(0,\sigma,\lambda)}{\partial\sigma^2} + \frac{\partial^2\phi(\mu,\sigma,0)}{\partial\sigma^2} - \frac{\partial^2\phi(0,\sigma,0)}{\partial\sigma^2} + \mu\left[\frac{\partial^3\phi(0,\sigma,\lambda)}{\partial\mu\partial\sigma^2} - \frac{\partial^3\phi(0,\sigma,0)}{\partial\mu\partial\sigma^2}\right] \\
&+ [\varUpsilon\, \varXi_{\mathcal{L}}(\lambda)]^T \varPhi[\varUpsilon^2 \varXi_{\mathcal{L}}(\mu)]\varXi_{\mathcal{L}}(\sigma).
\end{aligned}
\tag{7.5.49}
$$

Now, by taking the integral from 0 to σ relative to σ, the following equation is obtained

$$
\begin{aligned}
\frac{\partial\phi(\mu,\sigma,\lambda)}{\partial\sigma} &= \frac{\partial\phi(\mu,0,\lambda)}{\partial\sigma} + \frac{\partial\phi(0,\sigma,\lambda)}{\partial\sigma} - \frac{\partial\phi(0,0,\lambda)}{\partial\sigma} \\
&+ \frac{\partial\phi(\mu,\sigma,0)}{\partial\sigma} - \frac{\partial\phi(\mu,0,0)}{\partial\sigma} - \frac{\partial\phi(0,\sigma,0)}{\partial\sigma} + \frac{\partial\phi(0,0,0)}{\partial\sigma} \\
&+ \mu\left[\frac{\partial^2\phi(0,\sigma,\lambda)}{\partial\mu\partial\sigma} - \frac{\partial^2\phi(0,0,\lambda)}{\partial\mu\partial\sigma} - \frac{\partial^2\phi(0,\sigma,0)}{\partial\mu\partial\sigma} + \frac{\partial^2\phi(0,0,0)}{\partial\mu\partial\sigma}\right] \\
&+ \int_0^{\sigma} [\varUpsilon\, \varXi_{\mathcal{L}}(\lambda)]^T \varPhi[\varUpsilon^2 \varXi_{\mathcal{L}}(\mu)]\varXi_{\mathcal{L}}(\sigma)d\sigma \\
&= \frac{\partial\phi(\mu,0,\lambda)}{\partial\sigma} + \frac{\partial\phi(0,\sigma,\lambda)}{\partial\sigma} - \frac{\partial\phi(0,0,\lambda)}{\partial\sigma} + \frac{\partial\phi(\mu,\sigma,0)}{\partial\sigma} \\
&- \frac{\partial\phi(\mu,0,0)}{\partial\sigma} - \frac{\partial\phi(0,\sigma,0)}{\partial\sigma} + \frac{\partial\phi(0,0,0)}{\partial\sigma} \\
&+ \mu\left[\frac{\partial^2\phi(0,\sigma,\lambda)}{\partial\mu\partial\sigma} - \frac{\partial^2\phi(0,0,\lambda)}{\partial\mu\partial\sigma} - \frac{\partial^2\phi(0,\sigma,0)}{\partial\mu\partial\sigma} + \frac{\partial^2\phi(0,0,0)}{\partial\mu\partial\sigma}\right] \\
&+ [\varUpsilon\, \varXi_{\mathcal{L}}(\lambda)]^T \varPhi[\varUpsilon^2 \varXi_{\mathcal{L}}(\mu)][\varUpsilon\, \varXi_{\mathcal{L}}(\sigma)].
\end{aligned}
\tag{7.5.50}
$$

In the sequel, by taking the integral from Eq. (7.5.47) with respect to σ from 0 to σ, we have

$$\begin{aligned}\frac{\partial^3\phi(\mu,\sigma,\lambda)}{\partial\mu^2\partial\sigma} &= \frac{\partial^3\phi(\mu,0,\lambda)}{\partial\mu^2\partial\sigma} + \int_0^{\sigma}[\Upsilon\Xi_{\mathcal{L}}(\lambda)]^T\Phi\xi_{\mathcal{L}}(\mu)\xi_{\mathcal{L}}(\sigma)d\sigma \\ &= \frac{\partial^3\phi(\mu,0,\lambda)}{\partial\mu^2\partial\sigma} + \frac{\partial^3\phi(\mu,\sigma,0)}{\partial\mu^2\partial\sigma} - \frac{\partial^3\phi(\mu,0,0)}{\partial\mu^2\partial\sigma} + [\Upsilon\Xi_{\mathcal{L}}(\lambda)]^T\Phi\Xi_{\mathcal{L}}(\mu)[\Upsilon\Xi_{\mathcal{L}}(\sigma)].\end{aligned} \tag{7.5.51}$$

Again, by taking the integral from 0 to σ with respect to σ from Eq. (7.5.51), we have

$$\begin{aligned}&\frac{\partial^2\phi(\mu,\sigma,\lambda)}{\partial\mu^2} = \frac{\partial^2\phi(\mu,0,\lambda)}{\partial\mu^2} + \frac{\partial^2\phi(\mu,\sigma,0)}{\partial\mu^2} - \frac{\partial^2\phi(\mu,0,0)}{\partial\mu^2} \\ &+\sigma\left[\frac{\partial^3\phi(\mu,0,\lambda)}{\partial\mu^2\partial\sigma} + \frac{\partial^3 u(\xi,0,0)}{\partial\xi^2\partial y}\right] \\ &+[\Upsilon\Xi_{\mathcal{L}}(\lambda)]^T\Phi\Xi_{\mathcal{L}}(\mu)[\Upsilon^2\Xi_{\mathcal{L}}(\sigma)].\end{aligned} \tag{7.5.52}$$

To obtain $\frac{\partial\phi(\mu,\sigma,\lambda)}{\partial\mu}$, we take an integral from Eq. (7.5.52) with respect to μ, from 0 to μ. Then

$$\begin{aligned}&\frac{\partial\phi(\mu,\sigma,\lambda)}{\partial\mu} = \frac{\partial\phi(0,\sigma,\lambda)}{\partial\mu} + \frac{\partial\phi(\mu,0,\lambda)}{\partial\mu} - \frac{\partial\phi(0,0,\lambda)}{\partial\mu} \\ &+\frac{\partial\phi(\mu,\sigma,0)}{\partial\mu} - \frac{\partial\phi(0,\sigma,0)}{\partial\mu} - \frac{\partial\phi(\mu,0,0)}{\partial\mu} \\ &+\frac{\partial\phi(0,0,0)}{\partial\mu} + \sigma\left[\frac{\partial^2\phi(\mu,0,\lambda)}{\partial\mu\partial\sigma} - \frac{\partial^2\phi(0,0,\lambda)}{\partial\mu\partial\sigma} + \frac{\partial^2\phi(\mu,0,0)}{\partial\mu\partial\sigma} - \frac{\partial^2\phi(0,0,0)}{\partial\mu\partial\sigma}\right] \\ &+[\Upsilon\Xi_{\mathcal{L}}(\lambda)]^T\Phi[\Upsilon\Xi_{\mathcal{L}}(\mu)][\Upsilon^2\Xi_{\mathcal{L}}(\sigma)],\end{aligned} \tag{7.5.53}$$

we take integral with respect to μ

$$\begin{aligned}&\phi(\mu,\sigma,\lambda) = \phi(0,\sigma,\lambda) + \phi(\mu,0,\lambda) - \phi(0,0,\lambda) \\ &+\phi(\mu,\sigma,0) - \phi(0,\sigma,0) - \phi(\mu,0,0) \\ &+\mu\left[\frac{\partial\phi(0,\sigma,\lambda)}{\partial\mu} - \frac{\partial\phi(0,0,\lambda)}{\partial\mu} - \frac{\partial\phi(0,\sigma,0)}{\partial\mu} + \frac{\partial\phi(0,0,0)}{\partial\mu}\right] \\ &+\sigma\Bigg[\frac{\partial\phi(\mu,0,\lambda)}{\partial\sigma} - \frac{\partial\phi(0,0,\lambda)}{\partial\sigma} + \frac{\partial\phi(\mu,0,0)}{\partial\sigma} - \frac{\partial\phi(0,0,0)}{\partial\sigma} \\ &+\mu\left(-\frac{\partial^2\phi(0,0,\lambda)}{\partial\mu\partial\sigma} - \frac{\partial^2\phi(0,0,0)}{\partial\mu\partial\sigma}\right)\Bigg] \\ &+[\Upsilon\Xi_{\mathcal{L}}(\lambda)]^T\Phi[\Upsilon^2\Xi_{\mathcal{L}}(\mu)][\Upsilon^2\Xi_{\mathcal{L}}(\sigma)].\end{aligned} \tag{7.5.54}$$

Putting $\mu = j$ and $\sigma = t$ in the above equation, we have

$$\phi(j,t,\lambda) = \phi(0,t,\lambda) + \phi(j,0,\lambda) - \phi(0,0,\lambda) \tag{7.5.55}$$

$$
\begin{aligned}
&+ \phi(j,t,0) - \phi(0,t,0) - \phi(j,0,0) \\
&+ \mu\left[\frac{\partial\phi(0,\sigma,\lambda)}{\partial\mu} - \frac{\partial\phi(0,0,\lambda)}{\partial\mu} - \frac{\partial\phi(0,\sigma,0)}{\partial\mu} + \frac{\partial\phi(0,0,0)}{\partial\mu}\right] \\
&+ \sigma\left[\frac{\partial\phi(\mu,0,\lambda)}{\partial\sigma} - \frac{\partial\phi(0,0,\lambda)}{\partial\sigma} + \frac{\partial\phi(\mu,0,0)}{\partial\sigma} - \frac{\partial\phi(0,0,0)}{\partial\sigma}\right. \\
&\left.+ \mu\left(-\frac{\partial^2\phi(0,0,\lambda)}{\partial\mu\partial\sigma} - \frac{\partial^2\phi(0,0,0)}{\partial\mu\partial\sigma}\right)\right] \\
&+ ([\Upsilon\Xi_{\mathcal{L}}(\lambda)]^T\Phi[\Upsilon^2\Xi_{\mathcal{L}}(\mu)][\Upsilon^2\Xi_{\mathcal{L}}(\sigma)])|_{(\mu=j,\sigma=t)},
\end{aligned}
$$

as a result we have

$$
\begin{aligned}
&\left(\mu\left[\frac{\partial\phi(0,\sigma,\lambda)}{\partial\mu} - \frac{\partial\phi(0,0,\lambda)}{\partial\mu} - \frac{\partial\phi(0,\sigma,0)}{\partial\mu} + \frac{\partial\phi(0,0,0)}{\partial\mu}\right]\right. \\
&+ \sigma\left[\frac{\partial\phi(\mu,0,\lambda)}{\partial\sigma} - \frac{\partial\phi(0,0,\lambda)}{\partial\sigma} + \frac{\partial\phi(\mu,0,0)}{\partial\sigma} - \frac{\partial\phi(0,0,0)}{\partial\sigma}\right. \\
&\left.\left.+ \mu\left(-\frac{\partial^2\phi(0,0,\lambda)}{\partial\mu\partial\lambda} - \frac{\partial^2\phi(0,0,0)}{\partial\mu\partial\sigma}\right)\right]\right) \\
&= \phi(j,t,\lambda) - \phi(0,t,\lambda) - \phi(j,0,\lambda) + \phi(0,0,\lambda) - \phi(j,t,0) \\
&+ \phi(0,t,0) + \phi(j,0,0) \\
&- ([\Upsilon\Xi_{\mathcal{L}}(\lambda)]^T\Phi[\Upsilon^2\Xi_{\mathcal{L}}(\mu)][\Upsilon^2\Xi_{\mathcal{L}}(\sigma)])|_{(\mu=j,\sigma=t)}.
\end{aligned}
\tag{7.5.56}
$$

By putting the relation obtained in Eq. (7.5.54) we have

$$
\begin{aligned}
&\phi(\mu,\sigma,\lambda) = \phi(0,\sigma,\lambda) + \phi(\mu,0,\lambda) - \phi(0,0,\lambda) \\
&+ \phi(\mu,\sigma,0) - \phi(0,\sigma,0) - \phi(\mu,0,0) \\
&+ \phi(j,t,\lambda) - \phi(0,t,\lambda) - \phi(j,0,\lambda) \\
&+ \phi(0,0,\lambda) - \phi(j,t,0) + \phi(0,t,0) + \phi(j,0,0) \\
&+ [\Upsilon\Xi_{\mathcal{L}}(\lambda)]^T\Phi[\Upsilon^2\Xi_{\mathcal{L}}(\mu)][\Upsilon^2\Xi_{\mathcal{L}}(\sigma)] \\
&- ([\Upsilon\Xi_{\mathcal{L}}(\sigma)]^T\Phi[\Upsilon^2\Xi_{\mathcal{L}}(\mu)][\Upsilon^2\Xi_{\mathcal{L}}(\sigma)])|_{(\mu=j,\sigma=t)}.
\end{aligned}
\tag{7.5.57}
$$

Differentiating Eq. (7.5.54) three times concerning λ, we get

$$
\begin{aligned}
&\frac{\partial^3\phi(\mu,\sigma,\lambda)}{\partial\lambda^3} = \frac{\partial^3\phi(0,\sigma,\lambda)}{\partial\lambda^3} + \frac{\partial^3\phi(\mu,0,\lambda)}{\partial\lambda^3} \\
&+ \frac{\partial^3\phi(j,t,\lambda)}{\partial\lambda^3} - \frac{\partial^3\phi(0,t,\lambda)}{\partial^3\lambda} - \frac{\partial^3\phi(j,0,\lambda)}{\partial\lambda^3}
\end{aligned}
\tag{7.5.58}
$$

$$+ \frac{d^3}{d\lambda^3}\Bigg([\Upsilon \Xi_{\mathcal{L}}(\lambda)]^T \Phi[\Upsilon^2 \Xi_{\mathcal{L}}(\mu)][\Upsilon^2 \Xi_{\mathcal{L}}(\sigma)]$$
$$- ([\Upsilon \Xi_{\mathcal{L}}(\lambda)]^T \Phi[\Upsilon^2 \Xi_{\mathcal{L}}(\mu)][\Upsilon^2 \Xi_{\mathcal{L}}(\sigma)])|_{(\mu=j,\sigma=t)}\Bigg),$$

Similar to the process we did for ϕ, we also approximate θ. As a result, we obtain

$$\begin{aligned}\frac{\partial^2\theta(\mu,\sigma,\lambda)}{\partial\sigma^2} &= \frac{\partial^2\theta(0,\sigma,\lambda)}{\partial\sigma^2} + \frac{\partial^2\theta(\mu,\sigma,0)}{\partial\sigma^2} - \frac{\partial^2\theta(0,\sigma,0)}{\partial\sigma^2} \\ &+ \mu\left[\frac{\partial^3\theta(0,\sigma,\lambda)}{\partial\mu\partial\sigma^2} - \frac{\partial^3\theta(0,\sigma,0)}{\partial\mu\partial\sigma^2}\right] \\ &+ [\Upsilon \Xi_{\mathcal{L}}(\lambda)]^T \Theta[\Upsilon^2 \Xi_{\mathcal{L}}(\mu)]\Xi_{\mathcal{L}}(\sigma),\end{aligned} \tag{7.5.59}$$

$$\begin{aligned}\frac{\partial\theta(\mu,\sigma,\lambda)}{\partial\sigma} &= \frac{\partial\theta(\mu,0,\lambda)}{\partial\sigma} + \frac{\partial\theta(0,\sigma,\lambda)}{\partial\sigma} - \frac{\partial\theta(0,0,\lambda)}{\partial\sigma} \\ &+ \frac{\partial\theta(\mu,\sigma,0)}{\partial\sigma} - \frac{\partial\theta(\mu,0,0)}{\partial\sigma} - \frac{\partial\theta(0,\sigma,0)}{\partial\sigma} + \frac{\partial\theta(0,0,0)}{\partial\theta} \\ &+ \mu\left[\frac{\partial^2\theta(0,\sigma,\lambda)}{\partial\mu\partial\sigma} - \frac{\partial^2\theta(0,0,\lambda)}{\partial\mu\partial\sigma} - \frac{\partial^2\theta(0,\sigma,0)}{\partial\mu\partial\sigma} + \frac{\partial^2\theta(0,0,0)}{\partial\mu\partial\sigma}\right] \\ &+ [\Upsilon \Xi_{\mathcal{L}}(\lambda)]^T \Theta[\Upsilon^2 \Xi_{\mathcal{L}}(\mu)][\Upsilon \Xi_{\mathcal{L}}(\sigma)],\end{aligned} \tag{7.5.60}$$

$$\begin{aligned}\frac{\partial^2\theta(\mu,\sigma,\lambda)}{\partial\mu^2} &= \frac{\partial^2\theta(\mu,0,\lambda)}{\partial\mu^2} + \frac{\partial^2\theta(\mu,\sigma,0)}{\partial\mu^2} - \frac{\partial^2\theta(\mu,0,0)}{\partial\mu^2} \\ &+ \sigma\left[\frac{\partial^3\theta(\mu,0,\lambda)}{\partial\mu^2\partial\sigma} + \frac{\partial^3\theta(\mu,0,0)}{\partial\mu^2\partial\sigma}\right] \\ &+ [\Upsilon \Xi_{\mathcal{L}}(\lambda)]^T \Theta \Xi_{\mathcal{L}}(\mu)[\Upsilon^2 \Xi_{\mathcal{L}}(\sigma)],\end{aligned} \tag{7.5.61}$$

$$\begin{aligned}\frac{\partial\theta(\mu,\sigma,\lambda)}{\partial\mu} &= \frac{\partial\theta(0,\sigma,\lambda)}{\partial\mu} + \frac{\partial\theta(\mu,0,\lambda)}{\partial\mu} - \frac{\partial\theta(0,0,\lambda)}{\partial\mu} \\ &+ \frac{\partial\theta(\mu,\sigma,0)}{\partial\mu} - \frac{\partial\theta(0,\sigma,0)}{\partial\mu} - \frac{\partial\theta(\mu,0,0)}{\partial\mu} \\ &+ \frac{\partial\theta(0,0,0)}{\partial\mu} + \sigma\left[\frac{\partial^2\theta(\mu,0,\lambda)}{\partial\mu\partial\sigma} - \frac{\partial^2\theta(0,0,\lambda)}{\partial\mu\partial\sigma} + \frac{\partial^2\theta(\mu,0,0)}{\partial\mu\partial\sigma} - \frac{\partial^2\theta(0,0,0)}{\partial\mu\partial\sigma}\right] \\ &+ [\Upsilon \Xi_{\mathcal{L}}(\lambda)]^T \Theta[\Upsilon \Xi_{\mathcal{L}}(\mu)][\Upsilon^2 \Xi_{\mathcal{L}}(\sigma)],\end{aligned} \tag{7.5.62}$$

and

$$\begin{aligned}\theta(\mu,\sigma,\lambda) &= \theta(0,\sigma,\lambda)+\theta(\mu,0,\lambda)-\theta(0,0,\lambda)\\&+\theta(\mu,\sigma,0)-\theta(0,\sigma,0)-\theta(\mu,0,0)\\&+\theta(j,t,\lambda)-\theta(0,t,\lambda)-\theta(j,0,\lambda)\\&+\theta(0,0,\lambda)-\theta(j,t,0)+\theta(0,t,0)+\theta(j,0,0)\\&+[\Upsilon\Xi_{\mathcal{L}}(\lambda)]^T\Theta[\Upsilon^2\Xi_{\mathcal{L}}(\mu)][\Upsilon^2\Xi_{\mathcal{L}}(\sigma)]\\&-([\Upsilon\Xi_{\mathcal{L}}(\lambda)]^T\Theta[\Upsilon^2\Xi_{\mathcal{L}}(\mu)][\Upsilon^2\Xi_{\mathcal{L}}(\sigma)])|_{(\mu=j,\sigma=t)},\end{aligned}\tag{7.5.63}$$

and

$$\begin{aligned}\frac{\partial^3\theta(\mu,\sigma,\lambda)}{\partial\lambda^3} &= \frac{\partial^3\theta(0,\sigma,\lambda)}{\partial\lambda^3}+\frac{\partial^3\theta(\mu,0,\lambda)}{\partial\lambda^3}+\frac{\partial^3\theta(j,t,\lambda)}{\partial\lambda^3}-\frac{\partial^3\theta(0,t,\lambda)}{\partial^3\lambda}-\frac{\partial^3\theta(j,0,\lambda)}{\partial\lambda^3}\\&+\frac{d^3}{d\lambda^3}\Big([\Upsilon\Xi_{\mathcal{L}}(\lambda)]^T\Theta[\Upsilon^2\Xi_{\mathcal{L}}(\mu)][\Upsilon^2\xi_{\mathcal{L}}(\sigma)]-([\Upsilon\xi_{\mathcal{L}}(\lambda)]^T\Theta[\Upsilon^2\xi_{\mathcal{L}}(\mu)][\Upsilon^2\xi_{\mathcal{L}}(\sigma)])|_{(\mu=j,\sigma=t)}\Big).\end{aligned}\tag{7.5.64}$$

Now, by putting (7.5.64), (7.5.63), (7.5.62), (7.5.61), (7.5.60), (7.5.59), (7.5.58), (7.5.57), (7.5.53), (7.5.52) (7.5.50), (7.5.49) in Eq. (7.5.44), we obtain

$$\begin{aligned}&\frac{\partial^3\phi(0,\sigma,\lambda)}{\partial\lambda^3}+\frac{\partial^3\phi(\mu,0,\lambda)}{\partial\lambda^3}+\frac{\partial^3\phi(j,t,\lambda)}{\partial\lambda^3}-\frac{\partial^3\phi(0,t,\lambda)}{\partial^3\lambda}-\frac{\partial^3\phi(j,0,\lambda)}{\partial\lambda^3}\\&+\frac{d^3}{d\lambda^3}\Big([\Upsilon\Xi_{\mathcal{L}}(\lambda)]^T\Phi[\Upsilon^2\Xi_{\mathcal{L}}(\mu)][\Upsilon^2\Xi_{\mathcal{L}}(\sigma)]-([\Upsilon\Xi_{\mathcal{L}}(\lambda)]^T\Phi[\Upsilon^2\Xi_{\mathcal{L}}(\mu)][\Upsilon^2\Xi_{\mathcal{L}}(\sigma)])|_{(\mu=j,\sigma=t)}\Big)\\&\phi(0,\sigma,\lambda)+\phi(\mu,0,\lambda)-\phi(0,0,\lambda)+\phi(\mu,\sigma,0)-\phi(0,\sigma,0)-\phi(\mu,0,0)\\&+\phi(j,t,\lambda)-\phi(0,t,\lambda)-\phi(j,0,\lambda)+\phi(0,0,\lambda)-\phi(j,t,0)+\phi(0,t,0)+\phi(j,0,0)\\&+[\Upsilon\Xi_{\mathcal{L}}(\lambda)]^T\Phi[\Upsilon^2\Xi_{\mathcal{L}}(\mu)][\Upsilon^2\Xi_{\mathcal{L}}(\sigma)]-([\Upsilon\Xi_{\mathcal{L}}(\lambda)]^T\Phi[\Upsilon^2\Xi_{\mathcal{L}}(\mu)][\Upsilon^2\Xi_{\mathcal{L}}(\sigma)])|_{(\mu=j,\sigma=t))}\\&\Big(\frac{\partial\phi(0,\sigma,\lambda)}{\partial\mu}+\frac{\partial\phi(\mu,0,\lambda)}{\partial\mu}-\frac{\partial\phi(0,0,\lambda)}{\partial\mu}+\frac{\partial\phi(\mu,\sigma,0)}{\partial\mu}-\frac{\partial\phi(0,\sigma,0)}{\partial\mu}-\frac{\partial\phi(\mu,0,0)}{\partial\mu}\\&+\frac{\partial\phi(0,0,0)}{\partial\mu}+\sigma\Big[\frac{\partial^2\phi(\mu,0,\lambda)}{\partial\mu\partial\sigma}-\frac{\partial^2\phi(0,0,\lambda)}{\partial\mu\partial\sigma}+\frac{\partial^2\phi(\mu,0,0)}{\partial\mu\partial\sigma}-\frac{\partial^2\phi(0,0,0)}{\partial\mu\partial\sigma}\Big]\\&+[\Upsilon\Xi_{\mathcal{L}}(\lambda)]^T\Phi[\Upsilon\Xi_{\mathcal{L}}(\mu)][\Upsilon^2\Xi_{\mathcal{L}}(\sigma)]\Big)\\&+\phi(0,\sigma,\lambda)+\phi(\mu,0,\lambda)-\phi(0,0,\lambda)+\phi(\mu,\sigma,0)-\phi(0,\sigma,0)-\phi(\mu,0,0)\\&+\phi(j,t,\lambda)-\phi(0,t,\lambda)-\phi(j,0,\lambda)+\phi(0,0,\lambda)-\phi(j,t,0)+\phi(0,t,0)+\phi(j,0,0)\\&+[\Upsilon\Xi_{\mathcal{L}}(\lambda)]^T\Phi[\Upsilon^2\Xi_{\mathcal{L}}(\mu)][\Upsilon^2\Xi_{\mathcal{L}}(\sigma)]-([\Upsilon\Xi_{\mathcal{L}}(\lambda)]^T\Phi[\Upsilon^2\Xi_{\mathcal{L}}(\mu)][\Upsilon^2\Xi_{\mathcal{L}}(\sigma)])|_{(\mu=j,\sigma=t)}\\&\Big(\frac{\partial\theta(\mu,0,\lambda)}{\partial\sigma}+\frac{\partial\theta(0,\sigma,\lambda)}{\partial\sigma}-\frac{\partial\theta(0,0,\lambda)}{\partial\sigma}\\&+\frac{\partial\theta(\mu,\sigma,0)}{\partial\sigma}-\frac{\partial\theta(\mu,0,0)}{\partial\sigma}-\frac{\partial\theta(0,\sigma,0)}{\partial\sigma}+\frac{\partial\theta(0,0,0)}{\partial\sigma}\\&+\mu\Big[\frac{\partial^2\theta(0,\sigma,\lambda)}{\partial\mu\partial\sigma}-\frac{\partial^2\theta(0,0,\lambda)}{\partial\mu\partial\sigma}-\frac{\partial^2\theta(0,\sigma,0)}{\partial\mu\partial\sigma}+\frac{\partial^2\theta(0,0,0)}{\partial\mu\partial\sigma}\Big]+[\Upsilon\Xi_{\mathcal{L}}(\lambda)]^T\Theta[\Upsilon^2\Xi_{\mathcal{L}}(\mu)][\Upsilon\Xi_{\mathcal{L}}(\sigma)]\Big)\\&-\delta_0\Big(\frac{\partial^2\phi(\mu,0,\lambda)}{\partial\mu^2}+\frac{\partial^2\phi(\mu,\sigma,0)}{\partial\mu^2}-\frac{\partial^2\phi(\mu,0,0)}{\partial\mu^2}+\sigma\Big[\frac{\partial^3\phi(\mu,0,\lambda)}{\partial\mu^2\partial\sigma}+\frac{\partial^3\phi(\mu,0,0)}{\partial\mu^2\partial\sigma}\Big]\\&+[\Upsilon\Xi_{\mathcal{L}}(\lambda)]^T\Phi\Xi_{\mathcal{L}}(\mu)[\Upsilon^2\Xi_{\mathcal{L}}(\sigma)]+\frac{\partial^2\phi(0,\sigma,\lambda)}{\partial\sigma^2}+\frac{\partial^2\phi(\mu,\sigma,0)}{\partial\sigma^2}-\frac{\partial^2\phi(0,\sigma,0)}{\partial\sigma^2}\end{aligned}\tag{7.5.65}$$

$$+\mu\left[\frac{\partial^3\phi(0,\sigma,\lambda)}{\partial\mu\partial\sigma^2}-\frac{\partial^3\phi(0,\sigma,0)}{\partial\mu\partial\sigma^2}\right]$$
$$\left.+[\Upsilon\Xi_{\mathcal{L}}(\lambda)]^T\Phi[\Upsilon^2\Xi_{\mathcal{L}}(\mu)]\Xi_{\mathcal{L}}(\sigma)\right)-\mathrm{f}=0$$

$$\frac{\partial^3\theta(0,\sigma,\lambda)}{\partial\lambda^3}+\frac{\partial^3\theta(\mu,0,\lambda)}{\partial\lambda^3}+\frac{\partial^3\theta(j,t,\lambda)}{\partial\lambda^3}-\frac{\partial^3\theta(0,t,\lambda)}{\partial^3\lambda}-\frac{\partial^3\theta(j,0,\lambda)}{\partial\lambda^3}$$
$$+\frac{d^3}{d\lambda^3}\Big([\Upsilon\Xi_{\mathcal{L}}(\lambda)]^T\Theta[\Upsilon^2 Xi_{\mathcal{L}}(\mu)][\Upsilon^2\Xi_{\mathcal{L}}(\sigma)]$$
$$-([\Upsilon\Xi_{\mathcal{L}}(\lambda)]^T\theta[\Upsilon^2\Xi_{\mathcal{L}}(\mu)][\Upsilon^2\Xi_{\mathcal{L}}(\sigma)])|_{(\mu=j,\sigma=t)}\Big)$$
$$+\phi(0,\sigma,\lambda)+\phi(\mu,0,\lambda)-\phi(0,0,\lambda)+\phi(\mu,\sigma,0)-\phi(0,\sigma,0)-\phi(\mu,0,0)$$
$$+\phi(j,t,\lambda)-\phi(0,t,\lambda)-\phi(j,0,\lambda)+\phi(0,0,\lambda)-\phi(j,t,0)+\phi(0,t,0)+\phi(j,0,0)$$
$$+[\Upsilon\Xi_{\mathcal{L}}(\lambda)]^T\Phi[\Upsilon^2\Xi_{\mathcal{L}}(\mu)][\Upsilon^2\Xi_{\mathcal{L}}(\sigma)]-([\Upsilon\Xi_{\mathcal{L}}(\lambda)]^T\Phi[\Upsilon^2\Xi_{\mathcal{L}}(\mu)][\Upsilon^2\Xi_{\mathcal{L}}(\sigma)])|_{(\mu=j,\sigma=t)}$$
$$\Big(\frac{\partial\theta(0,\sigma,\lambda)}{\partial\mu}+\frac{\partial\theta(\mu,0,\lambda)}{\partial\mu}-\frac{\partial\theta(0,0,\lambda)}{\partial\mu}$$
$$+\frac{\partial\theta(\mu,\sigma,0)}{\partial\mu}-\frac{\partial\theta(0,\sigma,0)}{\partial\mu}-\frac{\partial\theta(\mu,0,0)}{\partial\mu}$$
$$+\frac{\partial\theta(0,0,0)}{\partial\mu}+\sigma\left[\frac{\partial^2\theta(\mu,0,\lambda)}{\partial\mu\partial\sigma}-\frac{\partial^2\theta(0,0,\lambda)}{\partial\mu\partial\sigma}+\frac{\partial^2\theta(\mu,0,0)}{\partial\mu\partial\sigma}-\frac{\partial^2\theta(0,0,0)}{\partial\mu\partial\sigma}\right]$$
$$+[\Upsilon\Xi_{\mathcal{L}}(\lambda)]^T\Theta[\Upsilon\Xi_{\mathcal{L}}(\mu)][\Upsilon^2\Xi_{\mathcal{L}}(\sigma)]\Big)$$
$$+\theta(0,\sigma,\lambda)+\theta(\mu,0,\lambda)-\theta(0,0,\lambda)+\theta(\mu,\sigma,0)-\theta(0,\sigma,0)-\theta(\mu,0,0)$$
$$+\theta(j,t,\lambda)-\theta(0,t,\lambda)-\theta(j,0,\lambda)+\theta(0,0,\lambda)-\theta(j,t,0)+\theta(0,t,0)+\theta(j,0,0)$$
$$+[\Upsilon\Xi_{\mathcal{L}}(\lambda)]^T\Theta[\Upsilon^2\Xi_{\mathcal{L}}(\mu)][\Upsilon^2\Xi_{\mathcal{L}}(\sigma)]-([\Upsilon\Xi_{\mathcal{L}}(\lambda)]^T\Theta[\Upsilon^2\Xi_{\mathcal{L}}(\mu)][\Upsilon^2\Xi_{\mathcal{L}}(\sigma)])|_{(\mu=j,\sigma=t)}\cdot$$
$$\Big(\frac{\partial\theta(\mu,0,\lambda)}{\partial\sigma}+\frac{\partial\theta(0,\sigma,\lambda)}{\partial\sigma}-\frac{\partial\theta(0,0,\lambda)}{\partial\sigma}$$
$$+\frac{\partial\theta(\mu,\sigma,0)}{\partial\sigma}-\frac{\partial\theta(\mu,0,0)}{\partial\sigma}-\frac{\partial\theta(0,\sigma,0)}{\partial\sigma}+\frac{\partial\theta(0,0,0)}{\partial\sigma}$$
$$+\mu\left[\frac{\partial^2\theta(0,\sigma,\lambda)}{\partial\mu\partial\sigma}-\frac{\partial^2\theta(0,0,\lambda)}{\partial\mu\partial\sigma}-\frac{\partial^2\theta(0,\sigma,0)}{\partial\mu\partial\sigma}+\frac{\partial^2\theta(0,0,0)}{\partial\mu\partial\sigma}\right]$$
$$+[\Upsilon\Xi_{\mathcal{L}}(\lambda)]^T\Theta[\Upsilon^2\Xi_{\mathcal{L}}(\mu)][\Upsilon\Xi_{\mathcal{L}}(\sigma)]\Big)$$
$$-\delta_0\Big(\frac{\partial^2\theta(\mu,0,\lambda)}{\partial\mu^2}+\frac{\partial^2\theta(\mu,\sigma,0)}{\partial\mu^2}-\frac{\partial^2\theta(\mu,0,0)}{\partial\mu^2}+\sigma\left[\frac{\partial^3\theta(\mu,0,\lambda)}{\partial\mu^2\partial\sigma}+\frac{\partial^3\theta(\mu,0,0)}{\partial\mu^2\partial\sigma}\right]$$
$$+[\Upsilon\Xi_{\mathcal{L}}(\lambda)]^T\Theta\Xi_{\mathcal{L}}(\mu)[\Upsilon^2\Xi_{\mathcal{L}}(\sigma)]$$
$$+\frac{\partial^2\theta(0,\sigma,\lambda)}{\partial\sigma^2}+\frac{\partial^2\theta(\mu,\sigma,0)}{\partial\sigma^2}-\frac{\partial^2\theta(0,\sigma,0)}{\partial\sigma^2}+\mu\left[\frac{\partial^3\theta(0,\sigma,\lambda)}{\partial\mu\partial\sigma^2}-\frac{\partial^3\theta(0,\sigma,0)}{\partial\mu\partial\sigma^2}\right]$$
$$+[\Upsilon\Xi_{\mathcal{L}}(\lambda)]^T\Theta[\Upsilon^2\Xi_{\mathcal{L}}(\mu)]\Xi_{\mathcal{L}}(\sigma)\Big)+\mathrm{f}=0.$$

By solving the above equation at the local points given in the previous section, the matrix of coefficients is obtained and by placing the matrices obtained in Eqs. (7.5.63) and (7.5.57), approximate solutions are obtained.

Example 7.8 We generalize Eq. (7.5.44), which is of the integer order, to the fractional-order equation, and obtain the approximate solution of the equation based on cliqUe polynomials. The generalization of the Navier–Stokes equation (NSE) to the fraction order is as follows

$$
\begin{aligned}
&{}^{C}D_{\lambda}^{\mathrm{m}}\phi(\mu,\sigma,\lambda)+\phi(\mu,\sigma,\lambda)D_{\mu}^{\mathrm{m}}\phi(\mu,\sigma,\lambda)+\phi(\mu,\sigma,\lambda)D_{\theta}^{\mathrm{m}}\theta(\mu,\sigma,\lambda) \qquad (7.5.66)\\
&=\delta_0\left(D_{\mu\mu}^{\mathrm{m}}\phi(\mu,\sigma,\lambda)+D_{\sigma\sigma}^{\mathrm{m}}\phi(\mu,\sigma,\lambda)\right)+\mathrm{f},\\
&{}^{C}D_{\lambda}^{\mathrm{m}}\theta(\mu,\sigma,\lambda)+\phi(\mu,\sigma,\lambda)D_{\mu}^{\mathrm{m}}\theta(\mu,\sigma,\lambda)+\theta(\mu,\sigma,\lambda)D_{\sigma}^{\beta}\theta(\mu,\sigma,\lambda)\\
&=\delta_0\left(D_{\mu\mu}^{\mathrm{m}}\theta(\mu,\sigma,\lambda)+D_{\sigma\sigma}^{\mathrm{m}}\theta(\mu,\sigma,\lambda)\right)+\mathrm{f},
\end{aligned}
$$

where $1<\mathrm{m}<2$, $(\mu,\sigma,\lambda)\in[i,j]\times[k,t]\times[0,\mathcal{F}]$ and with the initial conditions

$$
\begin{aligned}
\phi(\mu,\sigma,0)&=\varpi_1(\mu,\sigma)=-\sin(\mu+\sigma), \qquad D_{\lambda}\phi(\mu,\sigma,0)=\varpi_2(\mu,\sigma) \qquad (7.5.67)\\
\theta(\mu,\sigma,0)&=\varpi_1'(\mu,\sigma)=\sin(\mu+\sigma), \qquad D_{\lambda}\theta(\mu,\sigma,0)=\varpi_2'(\mu,\sigma)
\end{aligned}
$$

and boundary conditions

$$
\begin{aligned}
&\phi(i,\sigma,\lambda)=\upsilon_1(\lambda), \quad \phi(j,\sigma,\lambda)=\upsilon_2(\lambda), \quad \phi(\mu,k,\lambda)=\upsilon_3(\lambda), \quad \phi(\mu,t,\lambda)=\upsilon_4(\lambda) \qquad (7.5.68)\\
&\theta(\mu,k,\lambda)=\upsilon_1'(\lambda), \quad \theta(\mu,t,\lambda)=\upsilon_2'(\lambda), \quad \theta(i,\sigma,\lambda)=\upsilon_3'(\lambda), \quad \theta(j,\sigma,\lambda)=\upsilon_4'(\lambda),
\end{aligned}
$$

Now, using the clique polynomial method presented, we obtain the approximate solution of the above fractional equation. To do this, we first approximate $D_{\mu\mu}^{\mathrm{m}}\phi(\mu,\sigma,\lambda)$:

$$
\begin{aligned}
D_{\mu\mu}^{\mathrm{m}}\phi(\mu,\sigma,\lambda)&\simeq\sum_{\jmath=0}^{\mathcal{L}-1}\sum_{\iota=1}^{\mathcal{L}-1}\sum_{\epsilon=1}^{\mathcal{L}-1}\phi_{\jmath\iota\epsilon}\mathcal{W}(S_{\jmath},\mu)\mathcal{W}(S_{\iota},\sigma)\mathcal{W}(S_{\epsilon},\lambda) \qquad (7.5.69)\\
&=\Xi_{\mathcal{L}}(\lambda)^{T}\Phi\,\Xi_{\mathcal{L}}(\mu)\Xi_{\mathcal{L}}(\sigma)
\end{aligned}
$$

Now, we take integrals from Eq. (7.5.69) twice regard to μ and considering $i=0$,

$$
D_{\mu}^{m}\phi(\mu,\sigma,\lambda)\simeq D_{\mu}^{\mathrm{m}}\phi(0,\sigma,\lambda)+[\Upsilon\,\Xi_{\mathcal{L}}(\mu)]^{T}\Phi\,\Xi_{\mathcal{L}}(\sigma)\Xi_{\mathcal{L}}(\lambda), \qquad (7.5.70)
$$

and

$$
\phi(\mu,\sigma,\lambda)\simeq\mu D_{\mu}^{\mathrm{m}}\phi(0,\sigma,\lambda)+\upsilon_1(\lambda)+[\Upsilon^{2}\Xi_{\mathcal{L}}(\mu)]^{T}\Phi\,\Xi_{\mathcal{L}}(\sigma)\Xi_{\mathcal{L}}(\lambda). \qquad (7.5.71)
$$

By placing $\mu=j$ in (7.5.71) we have:

$$
D_{\mu}^{m}\phi(0,\sigma,\lambda)=\frac{1}{j}\left[\upsilon_2(\lambda)-\upsilon_1(\lambda)-[\Upsilon^{2}\Xi_{\mathcal{L}}(j)]^{T}\Phi\,\Xi_{\mathcal{L}}(\sigma)\Xi_{\mathcal{L}}(\lambda)\right]. \qquad (7.5.72)
$$

By putting Eq. (7.5.72) in Eqs. (7.5.70) and (7.5.71) we have

$$D_{\mu}^{\mathrm{m}}\phi(\mu,\sigma,\lambda) \simeq \frac{1}{j}\Big[\upsilon_2(\lambda) - \upsilon_1(\lambda) - [\Upsilon^2 \Xi_{\mathcal{L}}(j)]^T \Phi \Xi_{\mathcal{L}}(\sigma)\Xi_{\mathcal{L}}(\lambda)\Big] \tag{7.5.73}$$
$$+ [\Upsilon \Xi_{\mathcal{L}}(\mu)]^T \Phi \Xi_{\mathcal{L}}(\sigma)\Xi_{\mathcal{L}}(\lambda),$$

and also

$$\phi(\mu,\sigma,\lambda) \simeq \frac{\mu}{j}\Big[\upsilon_2(\lambda) - \upsilon_1(\lambda) - [\Upsilon^2 Xi_{\mathcal{L}}(j)]^T \Phi \Xi_{\mathcal{L}}(\sigma)\Xi_{\mathcal{L}}(\lambda)\Big] \tag{7.5.74}$$
$$+ [\Upsilon^2 \Xi_{\mathcal{L}}(\mu)]^T \Phi \Xi_{\mathcal{L}}(\sigma)\Xi_{\mathcal{L}}(\sigma) + \upsilon_1(\lambda).$$

Exactly the same as the above case, we also approximate $D_{\sigma\sigma}^{\mathrm{m}}\theta(\mu,\sigma,\lambda)$. Then, assuming

$$D_{\sigma\sigma}^{\mathrm{m}}\theta(\mu,\sigma,\lambda) \simeq \sum_{\jmath=0}^{\mathcal{L}-1}\sum_{\iota=1}^{\mathcal{L}-1}\sum_{\epsilon=1}^{\mathcal{L}-1} \phi_{\jmath\iota\epsilon}\mathcal{W}(\mathcal{S}_{\jmath},\mu)\mathcal{W}(\mathcal{S}_{\iota},\sigma)\mathcal{W}(\mathcal{S}_{\epsilon},\lambda) \tag{7.5.75}$$
$$= \Xi_{\mathcal{L}}(\lambda)^T \Theta \Xi_{\mathcal{L}}(\mu)\Xi_{\mathcal{L}}(\sigma),$$

again by taking the integral from Eq. (7.5.75) twice regard to σ and considering $k = 0$

$$D_{\sigma}^{\mathrm{m}}\theta(\mu,\sigma,\lambda) \simeq D_{\sigma}^{\mathrm{m}}\theta(\mu,0,\lambda) + [\Upsilon \Xi_{\mathcal{L}}(\sigma)]^T \Theta \Xi_{\mathcal{L}}(\mu)\Xi_{\mathcal{L}}(\lambda), \tag{7.5.76}$$

and

$$\theta(\mu,\sigma,\lambda) \simeq \sigma D_{\sigma}^{\mathrm{m}}\theta(\mu,0,\lambda) + \upsilon_1'(\lambda) + [\Upsilon^2 \Xi_{\mathcal{L}}(\sigma)]^T \Theta \Xi_{\mathcal{L}}(\mu)\Xi_{\mathcal{L}}(\lambda). \tag{7.5.77}$$

By placing $\sigma = t$ in (7.5.77) we will have:

$$D_{\sigma}^{\mathrm{m}}\theta(\mu,0,\lambda) = \frac{1}{t}\Big[\upsilon_2'(\lambda) - \upsilon_1'(\lambda) - [\Upsilon^2 \Xi_{\mathcal{L}}(t)]^T \Theta \Xi_{\mathcal{L}}(\mu)\Xi_{\mathcal{L}}(\lambda)\Big]. \tag{7.5.78}$$

By putting Eq. (7.5.78) in Eqs. (7.5.76) and (7.5.77) we have

$$D_{\sigma}^{\mathrm{m}}\theta(\mu,\sigma,\lambda) \simeq \frac{1}{t}\Big[\upsilon_2'(\lambda) - \upsilon_1'(\lambda) - [\Upsilon^2 \Xi_{\mathcal{L}}(t)]^T \Theta Xi_{\mathcal{L}}(\mu)Xi_{\mathcal{L}}(\lambda)\Big] \tag{7.5.79}$$
$$+ [\Upsilon \Xi_{\mathcal{L}}(\sigma)]^T \Theta \Xi_{\mathcal{L}}(\mu)\Xi_{\mathcal{L}}(\lambda),$$

also

$$\theta(\mu, \sigma, \lambda) \simeq \frac{\sigma}{t}\Big[\upsilon_2'(\lambda) - \upsilon_1'(\lambda) - [\Upsilon^2 \Xi_{\mathcal{L}}(t)]^T \Theta\, \Xi_{\mathcal{L}}(\mu)\Xi_{\mathcal{L}}(\lambda)\Big] \tag{7.5.80}$$
$$+ [\Upsilon^2 \Xi_{\mathcal{L}}(\sigma)]^T \Theta\, \Xi_{\mathcal{L}}(\mu)\Xi_{\mathcal{L}}(\lambda) + \upsilon_1'(\lambda).$$

To complete the process, we approximate $D^{\mathrm{m}}_{\mu\mu}\theta(\mu, \sigma, \lambda)$. Therefore, we assume

$$D^{\mathrm{m}}_{\mu\mu}\theta(\mu, \sigma, \lambda) \simeq \sum_{\jmath=0}^{\mathcal{L}-1}\sum_{\iota=1}^{\mathcal{L}-1}\sum_{\epsilon=1}^{\mathcal{L}-1} \phi_{\jmath\iota\epsilon}\mathcal{W}(\mathcal{S}_\jmath, \mu)\mathcal{W}(\mathcal{S}_\iota, \sigma)\mathcal{W}(\mathcal{S}_\epsilon, \lambda) \tag{7.5.81}$$
$$= \Xi_{\mathcal{L}}(\lambda)^T \Theta\, \Xi_{\mathcal{L}}(\mu)\Xi_{\mathcal{L}}(\sigma).$$

By taking the integral from Eq. (7.5.81) regard to μ and considering $i = 0$

$$D^{\mathrm{m}}_{\mu}\theta(\mu, \sigma, \lambda) = D^{\mathrm{m}}_{\mu}\theta(0, \sigma, \lambda) + [\Upsilon\, \Xi_{\mathcal{L}}(\mu)]^T \Theta\, \Xi_{\mathcal{L}}(\sigma)\Xi_{\mathcal{L}}(\lambda), \tag{7.5.82}$$

again by taking the integral, we will have

$$\theta(\mu, \sigma, \lambda) = [\Upsilon^2 \Xi_{\mathcal{L}}(\mu)]^T \Theta\, \Xi_{\mathcal{L}}(\sigma)\Xi_{\mathcal{L}}(\lambda) + \mu D^{\mathrm{m}}_{\mu}\theta(0, \sigma, \lambda) + \upsilon_3'(\lambda), \tag{7.5.83}$$

By putting $\mu = j$ in (7.5.83) we obtain

$$D^{\mathrm{m}}_{\mu}\theta(0, \sigma, \lambda) = \frac{1}{j}\Big[\upsilon_4'(\lambda) - \upsilon_3'(\lambda) - [\Upsilon^2 \Xi_{\mathcal{L}}(\mu)]^T \Theta\, \Xi_{\mathcal{L}}(\sigma)\Xi_{\mathcal{L}}(\lambda)\Big], \tag{7.5.84}$$

by putting Eq. (7.5.84) in Eq. (7.5.82) we have

$$D^{\mathrm{m}}_{\mu}\theta(\mu, \sigma, \lambda) = \frac{1}{j}\Big[\upsilon_4'(\lambda) - \upsilon_3'(\lambda) - [\Upsilon^2 \Xi_{\mathcal{L}}(\mu)]^T \Theta\, \Xi_{\mathcal{L}}(\sigma)\Xi_{\mathcal{L}}(\lambda)\Big] + \tag{7.5.85}$$
$$[\Upsilon\, \Xi_{\mathcal{L}}(\mu)]^T \Theta\, \Xi_{\mathcal{L}}(\sigma)\Xi_{\mathcal{L}}(\lambda)$$

By putting (7.5.85), (7.5.81), (7.5.80), (7.5.79), (7.5.75), (7.5.74), (7.5.73) and (7.5.69) in Eq. (7.5.66) and taking Riemann–Liouville fractional integral (RL-integral) of order m, we have

$$\frac{\mu}{j}\Big[\upsilon_2(\lambda) - \upsilon_1(\lambda) - [\Upsilon^2 \Xi_{\mathcal{L}}(j)]^T \Phi\, \Xi_{\mathcal{L}}(\sigma)\Xi_{\mathcal{L}}(\sigma)\Big] \tag{7.5.86}$$
$$+ [\Upsilon^2 \Xi_{\mathcal{L}}(\mu)]^T \Phi\, \Xi_{\mathcal{L}}(\sigma)\Xi_{\mathcal{L}}(\lambda) + \upsilon_1(\lambda) - \varpi_1(\mu, \sigma) - \lambda\varpi_2(\mu, \sigma) +$$
$$\Bigg(\bigg(\frac{\mu}{j}\Big[\upsilon_2(\lambda) - \upsilon_1(\lambda) - [\Upsilon^2 \Xi_{\mathcal{L}}(j)]^T \Phi\, \Xi_{\mathcal{L}}(\sigma)\Xi_{\mathcal{L}}(\lambda)\Big] + [\Upsilon^2 \Xi_{\mathcal{L}}(\mu)]^T \Phi\, \Xi_{\mathcal{L}}(\sigma)\Xi_{\mathcal{L}}(\sigma) + \upsilon_1(\lambda)\bigg)$$
$$\bigg[\frac{1}{j}\Big[\upsilon_2(\lambda) - \upsilon_1(\lambda) - [\Upsilon^2 \Xi_{\mathcal{L}}(j)]^T \Phi\, \Xi_{\mathcal{L}}(\sigma)\Xi_{\mathcal{L}}(\lambda)\Big] + [\Upsilon\, \Xi_{\mathcal{L}}(\mu)]^T \Phi\, \Xi_{\mathcal{L}}(\sigma)\Xi_{\mathcal{L}}(\lambda)\bigg]\Bigg) +$$
$$\Bigg(\bigg(\frac{\mu}{j}\Big[\upsilon_2(\lambda) - \upsilon_1(\lambda) - [\Upsilon^2 \Xi_{\mathcal{L}}(j)]^T \Phi\, \Xi_{\mathcal{L}}(\sigma)\Xi_{\mathcal{L}}(\lambda)\Big] + [\Upsilon^2 \Xi_{\mathcal{L}}(\nu)]^T \Phi\, \Xi_{\mathcal{L}}(\sigma)\Xi_{\mathcal{L}}(\lambda) + \upsilon_1(\lambda)\bigg)$$
$$\bigg[\frac{1}{t}\Big[\upsilon_2'(\lambda) - \upsilon_1'(\lambda) - [\Upsilon^2 \Xi_{\mathcal{L}}(t)]^T \Theta\, \Xi_{\mathcal{L}}(\mu)\Xi_{\mathcal{L}}(\lambda)\Big] + [\Upsilon\, \Xi_{\mathcal{L}}(\sigma)]^T \Theta\, \Xi_{\mathcal{L}}(\mu)\Xi_{\mathcal{L}}(\lambda)\bigg]\Bigg) -$$

$$
\begin{aligned}
&\delta_0\Big(\mathcal{R}^{\mathrm{m}}\,\Xi_{\mathcal{L}}(\lambda)^T\,\Phi\,\Xi_{\mathcal{L}}(\mu)\Xi_{\mathcal{L}}(\sigma)+\mathcal{R}^{\mathrm{m}}\,\Xi_{\mathcal{L}}(\lambda)^T\,\Phi\,\Xi_{\mathcal{L}}(\mu)\Xi_{\mathcal{L}}(\sigma)\Big)-\mathrm{f}=0\\
&\frac{\sigma}{t}\Big[\upsilon_2'(\lambda)-\upsilon_1'(\lambda)-[\Upsilon^2\,\Xi_{\mathcal{L}}(t)]^T\,\Theta\,\Xi_{\mathcal{L}}(\mu)\Xi_{\mathcal{L}}(\lambda)\Big]\\
&+[\Upsilon^2\,\Xi_{\mathcal{L}}(\sigma)]^T\,\Theta\,\Xi_{\mathcal{L}}(\mu)\Xi_{\mathcal{L}}(\lambda)+\upsilon_1'(\lambda)-\varpi_1'(\mu,\sigma)-\lambda\varpi_2'(\mu,\sigma)+\\
&\Big(\Big(\frac{\xi}{b}\Big[n_2(t)-n_1(t)-[P^2\,\Xi_{\mathcal{L}}(b)]^T\,U\,\Xi_{\mathcal{L}}(y)\Xi_{\mathcal{L}}(\lambda)\Big]+[\Upsilon^2\,\Xi_{\mathcal{L}}(\mu)]^T\,\Phi\,\Xi_{\mathcal{L}}(\sigma)\Xi_{\mathcal{L}}(\lambda)+\upsilon_1(\lambda)\Big)\\
&\Big[\frac{1}{j}\Big[\upsilon_4'(\lambda)-\upsilon_3'(\lambda)-[\Upsilon^2\,\Xi_{\mathcal{L}}(\mu)]^T\,\Theta\,\Xi_{\mathcal{L}}(\sigma)\Xi_{\mathcal{L}}(\lambda)\Big]+[\Upsilon\,\Xi_{\mathcal{L}}(\mu)]^T\,\Theta\,\Xi_{\mathcal{L}}(\sigma)\Xi_{\mathcal{L}}(\lambda)\Big]\Big)+\\
&\Big(\Big(\frac{\sigma}{t}\Big[\upsilon_2'(\lambda)-\upsilon_1'(\lambda)-[\Upsilon^2\,\Xi_{\mathcal{L}}(t)]^T\,\Theta\,\Xi_{\mathcal{L}}(\mu)\Xi_{\mathcal{L}}(\lambda)\Big]+[\Upsilon^2\,\Xi_{\mathcal{L}}(\sigma)]^T\,\Theta\,\Xi_{\mathcal{L}}(\mu)\Xi_{\mathcal{L}}(\lambda)+\upsilon_1'(\lambda)\Big)\\
&\Big[\frac{1}{t}\Big[\upsilon_2'(\lambda)-\upsilon_1'(\lambda)-[\Upsilon^2\,\Xi_{\mathcal{L}}(t)]^T\,\Theta\,\Xi_{\mathcal{L}}(\mu)\Xi_{\mathcal{L}}(\lambda)\Big]+[\Upsilon\,\Xi_{\mathcal{L}}(\sigma)]^T\,\Theta\,\Xi_{\mathcal{L}}(\mu)\Xi_{\mathcal{L}}(\lambda)\Big]\Big)-\\
&\delta_0\Big(\mathcal{R}^{\mathrm{m}}\,\Xi_{\mathcal{L}}(\lambda)^T\,\Theta\,\Xi_{\mathcal{L}}(\mu)\Xi_{\mathcal{L}}(\sigma)+\mathcal{R}^{\mathrm{m}}\,\Xi_{\mathcal{L}}(\lambda)^T\,\Theta\,\Xi_{\mathcal{L}}(\mu)\Xi_{\mathcal{L}}(\sigma)\Big)+\mathrm{f}=0
\end{aligned}
$$

By placing the local points $\mu_\jmath = \frac{\jmath(j-i)}{\mathcal{L}+2}$, $\sigma_\iota = \frac{\iota(j-i)}{\mathcal{L}+2}$ and $\lambda_\epsilon = \frac{\epsilon}{\mathcal{L}+2}\mathcal{F}$ in Eq. (7.5.86) and solving the set of equations, $\phi_\jmath\,\iota\epsilon$ and $\therefore_\jmath\,\iota\epsilon$ are obtained for $\jmath, \iota, \epsilon = 1, 2, \ldots, \mathcal{L}$. By placing the obtained matrices in (7.5.74) and (7.5.80), the approximate values of $\phi(\mu, \sigma, \lambda)$ and $\theta(\mu, \sigma, \lambda)$ are obtained.

Example 7.9 Consider the following Navier–Stokes equation (N-SE) (Fig. 7.19).

$$
\begin{aligned}
&\frac{\partial^3\phi(\mu,\sigma,\lambda)}{\partial\lambda^3}+\phi(\mu,\sigma,\lambda)\frac{\partial\phi(\mu,\sigma,\lambda)}{\partial\mu}+\phi(\mu,\sigma,\lambda)\frac{\partial\theta(\mu,\sigma,\lambda)}{\partial\sigma}=\delta_0\Big(\frac{\partial^2\phi(\mu,\sigma,\lambda)}{\partial\mu^2}+\frac{\partial^2\phi(\mu,\sigma,\lambda)}{\partial\sigma^2}\Big)+\mathrm{f}\\
&\frac{\partial^3\theta(\mu,\sigma,\lambda)}{\partial\lambda^3}+\phi(\mu,\sigma,\lambda)\frac{\partial\theta(\mu,\sigma,\lambda)}{\partial\mu}+\theta(\mu,\sigma,\lambda)\frac{\partial\theta(\mu,\sigma,\lambda)}{\partial\sigma}=\delta_0\Big(\frac{\partial^2\theta(\mu,\sigma,\lambda)}{\partial\mu^2}+\frac{\partial^2\theta(\mu,\sigma,\lambda)}{\partial\sigma^2}\Big)-f,
\end{aligned}
\tag{7.5.87}
$$

for $\mathrm{m} = 3$ and $(\mu, \sigma, \lambda) \in [i, j] \times [k, t] \times [0, \mathcal{F}]$. The initial conditions and boundary conditions are considered as follows

$$\phi(\mu, \sigma, 0) = -\exp(\mu + \sigma) \tag{7.5.88}$$
$$u(\mu, \sigma, 0) = \exp(\mu + \sigma), \tag{7.5.89}$$

and

$$
\begin{aligned}
&\phi(j, \sigma, \lambda) = \zeta_1(\lambda), \quad \phi(\mu, t, \lambda) = \zeta_2(\lambda), \quad \phi(j, t, \lambda) = \zeta_3(\lambda)\\
&\theta(j, \sigma, \lambda) = \zeta_1'(\lambda), \quad \phi(\mu, t, \lambda) = \zeta_2'(\lambda), \quad \phi(j, t, \lambda) = \zeta_3'(\lambda),
\end{aligned}
\tag{7.5.90}
$$

Similar to the process performed in Example 7.7, in this example we obtain numerical results based on the proposed method for this equation with different initial and boundary conditions.

In Tables 7.15 and 7.16, according to the absolute error, a comparison between L_2-error, L_∞-error, and RMS-error for N-SEs has been made using the clique polynomial

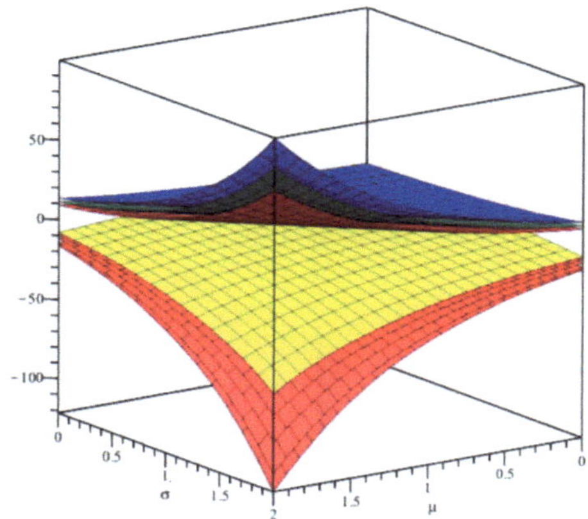

(a) Exact solution for example (7.9) for $\lambda = 0.1, 0.2, 0.3, 0.4, 0.5, 0.6, 0.7, 0.8, \mu \in (0, 2), \sigma \in (0, 2)$.

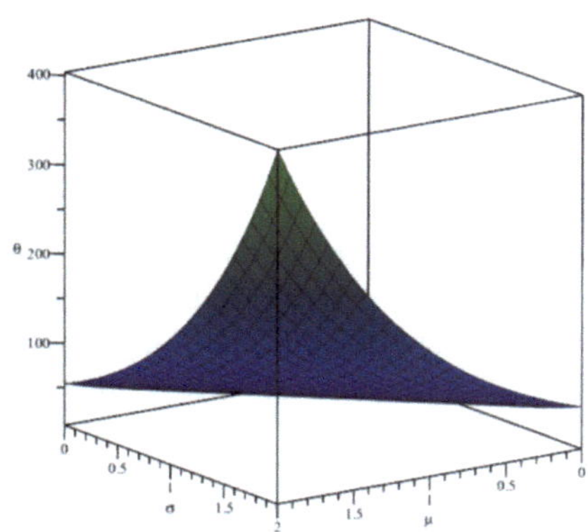

(b)

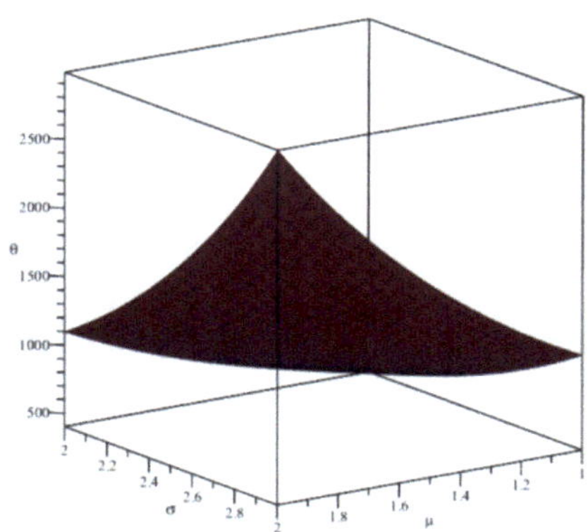

(c)

Fig. 7.19 Exact solution for Example 7.9. **a** $\lambda = 2$, $\mu \in (0, 2)$, $\sigma \in (0, 2)$, **b** $\lambda = 3$, $\mu \in (1, 2)$, $w \in (2, 3)$, **c** $\lambda = 5$, $u \in (0, \frac{1}{8})$, $w \in (0, \frac{9}{5})$, **d** $\lambda = 4$, $u \in (\frac{1}{2}, 1)$, $w \in (0, \frac{3}{7})$, **e** $\lambda = 2$, $\mu \in (\frac{1}{4}, \frac{1}{2})$, $w \in (0, \frac{7}{5})$, **f** $\lambda = 3$, $\mu \in (1, \frac{3}{2})$, $w \in (2, \frac{5}{2})$

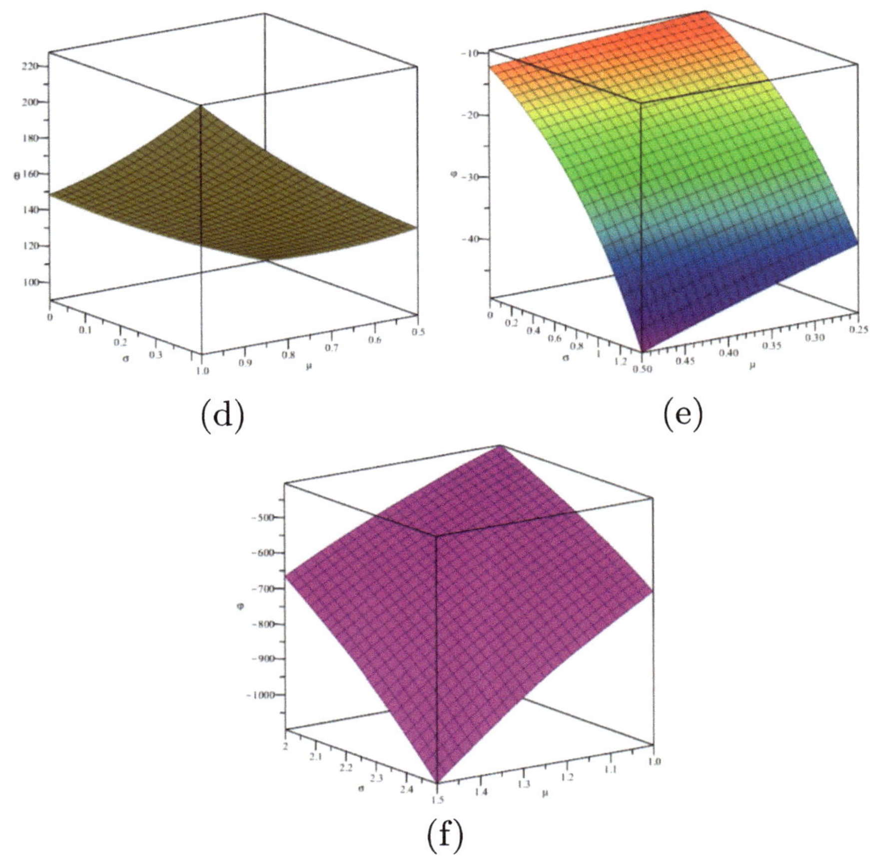

Fig. 7.19 (continued)

technique and using the FRDT method. FRDT method is another method of solving this type of equations. It is clear from the table that the error of the clique polynomial method is less than the FRDT method. Therefore, considering the many applications that N-SEs have, we propose the clique polynomial method to solve these equations.

Table 7.15 Compare the errors of our proposed method and other methods for Example 7.8

$\mathcal{F}$	$L_2 - error$	$RMS - error$	$L_\infty - error$
Numerical results using FRDT method			
–	2.1146	0.0552	0.7048
–	2.2115	0.2457	0.7371
–	2.2862	1.1971	0.7620
–	2.3382	1.3230	0.7794
–	2.3669	1.4622	0.7889
–	2.3718	1.6159	0.7906
–	2.3530	1.7859	0.7834
–	2.3107	5.9213	0.7702
–	2.2453	2.1813	0.7484
–	2.1575	2.4107	0.7191
Numerical results using our method			
1	0.0013	0.00010	0.0003
2	0.0015	0.00015	0.0004
3	0.0017	0.00018	0.0005
4	0.0019	0.00021	0.0006
5	0.0021	0.00024	0.0060
6	0.0023	0.00026	0.0007
7	0.0025	0.00028	0.0008
8	0.0027	0.00030	0.0009
9	0.0029	0.00032	0.0009
10	0.0030	0.00034	0.0010

Table 7.16 Compare the errors of our proposed method and other methods for Example 7.8

$\mathcal{F}$	$L_2 - error$	$RMS - error$	$L_\infty - error$
Numerical results using FRDT method			
–	2.772	0.9801	0.3099
–	3.249	1.0832	0.3425
–	3.591	1.1971	0.3785
–	3.969	1.3230	0.4183
–	4.386	1.4622	0.4623
–	4.847	1.6159	0.5110
–	5.357	1.7859	0.5647
–	5.921	5.9213	0.6241
–	6.544	2.1813	0.6898
–	7.232	2.4107	0.7623
Numerical results using our method			
1	0.0164	0.0016	0.0052
2	0.0005	0.0023	0.0073
3	0.0285	0.0028	0.0090
4	0.0329	0.0032	0.0104
5	0.0368	0.0036	0.0116
6	0.0403	0.0040	0.0127
7	0.0435	0.0043	0.0137
8	0.0465	0.0046	0.147
9	0.0494	0.0049	0.0156
10	0.0520	0.0052	0.0164

References

1. Ahmad, B., Ntouyas, S.K., Tariboon, J.: On hybrid Caputo fractional integro-differential inclusions with nonlocal conditions. J. Nonlinear Sci. Appl. **9**(6), 4235–4246 (2016)
2. Ahmadinia, M., Safari, Z.: Analysis of local discontinuous Galerkin method for time-space fractional sine-Gordon equations. Appl. Numer. Math. **148**, 1–17 (2020)
3. Argyris, J., Haase, M., Heinrich, J.C.: Finite element approximation to two-dimensional sine-Gordon solitons. Comput. Methods Appl. Mech. Engrg. **86**(1), 1–26 (1991)
4. Atangana, A.: Modelling the spread of COVID-19 with new fractal-fractional operators: can the lockdown save mankind before vaccination? Chaos Solitons Fractals **136**, 109860 (2020)
5. Atangana, A., Araz, S.I.: New numerical method for ordinary differential equations: Newton polynomial. J. Comput. Appl. Math. **372**(112622), 22 (2020)
6. Baleanu, D., Shiri, B.: Collocation methods for fractional differential equations involving non-singular kernel. Chaos Solitons Fractals **116**, 136–145 (2018)
7. Bekir, A., Güner, Ö., Cevikel, A.C.: Fractional complex transform and exp-function methods for fractional differential equations. Abstr. Appl. Anal. Art. ID 426462, 8 pp (2013)
8. Belmekki, M., Nieto, J.J., Rodriguez-Lopez, R.: Existence of periodic solution for a nonlinear fractional differential equation. Bound. Value Probl. Art. ID 324561, 18 pp (2009)
9. Bratsos, A.G.: A third order numerical scheme for the two-dimensional sine-Gordon equation. Math. Comput. Simulation **76**(4), 271–282 (2007)
10. Cao, Y., Yang, H., Yin, L.: Finite element methods for semilinear elliptic stochastic partial differential equations. Numer. Math. **106**(2), 181–198 (2007)
11. Carpinteri, A., Mainardi, F.: Fractals and fractional calculus in continuum mechanics. Papers from the CISM Course on Scaling Laws and Fractality in Continuum Mechanics held in Udine, September 23–27, 1996. CISM International Centre for Mechanical Sciences. Courses and Lectures, 378. Springer, Vienna (1997). vi+348 pp. ISBN: 3-211-82913-X
12. Chen, T.M., Rui, J., Wang, Q.P., Zhao, Z.Y., Cui, J.A., Yin, L.: A mathematical model for simulating the phase-based transmissibility of a novel Coronavirus. Infectious Diseases Poverty **9**(1), 1–8 (2020)
13. Chu, Y.M., Inc, M., Hashemi, M.S., Eshaghi, S.: Comput. Appl. Math. **41**(6), 1–17 (2022)
14. Cialenco, I., Fasshauer, G.E., Ye, Q.: Approximation of stochastic partial differential equations by a kernel-based collocation method. Int. J. Comput. Math. **89**(18), 2543–2561 (2012)
15. Cui, M., Lin, Y.: Nonlinear numerical analysis in the reproducing kernel space. Nova Science Publishers, Inc., New York (2009). xiv+226 pp. ISBN: 978-1-60456-468-6; 1-60456-468-7
16. Dehghan, M., Abbaszadeh, M.: Two meshless procedures: moving Kriging interpolation and element-free Galerkin for fractional PDEs. Appl. Anal. **96**(6), 936–969 (2017)
17. Dehghan, M., Abbaszadeh, M., Mohebbi, A.: An implicit RBF meshless approach for solving the time fractional nonlinear sine-Gordon and Klein-Gordon equations. Eng. Anal. Bound. Elem. **50**, 412–434 (2015)
18. Dhage, B.C., Lakshmikantham, V.: Basic results on hybrid differential equations. Nonlinear Anal. Hybrid Syst. **4**(3), 414–424 (2010)
19. Din, A., Yongjin, L., Abdullahi, Y., Aliyu, I.A.: Caputo type fractional operator applied to Hepatitis B system. Fractals 2240023 (2021)
20. Din, A., Yongjin, L., Faiz, M.K., Zia, U.K., Peijiang, L.: On Analysis of fractional order mathematical model of Hepatitis B using Atangana-Baleanu Caputo (ABC) derivative. Fractals **30**(01), 2240017 (2022)
21. Dodd, R.K., Eilbeck, J.C., Gibbon, J.D., Morris, H.C.: Solitons and nonlinear wave equations. Academic Press, Inc. [Harcourt Brace Jovanovich, Publishers], London-New York, 1982. x+630 pp. ISBN: 0-12-219120-X
22. Duffy, B.R., Parkes, E.J.: Phys. Lett. A **214**(5–6), 271 (1996)
23. Duffy, B.R., Parkes E.: J. Comput. Phys. Commun. **98**(3), 288 (1996)
24. Eidinejad, Z; Saadati, R; De La Sen, M. Picard method for existence, uniqueness, and Gauss Hypergeomatric stability of the fractional-order differential equations. Mathematical Problems in Engineering 2021 (2021)

25. Eidinejad, Z., Saadati, R., Li, C., Inc, M., Vahidi, J.: The multiple exp-function method to obtain soliton solutions of the conformable Date–Jimbo–Kashiwara–Miwa equations
26. Eidinejad, Z., Saadati, R., Li, C., Inc, M., Vahidi, J.: The multiple exp-function method to obtain soliton solutions of the conformable Date–Jimbo–Kashiwara–Miwa equations. Int. J. Modern Phys. B 2450043 (2023)
27. Eidinejad, Z., Saadati, R., Vahidi, J., Li, C.: Numerical solutions of 2D stochastic time-fractional Sine–Gordon equation in the Caputo sense. Int. J. Numer. Modell.: Electronic Netw. Dev. Fields p. e3121
28. Eidinejad, Z., Saadati, R.: Hyers-Ulam-Rassias-Wright stability for fractional oscillation equation. Discrete Dynamics in Nature and Society 2022 (2022)
29. Eidinejad, Z., Saadati, R.: Hyers-Ulam-Rassias-Kummer stability of the fractional integro-differential equations. Math. Biosci. Eng. **19**(7), 6536–6550 (2022)
30. Eidinejad, Z., Saadati, R., de la Sen, M.: Radu-Mihet method for the existence, uniqueness, and approximation of the ψ-Hilfer fractional equations by matrix-valued fuzzy controllers. Axioms **10**(2), 63 (2021)
31. El-Shahed, M., Salem, A.: On the generalized Navier-Stokes equations. Appl. Math. Comput. **156**(1), 287–293 (2004)
32. Fan, E., Hongqing, Z.: Phys. Lett. A **246**(5), 403 (1998)
33. Fasshauer, G.E., Ye, Q.: Kernel-based collocation methods versus Galerkin finite element methods for approximating elliptic stochastic partial differential equations. Meshfree methods for partial differential equations VI, 155–170, Lecture Notes in Computational Science and Engineering, 89. Springer, Heidelberg (2013)
34. Fukunaga, M.: A new method for laplace transforms of multiterm fractional differential equations of the caputo type. J. Comput. Nonlinear Dyn. **16**(10) (2021)
35. Ganji, Z.Z., Ganji, D.D., Ganji, A.D., Rostamian, M.: Analytical solution of time-fractional Navier-Stokes equation in polar coordinate by homotopy perturbation method. Numer. Methods Partial Diff. Equ. **26**(1), 117–124 (2010)
36. Ganji, R.M., Jafari, H., Kgarose, M., Mohammadi, A.: Numerical solutions of time-fractional Klein-Gordon equations by clique polynomials. Alexandria Eng. J. **60**(5), 4563–4571 (2021)
37. Ganji, R.M., Jafari, H., Adem, A.R.: A numerical scheme to solve variable order diffusion-wave equations. Thermal Sci. **23**(6), 2063–2071 (2019)
38. Ganji, R.M., Jafari, H., Kgarose, M., Mohammadi, A.: Numerical solutions of time-fractional Klein-Gordon equations by clique polynomials. Alexandria Eng. J **60**(5), 4563–4571 (2021)
39. Geng, F., Cui, M.: A reproducing kernel method for solving nonlocal fractional boundary value problems. Appl. Math. Lett. **25**(5), 818–823 (2012)
40. Gepreel, K.A., Saleh, O.: Exact solutions for nonlinear partial fractional differential equations. Chin. Phys. B **21**(11), 110204 (2012)
41. Guo, B., Pu, X., Huang, F.: Fractional partial differential equations and their numerical solutions. Originally published by Science Press in 2011. World Scientific Publishing Co. Pte. Ltd., Hackensack, NJ (2015). x+336 pp. ISBN: 978-981-4667-04-3
42. Guo, P.: The Adomian decomposition method for a type of fractional differential equations. J. Appl. Math. Phys. **7**(10), 2459–2466 (2019)
43. Guo, F., Lin, J.: Nonlinear Dyn **96**, 1233–1241 (2019)
44. Guo, S., Mei, L.: The fractional variational iteration method using He's polynomials. Phys. Lett. A **375**(3), 309–313 (2011)
45. Guo, X., Lexin, L., Qiang, W.: Modeling interactive components by coordinate kernel polynomial models. Math. Found. Comput. **3**(4), 263–277 (2020)
46. Hajira, K.H., Khan, A., Kumam, P., Baleanu, D., Arif, M.: An approximate analytical solution of the Navier-Stokes equations within Caputo operator and Elzaki transform decomposition method. Adv. Difference Equ. Paper No. 622, 23 pp (2020)
47. Hammachukiattikul, P., Mohanapriya, A., Ganesh, A., Rajchakit, G., Govindan, V., Gunasekaran, N., Lim, C.P.: A study on fractional differential equations using the fractional Fourier transform. Adv. Diff. Equ. Paper No. 691, 22 pp (2020)
48. Hashemi, M.S.: Chaos, Solitons Fractals **152**, 111367 (2021)

49. Hashemi, M.S., Bahrami, F., Najafi, R.: Optik **138**, 240–249 (2017)
50. He, J.-H., Wu, X.-H.: Chaos Solitons Fractals **30**(3), 700 (2006)
51. Herzallah, M.A.E.; Baleanu, D. On fractional order hybrid differential equations. Abstr. Appl. Anal. Art. ID 389386, 7 pp (2014)
52. Hirota, R.: Exact three-soliton solution of the two-dimensional sine-Gordon equation. J. Physical Soc. Jpn. **35**(5), 1566–1566 (1973)
53. Hirota, R.: The Direct Method in Soliton Theory, no. 155. Cambridge University Press (2004)
54. Hoede, C., Li, X.L.: Clique polynomials and independent set polynomials of graphs. In: 13th British Combinatorial Conference (Guildford, 1991). Discrete Math. **125**(1–3), 219–228 (1994)
55. Hoede, C.: Hard graphs for the maximum clique problem. Proceedings of the First Japan Conference on Graph Theory and Applications (Hakone, 1986). Discrete Math. **72**(1–3), 175–179 (1988)
56. Hu, T., Fan, J., Wu, Q., Zhou, D.-X.: Regularization schemes for minimum error entropy principle. Anal. Appl. (Singap.) **13**(4), 437–455 (2015)
57. Inc, M.; Miah, M., Chowdhury, A., Ali, S., Rezazadeh, H., Akinlar, M.A., Chu, Y.-M.: AIMS Math. **5**(6), 6726 (2020)
58. Iqbal, Md.A., Wang, Y., Miah, Md.M., Osman, M.S.: Fractal and Fractional **6**(1), 4 (2021)
59. Jentzen, A., Kloeden, P.E.: Taylor approximations for stochastic partial differential equations. CBMS-NSF Regional Conference Series in Applied Mathematics, 83. Society for Industrial and Applied Mathematics (SIAM), Philadelphia, PA, 2011. xiv+220 pp. ISBN: 978-1-611972-00-9
60. Jiang, W., Cui, M., Lin, Y.: Anti-periodic solutions for Rayleigh-type equations via the reproducing kernel Hilbert space method. Commun. Nonlinear Sci. Numer. Simul. **15**(7), 1754–1758 (2010)
61. Kaya, D.: A numerical solution of the sine-Gordon equation using the modified decomposition method. Appl. Math. Comput. **143**(2–3), 309–317 (2003)
62. Khan, M.A., Atangana, A.: Mathematical modeling and analysis of COVID-19: a study of new variant Omicron. Phys. A **599**, Paper No. 127452 (2022)
63. Khan, H., Shah, R., Kumam, P., Arif, M.: Analytical solutions of fractional-order heat and wave equations by the natural transform decomposition method. Entropy **21**(6), Paper No. 597, 21 pp (2019)
64. Khan, M.A., Atangana, A.: Modeling the dynamics of novel Coronavirus (2019-nCov) with fractional derivative. Alexandria Eng. J. **59**(4), 2379–2389 (2020)
65. Khan, H., Shah, R., Kumam, P., Arif, M.: Analytical solutions of fractional-order heat and wave equations by the natural transform decomposition method. Entropy **21**(6), 21 (2019)
66. Khater, M.M.A., Dianchen, L., Raghda, A.M.A.: Dispersive long wave of nonlinear fractional Wu-Zhang system via a modified auxiliary equation method. AIP Adv. **9**(2), 025003 (2019)
67. Kilbas, A.A., Srivastava, H.M., Trujillo, J.J.: Theory and applications of fractional differential equations. North-Holland Mathematics Studies, 204. Elsevier Science B.V., Amsterdam (2006). xvi+523 pp. ISBN: 978-0-444-51832-3; 0-444-51832-0
68. Kudryashov, N.A.: Commun. Nonlinear Sci. Numer. Simul. **14**(9–10), 3507 (2009)
69. Kumar, A., Ilhan, E., Ciancio, A., Yel, G., Baskonus, H.M.: AIMS Math. **6**(5), 4238 (2021)
70. Kumbinarasaiah, S., Ramane, H.S., Pise, K.S., Hariharan, G.: Numerical-solution-for-nonlinear-Klein-Gordon equation via operational-matrix by clique polynomial of complete graphs. Int. J. Appl. Comput. Math. **7**(1), Paper No. 12, 19 pp (2021)
71. Laadjal, Z., Jarad, F.: Existence, uniqueness and stability of solutions for generalized proportional fractional hybrid integro-differential equations with Dirichlet boundary conditions. AIMS Math. **8**(1), 1172–1194 (2023)
72. Lan, H.B., Wang, K.L.: I. J. Phys. A **23**(17), 3923 (1990)
73. Li, L.-X., Li, E.-Q., Wang, M.-L.: Appl. Math. J. Chinese Univ. Ser. B **25**(4), 454 (2010)
74. Li, Z., Wang, M., Wang, Y., Pang, J.: Using reproducing kernel for solving a class of fractional order integral differential equations. Adv. Math. Phys. Art. ID 8101843, 12 pp (2020)

75. Li, X., Zhang, S., Wang, Y., Chen, H.: Analysis and application of the element-free Galerkin method for nonlinear sine-Gordon and generalized sinh-Gordon equations. Comput. Math. Appl. **71**(8), 1655–1678 (2016)
76. Liu, P., Mati ur, R., Anwarud, D.: Fractal fractional based transmission dynamics of COVID-19 epidemic model. Comput. Methods Biomech. Biomed. Eng. 1–18 (2022)
77. Liu, W., Kewang, C.: The functional variable method for finding exact solutions of some nonlinear time-fractional differential equations. Pramana **81**(3), 377–384 (2013)
78. Liu, S., Fu, Z., Liu, S.: Exact solutions to sine-Gordon-type equations. Phys. Lett. A **351**(1–2), 59–63 (2006)
79. Loh, J.R., Isah, A., Phang, C., Toh, Y.T.: On the new properties of Caputo-Fabrizio operator and its application in deriving shifted Legendre operational matrix. Appl. Numer. Math. **132**(4), 138–153 (2018)
80. Lu, D., Seadawy, A.R., Khater, M.M.A.: Structure of solitary wave solutions of the nonlinear complex fractional generalized Zakharov dynamical system. Adv. Difference Equ. Paper No. 266, 18 pp (2018)
81. Ma, W.-X., Wu, H., He, J.: Phys. Lett. A **364**(1), 29 (2007)
82. Ma, W.-X., Zhou, R., Gao, L.: Modern Phys. Lett. A **24**(21), 1677 (2009)
83. Ma, W.-X.: Phys. Lett. A **180**(3), 221 (1993)
84. Ma, W.-X.; Lee, J.-H.: Chaos Solitons Fractals **42**(3), 1356 (2009)
85. Magin, R.: Fractional calculus in bioengineering, part3. Crit. Rev. Biomed. Eng. **32**(3–4) (2004)
86. Mainardi, F.: Fractional calculus and waves in linear viscoelasticity. An Introduction to Mathematical Models. Imperial College Press, London (2010). xx+347 pp. ISBN: 978-1-84816-329-4; 1-84816-329-0
87. Malfliet, W.: Amer. J. Phys. **60**(7), 650 (1992)
88. Miller, K.S., Ross, B.: An introduction to the fractional calculus and fractional differential equations. A Wiley-Interscience Publication. Wiley, New York (1993). xvi+366 pp. ISBN: 0-471-58884-9
89. Mirzaee, Farshid, Samadyar, Nasrin: On the numerical solution of fractional stochastic integro-differential equations via meshless discrete collocation method based on radial basis functions. Eng. Anal. Bound. Elem. **100**, 246–255 (2019)
90. Mirzaee, F., Rezaei, S., Samadyar, N.: Numerical solution of two-dimensional stochastic time-fractional Sine-Gordon equation on non-rectangular domains using finite difference and meshfree methods. Eng. Anal. Bound. Elements **127**, 53–63 (2021)
91. Mirzaee, Farshid, Solhi, Erfan, Samadyar, Nasrin: Moving least squares and spectral collocation method to approximate the solution of stochastic Volterra-Fredholm integral equations. Appl. Numer. Math. **161**, 275–285 (2021)
92. Momani, S., Odibat, Z.: Analytical solution of a time-fractional Navier-Stokes equation by Adomian decomposition method. Appl. Math. Comput. **177**(2), 488–494 (2006)
93. Navier, C.L.M.H.: Mémoire sur les lois du mouvement des fluides. Mémoires de l'Académie Royale des Sciences de l'Institut de France **6**(1823), 389–440 (1823)
94. Nawaz Khan, M., Imtiaz, A., Hijaz, A.: A radial basis function collocation method for space-dependent inverse heat problems. J. Appl. Comput. Mech. (2020)
95. Niu, J., Sun, L., Xu, M., Hou, J.: A reproducing kernel method for solving heat conduction equations with delay. Appl. Math. Lett. **100**, 106036, 7 pp (2020)
96. Novel, C.P.: The epidemiological characteristics of an outbreak of 2019 novel Coronavirus diseases (COVID-19) in China. Chinese J. Epidemiol. **41**(2), 145–151 (2020)
97. Oldham, K.B., Spanier, J.: The fractional calculus. Theory and applications of differentiation and integration to arbitrary order. With an annotated chronological bibliography by Bertram Ross. Mathematics in Science and Engineering, vol. 111. Academic [Harcourt Brace Jovanovich, Publishers], New York-London, 1974. xiii+234 pp
98. Owolabi, K.M., Hammouch, Z.: Mathematical modeling and analysis of two-variable system with noninteger-order derivative. Chaos **29**(1), 013145, 15 pp (2019)
99. Parkes, E.J., Duffy, B.R.: Comput. Phys. Commun. **98**(3), 288 (1996)

100. Podlubny, I.: Fractional differential equations, mathematics in science and engineering 340 (1999)
101. Podlubny, I.: Fractional differential equations. An introduction to fractional derivatives, fractional differential equations, to methods of their solution and some of their applications. Mathematics in Science and Engineering, 198. Academic, Inc., San Diego, CA, 1999. xxiv+340, pp. ISBN: 0-12-558840-2
102. Podlubny, I.: Fractional differential equations. An introduction to fractional derivatives, fractional differential equations, to methods of their solution and some of their applications. Mathematics in Science and Engineering, 198. Academic, San Diego, CA (1999). xxiv+340 pp. ISBN: 0-12-558840-2
103. Podlubny, I.: Fractional differential equations. An introduction to fractional derivatives, fractional differential equations, to methods of their solution and some of their applications. Mathematics in Science and Engineering, 198. Academic, San Diego, CA (1999). xxiv+340 pp. ISBN: 0-12-558840-2
104. Qiu, W., Xu, D., Guo, J.: Numerical solution of the fourth-order partial integro-differential equation with multi-term kernels by the Sinc-collocation method based on the double exponential transformation. Appl. Math. Comput. **392**, Paper No. 125693, 16 pp (2021)
105. Ragab, A.A., Hemida, K.M., Mohamed, M.S., Abd El Salam, M.A.: Solution of time-fractional Navier–Stokes equation by using homotopy analysis method. Gen. Math. Notes **13**(2), 13–21 (2012)
106. Ray, S.S.: A new coupled fractional reduced differential transform method for the numerical solutions of-dimensional time fractional coupled Burger equations. Modelling and Simulation in Engineering 2014 (2014)
107. Rezazadeh, H., Zafar, A., Hashemi, M.S., Tala-Tebue, E.: Int. J. Modern Phys. B **34**(18), 2050161 (2020)
108. Robler, A., Seaid, M., Zahri, M.: Method of lines for stochastic boundary-value problems with additive noise. Appl. Math. Comput. **199**(1), 301–314 (2008)
109. Saad, K.M., Srivastava, H.M., Gomez-Aguilar, J.F.: A fractional quadratic autocatalysis associated with chemical clock reactions involving linear inhibition. Chaos Solitons Fractals **132**, 109557, 9 pp (2020)
110. Saedshoar Heris, M., Javidi, M.: On fractional backward differential formulas for fractional delay differential equations with periodic and anti-periodic conditions. Appl. Numer. Math. **118**, 203–220 (2017)
111. Samadyar, Nasrin, Mirzaee, Farshid: Numerical solution of two-dimensional weakly singular stochastic integral equations on non-rectangular domains via radial basis functions. Eng. Anal. Bound. Elem. **101**, 27–36 (2019)
112. Shah, R., Hassan, K., Kumam, P., Muhammad, A.: An analytical technique to solve the system of nonlinear fractional partial differential equations. Mathematics **7**(6), 505 (2019)
113. Shah, R., Hassan, K., Poom, K., Muhammad, A.: An analytical technique to solve the system of nonlinear fractional partial differential equations. Mathematics **7**(6), 505 (2019)
114. Shah, N.A., Xiaoping, W., Haitao, Q., Shaowei, W., Ahmad, H.: Transient electro-osmotic slip flow of an oldroyd-B fluid with time-fractional Caputo-Fabrizio derivative. J. Appl. Comput. Mech. **5**(4), 779–790 (2019)
115. Shah, R., Hassan, K., Kumam, P., Muhammad, A.: An analytical technique to solve the system of nonlinear fractional partial differential equations. Mathematics **7**(6), 505 (2019)
116. Shan, Y., Liu, W., Wu, B.: Space-time Legendre-Gauss-Lobatto collocation method for two-dimensional generalized sine-Gordon equation. Appl. Numer. Math. **122**, 92–107 (2017)
117. Singh, M., Gupta, R.K.: Int. J. Appl. Comput. Math. **4**(3), Paper No. 88, 15 pp (2018)
118. Singh, B.K., Kumar, P.: FRDTM for numerical simulation of multi-dimensional, time-fractional model of Navier–Stokes equation. Ain Shams Eng. J. **9**(4), 827–834 (2018)
119. Tanwar, D.V.: Lie symmetry reductions and generalized exact solutions of Date-Jimbo-Kashiwara-Miwa equation. Chaos, Solitons Fractals **162**, 112414 (2022)
120. This site shows Coronavirus cases online. https://www.worldometers.info Coronavirus

121. Tidke, H.L.: Existence of global solutions to nonlinear mixed Volterra-Fredholm integrodifferential equations with nonlocal conditions. Electron. J. Differential Eq. (55), 7 pp (2009)
122. Tuan, N.H., Ganji, R.M., Jafari, H.: A numerical study of fractional rheological models and fractional Newell-Whitehead-Segel equation with non-local and non-singular kernel. Chinese J. Phys. **68**(5), 308–320 (2020)
123. Ullah, S., Khan, M.A.: Modeling the impact of non-pharmaceutical interventions on the dynamics of novel Coronavirus with optimal control analysis with a case study. Chaos, Solitons Fractals **139**, 110075 (2020)
124. Vargas, A.M.: Finite difference method for solving fractional differential equations at irregular meshes. Math. Comput. Simul. **193**, 204–216 (2022)
125. Wang, S.-N., Hu, J.: Modern Phys. Lett. B **33**(10), 1950119, 19 pp (2019)
126. Wang, M.L.: Phys. Lett. A **199**(3–4), 169 (1995)
127. Wazwaz, A.M.: Exact solutions for the generalized sine-Gordon and the generalized sinh-Gordon equations. Chaos Solitons Fractals **28**(1), 127–135 (2006)
128. Wei, Z., Dong, W.: Periodic boundary value problems for Riemann-Liouville sequential fractional differential equations. Electron. J. Qual. Theory Differ. Equ. (87), 13 pp (2011)
129. Wei, G.W.: Discrete singular convolution for the sine-Gordon equation. Phys. D **137**(3–4), 247–259 (2000)
130. Wei, Z., Dong, W., Che, J.: Periodic boundary value problems for fractional differential equations involving a Riemann-Liouville fractional derivative. Nonlinear Anal. **73**(10), 3232–3238 (2010)
131. Wenxiu, M., Zhou, D.: Acta Math. Scientia. **17**, 168 (1997)
132. World Health Organization. Coronavirus. World Health Organization. https://www.who.int/health-topics Coronavirus
133. Wu, G.-C., Lee, E.W.M.: Fractional variational iteration method and its application. Phys. Lett. A **374**(25), 2506–2509 (2010)
134. Wu, Q.: The maximum clique problems with applications to graph coloring. PhD Thesis: Universiti d'Angers (2013)
135. Xia, F.L., Jarad, F., Hashemi, M.S., Riaz, M.B.: Results Phys. 105512 (2022)
136. Xu, D., Qiu, W., Guo, J.: A compact finite difference scheme for the fourth-order time-fractional integro-differential equation with a weakly singular kernel. Numer. Methods Partial Diff. Equ. **36**(2), 439–458 (2020)
137. Yepez-Martinez, H., Gomez-Aguilar, J.F.: Laplace variational iteration method for modified fractional derivatives with non-singular kernel. J. Appl. Comput. Mech. **6**(3), 684–698 (2020)
138. Yin, F., Tian, T., Song, J., Zhu, M.: Spectral methods using Legendre wavelets for nonlinear Klein\Sine-Gordon equations. J. Comput. Appl. Math. **275**, 321–334 (2015)
139. Yuan, Y.-Q., Tian, B., Sun, W.-R., Chai, J., Liu, L.: Comput. Math. Appl. **74**(4), 873 (2017)
140. Yuan, Y.-Q., Tian, B., Sun, W.-R., Chai, J., Liu, L.: Comput. Math. Appl. **74**(4), 873 (2017)
141. Zagrodzinski, J.: Particular solutions of the sine-Gordon equation in $2 + 1$ dimensions. Phys. Lett. A **72**(4–5), 284–286 (1979)
142. Zeb, A., Abdon, A., Zareen, A.K., Salih, D.: A robust study of a piecewise fractional order COVID-19 mathematical model. Alexandria Eng. J. **61**(7), 5649–5665 (2022)
143. Zhang, Y., Ye, L.-Y., Lv, Y.-N., Zhao, H.-Q.: J. Phys. A **40**(21), 5539 (2007)
144. Zheng, B.: (G'/G)-expansion method for solving fractional partial differential equations in the theory of mathematical physics. Commun. Theor. Phys. (Beijing) **58**(5), 623–630 (2012)
145. Zhou, D.-X.: Deep distributed convolutional neural networks: universality. Anal. Appl. (Singap.) **16**(6), 895–919 (2018)
146. Zhou, Y., Cui, M., Lin, Y.: Numerical algorithm for parabolic problems with non-classical conditions. J. Comput. Appl. Math. **230**(2), 770–780 (2009)

MIX
Papier aus verantwortungsvollen Quellen
Paper from responsible sources
FSC® C105338

If you have any concerns about our products,
you can contact us on
ProductSafety@springernature.com

In case Publisher is established outside the EU,
the EU authorized representative is:
Springer Nature Customer Service Center GmbH
Europaplatz 3, 69115 Heidelberg, Germany

Printed by Libri Plureos GmbH
in Hamburg, Germany